# Handbook of
# Electronics Calculations

# OTHER McGRAW-HILL HANDBOOKS OF INTEREST

*American Institute of Physics* • AMERICAN INSTITUTE OF PHYSICS HANDBOOK
*Baumeister* • MARKS' STANDARD HANDBOOK FOR MECHANICAL ENGINEERS
*Beeman* • INDUSTRIAL POWER SYSTEMS HANDBOOK
*Bourns, Inc.* • THE POTENTIOMETER HANDBOOK
*Burington and May* • HANDBOOK OF PROBABILITY AND STATISTICS WITH TABLES
*Condon and Odishaw* • HANDBOOK OF PHYSICS
*Coombs* • BASIC ELECTRONIC INSTRUMENT HANDBOOK
*Coombs* • PRINTED CIRCUITS HANDBOOK
*Croft, Carr, and Watt* • AMERICAN ELECTRICIANS' HANDBOOK
*Fink* • ELECTRONICS ENGINEERS' HANDBOOK
*Fink and Beaty* • STANDARD HANDBOOK FOR ELECTRICAL ENGINEERS
*Hamsher* • COMMUNICATION SYSTEM ENGINEERING HANDBOOK
*Harper* • HANDBOOK OF ELECTRONIC PACKAGING
*Harper* • HANDBOOK OF ELECTRONICS SYSTEM DESIGN
*Harper* • HANDBOOK OF MATERIALS AND PROCESSES FOR ELECTRONICS
*Harper* • HANDBOOK OF THICK FILM HYBRID MICROELECTRONICS
*Harper* • HANDBOOK OF WIRING, CABLING, AND INTERCONNECTING FOR ELECTRONICS
*Henney* • RADIO ENGINEERING HANDBOOK
*Hicks* • STANDARD HANDBOOK OF ENGINEERING CALCULATIONS
*Hunter* • HANDBOOK OF SEMICONDUCTOR ELECTRONICS
*Huskey and Korn* • COMPUTER HANDBOOK
*Ireson* • RELIABILITY HANDBOOK
*Jasik* • ANTENNA ENGINEERING HANDBOOK
*Juran* • QUALITY CONTROL HANDBOOK
*Kaufman and Seidman* • HANDBOOK FOR ELECTRONICS ENGINEERING TECHNICIANS
*Giacoletto* • ELECTRONIC DESIGNER'S HANDBOOK
*Machol* • SYSTEM ENGINEERING HANDBOOK
*Maissel and Glang* • HANDBOOK OF THIN FILM TECHNOLOGY
*Markus* • ELECTRONICS DICTIONARY
*Perry* • ENGINEERING MANUAL
*Smeaton* • MOTOR APPLICATION AND MAINTENANCE HANDBOOK
*Smeaton* • SWITCHGEAR AND CONTROL HANDBOOK
*Stout and Kaufman* • HANDBOOK OF OPERATIONAL AMPLIFIER CIRCUIT DESIGN
*Terman* • RADIO ENGINEERS' HANDBOOK
*Truxal* • CONTROL ENGINEERS' HANDBOOK

# Handbook of
# Electronics Calculations

## FOR ENGINEERS AND TECHNICIANS

*Editors*

## MILTON KAUFMAN

President, Electronic Writers and Editors, Inc.

## ARTHUR H. SEIDMAN

Professor of Electrical Engineering
Pratt Institute, School of Engineering

**McGRAW-HILL BOOK COMPANY**
New York   St. Louis   San Francisco   Auckland   Bogotá
Düsseldorf   Johannesburg   London   Madrid
Mexico   Montreal   New Delhi   Panama
Paris   São Paulo   Singapore
Sydney   Tokyo   Toronto

Library of Congress Cataloging in Publication Data

Main entry under title:

Handbook of electronics calculations for engineers
    technicians.
    Includes bibliographies and index.
    1. Electronics—Handbooks, manuals, etc.
I. Kaufman, Milton.   II.  Seidman, Arthur H.
TK7825.H36      621.381'02'02      79-12387
ISBN 0-07-033392-0

1234567890   KPKP   7865432109

*The editors for this book were Harold B. Crawford and
Joseph Williams, the designer was Naomi Auerbach, and the
production supervisor was Thomas G. Kowalczyk. It was
set in Caledonia by University Graphics, Inc.*

*Printed and bound by The Kingsport Press.*

*Dedicated to our children,*
*Elissa Bishop, Richard Kaufman, and granddaughter Erica Bishop*
*and Ben and Rebecca Seidman*

# Contents

# 4. SELECTING *R*, *L*, AND *C* COMPONENTS

# 5. SELECTING SEMICONDUCTOR DEVICES

# 6. AUDIO AMPLIFIERS

## 7. TUNED AMPLIFIERS

## 8. FEEDBACK

# 9. OSCILLATORS

# 10. POWER SUPPLIES

# 11. BATTERY USES AND SPECIAL CELLS

## 12. OP AMP APPLICATIONS

## 13. DIGITAL LOGIC

## 14. COMPUTER-AIDED CIRCUIT DESIGN

## 15. ANALOG-DIGITAL CONVERSION

## 16. VIDEO AMPLIFIERS

## 17. THE MICROPROCESSOR

## 18. TRANSMISSION LINES

# 19. FILTERS

# 20. ANTENNAS

## 21. MICROWAVES

## 22. COMMUNICATIONS SYSTEMS

## 23. MEASUREMENTS

## 24. THICK-FILM TECHNOLOGY

## Appendix

**Index follows Appendixes.**

# Preface

This is the first comprehensive electronics handbook which emphasizes worked-out practical problems rather than theory. As such, it serves as a "cookbook" for engineers and technicians to help them solve their everyday on-the-job problems.

Hundreds of worked-out problems are presented, covering a diversity of topics in analog and digital circuits. The topics have been carefully chosen to serve the needs of the modern engineer and technician. They range from Selecting Semiconductor Devices, Feedback, Microwaves, and Communications Systems to Microprocessors, Computer-Aided Design, Thick Film Technology, and D/A and A/D Converters. In all, there are 24 chapters encompassing the most important topics in electronics. Every effort has been made to avoid purely theoretical problems and to concentrate on meaningful and practical problems.

The format of this handbook is such that approximately 75% of the contents consists of worked-out problems. The remainder is devoted to brief, but important, theory applicable to each specific problem.

To ensure maximum usefulness and ease of interpretation, many schematic and block diagrams have been included. In addition, numerous curves, tables, and graphs are presented.

Each of the 24 in-depth chapters follows the same basic format. This format has been carefully developed to help readers make maximum utilization of the material. The general approach to each worked-out problem is:

1. Statement of the problem.
2. Presentation of essential theory that relates to the problem.
3. Detailed solution of the problem.

This relatively simple, yet comprehensive, format adds greatly to the value of the handbook. All mathematics has been kept as simple as possible. Arithmetic and algebra are used in the solution of the majority of problems.

The handbook not only provides detailed solutions to hundreds of problems, but it also assists the reader in making the proper choice of components for a particular application. This is particularly true of Chapter 4, Selecting $R$, $L$, and $C$ Components, and Chapter 5, Selecting Semiconductor Devices

Each chapter contains a bibliography. The referenced books and publications provide practical information that is pertinent to the chapter subject matter.

Great care has been exercised in making this handbook as accurate as possible. It is inevitable, however, that in a first edition of this size and coverage some errors may remain. The editors will appreciate these being brought to their attention.

The editors wish to gratefully acknowledge the substantial effort and cooperation of all the contributors. They have covered their special fields in a comprehensive and practical manner.

*Milton Kaufman*
*Arthur H. Seidman*

# Chapter 1

# Technical Math Review

### ERNEST A. JOERG

**Professor and Chairperson, Electrical Technology Department,
Westchester Community College, Valhalla, N.Y.**

## 1.1  INTRODUCTION

Basic mathematics, so vital to engineers and technicians, are reviewed in this chapter.
Topics covered are complex numbers, elementary algebra, exponentials, logarithms,
trigonometry, and complex algebra. Some 38 worked-out problems illustrate the application of mathematical techniques in their solution.

## 1.2  COMPLEX NUMBERS

Any number ever used can be classified as a complex number. It is defined as a number
consisting of two parts: one real and the other imaginary. It is generally written as a $a + bi$
or, in engineering, as $a + jb$, where $i = j = \sqrt{-1}$. Term $j$ (or $i$) is called the *imaginary
unit*. In *Fig. 1-1*, $a$ is the real part and $jb$ the imaginary part of the complex number. Term
$j$ is treated as a variable, and $j^2 = -1$.

Complex numbers containing $a$ and $b$ other than zero may be added or subtracted in
the same manner as real numbers. However, only real quantities are combined with reals
and imaginaries with imaginaries.

**problem 1.1**  Add $2 + j6$ to $5 - j8$ and label the sum **Y**.

**solution**  $Y = 2 + j6 + (5 - j8) = (2 + 5) + j(6 - 8)$
$= 7 - j2$

**problem 1.2**  From $-1 + j8$ subtract $3 - j6$ and call the result **Z**.

**solution**  $Z = -1 + j8 - (3 - j6) = -1 + j8 - 3 + j6 = (-1 - 3) + j(8 + 6)$
$= -4 + j14$

**problem 1.3**  Multiply $7 - j2$ by $-4 - j14$ and label the product **L**.

**solution**    $L = (Y)(Z) = (7 - j2)(-4 + j14)$

$$= (7)(-4) + (7)(j14) + (-j2)(-4) + (-j2)(j14)$$
$$= -28 + j98 + j8 - j^2 \times 28$$
$$= -28 + j98 + j8 + 28 = (-28 + 28) + j(98 + 8)$$
$$= 0 + j106 = j106 \quad \text{(pure imaginary number)}$$

**problem 1.4**    Find the inverse of $Y$ and call it $Q$.

**solution**    $Q = 1/Y = 1/(7 - j2)$. Multiply numerator and denominator by the complex conjugate of the denominator. The *complex conjugate* of a complex number is found by changing only the sign of the imaginary part. For this particular problem it is $7 + j2$. Thus,

$$Q = \frac{1}{7 - j2} = \frac{1}{7 - j2}\left(\frac{7 + j2}{7 + j2}\right) = \frac{7 + j2}{(7)(7) + (-j2)(j2) + j14 - j14}$$

$$= \frac{7 + j2}{49 + 4} = \frac{7 + j2}{53} = \left(\frac{7}{53}\right) + j\left(\frac{2}{53}\right) \approx 0.132 + j0.038$$

A complex number $(a + jb)$ multiplied by its complex conjugate $(a - jb)$ always yields a real number equal to $a^2 + b^2$.

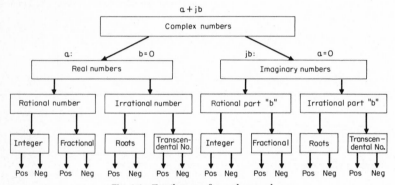

**Fig. 1-1**    Family tree of complex numbers.

## Graphing

The complex plane (Argand plot) is employed when graphing complex numbers. The vertical axis is called the *axis of imaginaries* and the horizontal axis is called the *axis of reals* with + and − in the conventional directions.

**problem 1.5**    Graph (a) $A = 2 + j2$; (b) $B = 3 + j4$; (c) $C = 5 - j2$; (d) $D = A + B$; (e) $E = -2 + j4$; (f) $F = -3 - j5$.

**solutions**    (a) See Fig. 1-2a; (b) see Fig. 1-2b; (c) see Fig. 1-2cc; (d) see Fig. 1-2d; (e) see Fig. 1-2e; (f) see Fig. 1-2f.

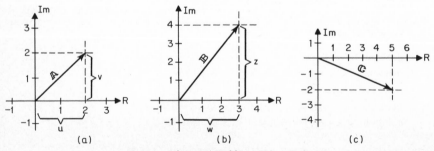

**Fig. 1-2**    Solution to Problem 1.5 (see text).

## 1.3   ELEMENTARY ALGEBRA

In algebra a letter represents a particular amount, or quantity, of something in question. For example, $V = RI$ (Ohm's law) states that voltage in volts $(V)$ equals the product of resistance in ohms $(R)$ and current in amperes $(I)$.

An algebraic expression is a combination of symbols representing operations, variables, and coefficients. In general, it has the form:

$$a_0 x^n + a_1 x^{n-1} + \cdots + a_{n-2}x^2 + a_{n-1}x^1 + a_n x^0 = 0 \qquad (1\text{-}1)$$

where the $a$'s are coefficients of variable $x$. Expression (1-1) is termed a polynomial in $x$ of the $n$th degree. A polynomial can consist of one or more terms; the terms are separated by the $+$ or $-$ operations. A one-termed polynomial is called a *monomial,* a three-termed a *trinomial,* etc. The coefficient of the $x^0$ term, which is $a_n$, is generally referred to as a *constant* term.

### Addition and Subtraction of Polynomials

**problem 1.6**   Add $5x^2 - 8x - 3$, $2x^2 + 9x - 2$, and $4x^2 + 6x - 7$.

**solution**

$$
\begin{array}{r}
5x^2 - 8x - 3 \\
2x^2 + 9x - 2 \\
4x^2 + 6x - 7 \\
\hline
11x^2 + 7x - 12
\end{array}
$$

**problem 1.7**   Subtract $5x^2 - 8x + 2$ (subtrahend) from $-2x + 8x^2 + 9$ (minuend).

**solution**

$$
\begin{array}{ll}
8x^2 - 2x + 9 & \quad 8x^2 - 2x + 9 \\
- (5x^2 - 8x + 2) = & -5x^2 + 8x - 2 \\
\cline{1-1}
& \quad \overline{3x^2 + 6x + 7}
\end{array}
$$

Note that each and every term in the subtrahend was changed in sign and then added to the minuend. Also, the minuend was rewritten in descending powers of $x$ prior to solving the given problem. In both addition and subtraction, only like-type terms are combined: the $x^2$ terms with the $x^2$ terms, the $x$- terms with the $x$- terms, and the constant terms with the constants.

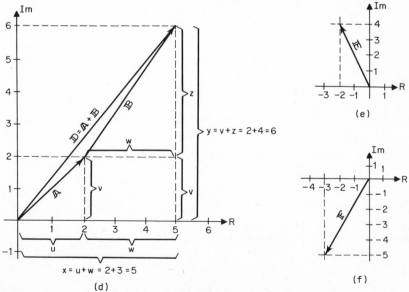

Fig. 1-2 (Cont.)

## Multiplication and Division of Polynomials

**problem 1.8**    Perform each of the following: (a) Multiply $x + 3$ by $x - 4$. (b) Multiply $4x^3$ by $-6x^2 + 5$.

**solution**

(a)  $(x + 3)(x - 4) = xx - 4x + 3x + (3)(-4)$
$= x^2 + x(-4 + 3) - 12$
$= x^2 - x - 12$

(b)  $(4x^3)(-6x^2 + 5) = (4)(-6)(x^3)(x^2) + (4)(5)(x^3)$
$= -24x^5 + 20x^3$

In multiplication of a base symbol, say $x$, raised to some power $m$ by the same base symbol, $x$, raised to some other power $n$, the base is kept and the powers added.

**problem 1.9**    Divide each of the following: (a) $-15x^4$ by $3x^3$; (b) $x^2 + 6x + 8$ by $x + 2$; (c) $x^2 - 30 + 5x$ by $8 + x$.

**solution**

(a)  $\dfrac{-15x^4}{3x^3} = -5x^1 = -5x$

When dividing a base symbol, say $x$, raised to power $p$ (the numerator) by the same base symbol, $x$, raised to some other power $q$ (the denominator), the base is kept and the power of the denominator is subtracted from the power of the numerator: $x^p/x^q = x^{p-q}$.

(b)  $\dfrac{x^2 + 6x + 8}{x + 2} =$

$$
\begin{array}{r}
x + 4 \\
x + 2 \overline{) x^2 + 6x + 8} \\
\oplus\ x^2 \oplus 2x \\
\hline
4x + 8 \\
\oplus\ 4x \oplus 8 \\
\hline
0
\end{array}
$$

In division, write the dividend and divisor in descending powers of the given variable. Successively, then:

1. Divide first term of the divisor into the first term of dividend to obtain the first term of the quotient.
2. Multiply through the complete divisor.
3. Subtract from the dividend.
4. Bring down the next remaining term from the dividend.
5. Repeat the process until all terms are accounted for.

If a quantity exists after the last subtraction, it is called the remainder.

(c)  $\dfrac{x^2 - 30 + 5x}{8 + x} =$

$$
\begin{array}{r}
x - 3 \\
x + 8 \overline{) x^2 + 5x - 30} \\
\oplus\ x^2 \oplus 8x \\
\hline
-3x - 30 \\
\overset{-}{\ominus}3x \overset{+}{\ominus} 24 \\
\hline
-6 \quad \text{remainder}
\end{array}
$$

Note that quotient $Q(x) = x - 3$ and remainder $R = -6$. The complete answer is:

$$x - 3 + \frac{-6}{x + 8} = Q(x) + \frac{R}{\text{divisor}}$$

## Algebraic Equations and Formulas

An algebraic equation is a mathematical statement that indicates the equality of two algebraic expressions. For example, $3x - 7 = x + 1$ states that the binomial (a two-termed polynomial) on the left-hand side (LHS) of the equation equals the binomial on the right-hand side (RHS) of the equation.

## Solution of First-Degree Equations

Here $x$, or any other chosen variable, is raised to power unity (1). A quantity can be added or subtracted from both sides of an equation. Each and every term on both sides of an equation can be multiplied or divided by the same quantity (excluding zero).

**problem 1.10**    Solve for $x$ or $y$ in each of the given equations: (a) $x + 2 = 9$; (b) $2x - 1 = 15$; (c) $(y - 9)/2 = (y + 12)/9$; (d) $(2x + 5)/6 = (2x)/3 - \frac{1}{2}$.

## solution

(a)   $x + 2 = 9$ Subtract 2 from both sides of the equation:
$$x + 2 - 2 = 9 - 2$$
$$x = 7$$

(b)   $2x - 1 = 15$ Add 1 to both sides:
$$2x - 1 + 1 = 15 + 1$$
$$2x = 16$$ Divide both sides by 2:
$$\frac{2x}{2} = \frac{16}{2}$$
$$x = 8$$

The solution can be verified by inserting the found value of $x$ back into the original given equation. For example,

$$2(8) - 1 = 15$$
$$16 - 1 = 15$$
$$15 = 15$$

(c)   $\dfrac{y - 9}{2} = \dfrac{y + 12}{9}$ Multiply both sides by 18:
$$\left(\frac{y - 9}{2}\right)(18) = \left(\frac{y + 12}{9}\right)(18)$$
$$(y - 9)(9) = (y + 12)(2)$$ Clear parentheses:
$$9y - 81 = 2y + 24$$ Add 81 to both sides:
$$9y - 81 + 81 = 2y + 24 + 81$$ Combine like terms:
$$9y = 2y + 105$$ Subtract $2y$ from both sides:
$$9y - 2y = 2y - 2y + 105$$ Combine like terms:
$$7y = 105$$ Divide both sides by 7:
$$7y/7 = 105/7$$
$$y = 15$$

The equation in this problem is also classified as a *fractional equation*. Initially, both sides of such an equation are multiplied through by the lowest common denominator (LCD) or the lowest common multiple (LCM) of the denominator.

(d)   $\dfrac{2x + 5}{6} = \dfrac{2x}{3} - \dfrac{1}{2}$ Multiply through by LCD = 6:
$$\left(\frac{2x + 5}{6}\right)(6) = \left(\frac{2x}{3}\right)(6) - \frac{1}{2}(6)$$
$$2x + 5 = 2x(2) - 3$$ Clear parenthesis:
$$2x + 5 = 4x - 3$$ Subtract 5 from both sides:
$$2x + 5 - 5 = 4x - 3 - 5$$ Combine like terms:
$$2x = 4x - 8$$ Subtract $4x$ from both sides:
$$2x - 4x = 4x - 4x - 8$$ Combine like terms:
$$-2x = -8$$ Divide both sides by $-2$:
$$\frac{-2x}{-2} = \frac{-8}{-2}$$ Like signs yield a positive number:
$$x = +4$$

## Literal Equations

A literal equation is one in which the constants are represented by letters of the alphabet, called *literals*. Sometimes one wishes to rearrange an equation and solve for a particular literal.

**problem 1.11**   Given $V = ZI$, solve for $I$.

**solution**   Dividing both sides by $Z$ yields:

$$\frac{V}{Z} = \frac{ZI}{Z}$$
$$= I$$

Quantity $1/Z$ is an *admittance* and is denoted by symbol $Y$. Therefore, the final solution can be written as $I = YV$. This is sometimes referred to as the dual form of Ohm's law.

**problem 1.12**   Given $°F = (9/5)°C + 32$, solve for $°C$.

**solution**   Subtract 32 from both sides: $°F - 32 = (9/5)°C + 32 - 32 = (9/5)°C$. Multiply both sides by 5/9: $5(°F - 32)/9 = °C$.

## Graphing Linear Equations

A linear equation has $x$ and $y$ to the first degree. It has an infinite number of solutions, or pairs of values, sometimes referred to as *coordinate pairs (x,y)*. (Note that the word *line* appears within the word *linear*.) The equation is plotted on rectangular (cartesian) coordinates.

**"Brute Force" Technique**   This technique calls for choosing several values of $x$ (the independent variable) and then determining the respective $y$ values (the dependent variable). Two points are necessary to graph the straight line; however, additional insurance points are sometimes desirable.

**problem 1.13**   Graph $-4x + 6y = 12$.

**solution**   Prior to making a table of values, the equation can be simplified by dividing both sides by 2 yielding $-2x + 3y = 6$; this will be used for the plot. Values of $x$ chosen (since it is the independent variable) are 4,3,2,1,0,−1,−2,−3, and −4. Each value is inserted into the equation, and then it is solved for $y$. The results are summarized below:

| $x$ | 4 | 3 | 2 | 1 | 0 | −1 | −2 | −3 | −4 |
|---|---|---|---|---|---|---|---|---|---|
| $y$ | 14/3 | 4 | 10/3 | 8/3 | 2 | 4/3 | 2/3 | 0 | −2/3 |

The graph is plotted in Fig. 1-3. Point (0,2) is called the $y$ intercept, and point (−3,0) is called the $x$ intercept.

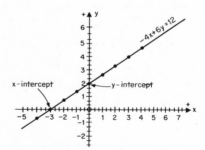

**Fig. 1-3**   Graph of $-4x + 6y = 12$ (Problem 1.13).

**The Slope-Intercept Method**   The slope-intercept method permits sketching of linear equations more rapidly. All linear equations can be written in the form:

$$y = mx + b \qquad (1\text{-}2)$$

where $m$ is the slope ($\Delta y/\Delta x$), or the vertical change divided by the horizontal change, and $b$ is the $y$ intercept $(0,b)$. Equation (1-2) is referred to as an equation of a straight line.

**problem 1.14**   Using the slope-intercept method, graph $3y - 6x = 12$.

**solution**   Dividing by 3 (common factor) yields:

$$y = \frac{6x + 12}{3}$$

Comparing with Eq. (1-2)       $y = 2x + 4$
$$y = mx + b$$

It is seen that $m = 2$, $b = 4$. Use $P_1 = (0,4)$ as the start point for plotting. Slope $m$ is + 2, or +2/+1. From the start point, proceed two units up in the $y$ direction and one unit over to the right in the $x$ direction to the new point $(1,6) = P_2$ (Fig. 1-4$a$). Because two points determine a straight line, an accurate graph of the given straight line is obtained in Fig. 1-4$b$.

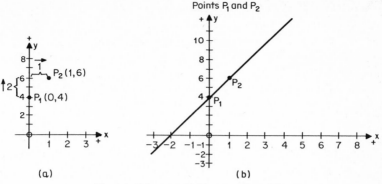

**Fig. 1-4**   Solution to Problem 1.14 (see text).

## Quadratic Equations

A quadratic equation is an equation involving the square and no higher power of the unknown quantity. *Quadratus* in Latin means "made square," or the variable squared. It should not be confused with the prefix *quad* meaning *four*. The general quadratic equation can be written:

$$ax^2 + bx + c = 0 \qquad (1\text{-}3)$$

A quadratic equation always has two roots; quadratics may be solved by factoring, by the quadratic formula, or by completing the square. The first two methods will be illustrated.

**Factoring Method**   The following steps should be taken: 1. Shuttle, or transpose, all terms to the LHS of the equation; then the RHS equals zero. 2. Factor the LHS (if it is not easily factorable, use the alternate quadratic formula method). 3. Set each factor containing the unknown quantity equal to zero. 4. Solve the two resulting equations and obtain the desired roots.

**problem 1.15**   Given $3x^2 - 6x = 0$, solve for $x$.

**solution**   Factor out the $3x$ term:

$$(3x)(x - 2) = 0$$

Since the product of the two quantities is zero, then either factor can equal zero. Thus,

$$3x = 0 \qquad x - 2 = 0$$
$$x = 0 \qquad x = 2$$

The two roots are real and different.

**problem 1.16**   Find the roots of $y^2 - y - 6 = 0$.

**solution**   Factoring the LHS,

$$(y \overset{-}{\phantom{.}} 3)(y + 2) = 0$$

Solving each separately yields:

$$y - 3 = 0 \qquad y + 2 = 0$$
$$y = 3 \qquad y = -2$$

The two roots are real and different.

**Quadratic Formula**   Given a quadratic equation, $ax^2 + bx + c = 0$. The two roots of the equation may be found from the quadratic formula:

$$x_{1,2} = \frac{-b \pm \sqrt{b^2 - 4ac}}{2a} \qquad (1\text{-}4)$$

This method of solution can be used for all quadratic equations.

**problem 1.17**   Given $3w^2 + 24w - 33 = 0$, find $w$.

**solution**   Divide through by 3, yielding:

$$w^2 + 8w - 11 = 0$$

Comparing to general form:

$$a = 1, b = 8, c = -11$$

Applying Eq. (1-4), let $x = w$:

$$w = \frac{-8 \pm \sqrt{(8)^2 - 4(1)(-11)}}{2(1)}$$

$$= \frac{-8}{2} \pm \frac{\sqrt{108}}{2}$$

$$= -4 \pm \frac{6\sqrt{3}}{2} = -4 \pm 3\sqrt{3}$$

Then $w_1 \approx -9.196$ and $w_2 \approx +1.196$. Both roots are real and different.

**problem 1.18**   Find the roots of $z^2 + 8z + 25 = 0$.

**solution**   $a = 1, b = 8$, and $c = 25$. Hence,

$$z = \frac{-8 \pm \sqrt{8^2 - 4 \times 1 \times 25}}{2 \times 1}$$

$$= \frac{-8 \pm \sqrt{-36}}{2} = -4 \pm \frac{\sqrt{36}\sqrt{-1}}{2}$$

Therefore, $z_1 = -4 + j3$ and $z_2 = -4 - j3$. The two roots are complex. Recall that if $z = a + jb$, its complex conjugate is $a - jb$. This is the case for $z_1$ and $z_2$ in the problem. They are complex conjugates of each other. Mathematically it can be indicated by $z_1 = z_2{}^*$ or $z_2 = z_1{}^*$. This problem illustrates the third possible solution for a quadratic, a pair of complex roots.

The *discriminant* of Eq. (1-4), $b^2 - 4ac$, indicates the nature of the roots:
1. Roots real and different: discriminant $> 0$.
2. Roots real and alike: discriminant $= 0$.
3. Roots complex: discriminant $< 0$.

## 1.4  EXPONENTS AND SCIENTIFIC NOTATION

An exponent indicates the number of times a factor is used. For example, consider $y \cdot y \cdot y \cdot y = y^4$. Exponent 4 tells us that $y$ is multiplied by itself four times. The rules governing exponents are:

$$a^m \cdot a^n = a^{m+n} \tag{1-5}$$

$$\frac{a^p}{a^q} = a^{p-q} = \frac{1}{a^{q-p}} \qquad \text{where } a \neq 0 \tag{1-6}$$

$$a^{+u} = \frac{1}{a^{-u}} \qquad a^{-k} = \frac{1}{a^{+k}} \qquad \text{where } a \neq 0 \tag{1-7}$$

$$a^0 \equiv 1 \qquad \text{where } a \neq 0 \tag{1-8}$$

$$a^w \cdot b^w = (a \cdot b)^w \tag{1-9}$$

$$\frac{a^z}{b^z} = \left(\frac{a}{b}\right)^z \qquad \text{where } b \neq 0 \tag{1-10}$$

$$(a^u)^v = a^{uv} \tag{1-11}$$

$$a^{P/R} = \sqrt[R]{a^P} = (\sqrt[R]{a})^P \qquad \text{where } a > 0 \text{ unless } R \text{ is odd} \tag{1-12}$$

$$a^{1/R} = \sqrt[R]{a} \tag{1-13}$$

**problem 1.19** Simplify $27^{2/3}$,

**solution** By Eq. (1-12): $27^{2/3} = (\sqrt[3]{27})^2 = (3)^2 = 9$.

**problem 1.20** Simplify $(c^3 d^2)/(c^{-2} d^{-1} f)$.

**solution** By Eqs. (1-7) and (1-5):

$$\frac{c^3 d^2}{c^{-2} d^{-1} f} = \frac{c^3 d^2 c^2 d^1}{f} = \frac{c^5 d^3}{f}$$

**problem 1.21** Simplify $(\sqrt{5a^3 b^7})^0 \cdot 7(a^2 b^4)^0$.

**solution** By Eq. (1-8), $(1) \cdot (7)(1) = 7$.

**problem 1.22** Simplify

$$\left(\frac{3x^2\, y^4}{5z^5}\right)^3$$

**solution** By Eqs. (1-9) and (1-11):

$$\left(\frac{3x^2 y^4}{5z^5}\right)^3 = \frac{3^3 \cdot (x^2)^3 \cdot (y^4)^3}{5^3 \cdot (z^5)^3} = \frac{27 x^6\, y^{12}}{125\, z^{15}}$$

**problem 1.23** Simplify $\sqrt[-5]{32}$.

**solution** By Eqs. (1-7) and (1-13): $\sqrt[-5]{32} = 32^{-1/5} = 1/32^{1/5} = 1/\sqrt[5]{32} = 1/2$.

## Standard Scientific Notation

Scientific notation enables the engineer and scientist to express very small or very large numbers in a compact form which can also be used in computations. For example, sixty-eight million can be written as 68,000,000 or $68 \times 10^6$, where $10^6$ represents 1 million; this is scientific notation. However, in *standard scientific notation* (SSN), the decimal point is placed to the right of the first significant nonzero digit. Thus $6.8 \times 10^7$ is SSN for 68,000,000. Handy prefix forms for decimal multiples or submultiples are:

| | | |
|---|---|---|
| $10^1 \rightarrow$ deka, deca | $10^{12} \rightarrow$ tera | $10^{-9} \rightarrow$ nano |
| $10^2 \rightarrow$ hecto | $10^{-1} \rightarrow$ deci | $10^{-12} \rightarrow$ pico |
| $10^3 \rightarrow$ kilo | $10^{-2} \rightarrow$ centi | $10^{-15} \rightarrow$ femto |
| $10^6 \rightarrow$ mega | $10^{-3} \rightarrow$ milli | $10^{-18} \rightarrow$ atto |
| $10^9 \rightarrow$ giga | $10^{-6} \rightarrow$ micro | |

**problem 1.24** Express each of the following statements in SSN: (*a*) The angstrom is one ten-billionth of a meter. (*b*) The speed of light $c$ is approximately thirty gigacentimeters per second. (*c*) The wavelength $\lambda$ of yellow light is about 0.0000228 in. (*d*) One coulomb of charge contains about 6,250,000,000,000,000,000 electrons. (*e*) Rock and roll music is heard at 99X or 99 megahertz on the fM dial.

**solution** (*a*) $1 \text{ Å} = 1/(10 \times 10^9) = 1/10^{10} = 1 \times 10^{-10}$ m. (*b*) $c = 30 \times 10^9$ cm/s $= 3 \times 10^{10}$ cm/s $= 3 \times 10^{10} \times 10^{-2}$ m/s $= 3 \times 10^8$ m/s. (*c*) $\lambda \approx 2.28 \times 10^{-5}$ in; in nonmetric notation, this quantity can be expressed as $22.8 \times 10^{-6}$ in or 22.8 $\mu$in. (*d*) If $N_e$ represents the number of electrons, then $N_e \approx 6.25 \times 10^{18}$ electrons per coulomb. (*e*) $99 \times 10^6 = 9.9 \times 10^7$ Hz.

## 1.5 LOGARITHMS

There are two generally used logarithms: the *common* and the *natural* logarithm. The common logarithm of a number $N$ is the exponent or power to which the base number 10 must be raised to give that number. Let symbol $x$ be used for the exponent. Then

$$N = 10^x \tag{1-14}$$
$$\log_{10} N = \log N = x \tag{1-15}$$

The natural logarithm of a number $M$ is the exponent or power to which the base number $\epsilon$ must be raised to give that number. Let symbol $y$ be used for the exponent.

Then

$$M = \epsilon^y \tag{1-16}$$
$$\log_\epsilon M = \ln M = y \tag{1-17}$$

Base number $\epsilon$ is approximately equal to 2.72. The natural logarithm is also called the *naperian* logarithm.

The laws of logarithms, applicable to any base, are:

$$\log AB = \log A + \log B \tag{1-18}$$

$$\log \frac{C}{D} = \log C - \log D \tag{1-19}$$

$$\log F^P = P \cdot \log F \tag{1-20}$$

$$\log \sqrt[n]{G} = \frac{1}{n} \cdot \log G \tag{1-21}$$

$$\log H^{P/Q} = \frac{P}{Q} \cdot \log H \tag{1-22}$$

$$\log 1 = 0 \tag{1-23}$$

## Common or Briggsian Logarithms

If a number $N$ is written in SSN, then the power of 10 is its *characteristic*. The *mantissa* is found in the tables of logarithms. *The complete logarithm is the characteristic plus mantissa.* Characteristics for common logs never appear in tables. The quickest way to find a log of a number is by use of an electronic hand calculator.

**problem 1.25**    (*a*) Find the characteristic for $N = 240,000$. (*b*) Find the complete common log for $P = 8,230,000$. (*c*) Find the common log for $Q = 0.00036$. (*d*) If $\log R = 4.8825$, find $R$. Using logs, find (*e*) $S = (246)(583)$ and (*f*) $\sqrt{3.56} = T$.

**solution**    (*a*) Rewriting $N$ in SSN, $N = 2.4 \times 10^5$; thus the characteristic is 5. (*b*) In SSN, $P = 8.23 \times 10^6$; thus the characteristic is 6. The mantissa found in the table (Appendix) is 0.9154. The complete log is 6.9154. (*c*) Rewriting $Q$ in SSN, $Q = 3.6 \times 10^{-4}$; hence the characteristic is $-4$. The mantissa found in the table is 0.5563. In this case, the answer is given as $\log Q = 6.5563 - 10$. (*d*) $R = $ antilog $(4.8825)$; look for 0.8825 in the tables (row 76, col 3). Since the characteristic given was 4, there must be five places to the left of the decimal point in the original number $R$. Therefore $R \approx 76,300$. (*e*) $\log S = \log 246 + \log 583 = 2.3909 + 2.7657 = 5.1566$. Then $S = $ antilog $(5.1566) \approx 143,000$ excluding interpolation. (*f*) $\log T = \log (3.56)^{1/2} = \frac{1}{2} \log 3.56 = \frac{1}{2}(0.5514) = 0.2757$. Then $T = $ antilog $(0.2757) \approx 1.89$.

**problem 1.26**    Find the outside diameter of the inner conductor of an air-insulated coaxial transmission line if the line impedance is 50 $\Omega$ and the inside diameter of the outer conductor is 1.02 cm. It is known that the characteristic impedance $Z_o = 138 \cdot \log (d_o /d_i)$, where $d_o$ and $d_i$ are the outer and inner diameters of the line, respectively.

**solution**    Inserting values

$$50 = 138 \cdot \log 1.02/d_i$$

$$\log \frac{1.02}{d_i} = 0.3623$$

$$\frac{1.02}{d_i} = \text{antilog} (0.3623)$$

$$\frac{1.02}{d_i} = 2.303$$

Solving for $d_i$ yields:

$$d_i = 0.443 \text{ cm}$$

## The Exponential Number

Exponential number $\epsilon$, when used as a base number, follows the same rules of Eqs. (1-5) to (1-13). For example, $\epsilon^0 = 1$, $\epsilon^A \epsilon^B = \epsilon^{A+B}$. The exponential number is sometimes referred to as the naperian base number and as Euler's number. Euler's relationship connects exponentials to trigonometric functions:

$$\epsilon^{\pm j\theta} = \cos\theta \pm j\sin\theta = a + jb \tag{1-24}$$

where $a = \cos\theta$ and $b = \sin\theta$.

Base number $\epsilon$ can be expressed by the following series:

$$\epsilon = \frac{1}{0!} + \frac{1}{1!} + \frac{1}{2!} + \frac{1}{3!} + \frac{1}{4!} + \cdots \tag{1-25}$$

where $0! = 1$, $1! = 1$, $2! = 2 \times 1$, $3! = 3 \times 2 \times 1 = 6$, etc. A number such as $3!$ is called $3$ *factorial*.

### The Natural Logarithm

When using tables of logarithms, the complete natural logarithm is found therein. To convert from $\log N$ to $\ln M$, multiply $\log N$ by 2.303; to convert from $\ln M$ to $\log N$, multiply $\ln M$ by 0.434. Natural logarithms obey the laws specified in Eqs. (1-18) to (1-23).

## 1.6  FUNDAMENTAL TRIGONOMETRIC FUNCTIONS

An angle is generated when a line is rotated about a point. The initial position of the line is called the *reference ray position,* and the final position is called the *terminal ray position* of the line. Both line positions are called the *sides* of the angle, and the point of intersection is called the *vertex* (Fig. 1-5). Angles are measured in degrees, radians, and grads.

A triangle which has one right angle (an angle of 90°) is called a *right* triangle (Fig. 1-6). From the figure, the sum of the angles in a triangle equals 180°, or

$$\alpha + \beta + \gamma = 180° \tag{1-26}$$

Since $\gamma = 90°$,

$$\alpha + \beta = 90° \tag{1-27}$$

The pythagorean theorem states that the sum of the squares of both sides of the triangle equals the square of its hypotenuse. By reference to Fig. 1-6,

$$a^2 + b^2 = h^2 \tag{1-28}$$

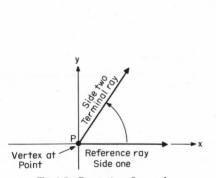

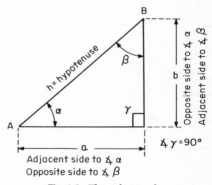

**Fig. 1-5**  Generation of an angle.            **Fig. 1-6**  The right triangle.

### Sine, Cosine, and Tangent

Referring to Fig. 1-6,

$$\sin\alpha = \frac{\text{opposite side}}{\text{hypotenuse}}$$

$$= \frac{b}{h} \tag{1-29}$$

$$\cos \alpha = \frac{\text{adjacent side}}{\text{hypotenuse}}$$

$$= \frac{a}{h} \tag{1-30}$$

$$\tan \alpha = \frac{\text{opposite side}}{\text{adjacent side}} \tag{1-31a}$$

$$= \frac{\sin \alpha}{\cos \alpha} \tag{1-31b}$$

## Other Circular Trigonometric Functions

Trigonometric functions of secondary importance are the secant (sec), cotangent (cot), and cosecant (csc):

$$\sec = \frac{\text{hypotenuse}}{\text{adjacent side}} \tag{1-32a}$$

$$\cot = \frac{\text{adjacent side}}{\text{opposite site}} \tag{1-32b}$$

$$\csc = \frac{\text{hypotenuse}}{\text{opposite side}} \tag{1-32c}$$

These functions are reciprocals of the three basic trigonometric functions:

$$\sec \alpha = \frac{1}{\cos \alpha} \tag{1-33a}$$

$$\cot \alpha = \frac{1}{\tan \alpha} \tag{1-33b}$$

$$\csc \alpha = \frac{1}{\sin \alpha} \tag{1-33c}$$

If angles $\alpha$ and $\beta$ are complementary angles, then the following equalities exist. By reference to Fig. 1-6,

$$\text{Sin } \alpha = \cos \beta \tag{1-34a}$$
$$\text{Tan } \alpha = \cot \beta \tag{1-34b}$$
$$\text{Sec } \alpha = \csc \beta \tag{1-34c}$$

## Inverse Trigonometric Functions

An inverse function exists when the roles of the independent and dependent variables are inverted or interchanged. A typical inverse statement would be "30° is an angle whose sine is ½"; the original direct statement was "sin 30° equals ½." The following statements are desired:

1. $\alpha$ equals an angle whose sine is $b/h$.
2. $\alpha$ equals an angle whose cosine is $a/h$.
3. $\alpha$ equals an angle whose tangent is $b/a$.

Mathematically a phrase such as "an angle whose sine is" can be abbreviated in two ways as *arcsin* or $sin^{-1}$. If the basic trigonometric functions are written as inverse functions, then,

$$\alpha = \arcsin \frac{b}{h} = \sin^{-1} \frac{b}{h} \tag{1-35a}$$

$$\alpha = \arccos \frac{a}{h} = \cos^{-1} \frac{a}{h} \tag{1-35b}$$

$$\alpha = \arctan \frac{b}{a} = \tan^{-1} \frac{b}{a} \tag{1-35c}$$

The secondary trigonometric functions also have inverse functions:

$$\alpha = \text{arcsec}\ \frac{h}{a} \tag{1-36a}$$

$$\alpha = \text{arccot}\ \frac{a}{b} \tag{1-36b}$$

$$\alpha = \text{arccsc}\ \frac{h}{b} \tag{1-36c}$$

Values of trigonometric functions are provided in the Appendix.

### Angles and Quadrants

Angles are positive if measured counterclockwise (ccw) from the reference ray and negative if measured clockwise (cw) from the reference ray. The positive $x$ direction (horizontal and to the right) is generally the reference ray. *Acute angles* are angles measuring between 0 and 90°; *right angles* measure exactly 90°; angles between 90 and 180° are referred to as *obtuse angles;* a *straight angle* measures exactly 180°; and a *reflex angle* measures greater than 180°.

One degree, written 1°, is an angle having its vertex at the center of the circle and subtending an arc equal to 1/360 of the circumference (Fig. 1-7a). There are two ways in which to subdivide the degree. One method employs the sexagesimal system whereby the degree is divided into 60 minutes and each minute is divided into 60 seconds. The other method employs the decimal system in which the degree is divided and subdivided into tenths, hundredths, thousandths, etc.

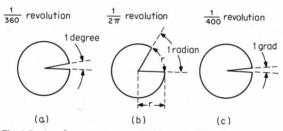

**Fig. 1-7**   Angular measurement. (*a*) Degrees. (*b*) Radians. (*c*) Grads.

Angles also can be measured by employing the *radian unit.* This method divides one complete revolution into $2\pi$ or approximately 6.28 equal parts (Fig. 1-7b). One radian is an angle which subtends an arc equal in length to the radius of the circle and which has its vertex at the center of the circle. One additional method of angular measurement uses a *grad* in which one revolution is divided into 400 *grad*ients (Fig. 1-7c). Conversions for the three systems of angular measurement are provided in Table 1.1.

**TABLE 1-1   Conversions for Angular Measurements**

| Degrees | Radians | Grads |
|---|---|---|
| 1° | $17.45 \times 10^{-3}$ | 1.11 |
| 57.296° or 57°17′45″ | 1 | 63.662 |
| 0.9° or 0°54′00″ | $15.71 \times 10^{-3}$ | 1 |

A circle can be divided into four parts called *quadrants* (quads). Quad I is between 0 and 90°; quad II is between 90 and 180°; quad III is between 180 and 270°; and quad IV is between 270 and 360°. Angles 0°, 90°, 180°, and 270° are also referred to as *quadrantal* angles. Signs of the trigonometric functions in all quadrants are shown in Fig. 1-8.

## Graphs of Primary Trigonometric Functions

**The Sine Waveform**    A graph of the sine function, $y = A \sin \theta$, is illustrated in Fig. 1-9a where $A = 1$.

**The Cosine Waveform**    A graph of the cosine function, $y = B \cos \theta$, where $B = 1$, is shown in Fig. 1-9b.

**The Tangent Waveform**    A graph of the tangent function, $y = D \tan \theta$, where $D = 1$, is shown in Fig. 1-9c.

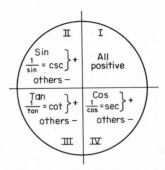

**Fig. 1-8**   Quadrants and signs of trigonometric functions.

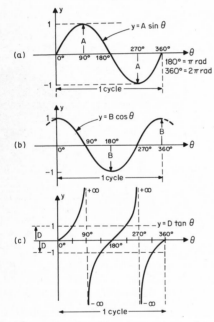

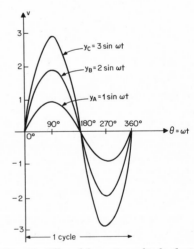

**Fig. 1-9**   Graphs of basic trigonometric functions. (a) Sine. (b) Cosine. (c) Tangent.

**Fig. 1-10**   Effect of change in amplitude of a sine wave.

**Amplitude, Frequency, and Phase**    Different amplitudes are shown in Fig. 1-10 in which $y_A = \sin \omega t$, $y_B = 2 \sin \omega t$, and $y_C = 3 \sin \omega t$ are graphed simultaneously. Figure 1-11 illustrates the effect of frequency changes (amplitude constant) by graphs of $v_A = V_{max} \sin \omega t$ and $v_B = V_{max} \sin 2\omega t$. In Fig. 1-12 the effect of a phase angle $\phi$, other than $0°$, is shown using $v = V_{max} \sin (\omega t + \phi°)$ as reference waveform.

A general sinusoidal equation (choosing the variable $v$) may be written as:

$$v = V_{max} \sin[\omega t - (\pm \phi)] \qquad (1\text{-}37)$$

where $V_{max}$ = peak amplitude
$\omega$ = angular frequency, rad/s
$t$ = time
$\phi$ = phase angle

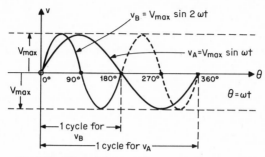

**Fig. 1-11**   Effect of change in frequency of a sine wave.

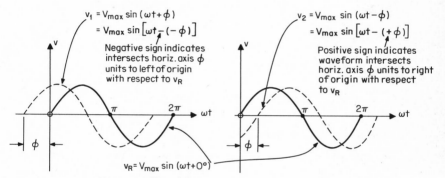

**Fig. 1-12**   Effect of change in phase angle of a sine wave.

Other useful expressions for sinusoidal waveforms are:

$$\theta = \omega t = 2\pi f \qquad (1\text{-}38a)$$

$$T = \frac{1}{f} = \frac{2\pi}{\omega} \qquad (1\text{-}38b)$$

$$\pi \text{ rad} = 180° \qquad (1\text{-}38c)$$

**problem 1.27**   (*a*) Calculate the radian frequency for a 60-Hz sine wave. (*b*) What angle $\theta$ is represented at a time of 10 ms?

**solution**

(*a*) $\omega = 2\pi f = 6.28 \times 60 = 377$ rad/s
(*b*) $\theta = \omega t = 377 \times 10 \times 10^{-3} = 377 \times 10 \times 10^{-3}$
$= 3.77$ rad $= 216°$

**problem 1.28**   Find the angle $\theta$ for a 400-Hz sine wave at $t = 2$ *ms* in (*a*) radians and (*b*) degrees.

**solution**   (*a*) $\theta = 2\pi f t = 6.28 \times 400 \times 2 \times 10^{-3} = 5.024$ rad (*b*) $\theta = 5.024$ rad $\times 180°/(\pi$ rad$) = 288°$.

**problem 1.29**   A current varies according to $i(t) = 3 \sin (400\pi t + \pi/4)$. Find (a) frequency $f$ and (b) the value of $i(t)$ at $t = 5$ ms.

**solution**

(a) $\omega = 2\pi f = 400\pi$; therefore, $f = 400\pi/2\pi = 200$ Hz.

(b) $i(5 \text{ ms}) = 3 \sin \left( 400\pi \times 5 \times 10^{-3} + \dfrac{\pi}{4} \right)$

$\qquad = 3 \sin \left( 2\pi + \dfrac{\pi}{4} \right) = 3 \sin \left( \dfrac{9\pi}{4} \right)$

$\qquad = 3 \sin (405°) = 3 \sin (45°) = 2.12$ A

**problem 1.30**   If the period of a cosinusoidal waveform is 800 $\mu$s, find its (a) frequency, (b) angular frequency, and (c) instantaneous value if the amplitude is 50 mA and phase angle is 0° at t = 66.7 $\mu$s.

**solution**   (a) $f = 1/T = 1/(800 \times 10^{-6}) = 1.25$ kHz. (b) $\omega = 6.28 \times 1.25 \times 10^3 = 7.85$ krad/s. (c) $i(t = 66.7 \ \mu s) = 50 \times 10^{-3} \cos (7.85 \times 10^3 \times 66.7 \times 10^{-6}) = 50 \times 10^{-3} \cos (0.523 \text{ rad}) = 0.05 \cos (30°) = 43.3$ mA.

**problem 1.31**   When a Lissajous pattern was viewed on an oscilloscope (Fig. 1-13), $a$ was measured to be 3 cm and $b$ to be 3.75 cm. Determine the value of the phase angle.

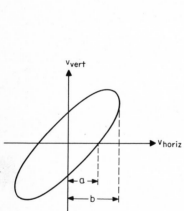

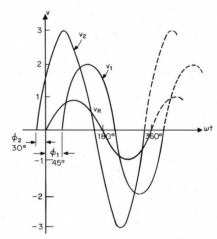

**Fig. 1-13**   Measurement pf phase angle from a Lissajous pattern (Problem 1.31).

**Fig. 1-14**   Solution to Problem 1.32 (see text).

**solution**   The phase angle $\phi$ is given by:

$$\phi = \sin^{-1} \frac{a}{b} \qquad (1\text{-}39)$$

Hence, $\phi = \sin^{-1} (3/3.75) = \sin^{-1}(0.8)$. From the trigonometric tables in the Appendix, $\phi = 53.1°$.

**problem 1.32**   Sketch $v_1 = 2 \sin (\omega t - \pi/4)$ and $v_2 = 3 \cos (\omega t - 60°)$. Use $v_1$ as the reference waveform.

**solution**   Inspection of $v_1$ indicates that the phase angle $\pi/4$ rad = 45° is to the right of the origin (Fig. 1-14). For a comparative sketch, it is preferable to express $v_2$ as a sine wave. A useful relationship between sine and cosine waves is:

$$\cos (\omega t + 90°) = \sin \omega t \qquad (1\text{-}40a)$$
$$\sin (\omega t - 90°) = \cos \omega t \qquad (1\text{-}40b)$$

By Eq. (1-40a), $v_2 = 3 \sin (\omega t + 30°) = 3 \sin [\omega t - (-30°)]$. Therefore, the phase angle is to the left of the origin, as shown in Fig. 1-14.

## 1.7   COMPLEX ALGEBRA

In a series circuit, the impedance **Z**, measured in ohms, is:

$$\mathbf{Z} = R \pm jX \tag{1-41}$$

where $R$ is the resistance and $X$ the reactance in the circuit. If $X$ is inductive reactance $(X_L = \omega L)$, the positive sign is used. If $X$ is capacitive reactance $[X_C = 1/(\omega C)]$, the negative sign is used.

In a parallel circuit, the admittance **Y**, measured in siemens (S), is:

$$\mathbf{Y} = G \pm jB \tag{1-42}$$

where $G$ is the conductance and $B$ is the *susceptance* in the circuit. If $B$ is inductive susceptance $[B_L = 1/(\omega L)]$, the negative sign is used. If $B$ is capacitive susceptance ($B_C = \omega C$), the positive sign is used.

An impedance or admittance triangle can be drawn if **Z** or **Y** is plotted on the complex plane. The vertical axis becomes the axis of imaginaries, and the horizontal axis the axis of reals.

Impedance and admittance are reciprocals of each other:

$$\mathbf{Z} = \frac{1}{\mathbf{Y}} \tag{1-43a}$$

$$\mathbf{Y} = \frac{1}{\mathbf{Z}} \tag{1-43b}$$

**problem 1.33**    Find the impedance of a series circuit of $R = 600\ \Omega$ and $X_L = 800\ \Omega$. Draw the impedance triangle of the circuit.

**solution**    The impedance triangle is drawn in Fig. 1-15. The magnitude of the impedance $|\mathbf{Z}|$ is:

$$|\mathbf{Z}| = \sqrt{600^2 + 800^2} = 1000\ \Omega$$

Angle $\theta$ is equal to $\tan^{-1}(X_L/R) = \tan^{-1}(800/600) = \tan^{-1}(1.33)$. From the trigonometric tables in the Appendix, $\theta = 53$.

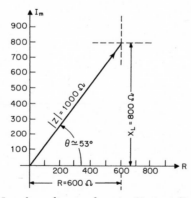

**Fig. 1-15**   Impedance diagram of a series $RL$ circuit (Problem 1.33).

**problem 1.34**    If the magnitude of impedance of an $RC$ series circuit is $1300\ \Omega$ and the reactive component is $1200\ \Omega$, find $R$.

**solution**    $Z^2 = R^2 + X^2$; hence, $R^2 = Z^2 - X^2$ or $R = \sqrt{Z^2 - X^2}$. By substitution of values in the last equation, $R = \sqrt{1300^2 - 1200^2} = 500\ \Omega$.

**problem 1.35**    The impedance across $a$-$b$ of a series $RL$ circuit (Fig. 1-16a) is $50 + j50\ \Omega$. Find the admittance across the terminals.

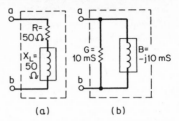

(a)                    (b)

**Fig. 1-16**   Solution to Problem 1.35 (see text).

**solution**   $Y_{ab} = 1/Z_{ab} = 1/(50 + j50)$. Multiplying the numerator and denominator by the complex conjugate of the denominator $(50 - j50)$ yields:

$$Y_{ab} = \frac{50 - j50}{2500 + 2500}$$
$$= 10 - j10 \text{ mS}$$

(See Fig. 1-16b.)

## Polar and Exponential Forms of Complex Numbers

Although one can multiply or divide a complex number, as well as take a root or raise it to a power, in rectangular form (for example, $Z = R + jX$), it is simpler to perform these operations when the complex number is expressed in *polar* or *exponential* form. In polar form, $Z$ is expressed as:

$$Z = |Z|\underline{/\theta°}$$

where $|Z|$ is the magnitude and $\theta$ is equal to $\tan^{-1}(X/R)$. The value of $R$ is:

$$R = |Z| \cos \theta \qquad (1\text{-}44a)$$

and the value of $X$ is:

$$X = |Z| \sin \theta \qquad (1\text{-}44b)$$

In exponential form, $Z$ is expressed as:

$$Z = |Z|e^{j\theta}$$

In multiplication of complex numbers expressed in polar (or exponential) forms, their magnitudes are multiplied and their angles are added algebraically. For division, their magnitudes are divided and their angles are subtracted algebraically.

**problem 1.36**   (a) Multiply $10\underline{/30°}$ by $2\underline{/60°}$. (b) Divide $10\underline{/30°}$ by $2\underline{/60°}$. Express your answer in polar and exponential forms.

**solution**   (a) $10\underline{/30°} \times 2\underline{/60°} = (10 \times 2)\underline{/30° + 60°} = 20\underline{/90°} = 20e^{j90°}$
(b) $10\underline{/30°}/2\underline{/60°} = (10/2)\underline{/30° - 60°} = 5\underline{/-30°} = 5e^{-j30°}$

## Phasor Transformation

A sinusoidal function $x$ may be expressed in exponential or polar form as:

$$x = Ae^{j\phi} \qquad (1\text{-}45a)$$
$$= A\underline{/\phi} \qquad (1\text{-}45b)$$

where $A$ is the amplitude and $\phi$ the phase angle of $x$. For currents and voltages, amplitudes $V$ or $I$ are generally expressed by their effective (rms) values. The effective value is equal to the maximum amplitude multiplied by 0.707 or divided by $\sqrt{2} \cong 1.414$. Voltage and current expressed by Eq. (1-45) are called *phasors*.

In the *inverse* phasor expansion, the phasor is converted to an expression which is a function of time. For example, if $V = 70.7e^{j30°}$ and $\omega$ is 1000 rad/s, then $v(t) = \sqrt{2} \times 70.7 \sin (1000t + 30°) = 100 \sin (1000t + 30°)$.

A useful relationship is Euler's equation:

$$A\underline{/\phi°} = Ae^{j\phi°} = A (\cos \phi° + j \sin \phi°) \qquad (1\text{-}46a)$$
$$A\underline{/-\phi°} = Ae^{-j\phi°} = A (\cos \phi° - j \sin \phi°) \qquad (1\text{-}46b)$$

**problem 1.37** Express $v(t) = 28.28 \sin (1000t + 45°)$ as a phasor.
**solution** $\mathbf{V} = (0.707 \times 28.28)e^{j45°} = 20e^{j45°} = 20\underline{/45°}$

**problem 1.38** The voltage across two impedances in series is $v_A + v_B = 75 \sin (377t + 45°) + 25 \sin (377t + 60°)$. Express the total voltage as a phasor.

**solution**

$$\mathbf{V}_A = (0.707 \times 75)\underline{/45°} = 53\underline{/45°} \text{ V}$$
$$\mathbf{V}_B = (0.707 \times 25)\underline{/60°} = 17.7\underline{/60°} \text{ V}$$

Because $v_A$ and $\mathbf{V}_B$ have the *same* frequency, their phasors can be added. (Addition of phasors of different frequencies is not permitted.)

Applying Euler's equation yields:

$$53 (\cos 45° + j \sin 45°) = 37.5 + j37.5$$
$$17.7 (\cos 60° + j \sin 60°) = 8.85 + j15.3$$

Hence, $\mathbf{V}_A + \mathbf{V}_B = (37.5 + 8.85) + j(37.5 + 15.3) = 46.35 + j52.8$. The magnitude of $\mathbf{V}_A + \mathbf{V}_B = \sqrt{46.35^2 + 52.8^2} = 70.2$ V. Phase angle $\phi$ is $\tan^{-1} (52.8/46.35) = 48.72°$. Hence, $\mathbf{V}_A + \mathbf{V}_B = 70.2 \underline{/48.72°} = 70.2e^{j48.72°}$ V.

## 1.8 BIBLIOGRAPHY

Barnett, Raymond: *College Algebra with Trigonometry,* McGraw-Hill, New York, 1974.

Boylestad, Robert: *Introductory Circuit Analysis,* 3d ed., Charles E. Merrill, Columbus, Ohio, 1977.

Cameron, E.: *Algebra and Trigonometry,* rev. ed., Holt, New York, 1965.

Chew, A. Little, Richard, and Little, Sherry: *Technical Mathematics,* Houghton Mifflin, Boston, 1976.

Drooyan and Wooton: *Elementary Algebra for College Students,* 3d ed., Wiley, New York, 1972.

Joerg, A. E.: Class notes on *College Algebra, Algebra and Trigonometry, and Computational Skills,* Westchester Community College, 1973–1976.

Joerg, E. A.: Class notes on *Survey of Mathematics, Algebra and Trigonometry, Electronic Computations I & II, Passive and Active Circuit Analysis, and Electronics,* Westchester Community College, Valhalla, N.Y., 1965–1976.

Millman, Jacob, and Halkias, Christopher: *Electronic Devices and Circuits,* McGraw-Hill, New York, 1967.

Mustafa, Munem, Tschirtart, W. and Yizze, James: *College Algebra,* Worth Publishers, Inc., New York, 1974.

Oppenheimer, Samuel: *Semiconductor Logic and Switching Circuits,* 2d ed. Charles E. Merrill Books, Inc., Columbus, Ohio, 1973.

Person, Russel V.: *Essentials of Mathematics,* 2d ed., Wiley, New York, 1968.

Pierce, F., and Paulus, T. J.: *Applied Electronics,* Charles E. Merrill Books, Inc., Columbus Ohio, 1972.

Rees, Paul, Sparks, Fred, and Rees, Chas.: *Algebra and Trigonometry,* 3d ed., McGraw-Hill, New York, 1975.

Thomson, C.: *Mathematics for Electronics,* Prentice-Hall, Englewood Cliffs, N.J., 1976.

Westlake, John, and Noden, G.: *Applied Mathematics for Electronics,* Prentice-Hall, Englewood Cliffs, N.J., 1968.

Zurflieh, Thomas: *Basic Technical Mathematics Explained,* McGraw-Hill, New York, 1974.

# Chapter 2

# DC Circuit Analysis

## ERNEST A. JOERG

Professor and Chairperson, Electrical Technology Department,
Westchester Community College, Valhalla, N.Y.

## 2.1 INTRODUCTION

The electron possesses the smallest amount of negative charge; it is equal to $-1.602 \times 10^{-19}$ coulomb. The unit of charge in the International System of Units (SI) is the *coulomb* (C). One coulomb of charge represents approximately $6.24 \times 10^{18}$ electrons.

### Current

In a given conductor (Fig. 2-1), electrons passing by a given point $x$ result in a flow of *charge Q*, measured in coulombs. For a given period of time that has elapsed during the migration of electrons, current $i$ is defined as the rate of change of charge with respect to time:

$$i = \frac{\Delta Q}{\Delta t} \tag{2-1}$$

where $\Delta$ (delta) here means "a change in."

The unit of current is the *ampere* ($A$). As noted in Fig. 2-1, conventional current flow is opposite in direction to electron flow. Conventional flow is adopted in this chapter and most of the succeeding chapters in the Handbook. If the flow of charge is unchanging and unidirectional, then it is classified as *direct current*, or *dc*. It is given the symbol $I$.

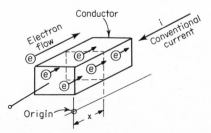

**Fig. 2-1**   Electron and conventional current flow in a conductor.

## Voltage

Energy is often defined as the capacity to do work. The SI unit for work is the *joule* (J), and the letter symbol is W. If 1 joule of energy is required to move 1 coulomb of charge from one point in a conductor to another, then a *potential difference*, or *voltage*, of 1 volt exists between the two points. The SI unit is the *volt*, and the letter symbol for voltage is V:

$$V_{xy} = \frac{\Delta W}{\Delta Q} \tag{2-2}$$

where $V_{xy} = V_x - V_y$

## Power

Power (P) is defined as the rate of change of energy or work with respect to time:

$$P = \frac{\Delta W}{\Delta t} \tag{2-3}$$

It is measured in *watts* (W) (joules/second). In terms of voltage and current, power is equal to their product:

$$P = V I \tag{2-4}$$

**problem 2.1**   If $0.784 \times 10^{18}$ electrons pass through a wire at a given point in 643 ms, what is the current?

**solution**   By Eq. (2-1),

$$I = \frac{[(0.784 \times 10^{18} \text{ electrons})(1.6 \times 10^{-19} \text{ C/electron})]}{643 \times 10^{-3} \text{ s}}$$

$$= 1.95 \times 10^{-1} \text{ A} = 195 \text{ mA}.$$

**problem 2.2**   If the potential difference between two points is 36 V, how much work is required to bring 4 $\mu$C of charge from one point to the other?

**solution**   By Eq. (2-2), $W = (36 \text{ V}) \times (4 \times 10^{-6} \text{ C}) = 1.44 \times 10^{-4}$ J.

## 2.2   RESISTANCE

When charge flows through a given material, it experiences an opposition to the flow. This opposition is called the *resistance* of the material. Resistance depends on the length, cross-sectional area, type, and operating temperature of the material. For a constant temperature, the resistance of a material is:

$$R = \frac{\rho l}{A} \tag{2-5}$$

where $R$ is the resistance (SI unit is the ohm, $\Omega$), $\rho$ is the *resistivity* of the material in ohm-meters ($\Omega \cdot$ m) or ohm-centimeters ($\Omega \cdot$ cm), $l$ is the length in meters or centimeters, and $A$ is the cross-sectional area in square meters or square centimeters. Resistance of a material is directly proportional to its length and inversely proportional to its cross-sectional area.

**problem 2.3**   An intrinsic germanium slab (Fig. 2-2) is used as a semiconductor resistor at 20°C. Find its resistance.

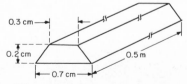

**Fig. 2-2**   Calculating the resistance of a semiconductor resistor (Problem 2.3).

**solution**   Form Table 2-1, $\rho = 0.45\ \Omega \cdot$ m. For a trapezoidal cross section, area $A = (\frac{1}{2})h(b_1 + b_2)$ where $h = 0.2$ cm, $b_1 = 0.3$ cm, and $b_2 = 0.7$ cm. By Eq. (2-5),

$$R = \frac{0.45 \times 0.5}{\frac{1}{2} \times 0.002(0.003 + 0.007)} = 22.5\ k\Omega$$

**TABLE 2-1   Typical Resistivities of Materials at 20°**

| Material | Resistivity $\rho,\ \Omega \cdot$ m* | Material | Resistivity $\rho,\ \Omega \cdot$ m* |
|---|---|---|---|
| Aluminum | $2.83 \times 10^{-8}$ | Graphite | $8 \times 10^{-6}$ |
| Antimony | $4.17 \times 10^{-7}$ | Iron | $9.8 \times 10^{-7}$ |
| Bakelite plastic resins | $1 \times 10^{+10}$ | Lead | $2.2 \times 10^{-7}$ |
| and compounds | | Manganin alloy | $4.4 \times 10^{-7}$ |
| Brass | $7 \times 10^{-8}$ | Mercury | $9.6 \times 10^{-7}$ |
| Carbon | $3.5 \times 10^{-5}$ | Mica | $1 \times 10^{+14}$ |
| Constantan alloy | $4.9 \times 10^{-7}$ | Nichrome | $1 \times 10^{-6}$ |
| Copper | $1.7 \times 10^{-8}$ | Rubber (hard) | $1 \times 10^{+16}$ |
| German silver | $3.3 \times 10^{-7}$ | Silicon | $1.3 \times 10^{+3}$ |
| Germanium | 0.45 | Silver | $1.6 \times 10^{-8}$ |
| Glass | $1 \times 10^{+12}$ | Steel (4% Si) | $5 \times 10^{-7}$ |
| Gold | $2.4 \times 10^{-8}$ | Tungsten | $5.2 \times 10^{-8}$ |

*Values may vary with purity of material.

## Conductance

Conductance is the inverse (reciprocal) of resistance and is designated by the letter $G$:

$$G = \frac{1}{R} = R^{-1} \tag{2-6}$$

Conductance indicates the ability of a material to conduct electricity. The unit is the *siemen* (formerly mho), abbreviated S. By letting $\sigma$ equal $1/\rho$, Eq. (2-5) may be expressed by:

$$R = \frac{l}{\sigma A} \tag{2-7}$$

where $\sigma$ is the *conductivity* of the material. Its unit is siemens/meter (S/m). Since $G = 1/R$, Eq. (2-7) may be written as:

$$G = \frac{\sigma A}{l} \tag{2-8}$$

**problem 2.4**   Find the conductance of nichrome wire whose resistance is 721.8 $\Omega$.

**solution**

$$G = 1/721.8 = 1.385 \times 10^{-3}\ S = 1.385\ mS.$$

**problem 2.5**   Given a gold bullion 5 cm on the square and 0.75 meter long. Specify its conductivity and conductance.

**solution**   From Table 2-1, $\sigma = 1/(2.4 \times 10^{-8}) = 4.167 \times 10^7$ S/m. By Eq. (2-8), $G = [4.167 \times 10^7 \times (5 \times 10^{-2})^2]/0.8 = 130.22$ kS.

## SI Unit for Temperature

The kelvin (K) is the SI unit for temperature. Expressing temperature in degrees Celsius (°C), however, is permissible because there is a 100-degree differential between the boiling and freezing points of water on both scales. Thus a 1° change on the Celsius scale is equivalent to a 1° change on the Kelvin scale. Conversion formulas for the kelvin, Celsius, Fahrenheit, and Rankine scales are given in Table 2-2. The Fahrenheit and Rankine scales have a 180-degree differential between freezing and boiling points; they are, however, not used in the International System of Units.

**TABLE 2-2    Temperature System Conversion Chart**

| Given ↓ | Desired temperature system | | | |
| | °C (Celsius) | K (kelvin) | °F (Fahrenheit) | °R (Rankine) |
| --- | --- | --- | --- | --- |
| °C | 1 | +273.15 | ($\frac{9}{5}$)°C + 32 | ($\frac{9}{5}$)°C + 491.67 |
| K | −273.15 | 1 | ($\frac{9}{5}$)K − 459.67 | ($\frac{9}{5}$)K |
| °F | ($\frac{5}{9}$)(°F-32) | ($\frac{5}{9}$)°F + 255.38 | 1 | +459.67 |
| °R | ($\frac{5}{9}$)°R-273.15 | ($\frac{5}{9}$)°R | −459.67 | 1 |

## 2.3  OHM'S LAW

Under fixed environmental conditions, Ohm's law states that the voltage $V$ across resistance $R$ is directly proportional to the current $I$ flowing through it:

$$V = R I \tag{2-9}$$

where $V$ is in volts, $R$ in ohms, and $I$ in amperes. If one divides both sides of Eq. (2-9) by $R$, then

$$I = \frac{V}{R} \tag{2-10a}$$

Because $1/R = G$,

$$I = V G \tag{2-10b}$$

Dividing Eq. (2-9) by $I$ yields:

$$R = \frac{V}{I} \tag{2-11a}$$

or

$$G = \frac{I}{V} \tag{2-11b}$$

### Power Dissipation

By Eq. (2-4), $P = V I$. If Eq. (2-10) is substituted in Eq. (2-4), the dissipated power in a resistor is found in terms of current $I$:

$$P = I^2R \tag{2-12a}$$
$$= \frac{I^2}{G} \tag{2-12b}$$

Similarly, in terms of voltage $V$:

$$P = \frac{V^2}{R} \tag{2-13a}$$
$$= V^2G \tag{2-13b}$$

**problem 2.6**  A television set has a connecting cable to its outdoor antenna whose length is 25 m and linear resistance is 0.11 Ω/m. Determine the resistance and conductance of the cable.

**solution**

$$R = 0.11 \, \Omega/m \times 25 \, m = 2.75 \, \Omega$$
$$G = \frac{1}{2.75} = 0.364 \, S$$

**problem 2.7**  Referring to Fig. 2-3, find $R_L$, $G_L$, and $P$.

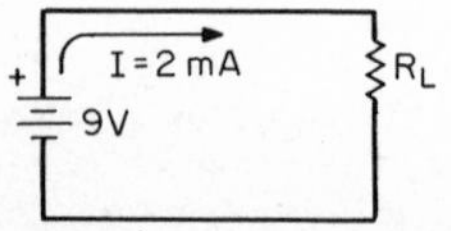

**Fig. 2-3**  Calculating the values of $R_I$, $G_I$, and $P$ in a simple dc circuit (Problem 2.7).

**solution**

$$R_L = \frac{V}{I} = \frac{9}{2 \times 10^{-3}} = 4.5 \text{ k}\Omega$$

$$G_L = \frac{1}{R_L} = 0.222 \text{ mS}$$

$$P = I\,V = 2 \times 10^{-3} \times 9 = 18 \text{ mW}$$

## 2.4  EQUIVALENT RESISTANCE

### Series Connection

If two or more elements in a circuit are connected so that each element has the same current flowing through it, they are in *series*. In Fig. 2-4, the elements are connected end to end in a chain fashion. This configuration presents only one path to the current flowing through the circuit.

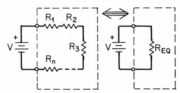

**Fig. 2-4**  Determining the equivalent resistance of a series $R$ circuit.

**Equivalent Series Resistance**  In a series circuit, the total or equivalent resistance $R_{EQ}$ seen by the source is equal to the sum of the values of the individual resistors:

$$R_{EQ} = R_1 + R_2 + R_3 + \cdots + R_n \qquad (2\text{-}14)$$

### Parallel Connection

If elements are connected such that the same voltage is across all of them, the elements are said to be connected in parallel (Fig. 2-5).

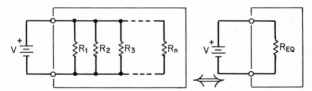

**Fig. 2-5**  Determining the equivalent resistance of a parallel $R$ circuit.

**Equivalent Parallel Resistance**  In a parallel circuit, the equivalent resistance is equal to the inverse of the sum of the reciprocals of the individual resistance values:

$$R_{EQ} = \frac{1}{1/R_1 + 1/R_2 + 1/R_3 + \cdots + 1/R_n} \qquad (2\text{-}15)$$

For the case of two resistors $R_A$ and $R_B$, then

$$R_{EQ} = R_A \,\|\, R_B = \frac{R_A R_B}{R_A + R_B} \qquad (2\text{-}16)$$

where $\|$ stands for "in parallel with." The equivalent parallel resistance is always less than the smallest valued resistor in the parallel combination.

### Joerg Equivalence Corollary

If a resistance $R$ is in parallel with another resistance of value $1/n$ of $R$, then the equivalent resistance is:

$$R_{EQ} = \frac{R}{1 + n} \qquad (2\text{-}17)$$

## Equivalent Conductance

Conductances in parallel combine in the same manner as resistances in series, and conductances in series combine in the same manner as resistances in parallel. The Joerg corollary applies to a conductance $G$ in series with another conductance whose value is 1/ $n$ of $G$.

**problem 2.8**    Find the equivalent resistance and conductance in Fig. 2-6.

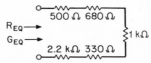

**Fig. 2-6**    Calculating the equivalent resistance and conductance of a series $R$ circuit (Problem 2.8).

**solution**    Because the resistors are in series, by Eq. (2-14),

$$R_{EQ} = 0.5 + 0.68 + 1 + 0.33 + 2.2 = 4.71 \text{ k}\Omega$$
$$G_{EQ} = \frac{1}{R_{EQ}} = \frac{1}{4.71} = 0.212 \text{ mS}$$

**problem 2.9**    Find the equivalent conductance and resistance in Fig. 2-7.

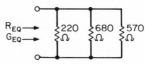

**Fig. 2-7**    Calculating the equivalent conductance and resistance of a parallel $R$ circuit (Problem 2.9).

**solution**    Because the resistors are in parallel, by Eq. (2-15),

$$G_{EQ} = 1/220 + 1/680 + 1/570 = 7.8 \text{ mS}$$
$$R_{EQ} = 1/(7.8 \times 10^{-3}) = 128.2 \ \Omega$$

**problem 2.10**    Referring to Fig. 2-8, find the equivalent resistance in ($a$) and the equivalent conductance in ($b$).

(a)          (b)

**Fig. 2-8**    Finding the equivalent resistance in ($a$) and the equivalent conductance in ($b$) (Problem 2.10).

**solution**    In Fig. 2-8$a$, the 4.5-k$\Omega$ resistance is ½ of the 9-k$\Omega$ resistance and they are in parallel. By Eq. (2-17), $R_{EQ} = 9/(1 + 2) = 3 \text{ k}\Omega$.
   In Fig. 2-8$b$, the conductances are in series. Hence, $G_{EQ} = 12/(1 + 3) = 3 \text{ mS}$. (The 4-mS conductance is one-third of the 12-mS conductance.)

## Series-Parallel Connection

For this case, the approach is:
   1. Label all junctions (nodes).
   2. Isolate the series sections and isolate the parallel sections.
   3. Simplify each isolated section using Eq. (2-14), (2-15), (2-16), or (2-17).
   4. Insert the equivalent elements between the appropriate nodes.
   5. Reduce to completion.

**problem 2.11**   Find $R_{EQ}$ and $G_{EQ}$ in Fig. 2-9.

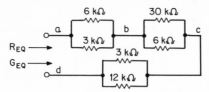

**Fig. 2-9**   Finding the equivalent resistance and conductance of a series-parallel circuit (Problem 2.11).

**solution**   $R_{EQ} = R_{ad} = R_{ab} + R_{bc} + R_{cd}$. All sections are in parallel, and each section is in series. By Eq. (2-17), $R_{ab} = 2$ kΩ, $R_{bc} = 5$ kΩ, and $R_{cd} = 2.4$ kΩ. Then $R_{EQ} = 2 + 5 + 2.4 = 9.4$ kΩ and $G_{EQ}$ = 1/9.4 = 0.109 mS.

**problem 2.12**   Find $R_{EQ}$ and $G_{EQ}$ in Fig. 2-10.

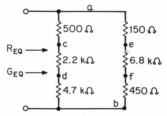

**Fig. 2-10**   Finding the equivalent resistance and conductance of a parallel-series circuit (Problem 2.12).

**solution**   It is noted that $R_{acdb}|R_{aefb}$ and each section is in series. Then, $R_{acdb} = R_{ac} + R_{cd} + R_{db}$ = 500 + 2200 + 4700 = 7.4 kΩ; $R_{aefb} = R_{ae} + R_{ef} + R_{fb}$ = 150 + 6800 + 450 = 7.4 kΩ. $R_{EQ}$ = 7.4|7.4 = 3.7 kΩ, and $G_{EQ}$ = 0.27 mS.

**problem 2.13**   Rind $R_{EQ}$ and $G_{EQ}$ in Fig. 2-11.

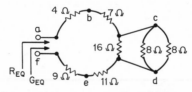

**Fig. 2-11**   Calculating the equivalent resistance and conductance of a series-parallel circuit (Problem 2.13).

**solution**   $R_{cd} = 16 | 8 | 8 = 3.2$ Ω. Inserting 3.2 Ω between nodes $c$ and $d$, $R_{EQ} = R_{af} = 4 + 7$ + 3.2 + 11 + 9 = 34.2 Ω; $G_{EQ}$ = 1/34.2 = 29.24 mS.

## Delta-to-Wye (Δ → Y) and Wye-to-Delta (Y → Δ) Conversions

The $\pi \to T$ and $T \to \pi$ conversions are alternate names. By reference to Fig. 2-12, the conversion formulas are:

$$R_1 = \frac{R_a R_c}{R_a + R_b + R_c} \tag{2-18a}$$

$$R_2 = \frac{R_b R_c}{R_a + R_b + R_c} \tag{2-18b}$$

$$R_3 = \frac{R_b R_a}{R_a + R_b + R_c} \tag{2-18c}$$

and

$$R_a = \frac{R_1 R_2 + R_1 R_3 + R_2 R_3}{R_2} \qquad (2\text{-}19a)$$

$$R_b = \frac{R_1 R_2 + R_1 R_3 + R_2 R_3}{R_1} \qquad (2\text{-}19b)$$

$$R_c = \frac{R_1 R_2 + R_1 R_3 + R_2 R_3}{R_3} \qquad (2\text{-}19c)$$

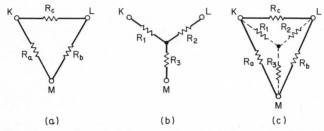

(a)        (b)        (c)

**Fig. 2-12** Delta-to-wye and wye-to-delta conversions. (a) Delta circuit. (b) Wye circuit. (c) Conversions.

**problem 2.14**  Find $R_{EQ}$ and $G_{EQ}$ for the bridge circuit of Fig. 2-13a.

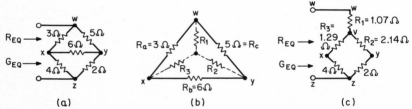

(a)        (b)        (c)

**Fig. 2-13**  Calculating the equivalent resistance and conductance of a bridge circuit. (a) Given circuit. (b) Converting the upper delta configuration to a wye circuit. (c) Simplified circuit (Problem 2.14).

**solution**  The upper $\Delta$ circuit across nodes $wxy$ is converted to a $Y$ circuit. By Eq. 2-18, $R_1 = (3 \times 5)/14 = 1.07$ kΩ, $R_2 = (6 \times 5)/14 = 2.14$ kΩ, and $R_3 = (3 \times 6)/14 = 1.29$ kΩ (Fig. 2-13b). If the internal node of the $Y$ circuit is called $v$ and the equivalent $Y$ is inserted across nodes $wxy$ (Fig. 2.13c), then $R_{vx}$ is in series with $R_{xz}$ and $R_{vy}$ is in series with $R_{yz}$. Hence, $R_{vxz} = 5.29$ kΩ and $R_{vyz} = 4.14$ kΩ. Also, $R_{vz} = R_{vxz} | R_{vyz} = 5.29 | 4.14 = 2.32$ kΩ. Thus, $R_{EQ} = R_1 + R_{vz} = 1.07 + 2.32 = 3.39$ kΩ and $G_{EQ} = 1/R_{EQ} = 0.295$ mS.

## 2.5  KIRCHHOFF'S LAWS

### Kirchhoff's Current Law (KCL)

At any node or junction of a circuit, the algebraic sum of the entering current $I_{ent}$ is zero:

$$\Sigma I_{ent} = 0 \qquad (2\text{-}20a)$$

The exiting currents $I_{exit}$ are treated as negative entering currents. An alternate form of KCL is:

$$\Sigma I_{ent} = \Sigma I_{exit} \qquad (2\text{-}20b)$$

### Kirchhoff's Voltage Law (KVL)

Around any closed pathway in either a clockwise or counterclockwise direction, the algebraic sum of the voltage drops $V_{drop}$ is zero:

$$\Sigma V_{drop} = 0 \qquad (2\text{-}21a)$$

A drop occurs when voltage passes through an element from + to −; a rise occurs when voltage passes from − to +. The voltage rise $V_{rise}$ is treated as a negative drop in Eq. (2-21$a$). An alternate form of KVL is:

$$\Sigma V_{drop} = \Sigma V_{rise} \qquad (2\text{-}21b)$$

**problem 2.15**    Find the value of $I_C$ in Fig. 2-14.

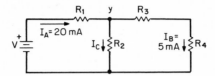

**Fig. 2-14**    Applying KCL to determine $I_C$ (Problem 2.15).

**solution**    From the figure, $I_A = 20$ mA and $I_B = 5$ mA are given. Current $I_B$ also flows in $R_3$ because $R_3$ is in series with $R_4$. Applying KCL at node $y$, 20 mA − 5 mA − $I_C$ = 0; therefore, $I_C$ = 15 mA downward, as indicated in the figure.

**problem 2.16**    Find the value of $R$ in Fig. 2-15.

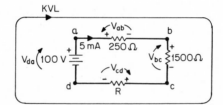

**Fig. 2-15**    Applying KVL to determine $R$ (Problem 2.16).

**solution**    If KVL is applied to the circuit, $-100 + V_{ab} + V_{bc} + V_{cd} = 0$; thus, $V_{cd} = 100 - V_{ab} - V_{bc}$. Also, $V_{ab} = 250 \times 5 \times 10^{-3} = 1.25$ V, and $V_{bc} = 1500 \times 5 \times 10^{-3} = 7.5$ V. Therefore, $V_{cd} = 100 - 1.25 - 7.5 = 91.25$ V, and $R = 91.25/(5 \times 10^{-3}) = 18.25$ kΩ.

## 2.6   VOLTAGE AND CURRENT DIVIDERS

### Voltage-Divider Principle

It is often desired to know the value of the voltage across an individual resistor in a series circuit (Fig. 2-16). Voltage $V_n$ across any resistor $R_n$ is equal to the product of the applied voltage $V_{tot}$ and $R_n$ divided by the sum of the resistances $R_{tot}$, in the series circuit:

$$V_n = V_{tot} \frac{R_n}{R_{tot}} \qquad (2\text{-}22)$$

From Eq. (2-22) it is seen that $V_n$ is directly proportional to $R_n$. The voltage-divider principle applies only to a circuit, such as the series circuit, where the current is the *same* in each element.

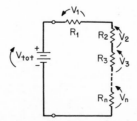

**Fig. 2-16**    Voltage-divider principle (see text).

## Current-Divider Principle

In a circuit containing $n$ parallel branches (Fig. 2-17), it is often desired to know the value of a current in some particular branch. The current $I_n$ in a particular branch $R_n$ is equal to the product of the applied current $I_{tot}$ and the equivalent resistance $R_{EQ}$ of the parallel circuit divided by $R_n$:

$$I_n = I_{tot} \frac{R_{EQ}}{R_n} \tag{2-23}$$

The current-divider principle applies only to a circuit, such as the parallel circuit, where the voltage across each element is the *same*.

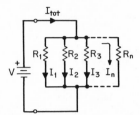

**Fig. 2-17**   Current-divider principle (see text).

Where there are two resistors $R_A$ and $R_B$ in parallel, the current $I_A$ in $R_A$ is:

$$I_A = I_{tot} \frac{R_B}{R_A + R_B} \tag{2-24a}$$

The current $I_B$ in $R_B$ is:

$$I_B = I_{tot} \frac{R_A}{R_A + R_B} \tag{2-24b}$$

The branch current is *inversely proportional* to the resistance of the branch. The branch containing the larger resistance receives the smaller current, and vice versa.

**problem 2.17**   Using voltage division, find the voltages shown in Fig. 2-18.

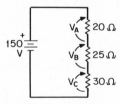

**Fig. 2-18**   Employing voltage division to find $V_A$, $V_B$, and $V_C$ (Problem 2.17).

**solution**   By Eq. (2-22), $V_A = 150 \times 20/(20 + 25 + 30) = 150 \times 20/75 = 40$ V; $V_B = 150 \times 25/75$ = 50 V; $V_C = 150 \times 30/75 = 60$ V.

**problem 2.18**   Using voltage division, find $V_0$ in Fig. 2-19.

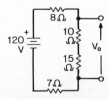

**Fig. 2-19**   Employing voltage division to find $V_o$ (Problem 2.18).

**solution**    By Eq. (2-22), $V_0 = 120 \times (10 + 15)/(8 + 10 + 15 + 7)$ 75 V.

**problem 2.19**    Using current division, find the currents indicated in Fig. 2-20.

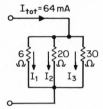

**Fig. 2-20**    Employing current division to find $I_1$, $I_2$, and $I_3$ (Problem 2.19).

**solution**    $R_{EQ} = 6|20|30 = 4 \, \Omega$. By Eq. (2-23), $I_1 = 64 \text{ mA} \times \frac{4}{6} = 42.67 \text{ mA}$; $I_2 = 64 \text{ mA} \times \frac{4}{20} = 12.8 \text{ mA}$; $I_3 = 64 \text{ mA} \times \frac{4}{30} = 8.53 \text{ mA}$.

## 2.7    MESH ANALYSIS

A *mesh* is a closed pathway with no other closed pathways within it. A *loop* is also a closed pathway but may have other closed pathways within it. Therefore, all meshes are loops; however, all loops are not meshes. If the schematic of a circuit is neatly drawn, one can view the circuit as a "window frame" with the meshes as the "windows."

To solve a circuit using mesh analysis:

1. Assume mesh currents in any direction.

2. Insert voltage sense arrows across each element due to the mesh current. The voltage senses are fixed across voltage sources. If an element has more than one sensing arrow owing to two mesh currents, label each separately within the respective mesh.

3. Write KVL around each mesh in any direction. It is convenient to employ the same direction as that of the mesh current.

4. Rewrite the resulting equations in block form.

**problem 2.20**    Write the mesh equations for the circuit of Fig. 2-21a.

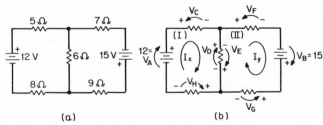

**Fig. 2-21**    Mesh analysis of a circuit. (a) Given circuit. (b) Assumed directions of currents and voltages across elements (Problem 2.20).

**solution**

1. In Fig. 2-21b, mesh currents $I_x$ and $I_y$ are assumed to flow clockwise.

2. Voltage sense arrows $V_A$ through $V_H$ are indicated in Fig. 2-21b.

3. Applying KVL yields for mesh I:

$$12 + V_E = V_C + V_D + V_H$$

or

$$12 + 6I_y = (5 + 6 + 8)I_x$$

For mesh II:

$$15 + V_D = V_F + V_G + V_E$$

or

$$15 + 6I_x = (7 + 9 + 6)I_y$$

4. Rewriting the equations in block form:

$$19I_x - 6I_y = 12$$
$$-6I_x + 22I_y = 15$$

## Joerg Mesh Equation*

The Joerg mesh equation (JME) is a general equation that enables one to write mesh equations by inspection. For dc circuits:

$$\Sigma V_{\text{rises}} \text{ (due to voltage sources)} = \Sigma(\text{self-resistance voltage drops})$$
$$\pm \Sigma(\text{common resistance voltage drops}) \quad (2\text{-}25)$$

The Joerg mesh equation calls for the algebraic sum of voltage rises. If a voltage drop owing to a voltage source is encountered, it is treated as a negative rise. The plus sign preceding the common-resistance voltage drops is used if both mesh currents in a particular element are in the same direction. A minus sign is used if the currents are in the opposite direction.

Self-resistance is equal to the total resistance encountered when traversing a complete mesh. Common resistance is the resistance encountered by two different mesh currents.

The self-resistance voltage drop is equal to the product of self-resistance of the mesh and the self-current. Self-current is the current in the mesh being analyzed. The common-resistance voltage drop is equal to the product of the common resistance and neighboring current. The neighboring, or common, current is the current flowing in the common resistor.

**problem 2.21**    By the JME, write the mesh equations for the circuit of Fig. 2-22.

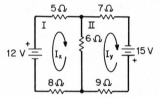

**Fig. 2-22**    Analyzing a two-mesh circuit (Problem 2.21).

**solution**    Applying Eq. (2-25) to mesh I, we have: Algebraic sum of voltage rises due to sources is 12 V. Self-resistance voltage drop = $(5 + 6 + 8)I_x$ V. Common-resistance voltage drop = $6I_y$. Hence,

$$12 = 19I_x - 6I_y$$

The negative sign is chosen because the mesh currents oppose each other in the 6-$\Omega$ resistor.
Applying Eq. (2-25) to mesh II yields:

$$15 = -6I_x + 22I_y$$

## Determinants

A square array of numbers enclosed by vertical lines on each side of them is called a *determinant*. A second-order (2 × 2) determinant is expressed by:

$$\Delta = \begin{vmatrix} a_{11} & a_{12} \\ a_{21} & a_{22} \end{vmatrix} \quad (2\text{-}26)$$

where an element, in general, is designated as $a_{ij}$. Subscript $i$ is the row number and $j$ the column number of element $a_{ij}$. For example, in Eq. (2-26) $a_{21}$ is in the second row and first column of the determinant $\Delta$.

A third-order (3 × 3) determinant is expressed by:

$$\Delta = \begin{vmatrix} a_{11} & a_{12} & a_{13} \\ a_{21} & a_{22} & a_{23} \\ a_{31} & a_{32} & a_{33} \end{vmatrix} \quad (2\text{-}27)$$

**problem 2.22**    Given: $19I_x - 6I_y = 12$ and $-6I_x + 22I_y = 15$. Using determinants, solve for $I_x$ and $I_y$.

*Copyrighted; permission to use granted by author.

**solution** The coefficient determinant is:

$$\Delta = \begin{vmatrix} 19 & -6 \\ -6 & 22 \end{vmatrix}$$

where the *primary diagonal P* and the *secondary diagonal S* are shown. The value of $\Delta$ is equal to the product of the primary diagonal elements minus the product of the secondary elements. Hence, $\Delta = 19 \times 22 - (-6)(-6) = 382$.

If the value of an unknown, such as $I_x$, is desired, the coefficients of $I_x$ in the coefficient determinant are replaced by the forcing functions of the circuit. Then by *Cramer's rule*,

$$I_x = \frac{\Delta_x}{\Delta} \quad \text{and} \quad I_y = \frac{\Delta_y}{\Delta} \tag{2-28}$$

The forcing functions in this problem are 12 and 15 V. By Eq. (2-28),

$$I_x = \frac{\begin{vmatrix} 12 & -6 \\ 15 & 22 \end{vmatrix}}{382} = \frac{354}{382} = 0.927 \text{ A}$$

$$I_y = \frac{\begin{vmatrix} 19 & 12 \\ -6 & 15 \end{vmatrix}}{382} = \frac{357}{382} = 0.935 \text{ A}$$

**problem 2.23** Given $5I_A + 0I_B + 4I_C = 40$; $0I_A + I_B - 4I_C = -5$; and $-I_A + I_B + I_C = 0$. Solve for $I_A$, $I_B$, and $I_C$ using determinants.

**solution** The coefficient determinant is:

$$\Delta = \begin{vmatrix} 5 & 0 & 4 \\ 0 & 1 & -4 \\ -1 & 1 & 1 \end{vmatrix}$$

If the left two columns are repeated to the right of $\Delta$, then three primaries and three secondaries can be formed. The value of $\Delta$ is now equal to the sum of products of the primary diagonals minus the sum of products of the secondary diagonals. (A 3 × 3 determinant is the highest-order determinant in which the diagonal method of solution can be used.) Hence,

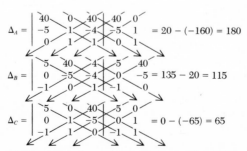

$$\Delta = \begin{vmatrix} 5 & 0 & 4 \\ 0 & 1 & -4 \\ -1 & 1 & 1 \end{vmatrix} \begin{matrix} 5 & 0 \\ 0 & 1 \\ -1 & 1 \end{matrix} = 5 - (-24) = 29$$

$$S_1 = -4 \; S_2 = -20 \; S_3 = 0 \; P_1 = 5 \; P_2 = 0 \; P_3 = 0$$

By Cramer's rule,

$$I_A = \frac{\Delta_A}{\Delta} \qquad I_B = \frac{\Delta_B}{\Delta} \qquad I_C = \frac{\Delta_C}{\Delta} \tag{2-29}$$

Hence,

$$\Delta_A = \begin{vmatrix} 40 & 0 & 4 \\ -5 & 1 & -4 \\ 0 & 1 & 1 \end{vmatrix} \begin{matrix} 40 & 0 \\ -5 & 1 \\ 0 & 1 \end{matrix} = 20 - (-160) = 180$$

$$\Delta_B = \begin{vmatrix} 5 & 40 & 4 \\ 0 & -5 & -4 \\ -1 & 0 & 1 \end{vmatrix} \begin{matrix} 5 & 40 \\ 0 & -5 \\ -1 & 0 \end{matrix} = 135 - 20 = 115$$

$$\Delta_C = \begin{vmatrix} 5 & 0 & 40 \\ 0 & 1 & -5 \\ -1 & 1 & 0 \end{vmatrix} \begin{matrix} 5 & 0 \\ 0 & 1 \\ -1 & 1 \end{matrix} = 0 - (-65) = 65$$

By Eq. (2-29), $I_A = 180/29 = 6.21$ A; $I_B = 115/29 = 3.97$ A; and $I_C = 65/29 = 2.24$ A.

## Laplace Expansion

This method is for a fourth- or higher-order determinant. Although it can be used for a second- or third-order determinant, for these Cramer's rule is preferred. The Laplace expansion should not be confused with the Laplace transform.

The procedure for applying Laplace expansion to a $4 \times 4$ determinant is:

1. Position value $PV$ of an element in the determinant is expressed by:

$$PV = (-1)^{i+j} \tag{2-30}$$

If $i + j$ is odd, than $PV = -1$; if $i + j$ is even, $PV = +1$.

2. *Any* row *or* column may be expanded.

3. Each term in the expansion contains the following product: (element of determinant)($PV$ of element)(minor determinant of element): $(a_{ij})(-1)^{i+j}(\Delta_{ij})$.

4. *Minor determinant* $\Delta_{ij}$ is formed from the given determinant by deleting row $i$ and column $j$ of element $a_{ij}$.

5. Product $(-1)^{i+j}(\Delta_{ij})$ is called the *cofactor* of element $a_{ij}$ and is designated by $A_{ij}$.

6. A fourth-order determinant contains four complete terms. For example, if an expansion is formed across the third row of a fourth-order determinant:

$$\Delta = \begin{vmatrix} a_{11} & a_{12} & a_{13} & a_{14} \\ a_{21} & a_{22} & a_{23} & a_{24} \\ a_{31} & a_{32} & a_{33} & a_{34} \\ a_{41} & a_{42} & a_{43} & a_{44} \end{vmatrix}$$

then $\Delta = a_{31}A_{31} + a_{32}A_{32} + a_{33}A_{33} + a_{34}A_{34}$

$= a_{31}(-1)^4\Delta_{31} + a_{32}(-1)^5\Delta_{32} + a_{33}(-1)^6\Delta_{33} + a_{34}(-1)^7\Delta_{34}$

$= a_{31}\Delta_{31} - a_{32}\Delta_{32} + a_{33}\Delta_{33} - a_{34}\Delta_{34} \tag{2-31}$

Note that there are three other rows and four different columns to choose for expansion. Hence, eight choices exist, each one of which results in the same answer.

**problem 2.24**   Given: $50 = 56I_A - 7I_B - 9I_C - 0I_D$; $30 = -7I_A + 21I_B - 6I_C + 8I_D$; $30 = -9I_A - 6I_B + 26I_C + 10I_D$; and $80 = 0I_A + 8I_B + 10I_C + 53I_D$. Apply the Laplace expansion to evaluate the coefficient determinant.

**solution**   The coefficient determinant is:

$$\Delta = \begin{vmatrix} 56 & -7 & -9 & -0 \\ -7 & 21 & -6 & 8 \\ -9 & -6 & 26 & 10 \\ 0 & 8 & 10 & 53 \end{vmatrix}$$

For minimum calculation, the row or column containing the most zeros is selected for expansion. Choosing row four yields:

$$\Delta = a_{41}A_{41} + a_{42}A_{42} + a_{43}A_{43} + a_{44}A_{44}$$
$$= -a_{41}\Delta_{41} + a_{42}\Delta_{42} - a_{43}\Delta_{43} + a_{44}\Delta_{44}$$

$$= -(0)\Delta_{41} + 8 \begin{vmatrix} 56 & -9 & -0 \\ -7 & -6 & 8 \\ -9 & 26 & 10 \end{vmatrix} - 10 \begin{vmatrix} 56 & -7 & -0 \\ -7 & 21 & 8 \\ -9 & -6 & 10 \end{vmatrix}$$

$$+ 53 \begin{vmatrix} 56 & -7 & -9 \\ -7 & 21 & -6 \\ -9 & -6 & 26 \end{vmatrix}$$

Because element $a_{41} = 0$, $\Delta_{41}$ is not needed; the first term of $\Delta$ is, therefore, zero. Each minor determinant is a $3 \times 3$. Their solution is determined by the application of Cramer's rule, as explained in the previous section. Hence,

$$\Delta = 8(-14,990) - 10(14,462) + 53(26,341) = 1,131,533$$

If, for example, current $I_A$ is to be found, the preceding steps are repeated with the forcing functions replacing the elements of the first column in the coefficient determinant. Then current $I_A$ is equal to the value of the resultant determinant divided by the coefficient determinant. Because of the lengthy calculations, evaluation of high-order determinants is best done on a computer.

## 2.8  NODAL ANALYSIS

Another method of circuit analysis is *nodal analysis*. This method is useful for finding unknown voltages when the values of the node currents are known in a given circuit. To solve a circuit using nodal analysis:

1. Designate all nodes by a letter or a numeral.
2. Choose one of the nodes as the *reference node* (zero volts, or ground).

3. Assume all other nodes at potentials above the reference node.

4. Direct currents through each element away from the node being analyzed toward the reference node. Assume currents between nodes, neither of which is a reference node, in a given direction. The voltage sense accommodates the assumed current direction (Fig. 2-23).

5. Apply KCL at each node, except the reference node.

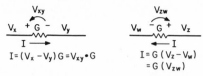

**Fig. 2-23**   Voltage senses for assumed directions of current $I$.

**problem 2.25**    For the circuit of Fig. 2-24, express the voltages in determinant form.

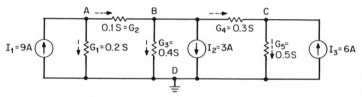

**Fig. 2-24**   Nodal analysis of a circuit (Problem 2.25).

**solution**    The nodes are labeled as $A$, $B$, $C$, and $D$. Node $D$ is chosen as the reference node. Currents through $G_1$, $G_3$, and $G_5$ flow away from their nodes toward the reference node. Current through $G_2$ is assumed to flow from node $A$ to node $B$; current through $G_4$ is assumed to flow from node $B$ to node $C$. Applying KCL to each node (except reference node $D$), we have:

Node $A$:         $9 = 0.3V_A - 0.1V_B - 0V_C$
Node $B$:         $3 = -0.1V_A + 0.8V_B - 0.3V_C$
Node $C$:         $6 = 0V_A - 0.3V_B + 0.8V_C$

Hence,

$$\Delta = \begin{vmatrix} 0.3 & -0.1 & -0 \\ -0.1 & 0.8 & -0.3 \\ 0 & -0.3 & 0.8 \end{vmatrix} \quad \Delta_A = \begin{vmatrix} 9 & -0.1 & -0 \\ -3 & 0.8 & -0.3 \\ 6 & -0.3 & 0.8 \end{vmatrix}$$

$$\Delta_B = \begin{vmatrix} 0.3 & 9 & -0 \\ -0.1 & -3 & -0.3 \\ 0 & 6 & 0.8 \end{vmatrix} \quad \Delta_C = \begin{vmatrix} 0.3 & -0.1 & 9 \\ -0.1 & 0.8 & -3 \\ 0 & -0.3 & 6 \end{vmatrix}$$

and $V_A = \Delta_A/\Delta$; $V_B = \Delta_B/\Delta$; $V_C = \Delta_C/\Delta$.

## Joerg Nodal Equation*

The Joerg nodal equation (JNE) is a general equation that enables one to write nodal equations by inspection. For dc circuits:

$$\Sigma(I_{\text{ent}} \text{ due to current sources}) = \Sigma(\text{self-conductance currents})$$
$$- \Sigma(\text{common-conductance currents}) \quad (2\text{-}32)$$

The Joerg nodal equation calls for the algebraic sum of entering currents ($I_{\text{ent}}$) due to sources. Current leaving a node owing to a source is treated as a negative entering current. Self-conductance current is the product of the total conductance connected to a node and the node voltage. The common conductance current is the product of the conductance straddling the node with another neighboring node (which is not the reference node) and the neighboring node voltage.

*Copyrighted; permission to use granted by author.

**problem 2.26**   Using the JNE method, write the nodal equations for Fig. 2-25.

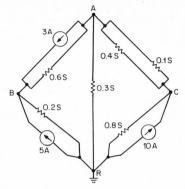

**Fig. 2-25**   Nodal analysis of a bridge circuit (Problem 2.26).

**solution**   The nodes are labeled $A$, $B$, $C$, and $R$ (reference node). At node $A$: $I_{ent}$ due to current sources $= -3$ A. Self-conductance currents $= (0.4 + 0.1 + 0.3 + 0.6)V_A = 1.4V_A$ A. Common-conductance currents $= 0.6\,V_B$ and $0.5\,V_C$ A. Therefore, $-3 = 1.4\,V_A - 0.6V_B - 0.5V_C$. In a similar manner, we obtain at nodes $B$ and $C$:

$$5 = -0.6V_A + 0.8V_B - 0V_C \text{ A}$$
$$10 = -0.5V_A - 0V_B + 1.3\,V_C$$

## 2.9  THÉVENIN AND NORTON THEOREMS

The Thévenin theorem (sometimes referred to as the Helmholtz theorem) states that a linear circuit with dc sources, regardless of its complexity, can be replaced by a simple circuit of a voltage source $V_{TH}$ in series with a resistance $R_{TH}$ (Fig. 2-26). Both circuits are identical with respect to their terminal behavior.

In the ensuing discussion, independent dc sources are assumed. The value of an *independent* voltage or current source does not depend on the circuit to which it is connected. For example, a battery is an independent source. The value of a dependent source, however, is not independent of the circuit to which it is connected. Dependent sources arise, for example, in the models of transistors.

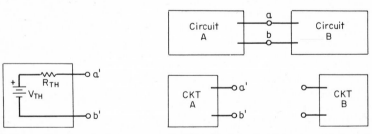

**Fig. 2-26**   Thévenin equivalent of a dc resistive circuit.

**Fig. 2-27**   Finding the Thévenin equivalent of circuit $A$. Circuit $B$ is the load for circuit $A$.

The procedure for applying Thévenin's theorem follows:

1. If the response (voltage or current) of a particular circuit element is desired, the circuit can be broken into two parts (Fig. 2-27). Labeling the terminals of circuit $A$ $a'$-$b'$, circuit $A$ is "killed." This means that the independent sources in the circuit are set to zero: voltage sources are short-circuited and current sources are open-circuited.

2. The *equivalent resistance* looking into points $a'$-$b'$, called the Thévenin resistance ($R_{TH}$), is found using appropriate circuit analysis discussed earlier in the chapter.

3. The sources in circuit A are restored and, by analysis, the *open-circuit voltage* across *a'-b'*, called the *Thévenin voltage* ($V_{TH}$), is determined.

4. Circuit A is redrawn as a voltage source $V_{TH}$ in series with a resistance $R_{TH}$ as illustrated in Fig. 2-26. As mentioned earlier, with respect to terminals *a'b'*, the electrical behavior of the Thévenin equivalent circuit is *identical* to circuit A.

In applying Norton's theorem:

1. After setting the sources to zero in circuit A, find the conductance looking into *a' – b'*. Label the conductance $G_N$, where $G_N = 1/R_{TH} = 1/R_N$.

2. Restore the sources in circuit A and short-circuit terminals *a'-b'*. By suitable analysis, find the short-circuit current $I_N$ flowing between terminals *a'-b'*.

3. Circuit A is redrawn as a current source $I_N$ in parallel with a conductance $G_N$ as illustrated in Fig. 2-28. With respect to terminals *a'-b'*, the electrical behavior of the Norton equivalent circuit is identical to circuit A.

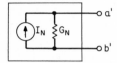

**Fig. 2-28** Norton equivalent of a dc resistive circuit.

The Thévenin and Norton equivalent circuits are linked by the following expressions:

$$V_{TH} = R_{TH} I_N \qquad (2\text{-}33a)$$
$$I_N = G_N V_{TH} \qquad (2\text{-}33b)$$

## Source Transformation

By Eq. (2-33), a circuit containing a voltage source in series with a resistance may be transformed into a circuit containing a current source in parallel with a conductance (or resistance), and vice versa. This is illustrated in Fig. 2-29).

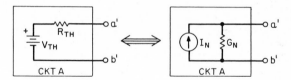

**Fig. 2-29** Source transformation.

**problem 2.27** Applying Thévenin's theorem, find the load current $I_L$ in Fig. 2-30a.

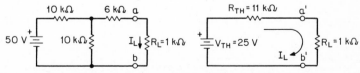

**Fig. 2-30** Applying Thévenin's theorem to find $I_L$. (*a*) Given circuit. (*b*) Thévenin equivalent with $R_L$ connected across terminals *a'-b'* (Problem 2.27).

**solution** Setting the 50-V source to zero, $R_{TH} = 6 + 10 \| 10 = 11$ kΩ. By voltage division, $V_{TH} = 50 \times 10/(10 + 10) = 25$ V. (When *a'-b'* are open-circuited, no current flows in the 6-kΩ resistance.) The Thévenin circuit is shown in Fig. 2-30b. Load current $I_L = 25/(11 + 1) = 2.08$ mA.

**problem 2.28** For the circuit of Fig. 2-31a, (*a*) draw its Thévenin equivalent circuit and determine $I_L$. (*b*) By source transformation, draw its Norton equivalent circuit.

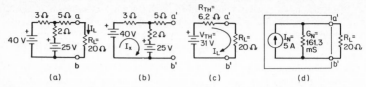

(a)    (b)    (c)    (d)

**Fig. 2-31** Application of Thévenin's theorem. (*a*) Given circuit. (*b*) Finding $V_{TH}$. (*c*) Thévenin equivalent. (*d*) Norton equivalent (Problem 2.28).

**solution**    (*a*) Current $I_x$ in Fig. 2-31*b* equals $(40 - 25)/5 = 3$ A. Then, $V_{TH} = 2 \times 3 + 25 = 31$ V. Setting the sources to zero, $R_{TH} = 3\|2 + 5 = 6.2\ \Omega$. The Thévenin circuit connected to $R_L$ is shown in Fig. 2-31*c*: $I_L = 31/(6.2 + 20) = 1.18$ A.
   (*b*) The Norton circuit is shown in Fig. 2-31*d* where $I_N = 31/6.2 = 5$ A and $G_N = 1/6.2 = 161.3$ mS.

**problem 2.29**    Applying Thévenin's theorem, find $I_L$ in the bridge circuit of Fig. 2-32*a*.

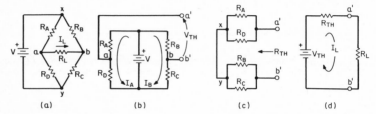

(a)    (b)    (c)    (d)

**Fig. 2-32** Finding $I_L$ in a bridge circuit. (*a*) Given circuit. (*b*) Determining $V_{TH}$. (*c*) Determining $R_{TH}$. (*d*) Load $R_L$ connected across the Thévenin equivalent (Problem 2.29).

**solution**    Removing $R_L$ from terminals *a-b*, from Fig. 2-32*b* $I_A = V/(R_A + R_D)$ and $I_B = V/(R_B + R_C)$. Hence, $V_{TH} = -I_A R_A + I_B R_B = V(R_B R_D - R_A R_C)/[(R_A + R_D)(R_B + R_C)]$ V.
   From Fig. 2-32*c*, $R_{TH} = R_A R_D/(R_A + R_D) + R_B R_C/(R_B + R_C)\ \Omega$. Hence, from Fig. 2-32*d*, $I_L = V_{TH}/(R_{TH} + R_L)$ A.

**problem 2.30**    Find $V_L$ in Fig. 2-33*a* using Norton's theorem.

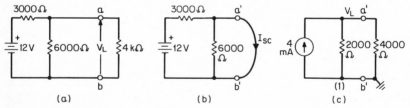

(a)    (b)    (c)

**Fig. 2-33** Finding $V_L$ by Norton's theorem. (*a*) Given circuit. (*b*) Determining $I_{SC}$. (*c*) Norton circuit (Problem 2.30).

**solution**    Removing the 4-kΩ load and shorting *a'-b'* in Fig. 2-33*b*, $I_{SC} = I_N = 12/3000 = 4$ mA. Setting the 12-V source to zero yields $R_N = 3000\|6000 = 2$ kΩ. The Norton circuit is drawn in Fig. 2-33*c*. Load voltage $V_L$ equals $4 \times 2\|4 = 5.33$ V.

## 2.10   OTHER NETWORK THEOREMS

Additional theorems useful in circuit analysis are presented in this section.

### Superposition Theorem

The superposition theorem states that when a linear circuit is driven by several sources, the total response can be found by finding the individual response to each source separately and then algebraically summing them.

**problem 2.31** Using superposition, find the current in the 22-kΩ resistor of Fig. 2-34a.

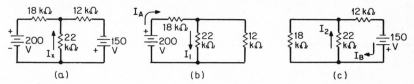

(a)                           (b)                          (c)

**Fig. 2-34** Using superposition to find an unknown current $I_x$. (a) Given circuit. (b) Determining $I_1$ due to the 200-V source. (c) Determining $I_2$ due to the 150-V source (Problem 2.31).

**solution** Let $I_x$, directed upward, be the total current in the 22-kΩ resistor. Setting the 150-V source to zero, Fig. 2-34b is obtained. Current $I_1$ is the current due to the 200-V source only. Source current $I_A = 200/(18 + 22\|12) = 7.76$ mA. By current division, $I_1 = 7.76 \times 12/(12 + 22) = 2.74$ mA.

Now, setting the 200-V source to zero, one obtains Fig. 2-34c. Source current $I_B = 150/(12 + 18\|22) = 6.85$ mA. Again, by current division, $I_2 = 6.85 \times 18/(18 + 22) = 3.08$ mA. Current $I_x$ is equal to the algebraic sum of currents $I_1$ and $I_2$, or $I_x = 3.08 - 2.74 = 0.34$ mA upward, as assumed.

## Maximum Power Transfer Theorem

Maximum power is transferred to a given load from a linear dc circuit if the load resistance is equal to the Thévenin resistance (or the inverse of the Norton conductance) of the driving circuit. Under this condition, the maximum power delivered to the load $P_{L,\max}$ is:

$$P_{L,\max} = \frac{V_{TH}^2}{4R_{TH}} \tag{2-34a}$$

$$= \frac{I_N^2}{4G_N} \tag{2-34b}$$

**problem 2.32** (a) In Fig. 2-35 find $R_L$ for maximum power transfer and the value of $P_{L,\max}$. (b) If $R_L = 2$ kΩ, what is the value of $P_L$?

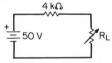

**Fig. 2-35** Application of power transfer theorem (Problem 2.32).

**solution** (a) For maximum power transfer, $R_L = R_{TH} = 4$ kΩ. By Eq. (2-34a), $P_{L,\max} = 50^2/(4 \times 4 \times 10^3) = 156$ mW. (b) $P_L = I^2R = [50/(4 + 2)]^2 \times 2 = 139$ mW.

## Millman's Theorem

The equivalent voltage $V_{EQ}$ for several parallel voltage sources each individually in series with a given resistor is:

$$V_{EQ} = \frac{V_1G_1 + V_2G_2 + V_3G_3 + \cdots}{G_1 + G_2 + G_3 + \cdots} \tag{2-35}$$

where $G = 1/R$.

The equivalent current $I_{EQ}$ for several series current sources with each source in parallel with a resistor is:

$$I_{EQ} = \frac{I_1R_1 + I_2R_2 + I_3R_3 + \cdots}{R_1 + R_2 + R_3 + \cdots} \tag{2-36}$$

**problem 2.33** Using Millman's theorem, find $I_L$ in Fig. 2-36a.

**solution** By Eq. (2-35),

$$V_{EQ} = \frac{20(1/5) + 50(1/10) + 90(1/15)}{1/5 + 1/10 + 1/15} = 40.91 \text{ V}$$

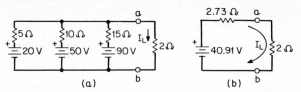

**Fig. 2-36**  Using Millman's theorem to find an unknown current. (*a*) Given circuit. (*b*) Resulting circuit for calculating $I_L$ (Problem 2.33).

$R_{EQ} = 1/G_{EQ} = 1/(11/30) = 2.73 \, \Omega$. Referring to Fig. 2-36*b*, it shows that $I_L = 40.91/(2.73 + 2) = 8.65$ A.

**problem 2.34**   Using Millman's theorem, find $I_L$ in Fig. 2-37*a*.

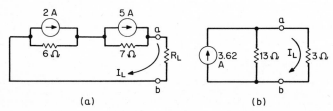

**Fig. 2-37**  Application of Millman's theorem to a circuit containing current sources in series. (*a*) Given circuit. (*b*) Resulting circuit for calculating $I_L$ (Problem 2.34).

**solution**   By Eq. (2-36),

$$I_{EQ} = \frac{2 \times 6 + 5 \times 7}{6 + 7} = 3.62 \text{ A}$$

$R_{EQ} = 6 + 7 = 13 \, \Omega$. Referring to Fig. 2-37*b*, yields by current division, $I_L = 3.62 \times 13/(13 + 3) = 2.94$ A.

## Compensation Theorem

Assume that the current in, and the voltage across, a given branch are known. Then, in a dc circuit, any combination of elements can be inserted into that branch provided that they retain the same current and voltage values originally specified.

**problem 2.35**   In Fig. 2-38*a*, branch *b-c* has a current of 2 A and a voltage across it of 40 V. Use the compensation theorem to replace the 20-$\Omega$ resistor.

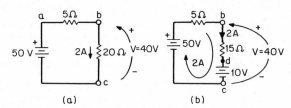

**Fig. 2-38**  Application of the compensation theorem. (*a*) Given circuit. (*b*) A possible solution (Problem 2.35).

**solution**   One possibility is shown in Fig. 2-38*b*. Current $I_{bc} = (50 - 40)/5 = 2$ A and $V_{bc} = 2 \times 15 + 10 = 40$ V, as required.

## Tellegen's Theorem

In a given circuit, a set of branch voltages is selected that satisfies the KVL around any closed pathway. In addition, a set of branch currents is selected that satisfies the KCL at each node. Then the algebraic sum of the products of the branch voltage $V_k$ and the branch current $I_k$ is equal to zero:

$$\sum_{k=1}^{n} V_k I_k = 0 \qquad (2\text{-}37)$$

**problem 2.36**   The circuit of Fig. 2-39 contains six elements. Let $V_1 = 5$ V, $V_2 = 2$ V, $V_6 = 12$ V, $I_1 = 3$ A, $I_2 = 2$ A, and $I_3 = 0.5$ A. Show that Eq. (2-37) is valid.

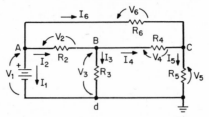

**Fig. 2-39**   Application of Tellegen's theorem (Problem 2.36).

**solution**   In this problem, $k = 6$. Since $V_3 = V_1 - V_2$, $V_3 = 5 - 2 = 3$ V. Also, $V_4 = V_6 - V_2 = 12 - 2 = 10$V, and $V_5 = V_3 - V_4 = 3 - 10 = -7$V.
   At node A, $I_6 = -(I_2 + I_3) = -(3 + 2) = -5$ A; at node B, $I_4 = I_2 - I_3 = 2 - 0.5 = 1.5$ A; and at node C, $I_5 = I_4 + I_6 = 1.5 - 5 = -3.5$ A.
   By Eq. (2-37), $5 \times 3 + 2 \times 2 + 3 \times 0.5 + 10 \times 1.5 + (-7) \times (-3.5) + 12 \times (-5) = 15 + 4 + 1.5 + 15 + 24.5 - 60 = 0$.

## 2.11   NETWORK TOPOLOGY

In network topology, one is concerned with the geometrical form of the circuit. Once the form is known, one can determine whether mesh or nodal analysis will yield the fewest equations for the solution of the circuit. The application of network topology to complex circuits, where the choice of analysis method may not be obvious, is especially useful.

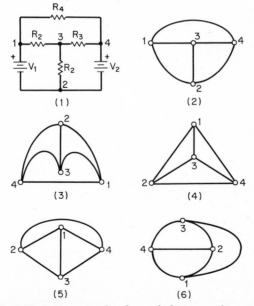

**Fig. 2-40**   A given network and several of its isomorphic graphs.

Important definitions used in topology include:

GRAPH, OR SKELETON.   The geometrical structure (skeletal outline) of a network. To obtain the graph, the given network is redrawn by replacing each element with a line segment.

ISOMORPHIC GRAPHS.   Graphs of a given network that are identical circuitwise. The length or curvature of a line segment is inconsequential. Graphs that look different to the eye may be identical, or isomorphic. For example, Fig. 2-40 shows a given network and several of its isomorphic graphs.

EDGE.   A line segment that represents a circuit element with two isolated endpoints.

DEGREE OF NODE.   Also termed vertex, it is equal to the number of edges incident to the node.

PLANAR GRAPH.   A graph is planar if it can be drawn in a plane with no two edges intersecting except at a node; otherwise, the graph is nonplanar. Examples of a planar and nonplanar graph are provided in Fig. 2-41.

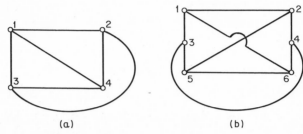

**Fig. 2-41**   Examples of ($a$) planar and ($b$) nonplanar graphs.

ORIENTED GRAPH.   A graph in which the edges have assumed directions.

CONNECTED GRAPH.   A graph in which a path exists between every pair of nodes.

SUBGRAPH.   A portion of a given graph. For example, the path between two nodes in a graph is a subgraph.

COMPLEMENT OF A SUBGRAPH.   The remainder of a graph once a particular subgraph has been chosen. Figure 2-42 shows a graph, the selected subgraph, and the complement of the subgraph.

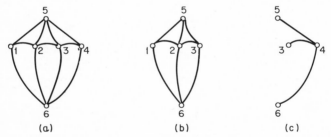

**Fig. 2-42**   Example of ($a$) a graph; ($b$) its subgraph; and ($c$) the complement of the subgraph.

CIRCUIT LOOP.   The resulting closed path obtained in traversing through edges and returning to the starting node. Every node in such a loop is of second degree.

TREE.   A tree contains all the nodes of a given graph but no circuit loops. Figure 2-43 illustrates a given graph and some possible trees.

BRANCH.   The edge of a tree.

COTREE.   The complement of a given tree.

LINKS, OR CHORDS.   The edges of a cotree.

WINDOW PANE.   A circuit loop that contains no other loops inside it.

INCIDENCE SET.   The set of all edges common to a given node.

## Relationships

The number of branches $b$ is equal to 1 less the number of nodes $n$ in a tree:

$$b = n - 1 \tag{2-38}$$

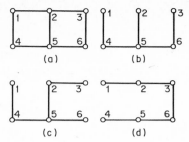

**Fig. 2-43** (a) A graph and (b), (c), (d) possible trees.

The number of branches in a tree plus the number of links $l$ in its cotree is equal to the number of edges $e$ in the graph:

$$b + l = e \tag{2-39}$$

The number of links is equal to 1 plus the difference between the number of edges and nodes:

$$l = 1 + (e - n) \tag{2-40}$$

The number of independent nodal equations $I_E$ needed to find the branch voltages is:

$$I_E = n - 1 - v_s \tag{2-41}$$

where $v_s$ is the number of voltage sources chosen as tree branches.

The number of independent mesh equations $V_E$ needed to find mesh currents is:

$$V_E = e - n + 1 - i_s \tag{2-42}$$

where $i_s$ is the number of current sources chosen as link branches.

**problem 2.37**  Determine the number of independent nodal equations for the circuit of Fig. 2-44a.

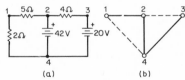

**Fig. 2-44**  An example of (a) a given circuit and (b) its graph (Problem 2.37).

**solution**  The graph of the circuit is given in Fig. 2-44b where the tree is shown in solid lines and the cotree in dashed lines. Two voltage sources were chosen as tree branches, hence $v_s = 2$. The number of nodes is 4. By Eq. (2-41), $I_E = 4 - 1 - 2 = 1$. One nodal equation is required.

**problem 2.38**  Determine the number of independent mesh equations for the circuit of Fig. 2-45a.

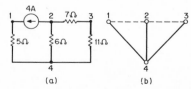

**Fig. 2-45**  Another example of (a) a given circuit and (b) its graph (Problem 2.38).

**solution**  The graph of the circuit is given in Fig. 2-45b where the tree is shown in solid lines and the cotree in dashed lines. Because the current source was chosen as a link branch, $i_s = 1$. The number of edges is 5, and the number of nodes is 4. By Eq. (2-42), $V_E = 5 - 4 + 1 - 1 = 1$. One mesh equation is required.

## 2.12   BIBLIOGRAPHY

Babb, Daniel: *Resistive Circuits,* International Textbook, Scranton, Pa., 1968.

Boylestad, Robert: *Introductory Circuit Analysis,* 3d ed., Charles E. Merrill, Columbus, Ohio, 1977.

Churchman, Lee: *Introduction to Circuits,* Holt, New York, 1976.

Close, C. M.: *Analysis of Linear Circuits,* Harcourt, Brace, New York, 1966.

Hayt, W., and Kemmerly, J.: *Engineering Circuit Analysis,* 3d ed., McGraw-Hill, New York, 1978.

Jackson, Herbert: Introduction to Electric Circuits, 4th ed., Prentice-Hall, Englewood Cliffs, N.J., 1976.

Joerg, A. E.: Lecture notes on *College Algebra II* and *Algebra and Trigonometry,* Bridgeport Engineering Institute, Bridgeport, Conn., and Westchester Community College, Valhalla, N.Y., 1967–1976.

Joerg, E. A.: Lecture notes on *Introductory Networks, Passive and Active Network Analysis, Advanced Circuit Analysis,* and *Physics III—Electricity and Magnetism,* Hofstra University, Hempstead, N.Y., Bridgeport Engineering Institute, Bridgeport, Conn., and Westchester Community College, Valhalla, N.Y., 1965–1976.

Karni, Shlomo: *Intermediate Network Analysis,* Allyn & Bacon, Boston, 1971.

Klayton, Marwin: *Fundamental Electrical Technology,* Addison-Wesley, Reading, Mass., 1977.

Leach, Donald: *Basic Electric Circuits,* Wiley, New York, 1969.

Risdale, R.: *Electrical Circuits for Electrical Technology,* McGraw-Hill, New York, 1976.

Romanowitz: *Introduction to Electric Circuits,* Wiley, New York, 1971.

Seshu, S., and Reed, M.: *Linear Graphs and Electric Networks,* Addison-Wesley, Reading, Mass., 1961.

Tocci, Ronald, *Introduction to Electric Circuit Analysis,* Charles E. Merrill Books, Inc., Columbus, Ohio, 1974.

Van Valkenburg, M. E.: *Network Analysis,* 3d ed., Prentice-Hall, Englewood Cliffs, N.J., 1974.

Vitrogen, David: *Elements of Electrical and Magnetic Circuits,* Rinehart Press, San Francisco, 1971.

# Chapter 3

# AC Circuit Analysis

## ERNEST A. JOERG

**Professor and Chairperson, Electrical Technology Department,
Westchester Community College, Valhalla, N.Y.**

## 3.1 INTRODUCTION

A generator whose voltage changes its polarity periodically generates an alternating current (ac) as shown in Fig. 3-1. The waveshape is *sinusoidal,* and it may be expressed, in general, by:

$$i(t) = I_m \sin(\omega t \pm \phi) \tag{3-1}$$

where $i(t)$ = current as a function of time
$I_m$ = maximum, or peak, amplitude of the current
$\omega$ = angular frequency, rad/s
$\phi$ = phase angle measured in radians or degrees
Angle $\phi$ specifies the shift of the waveform to the left or right from the origin.

A similar expression can be written for an ac voltage:

$$v(t) = V_m \sin(\omega t \pm \phi) \tag{3-2}$$

where $v(t)$ is the voltage as a function of time and $V_m$ is the maximum, or peak, amplitude of the voltage.

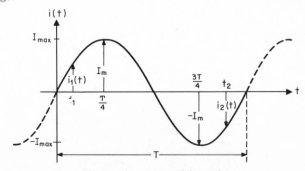

**Fig. 3-1** An ac sinusoidal waveform.

The angular frequency is related to frequency $f$ (in Hertz) by:

$$\omega = 2\pi f \tag{3-3a}$$

or

$$f = \frac{\omega}{2\pi} \tag{3-3b}$$

Period $T$, the time span for one complete cycle (in seconds), is:

$$T = \frac{1}{f} \tag{3-4a}$$

or

$$T = \frac{2\pi}{\omega} \tag{3-4b}$$

## Average Value

The *average* (dc) value of a periodic waveform is equal to the *net* area of the waveform divided by the period $T$ or $2\pi$ rad. Because the positive and negative half cycles of a sine wave are equal, its dc value is zero.

In terms of current, the general expression for the average value of any periodic waveform is:

$$I_{av} = I_{dc} = \frac{1}{T}\int_0^T i\, dt \tag{3-5a}$$

$$= \frac{1}{2\pi}\int_0^{2\pi} i\, d(\omega t) \tag{3-5b}$$

## Effective (RMS) Value

The *effective*, or *root-mean-square* (rms), value of a periodic waveform is equal to the direct current which dissipates the same energy in a given resistor. For a sine wave:

$$I_{rms} = 0.707I_m = \frac{I_m}{\sqrt{2}} \tag{3-6a}$$

$$V_{rms} = 0.707V_m = \frac{V_m}{\sqrt{2}} \tag{3-6b}$$

In general (in terms of *current*):

$$I_{rms} = \sqrt{\frac{1}{T}\int_0^T i^2\, dt} \tag{3-7a}$$

$$= \sqrt{\frac{1}{2\pi}\int_0^{2\pi} i^2\, d(\omega t)} \tag{3-7b}$$

In the ensuing discussion, rms voltage and currents will be simply designated as $V$ and $I$, respectively.

**problem 3.1**   Given $i(t) = 100 \sin (2000t + 45°)$A. Determine $I_m, I_{dc}, I, \omega, f, T, \phi$, and the peak-to-peak value $I_{pp}$.

**solution**   $I_m = 100$ A; $I_{dc} = 0$; $I = 0.707 \times 100 = 70.7$ A; $\omega = 2000$ rad/s; $f = 2000/6.28 = 318.3$ Hz; $T = 1/318.3 = 3.14$ ms; $\phi = 45°$; $I_{pp} = 2I_m = 2 \times 100 = 200$ A.

**problem 3.2**   Find the average (dc) value of the periodic waveform ($T = 23$ s) of Fig. 3-2.

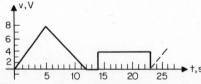

**Fig. 3-2**   Finding the average (dc) value of a periodic waveform (Problem 3.2).

**solution**    The net area for 1 cycle is ½ × 5 × 8 + ½ × 7 × 8 + 0 + 9 × 4 = 84. Hence, $V_{dc}$ = 84/23 = 3.65 V.

**problem 3.3**    Find the rms value of the periodic waveform ($T$ = 8s) of Fig. 3-3.

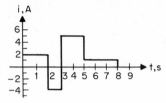

**Fig. 3-3**    Finding the rms (effective) value of a periodic waveform (Problem 3.3).

**solution**

$$I = \sqrt{\frac{2^2 \times 2 + (-4)^2 \times 1 + 5^2 \times 2 + 1^2 \times 3}{8}} = 3.1 \text{ A rms}$$

## 3.2   THE CAPACITOR AND INDUCTOR

In addition to resistors (discussed in Chap. 2) capacitors and inductors are two other important elements found in circuits. Resistors, capacitors, and inductors are *passive elements*. They are capable only of absorbing electric energy. Unlike a resistor, which dissipates energy, capacitors and inductors store energy and return it to the circuit in which they are connected.

### Capacitors

A capacitor is basically two parallel conducting plates separated by an insulating material called a *dielectric* (Fig. 3-4). (Refer to Chap. 4 for a complete discussion of the various types of capacitors available.) If the area of the plates is $A$ and the thickness of the dielectric $d$, the capacitance $C$ is:

$$C = \frac{\epsilon A}{D} \qquad \text{farads (F)} \tag{3-8}$$

where $\epsilon$ is the *overall permittivity* in farads/meter (F/m). It is equal to the product of the *dielectric constant* $k$ and the *permittivity of a vacuum* $\epsilon_0$:

$$\epsilon = k\epsilon_0 \tag{3-9}$$

The value of $\epsilon_0$ is 8.854 × 10$^{-12}$ F/m. A listing of dielectric constants of various materials is provided in Table 3-1.

The charge $Q$ (in coulombs) stored in a capacitor is:

$$q = Cv \tag{3-10}$$

where $v$ is the voltage (in volts) impressed across the capacitor and $C$ is in farads.

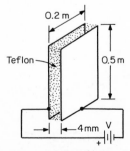

**Fig. 3-4**    Basic construction of a parallel-plate capacitor (Problem 3.4).

**Capacitors Connected in Series**   For capacitors connected in series, the equivalent capacitance $C_{EQ}$ is:

$$C_{EQ} = \frac{1}{1/C_1 + 1/C_2 + 1/C_3 + \cdots} \tag{3-11}$$

For two capacitors $C_1$ and $C_2$ connected in series:

$$C_{EQ} = \frac{C_1 C_2}{C_1 + C_2} \tag{3-12}$$

If a capacitor $C$ is in series with another capacitor whose value is $1/n$ of $C$, then:

$$C_{EQ} = \frac{C}{n+1} \tag{3-13}$$

**TABLE 3-1    Dielectric Constants: Typical Values***
($T = 20°C$; atmospheric pressure; $f < 1$ MHz)

| Dielectric type | $k$ |
| --- | --- |
| Air | 1.0059 |
| Amber | 2.9 |
| Asphalt | 2.7 |
| Bakelite | 6 (3½ to 8½) |
| Beeswax | 2.7 |
| Celluloid | 6.2 |
| Ceramic | 5½ × 10³ (4k to 7k) |
| Distilled water | 78 |
| Ebonite | 2.8 |
| Ethyl | 26 |
| Glass (window) | 6 |
| Glycerin | 56 |
| Mica | 5 (6 to 7½) |
| Mylar | 3 |
| Paper | 2.5 (2 to 4) |
| Paraffin | 4 (3 to 5) |
| Petroleum | 4 (2 to 6) |
| Polyethylene | 2.3 |
| Polystyrene | 2.6 |
| Porcelain | 6.5 (6 to 7½) |
| Quartz | 3.8 |
| Pyrex (glass) | 4.8 |
| Rubber | 3 (2 to 3½) |
| Slate | 6.8 |
| Soil | 2.9 |
| Teflon | 2 |
| Vacuum | 1.0 |
| Vaseline | 2.2 |
| Water | 81 |
| Wood | 5.5 (2½ to 8½) |

*Approximate values are given.

**Capacitors Connected in Parallel**   For capacitors connected in parallel, the equivalent capacitance $C_{EQ}$ is:

$$C_{EQ} = C_1 + C_2 + C_3 + \cdots \tag{3-14}$$

**Current and Voltage in a Capacitor**   The instantaneous current in a capacitor $i_C$ is:

$$i_C = \frac{C \, dv_C}{dt} \simeq \frac{C \, \Delta v_C}{\Delta t} \tag{3-15}$$

The instantaneous voltage across a capacitor is:

$$v_C = \frac{1}{C} \int i_C \, dt \tag{3-16}$$

Under steady-state conditions, the capacitor can be replaced by an *open circuit* in a *dc* circuit.

**Energy Stored in a Capacitor**   The instantaneous energy stored in a capacitor $w_C$ is:

$$w_C = \tfrac{1}{2}Cv_C^2 \qquad \text{joules (J)} \tag{3-17}$$

where $C$ is in farads and $v_C$ in volts.

**problem 3.4**   For the parallel-plate capacitor of Fig. 3-4 find (a) its capacitance and (b) the charge on each plate $v = 350$ V.

**solution**   (a) Area $A$ of a plate is $0.5 \times 0.2 = 0.1$ m². From Table 3-1, $k = 2$ for Teflon fluorocarbon resins. By Eq. (3-8), $C = (8.85 \times 10^{-12} \times 2 \times 0.1)/0.004 = 4.425 \times 10^{-10}$ F $= 442.5$ pF.
   (b) By Eq. (3-10), $q = 4.425 \times 10^{-10} \times 350 = 1.549 \times 10^{-7}$ C.

**problem 3.5**   Find the equivalent capacitance of $6$-$\mu$F and $3$-$\mu$F capacitors connected in (a) series and (b) parallel.

**solution**   (a) By Eq. (3-12), $C_{EQ} = (6 \times 3)/(6 + 3) = 2\ \mu$F. In a series circuit of capacitors, the equivalent capacitance is always less than the smallest value of capacitance in series.
   (b) By Eq. (3-14), $C_{EQ} = 6 + 3 = 9\ \mu$F.

**problem 3.6**   Find $C_{EQ}$ in the series-parallel circuit of Fig. 3-5.

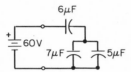

**Fig. 3-5**   Finding the equivalent capacitance of capacitors connected in a series-parallel circuit (Problem 3.6).

**solution**   The 5- and 7-$\mu$F capacitors in parallel yields $5 + 7 = 12\ \mu$F. By Eq. (3-12), $C_{EQ} = (6 \times 12)/(6 + 12) = 4\ \mu$F.

**problem 3.7**   The voltage $v_C$ across a $3$-$\mu$F capacitor is plotted in Fig. 3-6a. Sketch the current $i_C$ in the capacitor.

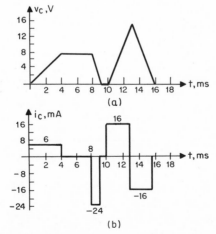

**Fig. 3-6**   Voltage and current waveforms of a capacitor (Problem 3.7). (a) Voltage across capacitor. (b) Resulting current in capacitor.

**solution**   The current waveform is shown in Fig. 3-6b. By Eq. (3-15), for the interval from 0 to 4 ms, $\Delta v_C/\Delta t = 8/0.004 = 2000$ V/s; hence, $i_C = 3 \times 10^{-6} \times 2000 = 6$ mA. For the interval of 4 to 8 ms, $v_C$ is constant; hence $\Delta v_C/\Delta t = 0$ and $i_C = 0$. For the interval of 8 to 9 ms, $\Delta v_C/\Delta t = -8/0.001 = -8000$ V/s; hence, $i_C = -8000 \times 3 \times 10^{-6} = -24$ mA. For 9 to 10 ms, $v_C = i_C = 0$. For the interval of 10 to 13 ms, $\Delta v_C/\Delta t = 16/0.003 = 5330$ V/s and $i_C = 5330 \times 3 \times 10^{-6} = 16$ mA. In a similar manner, for the interval from 13 to 16 ms, $i_C = -16$ mA.

## Inductors

A practical inductor is formed when wire is wrapped around a core of a magnetic material or of air (Fig. 3-7). The ability of an inductor (also called a *coil* or *choke*) to oppose a change in current through it determines its *self-inductance L*. For the simple geometries of Fig. 3-7, inductance $L$ is:

$$L = \frac{N^2 \mu A}{l} \qquad \text{henrys (H)} \qquad (3\text{-}18)$$

where $N$ = number of turns
$A$ = cross-sectional area
$l$ = length of core
$\mu$ = permeability, H/m
(Refer to Chap. 4 for a complete discussion of inductors having various geometries.)

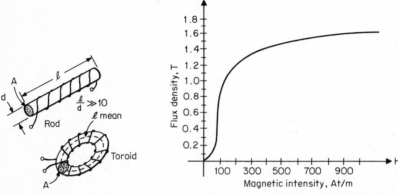

**Fig. 3-7**  Examples of inductors.          **Fig. 3-8**  A typical magnetization curve.

The *permeability* is equal to the product of the *relative permeability* $\mu_r$ and the *permeability of free space* $\mu_0 = 4 \times 10^{-7}$ H/m:

$$\mu = \mu_r \mu_0 \qquad (3\text{-}19)$$

A typical magnetization curve for annealed iron is given in Fig. 3-8. Plotted is *flux density B* (magnetic lines per unit area) in teslas as a function of *magnetic intensity H* in ampere-turns/meter. Values of relative permeability for different materials are provided in Table 3-2. The designation $\mu_{ri}$ is equal to the tangent to the magnetization curve for H =

**TABLE 3-2   Relative Permeability $\mu_{ri}$ and $\mu_{rm}$**

| Material type | $\mu_{ri}$, H/m | $\mu_{rm}$, H/m |
|---|---|---|
| Cast iron | 125 | 500 |
| Cast steel | 175 | 1,500 |
| Cold-rolled steel | 180 | 2,000 |
| Iron | 200 | 5,000 |
| Low-carbon steel | 250 | 2,500 |
| Silicon steel (4%) | 500 | 7,000 |
| Special iron (99½% purity) | 5,000 | 180,000 |
| Permalloy (78½% Ni, 21.2% Fe, 0.3% Mn) | 2,500 | 25,000 |
| Parmalloy 4-79 (4% Mo, 79% Ni, 17% Fe) | 20,000 | 100,000 |
| Supermalloy (79% Ni, 5% Mo, 0.3% Mn, 15.7% Fe) | 100,000 | 800,000 |

0. Maximum permeability $\mu_{rm}$ occurs at the maximum slope point on the magnetization curve (see Fig. 3-9). For air, $\mu_r = 1$.

**Inductors Connected in Series**    For inductors connected in series (assuming no magnetic coupling between them), the equivalent inductance $L_{EQ}$ is:

$$L_{EQ} = L_1 + L_2 + L_3 + \cdots \tag{3-20}$$

(The effect of magnetic coupling between inductors is considered in Chap. 4.)

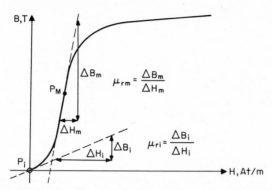

**Fig. 3-9**    Illustrating initial permeability ($\mu_{ri}$) and maximum permeability ($\mu_{rm}$).

**Inductors Connected in Parallel**    For inductors connected in parallel (assuming no magnetic coupling between them), the equivalent inductance $L_{EQ}$ is:

$$L_{EQ} = \frac{1}{1/L_1 + 1/L_2 + 1/L_3 + \cdots} \tag{3-21a}$$

For two inductors $L_1$ and $L_2$ connected in parallel:

$$L_{EQ} = \frac{L_1 L_2}{L_1 + L_2} \tag{3-21b}$$

If an inductor $L$ is in parallel with another inductor whose value is $1/n$ of $L$, then:

$$L_{EQ} = \frac{L}{n + 1} \tag{3-22}$$

**Voltage and Current in an Inductor**    The instantaneous voltage across an inductor $v_L$ is:

$$v_L = L \frac{di_L}{dt} \simeq L \frac{\Delta i_L}{\Delta t} \tag{3-23}$$

The instantaneous current in an inductor is:

$$i_L = \frac{1}{L} \int v_L \, dt \tag{3-24}$$

Under steady-state conditions, an inductor can be replaced by a *short circuit* in a dc circuit.

**Energy Stored in an Inductor**    The instantaneous energy stored in an inductor $w_L$ is:

$$w_L = \tfrac{1}{2} L i_L^2 \quad \text{joules (J)} \tag{3-25}$$

**problem 3.8**    The current in a 3-mH inductor is plotted in Fig. 3-10$a$. Sketch the voltage $v_L$ across the inductor.

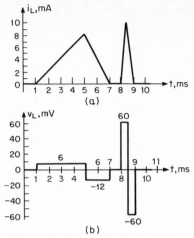

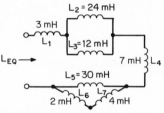

**Fig. 3-10** Voltage and current waveforms in an inductor (Problem 3.8). (*a*) Current in an inductor. (*b*) Resulting voltage across the inductor.

**solution**    The voltage waveform is plotted in Fig. 3-10*b*. For the interval from 0 to 1 ms, $i_L$ and, hence, $v_L = 0$. From 1 to 5 ms, $\Delta i_L/\Delta t = 0.008/0.004 = 2$ A/s. By Eq. (3-23), $v_L = 2 \times 0.003 = 6$ mV. The remainder of the plot is obtained in a similar manner.

**problem 3.9**    Find $L_{EQ}$ in Fig. 3-11.

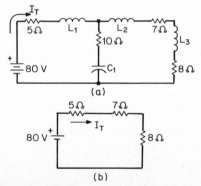

**Fig. 3-11**    Finding the equivalent inductance of inductors connected in a series-parallel circuit (Problem 3.9). No magnetic coupling between inductors exists.

**solution**    The equivalent inductance $L_{EQ} = L_1 + L_2\|L_3 + L_4 + L_5\|(L_6 + L_7) = 3 + 24\|12 + 7 + 30\|(2 + 4) = 3 + 8 + 7 + 5 = 23$ mH.

**problem 3.10**    Find $I_T$ in Fig. 3-12*a* under steady-state conditions.

**Fig. 3-12**    Finding $I_T$ under dc steady-state conditions (Problem 3.10). (*a*) Given circuit. (*b*) Resulting simplified circuit for finding $I_T$.

**solution**   In dc steady state, a capacitor acts as an open circuit and an inductor acts like a short circuit. Redrawing the circuit to reflect these conditions, Fig. 3-12$b$ is obtained. Current $I_T = 80/(5 + 7 + 8) = 80/20 = 4$ A.

## 3.3  VOLTAGE-CURRENT (*V-I*) RELATIONSHIPS AND PHASOR DIAGRAMS

### Pure Resistance

An ac voltage source, $v = V_m \sin \omega t$, is connected across a resistive element $R$ in Fig. 3.13$a$. By Ohm's law:

$$i = \frac{v}{R}$$

$$= \frac{V_m}{R} \sin \omega t \tag{3-26}$$

$$= I_m \sin \omega t$$

where $I_m = V_m/R$. By plotting $v$ and $i$ as a function of time in Fig. 3-13$b$, it is seen that the voltage across, and the current in, a pure resistance are in *phase*.

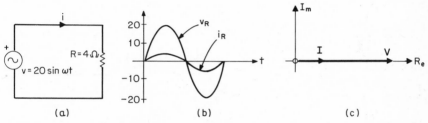

**Fig. 3-13**  Current and voltage in a pure resistive element. (*a*) Circuit. (*b*) Voltage and current waveforms. (*c*) Phasor diagram.

In terms of phasors (refer to Chap. 1), we can show this in the *frequency domain* (Fig. 3-13$c$) where:

$$\mathbf{I} = I\underline{/0°} \tag{3-27a}$$
$$\mathbf{V} = V\underline{/0°} \tag{3-27b}$$

The impedance $\mathbf{Z}$ is:

$$\mathbf{Z} = \frac{\mathbf{V}}{\mathbf{I}} \tag{3-28}$$

In a purely resistive circuit, the impedance is real and is equal to $R$.

### Pure Inductance

In Fig. 3-14$a$, the current flowing in a pure inductance $L$ is $i = I_m \sin \omega t$. By Eq. (3-23),

$$v = \omega L I_m \cos \omega t$$
$$= V_m \sin (\omega t + 90°) \tag{3-29}$$

where $V_m = \omega L I_m$ and $\cos \omega t = \sin (\omega t + 90°)$. By plotting $v$ and $i$ as a function of time in

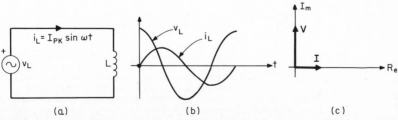

**Fig. 3-14**  Current and voltage in a pure inductive element. (*a*) Circuit. (*b*) Voltage and current waveforms. (*c*) Phasor diagram.

Fig. 3-14*b*, it is seen that the voltage *leads* the current by 90° (or the current *lags* the voltage by 90°) in a pure inductance.

In terms of phasors (Fig. 3-14*c*):

$$\mathbf{I} = I\underline{/0°} \qquad (3\text{-}30a)$$
$$\mathbf{V} = V\underline{/90°} \qquad (3\text{-}30b)$$

The impedance of a purely inductive circuit is imaginary and is equal to:

$$\mathbf{Z} = jX_L \qquad (3\text{-}31)$$

where $X_L = 2\pi fL$ and is called the *inductive reactance*.

## Pure Capacitance

In Fig. 3-15*a*, the voltage across a pure capacitance $C$ is $v = V_m \sin \omega t$. By Eq. (3-15),

$$i = CV_m \cos \omega t$$
$$= I_m \sin(\omega t + 90°) \qquad (3\text{-}32)$$

where $I_m = \omega CV_m$ and $\cos \omega t = \sin(\omega t + 90°)$. By plotting $v$ and $i$ as a function of time in Fig. 3-15*b*, it is seen that the current *leads* the voltage by 90° (or the voltage *lags* the current by 90°) in a pure capacitance.

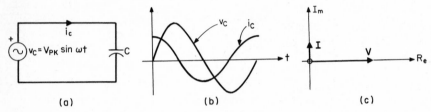

**Fig. 3-15** Current and voltage in a pure capacitive element. (*a*) Circuit. (*b*) Voltage and current waveforms. (*c*) Phasor diagram.

In terms of phasors (Fig. 3-15*c*):

$$\mathbf{I} = I\underline{/90°} \qquad (3\text{-}33a)$$
$$\mathbf{V} = V\underline{/0°} \qquad (3\text{-}33b)$$

The impedance of a purely capacitive circuit is imaginary and is equal to:

$$\mathbf{Z} = -jX_C \qquad (3\text{-}34)$$

where $X_C = 1/(2\pi fC)$ and is called the *capacitive reactance*.

**problem 3.11**    Given a black box with the following terminal current and voltage. In each case determine what ideal passive element is inside the box and its value.
(*a*) $i = 4 \sin(754t - 70°)$; $v = 1600 \sin(754t + 20°)$
(*b*) $i = 7 \sin(377t + 120°)$; $v = 70 \cos(377t + 30°)$
(*c*) $i = 60 \sin(1000t + 45°)$; $v = 300 \sin(1000t + \pi/4)$
(*d*) $i = 0.5 \sin(200t + 140°)$; $v = 250 \sin(200t + 50°)$
(*e*) $i = 4 \cos(400t + 60°)$; $v = 3800 \sin(400t + 60°)$

**solution**    (*a*) The voltage leads the current by $20° - (-70°) = 90°$; hence, the element is an inductor. The inductive reactance $X_L = 1600/4 = 400\ \Omega$. Then, $L = X_L/\omega = 400/754 = 0.53$ H.
(*b*) Current $i = 7 \sin(377t + 120°)$ also may be expressed by $i = 7 \cos(377t + 30°)$. Because the current and voltage are actually in phase, the element is resistive. The value of the resistance is $70/7 = 10\ \Omega$.
(*c*) The $\pi/4$ rad is equal to 45°; hence, the current and voltage are in phase, and the element is resistive. The value of the resistance is $300/60 = 5\ \Omega$.
(*d*) The current leads the voltage by $140° - 50° = 90°$; hence, the element is a capacitor. The capacitive reactance $X_C = 250/0.5 = 500\ \Omega$. Then, $C = 1/(\omega X_C) = 1/(200 \times 500) = 10\ \mu$F.
(*e*) The current may be rewritten as $i = 4 \sin(400t + 150°)$. The current leads the voltage by $150° - 60° = 90°$; hence, the element is a capacitor. The capacitive reactance $X_C = 3800/4 = 950\ \Omega$. Then, $C = 1/(400 \times 950) = 2.63\ \mu$F.

### Reactance Values at Zero and High Frequencies

Capacitors and inductors are frequency-sensitive elements. Under certain conditions, they can be replaced by a short or an open circuit.

   **Capacitors**   Becaue $X_C = 1/(\omega C)$, for direct current (zero frequency) $X_C$ approaches infinity, and a capacitor may be replaced by an open circuit. At very high frequencies, $X_C$ approaches 0, and a capacitor may be replaced by a short circuit.

   **Inductors**   Because $X_L = \omega L$, for direct current $X_L = 0$, and an inductor may be replaced by a short circuit. At very high frequencies, $X_L$ approaches infinity and may be replaced by an open circuit.

### 3.4   IMPEDANCE AND THE IMPEDANCE TRIANGLE

In general, impedance **Z** may be expressed as:

$$\mathbf{Z} = R \pm jX \tag{3-35}$$

and is a complex number. It it not, however, a phasor. The real part of **Z** is designated by $R$ (resistance) and the imaginary part by $X$ (reactance). The plus sign is associated with inductive reactance and the minus sign with capacitive reactance. Because the dimension of **Z** is ohms, $R$ and $X$ also are in ohms.

   In a series $RL, RC$, or $RLC$ circuit, an *impedance triangle* can be plotted in the complex plane. Because current is the same in each element in a series circuit, a voltage triangle also can be drawn. The impedance and voltage triangles will be similar for a given circuit.

   **Impedance Triangle for a Series RL Circuit**   In a series $RL$ circuit (Fig. 3-16a), $\mathbf{Z} = R + jX_L$, Angle $\theta$ associated with the impedance is $\theta = \tan^{-1}(X_L/R)$ (see Chap. 1). The impedance and voltage triangles are drawn in Fig. 3-16b. Because **I** is the same in each element, $\mathbf{V_R} = \mathbf{I}R$ and $\mathbf{V_L} = \mathbf{j}\mathbf{I}X_L$, Values of $R$ and $\mathbf{V_R}$ are plotted on the real axis, and $X_L$ and $\mathbf{V_L}$ are plotted on the imaginary axis of their respective triangles. Impedance **Z** and $\mathbf{V_{tot}}$ are the hypotenuse of their respective triangles. Angle $\theta$ is the same because the triangles are similar.

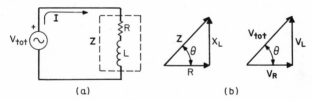

        (a)                    (b)

**Fig. 3-16**   Series $RL$ (*a*) circuit and (*b*) impedance and voltage triangles.

### Impedance Triangle for a Series RC Circuit

In a series $RC$ circuit (Fig. 3-17a), $\mathbf{Z} = R - jX_C$ and $\theta = -\tan^{-1}(X_C/R)$. The impedance and voltage triangles are drawn in Fig. 3-17b.

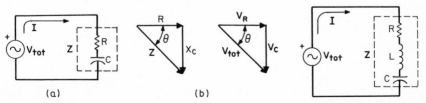

   (a)                     (b)

**Fig. 3-17**   Series $RC$ (*a*) circuit and (*b*) impedance and voltage triangles.       **Fig. 3-18**   Series $RLC$ circuit.

### Impedance Triangle for a Series RLC Circuit

In a series $RLC$ circuit (Fig. 3-18), $\mathbf{Z} = R + j(X_L - X_C)$. If $X_L > X_C$, the circuit appears inductive; if $X_L < X_C$, the circuit appears capacitive. For these conditions, the impedance and voltage triangles appear as shown in Figs. 3-16b and 3-17b, respectively. Angle $\theta = \tan^{-1}[(X_L - X_C)]/R$.

If $X_L = X_C$, the circuit is resistive, and maximum current flows in a series circuit. The frequency $f_0$ for which $X_L = X_C$ is called the *resonant frequency* of the series circuit and is expressed by:

$$f_0 = \frac{1}{2\pi\sqrt{LC}} \tag{3-36}$$

**problem 3.12**    The impedance of a series circuit is 1000 $\Omega$ and its phase angle is 35°. If $\omega = 2500$ rad/s, find the resistive and reactive components in the circuit.

**solution**    Because the phase angle is positive, the reactance is inductive. Referring to the impedance triangle in Fig. 3-16b shows that

$$R = |Z| \cos \theta = 1000 \cos 35° = 819.2 \ \Omega$$
$$X_L = |Z| \sin \theta = 1000 \sin 35° = 573.6 \ \Omega$$
$$L = \frac{X_L}{\omega} = \frac{573.6}{2500} = 0.229 \ \text{H}$$

**problem 3.13**    In a series circuit, the resistance value is 520 $\Omega$ and the phase angle is −70°. If the frequency is 60 Hz, find the (a) reactance and (b) the impedance of the circuit.

**solution**    (a) Because the phase angle is negative, an RC series circuit is implied. By reference to the impedance triangle of Fig. 3-17b, $X_C = 520 \tan 70° = 1428.7 \ \Omega$.
   (b) Impedance $|Z| = 520/\cos 70° = 1520.4 \ \Omega$.

**problem 3.14**    In a series RLC circuit, $X_C = 15$ k$\Omega$, $X_L = 22$ k$\Omega$, and $R = 12$ k$\Omega$. Determine the magnitude and phase angle of **Z**.

**solution**    $Z = 12,000 + j(22,000 - 15,000) = 12,000 + j7000$. $|Z| = \sqrt{12,000^2 + 7000^2} = 13,892.4 \ \Omega$. Phase angle $\theta = \tan^{-1}(7000/12,000) = 30.26°$.

**problem 3.15**    Find impedance **Z** in Fig. 3-19a.

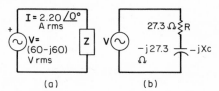

(a)                    (b)

**Fig. 3-19**    Finding impedance **Z** (Problem 3.15). (a) Circuit. (b) Impedance **Z** is equal to a resistor in series with a capacitor.

**solution**    In polar form, $V = \sqrt{60^2 + 60^2}/\underline{-\tan^{-1}(60/60)} = 84.85/\underline{-45°}$. Impedance $Z = V/I = (84.85/\underline{-45°})(2.2/\underline{0°}) = 38.57/\underline{-45°} \ \Omega$. In rectangular form, $Z = 38.75[\cos(-45°) + j \sin(-45°)] = 27.3 - j27.3 \ \Omega$. By reference to Fig. 3-19b, **Z** is shown as a resistance of 27.3 $\Omega$ in series with a capacitive reactance of 27.3 $\Omega$.

## 3.5    ADMITTANCE AND THE ADMITTANCE TRIANGLE

Admittance **Y** is the ratio of a current to a voltage:

$$Y = \frac{I}{V} \tag{3-37}$$

Its unit is the siemen (S). Admittance is also the reciprocal of impedance:

$$Y = \frac{1}{Z} \tag{3-38}$$

Since impedance is not a phasor, neither is admittance.
   The general form of admittance may be expressed by:

$$Y = G \pm jB \tag{3-39}$$

where $G$ is the conductance and $B$ is the *susceptance*. The plus sign is associated with a capacitive susceptance and the negative sign with an inductive susceptance.

A common error is to think that $G = 1/R$ and $B = 1/X$. The correct result is:

$$\mathbf{Y} = \frac{1}{\mathbf{Z}} = \frac{1}{R \pm jX} = \frac{R \mp jX}{R^2 + X^2}$$

where

$$G = \frac{R}{R^2 + X^2} \tag{3-40a}$$

$$B = \frac{X}{R^2 + X^2} \tag{3-40b}$$

Also,

$$R = \frac{G}{G^2 + B^2} \tag{3-41a}$$

$$X = \frac{B}{G^2 + B^2} \tag{3-41b}$$

In a parallel $RC$, $RL$, or $RLC$ circuit, an *admittance triangle* can be plotted on the complex plane, similar to the impedance triangle. Because the voltage is the same across each element in a parallel circuit, a current triangle also can be drawn. The admittance and current triangles are similar triangles.

### Admittance Triangle for a Parallel *RC* Circuit

In a parallel $RC$ circuit (Fig. 3-20a), $\mathbf{Y} = G + j\omega C$. Angle $\theta = \tan^{-1} (\omega C/G)$. The admittance and current triangles are drawn in Fig. 3-20b.

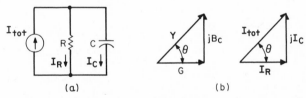

**(a)**                    **(b)**

**Fig. 3-20**    Parallel $RC$ (a) circuit and (b) admittance and current triangles.

### Admittance Triangle for a Parallel *RL* Circuit

In a parallel $RL$ circuit (Fig. 3-21a), $\mathbf{Y} = G - j/(\omega L)$. Angle $\theta = -\tan^{-1} (1/\omega LG)$. The admittance and current triangles are drawn in Fig. 3-21b.

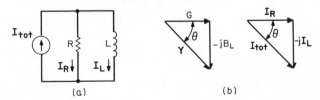

**(a)**                    **(b)**

**Fig. 3-21**    Parallel $RL$ (a) circuit and (b) admittance and current triangles.

### Admittance Triangle for a Parallel *RLC* Circuit

In a parallel $RLC$ circuit (Fig. 3-22), $\mathbf{Y} = G + j(B_C - B_L)$ where $B_C = \omega C$ and $B_L = 1/\omega L$. If $B_C > B_L$, the circuit appears capacitive; if $B_C < B_L$, the circuit is inductive. For these conditions, the admittance and current triangles appear as shown in Figs. 3-20b and 3-21b, respectively. Angle $\theta = \tan^{-1}[(B_C - B_L)/G]$.

If $B_C = B_L$, $\mathbf{Y} = G$, and the impedance is maximum. The resonant frequency at which this occurs is given by Eq. (3-36).

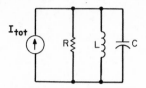

**Fig. 3-22**   Parallel $RLC$ circuit.

**problem 3.16**   For the parallel $RC$ circuit of Fig. 3-23$a$, find $(a)$ $G$, $B_C$, $\mathbf{I}_R$, $\mathbf{I}_C$, $\mathbf{I}$, $\mathbf{Y}$ and $(b)$ draw the admittance and current triangles.

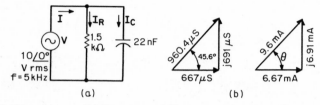

**Fig. 3-23**   Given $(a)$ a parallel $RC$ circuit and $(b)$ resulting admittance and current triangles (Problem 3.16).

**solution**   $(a)$ $G = 1/1500 = 0.667$ mS; $B_C = 6.28 \times 5000 \times 22 \times 10^{-9} = 0.691$ mS; $\mathbf{I}_R = 0.667 \times 10^{-3} \times 10\underline{/0^\circ} = 6.67\underline{/0^\circ}$ mA; $\mathbf{I}_C = (0.691 \times 10^{-3})\underline{/90^\circ} \times 10\underline{/0^\circ} = 6.91\underline{/90^\circ}$ mA; $\mathbf{I} = \mathbf{I}_R + j\mathbf{I}_C = 6.67 + j6.91$ mA. Angle $\theta = \tan^{-1}(0.691/0.667) = 45.6^\circ$; hence, $\mathbf{Y} = \sqrt{0.667^2 + 0.691^2} \ \underline{/45.6^\circ} = 0.96\underline{/45.6^\circ}$ mS.

$(b)$ The admittance and current triangles are drawn in Fig. 3-23$b$.

**problem 3.17**   For a parallel $RLC$ circuit, $G = 8$ mS, $B_C = 5$ mS, and $B_L = 10$ mS. $(a)$ Find $\mathbf{Y}$. $(b)$ If 25 V is impressed across the circuit, determine $\mathbf{I}$.

**solution**   $(a)$ $\mathbf{Y} = 8 + j(5 - 10) = 8 - j5$ mS $= 9.43\underline{/-32^\circ}$ mS. The minus sign indicates a net inductive susceptance for the parallel circuit.

$(b)$ $\mathbf{I} = 9.43\underline{/-32^\circ} \times 25\underline{/0^\circ} = 0.235.75\underline{/-32^\circ}$ A

**problem 3.18**   The terminal voltage and current of a parallel $RLC$ circuit are 100 53° V and 5 0° A. $(a)$ Determine the value of $\mathbf{Z}$ and $\mathbf{Y}$. $(b)$ If $B_C = 0.01$ S, determine the value of $B_L$. $(c)$ Find the branch currents.

**solution**

$$(a)\ \mathbf{Z} = (100\underline{/53^\circ})/5\underline{/0^\circ} = 20\underline{/53^\circ} = 12 + j16\Omega$$
$$\mathbf{Y} = 1/(12 + j16) = 0.05\underline{/-53^\circ}\ \text{S} = 0.03 - j0.04\ \text{S}$$
$$(b)\ -0.04 = 0.01 - B_L;\ \text{solving,}\ B_L = 0.05\ \text{S}$$
$$(c)\ \mathbf{I}_R = 100\underline{/53^\circ} \times 0.03 = 3\underline{/53^\circ}\ \text{A}$$
$$\mathbf{I}_C = 100\underline{/53^\circ} \times 0.01\underline{/90^\circ} = 1\underline{/143^\circ}\ \text{A}$$
$$\mathbf{I}_L = 100\underline{/53^\circ} \times 0.05\underline{/-90^\circ} = 5\underline{/-37^\circ}\ \text{A}$$

## 3.6   EQUIVALENT IMPEDANCE AND ADMITTANCE

Like resistances, impedances and admittances may be connected in series, parallel, or in series-parallel. Whereas real numbers are associated with resistances, complex numbers, in general, are associated with impedances and admittances.

**Equivalent Series Impedance**   For impedances in series, the equivalent impedance $Z_{EQ}$ is:

$$\mathbf{Z}_{EQ} = \mathbf{Z}_1 + \mathbf{Z}_2 + \mathbf{Z}_3 + \cdots \tag{3-42}$$

Impedances in series combine like resistors in series.

**Equivalent Series Admittance**    For admittances in series, the equivalent admittance $Y_{EQ}$ is:

$$Y_{EQ} = \frac{1}{1/Y_1 + 1/Y_2 + 1/Y_3 + \cdots} \tag{3-43}$$

Admittances in series combine like conductances in series.

**Equivalent Parallel Impedance**    For impedances in parallel, the equivalent impedance $Z_{EQ}$ is:

$$Z_{EQ} = \frac{1}{1/Z_1 + 1/Z_2 + 1/Z_3 + \cdots} \tag{3-44}$$

Impedances in parallel combine like resistors in parallel.

**Equivalent Parallel Admittance**    For admittances in parallel, the equivalent admittance $Y_{EQ}$ is:

$$Y_{EQ} = Y_1 + Y_2 + Y_3 + \cdots \tag{3-45}$$

Admittances in parallel combine like conductances in parallel.

### Series-Parallel and Parallel-Series Conversions

By the application of Eqs. (3-40) and (3-41) it becomes possible to convert a series $RL$ or $RC$ circuit to a parallel $RL$ or $RC$ circuit, and vice versa. To distinguish the series and parallel elements, subscript $s$ is used for a series element and subscript $p$ for a parallel element.

**problem 3.19**    Convert the circuit of Fig. 3-24 to an all-parallel circuit. Frequency $f = 750$ kHz.

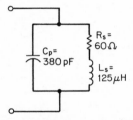

**Fig. 3-24**    Converting the given circuit to an all-parallel circuit (Problem 3.19).

**solution**    The reactance of $L_s$ is $X_{Ls} = 6.28 \times 0.750 \times 10^6 \times 125 \times 10^{-6} = 589$ Ω. By Eq. (3-40a), $G_p = 60/(60^2 + 589^2) = 0.17$ mS. Hence, $R_p = 1/(0.17 \times 10^3) = 5882$ Ω.
By Eq. (3-40-b-), $B_p = 589/(60^2 + 589^2) = 1.68$ mS. Hence, $L_p = 1/(6.28 \times 0.750 \times 10^6 \times 1.68 \times 10^{-3}) = 0.127$ mH. The resulting parallel circuit consists of (380 pF)‖(5882 Ω)‖(0.127 mH).

**problem 3.20**    Convert the circuit of Fig. 3-25 to an all-series circuit. Frequency $f = 400$ Hz.

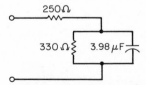

**Fig. 3-25**    Converting the given circuit to an all-series circuit (Problem 3.20).

**solution**    $G = 1/330 = 3$ mS; $B = 6.28 \times 400 \times 3.98 \times 10^{-6} = 10$ mS. By Eq. (3-41a), $R_s = (3 \times 10^{-3})/(9 \times 10^{-6} + 100 \times 10^{-6}) = 27.7$ Ω. By Eq. (3-41b), $X_s = (10 \times 10^{-3})/(9 \times 10^{-6} + 100 \times 10^{-6}) = 91.6$ Ω. The value of $C_s = 1/(6.28 \times 400 \times 91.6) = 4.34$ μF. The total series resistance is $250 + 27.7 = 277.7$ Ω. Hence, the resulting circuit consists of a resistance of 277.7 Ω in series with a 4.34-μF capacitor.

## 3.7   KIRCHHOFF'S LAWS FOR AC CIRCUITS

Kirchhoff's voltage and current laws apply in the same manner (see Chap. 2) to ac circuits in steady state as to dc circuits. The only difference is that, in general, complex numbers and phasors are associated with an ac circuit.

   **problem 3.21**   Find the unknown branch current $I_U$ and the unknown impedance $Z_U$ in the circuit of Fig. 3-26.

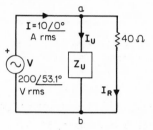

**Fig. 3-26**   Determining current $I_U$ and impedance $Z_U$ (Problem 3.21).

**solution**   By application of the KCL to node $a$, $I = I_U + I_R$. Current $I_R = (200\underline{/53.1°})/40\underline{/0°} = 5\underline{/53.1°}$ A. Then, $I_U = I - I_R = 10\underline{/0°} - 5\underline{/53.1°} = 10 + j0 - (3 + j4) = 8.06\underline{/-29.7°}$ A.
   Impedance $Z_U = V/I_U = (200\underline{/53.1°})/(8.06\underline{/-29.7°}) = 24.8\underline{/82.8°}$ $\Omega$. In rectangular form, $Z_U = 3.11 + j24.6$ $\Omega$, which is equivalent to a 3.11-$\Omega$ resistor in series with an inductive reactance of 24.6 $\Omega$.

   **problem 3.22**   Referring to Fig. 3-27, (a) find voltage $V_B$. (b) If $I = 0.548\underline{/43.2°}$ A, determine the values of $Z_A$ and $Z_B$.

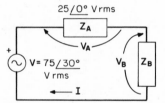

**Fig. 3-27**   Determining voltage $V_B$ and impedances $Z_A$, $Z_B$ (Problem 3.22).

**solution**   (a) Applying the KVL around the closed loop, $V_B = V - V_A = 75\underline{/30°} - 25\underline{/0°} = 64.95 + j37.5 - (25 + j0) = 54.79\underline{/43.2°}$ V.
   (b) $Z_A = V_A/I = (25\underline{/0°})/(0.548\underline{/43.2°}) = 45.6\underline{/-43.2°}$ $\Omega$.

$$Z_B = V_B/I = (54.79\underline{/43.2°})/(0.548\underline{/43.2°}) = 100 \ \Omega.$$

## 3.8   VOLTAGE AND CURRENT DIVIDERS FOR AC CIRCUITS

Voltage and current dividers (Chap. 2) apply equally to ac circuits. The only difference is that, in general, complex quantities and phasors are associated with ac circuits.
   **Voltage-Driver Principle**   For a series circuit of impedances, the voltage $V_n$ across any impedance $Z_n$ is equal to the product of the applied voltage $V_{tot}$ and $Z_n$ divided by the sum of the impedances $Z_{tot}$:

$$V_n = V_{tot} \frac{Z_n}{Z_{tot}} \tag{3-46}$$

   **Current-Divider Principle**   For a parallel circuit of impedances, the current $I_n$ in a particular branch $Z_n$ is equal to the product of the applied current $I_{tot}$ and the equivalent impedance $Z_{EQ}$ divided by $Z_n$:

$$I_n = I_{tot} \frac{Z_{EQ}}{Z_n} \tag{3-47}$$

For two impedances $\mathbf{Z}_1$ and $\mathbf{Z}_2$ in parallel, the current $\mathbf{I}_1$ in $\mathbf{Z}_1$ *is:*

$$\mathbf{I}_1 = \mathbf{I}_{tot} \frac{\mathbf{Z}_2}{\mathbf{Z}_1 + \mathbf{Z}_2} \qquad (3\text{-}48a)$$

The current in $\mathbf{Z}_2$ is:

$$\mathbf{I}_2 = \mathbf{I}_{tot} \frac{\mathbf{Z}_1}{\mathbf{Z}_1 + \mathbf{Z}_2} \qquad (3\text{-}48b)$$

**problem 3.23**    Using voltage division, find the voltage across the inductor in Fig. 3-28.

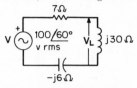

**Fig. 3-28**    Finding voltage $V_L$ by voltage division (Problem 3.23).

**solution**    $\mathbf{Z}_{tot} = 7 + j(30 - 6) = 7 + j24 = 25\underline{/73.74°}$ Ω. Also, $j30 = 30\underline{/90°}$ Ω. Hence, by Eq. (3-46), $\mathbf{V}_L = (30\underline{/90°}) \times (100\underline{/60°})/(25\underline{/73.74°}) = 120\underline{/76.25°}$ V.

**problem 3.24**    Using current division, find current $I_R$ in the 5-Ω resistance of Fig. 3-29.

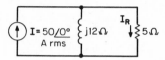

**Fig. 3-29**    Finding current $I_R$ by current division (Problem 3.24).

**solution**    By Eq. (3-48b), $I_R = 50\underline{/0°} \times j12/(5 + j12) = 46.15\underline{/22.6°}$ A.

## 3.9    MESH AND NODAL ANALYSIS OF AC CIRCUITS

The methods of mesh and nodal analysis described in Chap. 2 for dc circuits also apply to ac circuits in steady state.

**problem 3.25**    Find the mesh currents in the circuit of Fig. 3-30.

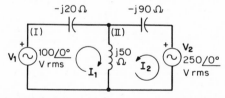

**Fig. 3-30**    Calculating mesh currents $I_1$ and $I_2$ (Problem 3.25).

**solution**    The voltage rise due to voltage sources in mesh $I$ is $100\underline{/0°}$ V, and in mesh II it is $250\underline{/0°}$ V. The self-impedance of mesh $I$ is $-j20 + j50 = j30$ Ω. The common impedance between meshes I and II is $j50$ Ω. The self-impedance of mesh II is $j50 - j90 = -j40$ Ω. The resulting mesh equations are:

$$100 = j30\mathbf{I}_1 + j50\mathbf{I}_2$$
$$250 = j50\mathbf{I}_1 - j40\mathbf{I}_2$$

Solving by determinants, we have

$$\Delta = \begin{vmatrix} j30 & j50 \\ j50 & -j40 \end{vmatrix} = -j^2 \times 1200 - j^2 \times 2500 = 3700$$

Applying Cramer's rule yields:

$$I_1 = \frac{\Delta_1}{\Delta} \quad \text{and} \quad I_2 = \frac{\Delta_2}{\Delta}$$

$$\Delta_1 = \begin{vmatrix} 100 & j50 \\ 250 & -j40 \end{vmatrix} = -j400 - j12{,}500 = -j16{,}500$$

$$\Delta_2 = \begin{vmatrix} j30 & 100 \\ j50 & 250 \end{vmatrix} = j7500 - j5000 = j2500$$

Hence, $I_1 = -j16{,}500/3700 = -j4.46 = 4.46\underline{/-90°}$ A and $I_2 = j2500/3700 = j0.676 = 0.676\underline{/90°}$ A

**problem 3.26**    Write the nodal equations for the circuit of Fig. 3-31.

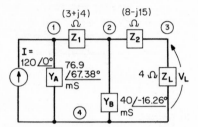

**Fig. 3-31**    Writing nodal equations for the given circuit (Problem 3.26).

**solution**    Applying KCL to:

Node 1:    $120\underline{/0°} = (76.9 \times 10^{-3})\underline{/67.38°}\,V_1 - [1/(3 + j4)](V_1 - V_2) - 0V_3$

Node 2:    $0 = -V_1/(3 + j4) + [1/(3 + j4) + (40 \times 10^{-3}\underline{/-16.26°}$
$\phantom{0 =} 1/(8 - j15)]V_2 - V_3/(8 - j15)$

Node 3:    $0 = -0V_1 - V_2/(8 - j15) + [1/4 + 1/(8 - j15)]\,V_3$

## 3.10    THÉVENIN AND NORTON THEOREMS APPLIED TO AC CIRCUITS

Thévenin and Norton theorems described in Chap. 2 for dc circuits also apply to ac circuits. As for dc circuits, only independent voltage and current sources may be set to zero.

**problem 3.27**    Determine the Thévenin equivalent circuit across terminals $x$-$y$ for the circuit of Fig. 3-32$a$.

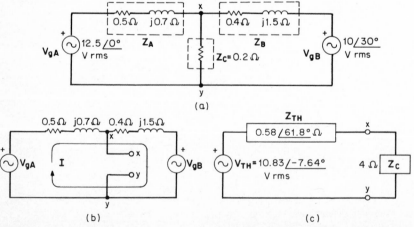

**Fig. 3-32**    Determining a Thévenin equivalent circuit (Problem 3.27). ($a$) Given circuit. ($b$) Impedance $Z_C$ disconnected across terminals $x$-$y$. ($c$) Resulting Thévenin equivalent circuit.

**solution** Impedance $Z_C$ is disconnected across terminals $x$-$y$, resulting in Fig. 3-32$b$. Current $\mathbf{I}$ = $(12.5/0° - 10/30°)/(0.5 + j0.7 + 0.4 + j1.5) = 2.65/-15.22°$ A. $\mathbf{V}_{TH} = \mathbf{V}_{gA} - \mathbf{IZ}_A = 12.5/0° - (0.5 + j0.7)(2.65/-15.22°) = 10.83/-7.64°$ V.

By setting the sources to zero, $\mathbf{Z}_{TH} = \mathbf{Z}_A \| \mathbf{Z}_B = (0.5 + j0.7)(0.4 + j1.5)/(0.5 + j0.7 + 0.4 + j1.5) = 0.58/61.78°$ $\Omega$. The Thévenin equivalent is drawn in Fig. 3-32$c$.

## Source Transformation

For ac circuits:

$$\mathbf{I}_N = \mathbf{Y}_N \mathbf{V}_{TH} \tag{3-49a}$$

$$\mathbf{Y}_N = \frac{1}{\mathbf{Z}_{TH}} \tag{3-49b}$$

**problem 3.28** Using the results of Problem 3.27, draw the Norton equivalent for the circuit of Fig. 3-32$a$.

**solution** By Eq. (3-49$b$), $\mathbf{Y}_N = 1/(0.58/61.78°) = 1.7/-61.78°$ S. Substituting the value of $\mathbf{Y}_N$ in Eq. (3-49$a$), $I_N = (1.72/-61.78°)(10.83/-7.64°) = 18.63/-69.42°$ A. The Norton equivalent circuit is given in Fig. 3-33.

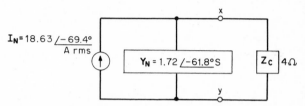

**Fig. 3-33** Norton equivalent circuit (Problem 3.28).

**problem 3.29** Find the Norton equivalent to the left of terminals $x$-$y$ for the circuit of Fig. 3-34$a$

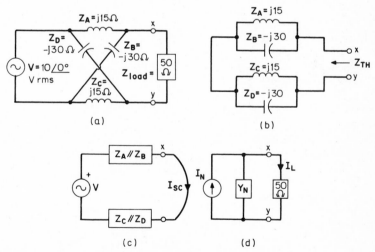

**Fig. 3-34** Determining a Norton equivalent circuit (Problem 3.29). ($a$) Given circuit. ($b$) Finding $\mathbf{Z}_{TH}$. ($c$) Finding $\mathbf{I}_{SC}$. ($d$) Resulting Norton circuit.

**solution** With the load removed across terminals $x$-$y$ and setting the sources to zero, one obtains two identical LC parallel circuits in series (Fig. 3-34$b$). Then, $\mathbf{Z}_{TH} = \mathbf{Z}_A \| \mathbf{Z}_B + \mathbf{Z}_C \| \mathbf{Z}_D = 2 \times (j15)(-j30)/(j15 - j30) = 60/90°$ $\Omega$.

Shorting terminals $x$-$y$, from Fig. 3-34$c$, $\mathbf{I}_{SC} = \mathbf{I}_N = (120/0°)/(60/90°) = 2/-90°$ A. $\mathbf{Y}_N = 1/\mathbf{Z}_{TH} = 1/(60/90°) = 16.67/-90°$ S. The Norton equivalent circuit is shown in Fig. 3-34$d$.

## Thévenin and Norton Applications with Dependent Sources

For circuits containing dependent sources, $\mathbf{Z}_{TH}$ *cannot* be found by setting the sources to zero. (This is permitted only if the sources are independent.) Instead, the terminals are opened to find the open-circuit voltage $\mathbf{V}_{OC}$ ($\mathbf{V}_{TH}$) and then shorted to find the short-circuit current $\mathbf{I}_{SC}$. The Thévenin impedance is then found by:

$$\mathbf{Z}_{TH} = \frac{\mathbf{V}_{OC}}{\mathbf{I}_{SC}} \qquad (3\text{-}50)$$

Another method used for determining $\mathbf{Z}_{TH}$ (or $\mathbf{Y}_N$) is illustrated in the next two problems.

**problem 3.30**    A hybrid model of a common-emitter amplifier stage is illustrated in Fig. 3-35$a$. Find the Norton admittance $\mathbf{Y}_N$ across output terminals 2-2'.

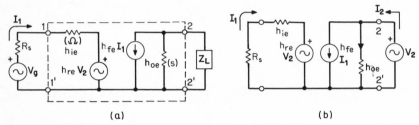

(a)                                            (b)

**Fig. 3-35**    Finding the output admittance $\mathbf{Y}_N$ of an amplifier (Problem 3.30). ($a$) Given circuit containing two dependent sources. ($b$) $\mathbf{Y}_N = \mathbf{I}_2/\mathbf{V}_2$.

**solution**    Note that the model has two dependent sources: a current-dependent source $h_{fe}\mathbf{I}_1$ and a voltage-dependent source $h_{re}\mathbf{V}_2$. A voltage source $\mathbf{V}_2$ is impressed across the output terminals 2-2'' (Fig. 3-35$b$). (In finding the output admittance or impedance of an amplifier, the signal source $\mathbf{V}_g$ is set to zero.) The output admittance is equal to $\mathbf{I}_2/\mathbf{V}_2$.

Applying KVL to the input side, we obtain $-h_{re}\mathbf{V}_2 = R_s\mathbf{I}_1 + h_{ie}\mathbf{I}_1$. Solving for $\mathbf{I}_1/\mathbf{V}_2$ yields $\mathbf{I}_1/\mathbf{V}_2 = -h_{re}/(h_{ie} + R_s)$. Applying KCL at the output side results in $-h_{fe}\mathbf{I}_1 + \mathbf{I}_2 = h_{oe}\mathbf{V}_2$. Solving for $\mathbf{I}_2/\mathbf{V}_2 = \mathbf{Y}_N$, we get $\mathbf{Y}_N = h_{fe}(\mathbf{I}_1/\mathbf{V}_2) + h_{oe}$. Substituting the expression for $\mathbf{I}_1/\mathbf{V}_2$ yields $\mathbf{Y}_N = h_{oe} - h_{fe}h_{re}/(h_{ie} + R_s)$ S.

**problem 3.31**    A small-signal model of a source follower is shown in Fig. 3-36$a$. Determine the Thévenin (output) impedance of the circuit across terminals $S$-$D$.

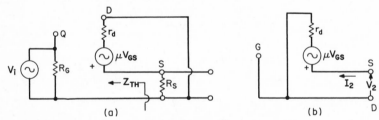

(a)                                            (b)

**Fig. 3-36**    Finding the output impedance $\mathbf{Z}_{TH}$, of a source follower (Problem 3.31). ($a$) Given circuit containing one dependent source. ($b$) $\mathbf{Z}_{TH} = \mathbf{V}_2/\mathbf{I}_2$.

**solution**    The model has a single dependent source $\mu\mathbf{V}_{GS}$. By setting $\mathbf{V}_i$ to zero, removing resistance $R_S$, and impressing voltage $\mathbf{V}_2$ across the output ($S$-$D$) terminals, Fig. 3-36$b$ is obtained. From the figure, $\mathbf{V}_2 = \mu\mathbf{V}_{GS} + r_d\mathbf{I}_2$. Voltage $\mathbf{V}_{GS}$ is the voltage across the gate ($G$) and source ($S$) terminals. It is seen that $\mathbf{V}_{GS} = -\mathbf{V}_2$. Hence, $\mathbf{V}_2 = -\mu\mathbf{V}_2 + r_d\mathbf{I}_2$. Solving for $\mathbf{V}_2/\mathbf{I}_2$,

$$\mathbf{Z}_{TH} = \frac{\mathbf{V}_2}{\mathbf{I}_2} = \frac{r_d}{1 + \mu} \qquad \Omega$$

## 3.11 OTHER NETWORK THEOREMS FOR AC CIRCUITS

The superposition, maximum power transfer, Millman's, compensation, and Tellegen's theorems, as well as topology, discussed in Chap. 2 also apply to ac circuits. Some remarks, however, concerning superposition, maximum power transfer, and Millman's theorem are appropriate here.

**Superposition** Superposition can be extended to circuits containing both dc and ac sources of different frequencies.

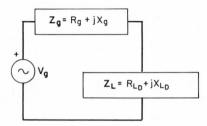

**Fig. 3-37** Illustrating the maximum power transfer theorem for an ac circuit.

**Maximum Power Transfer** As reference to Fig. 3-37 shows, in general the source impedance $Z_g = R_g + jX_g$ and the load impedance $Z_L = R_L + jX_L$. The power delivered to the load $P_L$ may be expressed by:

$$P_L = \frac{|V_g|^2 R_L}{(R_g + R_L)^2 + (X_g + X_L)^2} \tag{3-51}$$

For maximum power to the load:

$$R_L = \sqrt{R_g^2 + (X_g + X_L)^2} \tag{3-52a}$$

and
$$X_L = -X_g \tag{3-52b}$$

Maximum power transferred from the source to the load, therefore, occurs when the load impedance is the *complex conjugate* of the source impedance. Substituting $R_L = R_g$ and $X_L = -X_g$ in Eq. (3-51), an expression for maximum power, $P_{Lmax}$, is obtained:

$$P_{Lmax} = \frac{|V_g|^2}{4R_L} \tag{3-53}$$

**Millman's Theorem** For several sources of the same frequency, $G$'s become $Y$'s and the voltages are phasors in Eq. (2-35). Similarly, referring back to Eq. (2-36), all $R$'s become $Z$'s and the currents are phasor quantities.

**problem 3.32** A half-wave rectified sine wave having a maximum amplitude of 200 V is applied to the series circuit of Fig. 3-38. The first three terms of $v(t)$ are:

$$v(t) = \frac{200}{\pi} + 100 \sin 377t - 40 \cos 754t$$

Using superposition, determine the output voltage $v_o(t)$ when the input voltage is at its maximum value.

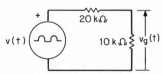

**Fig. 3-38** Using superposition to find $v_o(t)$ (Problem 3.32).

**solution** For the dc component, by voltage division, $v_o(t)_0 = (200/\pi)(10/30) = 21.22$ V. For the fundamental term, $v_o(t)_1 = (100 \sin 377t)(10/30) = 33.33 \sin 377t$ V. And last, for the second harmonic term, $v_o(t)_2 = (-40 \cos 754t)(10/30) = -13.33 \cos 754t$ V. The total response is equal to the sum of the individual responses:

$$v_o(t) = 21.22 + 33.33 \sin 377t - 13.33 \cos 754t \text{ V}$$

**problem 3.33**    A source having an impedance $\mathbf{Z}_g = 63 + j16\ \Omega$ delivers power to a variable resistive load. Find the value of $R_L$ for most power under these conditions.

**solution**    We cannot achieve maximum power given by Eq. (3-53) because only a variable resistance is available. The most power obtained is realized by letting $R_L$ [by Eq. (3-52a)] equal to $\sqrt{63^2 + 16^2} = 65\ \Omega$.

**problem 3.34**    Referring to Fig. 3-39, (a) calculate the value of $\mathbf{Z}_L$ for maximum power transfer and (b) determine $P_{L,\text{max}}$.

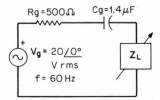

**Fig. 3-39**    Finding $\mathbf{Z}_L$ for maximum power transfer (Problem 3.34).

**solution**    (a) $R_L = 500\ \Omega$ and $X_L = 1/\omega C_g$; hence, $L = 1/(\omega^2 C_g) = 1/(377^2 \times 1.4 \times 10^{-6}) = 5.03$ H. (b) By Eq. (3-53), $P_{L,\text{max}} = 20^2/(4 \times 500) = 0.2$ W.

**problem 3.35**    Using Millman's and Thévenin's theorems, find the power dissipated in $R_L$ in the circuit of Fig. 3-40a.

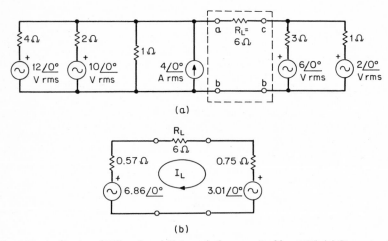

**Fig. 3-40**    Application of Millman's and Thévenin's theorems (Problem 3.35). (a) Given circuit. (b) Thévenin circuit.

**solution**    Removing the 6-$\Omega$ load across terminals $a$-$c$ and looking to the left of terminals $a$-$b$, by Millman's theorem, $\mathbf{V}_{\text{TH}} = (12/4 + 10/2 + 4/1)/(1/4 + 1/2 + 1) = 6.86$ V. If the sources are set to zero, $\mathbf{Z}_{\text{TH}} = 1/(1/4 + 1/2 + 1) = 0.57\ \Omega$.

By repetition of the preceding steps for the circuit to the right of terminals $c$-$b$, $\mathbf{V}_{\text{TH}} = (6/3 + 2/1)/(1/3 + 1) = 3.01$ V, and $\mathbf{Z}_{\text{TH}} = 1/(1/3 + 1) = 0.75\ \Omega$. The resulting Thévenin circuit is shown in Fig. 3-40b. Load current $\mathbf{I}_L = (6.86 - 3.01)/(6 + 0.75 + 0.57) = 0.526$ A. The power dissipated in the load is $P_L = I^2 R = (0.526)^2 \times 6 = 1.66$ W.

### 3.12  AC POWER AND THE POWER TRIANGLE

Instantaneous power $p$ in an ac circuit may be expressed by:

$$p = vi$$
$$= (V_m \sin \omega t)[I_m \sin (\omega t + \phi)] \tag{3-54}$$

where $\phi$ is the phase angle between the voltage and current waveforms.

Average (real) power $P_{av}$ is:

$$P_{av} = \tfrac{1}{2}V_m I_m \cos \phi \tag{3-55a}$$
$$= VI \cos \phi \tag{3-55b}$$

If $\phi \leq 90°$ or $\phi \geq -90°$, $P_{av}$ is positive and the circuit absorbs power. If $\phi > 90°$ or $\phi < 270°$, then $P_{av}$ is negative and the circuit gives up power.

#### Power in a Pure Resistive Circuit

For a pure resistive circuit, $\phi = 0°$ and

$$P_{av} = VI \tag{3-56a}$$
$$= I^2 R \tag{3-56b}$$
$$= \frac{V^2}{R} \tag{3-56c}$$
$$= V^2 G \tag{3-56d}$$

Referring to Fig. 3-41, one sees that at any instant of time the voltage and current are simultaneously positive or negative. Because power is the product of voltage and current, the power in a resistor is always positive. Power is dissipated in a resistor and not returned to the circuit.

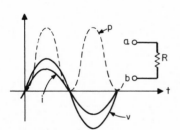

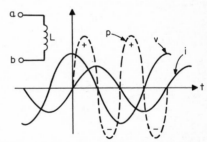

**Fig. 3-41** Voltage, current, and power waveforms in a pure resistive circuit. (Voltage and current are in phase.)

**Fig. 3-42** Voltage, current, and power waveforms in a pure inductive circuit. (Current lags the voltage by 90°.)

#### Power in a Pure Inductive Circuit

Reference to Fig. 3-42 shows that the voltage leads the current by 90° in a pure inductive circuit. The positive power pulses indicate that energy is imparted to the inductance and stored. During the negative power pulse, the energy is returned to the circuit. The average power dissipated in a pure inductive circuit, therefore, is zero.

#### Power in a Pure Capacitive Circuit

Referring to Fig. 3-43, one sees that current leads the voltage by 90° in a pure capacitive circuit. Power is imparted to the capacitor during the positive power pulses and returned to the circuit during the negative power pulses. The average power dissipated in a pure capacitive circuit, therefore, is zero.

In summary, power is dissipated only in resistance; the average power dissipated in an inductance or capacitance is zero. For the general case of $\mathbf{Z} = R \pm jX$, power is dissipated only in $R$.

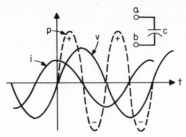

**Fig. 3-43**  Voltage, current, and power waveforms in a pure capacitive circuit. (Current leads the voltage by 90°.)

### Apparent, Real, and Reactive Power

The product of the total voltage and current in a circuit is called the *apparent power* $P_a$:

$$P_a = VI \tag{3-57}$$

If the ratio of real to apparent power is taken, then:

$$\frac{P_{av}}{P_a} = \frac{VI \cos \phi}{VI}$$
$$= \cos \phi \tag{3-58}$$

Because this ratio is a cosine function, a right triangle, called a *power triangle,* may be constructed (Fig. 3-44). The average power is the adjacent side and the apparent power the hypotenuse of the triangle, in accordance with the definition of cosine.

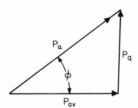

**Fig. 3-44**  A power triangle.

The opposite side of the triangle divided by the hypotenuse is $\sin \phi$. The opposite side represents *reactive power* $P_q$. Reactive power can be thought of as the rate at which energy is alternately stored and released from a reactive element.

From the right triangle of Fig. 3-44,

$$P_a^2 = P_{av}^2 + P_q^2 \tag{3-59}$$

Dimensionally, $P_a$ is in *voltamperes* (VA), $P_q$ is in *voltamperes reactive* (var) and, of course, $P_{av}$ is in watts (W). The cosine of $\phi$ is called the *power factor* (PF) of the circuit:

$$PF = \cos \phi \tag{3-60}$$

The power factor in an $RL$ circuit is a *lagging* power factor; in an $RC$ circuit, it is a *leading* power factor. The ratio $P_q/P_a = \sin \phi$ is called the *reactive factor* RF:

$$RF = \sin \phi \tag{3-61}$$

**problem 3.36**    A circuit connected to a 40-V source dissipates 30 W. If the current flowing is 3 A, find $P_a$, $P_{av}$, $P_q$, PF, $\phi$, and Z.

**solution**

$$P_a = 40 \times 3 = 120 \text{ VA.}$$
$$P_{av} = 30 \text{ W (given).}$$

By Eq. (3-59),

$P_q = \sqrt{120^2 - 30^2} = 116.2$ var.

PF $= \cos \phi = 30/120 = 0.25$. Because of insufficient information, it can be either a lagging or leading power factor.

$$\phi = \cos^{-1}(0.25) = 75.52°.$$
$$|\mathbf{Z}| = 40/3 = 13.34\ \Omega \quad \text{and} \quad \mathbf{Z} = 13.34/\underline{+75.52°}\ \Omega.$$

**problem 3.37** For the series $RLC$ circuit of Fig. 3-45$a$, find the voltage across each element and the total $P_a$, $P_{av}$, $P_q$, and $\phi$ of the circuit.

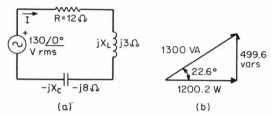

(a)   (b)

**Fig. 3-45** Finding power in an ac circuit (Problem 3.37) ($a$) Given circuit. ($b$) Power triangle.

**solution** Current $\mathbf{I} = (130/\underline{0°})/(12 + j3 - j8) = 10/\underline{22.6°}$ A. Then, $\mathbf{V}_R = 10/\underline{22.6°} \times 10 = 120/\underline{22.6°}$ V; $\mathbf{V}_L = 10/\underline{22.6°} \times 3/\underline{90°} = 30/\underline{112.6°}$ V; and $\mathbf{V}_C = 10/\underline{22.6°} \times 8/\underline{-90°} = 80/\underline{-67.4°}$ V.

Phase angle $\theta = \tan^{-1}(-5/12) = -22.6°$. $P_a = 130 \times 10 = 1300$ VA; $P_{av} = 1300 \cos(-22.6°) = 1200.2$ W; $P_q = 1300 \sin(-22.6°) = -499.6$ var; and PF $= \cos(-22.6°) = 0.923$ leading. The power triangle is drawn in Fig. 3-45$b$.

**problem 3.38** A black box containing an admittance of $20/\underline{-60°}$ mS is connected across a 120-V, 60-Hz source. Find $P_{av}$, $P_a$, $P_q$, and PF.

**solution** $\mathbf{Z} = 1/(20/\underline{-60°}) = 50/\underline{60°} = 25 + j43.3\ \Omega$. This represents a 25-$\Omega$ resistor in series with an inductive reactance of 43.3 $\Omega$. Current $\mathbf{I} = (120/\underline{0°})/(50/\underline{60°}) = 2.4/\underline{-60°}$ A.

$P_{av} = 120 \times 2.4$ x $\cos(-60°) = 144$ W; $P_a = 120 \times 2.4 = 288$ VA; $P_q = 288 \sin(-60°) = -249.4$ var; and PF $= \cos(-60°) = 0.5$ lagging. The power triangle is drawn in Fig. 3-46.

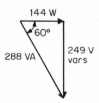

**Fig. 3-46** Power triangle for Problem 3.38.

## 3.13 BIBLIOGRAPHY

Bartowiak, Robert: Electric Circuits, IEP (Intext), New York, 1973.

Boylestad, Robert: Introduction to Circuit Analysis, Charles E. Merrill Books, Inc., Columbus, Ohio, 3d. ed., 1977.

Churchman, Lee: Introduction to Circuits, Holt, New York, 1976.

Fitzgerald, A. E., Higgenbotham, D. E., and Grabel, A. Basic Electrical Engineering, 4th ed., McGraw-Hill, New York, 1975.

Hayt, W. H., and Kemmerly, J.: Engineering Circuit Analysis, 3d ed., McGraw-Hill, New York, 1978.

Jackson, Herbert: Introduction to Electric *Circuits*, 4th ed., Prentice-Hall, Englewood Cliffs, N.J., 1976.

Joerg, E. A.: Lecture notes on *Introductory Networks, Passive and Active Network Analysis,* and *Advanced Circuit Analysis,* Hofstra University, Hempstead, N.Y., Bridgeport Engineering Institute, Bridgeport, Conn., and Westchester Community College, Valhalla, N.Y., 1965–1977.

Karni, Shlomo: Intermediate Network Analysis, Allyn and Bacon, Boston, 1971.

Klayton, Marwin: Fundamental Electrical Technology, Addison-Wesley, Reading, Mass., 1977.

Kinariwala, Kuo, and Tsao: Linear Circuits and Computations, Wiley, New York, 1973.
Oppenheimer and Borchers: Direct and Alternating Currents, McGraw-Hill, New York, 1973.
Risdale, R.: Electrical Circuits for Electrical Technology, McGraw-Hill, New York, 1976.
Taylor, D.: Electronic Engineering Mathematics, Wiley, Australia, 1972.
Tocci, Ronald: Introduction to Electric Circuit Analysis, Charles E. Merrill, Columbus, Ohio, 1974.
Van Valkenburg, M. E.: Network Analysis, 3d ed., Prentice-Hall, Englewood Cliffs, N.J., 1974.

# Chapter 4

# Selecting *R, L,* and *C* Components

## MORRIS E. LEVINE

**Professor of Electrical Technology, College of Staten Island, Staten Island, N.Y.**

## 4.1 INTRODUCTION

Passive components, resistance *(R)*, inductance *(L)*, and capacitance *(C)*, have many applications in electronic circuits and equipments. In this chapter we shall discuss the factors involved in the selection of these components.

## 4.2 RESISTORS

In selecting resistors, the following factors have to be considered:
1. Resistance value and limiting values
2. Dissipation
3. Current-carrying capacity
4. Maximum voltage limits
5. Tolerance or precision
6. Temperature coefficient and limitations
7. Voltage coefficient
8. Noise
9. Size and mounting requirements
10. Parasitic reactive effects: inductance and capacitance
11. Environmental stability with respect to solderability, shock, vibration, thermal cycling, humidity, altitude, insulation, mechanical strength, and color code durability
12. Drift
13. Frequency effects
14. Cost
15. Maximum temperature and temperature derating factor

### Fixed Resistor Tolerance

In terms of tolerance, resistors can be classified as:
1. General purpose: tolerance 5% or greater
2. Semiprecision: tolerance between 1 and 5%
3. Precision: tolerance between 0.5 and 1%
4. Ultraprecision: tolerance better than 0.5%

### Composition Resistors

Composition, or general purpose, resistors have the widest tolerance of the available resistors and are the least expensive. A cutaway view of this type of resistor is shown in Fig. 4-1. These resistors are made by embedding particles of carbon, the resistance material, in a binder in the form of a slug. The slug and each wire lead are molded under high pressure and temperature.

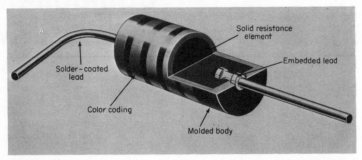

**Fig. 4-1** Cutaway view of a carbon-composition resistor. *(Courtesy Allen-Bradley.)*

Composition resistors are made in ⅛-, ¼-, ½-, 1- and 2-W sizes and in 5%, 10%, and 20% tolerance levels. Values from 1 Ω to 100 MΩ are available, and the nominal values of resistance and tolerance are indicated by means of a series of color bands (Fig. 4-2). The resistance value is indicated by four color bands (or three for 20% tolerance). When

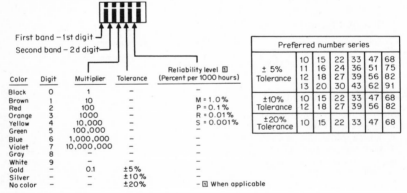

**Fig. 4-2** Color coding and preferred number series for carbon-composition resistors. *(Courtesy Allen-Bradley.)*

applicable an additional band is used to indicate the reliability level. These resistors are made in a preferred series whose values are given in Fig. 4-2. For example, in the decade between 1000 and 10,000 Ω, in the ±20% tolerance series, resistance values of 1000, 1500, 2200, 3300, 4700, 6800, and 10,000 Ω are available.

> **problem 4.1**  A resistor has the following color bands marked on it to indicate its nominal resistance value: gray, red, orange, silver. What is its nominal value of resistance and what are the limiting values of resistance to be expected?

**solution**    From Fig. 4-2

$$R = 82,000 \ \Omega \ \pm 10\% \ \text{nominal}$$
$$R_{max} = 82,000 \ \Omega + 8200 \ \Omega = 90,200 \ \Omega$$
$$R_{min} = 82,000 \ \Omega - 8200 \ \Omega = 73,800 \ \Omega$$

Although these resistors come in wide tolerance values, it is possible to make a resistance of narrower tolerance by combining two wider-tolerance resistors. This can be done by initially selecting a resistor whose nominal value is close to the desired value and combining it with another resistor whose value is determined by the difference between that of the initially selected resistor and the desired value. Resistors can be combined in series or in parallel, as required.

**problem 4-2**    Show how to make an improved-tolerance resistance by combining two wider-tolerance resistors when the nominally selected resistor is lower than the desired resistance value. The resistor is a nominally 27,000-$\Omega$ $\pm$ 20% resistor whose actual value is measured to be 23,200 $\Omega$. It is desired to make a resistor close to 27,000 $\Omega$.

**solution**    Since the resistance value is less than 27,000 $\Omega$, a series resistance is required.

$$27,000 \ \Omega - 23,200 \ \Omega = 3800 \Omega$$

Select a resistor closest to this value from the preferred series, in this case a 3300-$\Omega$ $\pm$20% value. (*Note:* A 3900-$\Omega$ $\pm$10% value could also have been selected.) The limiting expected values of the 3300-$\Omega$ $\pm$20% values are 3960 $\Omega$ and 2640 $\Omega$. The final series combination will fall between

$$R_{max} = 23,000 \ \Omega + 3960 \ \Omega = 27,160 \ \Omega \ \text{or} \ +0.6\%$$
$$R_{min} = 23,200 \ \Omega + 2640 \ \Omega = 25,840 \ \Omega \ \text{or} \ -4.3\%$$

**problem 4.3**    Show how to make an improved-tolerance resistance by combining two wider-tolerance resistors when the nominally selected resistor is higher than the required value. In this case the resistance is a nominal 27,000-$\Omega$ $\pm$20% resistor whose actual value is measured to be 30,600 $\Omega$. It is desired to make a resistor close to 27,000 $\Omega$.

**solution**    Since the resistance value is greater than the desired 27,000 $\Omega$, a parallel combination must be used. To obtain an idea of the value of the parallel resistor, the following ratio is taken:

$$\frac{30,600 \ \Omega}{30,600 \ \Omega - 27,000 \ \Omega} = 8.5$$

Use a parallel resistor whose value is close to 8.5 $\times$ 27,000 $\Omega$ = 230,000 $\Omega$. A 220,000-$\Omega$ $\pm$20% resistor meets this requirement. Its limiting expected values are $R_{max}$ = 264,000 $\Omega$ and $R_{min}$ = 176,000 $\Omega$. It is combined in parallel with the 30,600-$\Omega$ resistor. The final parallel combination will fall between

$$R_{max} = \frac{30,600 \ \Omega \times 264,000 \ \Omega}{30,600 \ \Omega + 264,000 \ \Omega} = 27,400 \ \Omega \ \text{or} \ 1.5\%$$

$$R_{min} = \frac{30,600 \ \Omega \times 176,000 \ \Omega}{30,600 \ \Omega + 176,000 \ \Omega} = 26,100 \ \Omega \ \text{or} \ -3.3\%$$

## Lower-Tolerance Resistors

Where resistances of lower tolerances are required, metal-film, carbon-film, and wire-wound types are available. Semiprecision resistors in the 0.1, 0.25, 0.5, 1 and 2% tolerance levels are supplied also in a set of standard resistance values (Table 4-1).

The resistance values may be indicated by means of a color code. The color code is the same as that shown in Fig. 4-2 with the following additions:

| *Tolerance* | | *Multiplier* |
|---|---|---|
| Brown | 1% | Silver $\times$ 0.01 |
| Red | 2% | |

Two percent tolerance resistors use the same four-band color coding scheme of Fig. 4-2. One percent tolerance resistors use the five-band color coding of Fig. 4-3.

Resistance values for lower-tolerance resistors are frequently indicated by means of a four-digit number. The first three digits indicate the significant digits and the last digit the number of zeros to follow. When no zero follows, a letter is used to indicate the decimal place. For example,

$$1271 = 1270 \ \Omega$$
$$12R7 = 12.7 \ \Omega$$

## TABLE 4-1 Standard EIA/MIL/IEC Resistance Decade Values
Subject to minimum and maximum limits for each type

| * | ±1% | ±2% | * | ±1% | ±2% | * | ±1% | ±2% | * | ±1% | ±2% | * | ±1% | ±2% | * | ±1% | ±2% | ±5% | ±10% | ±20% |
|---|---|---|---|---|---|---|---|---|---|---|---|---|---|---|---|---|---|---|---|---|
| 1.00 | 1.00 | 1.0 | 1.47 | 1.47 |  | 2.15 | 2.15 |  | 3.16 | 3.16 |  | 4.64 | 4.64 | 4.7 | 6.81 | 6.81 | 6.8 | 1.0 | 1.0 | 1.0 |
| 1.01 |  |  | 1.49 |  |  | 2.18 |  |  | 3.20 |  |  | 4.70 |  |  | 6.90 |  |  | 1.1 |  |  |
| 1.02 | 1.02 |  | 1.50 | 1.50 | 1.5 | 2.21 | 2.21 | 2.2 | 3.24 | 3.24 |  | 4.75 | 4.75 |  | 6.98 | 6.98 |  | 1.2 | 1.2 |  |
| 1.04 |  |  | 1.52 |  |  | 2.23 |  |  | 3.28 |  |  | 4.81 |  |  | 7.06 |  |  | 1.3 |  |  |
| 1.05 | 1.05 |  | 1.54 | 1.54 | 1.54 | 2.26 | 2.26 |  | 3.32 | 3.32 | 3.3 | 4.87 | 4.87 |  | 7.15 | 7.15 |  | 1.5 | 1.5 | 1.5 |
| 1.06 |  |  | 1.56 |  |  | 2.29 |  |  | 3.36 |  |  | 4.93 |  |  | 7.23 |  |  | 1.6 |  |  |
| 1.07 | 1.07 |  | 1.58 | 1.58 |  | 2.32 | 2.32 |  | 3.40 | 3.40 |  | 4.99 | 4.99 |  | 7.32 | 7.32 |  | 1.8 | 1.8 |  |
| 1.09 |  |  | 1.60 |  |  | 2.34 |  |  | 3.44 |  |  | 5.05 |  |  | 7.41 |  |  | 2.0 |  |  |
| 1.10 | 1.10 | 1.1 | 1.62 | 1.62 | 1.6 | 2.37 | 2.37 | 2.4 | 3.48 | 3.48 |  | 5.11 | 5.11 | 5.1 | 7.50 | 7.50 | 7.5 | 2.2 | 2.2 | 2.2 |
| 1.11 |  |  | 1.64 |  |  | 2.40 |  |  | 3.52 |  |  | 5.17 |  |  | 7.59 |  |  | 2.4 |  |  |
| 1.13 | 1.13 |  | 1.65 | 1.65 |  | 2.43 | 2.43 |  | 3.57 | 3.57 |  | 5.23 | 5.23 |  | 7.68 | 7.68 |  | 2.7 | 2.7 |  |
| 1.14 |  |  | 1.67 |  |  | 2.46 |  |  | 3.61 |  |  | 5.30 |  |  | 7.77 |  |  | 3.0 |  |  |
| 1.15 | 1.15 |  | 1.69 | 1.69 |  | 2.49 | 2.49 |  | 3.65 | 3.65 | 3.6 | 5.36 | 5.36 |  | 7.87 | 7.87 |  | 3.3 | 3.3 | 3.3 |
| 1.17 |  |  | 1.72 |  |  | 2.52 |  |  | 3.70 |  |  | 5.42 |  |  | 7.96 |  |  | 3.6 |  |  |
| 1.18 | 1.18 |  | 1.74 | 1.74 |  | 2.55 | 2.55 |  | 3.74 | 3.74 |  | 5.49 | 5.49 |  | 8.06 | 8.06 |  | 3.9 | 3.9 |  |
| 1.20 |  |  | 1.76 |  |  | 2.58 |  |  | 3.79 |  |  | 5.56 |  |  | 8.16 |  |  | 4.3 |  |  |
| 1.21 | 1.21 | 1.2 | 1.78 | 1.78 | 1.8 | 2.61 | 2.61 |  | 3.83 | 3.83 |  | 5.62 | 5.62 | 5.6 | 8.25 | 8.25 | 8.2 | 4.7 | 4.7 | 4.7 |
| 1.23 |  |  | 1.80 |  |  | 2.64 |  |  | 3.88 |  |  | 5.69 |  |  | 8.35 |  |  | 5.1 |  |  |
| 1.24 | 1.24 |  | 1.82 | 1.82 |  | 2.67 | 2.67 |  | 3.92 | 3.92 | 3.9 | 5.76 | 5.76 |  | 8.45 | 8.45 |  | 5.6 | 5.6 |  |
| 1.26 |  |  | 1.84 |  |  | 2.71 |  | 2.7 | 3.97 |  |  | 5.83 |  |  | 8.56 |  |  | 6.2 |  |  |
| 1.27 | 1.27 |  | 1.87 | 1.87 |  | 2.74 | 2.74 |  | 4.02 | 4.02 |  | 5.90 | 5.90 |  | 8.66 | 8.66 |  | 6.8 | 6.8 | 6.8 |
| 1.29 |  |  | 1.89 |  |  | 2.77 |  |  | 4.07 |  |  | 5.97 |  |  | 8.76 |  |  | 7.5 |  |  |
| 1.30 | 1.30 | 1.3 | 1.91 | 1.91 |  | 2.80 | 2.80 |  | 4.12 | 4.12 |  | 6.04 | 6.04 |  | 8.87 | 8.87 |  | 8.2 | 8.2 |  |
| 1.32 |  |  | 1.93 |  |  | 2.84 |  |  | 4.17 |  |  | 6.12 |  |  | 8.98 |  |  | 9.1 |  |  |
| 1.33 | 1.33 |  | 1.96 | 1.96 |  | 2.87 | 2.87 |  | 4.22 | 4.22 |  | 6.19 | 6.19 | 6.2 | 9.09 | 9.09 | 9.1 |  |  |  |
| 1.35 |  |  | 1.98 |  |  | 2.91 |  |  | 4.27 |  |  | 6.26 |  |  | 9.20 |  |  |  |  |  |
| 1.37 | 1.37 |  | 2.00 | 2.00 | 2.0 | 2.94 | 2.94 |  | 4.32 | 4.32 | 4.3 | 6.34 | 6.34 |  | 9.31 | 9.31 |  |  |  |  |
| 1.38 |  |  | 2.03 |  |  | 2.98 |  |  | 4.37 |  |  | 6.42 |  |  | 9.42 |  |  |  |  |  |
| 1.40 | 1.40 |  | 2.05 | 2.05 |  | 3.01 | 3.01 | 3.0 | 4.42 | 4.42 |  | 6.49 | 6.49 |  | 9.53 | 9.53 |  |  |  |  |
| 1.42 |  |  | 2.08 |  |  | 3.05 |  |  | 4.48 |  |  | 6.57 |  |  | 9.65 |  |  |  |  |  |
| 1.43 | 1.43 |  | 2.10 | 2.10 |  | 3.09 | 3.09 |  | 4.53 | 4.53 |  | 6.65 | 6.65 |  | 9.76 | 9.76 |  |  |  |  |
| 1.45 |  |  | 2.13 |  |  | 3.12 |  |  | 4.59 |  |  | 6.73 |  |  | 9.88 |  |  |  |  |  |

*±0.1%, ±0.25% and ±0.5%

SOURCE: Courtesy TRW/IRC Resistors. Philadelphia, Pa.

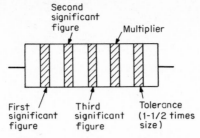

**Fig. 4-3** Color coding for 1% tolerance resistors.

Still another method of indicating the resistance value is to use three significant digits and a following letter to indicate a multiplier. The letters are: R = ohms, K = thousand ohms, and M = megohms. For example, 53.6 R = 53.6 Ω; 53.6 K = 53,600 Ω.

### Special Resistors

There are categories of resistors which do not fall within the classifications provided in the previous sections of the chapter. Some of these are:

1. *Ultraprecision (0.002%) and ultrastable.* These are made by carefully winding a resistance wire on a form.

2. *High voltage,* for kilovolt applications.

3. *High resistance: multi-megohm.* These are metal film or metal oxide on a ceramic core.

4. *Metal film networks.* These are being made available in dual-in-line DIP packages identical to those used for integrated circuits and in a similar single-in-line SIP package. They are used in pull-up and pull-down resistors with integrated circuits for driving LEDs and for D/A and A/D ladder networks. In addition to ease of insertion of many resistors simultaneously on PC boards, the proximity of the resistors which are on a common ceramic chip allows resistors to track and resistor ratios to be maintained with temperature variation.

5. *Noninductive wire-wound.* These use two windings wound in opposite directions so that their magnetic fields cancel each other.

6. *Power resistors.* These are wire wound with wire locked to a ceramic core by a vitreous enamel. They are frequently made with an adjustable slider to provide means for semivariable capability.

### Temperature Effects

An important consideration in the selection of a resistor is the way its resistance varies with temperature. Sometimes it is simply a question of how the resistance of an individual resistor varies. In voltage-divider applications, it is a question of matching the resistance variations of two resistors. In thermistor-sensing applications, the user desires a large but controlled and specified resistance variation over a required temperature range.

Specification of temperature resistance variation is usually expressed as a temperature coefficient of resistance (abbreviated TCR or tempco) in parts per million per degree Celsius (ppm/°C). It represents a percent change of the nominal resistance at 25°C. The TCR may be positive or negative.

**problem 4.4**   A resistor with a TCR of +4000 ppm/°C has a resistance of 5000 Ω at +25°C. What is its resistance at +75°C?

**solution**   The quantity 4000 ppm is equal to $4000/10^6$ = 0.004. The change in resistance is given by the product of this factor and the change in temperature. We can, therefore, write

$$R = 5000 \, [1 + 0.004 \, (75 - 25)] \ \Omega$$
$$= 6000 \ \Omega$$

A shortcoming of the previous method of determining the change in resistance with temperature is that the TCR is dependent upon the choice of temperature for the nominal resistance. A change in the reference temperature results in a change in TCR. A unifying method is to determine a theoretical temperature of zero resistivity, which can be done if the resistance variation is linear, as is implied by the concept of TCR. The resistance at any temperature can then be obtained by means of similar triangles.

**problem 4.5**   The theoretical temperature for zero resistance of a resistor is $-225°C$. If its resistance is 5000 Ω at $+25°C$, what will its resistance be at 75°C?

**solution**   From similar triangles (Fig. 4-4), we can write

$$\frac{R_1}{225 + 25} = \frac{R_2}{225 + 75}$$

For $R_1 = 5000$ Ω at 25°C we obtain

$$R_2 = 6000 \ \Omega$$

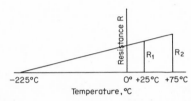

**Fig. 4-4**   Zero-resistance method for determination of resistance temperature variation (Problem 4.4).

**problem 4.6**   Determine the theoretical temperature for zero resistance for a resistor whose TCR at 25°C is 4000 ppm/°C.

**solution**   The general equation relating resistance, temperature, and TCR is

$$R = R_{25°} \ (1 + TCR \times \Delta T) \tag{4-1}$$

For $R$ to be equal to zero, we write

$$1 + TCR \times \Delta T = 0$$

and

$$\Delta T = -\frac{1}{TCR}$$

For the TCR = 0.004

$$\Delta T = -\frac{1}{TCR} = -250°C$$

Hence, the projected temperature for zero resistance is

$$T_0 = +25°C - 250°C = -225°C$$

## Rated Continuous Working Voltage

The rated continuous working voltage (RCWV) is the maximum voltage that can be safely applied to a resistor without exceeding its power rating.

**problem 4.7**   What is the maximum voltage that can be safely applied to a 500-Ω 2-W resistor?

**solution**

$$P = \frac{V^2}{R};$$

hence

$$V = \sqrt{2 \times 500} = 31.6 \ V$$

## Critical Resistance

The critical resistance $R_C$ of a series of resistors is that resistance at which the maximum allowed voltage and power dissipation occur simultaneously.

**problem 4.8**   The maximum allowable voltage that can be applied to ¼-W composition resistors is 250 V. What is the critical resistance?

**solution**

$$\frac{1}{4} = \frac{250^2}{R}$$

$$R = 250{,}000 \ \Omega$$

## Noise

In a resistor, at any temperature above 0° absolute, the random motion of electrons generates small noise currents and small noise terminal voltages. These voltages are determined by the absolute temperature, the value of resistance, and the bandwidth. The expression for the noise voltage $V_n$ generated in a resistor is:

$$V_n = \sqrt{4kTR \ \Delta f} \qquad \text{volts} \tag{4-2}$$

where $k$ = Boltzmann constant, $1.38 \times 10^{-23}$ J/K
$\quad T$ = absolute temperature, K
$\quad R$ = resistance, $\Omega$
$\quad \Delta f$ = bandwidth, Hz
This noise is known as *Johnson,* or *white,* noise.

**problem 4.9**   Determine the noise voltage generated in a 1-M$\Omega$ resistor at room temperature (27°C) over a 10 kHz bandwidth.

**solution**

$$T = 27° + 273° = 300 \text{ K. By Eq. (4-2)},$$
$$V_n = \sqrt{4 \times 1.38 \times 10^{-23} \times 300 \times 10^6 \times 10^4}$$
$$= 13 \times 10^{-6} \text{ V} = 13 \ \mu\text{V}$$

## High-Frequency Effects

Because of terminal lead and wiring inductance in wirewound resistors, capacitance between the terminal leads and between the carbon particles in composition resistors, and capacitance between turns in wirewound resistors, the high-frequency model of a resistor differs from its low-frequency resistance. Figure 4-5 shows a lumped model of a resistor operating at high frequency. Typical high-frequency characteristics of some fixed resistors are provided in Fig. 4-6.

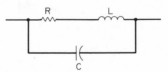

**Fig. 4-5**   A model of a resistor operating at high frequencies.

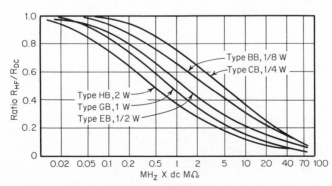

**Fig. 4-6**   Typical high-frequency characteristics of different-type fixed resistors. (*Courtesy Allen-Bradley.*)

**problem 4.10**   Using the appropriate curve of Fig. 4-6, determine the high-frequency resistance $R_{HF}$ of a 100,000-$\Omega$ ⅛-W resistor at a frequency of 5 MHz.

**solution**   The product of megahertz and direct current megohms is:

$$5 \text{ MHz} \times 0.1 \text{ M}\Omega = 0.5$$

From Fig. 4-6, for the ⅛-W curve,

$$\frac{R_{HF}}{R_{DC}} = 0.75$$

$$R_{HF} = 0.75 \times 100,000 \ \Omega = 75,000 \ \Omega$$

## Power Rating

Power rating is the maximum power in watts that a resistor can safely dissipate at ambient temperatures up to 70°C. At temperatures above that, the power rating is derated or reduced linearly. Figure 4-7 shows a typical derating curve.

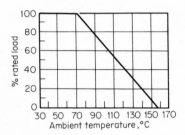

**Fig. 4-7**   Typical derating curve for a resistor. Ambient temperature refers to the temperature of the environment in which the resistor is operating.

**problem 4.11**   A 10-W resistor is used at an ambient temperature of 110°C. What is the maximum power that the resistor can safely dissipate?

**solution**   From Fig. 4-7, the derating factor at 110°C is 55%. The resistor can therefore safely dissipate $0.55 \times 10 = 5.5$ W.

## Variable Resistors—Potentiometers

In addition to the factors listed for fixed resistors, the following additional factors have to be considered in the selection of variable resistors.

1. Resolution. The smallest change in resistance that can be realized as the wiper arm is rotated. This is an important consideration for wirewound potentiometers.

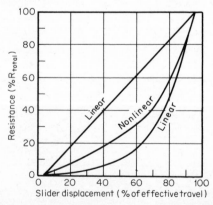

**Fig. 4-8**   Examples of taper curves for potentiometers. (*Courtesy MEPCO/ELECTRA, Inc.*)

2. Wiper current. The maximum current that can flow in and out of the wiper terminal.

3. End resistance. Resistance between wiper arm and end terminals with wiper positioned at the corresponding endpoint.

4. Setting stability. Repeatability of the resistance value as the resistor is reset to the same position.

5. Maximum number of allowed rotations.

6. Number of turns.

7. Number of coupled (ganged) potentiometers and their effect on temperature derating.

8. Ability to couple a potentiometer shaft to a switch

9. Taper and linearity. Typical taper curves are given in Fig. 4-8.

## Thermistors

Thermistors are semiconductor resistors whose resistance varies to a considerable degree with change in temperature. They can have a positive or negative temperature coefficient and are available in disk, washer bead, or bolted assembly packages. They have important applications, in measurement and control of temperature, in time delay applications, and in liquid level indicators. Figure 4-9 shows the electrical symbol of a thermistor, and Figs. 4-10 to 4-12 show the resistance characteristics of a thermistor.

**Fig. 4-9**   Electrical symbol for a thermistor.

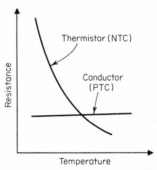

**Fig. 4-10**   The variation in resistance of a conductor (PTC) and a thermistor (NTC) with temperature.

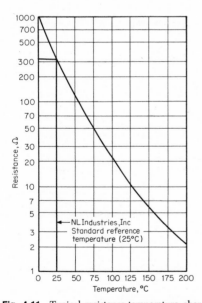

**Fig. 4-11**   Typical resistance-temperature characteristics of a thermistor. *(Courtesy N. L. Industries, Inc., Electronics Dept.)*

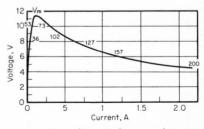

**Fig. 4-12**   Typical static voltampere characteristics of a thermistor. *(Courtesy N. L. Industries, Inc., Electronics Dept.)*

## Summary of Resistor Types

A summary of commercially available fixed and variable resistor types is given in Table 4-2. It presents their characteristics and properties and gives an indication of their major field of application.

**TABLE 4-2    Summary of Fixed and Variable Resistor Types**

| Category | Type | Key property | Power | Temperature coefficient, °C | Resistance range |
|---|---|---|---|---|---|
| General purpose | Carbon composition | Cost | ⅛–2 W | >500 | 1 Ω ⩾ 100 MΩ |
| ⩾5% tolerance | Molded wire-wound | Tempco | ½–2 W | ⩾200 | 0.1 Ω–2.4 kΩ |
| ⩾200 ppm | Ceramic wire-wound | Low voltage W | 2–50 W | ⩾200 | 0.1 Ω–30 kΩ |
| | Metal glaze | Flexibility | ⅛–5 W | 200 | 4.3 Ω–1.5 MΩ |
| | Tin oxide | Reliability | ⅛–20 W | 200 | 4.3 Ω–2.5 MΩ |
| | Carbon film (import) | Cost | ¼–2 W | >200 | 10 Ω ⩾ 1 MΩ |
| | Cermet film | Stability | ¼–3 W | 150 | 10 Ω–10 MΩ |
| Semiprecision | Metal glaze | Flexibility | ⅛–2 W | ⩽200 | 1 Ω–1,5 MΩ |
| >1 <5% ⩽200 ppm | Tin oxide | Stability | ⅛–2 W | ⩽200 | 4.3 Ω–1.5 MΩ |
| Power | Ceramic wire-wound | Cost | 2–50 W | ⩾200 | 0.1 Ω–30 kΩ |
| ⩾2 W | Axial lead coated WW | Auto insertion | ½–15 W | ⩽50 | 0.1 Ω–175 kΩ |
| | Tubular and flat WW | | 4–250 W | ⩽100 | 0.1 Ω ⩾ 1 mΩ |
| Precision | Metal film | Tolerance | 1/10–1 W | 20 | 0.1 Ω–1 MΩ |
| ⩽1% | Metal glaze | Environment | 1/10–1 W | ⩽100 | 1 Ω–1 MΩ |
| ⩽100 ppm | Tin oxide | Power | 1/10–½ W | ⩽100 | 10 Ω–1 MΩ |
| | Thin film | Size, networks | 1/20–5 W | ⩽100 | 10 Ω–100 MΩ |
| | Encaps, wire-wound | Power, Tempco | 1/20–1 W | ⩽20 | 0.1 Ω ⩾ 1 MΩ |
| Ultraprecision | Thin film | Flexibility | 1/20–½ W | ⩽25 | 20 Ω–1 MΩ |
| ⩽0.5% | Encaps, wire-wound | Noise | 1/20–1 W | ⩽20 | 0.1 Ω ⩾ 1 MΩ |
| ⩽25 ppm | | | | | |
| Variable devices | Wire-wound | Tempco | 5 at 70°C | ±20 | 10 Ω–100 kΩ |
| (pots, trimmers) | Conductive plastic | Rotational life | 2 at 70°C | ±250–500 | 1 kΩ–100 kΩ |
| | Cermet | Environmental | 12 at 70°C | ±250–500 | 500 Ω–2 MΩ |
| | Carbon | Cost | 5 at 70°C | ±300–2000 | 100 Ω–2 MΩ |
| Networks | Thick film | Cost | ⩽2 W/pkg. | ⩽200 | 10 Ω–10 MΩ |
| | Thin film | Performance | ⩽2 W/pkg. | ⩽100 | 10 Ω–1 MΩ |

SOURCE: Reprinted from *Electronics*, May 30, 1974; copyright McGraw-Hill, New York, 1974.

## 4.3  INDUCTORS

When current flows in a wire (or coil), a magnetic field is developed around the wire (or coil). When the current is increased, the magnetic flux increases. An increase in the magnetic flux generates a voltage in the wire or coil with a polarity to oppose the change in flux. The ability of a coil to oppose this change is called its *self-inductance* or, more commonly, *inductance;* coils are called *inductors.*

The larger the flux, the larger the inductance. Since iron-core coils develop more flux, their inductance is higher than air-core coils. Because the amount of flux in iron is determined by the region of the hysteresis loop being traversed, the inductance of magnetic-core coils depends on many factors, and it is variable.

When two coils are linked by a common magnetic field (transformer), the measure of the magnetic-flux interaction between the two coils is called the *mutual inductance.* The unit of inductance $(L)$ is called the henry (H); millihenry (mH) and microhenry ($\mu$H) inductors are quite common.

When alternating voltages are applied to inductors, they act as a reactance of magnitude:

$$X_L = j2\pi f L \qquad \text{ohms}$$

where $j = \sqrt{-1} \equiv 1\underline{/90°}$, a 90° rotator
$\quad L$ = inductance, H
$\quad f$ = frequency, Hz

Inductors in many styles and forms are commercially available. Small inductance values can be obtained in a size and shape approximating the 1-W composition resistor. Their inductance in microhenries is indicated by a color code identical to that used for composition resistors. For high values of inductance, powdered iron-core coils are available. For still larger values of inductance used at power line and low frequencies, coils with cores made of sheet-steel laminations are used.

Variable inductors have movable powdered-iron cores, which can be moved by a positioning tool or positioning arm. Despite the many commercially available coils, the wide variety of requirements make it necessary for the engineer and technician to design and wind coils. The design factors for some common air-core coils are discussed in the following sections.

## Choice of Inductors

In selection of inductors, the following factors have to be considered:
1. Inductance value
2. Size and mounting requirements
3. $Q$
4. Frequency range
5. Composition of core (air or iron)
6. Direct-current level and magnitude of alternating current in iron coils
7. Stray capacitance effects and self-resonant frequency
8. For coupled coils, turns ratio, mutual inductance, and capacitive coupling between windings
9. Environmental factors: Temperature, humidity, shock, vibration, insulation, thermal cycling altitude
10. Power dissipation
11. Shielding
12. Fixed or variable

## Design of Small Air-Core Coils

Design equations for some common air-core coils (Fig. 4-13) are given.
1. *Single-layer air-core coil* (solenoid) of Fig. 4-13a,

$$L = \frac{0.394r^2N^2}{9r + 10l} \quad \mu H \tag{4-3}$$

where the dimensions are in centimeters. This is accurate to within 1% for long coils in which $l > 2/3 \ r$.

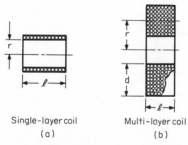

Single-layer coil          Multi-layer coil
        (a)                        (b)

**Fig. 4-13**   Computing the inductance of some typical air-core coils.

**problem 4.12**   What is the inductance of a 20-turn coil wound on a ½-in-diameter (1.27 cm) coil form with a coil length of 2 in (5.08 cm)?

**solution**   Substituting the given values in Eq. (4-3):

$$r = 0.5 \times 1.27 \text{ cm} = 0.64 \text{ cm}$$

$$L = \frac{0.394 \ (0.64)^2(20)^2}{9 \times 0.64 + 10 \times 5.08} = 1.14 \ \mu H$$

**problem 4.13**   How many turns are required for an inductance of 20 $\mu H$ to be wound on a 5-cm-diameter coil form with a winding length of 3.8 cm?

**solution**   Substituting the values in Eq. (4-3) yields:

$$r = 2.5 \text{ cm}$$

$$20 = \frac{0.394(2.5)^2 \ N^2}{9 \times 2.5 + 10 \times 3.8}$$

Solving for $N$ yields $N = 22$ turns.

2. Many single-layer coils are tight-wound, and the *winding pitch p* (turns per unit length) is determined by the wire size and the type of insulation. Length *l* in terms of N and p is:

$$l = \frac{N}{p} \qquad (4\text{-}4)$$

Substituting Eq. (4-4) in Eq. (4-3) and solving for N, we obtain

$$N = \frac{12.7L}{pr^2}\left[1 + \sqrt{1 + \frac{0.14r^3p^2}{L}}\right] \qquad (4\text{-}5)$$

**problem 4.14**    Determine the number of turns required for a single-layer coil of 120 $\mu$H to be tight-wound on a 2.5-cm-diameter coil form using No. 22 enamel-coated wire. What is *l*?

**solution**    From wire tables (Appendix) for No. 22 enamel wire, $p = 37$ turns/in (14.6 turns/cm). Substituting the values in Eq. (4-5) yields:

$$N = \frac{12.7(120)}{14.6(1.25)^2}\left[1 + \sqrt{1 + \frac{0.14(1.25)^3(14.6)^2}{120}}\right]$$

$$= 148 \text{ turns}$$

and

$$l = \frac{N}{p} = 148/14.6 = 10.14 \text{ cm}$$

3. *Short air-core coils.* For values of *l* between $\frac{2}{3}r$ and $r/10$, Eq. (4-3) is modified, as given in Eq. (4-6). This is accurate to within 2%.

$$L = \frac{0.394r^2N^2}{\left(9 - \dfrac{r}{5l}\right)r + 10l} \qquad \mu\text{H} \qquad (4\text{-}6)$$

**problem 4.15**    Determine the inductance of an air-core solenoid of 5 turns, wound on a 10-cm coil form whose length is 2.5 cm.

**solution**    By substituting in Eq. (4-6),

$$L = \frac{0.394(5)^2(5)^2}{[9 - (5/12.5)]5 + 10(2.5)} = 3.7 \ \mu\text{H}$$

4. *Multilayer coils.* Figure 4-13b shows a multilayer coil. The formula for computing its inductance is given by:

$$L = \frac{0.315 \, r^2N^2}{6r + 9l + 10d} \qquad \mu\text{H} \qquad (4\text{-}7)$$

where $r$ = mean radius, cm
     $l$ = length, cm
     $d$ = winding depth, cm
     $N$ = number of turns
     $L$ = inductance, $\mu$H

**problem 4.16**    Determine the inductance of a multilayer coil of 400 turns. The mean radius is 2.5 cm, the winding length is 2 cm, and the winding depth is 1.2 cm.

**solution**    Substituting the values in Eq. (4-7) yields:

$$L = \frac{0.315(2.5)^2(400)^2}{6 \times 2.5 + 9 \times 2 + 10 \times 1.2} \ \mu\text{H}$$

$$= 7000 \ \mu\text{H}$$

$$= 7 \text{ mH}$$

5. *Inductance of a straight wire.* At high frequencies the inductance of a straight wire is of importance since wires are used for interconnections or as terminal leads for

both passive and active components. The inductance of a straight nonmagnetic wire is given by:

$$L = 0.002 \ l \left[ 2.3 \log \left( \frac{4l}{d} - 0.75 \right) \right] \mu H \tag{4-8}$$

where $l$ and $d$ are the length and diameter in cm and $l \gg d$.

**problem 4.17**    Determine the inductance of a 1.25-cm length of No. 22 wire and its reactance at 50 MHz.

**solution**    No. 22 wire has a diameter of 0.0254 in (0.065 cm). Substituting the values in Eq. (4-8) gives:

$$L = 0.002 \times 1.25 \ [2.3 \log (4 \ 1.25/0.065 - 0.75)] \ \mu H$$
$$= 0.011 \ \mu H$$
$$X_L = 2\pi f L = 2\pi \times 50 \times 10^6 \times 0.011 \times 10^{-6}$$
$$= 3.5 \ \Omega$$

Although Fig. 4-13$b$ and Eq. (4-7) are for a winding in which each layer is tight-wound and one layer falls directly above the previous layer, most multilayer rectangular coils are of the *universal*, or *pie*-type, winding (Fig. 4-14). For this coil, the winding progresses both around the coil form and from side to side of the winding. The considerations involved in designing such a winding are mechanical stability, reduction in distributed capacity, and $Q$ (to be discussed later). For the same number of turns and coil form shown in Fig. 4-13$b$, the inductance will be approximately 10% higher for the universal winding than that given by Eq. (4-7). Other winding techniques are available for reducing the distributed capacitance.*

**Fig. 4-14**    Universal or pie-type winding.

### Mutual Inductance

When two coils are close to each other and there is current in one coil, the flux of the first coil links the second coil. If the current in the first coil is changed, a voltage is induced in the second coil. This effect is called *mutual inductance M* and its unit is the henry (H). The mutual inductance from the second coil to the first coil is identical to that from the first to the second:

$$M_{12} = M_{21} = M \tag{4-9}$$

The measure of the closeness of the coupling is given by the coefficient of coupling, $K$:

$$K = \frac{M}{\sqrt{L_1 L_2}} \tag{4-10}$$

where $L_1$ and $L_2$ are the inductances of the individual coils. If all the flux of the first coil enters the second coil and there is no leakage, $K = 1$. Coefficient of coupling $K = 1$ can be approached in iron-core coils, but large values of $K$ are difficult to attain with air-core coils. In many cases, as, for example in the intermediate-frequency (I-F) transformers used in radio receivers, values of $K = 0.01$ are quite common.

1. *Measurement of mutual inductance.* While there are impedance bridges that can

---

*M. Kaufman and A. H. Seidman, *Handbook for Electronics Engineering Technicians*, McGraw-Hill, New York, 1976, Fig. 3.8, p. 3–12.

measure $M$, a common method to determine $M$ is on an inductance bridge. If the two coils $L_1$ and $L_2$ are connected in series, they can be connected so that the mutual inductance aids the net inductance. By reversing the connections to the second coil, the mutual inductance will oppose, or subtract, from the net inductance. We have:

$$(1)\ L_A = L_1 + L_2 + 2M \qquad \text{for } M \text{ aiding}$$
$$(2)\ L_O = L_1 + L_2 - 2M \qquad \text{for } M \text{ opposing}$$

Subtracting (2) from (1) and solving for $M$ yields

$$M = \frac{L_A - L_O}{4} \tag{4-11}$$

**problem 4.18**  Determine the mutual inductance and coefficient of coupling $K$ for a transformer in which $L_1 = 1$ H, $L_2 = 0.25$ H, $L_A = 1.45$ H, and $L_O = 1.05$ H.

**solution**

$$M = \frac{L_A - L_O}{4} = \frac{1.45 - 1.05}{4}\,\text{H} = 0.1\,\text{H}$$

$$K = \frac{M}{\sqrt{L_1 L_2}} = \frac{0.1}{\sqrt{1 \times 0.25}} = 0.2$$

2. *Determination of M and K from physical dimensions.* In general, the determination of self-inductance from physical dimensions is complex, and the previous section gave equations for inductance of some relatively simple geometries only. The determination of mutual inductance from physical dimensions is much more difficult, and only in the case of coils and transformers based upon the single-layer solenoid can simple solutions to $M$ and $K$ be attained.

*Case 1.*  There is no gap in the winding of the single-layer solenoid, and a tap is brought out at a point along the winding (Fig. 4-15). We wish to determine the mutual

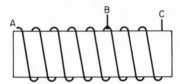

**Fig. 4-15**   Single-layer solenoid AC with tap at $B$.

inductance and coefficient of coupling between windings $AB$ and $BC$. The mutual inductance between the two windings aids in the total coil inductance, and we can define the following:

$$L = L_{AC} = L_{AB} + L_{BC} + 2M$$

Solving for $M$, we obtain

$$M = \frac{1}{2}(L - L_{AB} - L_{BC}) \tag{4-12}$$

**problem 4.19**  In Fig. 4-15, $AC$ is a 20-turn coil wound uniformly on a 1.27-cm-diameter coil form with a coil length of 5.08 cm. Tap $B$ is located 6 turns from the start of winding $A$. Determine $M$ and $K$ between windings $AB$ and $BC$.

**solution**   $L = 1.14\ \mu$H (Problem 4.12). For $L_{AB}$, $l = (6/20) \times 5.08 = 1.52$ cm. By Eq. (4-3) $L_{AB} = 0.28\ \mu$H.

For $L_{BC}$, $l = 3.6$ cm and $N = 14$ turns.

By Eq. (4-3), $L_{BC} = 0.76\ \mu$H. Substituting in Eq. (4-12),

$$M = \frac{1}{2}(1.14 - 0.28 - 0.76)\mu\text{H} = 0.05\ \mu\text{H}$$

$$K = \frac{0.05}{\sqrt{0.28 \times 0.76}} = 0.108$$

*Case 2.* There is a gap in the winding of the single-layer solenoid (Fig. 4-16), and there are two separate windings. Both windings are wound identically. In Fig. 4-16 the coils are windings $AB$ and $CD$. The dashed coil between $B$ and $C$ is an assumed winding needed to complete the total winding between $A$ and $D$. By following the same principles used in deriving Eq. (4-12), it can be shown that:

$$M_{AB-CD} = \frac{1}{2}(L_{AD} + L_{BC} - L_{AC} - L_{BD}) \tag{4-13}$$

**problem 4.20**    In Fig. 4-16 the coil is tight-wound using No. 22 wire on a 2.54-cm-diameter coil form. Winding $AB$ is 2.54 cm long, winding $CD$ is 5.08 cm long. The spacing between coils $AB$ and $CD$ is 1.27 cm long. Determine $M_{AB-CD}$ and $K$.

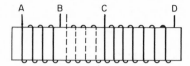

**Fig. 4-16**   Two separated coils wound in an identical manner on the same coil form.

**solution**   From Problem 4.14, the winding pitch for No. 22 enamel wire is 14.6 turns/cm. Tabulating and using Eq. (4-3), we obtain:

| Winding | Length, cm | Turns | L, μH |
|---------|-----------|-------|-------|
| AB | 2.54 | 37 | 23.7 |
| AC | 3.8 | 55.5 | 39.4 |
| BC | 1.27 | 18.5 | 9.0 |
| BD | 6.35 | 92.5 | 72.5 |
| CD | 5.08 | 74 | 55.9 |
| AD | 8.9 | 129.5 | 105.7 |

Substituting in Eq. (4-13),

$$M = \frac{1}{2}(105.7 + 9.0 - 39.4 - 72.5) \ \mu H$$

$$= 1.4 \ \mu H$$

$$K = \frac{1.4}{\sqrt{23.7 \times 55.9}} = 0.038$$

## Coupled Inductances in Parallel

Suppose coils $AB$ and $CD$ in Fig. 4-16 were connected in parallel. There are two possibilities:

*Case 1.*   Connect $A$ to $C$ and $B$ to $D$. In this case, the magnetic fields and the mutual inductance $M$ are aiding. Under these conditions,

$$L = \frac{1}{\dfrac{1}{L_{AB} + M} + \dfrac{1}{L_{CD} + M}} \tag{4-14}$$

**problem 4.21**   Determine the parallel inductance for the coils in Prob. 4.20 when the coils are connected for $M$ aiding.

**solution**   Substituting in Eq. (4-14),

$$L = \frac{1}{\dfrac{1}{23.7 + 1.4} + \dfrac{1}{55.9 + 1.4}}$$

$$= 19.6 \ \mu H$$

*Case 2.*    Connect A to D and B to C. In this case the magnetic fields and the mutual inductance $M$ are opposing. Under these conditions,

$$L = \frac{1}{\dfrac{1}{L_{AB} - M} + \dfrac{1}{L_{CD} - M}} \tag{4-15}$$

**problem 4.22**   Determine the parallel inductance for the coils in Problem 4-20 when the coils are connected for $M$ opposing.

**solution**    Substituting in Eq. (4-15),

$$L = \frac{1}{\dfrac{1}{23.7 - 1.4} + \dfrac{1}{55.9 - 1.4}}$$

$$= 15.8\ \mu\text{H}$$

## Q, or Figure of Merit, of a Coil (Loss Factor)

Factor $Q$ (figure of merit) of a coil is defined as

$$Q = \frac{X_L}{R_{\text{series}}} \tag{4-16}$$

where $R_{\text{series}}$ is the effective series resistance of the coil. In cases where the coil is shunted by an element having a resistance component, it is simpler to convert to a form in which the coil is in parallel with an equivalent resistance (Fig. 4-17). If the $Q$ of the coil is greater than 5, it is easy to show that

$$R_{\text{series}} \times R_{\text{parallel}} = X_L^2 \tag{4-17}$$

and

$$Q = \frac{X_L}{R_{\text{series}}} = \frac{R_{\text{parallel}}}{X_L} \tag{4-18}$$

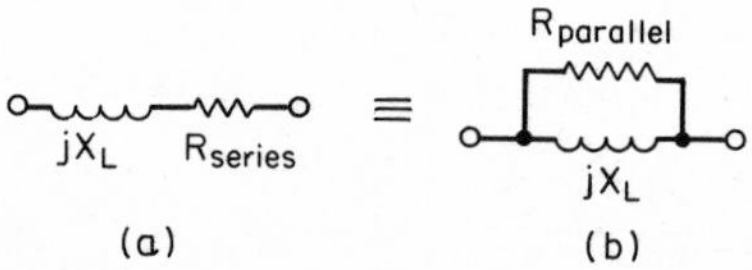

**Fig. 4-17**   Series RL circuit and equivalent parallel circuit. (*a*) Circuit. (*b*) Parallel equivalent.

**problem 4.23**   The coil in Fig. 4-17 has a reactance of 250 Ω and an effective $R_{\text{series}}$ of 2.5 Ω. Find $Q$ and the equivalent parallel resistance.

**solution**

$$Q = \frac{X_L}{R_{\text{series}}} = \frac{250}{2.5} = 100$$

$$= \frac{R_{\text{parallel}}}{X_L} = 100$$

$$R_{\text{parallel}} = 250(100) = 25{,}000\ \Omega$$

**problem 4.24**   The coil in Fig. 4.17*a* is shunted with a 50,000-Ω resistor. What is the effective $Q$ of the coil?

**solution**   *Method 1.*   Convert the series resistance to the parallel form (Fig. 4-17*b*), as was done in the previous problem. The equivalent parallel resistance of the 25,000- and 50,000-Ω resistors is 16,670 Ω. The net value of $Q$ is:

$$Q = \frac{16{,}670}{250} = 67$$

*Method 2.*   Convert the parallel 50,000-Ω resistor to a series form:

$$R_{\text{series}} = \frac{X_L^2}{R_{\text{parallel}}} = \frac{(250)^2}{50{,}000} = 1.25\ \Omega$$

The total net series resistance shown in Fig. 4-18 is 2.5 Ω + 1.25 Ω = 3.75 Ω.

$$Q = \frac{250 \ \Omega}{3.25 \ \Omega} = 67$$

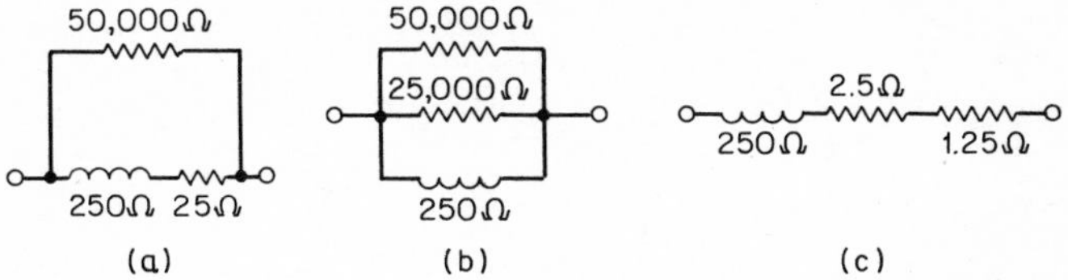

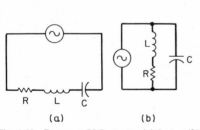

**Fig. 4-18** Circuits for Problem 4.23. (*a*) Circuit. (*b*) Parallel form. (*c*) Series form.

A major application of the *Q* factor is in resonant circuits, both series and parallel. In a series circuit of *R*, *L*, and *C* (Fig. 4-19*a*), at resonance the voltage developed across *C* and *R* is *Q* times the applied voltage. This is used in the commercial *Q* meter to measure *Q*. The impedance *Z* of an *R*, *L*, *C* series circuit may be expressed by:

$$Z = R_{\text{series}}\left(1 + j2\frac{\Delta f}{f_0}Q\right) \tag{4-19}$$

where $f_0 = 1/2\pi \sqrt{LC}$ is the series resonant frequency and $\Delta f$ is the change in frequency.

Figure 4-20 shows a family of resonance curves in which relative currents are plotted versus frequency for a constant applied voltage to a *series RLC* circuit for three different values of effective series resistance. In curve *a*, the effective resistance is low, and the *Q* is 150. In curve *b*, the resistance is higher, and in curve *c* still higher for *Q*'s of 75 and 30, respectively. The higher the value of *Q*, the narrower the resonance curve becomes.

**Fig. 4-19** Resonant *RLC* circuits. (*a*) Series. (*b*) Parallel.

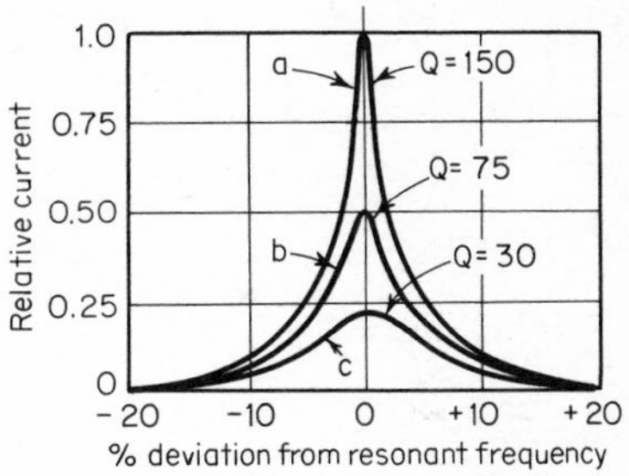

**Fig. 4-20** Selectivity curves of tuned circuits which show how selectivity depends on coil *Q*.

The bandwidth is defined as the change in frequency between the two points on the resonant curve where the current is reduced to a value of 0.707 times the maximum value. With this definition, bandwidth BW is given by:

$$\text{BW} = \frac{f_0}{Q} \tag{4-20}$$

where $f_o$ is the resonant frequency.

Identical concepts hold for the parallel resonant circuit of Fig. 4-19*b*. In this case the impedance is given by:

$$Z = \frac{R_{\text{parallel}}}{1 + j2\,(\Delta f/f_o)Q} \tag{4-21}$$

where $R_{\text{parallel}}$ is the effective parallel resistance (see Problem 4.24) at resonance. The bandwidth, also given by Eq. (4-20), is defined as the change in frequency between the two points where the impedance has been reduced to values 0.707 times the value at resonance.

**problem 4.25**   In the *RLC* series circuit of Fig. 4-21 determine: (*a*) the resonant frequency; (*b*) the current, and the voltage across $R_{\text{series}}$, $L$, and $C$ at resonance; (*c*) the bandwidth.

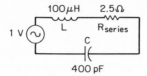

**Fig. 4-21**   Circuit for Problem 4.25.

**solution**

(*a*)
$$f_0 = \frac{1}{2\pi\sqrt{LC}} = \frac{1}{2\pi\sqrt{100 \times 10^{-6} \times 400 \times 10^{-12}}}$$

$$= 795 \text{ kHz}$$

(*b*) At resonance $X_L = -X_C$ and $R_{\text{series}} = 2.5 \ \Omega$.

$$I = \frac{1 \text{ V}}{2.5 \ \Omega} = 0.4 \text{ A}$$

$$Q = \frac{X_L}{R_{\text{series}}} = \frac{250}{2.5} = 100$$

$$V_L = V_C = Q \times V = 100 \times 1 = 100 \text{ V}$$

(*c*) $BW = \dfrac{f_o}{Q} = \dfrac{795 \times 10^3}{100} = 7950 \text{ Hz}$

### Distributed Capacitance of a Coil and Its Effects

In a coil, capacitance is developed between individual turns and terminal leads. The wires (conductors), separated by an insulator (dielectric), give rise to capacitance between turns. In addition, there is capacitance to ground. The total effect of the various capacitances is called *distributed capacitance*. The distributed capacitance has an apparent effect upon the coil parameters as expressed by the following equations:

Apparent inductance:
$$L_a = L\left(1 + \frac{C_d}{C}\right) \tag{4-22a}$$

Apparent resistance:
$$R_a = R\left(1 + \frac{C_d}{C}\right)^2 \tag{4-22b}$$

Apparent $Q$:
$$Q_a = \frac{Q}{1 + Cd/C} \tag{4-22c}$$

where $C_d$ = distributed capacitance of the coil
$\quad\quad C$ = external capacitance needed to tune $L$ to resonance
$\quad\quad L$ = true inductance of coil
$\quad\quad R$ = true resistance of coil
$\quad\quad Q$ = true $Q$ of coil

When the true inductance is known, the distributed capacitance of a coil can be determined experimentally by measuring its self-resonant frequency. This can be done with a grid-dip meter directly or on the $Q$ meter by determining the frequency at which the coil has no net reactance.

The external capacitance $C$ required to resonate the coil is determined from:

$$f = \frac{1}{2\pi\sqrt{L(C + C_d)}} \tag{4-23}$$

One can experimentally determine the external capacitance required to resonate the coil at two different frequencies. Substituting the obtained values in Eq. (4-23) and then solving the resultant equations simultaneously for $C_d$, we obtain:

$$C_d = \frac{(f_1/f_2)^2 C_1 - C_2}{1 - (f_1/f_2)^2} \tag{4-24}$$

where $C_d$ = distributed capacitance, pF
$\quad f_1, f_2$ = resonant frequencies
$\quad C_1, C_2$ = external resonant capacitance values, pF
If we select the measurement frequencies to be $f_2 = 2f_1$, then

$$C_d = \frac{C_1 - 4C_2}{3} \tag{4-25}$$

**problem 4.26**  The resonant frequency of a coil when shunted with a 100-pF capacitor is 800 kHz. At 1600 kHz, it must be shunted with a 10-pF capacitor to resonate it. Determine $C_d$.

**solution**   *Method 1*   By Eq. (4-24),

$$C_d = \frac{(800/1600)^2(100) - 10}{1 - (800/1600)^2} = 20 \text{ pF}$$

*Method 2*   By Eq. (4-25),

$$C_d = \frac{100 - 4(10)}{3} = 20 \text{ pF}$$

**problem 4.27**  The coil in Problem 4.26 has a measured value of $Q = 100$ at 1600 kHz. Determine the true inductance and true $Q$ of the coil at 1600 kHz.

**solution**   The apparent inductance is obtained from the resonance measurement:

$$1600 \times 10^3 = \frac{1}{2\pi \sqrt{L_a \times 10 \times 10^{-12}}}$$
$$L_a \cong 1000 \, \mu\text{H}$$

By Eq. (4-22*a*),

$$1000 \, \mu\text{H} = L\left(1 + \frac{20}{10}\right)$$
$$\text{True } L = 333 \, \mu\text{H}$$

By Eq. (4-22*c*),

$$Q_a = 100 = \frac{Q}{1 + (20/10)}$$
$$Q = 300$$

## Radio Frequency Chokes

Radio frequency (rf) chokes are coils wound so that distributed capacitance is minimized. The object is to obtain the highest possible self-resonant frequency independent of $Q$. This is accomplished by winding the coil as a series of pie-shaped sections.

## Iron-Core Coils

It is difficult to obtain high values of inductance with air-core coils and keep the physical size of the coil within reasonable dimensions. The inductance of a coil is determined by the flux, and the flux can be increased by using magnetic cores. At low frequencies steel-sheet laminations are used. As the frequency increases, the losses become greater, and iron-powder and ferrite cores are used. In addition, the use of movable cores provides coils of adjustable inductance.

## High-Frequency Effects

At low frequencies solid copper wires are used. As the frequency is increased, *skin effect* begins to occur, and the current in a solid conductor flows on its outer layer. Its effective resistance is thereby increased.

To overcome this, the wire is broken into small insulated enameled braided conductors. This is effective in the broadcast band i-f range of about 450 kHz. Above this, the skin effect continues to increase, and so at higher frequencies solid wire is again used.

## 4.4  CAPACITORS

A capacitor is formed when two conducting plates are separated by an insulator (dielectric). If a voltage is applied to the plates, lines of electric flux form in the dielectric between the plates. The amount of flux developed is a measure of the capacitance formed by the plates and the dielectric.

The basic unit of capacitance is the farad (F). This unit, however, is too large for practical applications and microfarads ($1 \mu F = 10^{-6}$ F) and picofarads ($1$ pF $= 10^{-12}$ F) are more practical units.

When a dc voltage is applied to the terminals of a capacitor, a charge $q$ is developed on the plates of the capacitor:

$$q = C \times v \qquad \text{coulombs (C)} \qquad (4\text{-}26)$$

where $C$ = capacitance, farads (F) and $v$ = applied voltage, volts (V).

When a capacitor is charged, it stores energy. The energy $w$ stored in a capacitor is given by:

$$w = \frac{1}{2} C v^2 \qquad \text{joules (J)} \qquad (4\text{-}27)$$

**problem 4.28**    Determine the charge and energy stored in a $4\text{-}\mu F$ capacitor when 200 V is applied to its terminals.

**solution**

$$\begin{aligned}
q &= (4 \times 10^{-6}) \times 200 \\
&= 0.8 \times 10^{-3} \text{ C} \\
w &= \frac{1}{2} \times (4 \times 10^{-6}) \times 200^2 \\
&= 0.08 \text{ J}
\end{aligned}$$

When an alternating voltage is applied to a capacitor, the capacitor is alternately charged and discharged. The reactance (impedance) $X_C$ of a capacitor is:

$$\begin{aligned}
X_C &= \frac{1}{j 2\pi f C} \\
&= \frac{1}{2\pi f C} \underline{/\!-90^\circ} \; \Omega
\end{aligned}$$

where $1/j = -j$, a $-90^\circ$ rotator
$\quad C$ = capacitance, F
$\quad f$ = frequency, Hz

When a dc voltage is applied to a capacitor through a resistor (Fig. 4-22a), time is required to charge the capacitor. The time is determined by the time constant $\tau = R \times C$

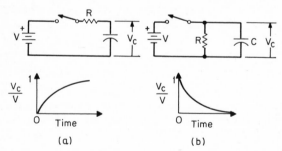

**Fig. 4-22**    *RC circuits. (a) Capacitor charging. (b) Capacitor discharging.*

($\tau$ in seconds, $R$ in ohms, and $C$ in farads). If the applied voltage is $V$, the voltage across the capacitor $v_C$ increases exponentially:

$$v_C = V(1 - \epsilon^{-t/\tau}) \qquad (4\text{-}28)$$

where $\epsilon = 2.72$ is the base of natural logarithms and $t$ is the time in seconds. From Eq. (4-28) we can develop Table 4-3a for the relative voltage $v_C/V$ as a function of the number of time constants.

There are times when it is more convenient to use a practical time constant. This is the time ($0.7RC$ or $0.7\tau$) it takes the capacitor voltage to rise to 50% of $V$. By use of the

**TABLE 4-3   Capacitor-Charging Voltages**

| (a) Using full time constants | | (b) Using practical time constants | |
|---|---|---|---|
| Time in time constants | $v_c/V$ | Time in time constants | $v_c/V$ |
| 0 | 0 | 0 | 0 |
| 1 | 0.632 | 0.7 | 0.5 |
| 2 | 0.865 | 1.4 | 0.75 |
| 3 | 0.951 | 2.1 | 0.875 |
| 4 | 0.981 | 2.8 | 0.937 |
| 5 | 0.993 | 3.5 | 0.969 |
| | | 4.2 | 0.984 |

practical time constant, Table 4-3b can be developed for $v_c/V$. Practically, the capacitor is fully charged after five times constants. When a charged capacitor discharges through a resistor (Fig. 4-22b), the voltage across the capacitor decays exponentially:

$$v_C = V\epsilon^{-t/\tau} \tag{4-29}$$

Table 4-4 gives the relative voltage $v_c/V$ as a function of the time in time constants. Practically the capacitor is fully discharged after five time constants.

**TABLE 4-4   Capacitor-Discharging Voltages**

| (a) Using full time constants | | (b) Using practical time constants | |
|---|---|---|---|
| Time in time constants | $v_c/V$ | Time in time constants | $v_c/V$ |
| 0 | 1 | 0 | 0 |
| 1 | 0.378 | 0.7 | 0.5 |
| 2 | 0.135 | 1.4 | 0.25 |
| 3 | 0.049 | 2.1 | 0.125 |
| 4 | 0.019 | 2.8 | 0.063 |
| 5 | 0.007 | 3.5 | 0.031 |
| | | 4.2 | 0.016 |

**problem 4.29**   In the circuit of Fig. 4-22a, $V = 100$ V, $R = 10$ MΩ, and $C = 0.01$ $\mu$F. Determine $v_C$, 0.2 s after the switch has closed.

**solution**   $\tau = RC = 10 \times 10^6 \times 0.01 \times 10^{-6} = 0.1$ s. $t/\tau = 0.2/0.1 = 2$. From Table 4-3a, at $t/\tau = 0.2$, $v_c/V = 0.865$. Hence, $v_C = 0.865(100) = 86.5$ V.

**problem 4.30**   In the circuit of Fig. 4-22b, $V = 100$ V, $R = 10$ MΩ, and $C = 0.01$ $\mu$F. The capacitor is fully charged. Determine voltage $v_C$ 0.21 s after the switch is opened.

**solution**   From Prob. 4.29, $\tau = 0.1$ s. $t/\tau = 0.21/0.1 = 2.1$. From Table 4-4b, $v_c/V = 0.125$. Hence, $v_C = 0.125(100) = 12.5$ V.

## Selection of Capacitors

The following factors are involved in the selection of capacitors.
1. Capacitance value and limiting values.
2. Voltage: dc, ac, peak, and surge.
3. Physical size and mounting requirements.
4. Temperature limits.
5. Temperature coefficient of capacitance.
6. Tolerance or precision.
7. Variation of capacitance with voltage.
8. Leakage.
9. Whether polarized or nonpolarized.
10. Value of $Q$.
11. Parasitic effects, series inductance, series resonance.
12. Whether fixed or variable. In variable capacitors, the maximum allowed number of variational adjustments.
13. Stability.

14. Environmental effects: Shock, vibration, temperature cycling, humidity, solderability, mechanical strength, altitude, insulation, color code durability.
15. Maximum ripple voltage.
16. Maximum ripple current.
17. Frequency range.
18. Cost.

## Capacitors in Series and in Parallel

1. *Capacitors in parallel.* When two or more capacitors are connected in parallel (Fig. 4-23), the equivalent capacitance $C_{EQ}$ is the sum of the individual capacitances:

$$C_{EQ} = C_1 + C_2 + C_3 + \cdots \tag{4-30}$$

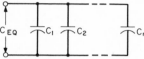

**Fig. 4-23** Capacitors in parallel.

**problem 4.31** What is the capacitance of a 0.001-$\mu$F capacitor and a 100-pF capacitor in parallel?

**solution** Both capacitance values must be expressed in the same units. To express both in picofarads, the conversion factor 1 $\mu$F = $10^6$ pF is used: 0.001 $\mu$F = 1000 pF. Therefore, $C_{EQ}$ = 1000 pF + 100 pF = 1100 pF.

2. *Capacitors in series.* When two or more capacitors are connected in series (Fig. 4-24), the resultant capacitance $C_{EQ}$ is:

$$\frac{1}{C_{EQ}} = \frac{1}{C_1} + \frac{1}{C_2} + \frac{1}{C_3} + \cdots \tag{4-31}$$

When there are two capacitors

$$C_{EQ} = \frac{C_1 C_2}{C_1 + C_2} \tag{4-32}$$

**Fig. 4-24** Capacitors in series.

**TABLE 4-5 Dielectric Constants of Some Commonly Used Materials**

| Dielectric | $k$ |
|---|---|
| Vacuum | 1 |
| Air | 1.0006 |
| Teflon | 2 |
| Polystyrene | 2.5 |
| Mylar | 3 |
| Paper, paraffin | 4 |
| Mica | 5 |
| Aluminum oxide | 7 |
| Tantalum oxide | 25 |
| Ceramic (low $k$) | 10 |
| Ceramic (high $k$) | 100–10 000 |

**problem 4.32**    What is the capacitance of a 2-$\mu$F and a 4-$\mu$F capacitor in series?

**solution**

$$C_{EQ} = \frac{(2\ \mu F)(4\ \mu F)}{2\ \mu F + 4\ \mu F} = 1.33\ \mu F$$

## Relative Dielectric Constant

The dielectric used in capacitors has considerable effect upon the flux and capacitance. The relative dielectric constant $k$ compares the flux in a vacuum ($k = 1$) with the flux in the dielectric material. Table 4-5 gives dielectric constant values for some commonly used materials.

## Calculation of Capacitance from Dimensions

1. *Parallel-plate capacitors* (see Fig. 4-25$a$).

$$C = \frac{kA}{11.3d} \quad pF \tag{4-33}$$

where $k$ = relative dielectric constant
$A$ = area, cm$^2$
$d$ = spacing between plates, cm

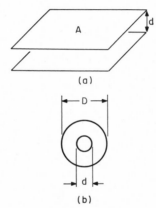

(a)

(b)

**Fig. 4-25**    Basic capacitors. ($a$) Parallel-plate. ($b$) Cylindrical.

**problem 4.33**    Determine the capacitance of a parallel-plate capacitor for which the plate dimensions are 2.5 cm by 100 cm, the distance between the plates is 0.025 cm, and the dielectric material is paper ($k = 4$).

**solution**    Substituting in Eq. (4-33),

$$C = \frac{4 \times 2.5 \times 100}{11.3 \times 0.025} = 3540\ pF$$

2. *Cylindrical capacitor* (concentric cable) (see Fig. 4-25$b$).

$$C = \frac{24.5\ k}{\log D/d} \quad pF/m \tag{4-34}$$

where $D$ = the outside diameter of the cable and $d$ = the inside diameter of the cable. Both diameters are in the same units.

**problem 4.34**    Determine the capacitance per meter of a coaxial cable. The inner conductor has a diameter of 0.06 cm and the outer cable a diameter of 0.4 cm; $k = 2.23$.

**solution**    Substituting in Eq. (4-34) yields

$$C = \frac{24.5(2.23)}{\log\left(\dfrac{0.4}{0.06}\right)} = 66.3\ pF/m$$

## Impedance of a Capacitor

The impedance of a capacitor is not a pure reactance but is modified by the series resistance of the leads and plates, by losses in the dielectric, and by parallel resistance and leakage effects. One way to handle this is to combine all these effects into an equivalent series resistance (ESR), measurable directly on impedance bridges or indirectly on instruments such as a $Q$ meter.

The impedance Z is given by:

$$Z = \sqrt{(ESR)^2 + X_C^2} \qquad (4\text{-}35)$$

**problem 4.35**   A 20-$\mu$F electrolyte capacitor has an ESR of 0.25 $\Omega$ at a frequency of 15,900 kHz. What is the impedance of the capacitor?

**solution**

$$X_C = \frac{1}{2\pi(15,900) \times 20 \times 10^{-6}} = 0.5\ \Omega$$

$$Z = \sqrt{(0.25)^2 + (0.5)^2} = 0.56\ \Omega$$

## Q and Dissipation Factor

The dissipation factor DF is given by:

$$DF = \frac{ESR}{X_C} \times 100\% \qquad (4\text{-}36)$$

The $Q$ factor, identical in interpretation to the $Q$ of a coil, is given by:

$$Q = \frac{1}{DF} = \frac{X_C}{ESR} \qquad (4\text{-}37)$$

In dealing with parallel circuits involving capacitors, it is easier to convert the ESR to an equivalent parallel resistance EPR. The specific EPR value is correct only at one frequency for two reasons; $X_C$ varies with frequency, and the resistance values also are frequency-dependent.

$$EPR = \frac{X_C^2}{ESR} \qquad (4\text{-}38a)$$

$$Q = \frac{EPR}{X_C} \qquad (4\text{-}38b)$$

**problem 4.36**   A 100-pF capacitor has an ESR equal to 20 $\Omega$ at a frequency of 1590 kHz. Determine its DF, $Q$, and EPR.

**solution**

$$X_C = \frac{1}{2\pi(1590 \times 10^3) \times 100 \times 10^{-12}}$$

$$= 1000\ \Omega$$

$$DF = \frac{20}{1000} \times 100\% = 2\%$$

$$Q = \frac{X_C}{R} = \frac{1000}{20} = 50$$

$$EPR = \frac{X_C^2}{ESR} = \frac{(10^3)^2}{20} = 50,000\ \Omega$$

or

$$EPR = Q \cdot X_C = 50(1000) = 50,000\ \Omega$$

## Temperature Coefficient

Capacitors, as well as resistors, are subject to variation in their value with temperature. Whereas for a resistor, except for the thermistor, temperature variation is undesirable, some capacitors are manufactured with specified temperature coefficients and are used for temperature compensation.

The temperature coefficient TC is expressed as the change in capacitance per degree Celsius change in temperature. It is generally expressed in parts per million per degree Celsius (ppm/°C). It may be positive (P precedes the coefficient), negative (N), or zero (NPO).

**problem 4.37**    A 220-pF ceramic capacitor has a TC of N1500. What is the change of capacitance when the temperature rises from 25 to 75°C?

**solution**

$$\Delta C = C \times TC \times \Delta T$$

$$= 220 \times \frac{-1500}{1,000,000} \times (75 - 25)$$

$$= -16.5 \text{ pF}$$

**problem 4.38**    A tuned circuit is designed to oscillate at 2 MHz. The inductance of the coil is 50 $\mu$H. It is found that the oscillator frequency drifts downward by 14,000 Hz as the ambient temperature changes from 25 to 60°C owing to an increase in coil inductance. Show how to compensate for this change by using a TC capacitor.

**solution**    For resonance, the nominal value of capacitance required is:

$$C_0 = \frac{1}{4\pi^2 f^2 L_0} = \frac{1}{4\pi^2 \times (2 \times 10^6)^2 \times 50 \times 10^{-6}}$$
$$= 126 \text{ pF}$$

For an uncompensated circuit, we can write:

$$f_0 - \Delta f = \frac{1}{2\pi \sqrt{(L_0 + \Delta L)C_0}}$$

or

$$f_0 \left(1 - \frac{\Delta f}{f_0}\right) = \frac{1}{2\pi} \frac{1}{\sqrt{L_0[1 + \Delta L/L_0]C_0}}$$

$$= \frac{1}{2\pi} \frac{1}{\sqrt{L_0 C_0}} \left(1 - \frac{1}{2} \frac{\Delta L}{L_0}\right)$$

Therefore

$$\frac{\Delta f}{f_0} = \frac{1}{2} \frac{\Delta L}{L_0}$$

and

$$\frac{\Delta L}{L_0} = \frac{2\Delta f}{f_0}$$

This can be compensated by a TC capacitor, and so

$$f_0 = \frac{1}{2\pi \sqrt{(L_0 + \Delta L)(C_0 - \Delta C)}}$$

To compensate,

$$\frac{\Delta C}{C_0} = \frac{\Delta L}{L_0} = 2 \frac{\Delta f}{f_0}$$

$$\Delta C = 126 \times 2 \times \frac{14,000}{2,000,000} \text{ pF} = 1.764 \text{ pF}$$

There are many possible ways to use a compensating capacitor. Let us make up the 126-pF tuning capacitor $C_0$ by using two capacitors in parallel. One of these is to be a 100-pF negative TC capacitor and the 26 pF is to be obtained from a variable capacitor (assumed zero TC), used for exact tuning.

To determine the TC, we have

$$\Delta C = C \times TC \times \Delta T$$
$$1.764 = 100 \times TC \times (60° - 25°)$$

TC as a fraction:

$$TC = \frac{1.764}{100 \times 35} \times 10^6 \text{ ppm/°C}$$

$$= 504 \text{ ppm/°C}$$

A 100-pF N470 ceramic capacitor can be selected to meet this requirement.

Since temperature-compensating capacitors come in many values of TC, another approach is to select the TC and use it to determine the fixed part of $C_0$. Assuming a TC of N2200,

$$\Delta C = 1.764 \, \text{pF} = C \times 2200 \times 10^{-6} \times (60° - 25°)$$
$$C = 22.9 \, \text{pF}$$

Select a 22-pF N2200 ceramic capacitor. The remaining 101 pF can be obtained from a variable-tuning capacitor.

## Series Resonance in a Capacitor

Capacitors must be connected in circuits by means of wire leads, and we have seen that a single wire has inductance. The equivalent circuit of a capacitor (assuming ESR = 0) is shown in Fig. 4-26. It is seen that there will be a frequency at which $X_L = X_C$. The capacitor is at series resonance, and the reactance is at a minimum. At higher frequencies, the capacitor acts as an inductor.

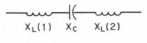

$X_C$ = capacitive reactance

$X$ = inductive reactance of leads =
    $X_L(1) + X_L(2)$

**Fig. 4-26**  Equivalent reactive circuit of a capacitor. $X_C$ = capacitive reactance. $X_L$ = inductive reactance of leads = $X_L(1) + X_L(2)$.

At high frequencies some circuits are designed to make use of this minimum reactance effect, and one must be careful to maintain the same terminal lead length if a capacitor has to be replaced. Some capacitors, such as the button feed-through type, are designed to minimize lead inductance effects.

**problem 4.39**  Determine the series resonant frequency of a 0.01-$\mu$F capacitor where each lead is 1.25 cm of No. 22 wire (the minimum practical length).

**solution**  From Prob. 4.17, the inductance of two 1.25-cm leads is $2 \times 0.011 \, \mu\text{H} = 0.022 \, \mu\text{H}$. For series resonance

$$f_0 = \frac{1}{2\pi\sqrt{LC}} = \frac{1}{2\pi} \frac{1}{\sqrt{0.022 \times 10^{-6} \times 0.01 \times 10^{-6}}}$$
$$= 10.7 \, \text{mHz}$$

## Fixed Capacitors

To meet the requirements and applications for fixed capacitors, many different styles and sizes have been developed. Table 4-6 is a listing of the characteristics of commonly used fixed capacitors.

## Variable Capacitors

There are many applications for variable capacitors, particularly in the field of communications. Variable capacitors can be divided into two basic types. In one type, the variable capacitor is made so that it can be adjusted continually, as is required for tuning a communications receiver over a wide band of frequencies. This is done with variable capacitors which may be ganged on the same shaft to simultaneously resonate several circuits.

A typical capacitor for such a function is the air-variable capacitor. This consists of two sets of aluminum plates which mesh with each other. One set of plates, the stator, is fixed. The other set, the rotor, is on a ball-bearing-supported shaft. It rotates and varies the capacitor area within the stator.

The second type is a partially variable capacitor, called a *trimmer*. The number of times trimmers can be adjusted is limited. Their function is to be adjustable to a desired capacitance value and then remain at this value. They are used for padders in communications receivers, for fine-precision tuning in fixed-frequency communications receivers, for crystal frequency adjustment, for tuning of microwave resonant circuits, for trimming of microstrips, and for adjustment of filter characteristics. Trimmers are made with mica, air,

**TABLE 4-6   Typical Characteristics of Commonly Used Fixed Capacitors**

| Type | Capacitance range | Maximum working voltage, V | Maximum operating temperature, °C | Tolerance, % | Insulation resistance, MΩ |
|---|---|---|---|---|---|
| Mica | 1 pF–0.1 $\mu$F | 50 000 | 150 | ±0.25 to ±5 | >100 000 |
| Silvered mica | 1 pF–0.1 $\mu$F | 75 000 | 125 | ±1 to ±20 | 1000 |
| Paper | 500 pF–50 $\mu$F | 100 000 | 125 | ±10 to ±20 | 100 |
| Polystyrene | 500 pF–10 $\mu$F | 1000 | 85 | ±0.5 | 10 000 |
| Polycarbonate | 0.001–1 $\mu$F | 600 | 140 | ±1 | 10 000 |
| Polyester | 5000 pF–10 $\mu$F | 600 | 125 | ±10 | 10 000 |
| Ceramic: | | | | | |
|   Low $k$ | 1 pF–0.001 $\mu$F | 6000 | 125 | ±5 to ±20 | 1000 |
|   High $k$ | 100 pF–2.2 $\mu$F | 100 | 85 | +100 to −20 | 100 |
| Glass | 10 pF–0.15 $\mu$F | 6000 | 125 | ±1 to ±20 | >100 000 |
| Vacuum | 1–5000 pF | 60 000 | 85 | ±5 | >100 000 |
| Energy storage | 0.5–250 $\mu$F | 50 000 | 100 | ±10 to ±20 | 100 |
| Electrolytic: | | | | | |
|   Aluminum | 1 $\mu$F–1 F | 700 | 85 | +100 to −20 | <1 |
|   Tantalum | 0.001–1000 $\mu$F | 100 | 125 | ±5 to ± 20 | >1 |

Teflon, ceramic, quartz, and glass dielectrics. For ultra-high-frequency applications trimmers with air, glass, and quartz dielectrics have very high values of $Q$. Mica units are of the compression type; ceramics rotate and vary the capacitance in a manner similar to that of the air-variable capacitor. In air, glass, and quartz dielectric capacitors the position of a piston is adjusted.

## 4.5  BIBLIOGRAPHY

Brotherton, M.: *Capacitors: Their Use in Electronics Circuits*, Van Nostrand, 1946.

Dummer, G. W., and Nordenberg, H. M.: *Fixed and Variable Capacitors*, McGraw-Hill, New York, 1960.

Henny, Keith: *Radio Engineers Handbook*, 5th ed., McGraw-Hill, New York, 1956.

International Telephone and Telegraph Corp.: *Reference Data for Radio Engineers*, 5th ed., Sams, Indianapolis, 1968.

Kaufman, Milton, and Seidman, A. H.: *Handbook for Electronics Engineering Technicians*, McGraw-Hill, New York, 1976.

Langford-Smith, F.: *Radiotron Designers Handbook*, 4th ed., Sydney, Australia, Radio Corporation of America, Harrison, N.J., 1957.

Mullin, W. F.: *ABCs of Capacitors*, 2d ed., Sams, Indianapolis, 1971.

Terman, F. E.: *Radio Engineering*, 3d ed., McGraw-Hill, New York, 1947.

———: *Radio Engineers Handbook*, McGraw-Hill, New York, 1943.

Turner, R. P.: *ABCs of Resistance and Resistors*, Sams, Indianapolis, 1974.

Chapter **5**

# Selecting Semiconductor Devices

## ROGER C. THIELKING

Chairperson and Associate Professor of Electrical Technology,
Onondaga Community College, Syracuse, N.Y.

## 5.1 INTRODUCTION

In selecting a semiconductor device, one is confronted with a huge number of options. Over 6000 types each of diodes and transistors have been registered with the Joint Electron Device Engineering Council (JEDEC), and thousands more have been marketed under manufacturers' type numbers. The catalog of a major manufacturer consists of, perhaps, several thick books. The task of researching such volumes of information appears to be very formidable. Fortunately, guidance and aids are available to simplify the procedure.

If large quantities are to be purchased, manufacturers' representatives are the best source of information. For smaller quantities, industrial electronics distributors should be contacted. In addition, manufacturers will usually respond promptly and helpfully to direct inquiry.

Other excellent sources of information are the D.A.T.A. books, published by Derivation and Tabulation Associates, Inc., Orange, New Jersey. These books are available only by subscription and are updated semiannually. They list virtually every device which is available from domestic and foreign sources, categorized in various ways, with condensed ratings and parameter data.

By one or more of these means, the choice can usually be narrowed to several economical devices that seem appropriate. At this point, manufacturers' data sheets giving complete specifications should be obtained for all the devices under consideration. The final selection is made by calculating device parameter requirements, as shown in this chapter, and choosing the least expensive type that meets these requirements.

There are two selection factors which are general in nature and need to be mentioned here:

*Hermetic vs. Plastic* In many cases, similar devices are available either in hermetically sealed cases of metal and glass or in plastic encapsulation. Hermetically sealed devices work better at high temperatures and may be more reliable under some conditions. Plastic devices, which historically have been less reliable, have been greatly improved in recent years. They are usually less expensive and have excellent reliability when operated within their ratings.

*JEDEC vs. Manufacturers' Type Numbers*    JEDEC-registered devices are numbered according to the familiar system consisting of N preceded by a digit (examples: 1N4148, 2N6222). Other factors being equal, JEDEC-numbered devices are preferred because they may be available from more than one manufacturer. Certain ratings and parameters are JEDEC-registered and will be the same for all suppliers, while others are not and may vary among companies. If nonregistered parameters are important, the specification sheet from the vending manufacturer must be checked.

## 5.2  SELECTING RECTIFIER DIODES FOR A DC POWER SUPPLY

Commonly used parameters for junction diodes are listed in Table 5-1.

**TABLE 5-1  Commonly Used Parameters for Junction Diodes**

| Parameter | Symbol | Meaning |
|---|---|---|
| Dc forward voltage | $V_F$ | Voltage across a forward-biased diode (anode positive with respect to cathode) |
| Dc forward current | $I_F$ | Direct current flowing in a forward-biased diode |
| Dc reverse voltage | $V_R$ | Voltage across a reverse-biased diode (anode negative with respect to cathode) |
| Dc reverse current | $I_R$ | Leakage current flowing in reverse-biased diode |
| Reverse breakdown voltage | $V_{br}$, $B_V$, PRV, PIV | Maximum reverse voltage across diode before it breaks down |
| Power dissipation | $P_D$ | Maximum power that may be dissipated in a diode |
| Operating junction temperature | $T_j$ | Temperature of the pn junction |
| Capacitance | $C$ | Capacitance across diode in its forward- or reverse-biased state |
| Reverse recovery time | $t_{rr}$ | Time required for reverse current or voltage to reach a specified value after switching diode from forward- to reverse-biased state |
| Forward recovery time | $t_{fr}$ | Time required for forward voltage or current to reach a specified value after switching diode from its reverse- to forward-biased state |
| Noise figure | $NF_o$ | Ratio of rms output noise power of receiver in which diode is used to that of an ideal receiver of same gain and bandwidth |
| Conversion loss | $L_C$ | Power lost in mixer diode when converting an rf signal to an i-f signal |
| Video resistance | $R_v$ | Low-level impedance of a detector diode |

SOURCE: M. Kaufman and A. H. Seidman, *Handbook for Electronics Engineering Technicians*, McGraw-Hill, New York, 1976.

**problem 5.1**    Rectifier diodes $D_1$, $D_2$, $D_3$, and $D_4$ must be selected for the bridge rectifier of Fig. 5-1. The load draws a maximum current of 800 mA, and the maximum ambient temperature is

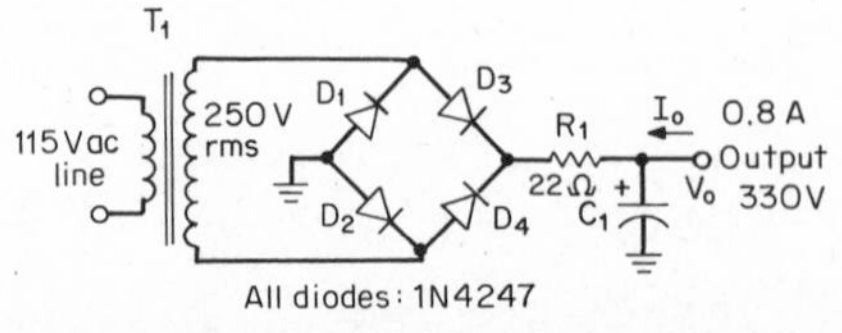

All diodes: 1N4247

**Fig. 5-1**  A bridge rectifier circuit.

60°C. Power transformer $T_1$ has a nominal output of 250 V rms at full load. Select a suitable rectifier diode.

**theory**   The key parameters for rectifier diode selection are reverse voltage and average, or direct, forward current. Where a capacitive input filter is used, as in this case, surge current is also important.

At each peak of applied voltage, current flow is through either $D_1$ and $D_4$ or $D_2$ and $D_3$, with the other diodes cut off. The peak reverse voltage $V_R$ is then

$$V_R = \sqrt{2}V_s - 0.7 \text{ V} \tag{5-1}$$

where $V_s$ is the rms value of the secondary of $T_1$, and 0.7 V is the approximate drop across a conducting diode.

---

The General Electric 1N4245-49 Series are A14 types, 2.5 ampere rated, axial-leaded, general-purpose rectifiers. Dual heat-sink construction provides rigid mechanical support for the pellet and excellent thermal characteristics. Passivation and protection of the silicon pellet's pn junction are provided by solid glass; no organic materials are present within the hermetically sealed package.

The 1N4245-49 series (A14's) are "Transient-Voltage Protected." These devices will dissipate up to 1000 W in the reverse direction without damage. Voltage transients generated by household or industrial power lines are dissipated.

Absolute maximum ratings: (25°C unless otherwise specified)

| | 1N4245 | 1N4246 | 1N4247 | 1N4248 | 1N4249 | |
|---|---|---|---|---|---|---|
| Reverse voltage ($-65$ to $+160°C$, $T_J$) | | | | | | |
| Working peak, $V_{RWM}$ | 200 | 400 | 600 | 800 | 1000 | V |
| DC, $V_R$ | 200 | 400 | 600 | 800 | 1000 | V |
| Average forward current, $I_O$ | | | | | | |
| 55°C ambient | ← | | 1.0 | | → | A |
| 25°C ambient | ← | | 2.5 | | → | A |
| Peak surge forward current, $I_{FSM}$ | | | | | | |
| Nonrepetitive, 0.0083 s | | | | | | |
| Half sine wave | ← | | 25 | | → | A |
| Full load JEDEC method | | | | | | |
| Peak surge forward current, $I_{FSM}$ | | | | | | |
| Nonrepetitive, .001 | | | | | | |
| Half sine wave | | | | | | |
| Full load 160°C, $T_J$ | ← | | 90 | | → | A |
| No load (25°C case) | ← | | 100 | | → | A |
| Junction operating temperature range, $T_J$ | ← | | $-65$ to $+160°C$ | | → | |
| Storage temperature range, $T_{STG}$ | ← | | $-65$ to $+200°C$ | | → | |
| $I^2t$, RMS for fusing, 0.001 to 0.01 s | ← | | 4.0 | | → | A |
| Peak nonrepetitive reverse | | | | | | |
| Power rating, $P_{RM}$ | ← | | 1000 | | → | W |
| (20 μs half sine wave, at max. $T_J$) | | | | | | |
| Mounting: Any position. Lead temperature 290°C maximum to ⅛″ from body for 5 seconds maximum during mounting. | | | | | | |

Electrical characteristics: (25°C unless otherwise specified)

| | | | | | | |
|---|---|---|---|---|---|---|
| Maximum forward voltage drop, $V_{FM}$ | ← | | 1.2 | | → | V |
| $I_F = 1.0A$, $T_A = +55°C$ | | | | | | |
| Maximum forward voltage drop, $V_{FM}$ | ← | | 1.25 | | → | V |
| $I_F = 2.5A$, $T_A = +25°C$ | | | | | | |
| Maximum reverse current, $I_{RM}$ | | | | | | |
| at rated $V_R$ | | | | | | |
| $T_J = +25°C$ | ← | | 1.0 | | → | μA |
| $T_J = +125°C$ | ← | | 25 | | → | μA |
| Typical reverse recovery time, $t_{rr}$ | ← | | 2.5 | | → | μs |
| Maximum reverse recovery time, $t_{rr}$ | ← | | 5.0 | | → | μs |
| (Recovery circuit per MIL-S-19500/286B) | | | | | | |

**Fig. 5-2**  Excerpts of specification, 1N4247. *(Courtesy General Electric Company.)*

Since each diode conducts on alternate half-cycles, the average direct current $I_F$ through each is:

$$I_F = \frac{I_o}{2} \tag{5-2}$$

where $I_o$ is the dc output of the circuit. Maximum surge current $I_S$ occurs upon turn-on when $C_1$ is discharged and peak input voltage is applied. There are two conducting diodes in series in the current path, so that:

$$I_S = \frac{\sqrt{2}V_s - 2 \times 0.7}{R_1} \tag{5-3}$$

which assumes negligible impedances for $T_1$ and the power source. If these values are significant, $I_S$ will have a smaller value.

**solution**   Select a 1N4247 device (Fig. 5-2). The key parameters previously mentioned should be determined first. Reverse voltage $V_R$, by Eq. (5-1), is $V_R = \sqrt{2} \times 250\text{ V} - 0.7\text{ V} \approx 353\text{V}$. At first glance, a 400-V device would seem appropriate. However, line voltage can usually vary by $\pm 10\%$, and the tolerance and load regulation factor of transformer $T_1$ could add another 10% or more, in the worst case. This could increase $V_R$ above 400 V, so that the next higher voltage device is needed.

Current $I_F$, by Eq. (5-2), is $I_F = 0.8\text{A}/2 = 0.4\text{A}$ through each diode, with the load current at its maximum. This compares with a maximum rating for the 1N4247, at the maximum ambient of $60^\circ$C, of about 1 A or more, depending upon the details of lead mounting. (With rectifiers of this type, most of the heat dissipation is through the leads.) But note that this rating is for a resistive load. With a capacitive load, the diode current should be *limited to about half* of the $I_F$ rating, or about 0.5 A for these devices. This is because the repeated, high-current surges of the capacitive circuit apply more stress to the diodes.

A few rectifier specifications give rating data for capacitive loads, which usually allow more than 50% of the current for resistive loads. In the absence of such data, the 50% derating is a safe approach.

Nonrecurrent surge current, by Eq. (5-3), is

$$I_S = \frac{\sqrt{2} \times 250\text{ V} - 2 \times 0.7\text{ V}}{22\ \Omega} = 16\text{ A}$$

However, this is subject to the same variations as the reverse voltage—a total of perhaps 20%, bringing the worst-case value to about 20 A. This is still within the 25-A rating given for the JEDEC measurement conditions.

Other items in the specification also should be reviewed. Some of these are self-explanatory or clearly not applicable, but others need some mention. The alternate surge current ratings of 90 and 100 A may be used if the indicated conditions are met. To determine this, measurements with an oscilloscope using a current probe may be easier than analytical calculations.

The reverse current of up to 25 $\mu$A has a negligible effect in rectifier circuits, and the forward voltage drop of 1.2 V also is insignificant when the output voltage is high. Reverse recovery time of 5 $\mu$s is important only at frequencies much greater than 60 Hz, or in certain switching applications.

The peak reverse power rating $P_{RM}$ of 1000 W is for unusual conditions of reverse breakdown which might be caused, for example, by a severe power line transient. If fusing is used, the $I^2t$ value of 4A²s indicates the maximum rating for a fuse which will protect the diode.

Finally, it must be noted that some of these parameter values are not JEDEC registered. If such values are important, the specifications of the particular vending company must be checked.

## 5.3  SELECTING A ZENER DIODE FOR A SHUNT REGULATOR

Commonly used parameters for zener diodes are listed in Table 5-2.

**TABLE 5-2   Commonly Used Parameters for Zener Diodes**

| Parameter | Symbol | Meaning |
| --- | --- | --- |
| Zener voltage | $V_Z$ | Nominal voltage at which zener diode regulates |
| Knee current | $I_{ZK}$ | Minimum current necessary for operation of zener diode |
| Maximum zener current | $I_{ZM}$ | Maximum current that can flow in zener diode |
| Zener impedance | $Z_Z$ | Indicates change in zener voltage for small changes in zener current with respect to a specified test current $I_{ZT}$ |

SOURCE: M. Kaufman and A. H. Seidman, *Handbook for Electronics Engineering Technicians*, McGraw-Hill, New York, 1976.

**problem 5.2**    The voltage regulator of Fig. 5-3 provides a 12-V dc output using a 25-V dc source. The output voltage must be within ±10%, and it must vary less than 100 mV with a 2-V change of the 25-V source. The load, represented by $R_L$, can vary between 400 Ω and 1 kΩ. Maximum ambient temperature is 40°C.

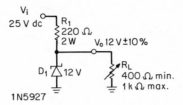

**Fig. 5-3**   Basic shunt regulator circuit.

**theory**    The performance of a zener diode is largely dependent on the zener current $I_Z$ flowing through it. In this circuit:

$$I_Z = \frac{V_i - V_o}{R_1} - \frac{V_o}{R_L} \tag{5-4}$$

An unvarying dc output voltage $V_o$ is usually desired. However, as $I_Z$ changes, there will be a slight change in $V_o$, given by

$$\Delta V_o = \Delta I_Z Z_Z \tag{5-5}$$

where $Z_Z$ is the zener impedance of the diode.

When $V_i$ varies, most of the voltage change occurs across $R_1$, and most of the resulting change of current flows through the diode (since the output voltage and current are relatively constant). Thus,

$$\Delta I_Z \cong \frac{\Delta V_i}{R_1} \tag{5-6}$$

and, by Eq. (5-5),

$$\Delta V_o \cong Z_Z \frac{\Delta V_i}{R_1} \tag{5-7}$$

**solution**    Select a 1N5927B device (Fig. 5-4). In general, zener diodes operate well at the zener test current $I_{ZT}$ at which they are tested. They also work well at increased currents, if the device is not overdissipated. Therefore, it is good practice to select a device whose $I_{ZT}$ is comparable with the minimum $I_Z$ of the circuit. This occurs when the load $R_L$ is at 400 Ω and, by Eq. (5-4)

$$I_Z = \frac{25\text{ V} - 12\text{ V}}{220\,\Omega} - \frac{12\text{ V}}{400\,\Omega} = 29\text{ mA} \tag{5-8}$$

which is close to the $I_{ZT}$ of 31.2 mA for the 1N5927B.

Maximum power dissipation must also be checked. This occurs at maximum $I_Z$, when $R_L = 1$ kΩ. Then,

$$I_{Z,\text{max}} = \frac{25\text{ V} - 12\text{ V}}{220\,\Omega} - \frac{12\text{ V}}{1\text{ k}\Omega} = 47\text{ mA} \tag{5-9}$$

Power dissipation $P_D$ is:

$$P_D = I_Z V_Z = 47\text{ mA} \times 12\text{ V} = 564\text{ mW}$$

which is well within the device rating of 1.5 W at $T_L$ (lead temperature) of 75°C.

The amount of output voltage change, owing to an input voltage change, is calculated from the specified zener impedance $Z_{ZT}$ of 6.5 Ω and, by Eq. (5-7):

$$\Delta V_o \cong 6.5\,\Omega \times \frac{2\text{ V}}{220\,\Omega} = 59\text{ mV}$$

which is within the circuit requirements. This value applies when $R_L = 400$ Ω, which causes $I_Z \cong I_{ZT}$. At larger values of $R_L$, $I_Z$ will be greater and $Z_Z$ less, so that $\Delta V_o$ would then be somewhat smaller.

The 10% tolerance on $V_o$ requires the use of the 1N5927B, which has a ±5% tolerance. The 10%

Maximum ratings*

| Rating | Symbol | Value | Unit |
|---|---|---|---|
| DC Power Dissipation @ $T_L$ = 75°C, Lead Length = ⅜″ | $P_D$ | 1.5 | W |
| Derate above 75°C | | 12 | mW/°C |
| Operating storage junction Temperature range | $T_J$, $T_{stg}$ | −55 to +200 | °C |

*Indicates JEDEC Registered Data.

Electrical characteristics ($T_L$ = 30°C unless otherwise noted.) ($V_F$ = 1.5 Volts Max @ $I_F$ = 200 mAdc for all types.)

| Motorola type number (Note 1) | Nominal Zener voltage $V_Z$ @ $I_{ZT}$, V | Test current $I_{ZT}$, mA | Max. zener impedance | | | | Max. reverse leakage current | | Maximum dc zener current $I_{ZM}$, mAdc |
|---|---|---|---|---|---|---|---|---|---|
| | | | $Z_{ZT}$ @ $I_{ZT}$ ohms | $Z_{ZK}$, @ ohms | $I_{ZK}$, mA | | $I_R$, @ μA | $V_R$, volts | |
| 1N5927 | 12 | 31.2 | 6.5 | 550 | 0.25 | | 1.0 | 9.1 | 125 |

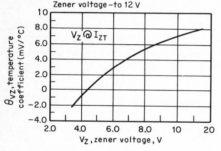
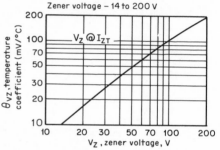

NOTE 1—TOLERANCE AND VOLTAGE DESIGNATION:
Tolerance designation—The type numbers listed indicate a tolerance of ±20%. Device tolerances of ±10% are indicated by an "A" suffix, ±5% by a "B" suffix, ±2% by a "C" suffix, ±1% by a "D" suffix.

**Fig. 5-4**   Excerpts of specification, 1N5927B. *(Courtesy Motorola, Inc.)*

version, 1N5927A, is not suitable because additional changes will result from circuit current and temperature variations. A rough indication of these changes may be estimated as follows.

When $R_L$ changes from 400 Ω to 1 kΩ, $I_Z$ increases by 18 mA as shown by Eqs. (5-8) and (5-9). This gives an output voltage change, by Eq. (5-5), of

$$\Delta V_o \simeq 6.5 \ \Omega \times 18 \ \text{mA} = 117 \ \text{mV} \tag{5-10}$$

(This is somewhat greater than the actual change, because $Z_Z$ becomes less than 6.5 Ω at the larger current.) The increase in $I_Z$ also causes increased power dissipation:

$$\Delta P_D = \Delta I_Z \times V_o = 18 \ \text{mA} \times 12 \ \text{V} = 216 \ \text{mW}$$

The resulting increase in temperature is calculated from the power-derating factor of 12 mW/°C: $\Delta T = \Delta P_D/(12 \ \text{mW/°C}) = 216 \ \text{mW}/(12 \ \text{mW/°C}) = 18°C$. The 15°C ambient temperature range from 25°C (room temperature) to 40°C, must be added, giving a total possible temperature change of 33°C. The temperature coefficient $\theta_{VZ}$ for the 12-V diode, from the curves shown in Fig. 5-4, is about + 9 mV/°C; then,

$$\Delta V_o = \Delta T \times \theta_{VZ} = 33°C \times 9 \ \text{mV/°C} = 297 \ \text{mV} \tag{5-11}$$

Adding Eqs. (5-10) and (5-11) yields a total possible output voltage change of +414 mV. This, added to the basic ±5% tolerance for the device, keeps the total variation within ±10% as required.

The knee and the reverse-current parameters do not apply in this circuit because the diode is not operated at low current levels.

## 5.4   SELECTING AN LED INDICATOR LIGHT

Commonly used parameters which apply to a light-emitting diode (LED) are listed in Table 5-3.

**TABLE 5-3   Commonly Used Parameters for the LED**

| Parameter | Symbol | Meaning |
|---|---|---|
| Forward voltage | $V_F$ | Dc forward voltage across an LED |
| Candlepower, candela | CP, cd | Measure of luminous intensity (brightness) of emitted light |
| Radian power output | $P_o$ | Light power, or brightness, of an LED |
| Peak spectral emission | $\lambda_{peak}$ | Wavelength of brightest emitted color. (Symbol $\lambda$ is the Greek letter *lambda*.) |
| Spectral bandwidth | $BW_\lambda$ | Indication of how concentrated is the emitted color |

SOURCE: M. Kaufman and A. H. Seidman, *Handbook for Electronics Engineering Technicians,* McGraw-Hill, New York, 1976.

**problem 5.3**   In the circuit of Fig. 5-5 transistor $Q_1$ is driven as a switch, so that it is either saturated or cut off. Diode $D_1$ is an LED. Ambient temperature may reach 40°C. Select a suitable LED for the circuit.

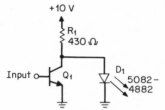

**Fig. 5-5**   A LED driver circuit.

**theory**   When $Q_1$ is saturated, its collector voltage is only a few tenths of a volt, which is insufficient to forward-bias $D_1$, causing it to be extinguished. With $Q_1$ cut off, $D_1$ is forward-biased with a current given by:

$$I_F = \frac{V_{CC} - V_F}{R_1} \tag{5-12}$$

where $V_{CC}$ is the supply voltage and $V_F$ the forward voltage drop across $D_1$.

**solution**   Select a 5082-4882 device (Fig. 5-6). When a red LED is used, $V_F$ is approximately 1.6 V. Then, by Eq. (5-12),

$$I_F = \frac{10 \text{ V} - 1.6 \text{ V}}{430 \text{ }\Omega} = 19.5 \text{ mA}$$

The 5082-4882 is characterized for 20 mA, indicating that it operates well for the $I_F$ of the circuit. Its relatively high luminous intensity of 1.8 mcd and the red diffused lens, giving a high-contrast ratio, improve visibility. At the maximum $V_F$ of 2 V, power dissipation is about 40 mW, which is well within the 100-mW rating.

LEDs usually have a low reverse-voltage breakdown, 3 V in this case. This is no problem if the circuit is designed so that reverse voltage is not applied.

Lamps shaped like the 5082-4882 are ideal for panel mounting, with convenient clips available for this purpose. Since these items are visual outputs, many factors in their selection call for subjective judgments. Appearance may, in some cases, be very important. In determining the adequacy of the brightness, several facts should be kept in mind:

1. Brightness may vary by a factor of 2:1 or more within a particular type number.

2. Brightness decreases, usually by about 1%/°C, as temperature increases. This indicates a brightness decrease of about 15% in this problem, at maximum ambient.

3. LEDs have high reliability and long life comparable with other good semiconductor devices. There is, however, a significant decline in brightness after aging.

These factors argue for selecting a device with more than adequate brightness, when a typical sample is evaluated.

**Lens Appearance**

The 5082-4880 series is available in three different lens configurations. These are red diffused, clear diffused, and clear.

The red diffused lens provides an excellent off/on contrast ratio. The clear lens is designed for applications where a point source is desired. It is particularly useful where the light must be focused or diffused with external optics. The clear diffused lens is useful in masking the red color in the off condition.

LED Selection Guide

| Minimum light output (mcd) | Lens type | | |
|---|---|---|---|
| | Red diffused lens | Clear plastic lens | Clear diffused lens |
| 0.5 | 5082-4880 | 5082-4883 | 5082-4886 |
| 1.0 | 5082-4881 | 5082-4884 | 5082-4887 |
| 1.6 | 5082-4882 | 5082-4885 | 5082-4888 |

Maximum Rating (25°C)

| | | | |
|---|---|---|---|
| DC power dissipation | 100 mW | Isolation voltage (between lead and case) | 300 V |
| DC forward current | 50 mA | Operating and storage | |
| Peak forward current | 1 A | temperature range | −55°C to +100°C |
| (1 μs pulse width, 300 pps) | | Lead soldering temperature | 230°C for 7 s |

Electrical Characteristics (25°C)

| Symbol | Parameters | 5082-4880 5082-4883 5082-4886 | | | 5082-4881 5082-4884 5082-4887 | | | 5082-4882 5082-4885 5082-4888 | | | Units | Test conditions |
|---|---|---|---|---|---|---|---|---|---|---|---|---|
| | | Min. | Typ. | Max. | Min. | Typ. | Max. | Min. | Typ. | Max. | | |
| $I$ | Luminous intensity | 0.5 | 0.8 | | 1.0 | 1.3 | | 1.6 | 1.8 | | mcd | $I_F = 20$ mA |
| $I$ | Luminous intensity | | | | | | | | 0.8 | | mcd | $I_F = 10$ mA |
| $\lambda_{pk}$ | Wavelength | | 655 | | | 655 | | | 655 | | nm | Measurement at Peak |
| $\tau_s$ | Speed of response | | 10 | | | 10 | | | 10 | | ns | |
| $C$ | Capacitance | | 200 | | | 200 | | | 200 | | pF | |
| $\theta_{JC}$ | Thermal resistance | | 270 | | | 270 | | | 270 | | °C/W | Junction to Cathode Lead |
| $V_F$ | Forward voltage | | 1.6 | 2.0 | | 1.6 | 2.0 | | 1.6 | 2.0 | V | $I_F = 20$ mA |
| $BV_R$ | Reverse breakdown voltage | 3 | 4 | | 3 | 4 | | 3 | 4 | | V | $I_R = 10$ μA |

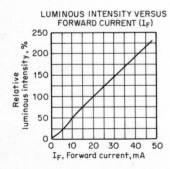

LUMINOUS INTENSITY VERSUS
FORWARD CURRENT ($I_F$)

Relative luminous intensity, %

$I_F$, Forward current, mA

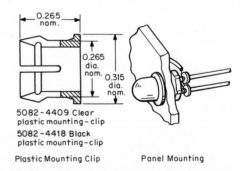

5082-4409 Clear
plastic mounting-clip

5082-4418 Black
plastic mounting-clip

Plastic Mounting Clip          Panel Mounting

**Fig. 5-6** Excerpts of specification, 5082-4882. *(Courtesy Hewlett-Packard, Inc.)*

Another characteristic of the LED is that its luminous efficiency increases with current. Increasing the direct current, however, increases junction temperature which reduces efficiency. By operating the device in a pulse mode, so as to maintain the same average current and power while increasing peak current, the perceived brightness can be increased by a factor of 2 or more.

## 5.5  LIQUID-CRYSTAL-DISPLAY (LCD) DEVICES

These devices are neither diodes nor semiconductors. They are discussed here briefly because of their function, which is similar to that of LEDs used for numeric or alphanumeric displays. A familiar example is a digital watch readout where both LEDs and LCDs are widely used.

The LCD does not emit light, but rather depends upon ambient light for its display (as does this printed page). It consists of a reflective backing behind the liquid crystal material, which is normally transparent but becomes opaque and black in appearance when voltage is applied across it. The liquid crystal is contained within two transparent and electrically conductive layers, to form a "sandwich," similar to the plates and dielectric of a capacitor. Its electrical characteristics are also like a low-value capacitor, since the liquid crystal material is an electrical insulator.

The only power consumed by an LCD is the minute amount needed to charge or discharge the small capacitance upon changing the display. This makes it especially attractive for displays using small batteries as the power source. Its main disadvantage is poor visibility in dim ambient light, which may require adding a light source.

## 5.6  SELECTING A VOLTAGE-VARIABLE CAPACITANCE DIODE FOR AN RF-TUNED CIRCUIT

Commonly used parameters which apply to voltage-variable capacitor diodes are listed in Table 5-4.

**TABLE 5-4  Commonly Used Parameters for Voltage-Variable Capacitors**

| Parameter | Symbol | Meaning |
|---|---|---|
| Series inductance | $L_S$ | Inductance, mostly of the leads, which is in series with the diode capacitance |
| Capacitance temperature coefficient | $TC_C$ | The change in capacitance caused by a temperature change |
| Figure of merit | $Q$ | Ratio of the capacitive reactance to the effective series resistance |
| Tuning ratio | $TR$ | Ratio of high to low capacitance values at specified reverse voltages |

**problem 5.4**   In Fig. 5-7, a voltage-variable capacitance diode $D_1$ must be selected for use with inductor $L_1$ to form a parallel-resonant tuned circuit. Inductor $L_1$ may be slug-tuned from 6.5 to 9.5 $\mu$H. The control voltage from potentiometer $R_1$ ranges from $+2$ to $+45$ V, which must tune the circuit from a minimum frequency of 16 MHz or less to a maximum of 20 MHz or more. The resonant frequency must change less than 1% over the ambient temperature of 25 to 50°C. Stray capacitance across the circuit of 2 pF is represented by $C_s$. The Q of $L_1$ is 125, and a minimum circuit Q of 100 must be maintained.

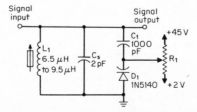

**Fig. 5-7**   A voltage-variable capacitor-tuned circuit.

**theory**   Capacitor $C_1$ blocks direct current. It is sufficiently large to act as an ac short circuit so that it does not affect the tuning. Then, the resonant frequency is given by

$$f = \frac{1}{2\pi\sqrt{L_1(C_D + C_S)}} \tag{5-13}$$

where $C_D$ is the capacitance of $D_1$. When the control voltage (which reverse biases $D_1$) is increased, the depletion layer at the $pn$ junction is thickened, and the capacitance is thereby reduced.

To determine capacitance $C_D$ and inductance $L_1$, from Eq. (5-13),

$$L_1 = \frac{1}{4\pi^2 f^2(C_D + C_S)} \tag{5-14}$$

and

$$C_D = \frac{1}{4\pi^2 f^2 L_1} - C_S \tag{5-15}$$

The $Q$ of the circuit, affected by the $Q_L$ of the inductor and $Q_C$ of the diode, is given by

$$Q = \frac{Q_L Q_C}{Q_L + Q_C} \tag{5-16}$$

(Blocking capacitor $C_1$ has a sufficient $Q$ so that it will not affect the circuit.)

**solution**   Select a 1N5140 device (Fig. 5-8). The 1N5140 family of diodes has a 60-V reverse-voltage rating, making it suitable for the control voltage of the circuit. The nominal capacitance needed may be found by Eq. (5-15), using midpoint values of 18 MHz for $f$ and 8 $\mu$H for $L_1$:

$$C_D = \frac{1}{4\pi^2(18 \text{ MHz})^2 8 \,\mu\text{H}} - 2 \text{ pF}$$

$$= 9.8 \text{ pF} - 2 \text{ pF} = 7.8 \text{ pF}$$

The 1N5140 has about the right value. It must be checked under limit conditions. Maximum capacitance at $V_R = 4$ V is 10 pF $\pm 10\%$, giving limits of 9 and 11 pF. (Somewhat larger capacitance is obtained at $V_R < 4$ V, but $Q$ decreases below its specified value.) The circuit must be tuned to the minimum frequency, 16 MHz, by means of the inductor for both limit values of capacitance. Inductance is determined by Eq. (5-14). When $C_D = 9$ pF,

$$L_1 = \frac{1}{4\pi^2(16 \text{ MHz})^2(9 \text{ pF} + 2 \text{ pF})} = 9.0 \,\mu\text{H} \tag{5-17a}$$

and when $C_D = 11$ pF,

$$L_1 = \frac{1}{4\pi^2(16 \text{ MHz})^2(11 \text{ pF} + 2 \text{ pF})} = 7.6 \,\mu\text{H} \tag{5-17b}$$

Both of these values are within the 6.5- to 9.5-$\mu$H range of $L_1$. The ability of $C_D$ to tune to the upper-frequency limit of 20 MHz by changing the control voltage may be checked with low-limit values of $L_1$, Eq. (5-17a), and $C_D$. The highest resonant frequency corresponds to a control voltage of 45 V. Specification sheet data show that $C_D$ then reduces from 9 pF (see Fig. 5-8) to approximately 3.6 pF. Frequency, calculated by Eq. (5-13), is:

$$f = \frac{1}{2\pi\sqrt{9.0 \,\mu\text{H}(3.6 \text{ pF} + 2 \text{ pF})}} = 22.4 \text{ MHz}$$

which is well above the circuit requirement.

Note that this calculation was made with low-limit $C_D$ values. For nominal or high-limit values, the tuning range would be even greater because the shunt combination of $C_D + C_S$ would change by a somewhat greater percentage.

The tuning ratio, TR, of 2.8 given in the specifications is unrealistic in practice. It requires that control voltage $V_R$ be brought to the maximum rating of the diode. Voltage $V_R$ must actually be kept appreciably below the specified maximum, to allow for tolerance and to provide a safety margin.

The $Q$ of the diode $Q_C$ is specified as 300 minimum at 50 MHz. The specifications show that it increases by a factor of about 2 at 20 MHz, making it 600. This is reduced by 10% at the 50°C high ambient limit, bringing it to 540. Then, by Eq. (5-16), circuit $Q$ will be:

$$Q = \frac{540 \times 125}{540 + 125} = 101$$

Hence, the $Q$ remains above 100 as required. This is a worst-case value (when $V_R = 4$ V) since $Q_C$ is much greater over most of the tuning range.

Silicon voltage-variable capacitance diodes, designed for electronic tuning and harmonic-generation applications, and providing solid-state reliability to replace mechanical tuning methods.
Maximum Ratings ($T_C = 25°C$ unless otherwise noted)

| Rating | Symbol | Value | Unit |
|---|---|---|---|
| Reverse voltage | $V_R$ | 60 | Vdc |
| Forward current | $I_F$ | 250 | mAdc |
| RF power input† | $P_{in}$† | 5.0 | Watts |
| Device dissipation @ $T_A = 25°C$ | $P_D$ | 400 | mW |
| Derate above 25°C | | 2.67 | mW/°C |
| Device dissipation @ $T_C = 25°C$ | $P_C$ | 2.0 | Watts |
| Derate above 25°C | | 13.3 | mW/°C |
| Junction temperature | $T_J$ | +175 | °C |
| Storage temperature range | $T_{stg}$ | −65 to +200 | °C |

†The RF power input rating assumes that an adequate heat sink is provided.

Electrical Characteristics ($T_A = 25°C$ unless otherwise noted)

| Characteristic—all types | Test conditions | Symbol | Min | Typ | Max | Unit |
|---|---|---|---|---|---|---|
| Reverse breakdown voltage | $I_R = 10\ \mu Adc$ | $B_{VR}$ | 60 | 70 | — | Vdc |
| Reverse voltage leakage current | $V_R = 55$ Vdc, $T_A = 25°C$ | $I_R$ | — | — | 0.02 | $\mu Adc$ |
| | $V_R = 55$ Vdc, $T_A = 150°C$ | | — | — | 20 | |
| Series inductance | $f = 250$ MHz, $L \approx \frac{1}{16}''$ | $L_S$ | — | 5.0 | — | nH |
| Case capacitance | $f = 1$ MHz, $L \approx \frac{1}{16}''$ | $C_C$ | — | 0.25 | — | pF |
| Diode capacitance temperature coefficient | $V_R = 4$ Vdc, $f \approx 1$ MHz | $TC_C$ | — | 200 | 300 | ppm/°C |

| | $C_T$, diode capacitance $V_R = 4$ Vdc, $f = 1$ MHz pF | | | $Q$, figure of merit $V_R = 4$ Vdc, $f = 50$ Mhz (min) | $\alpha\ V_R = 4$ Vdc, $f = 1$ MHz | | $TR$, tuning ratio $C_4/C_{60}$ $f = 1$ MHz | |
|---|---|---|---|---|---|---|---|---|
| Device | Min | Typ | Max | | Min | Typ | Min | Typ |
| 1N5139 | 6.1 | 6.8 | 7.5 | 350 | 0.37 | 0.40 | 2.7 | 2.9 |
| 1N5139A | 6.5 | 6.8 | 7.1 | 350 | 0.37 | 0.40 | 2.7 | 2.9 |
| 1N5140 | 9.0 | 10.0 | 11.0 | 300 | 0.38 | 0.41 | 2.8 | 3.0 |
| 1N5140A | 9.5 | 10.0 | 10.5 | 300 | 0.38 | 0.41 | 2.8 | 3.0 |

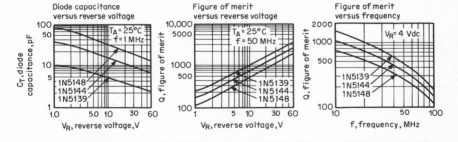

**Fig. 5-8**  Excerpts of specification, 1N5140. (*Courtesy Motorola, Inc.*)

The variation in resonant frequency with temperature results from the TC of the diode of up to 300 ppm/°C. With a 25°C change in ambient temperature, $\Delta C_D = 50°C \times 300$ ppm/°C = 15,000 ppm or 1.5%. Resonant frequency, however, is a function of the square root of $C_D$ (approximately) as indicated by Eq. (5-13). This reduces the variation by a factor of 2, to about 0.75%.

Power dissipation and forward-current parameters are not relevant because the device is always reverse-biased. Reverse current is insignificant unless the control voltage source impedance ($R_1$) is very large.

The 5% tolerance version, 1N5140A, is also suitable, but would probably cost more.

## 5.7  SELECTING MICROWAVE DIODES

Special parameters used for Schottky barrier and PIN diodes in microwave applications as mixers, detectors, and switches are listed in Table 5-5.

**TABLE 5-5    Special Parameters for Schottky and PIN Microwave Diodes**

| Parameter | Symbol | Meaning |
|---|---|---|
| Carrier lifetime | $\tau$ | Average time that a carrier will continue to exist after crossing the forward-biased junction; an indication of the switching speed of the diode |
| Standing wave ratio | SWR | Ratio of the maximum to minimum current; an indication of the amount of transmission line impedance mismatch caused by the diode |
| Intermediate frequency impedance | $Z_{IF}$ | Impedance of a diode, used as a mixer, at the if specified |
| Isolation | | Attenuation in decibels between input and output with PIN switch biased to isolate the signal |
| Insertion loss | | Attenuation in decibels of signal through a PIN switch biased for continuity |

**problem 5.5**    The balanced mixer of Fig. 5-9 is part of the front end of a microwave receiver having an rf input of 5.0 GHz, local oscillator (LO) of 4.97 GHz, and output frequency of 30 MHz. Stripline construction is used in the rf sections. The input impedance at the primary of transformer $T_1$ is 350 Ω. Select diodes for $D_1$ and $D_2$ that exhibit a good dynamic range so as to give low distortion.

(Stripline construction)

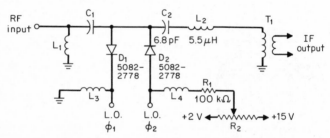

**Fig. 5-9**   A balanced mixer circuit.

**theory**    To obtain the difference frequency for the output, the LO frequencies are applied to both the mixer diodes. The two phases $\phi_1$ and $\phi_2$ of the LO are 180° apart so as to effectively cancel most of the oscillator noise at the output. Potentiometer $R_2$ adjusts the dc bias current between 20 and 150 $\mu$A as needed for best performance. Inductor $L_2$ and capacitor $C_2$ are series-resonant at 30 MHz, to couple out the i-f and provide rf isolation. Inductor $L_1$, together with the stripline impedance, provides impedance matching of the diodes to the rf line, and $C_1$ blocks the dc bias current. Inductors $L_3$ and $L_4$ are rf chokes.

**solution**    The Schottky barrier diode matched pair, 5082-2778 (Fig. 5-10), is selected. The only diode types suitable for gigahertz operation are point-contact and Schottky barrier types, which have minority carrier lifetimes of about 100 ps. The Schottky types, also known as "hot carrier" diodes, are superior for most applications. The first step in selection is to find those which are characterized for mixer application.

The next selection step is to find those mixer diodes which are made for the particular rf physical configuration being used. Stripline construction is widely used for this problem and consists of a conductor strip which is "sandwiched" between two ground planes and separated from them with thin layers of insulation. Final selection of the 5082-2778 is based on these factors:

1. The 5-GHz frequency is included in the recommended range for the devices.

2. The i-f impedance $Z_{IF}$ range of 250 to 500 Ω ensures a good match to the 400 Ω input impedance of $T_1$.

3. The maximum noise figure, NF, of 6.5 dB and standing-wave ratio, SWR, of 1.5 (for the specified conditions) are about as good as can be expected for this type of diode.

4. Use of a matched pair is important in a balanced detector, to provide nearly equal currents through each leg for optimum cancellation of the local oscillator noise.

Maximum ratings
Pulse power incident at $T_A = 25°C$                                          1 W
CW power dissipation at $T_A = 25°C$                                    300 mW
$T_{OPR}$—operating temperature range                           −60 to +150°C
$T_{STG}$—storage temperature range                              −60 to +150°C
Maximum pull on any lead                                                         2 g
Diode mounting temperature                                 220°C for 10 s max.

DC electrical specifications $T_A = 25°C$

| Parameter | Symbol | 5082-2709 -2768 -2778 | 5082-2716 -2769 -2779 | 5082-2767 | Units | Test conditions |
|---|---|---|---|---|---|---|
| Minimum breakdown voltage | $V_{SR}$ | 3 | 3 | 3 | V | $I_R = 10\mu A$ |
| Maximum forward voltage | $V_F$ | 1V @ 30mA | 1V @ 20mA | 1V @ 20mA | V | $I_F$—as shown |
| Maximum total capacitance | $C_T$ | 0.25 | 0.15 | 0.10 | pF | $V_R = 0V$, F = 1MHz |
| DC batch matched units* | — | 5082-2509 | 5082-2510 | — | — | $\Delta V_F \le 15mV$ @ 1mA |

*Minimum batch size 20 units or 20% of order, whichever is smaller.

RF electrical specifications at $T_A = 25°C$

| Parameter | Symbol | 5082-2768 | 5082-2769 | Units | Test conditions |
|---|---|---|---|---|---|
| Maximum noise figure | $NF_{SSB}$ | 6.5 @ 9.375GHz | 7.5 @ 16GHz | dB | |
| Maximum VSWR | VSWR | 1.5:1 | 1.5:1 | — | 1mW L.O. Power |
| IF impedance | $Z_{IF}$ | 250–500 | 250–500 | Ω | IF = 30MHz, 1.5dB NF |
| RF matched pairs | — | 5082-2778 | 5082-2779 | — | $\Delta NF \le 0.3dB$, $\Delta Z_{IF} < 25\Omega$ |

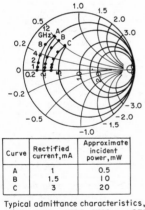

| Curve | Rectified current, mA | Approximate incident power, mW |
|---|---|---|
| A | 1 | 0.5 |
| B | 1.5 | 10 |
| C | 3 | 20 |

Typical admittance characteristics,
5082-2709, −2509, −2768 and −2778
with self bias.

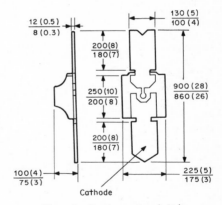

Cathode

Dimensions in micrometers (mils)

**Fig. 5-10**  Excerpts of specification, 5082-2778. *(Courtesy Hewlett-Packard, Inc.)*

The input admittance of the diode is shown in the Smith chart, Fig. 5-10. These data are needed for calculation of impedance-matching circuitry, consisting in this case of $L_1$ located at some fraction of a wavelength from $D_1$. The impedance-matching techniques involve the physical configuration of the stripline conductor and are beyond the scope of this chapter.

## 5.8  SELECTING AN AMPLIFIER BIPOLAR JUNCTION TRANSISTOR (BJT)

Parameters which apply to most transistor types are listed in Table 5-6.

**problem 5.6**   Transistor $Q_1$ is needed for the amplifier of Fig. 5-11. Maximum signal frequency is 25 kHz, minimum input impedance is 1 kΩ, and the dc-output voltage $V_o$ should not change by

**TABLE 5-6    Commonly Used Parameters for Bipolar Junction Transistors**

| Parameter | Symbol | Meaning |
|---|---|---|
| Collector-base voltage (emitter open) | $V_{CBO}$ | Maximum voltage that can be impressed across collector and base of a transistor with emitter open |
| Collector-emitter voltage (base short-circuited to emitter) | $V_{CES}$ | Maximum voltage that can be impressed across collector and emitter of a transistor with base short-circuited to emitter. Its value is in the order of one-half of $V_{CBO}$ |
| Collector-emitter voltage (with specified resistor between base and emitter) | $V_{CER}$ | For this condition, maximum collector-emitter voltage is greater than $V_{CES}$ but less than $V_{CBO}$ |
| Emitter-base voltage (collector open) | $V_{EBO}$ | Maximum voltage that can be impressed across emitter and base of a transistor with collector open |
| Collector saturation voltage | $V_{CE,sat}$ | Collector-emitter voltage of a transistor that is fully conducting, as in a transistor switch |
| Small-signal input resistance | $h_{ib}, h_{ie}$ | Input resistance of a transistor with output short-circuited for the signal. This and other $h$ parameters are called *hybrid parameters*. The second subscript refers to the transistor configuration: $b$ for common base and $e$ for common emitter. |
| Small-signal output admittance | $h_{ob}, h_{oe}$ | Output admittance of a transistor with input open-circuited for the signal |
| Small-signal reverse-voltage transfer ratio | $h_{rb}, h_{re}$ | Ratio of voltage developed across input of a transistor to voltage present at output, with input open-circuited for the signal |
| Small-signal forward-current gain | $h_{fb}, h_{fe}$ | Ratio of output to input signal currents with output short-circuited |
| Dc forward-current gain | $h_{FE}$ | Ratio of dc collector current to dc base current for transistor in common-emitter configuration |
| Collector dissipation | $P_C$ | Power dissipated in collector of a transistor. It is equal to the product of the dc collector current and dc collector-emitter voltage. |
| Gain-bandwidth product | $f_T$ | Frequency at which common-emitter forward current gain is unity |
| Cutoff frequency | $f_{hfb}, f_{hfe}$ | Frequency at which $h_{fb}$ (or $h_{fe}$) is 0.707 times its value at 1 kHz |
| Collector-cutoff current (emitter open) | $I_{CBO}, I_{CO}$ | Reverse saturation (leakage) current flowing between collector and base with emitter open |
| Collector-cutoff current (base open) | $I_{CEO}$ | Reverse saturation current flowing between collector and emitter with base open |
| Collector-base capacitance | $C_{ob}, C_{cb}$ | Transistor capacitance across collector and base. This capacitance influences, to a great degree, the high-frequency performance of a transistor amplifier |

SOURCE: M. Kaufman and A. H. Seidman, *Handbook for Electronics Engineering Technicians,* McGraw-Hill, New York, 1976.

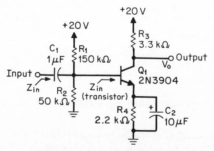

**Fig. 5-11**   A BJT amplifier circuit.

more than 1.5 V with transistor variations (when another transistor of the same type is substituted). Ambient temperature range is from 0 to 50°C. Select a suitable BJT for the circuit.

**theory**    In this type of amplifier circuit, the voltage gain of the stage is largely independent of transistor parameters. However, input impedance $Z_{in}$ is affected by the transistor and, because this is the load for the prior stage, it may affect the overall gain. Some minimum value of $Z_{in}$ must often be maintained, as indicated in this problem.

The input impedance is lowest over the mid- to high-frequency range when capacitors $C_1$ and $C_2$ have negligible impedance. Then, $Z_{in}$ of the transistor is given approximately by $h_{fe}Z_e$ where $Z_e$ is the impedance of the emitter circuit. In this case, $Z_e = r_e$, the emitter resistance of $Q_1$. For most transistor conditions, $r_e = I_E/26$ mV (from basic $pn$-junction theory). Thus,

$$Z_{in,transistor} = h_{fe}26 \text{ mV}/I_E \tag{5-18}$$

The input impedance of the entire circuit, $Z_{in}$, is equal to the transistor input impedance $Z_{in,transistor}$ shunted by bias resistors $R_1$ and $R_2$. The dc output voltage is given by

$$V_o = 20 \text{ V} - I_C R_3 \tag{5-19}$$

and

$$I_C \cong I_E \cong \frac{(20 \text{ V} \times R_2)/(R_1 + R_2) - 0.7 \text{ V}}{R_4} \tag{5-20}$$

This neglects the effect of voltage drop due to base current. This drop, however, is precisely what causes $V_o$ to vary with transistors, owing to different values of dc current gain $h_{FE}$.

The additional voltage changes $\Delta V_B$ and $\Delta V_o$ are given by:

$$\Delta V_B = \Delta V_E = \frac{R_1 R_2}{R_1 + R_2}\frac{I_C}{h_{FE}} \tag{5-21a}$$

$$\Delta V_o = \frac{R_3}{R_4}\frac{R_1 R_2}{R_1 + R_2}\frac{I_C}{h_{FE}} \tag{5-21b}$$

**solution**    Select a 2N3904 device (Fig. 5-12). There are many transistors that would be suitable for this application. For general-purpose applications of this type, transistors already being used or on hand should be examined for suitability, so as to minimize the number of types used. In selecting the 2N3904, it is assumed that another suitable device is not already on hand.

The selection is based mainly on these factors: good characteristics (40 to 60 V, current gain of about 100 or more, and gain bandwidth of 100 MHz or more), popularity, availability from several manufacturers, low cost, and characterization at or near the collector current of the circuit.

The approximate collector current is determined by Eq. (5-20):

$$I_C = \frac{[(20 \text{ V} \times 50 \text{ k}\Omega)/(150\text{k}\Omega + 50 \text{ k}\Omega)] - 0.7 \text{ V}}{2.2 \text{ k}\Omega} = 1.95 \text{ mA}$$

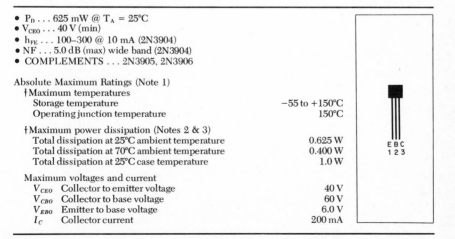

- $P_D$ ... 625 mW @ $T_A = 25$°C
- $V_{CEO}$ ... 40 V (min)
- $h_{FE}$ ... 100–300 @ 10 mA (2N3904)
- NF ... 5.0 dB (max) wide band (2N3904)
- COMPLEMENTS ... 2N3905, 2N3906

Absolute Maximum Ratings (Note 1)
†Maximum temperatures

| | |
|---|---|
| Storage temperature | −55 to +150°C |
| Operating junction temperature | 150°C |

†Maximum power dissipation (Notes 2 & 3)

| | |
|---|---|
| Total dissipation at 25°C ambient temperature | 0.625 W |
| Total dissipation at 70°C ambient temperature | 0.400 W |
| Total dissipation at 25°C case temperature | 1.0 W |

Maximum voltages and current

| | | |
|---|---|---|
| $V_{CEO}$ | Collector to emitter voltage | 40 V |
| $V_{CBO}$ | Collector to base voltage | 60 V |
| $V_{EBO}$ | Emitter to base voltage | 6.0 V |
| $I_C$ | Collector current | 200 mA |

E B C
1 2 3

**Fig. 5-12**    Excerpts of specification, 2N3904. *(Courtesy Fairchild Camera and Instrument Corp.)*

Electrical Characteristics (25°C ambient temperature unless otherwise noted)

| Symbol | Characteristic | 2N3903 Min. | 2N3903 Max. | 2N3904 Min. | 2N3904 Max. | Units | Test conditions |
|---|---|---|---|---|---|---|---|
| $BV_{CBO}$ | Collector-to-base breakdown voltage | 60 | | 60 | | V | $I_C = 10\ \mu A,\ I_E = 0$ |
| $BV_{CEO}$ | Collector-to-emitter breakdown voltage (Note 4) | 40 | | 40 | | V | $I_C = 1.0\ mA,\ I_B = 0$ |
| $BV_{EBO}$ | Emitter-to-base breakdown voltage | 6.0 | | 6.0 | | V | $I_E = 10\ \mu A,\ I_C = 0$ |
| $I_{CEX}$ | Collector cutoff current | | 50 | | 50 | nA | $V_{CE} = 30\ V,\ V_{EB} = 3.0\ V$ |
| $I_{BL}$ | Base cutoff current | | 50 | | 50 | nA | $V_{CE} = 30\ V,\ V_{EB} = 3.0\ V$ |
| $h_{FE}$ | DC current gain (Note 4) | 20 | 40 | | | | $I_C = 0.1\ mA,\ V_{CE} = 1.0\ V$ |
| | | 35 | 70 | | | | $I_C = 1.0\ mA,\ V_{CE} = 1.0\ V$ |
| | | 50 | 150 | 100 | 300 | | $I_C = 10\ mA,\ V_{CE} = 1.0\ V$ |
| | | 30 | 60 | | | | $I_C = 50\ mA,\ V_{CE} = 1.0\ V$ |
| | | 15 | 30 | | | | $I_C = 100\ mA,\ V_{CE} = 1.0\ V$ |
| $V_{CE(sat)}$ | Collector-to-emitter saturation voltage (Note 4) | | 0.2 | | 0.2 | V | $I_C = 10\ mA,\ I_B = 1.0\ mA$ |
| | | | 0.3 | | 0.3 | V | $I_C = 50\ mA,\ I_B = 5.0\ mA$ |
| $V_{BE(sat)}$ | Base-to-emitter saturation voltage (Note 4) | 0.65 | 0.85 | 0.65 | 0.85 | V | $I_C = 10\ mA,\ I_B = 1.0\ mA$ |
| | | | 0.95 | | 0.95 | V | $I_C = 50\ mA,\ I_B = 5.0\ mA$ |
| $f_T$ | Current gain bandwidth product | 250 | | 300 | | MHz | $I_C = 10\ mA,\ V_{CE} = 20\ V,$ $f = 100\ MHz$ |
| $C_{ob}$ | Output capacitance | | 4.0 | | 4.0 | pF | $I_E = 0,\ V_{CB} = 5.0\ V,\ f = 100\ kHz$ |
| $C_{ib}$ | Input capacitance | | 8.0 | | 8.0 | pF | $V_{BE} = 0.5\ V,\ I_C = 0,\ f = 100\ kHz$ |
| $h_{ie}$ | Input impedance | 1.0 | 8.0 | 1.0 | 10 | kΩ | $I_C = 1.0\ mA,\ V_{CE} = 10\ V,$ $f = 1.0\ kHz$ |
| $h_{re}$ | Voltage feedback ratio | 0.1 | 5.0 | 0.5 | 8.0 | X10⁻⁴ | $I_C = 1.0\ mA,\ V_{CE} = 10\ V,$ $f = 1.0\ kHz$ |
| $h_{fe}$ | Small signal current gain | 50 | 200 | 100 | 400 | | $I_C = 1.0\ mA,\ V_{CE} = 10\ V,$ $f = 1.0\ kHz$ |
| $h_{oe}$ | Output admittance | 1.0 | 40 | 1.0 | 40 | $\mu S$ | $I_C = 1.0\ mA,\ V_{CE} = 10\ V,$ $f = 1.0\ kHz$ |
| NF | Noise figure | | 6.0 | | 5.0 | dB | $I_C = 100\ \mu A,\ V_{CE} = 5.0\ V,$ $R_G = 1.0\ k\Omega,\ f = 10$ Hz to 15.7 kHz |

*Planar is a patented Fairchild process.
†Fairchild exceeds JEDEC registered value for this parameter.
NOTES:
1. These ratings are limiting values above which the serviceability of any individual semiconductor device may be impaired.
2. These are steady state limits. The factory should be consulted on applications involving pulsed or low duty cycle operations.
3. These ratings give a maximum junction temperature of 150°C and junction to case thermal resistance of 125°C/W (derating factor of 8.0 mW/°C); junction to ambient thermal resistance of 200°C/W (derating factor of 5.0 mW/°C).
4. Pulse conditions: Length = 300 μs; duty cycle = 2%.
5. For product family characteristic curves, refer to Section 5-SS14.

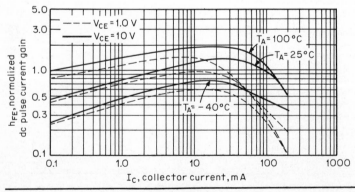

**Fig 5-12 (Cont.)**

The effect of $h_{FE}$ on output voltage is now calculated by Eq. (5-21$b$):

$$\Delta V_o = \left(\frac{3.3 \text{ k}\Omega}{2.2 \text{ k}\Omega}\right)\left(\frac{50 \text{ k}\Omega \times 150 \text{ k}\Omega}{50 \text{ k}\Omega + 150 \text{ k}\Omega}\right)\left(\frac{1.95 \text{ mA}}{h_{FE}}\right) = \frac{110 \text{ V}}{h_{FE}} \tag{5-22}$$

This is the change in output voltage compared with an idealized transistor with $h_{FE}$ of infinity. It will be greatest for a low-limit value of $h_{FE}$. From Fig. 5-12, the high limit is seen to be about three times this value. Therefore, the greatest change with transistors $\Delta(\Delta V_o)$max, which may not exceed 1.5 V, is:

$$\Delta(\Delta V_o)\text{max} = \frac{110 \text{ V}}{h_{FE,\text{min}}} - \frac{110 \text{ V}}{3h_{FE,\text{min}}} = \frac{73.3}{h_{FE,\text{min}}} = 1.5 \text{ V}$$

Solving,

$$h_{FE,\text{min}} = 48.9$$

The 2N3904 has an $h_{FE} = 70$ at $I_C = 1$ mA and at 25°C. Current gain $h_{FE}$ declines by about 20% at the low ambient temperature of 0°C. This is approximately offset by the increased value at the actual collector current of 1.95 mA.

The input impedance must also be checked. Here the low value of $h_{fe}$ the ac gain, should be used. As with $h_{FE}$, the specified value of 100 at $I_C = 1$ mA may be used. Then, by Eq. (5-18), $Z_{\text{in,transistor}} = 100 \times 26 \text{ mV}/1.95 \text{ mA} = 1.33$ k$\Omega$. This also equals the input impedance of the circuit $Z_{\text{in}}$, since bias resistors $R_1$ and $R_2$ are sufficiently large to have negligible shunting effect.

The device is being operated well within its maximum ratings in every respect. Its output capacitance of 4 pF has little effect at the 25-kHz maximum signal frequency, and the transistor gain-bandwidth product of 300 MHz is more than adequate.

## 5.9 SELECTING A SWITCHING BJT

Special parameters which apply to switching transistors are listed in Table 5-7.

**TABLE 5-7**    **Special Parameters for Switching Transistors**

| Parameter | Symbol | Meaning |
|---|---|---|
| Delay time | $t_d$ | Time delay between the beginning of the input and output signal as the transistor becomes conductive |
| Rise time | $t_r$ | Time for the output signal to increase from 10 to 90% of final value as the transistor becomes conductive |
| Storage time | $t_s$ | Time delay between the trailing edges of the input and output signal as the transistor becomes cut off |
| Fall time | $t_f$ | Time for the output signal to decrease from 90 to 10% of final value as the transistor becomes cut off |
| Turnon time | $t_{\text{on}}$ | Total time for the transistor to become fully conductive, equal approximately to $t_d + t_r$ |
| Turnoff time | $t_{\text{off}}$ | Total time for the transistor to become fully cut off, equal approximately to $t_s + t_f$ |
| Stored charge | $Q_T$ | Amount of charge of the carriers in transit in the transistor; an indication of the transistor turnoff time |

**problem 5.7**   By use of the circuit of Fig. 5-13$a$, a logic input signal is inverted and amplified to a 10-V switching level. The delay between input and output transitions, shown as $t_{\text{on}}$ and $t_{\text{off}}$ in Fig. 5-13$b$, must be less than 40 ns for the given input signal. Low level at the output should be 0.4 V or less. Select a transistor for $Q_1$.

**theory**   The "on" currents of $Q_1$ are given approximately by:

$$I_B = \frac{v_i - 0.7 \text{ V}}{R_1} \tag{5-23}$$

and

$$I_C = \frac{10 \text{ V}}{R_2} \tag{5-24}$$

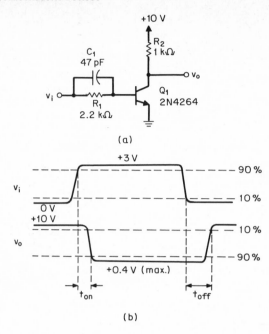

(a)

(b)

**Fig. 5-13** A transistor switching circuit. (*a*) Schematic. (*b*) Waveforms of input and output signals.

"Speed-up" capacitor $C_1$ is used to turn $Q_1$ on and off rapidly. Its principal function is to provide reverse current, upon turn-off, to remove the stored charge $Q_T$ in $Q_1$. The maximum amount of $Q_T$ which $C_1$ will quickly remove is:

$$Q_{T,\max} = C_1 \, \Delta V \qquad (5\text{-}25)$$

where $\Delta V$ is the voltage drop across $C_1$ just prior to the off transition.

Switching speed is improved if the transistor is driven less hard into saturation. However, in order to ensure that the transistor is always saturated under worst-case conditions with a reasonably low saturation voltage $V_{CE,\text{sat}}$ at the collector, signal levels near those given in the transistor specifications are usually needed.

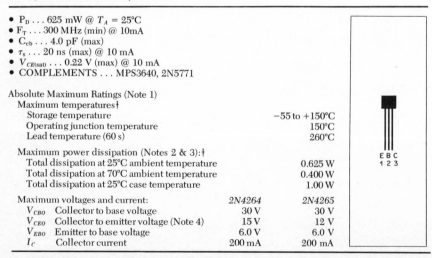

- $P_D$ ... 625 mW @ $T_A = 25°C$
- $F_T$ ... 300 MHz (min) @ 10mA
- $C_{cb}$ ... 4.0 pF (max)
- $\tau_s$ ... 20 ns (max) @ 10 mA
- $V_{CE(\text{sat})}$ ... 0.22 V (max) @ 10 mA
- COMPLEMENTS ... MPS3640, 2N5771

Absolute Maximum Ratings (Note 1)
Maximum temperatures†
| | |
|---|---|
| Storage temperature | $-55$ to $+150°C$ |
| Operating junction temperature | 150°C |
| Lead temperature (60 s) | 260°C |

Maximum power dissipation (Notes 2 & 3):†
| | |
|---|---|
| Total dissipation at 25°C ambient temperature | 0.625 W |
| Total dissipation at 70°C ambient temperature | 0.400 W |
| Total dissipation at 25°C case temperature | 1.00 W |

Maximum voltages and current:

| | | 2N4264 | 2N4265 |
|---|---|---|---|
| $V_{CBO}$ | Collector to base voltage | 30 V | 30 V |
| $V_{CEO}$ | Collector to emitter voltage (Note 4) | 15 V | 12 V |
| $V_{EBO}$ | Emitter to base voltage | 6.0 V | 6.0 V |
| $I_C$ | Collector current | 200 mA | 200 mA |

E B C
1 2 3

**Fig. 5-14** Excerpts of specification, 2N4264. (*Courtesy Fairchild Camera and Instrument Corp.*)

Electrical Characteristics (25°C ambient temperature unless otherwise noted)

| Symbol | Characteristic | 2N4264 | | 2N4265 | | Units | Test conditions |
|---|---|---|---|---|---|---|---|
| | | Min. | Max. | Min. | Max. | | |
| $h_{FE}$ | DC current gain | 25 | | 50 | | | $I_C = 1.0$ mA, $V_{CE} = 1.0$ V |
| | | 40 | 160 | 100 | 400 | | $I_C = 10$ mA, $V_{CE} = 1.0$ V |
| | | 20 | | 45 | | | $I_C = 10$ mA, $V_{CE} = 1.0$ V, $T_A = 55°C$ |
| | | 40 | | 90 | | | $I_C = 30$ mA, $V_{CE} = 1.0$ V |
| $h_{FE}$ | DC pulse current gain (Note 5) | 30 | | 55 | | | $I_C = 100$ mA, $V_{CE} = 1.0$ V |
| | | 20 | | 35 | | | $I_C = 200$ mA, $V_{CE} = 1.0$ V |
| $V_{CE(sat)}$ | Collector saturation voltage (Note 5) | | 0.22 | | 0.22 | V | $I_C = 10$ mA, $I_B = 1.0$ mA |
| | | | 0.35 | | 0.35 | V | $I_C = 100$ mA, $I_B = 10$ mA |
| $V_{BE(sat)}$ | Base saturation voltage (Note 5) | 0.65 | 0.80 | 0.65 | 0.80 | V | $I_C = 10$ mA, $I_B = 1.0$ mA |
| | | 0.75 | 0.95 | 0.75 | 0.95 | V | $I_C = 100$ mA, $I_B$ |
| $h_{fe}$ | Magnitude of common emitter Small signal current gain | 3.0 | | 3.0 | | | $V_{CE} = 10$ V, $I_C = 10$, $f = 100$ mHz |
| $C_{cb}$ | Collector base capacitance | | 4.0 | | 4.0 | pF | $I_E = 0$, $V_{CB} = 5.0$ V, $f = 100$ kHz |
| $C_{ib}$ | Input capacitance | | 8.0 | | 8.0 | pF | $V_{BC} = 0.5$ V, $I_C = 0$, $f = 100$ kHz |
| $I_{CEX}$ | Collector cutoff current | | 100 | | 100 | nA | $V_{CE} = 12$ V, $V_{EB} = 0.25$ V |
| $I_{BL}$ | Base cutoff current | | 100 | | 100 | nA | $V_{CE} = 12$ V, $V_{EB} = 0.25$ V |
| | | | 10 | | 10 | $\mu$A | $V_{CE} = 12$ V, $V_{EB} = 0.25$ V, $T_A = 100°C$ |
| $BV_{CBO}$ | Collector-to-base breakdown voltage | 30 | | 30 | | V | $I_C = 10$ $\mu$A, $I_E = 0$ |
| $BV_{CEO}$ | Collector-to-emitter breakdown voltage | 15 | | 12 | | V | $I_C = 1.0$ mA, $I_E = 0$ |
| $BV_{EBO}$ | Emitter-to-base breakdown voltage | 6.0 | | 6.0 | | V | $I_C = 0$, $I_E = 10$ $\mu$A |
| $\tau_s$ | Storage time (see test circuit no. 561) | | 20 | | 20 | ns | $I_C \cong I_{B1} \cong I_{B2} = 10$ mA |
| $t_{on}$ | Turn-on time | | 25 | | 25 | ns | $I_C \cong 10$ mA, $I_C = 3.0$ mA, $V_{CC} = 3.0$ V |
| $t_{off}$ | Turn-off time | | 35 | | 35 | ns | $I_C \cong 10$ mA, $I_{B1} \cong 3.0$ mA, $I_{B2} \cong 1.5$ mA, $V_{CC} = 3.0$ V |
| $t_d$ | Delay time | | 8.0 | | 8.0 | ns | $I_C = 100$ mA, $I_{B1} = 10$ mA, $V_{CC} = 10$ V |
| $t_r$ | Rise time | | 15 | | 15 | ns | $I_C = 100$ mA, $I_{B1} = 10$ mA, $V_{CC} = 10$ V |
| $t_s$ | Storage time | | 20 | | 20 | ns | $I_C = 100$ mA, $I_{B1} = I_{B2} = 10$ mA, $V_{CC} = 10$ V |
| $t_f$ | Fall time | | 15 | | 15 | ns | $I_C = 100$ mA, $I_{B1} = I_{B2} = 10$ mA, $V_{CC} = 10$ V |
| $Q_T$ | Total charge control | | 80 | | 80 | pC | $I_C = 10$ mA, $I_B = 1.0$ mA, $V_{CC} = 3.0$ V |

*Planar is a patented Fairchild process.
†Fairchild exceeds JEDEC registered values.
NOTES:
1. These ratings are limiting values above which the serviceability of any individual semiconductor device may be impaired.
2. These are steady state limits. The factory should be consulted on applications involving pulsed or low duty cycle operations.
3. These ratings give a maximum junction temperature of 150°C, and junction to case thermal resistance of 125°C/W (derating factor of 8.0 mW/°C) and junction to ambient thermal resistance of 200°C/W (derating factor of 5.0 mW/°C).
4. Rating refers to a high current point where collector to emitter voltage is lowest.
5. Pulse conditions: length = 300 $\mu$s; duty cycle = 2%.
6. For product family characteristic curves, refer to Section 5-SS5.

**Fig. 5-14 (Cont.)**

**solution**   Select a 2N4264 device (Fig. 5-14). Switching transistors should be selected which have switching speeds specified at currents close to the $I_C$ and $I_B$ of the circuit. These are determined by Eqs. (5-23) and (5-24):

$$I_B = \frac{3\text{ V} - 0.7\text{ V}}{2.2\text{ k}\Omega} = 1.05\text{ mA}$$

and

$$I_C = \frac{10\text{ V}}{1\text{ k}\Omega} = 10\text{ mA}$$

The 2N4264 is specified for this value of $I_C$ and has switching times $t_{on}$ and $t_{off}$ which are less than those required by the circuit.

Base current $I_{B1}$ and $I_{B2}$ are different for the circuit than in the specified test conditions. The current for turn-on, $I_{B1}$ is actually greater than the 3-mA test condition because of $C_1$. The same is true for the turn-off current, $I_{B2}$. Just before turn-off $I_B$ is only about 1 mA, which results in less saturation than in the specification. This further improves the turn-off action. The driving source impedance of the input voltage is assumed to be much less than the value of $R_1$, as is true for any fast-switching logic circuit source.

One further check should be made. The maximum stored charge $Q_T$ that the circuit discharges quickly is given by Eq. (5-25): $Q_{T,max} = 47\text{ pF} \times 2.3\text{ V} = 108\text{ pC}$. Since the maximum $Q_T$ from Fig. 5-14 is 80 pC, turn-off of $Q_1$ is very rapid. The exact amount is determined by the transient reverse-base current, which in turn depends upon the driving-source impedance.

The above considerations ensure that the circuit will switch in less than the specified $t_{on}$ and $t_{off}$ times of 25 ns and 35 ns, respectively. Saturation voltage $V_{CE,sat} = 0.22$ V is also within the problem requirement of $V_o < 0.4$ V at the low level. Transistor power dissipation is almost nil while the transistor is on or off and does not exceed 25 mW even during switching transients.


## 5.10   RADIO FREQUENCY (RF) TRANSISTORS

These devices are often selected according to the intended frequency range. *Megahertz-rated BJTs* are usually in conventional packages. They are, for the most part, characterized by having a gain-bandwidth product $f_T$ up to 2 GHz. Maximum power gain is usually specified for the intended frequency of operation, where a high value of gain is desirable. If used for very low signal levels (as in receiver front ends), devices with specified noise figures should be selected. In such uses the lowest values are best. Admittance or $y$ parameters are often provided to facilitate impedance matching of the device to the circuit.

Some devices are characterized with *scattering* or *s* parameters for the same purpose. The *s* parameters are a measure of reflected voltage levels related to transmission line operation. They are easier for the manufacturer to measure accurately, and they facilitate the use of Smith chart analysis techniques for complex admittance and impedance calculations.

*Gigahertz-rated BJTs* operate well up to 5 GHz. As with microwave diodes, they are usually packaged for use in the stripline or microstrip construction generally used at these frequencies. In characterization, emphasis is on power gain measurements. Conventional ac parameters are of little use at these frequencies, and so they are often not given. The newer devices, which have the best characteristics, are usually characterized with *s* parameters.

*Gigahertz-rated FETs* are now becoming the device of choice for frequencies of about 5 to 12 GHz. Constructed of gallium arsenide with a Schottky barrier gate, performance is available which is superior to that of silicon devices. Several handling precautions are needed to prevent damage, and biasing considerations are similar to those of conventional FETs. In other respects, the device characterization and selection factors are like those of gigahertz transistors.


## 5.11   SELECTING POWER TRANSISTORS

**problem 5.8**   In the circuit of Fig. 5-15, two power BJTs are used for a class B audio output stage driving an 8-$\Omega$ speaker. The transistors are mounted on heat sinks with a thermal resistance of 4°C/W. An ac input impedance of at least 100 $\Omega$ is needed for proper operation of the prior stage. The biasing circuit can supply up to 50 mA of average base current before appreciable signal limiting occurs. The maximum ambient temperature is 45°C. Select transistors for $Q_1$ and $Q_2$.

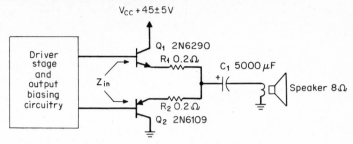

**Fig. 5-15** A class B audio power amplifier.

**theory**    In selecting power transistors, the amount of power they dissipate in the worst case must be determined. For this type of class B circuit exact calculations are difficult. A good approximation, however, may be made by assuming (1) that the output voltage $v_o$ can vary between $+V_{CC}/2$ and $-V_{CC}/2$ peak; (2) that all the power supplied to the stage by $V_{CC}$ which is not dissipated in the load is dissipated in the transistor; and (3) when either transistor is conducting, the other is cut off so that all its current flows through the speaker load. Then the load power $P_L$ is:

$$P_L = \frac{(v_{o,\text{rms}})^2}{R_L} \tag{5-26}$$

The total power $P_T$ supplied by $V_{CC}$ is:

$$P_T = \frac{v_{o,\text{rms}} V_{CC}}{2R_L} \tag{5-27}$$

If power dissipation in each transistor is represented by $P_Q$, then twice this value is equal to the difference between Eqs. (5-26) and (5-27):

$$2P_Q = \frac{v_{o,\text{rms}} V_{CC}}{2R_L} - \frac{(v_{o,\text{rms}})^2}{R_L} \tag{5-28}$$

The dissipated power $P_Q$ reaches a maximum when $v_{o,\text{rms}} = V_{CC}/4$. Substituting this value into Eq. (5-28) and solving for the maximum transistor dissipation $P_{Q\text{max}}$ yields:

$$P_{Q,\text{max}} = \frac{V_{CC}^2}{32\,R_L} \tag{5-29}$$

Power calculations are not completed when this value is determined. Power ratings of transistors are of limited use, except as a rough guide. The real limit to power dissipation is junction temperature $T_J$, which depends not only upon power dissipation but also upon thermal resistance (junction to case) of the transistor $\theta_{JC}$, thermal resistance (case to ambient) of the heat sink $\theta_{CA}$, and the ambient temperature $T_A$. These are related by the following equation:

$$T_J = P_Q\,(\theta_{JC} + \theta_{CA}) + T_A \tag{5-30}$$

Input ac impedance $Z_{\text{in}}$ is given by:

$$Z_{\text{in}} \cong h_{fe}R_L \tag{5-31}$$

Resistors $R_1$ and $R_2$ and the internal resistances of the transistors are small compared with the load. Since only one transistor conducts at a time, the input impedance for the combination is the same as for either one individually. Maximum average current delivered by each transistor is:

$$I_{E,\text{av}} = \tfrac{1}{2} \times 0.63\,\frac{V_{CC}}{2}\,\frac{1}{R_L} = 0.158 V_{CC} \tag{5-32}$$

**solution**    Select the 2N6290 and 2N6109 power transistors (Fig. 5-16). These are a complementary pair with similar ratings and characteristics.

To select power transistors, first calculate their maximum dissipation. By Eq. (5-29), using the worst-case value of $V_{CC} = 45\text{ V} + 5\text{ V} = 50\text{ V}$,

$$P_{Q,\text{max}} = \frac{(50\text{ V})^2}{32 \times 8\,\Omega} = 9.77\text{ W}$$

This calls, probably, for a transistor rated for about 20 W or more. (Wattage ratings are usually for a case temperature of 25°C, which is an unrealistic constraint.) The 2N6290 and 2N6109 are

Electrical Characteristics, *at case temperature* $(T_C)$ = 25°C, *unless otherwise specified*

| Characteristic | Symbol | Test Conditions♦ Voltage, V dc $V_{CE}$ | $V_{BE}$ | Current, A dc $I_C$ | $I_B$ | Limits 2N6292 2N6293 2N6106♦ 2N6107♦ Min. | Max. | 2N6290 2N6291 2N6108♦ 2N6109♦ Min. | Max. | |
|---|---|---|---|---|---|---|---|---|---|---|
| **Collector-cutoff current:** | | | | | | | | | | |
| With external base-to-emitter resistance ($R_{BE}$) = 100 Ω | $I_{CER}$ | 75<br>55 | | | | — | 0.1 | — | — | mA |
| With ($R_{BE}$) = 100 Ω and $T_C$ = 150°C | | 70<br>50 | | | | — | 2 | 0.1 | — | |
| With base-emitter junction reverse-biased | $I_{CEX}$ | 75<br>56 | −1.5<br>−1.5 | | | — | 2 | — | 2 | mA |
| With base-emitter junction reverse-biased and $T_C$ = 150°C | | 70<br>50 | −1.5<br>−1.5 | | | — | 0.1 | — | 0.1 | |
| With base open | $I_{CEO}$ | 40<br>60 | | | 0<br>0 | — | 2<br>1 | — | 2<br>1 | |
| Emitter-cutoff current | $I_{EBO}$ | | −5 | 0 | | — | 1 | — | 1 | mA |
| **Collector-to-emitter sustaining voltage:** | | | | | | | | | | |
| With base open | $V_{CEO(sus)}$ | | | 0.1[a] | 0 | 70 | — | 50 | — | V |
| With external base-to-emitter resistance ($R_{BE}$) = 100 Ω | $V_{CER(sus)}$ | | | 0.1 | | 80 | — | 60 | — | V |
| DC forward-current transfer ratio | $h_{FE}$ | 4<br>4<br>4 | | 2[a]<br>2.5[a]<br>7[a] | | 30<br>2.3 | 150 | 30<br>2.3 | 150 | |

| Characteristic | Symbol | | | | | | | Units |
|---|---|---|---|---|---|---|---|---|
| Base-to-emitter voltage: | $V_{BE}$ | | | | | | | V |
|   2N6292, 2N6293 | | 4 | 2[a] | — | 15 | — | — | |
|   2N6290, 2N6291 | | 4 | 2.5[a] | — | — | — | 1.5 | |
|   All Types | | 4 | 7[a] | — | 3 | — | 3 | |
| Collector-to-emitter saturation voltage | $V_{CE(sat)}$ | 4 | 2[a] | 0.2 | 1 | — | — | V |
| | | | 2.5[a] | 0.25 | — | — | 1 | |
| | | | 7[a] | 3[a] | 3.5 | — | 3.5 | |
| Common-emitter, small-signal, Forward current transfer ratio: $f = 50$ kHz | $h_{fe}$ | 4 | 0.5 | | 20 | | 20 | |
| Gain-bandwidth product: | $f_T$ | | | | | | | MHz |
|   2N6290–2N6293 | | 4 | 0.5 | | 4 | | 4 | |
|   2N6106–2N6109 | | −4 | −0.5 | | 10 | | 10 | |
| Magnitude of common-emitter, small-signal, forward-current transfer ratio: $f = 1$ MHz | $|h_{fe}|$ | | | | | | | |
|   2N6290–2N6293 | | 4 | 0.5 | | 4 | | 4 | |
|   2N6106–2N6109 | | −4 | −0.5 | | 10 | | 10 | |
| Collector-to-base capacitance: $f = $ MHz, $V_{CB} = 10$ V | $C_{obo}$ | 0 | 0 | | 250 | | 250 | pF |
| Thermal resistance: | | | | | | | | °C/W |
|   Junction-to-case | $R_{\theta JC}$ | | | | 3.125 | | 3.125 | |
|   Junction-to-ambient | $R_{\theta JA}$ | | | | 70 | | 70 | |

[a] Pulsed: Pulse duration = 300 μs, duty factor = 0.018.   ♦For pnp devices, voltage and current values are negative.

*In accordance with JEDEC registration data format (JS-6 RDF-2).

CAUTION: The sustaining voltage $V_{CER}$(sus) MUST NOT be measured on a curve tracer.

**Fig. 5-16**   Excerpts of specification, 2N6109 and 2N6290. (*Courtesy Radio Corporation of America.*)

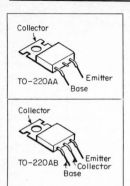

*Features:*

- Low saturation voltages
- VERSAWATT package (molded silicone plastic)
- Complementary npn and pnp types
- Thermal-cycling ratings
- Maximum safe-area-of-operation curves specified for dc operation

RCA-2N6106–2N6111, 2N6288–2N6293, and 2N6473–2N6476 are epitaxial-base silicon transistors supplied in a VERSAWATT package. The 2N6288–2N6293, 2N6473, and 2N6474[1] are n-p-n complements of p-n-p types 2N6106–2N6111, 2N6475, and 2N6476[2] respectively. All these transistors are intended for a wide variety of medium-power switching and amplifier applications, such as series and shunt regulators and driver and output stages of high-fidelity amplifiers.

The 2N6289, 2N6291, and 2N6293 n-p-n types and 2N6106, 2N6108, and 2N6110 p-n-p devices fit into TO-66 sockets. The remaining types are supplied in the JEDEC TO-220AB straight-lead version of the VERSAWATT package. All of these devices are also available on special order in a variety of leadform configurations. Detailed information on these and other VERSAWATT outlines is contained in "RCA's Lineup of Power Transistors" (PSP-704)

[1]Formerly RCA Dev. Nos. TA7784, TA8323, TA7783, TA8232, TA7782, TA8231, TA8444, and TA8723, respectively.
[2]Formerly RCA Dev. Nos. TA8210, TA7741, TA8211, TA7742, TA8212, TA7743, TA8445, and TA8722, respectively.

| | | 2N6288 | 2N6290 | 2N6292 | | | |
|---|---|---|---|---|---|---|---|
| | N-P-N | 2N6289 | 2N6291 | 2N6293 | 2N6473 | 2N6474 | |
| MAXIMUM RATINGS (Absolute-Maximum Values): | P-N-P | 2N6110♦ 2N6111♦ | 2N6108♦ 2N6109♦ | 2N6106♦ 2N6107♦ | 2N6475♦ | 2N6476♦ | |
| *Collector-to-base voltage | $V_{CBO}$ | 40 | 60 | 80 | 110 | 130 | V |
| *Collector-to-emitter voltage With external base-supply resistance ($R_{BB}$) = 100Ω, and base supply voltage ($V_{BB}$) = 0 | $V_{CEX}$ | 40 | 60 | 80 | 110 | 130 | V |
| With base open | $V_{CEO}$ | 30 | 50 | 70 | 100 | 120 | V |
| *Emitter-to-base voltage | $V_{EBO}$ | 5 | 5 | 5 | 5 | 5 | V |
| *Collector current (continuous) At case temperature ≤ 106°C | $I_C$ | 7 | 7 | 7 | 4 | 4 | A |
| *Base current (continuous) At case temperature ≤ 130°C | $I_B$ | 3 | 3 | 3 | 2 | 2 | A |
| Transistor dissipation: | $P_T$ | | | | | | |
| At case temperatures up to 25°C | | 40 | 40 | 40 | 40 | 40 | W |
| *At case temperatures up to 100°C | | 16 | 16 | 16 | 16 | 16 | W |
| At ambient temperature up to 25°C | | 1.8 | 1.8 | 1.8 | 1.8 | 1.8 | W |
| At case temperatures above 25°C | | | | Derate linearly at 0.32 W/°C | | | |
| *At case temperatures above 100°C | | | | Derate linearly at 0.32 W/°C | | | |
| At ambient temperatures above 25°C | | | | Derate linearly at 0.0144 W/°C | | | |
| *Temperature range: Storage and operating (junction) | | | ← | 65 to 150 | → | | °C |
| *Lead temperature (during soldering): At distance ≥ ⅛ in. (3.17 mm) from case for 10 s max. | | | ← | 235 | → | | °C |

*In accordance with JEDEC registration data format (JS-6, RDF-2). ♦For pnp devices, voltage and current values are negative.
Trademark(s) Registered®
Marca(s) Registrada(s)

**Fig. 5-16 (Cont.)**

economical devices with even higher ratings. Selection of them should result in lower junction temperature, better reliability, and perhaps a less massive heat sink.

To confirm the power-handling ability of the transistors, the maximum junction temperature should be calculated. Thermal resistance $\theta_{JC}$, from Fig. 5-16, is 3.125°C/W. Then, by Eq. (5-30), $T_{J,max} = 9.77$ W (3.125°C/W + 4°C/W) + 45°C = 115°C. This is well under the maximum rating of 150°C.

Maximum average emitter current per transistor, calculated by Eq. (5-32), is $I_{E,av} = 0.158 \times$ 50 V/8 Ω = 0.98 A. Dc current gain $h_{FE}$, from Fig. 5-16, is a minimum of 30; hence, the maximum average base current is $I_{B,av} = I_{E,av}/h_{FE} = 0.98$ A/30 = 32.7 mA, which meets the problem requirement. The alternating current gain $h_{fe}$ of 20 is used for the input impedance calculation by Eq. (5-31): $Z_{in} \cong 20 \times 8$ Ω = 160 Ω, which is greater than the 100 Ω minimum requirement.

Since these transistors have a gain-bandwidth product in the megahertz region, they have more than adequate response for the 20-kHz maximum frequency of an audio signal. The maximum voltage rating $V_{CEO}$ of 50 V cannot actually be reached across the cutoff device, because the conducting device always has a voltage drop across it. In addition, the device cannot be operated this way very long, as voltage limiting would cause severe audio distortion.

## 5.12   SELECTING AN AMPLIFIER JFET

Table 5-8 lists commonly used parameters of FETs.

**TABLE 5-8   Commonly Used Parameters for the JFET and MOSFET**

| Parameter | Symbol | Meaning |
|---|---|---|
| Drain current for zero bias | $I_{DSS}$ | Drain current flowing when gate is short-circuited to source ($V_{GS} = 0$) |
| Gate reverse current | $I_{GSS}$ | Leakage current flowing between gate and source for a specified reverse bias across gate and source terminals |
| Drain cutoff current | $I_{D,off}$ | Drain current flowing when device is biased in its OFF state |
| Gate-source breakdown voltage | $BV_{GSS}$ | Maximum reverse voltage that may be impressed across gate and source terminals without damaging the device |
| Gate-source pinchoff voltage | $V_P$ | Gate-to-source voltage which reduces $I_{DSS}$ to 1% or less of maximum value at a specified drain-to-source voltage |
| Small-signal forward transconductance | $g_{fs}, g_m,$ $y_{fs}$ | Ratio of a small change in signal drain current to a small change in signal gate-to-source voltage in common-source configuration. Parameter $g_{fs}$ is an indication of the gain for the device. |
| Dc drain-source ON resistance | $r_{DS}$ | Ratio of dc drain-source voltage to the dc drain current, generally measured at $V_{GS} = 0$ |
| Input capacitance | $C_{iss}$ | Small-signal input capacitance for device in the common-source configuration with $V_{DS} = 0$ |
| Reverse transfer capacitance | $C_{rss}$ | Capacitance between drain and gate in common-source configuration with $V_{DS} = 0$ |

SOURCE: M. Kaufman and A. H. Seidman, *Handbook for Electronics Engineering Technicians*, McGraw-Hill, New York, 1976.

**problem 5.9**   The circuit of Fig. 5-17 has an input impedance of 1 MΩ for a high impedance. The ac signal is a low-frequency source having a maximum amplitude of 200 mV peak to peak. A voltage gain of at least −4 is required. Capacitors $C_1$ and $C_2$ are for coupling and bypass, respectively, and have negligible ac impedance. Select a suitable N-channel JFET.

**theory**   To perform well as an amplifier, an FET must be operated in the pinch-off region where drain-to-source voltage $V_{DS}$ is greater than the pinch-off voltage $V_P$ (which is equal in magnitude to the gate-to-source cutoff voltage $V_{GS,off}$). This must be true for all devices over the wide range of parameter variations for the particular type used. Currents and voltages must be calculated for the worst cases.

Drain current $I_D$ is determined from the square-law equation that applies to all FETs:

$$I_D = I_{DSS}\left(1 - \frac{V_{GS}}{V_{GS,off}}\right)^2 \qquad (5\text{-}33)$$

where $V_{GS}$ is the gate-to-source voltage. When $I_D$ is known, drain-to-source voltage $V_{DS}$ may be calculated since:

$$V_{DS} = 25\text{ V} - I_D\,(R_4 + R_5) \tag{5-34}$$

When the peak value of the signal is subtracted from Eq. (5-34), the result must exceed $V_P$ if the device stays in pinch-off.

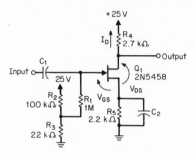

**Fig. 5-17**  A FET amplifier circuit.

Gain is calculated using the device transconductance $y_{fs}$, which is the ratio of output current to input voltage. Voltage gain of the stage $A_v$ is:

$$A_v = -y_{fs}R_4 \tag{5-35}$$

This approximation is good if the output impedance of the FET is high, as it will be if operated in pinch-off. Limits of $y_{fs}$ are usually specified, and other values can be calculated for a particular device with the following formula. The formula is derived from the definition of $y_{fs}$ and Eq. (5-33):

$$y_{fs} = \frac{2\,I_{DSS}}{V_{GS,\text{off}}}\left(1 - \frac{V_{GS}}{V_{GS,\text{off}}}\right) \tag{5-36}$$

Another relationship that may be useful is that between $I_D$ and $V_{GS}$ in the circuit. The voltage across $R_5$ is the bias voltage $25\text{V} \cdot R_3/(R_2 + R_3)$ less $V_{GS}$, so that

$$I_D = \frac{[(25\text{ V} \times R_3)/(R_2 + R_3)] - V_{GS}}{R_5} \tag{5-37}$$

The minimum $I_{DSS}$ of the transistor selected should be sufficient to provide $I_D$ given by Eq. (5-37) when $V_{GS} = 0$. This avoids forward bias on the gate junction. The result is:

$$I_{DSS,\text{min}} = \frac{25\text{ V} \times R_3}{(R_2 + R_3)R_5} \tag{5-38}$$

**solution**   Select a 2N5458 device (Fig. 5-18). The minimum $I_{DSS}$ needed, by Eq. (5-38), is

$$I_{DSS,\text{min}} = \frac{25\text{ V} \times 22\text{ k}\Omega}{(100\text{ k}\Omega + 22\text{ k}\Omega)2.2\text{ k}\Omega} = 2.05\text{ mA}$$

The minimum transconductance $y_{fs}$ is determined by Eq. (5-35) for the minimum required gain of 4: $4 = y_{fs,\text{min}}$ 2.7 k$\Omega$. Solving, $y_{fs,\text{min}} = 1480\ \mu$S.

These considerations lead to the selection of the 2N5458, which meets these requirements. A device with a slightly lower minimum $I_{DSS}$ may be used, as long as the forward bias on the gate junction is limited to about 0.3 V or less. This condition does not bias the junction enough to cause gate current. A higher $I_{DSS,\text{min}}$ device may also be used, but the resulting $V_{GS}$ bias causes a reduction in $y_{fs}$ below the specified value as indicated in Eq. (5-36).

To determine that $Q_1$ is always in pinch-off, we must first take a minimum $I_{DSS}$ device where $I_D \cong I_{DSS} = 2$ mA. Drain-to-source voltage $V_{DS}$ may be calculated by Eq. (5-34): $V_{DS} = 25\text{ V} - 2.0\text{ mA }(2.7\text{ k}\Omega + 2.2\text{ k}\Omega) = 15.2$ V. Since $V_{GS,\text{off}}$ is 7 V maximum, the transistor is operating well into pinch-off.

Now we check operation with a high $I_{DSS} = 9$ mA device. The worst case occurs when $V_{GS,\text{off}} = -7$ V, its largest value. Then, by Eq. (5-33),

$$I_D = 9\text{ mA}\left(1 - \frac{V_{GS}}{-7\text{V}}\right)^2 \tag{5-39}$$

and by Eq. (5-37),

$$I_D = \frac{[(25\text{ V} \times 22\text{ k}\Omega)/(100\text{ k}\Omega + 22\text{ k}\Omega)] - V_{GS}}{2.2\text{ k}\Omega} = 2.05\text{ mA} - \frac{V_{GS}}{2.2\text{ k}\Omega} \tag{5-40}$$

Silicon N-channel junction Field-effect transistors depletion mode (Type A) designed for general-purpose audio and switching applications.

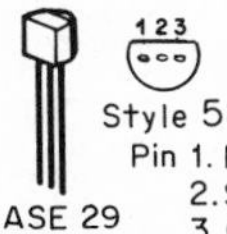

Style 5:
Pin 1. Drain
2. Source
CASE 29     3. Gate
(TO-92)
Drain and source
may be interchanged

Maximum Ratings

| Rating | Symbol | Value | Unit |
|---|---|---|---|
| Drain-source voltage | $V_{DS}$ | 25 | Vdc |
| Drain-gate voltage | $V_{DG}$ | 25 | Vdc |
| Reverse gate-source voltage | $V_{GS(r)}$ | 25 | Vdc |
| Gate current | $I_G$ | 10 | mAdc |
| Total device dissipation @ $T_A = 25°C$ | $P_D{}^{(2)}$ | 310 | mW |
| Derate above 25°C | | 2.82 | mW/°C |
| Operating junction temperature | $T_J{}^{(2)}$ | 135 | °C |
| Storage temperature range | $T_{stg}{}^{(2)}$ | −65 to +150 | °C |

Electrical Characteristics ($T_A = 25°C$ unless otherwise noted)

| Characteristic | Symbol | Min. | Typ. | Max. | Unit |
|---|---|---|---|---|---|
| **Off characteristics** | | | | | |
| Gate-source breakdown voltage | $BV_{GSS}$ | | | | Vdc |
| ($I_G = -10\ \mu Adc,\ V_{DS} = 0$) | | 25 | — | — | |
| Gate reverse current | $I_{GSS}$ | | | | nAdc |
| ($V_{GS} = -15\ Vdc,\ V_{DS} = 0$) | | — | — | 1.0 | |
| ($V_{GS} = -15\ Vdc,\ V_{DS} = 0,\ T_A = 100°C$) | | — | — | 200 | |
| Gate-source cutoff voltage | $V_{GS(off)}$ | | | | Vdc |
| ($V_{DS} = 15Vdc,\ I_D = 10\ nAdc$)  2N5457 | | 0.5 | — | 6.0 | |
| 2N5458 | | 1.0 | — | 7.0 | |
| 2N5459 | | 2.0 | — | 8.0 | |
| Gate-Source Voltage | $V_{GS}$ | | | | Vdc |
| ($V_{DS} = 15\ Vdc,\ I_D = 100\ \mu Adc$)  2N5457 | | — | 2.5 | — | |
| ($V_{DS} = 15\ Vdc,\ I_D = 200\ \mu Adc$)  2N5458 | | — | 3.5 | — | |
| ($V_{DS} = 15\ Vdc,\ I_D = 400\ \mu Adc$)  2N5459 | | — | 4.5 | — | |
| **On characteristics** | | | | | |
| Zero-gate-voltage drain current (1) | $I_{DSS}$ | | | | mAdc |
| ($V_{DS} = 15\ Vdc,\ V_{GS} = 0$)  2N5457 | | 1.0 | 3.0 | 5.0 | |
| 2N5458 | | 2.0 | 6.0 | 9.0 | |
| 2N5459 | | 4.0 | 9.0 | 16 | |
| **Dynamic characteristics** | | | | | |
| Forward transfer admittance (1) | $|y_{fs}|$ | | | | $\mu$mhos |
| ($V_{DS} = 15\ Vdc,\ V_{GS} = 0,\ f = 1\ kHz$)  2N5457 | | 1000 | 3000 | 5000 | |
| 2N5458 | | 1500 | 4000 | 5500 | |
| 2N5459 | | 2000 | 4500 | 6000 | |
| Output admittance (1) | $|y_{os}|$ | | | | $\mu$mhos |
| ($V_{DS} = 15\ Vdc,\ V_{GS} = 0,\ f = 1\ kHz$) | | — | 10 | 50 | |
| Input capacitance | $C_{iss}$ | | | | pF |
| ($V_{DS} = 15\ Vdc,\ V_{GS} = 0,\ f = 1\ MHz$) | | — | 4.5 | 7.0 | |
| Reverse transfer capacitance | $C_{rss}$ | | | | pF |
| ($V_{DS} = 15\ Vdc,\ V_{GS} = 0,\ f = 1\ MHz$) | | — | 1.5 | 3.0 | |

(1) Pulse Test: Pulse Width $\leq$ 630 ms; Duty Cycle $\leq$ 10%

(2) Continuous package improvements have enhanced these guaranteed Maximum Ratings as follows: $P_D = 1.0\ W$ @ $T_C = 25°C$. Derate above 25°C − 8.0 mW/°C, $T_J = -65$ to +150°C, $\theta_{JC} = 125°C/W$.

**Fig. 5-18** Specification, 2N5458. (*Courtesy Motorola, Inc.*)

Exact calculations would require simultaneous solution of Eqs. (5-39) and (5-40). This can be done but is difficult since it involves a quadratic equation. Often an easier approach is to estimate a value of $V_{GS}$. In a square-law relationship of Eq. (5-39), setting $V_{GS}$ to half of $V_{GS,\text{off}}$ reduces $I_D$ to ¼ of $I_{DSS}$, giving $I_D = 2.25$ mA and $V_{GS} = -3.5$ V. Drain current $I_D$ will probably be somewhat more, so we shall estimate $V_{GS} = -3$ V. Then, by Eq. (5-40),

$$I_D = 2.05 \text{ mA} + \frac{3 \text{ V}}{2.2 \text{ k}\Omega} = 3.41 \text{ mA}$$

Checking our estimate with Eq. (5-39), we have

$$3.41 \text{ mA} = 9 \text{ mA} \left(1 - \frac{V_{GS}}{-7 \text{ V}}\right)^2$$

Solving for $V_{GS}$ gives

$$V_{GS} = -\left[1 - \left(\frac{3.41 \text{ mA}}{9 \text{ mA}}\right)^{1/2}\right](7 \text{ V}) = -2.69 \text{ V}$$

It is, therefore, seen that $-3$ V was estimated a little high. By taking a new value between $-3$ and $-2.69$ V, repeating the calculations, accurate values can be determined with only a few tries. The result of this process yields $I_D = 3.30$ mA and $V_{GS} = -2.76$ V. Then, checking with Eq. (5-34), $V_{DS} = 25$ V $- 3.30$ mA $(2.7 \text{ k}\Omega + 2.2 \text{ k}\Omega) = 8.83$ V. Since this is not much above the $V_{DS} = 7$ V needed for pinch-off, the signal level should also be considered.

Transconductance is calculated by Eq. (5-36):

$$y_{fs} = \frac{2 \cdot 9 \text{ mA}}{-7 \text{ V}}\left[1 - \left(\frac{-2.76 \text{ V}}{-7 \text{ V}}\right)\right] = 1560 \ \mu\text{S}$$

and by Eq. (5-35),

$$A_v = -1560 \ \mu\text{S} \times 2.7 \text{ k}\Omega = -4.21$$

This value of gain multiplied by the 200 mV peak-to-peak (p-p) signal results in an output signal of 842 mV p-p. The instantaneous value of $V_{DS}$ is thereby reduced to 8.83 V $- 0.842$ V/2 = 8.41 V, still in pinch-off.

Although the supply voltage is 25 V, the biasing of the gate at +4.51 V limits $V_{DS}$ to about 20 V. The device is well within its voltage rating. Power dissipation is only about 32 mW maximum, a fraction of the power rating.

## 5.13  SELECTING A SILICON-CONTROLLED RECTIFIER (SCR)

Commonly used SCR and TRIAC parameters are summarized in Table 5-9.

**TABLE 5-9  Commonly Used Parameters for the SCR and TRIAC**

| | | |
|---|---|---|
| Forward breakdown voltage | $V_{(br)F}$ $V_{BO}$ | Forward voltage at which device fires |
| Reverse breakdown voltage | $V_{(br)R}$ | Maximum reverse voltage that causes device to go into avalanche |
| On-state voltage | $V_T, V_F$ | Voltage across device when it is conducting (ON state) |
| On-state current | $I_T, I_F$ | Current flowing between anode and cathode in ON state |
| Holding current | $I_H$ | Minimum current for device to be in ON state |
| Latching current | $I_L$ | Minimum current to maintain device in ON state, after switching from OFF to ON state with trigger removed |
| Gate trigger current | $I_{GT}$ | Minimum gate current for switching device from OFF to ON state |
| Gate trigger voltage | $V_{GT}$ | Gate voltage needed to produce required gate current |
| Gate turn-on time | $t_{on}$ | Time for device to turn on |
| Commutated turn-off time | $t_{off}, t_q$ | Time for device to turn off |
| Critical rate-of-rise | $dv/dt$ | Rate of voltage change applied to the anode or main terminal of a device in the OFF state which, if exceeded, may switch the device on |

SOURCE: Adapted from M. Kaufman and A. H. Seidman, *Handbook for Electronics Engineering Technicians*, McGraw-Hill, New York, 1976.

**problem 5.10** Select an SCR for the half-wave rectified control of a dc motor (Fig. 5-19). Triggering is from a phase controller, so that $Q_1$ is made conductive for any desired portion of a positive half-cycle of applied voltage. The motor and its load are controlled to limit the maximum rms motor current to 8 A. A triggering circuit produces a pulse, $v_T = 4$ V (open output value), for 10

$\mu s$; source impedance is 40 $\Omega$. Ambient temperature varies from 0 to +40°C. The SCR is mounted on a heat sink whose thermal resistance $\theta_{CA} = 4°$C/W

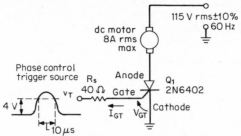

**Fig. 5-19** An SCR motor control circuit.

**theory**    SCRs and TRIACs are both often used with this type of phase control. The triggering circuit provides a pulse with a varying delay from the start of the positive half-cycle of applied voltage. With no delay, triggering is immediate and $Q_1$ conducts for the full half-cycle. If delayed, it conducts for only a portion of the half-cycle so that the power to the load is reduced.

The amount of trigger current $I_{GT}$ is determined by the required trigger voltage $V_{GT}$, and the resulting voltage drop across source resistance $R_S$ is as follows:

$$I_{GT} = \frac{v_T - V_{GT}}{R_S} \tag{5-41}$$

where $v_T$ is the trigger voltage. The peak anode current through the device $I_{A,\text{peak}}$ flows only on alternate half-cycles. Therefore, the rms current is half what it would be for a sine wave, assuming maximum conduction (with the SCR triggered at the beginning of each positive half-wave). Thus,

$$I_{A,\text{rms}} = \frac{\sqrt{2}}{4} I_{A,\text{peak}} \tag{5-42}$$

After the power dissipation is determined, maximum junction temperature can be calculated as in Sec. 5.11:

$$T_J = P_D (\theta_{JC} + \theta_{CA}) + T_A \tag{5-43}$$

where $\theta_{JC}$ = junction-to-case thermal resistance of $Q_1$
      $\theta_{CA}$ = case-to-ambient thermal resistance of the heat sink
      $T_A$ = ambient temperature

**solution**    Select a 2N6402 device (Fig. 5-20). There are many available SCRs that will do this job. The 2N6402 is selected mainly because it is an economical plastic device with a JEDEC-registered number that appears to have adequate characteristics.

The 8-A rms current is well within the 16-A rating, but actual junction temperature needs to be checked. Peak current and voltage are first determined. By Eq. (5-42), 8 A = $(\sqrt{2}/4)I_{A,\text{peak}}$. Solving, $I_{A,\text{peak}} = 4 \times 8$ A/$\sqrt{2}$ = 22.6 A.

The graph of maximum on-state instantaneous voltage vs. current will demonstrate that $V_F = 1.5$ V. While this is a maximum, current is near the peak level for a major fraction of the half-cycle. Assuming $V_F = 1.5$ V for the entire period of conduction is a conservative, but realistic, approach. Power dissipation is then equal to $V_F$ times $I_{A,\text{rms}}$, or 1.5 V $\times$ 8 A = 12 W. Junction temperature is determined by Eq. (5-43). Using 4°C/W for $\theta_{CA}$ for the heat sink, $T_{A,\text{max}} = 40$°C (from the problem statement), and $\theta_{JA} = 1.5$°C/W for the device (Fig. 5-20), then, $T_J = 12$ W (1.5°C/W + 4°C/W) + 40°C = 106°C, which is below the rated value for $T_{J,\text{max}} = 125$°C.

Triggering must also be considered. The triggering parameters of $I_{GT,\text{max}} = 30$ mA and $V_{GT,\text{max}} = 1.5$ V are for 25°C operation. But triggering becomes more difficult at lower temperatures, and graphical data show that, at 0°C, $I_{GT}$ increases by about 33% and $V_{GT}$ by 10%. Therefore, in this application $I_{GT,\text{max}} = 40$ mA and $V_{GT,\text{max}} = 1.65$ V.

To determine the actual available $I_{GT}$, the value $V_{GT,\text{max}} = 1.65$ V is substituted in Eq. (5-41):

$$I_{GT} = \frac{4 \text{ V} - 1.65 \text{ V}}{40\Omega} = 58.8 \text{ mA}$$

Since this is well over 40 mA, the device will trigger reliably in the circuit even at the low ambient temperature limit.

The peak reverse voltage applied is $\sqrt{2} \times 115$ V = 163 V. The +10% tolerance brings it to 179 V, still within the 200 V rating. Circuit values are well within the trigger power ratings of 20 W pulsed and 0.5 W average and the peak gate current rating of 2 A.

Silicon controlled rectifiers are designed primarily for half-wave ac control applications, such as motor controls, heating controls and power supplies; or wherever half-wave silicon gate controlled, solid-state devices are needed.

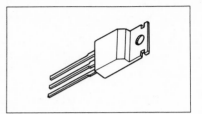

• Glass passivated junctions with center gate fire for greater parameter uniformity and stability

• Small, rugged, thermowatt construction for low thermal resistance, high heat dissipation, and durability

• Blocking voltage to 800 V

*Maximum Ratings

| Rating | Symbol | Value | Unit |
|---|---|---|---|
| Peak reverse blocking voltage[1] | $V_{RRM}$ | | V |
| 2N6400 | | 50 | |
| 2N6401 | | 100 | |
| 2N6402 | | 200 | |
| 2N6403 | | 400 | |
| 2N6404 | | 600 | |
| 2N6405 | | 800 | |
| Forward current rms ($T_C$ = 90°C) (all conduction angles) | $I_{T(rms)}$ | 16 | A |
| Peak forward surge current (½ cycle, sine wave, 60 Hz, $T_J$ = 125°C) | $I_{TSM}$ | 160 | A |
| Circuit fusing considerations ($T_J$ = −40 to +125°C, t = 1.0 to 8.3 ms) | $I^2t$ | 100 | A²s |
| Forward peak gate power | $P_{GM}$ | 20 | W |
| Forward average gate power | $P_{G(av)}$ | 0.5 | W |
| Forward peak gate current | $I_{GM}$ | 2.0 | A |
| Operating junction temperature range | $I_J$ | −40 to +125 | °C |
| Storage temperature range | $T_{stg}$ | −40 to +150 | °C |

Thermal Characteristics

| Characteristic | Symbol | Max | Unit |
|---|---|---|---|
| Thermal Resistance, Junction to Case | $R_{\theta JC}$ | 1.5 | °C/W |

[1]$V_{RRM}$ for all types can be applied on a continuous dc basis without incurring damage. Ratings apply for zero or negative gate voltage. Devices should not be tested for blocking capability in a manner such that the voltage supplied exceeds the rated blocking voltage.

*Indicates JEDEC Registered Data

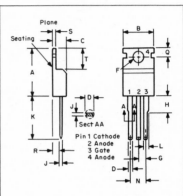

Pin 1 Cathode
2 Anode
3 Gate
4 Anode

| Dim | Millimeters | | Inches | |
|---|---|---|---|---|
| | Min | Max | Min | Max |
| A | 14.23 | 15.87 | 0.560 | 0.625 |
| B | 9.66 | 10.66 | 0.380 | 0.420 |
| C | 3.56 | 4.82 | 0.140 | 0.190 |
| D | 0.51 | 1.14 | 0.020 | 0.045 |
| F | 3.531 | 3.733 | 0.139 | 0.147 |
| G | 2.29 | 2.79 | 0.090 | 0.110 |
| H | — | 6.35 | — | 0.250 |
| J | 0.31 | 1.14 | 0.012 | 0.045 |
| K | 12.70 | 14.27 | 0.500 | 0.562 |
| L | 1.14 | 1.77 | 0.045 | 0.070 |
| N | 4.83 | 5.33 | 0.190 | 0.210 |
| Q | 2.54 | 3.04 | 0.100 | 0.120 |
| R | 2.04 | 2.92 | 0.080 | 0.115 |
| S | 0.51 | 1.39 | 0.020 | 0.055 |
| T | 5.85 | 6.85 | 0.230 | 0.270 |

CASE 221-02
TO 220 AB
All JEDEC dimensions and notes apply

**Fig. 5-20** Specification, 2N6402. (*Courtesy Motorola, Inc.*)

Electrical Characteristics ($T_C$ = 25°C unless otherwise noted.)

| Characteristic | Symbol | Min | Type | Max | Unit |
|---|---|---|---|---|---|
| *Peak forward blocking voltage | $V_{DRM}$ | | | | Volts |
| ($T_J$ = 125°C) | | | | | |
| 2N6400 | | 50 | — | — | |
| 2N6401 | | 100 | — | — | |
| 2N6402 | | 200 | — | — | |
| 2N6403 | | 400 | — | — | |
| 2N6404 | | 600 | — | — | |
| 2N6405 | | 800 | — | — | |
| *Peak forward blocking current | $I_{DRM}$ | — | — | 2.0 | mA |
| (rated $V_{DRM}$ @ $T_J$ = 125°C) | | | | | |
| *Peak reverse blocking current | $I_{RRM}$ | — | — | 2.0 | mA |
| (rated $V_{RRM}$ @ $T_J$ = 125°C) | | | | | |
| *Forward "on" voltage | $V_{TM}$ | — | — | 1.7 | Volts |
| ($I_{TM}$ = 32 A peak) | | | | | |
| *Gate trigger current (continuous dc) | $I_{GT}$ | — | 5.0 | 30 | mA |
| (anode voltage = 12 Vdc, $R_L$ = 100 Ohms) | | | | | |
| *Gate trigger voltage (continuous dc) | $V_{GT}$ | — | 0.7 | 1.5 | Volts |
| (anode voltage = 12 Vdc, $R_L$ = 100 Ohms) | | | | | |
| *Gate non-trigger voltage | $V_{GD}$ | 0.2 | — | — | Volts |
| (anode voltage = Rated $V_{DRM}$, $R_L$ = 100 Ω, $T_J$ = 125°C) | | | | | |
| *Holding current | $I_H$ | — | 6.0 | 40 | mA |
| (anode voltage = 12 Vdc) | | | | | |
| Turn-on time | $t_{gt}$ | — | 1.0 | — | μs |
| ($I_{TM}$ = 16 A, $I_{GT}$ = 40 mAdc) | | | | | |
| Turn-off time ($V_{DRM}$ = rated voltage) | $t_q$ | | | | μs |
| ($I_{TM}$ = 16 A, $I_R$ = 16 A) | | — | 15 | — | |
| ($I_{TM}$ = 16 A, $I_R$ = 16 A, $T_J$ = 125°C) | | — | 35 | — | |
| Forward voltage application rate | $dv/dt$ | — | 50 | — | V/μs |
| ($T_J$ = 125°C) | | | | | |

*Indicates JEDEC Registered Data.

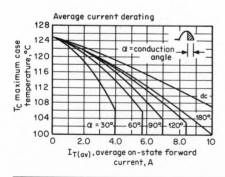

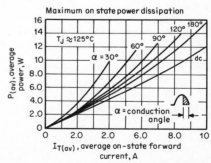

**Fig. 5-20 (Cont.)**

One other parameter needs some mention. The forward voltage application rate, $dv/dt$, of 50 V/$\mu$s is much greater than the rate of rise of a 115-V 60-Hz sine wave used here. However, in applications where the power source is direct current and is switched, high rates of voltage change can occur. They produce transients at the gate owing to the internal capacitance of the device and may cause unwanted self-triggering. This can be prevented by using SCRs with higher $dv/dt$ ratings or by using a snubber.*

## 5.14   SELECTING A TRIAC

Because it is a bidirectional device, a TRIAC has no cathode and anode as do most devices. Terminals for the main conductive path are designated Main Terminal 1 and Main Terminal 2, symbolized as MT1 and MT2 (Fig. 5-21). The gate is associated with MT1, with trigger levels and polarities stated with respect to MT1.

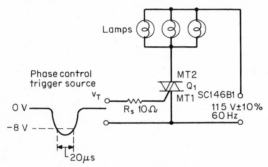

**Fig. 5-21**   A TRIAC lamp control circuit.

**problem 5.11**   Select a TRIAC for the phase-controlled lamp-dimming circuit of Fig. 5-21. The load consists of 800 W of incandescent lamps operating at 115 V and 60 Hz. Phase control triggering produces a negative 8-V, 20-$\mu$s pulse with 10-$\Omega$ source impedance, as indicated on the schematic. The trigger occurs during each half-cycle with a variable delay as in the previous SCR problem. Ambient temperature range is 10 to 40°C. A heat sink with thermal resistance $\theta_{CA} = 3$°C/W is used.

**theory**   The operation of the circuit is the same as that described in a previous problem, but with the switching action occurring during each half-cycle instead of during positive half-cycles only. Equations (5-41) and (5-43) apply in the same manner. If the trigger pulses are of positive polarity during positive half-cycles and of negative polarity for negative half-cycles (which is frequently used), the circuit works equally well and the device selection rationale is the same.

**solution**   Select a SC 146B1 device (Fig. 5-22). The selection factors are similar in many respects to those for selection of an SCR. The current drawn by the load is 800 W/115 V = 6.9 A. This must be increased by 10% for high line voltage, to 7.65 A. Figure 5-22 provides a graph of rms current vs. power dissipation, showing $P_D = 10$ W for 7.65 A. Junction temperature is calculated using $\theta_{JC,\text{ac}} = 1.5$°C/W, as recommended in the specification. Then, by Eq. (5-43), $T_J = 10$ W × (1.5°C/W + 3°C/W) + 40°C = 85°C. This is within the 100°C maximum rating for $T_J$.

Triggering current for the worst case is determined by Eq. (5-41), using $V_{GT} = 3.5$ V (for $T_C = -40$°C):

$$I_{GT} = \frac{8\text{ V} - 3.5\text{ V}}{10\ \Omega} = 450\text{ mA}$$

This is above the 80-mA minimum specified for $-40$°C. If the triggering were inadequate for $-40$°C, the circuit might still work at the actual low ambient temperature of $+10$°C. To determine this, the triggering conditions for $+10$°C are estimated as was done in the previous problem.

Gate power ratings also should be checked. Pulse power is 450 mA × 3.5 V = 1.58 W. For average power, this is multiplied by the duty cycle of approximately 20 $\mu$s/8.3 ms, giving 3.81 mW. These are well with the maximum ratings of 10 and 0.5 W, respectively.

The SC 146B1 is rated for 200 V, the right value for a power source of 115 V ac. If the application requires an electrically isolated tab, so that the heat sink is not connected to the circuit, the SC 147B1 may be used. Certain parameters are slightly different, resulting in a small change in $T_J$ from that calculated above.

*A snubber consists of a large capacitor and small resistor in series, connected between the SCR anode and cathode. See D. R. Grafham and J. C. Hey (eds.), *SCR Manual*, Chaps. 5 and 16, General Electric Company, Syracuse, N.Y., 1972, for more information.

Maximum Allowable Ratings

| Type | RMS on-state current, $I_{T(rms)}$ Amperes | Repetitive peak off-state voltage, $V_{DRM}$ | | | | Peak one full cycle surge (non-rep) on-state current, $I_{TSM}$ Amperes | | $I^2t$ for fusing for times $AT$ | |
|---|---|---|---|---|---|---|---|---|---|
| | | B Volts | D Volts | E Volts | M Volts | 50 Hz Amperes | 60 Hz Amperes | (RMS ampere)² seconds 1.0 millisecond | (RMS ampere)² seconds, 8.3 milliseconds |
| SC146 | 10 | 200 | 400 | 500 | 600 | 110 | 120 | 20 | 60 |

Peak gate power dissipation, $P_{GM}$          10 W for 10 $\mu$s
Average gate power dissipation, $P_{G(av)}$
Peak gate current, $I_{GM}$
Peak gate voltage, $V_{GM}$
Storage temperature, $T_{stg}$                     −40 to +125°C
Operating temperature, $T_J$              −40 to +100°C
Surge isolation voltage                     1600 V rms

Characteristics

| Test | Symbol | Min. | Typ. | Max. | Units | Test Conditions | | |
|---|---|---|---|---|---|---|---|---|
| Repetitive peak off-state current | $I_{DRM}$ | | | | mA | $V_{DRM}$ = Maximum Allowable Repetitive Off-State Voltage Rating Gate Open Circuited | | |
| | | — | — | 0.1 | | $T_C$ = +25°C | | |
| | | — | — | 0.5 | | $T_C$ = +100°C | | |
| Peak on-state voltage | $V_{TM}$ | | | | Volts | $T_C$ = +25°C, $I_{TM}$ = 1 ms, wide pulse, duty cycle ≤ 2% | | |
| SC146 | | — | — | 1.65 | | $I_{TM}$ = 14 A Peak | | |
| DC gate trigger current | $I_{GT}$ | | | | mAdc | $V_D$ = 12 Vdc | | |
| | | | | | | Trigger Mode | $R_L$ | $T_C$ |
| | | — | — | 50 | | MT2 + Gate + | 100 Ω | |
| | | — | — | 50 | | MT2 − Gate − | 100 Ω | +25°C |
| | | — | — | 50 | | MT2 + Gate − | 50 Ω | |
| | | — | — | 80 | | MT2 + Gate + | 50 Ω | |
| | | — | — | 80 | | MT2 − Gate − | 50 Ω | −40°C |
| | | — | — | 80 | | MT2 + Gate − | 25 Ω | |
| DC gate trigger voltage | $V_{GT}$ | | | | Vdc | $V_D$ = 12 Vdc | | |
| | | | | | | Trigger Mode | $R_L$ | $T_C$ |
| | | — | — | 2.5 | | MT2 + Gate + | 100 Ohms | |
| | | — | — | 2.5 | | MT2 − Gate − | 100 Ohms | +25°C |
| | | — | — | 2.5 | | MT2 + Gate − | 50 Ohms | |
| | | — | — | 3.5 | | MT2 + Gate + | 50 Ohms | |
| | | — | — | 3.5 | | MT2 − Gate − | 50 Ohms | −40°C |
| | | — | — | 3.5 | | MT2 + Gate − | 25 Ohms | |
| DC gate non-trigger voltage | $V_{GD}$ | 0.2 | — | — | Vdc | Trigger Mode | $R_L$ | $T_C$ |
| | | | | | | MT2 + Gate + | | |
| | | | | | | MT2 − Gate − | 1000 | +100°C |
| | | | | | | MT2 + Gate − | Ohms | |
| | | | | | | MT2 − Gate + | | |

**Fig. 5-22** Excerpts of specification, SC 146B1. (*Courtesy of General Electric Company.*)

Characteristics

| Test | Symbol | Min. | Typ. | Max. | Units | Test Conditions |
|------|--------|------|------|------|-------|-----------------|
| DC holding current | $I_H$ | | | | mAdc | Main terminal source voltage = 24 Vdc Peak initiating on-state current = 0.5 A, 0.1 to 10 milliseconds wide pulse, gate trigger source = 7 V, 20 Ω. |
| | | — | — | 50 | | $T_C = +25°C$ |
| | | — | — | 100 | | $T_C = -40°C$ |
| DC latching current | $I_L$ | | | | mAdc | Main terminal source voltage = 24 V dc Gate trigger source = 15 V, 100 Ω 50-$\mu$s pulse width, 5 $\mu$s rise and fall times maximum |

| | | | | | | Trigger Mode | $T_C$ |
|---|---|---|---|---|---|---|---|
| | | — | — | 100 | | MT2 + Gate + | |
| | | — | — | 100 | | MT2 − Gate − | +25°C |
| | | — | — | 200 | | MT2 + Gate − | |
| | | — | — | 200 | | MT2 + Gate + | |
| | | — | — | 200 | | MT2 − Gate − | −40°C |
| | | — | — | 400 | | MT2 + Gate − | |

| Test | Symbol | Min. | Typ. | Max. | Units | Test Conditions |
|------|--------|------|------|------|-------|-----------------|
| Steady state thermal resistance | $R\theta_{JA}$ | — | — | 75 | °C/W | Junction-to-ambient |
| Steady-state thermal resistance: | $R\theta_{JC}$ | | | | °C/W | Junction-to-case This characteristic is useful as an acceptance test at an incoming inspection station. |
| SC140 | | — | — | 3.1 | | |
| SC141 | | — | — | 3.0 | | |
| SC142 | | — | — | 3.3 | | |
| SC143 | | — | — | 3.2 | | |
| SC146 | | — | — | 2.2 | | |
| SC147 | | — | — | 2.5 | | |
| SC149 | | — | — | 2.0 | | |
| SC151 | | — | — | 2.0 | | |
| Apparent thermal resistance: | $R\theta_{JC(ac)}$ | | | | °C/W | Junction-to-case This characteristic is useful in the calculation of junction temperature rise above case temperature for AC current conduction. |
| SC140 | | — | — | 2.04 | | |
| SC141 | | — | — | 2.22 | | |
| SC142 | | — | — | 2.31 | | |
| SC143 | | — | — | 1.97 | | |
| SC146 | | — | — | 1.50 | | |
| SC147 | | — | — | 1.69 | | |
| SC149 | | — | — | 1.52 | | |
| SC151 | | — | — | 1.10 | | |

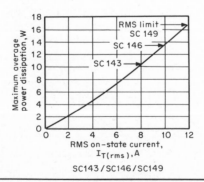

SC143 / SC146 / SC149

**Fig. 5-22 (Cont.)**

## 5.15   SELECTING A PROGRAMMABLE UNIJUNCTION TRANSISTOR

Commonly used parameters of a programmable unijunction transistor (PUT) are summarized in Table 5-10.

**TABLE 5-10   Commonly Used Parameters for the PUT**

| Parameter | Symbol | Meaning |
|---|---|---|
| Peak anode voltage | $V_P$ | Maximum anode voltage before the PUT switches on |
| Peak anode current | $I_P$ | Minimum anode current required to cause the PUT to switch on |
| Valley anode voltage | $V_V$ | Anode voltage at the valley point, with PUT switched on |
| Valley anode current | $I_V$ | Anode current at the valley point, with PUT switched on (if current reduces below this value, PUT will switch off) |
| Gate voltage | $V_s$ | Voltage at the gate determined by the programming resistors, with PUT switched off |
| Offset voltage | $V_T$ | Forward voltage on the gate junction, equal to $V_P - V_s$ |
| Pulse output voltage | $V_O$ | Pulse output amplitude with the PUT used in a specified circuit |
| Capacitive discharge energy | | Amount of energy which can be dissipated in the device, due to the discharge of the anode capacitor, with no current limiting |

**problem 5.12**   Figure 5-23 shows a relaxation oscillator using a PUT. The output at the cathode of $Q_1$ is used to gate an SCR (not shown). Select a PUT to operate over an ambient temperature range of +20 to +50°C.

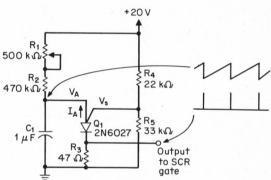

**Fig. 5-23**   A PUT relaxation oscillator and basic waveforms.

**theory**   Resistors $R_4$ and $R_5$ are for programming the device. They serve as a voltage divider to establish the off-state gate voltage $V_s$. The anode voltage $V_A$ must exceed $V_s$ by 0.7 V to turn on the PUT. A brief explanation of the circuit operation follows:

1. When voltage is first applied, $C_1$ is discharged, the gate is reverse-biased, and $Q_1$ is off.
2. Capacitor $C_1$ begins charging through $R_1$, $R_2$ and continues to charge until $V_A$ exceeds $V_s$ by 0.7 V. At this value, the gate becomes forward-biased, and $Q_1$ turns on provided that cathode current $I_A$ exceeds the maximum peak current $I_P$ required to fire.
3. When $Q_1$ conducts, $C_1$ discharges rapidly through $Q_1$ and $R_3$, developing a positive pulse across $R_3$. Voltage across $Q_1$, anode to cathode, is about 1 V during conduction.
4. Upon completion of the discharge of $C_1$, $I_A$ reduces to a low value determined by the circuit. This value must be less than the minimum valley current $I_V$ of the device for $Q_1$ to turn off so that the next cycle can begin. If $I_A$ exceeds $I_V$ at this time, $Q_1$ remains conducting indefinitely.
5. After discharge of $C_1$ and turnoff of $Q_1$, as described above, the events repeat themselves, producing a positive sawtooth waveform at $V_A$ and a periodic positive pulse (coinciding with the rapid negative-going excursion of the sawtooth) at the output.

From the preceding explanation it is seen that two values of $I_A$ are critical:
1. To bring $Q_1$ into conduction, $I_A \geq I_P$.
2. To turn $Q_1$ off, $I_A \leq I_V$.

When $Q_1$ begins to conduct, $V_A = V_s + 0.7$ V. Then,

$$I_A = \frac{20\text{ V} - V_s - 0.7\text{ V}}{R_1 + R_2} \qquad (5\text{-}44)$$

Voltage $V_s$ is established by voltage divider $R_4$ and $R_5$ so that:

$$V_s = \frac{20\text{ V} \times R_5}{R_4 + R_5} \qquad (5\text{-}45)$$

When $Q_1$ turns off, the voltage across $R_3$ is negligible and $V_A = 1$ V, the drop across $Q_1$. Thus,

$$I_A = \frac{20\text{ V} - 1\text{ V}}{R_1 - R_2} = \frac{19\text{ V}}{R_1 + R_2} \qquad (5\text{-}46)$$

The frequency of oscillation and the characteristics of the output pulse need not be calculated, as they are virtually independent of the parameters of $Q_1$ (as long as switching action occurs and the circuit oscillates). The actual frequency is approximately 1 Hz, variable over a 2:1 range with $R_1$.

The value of $I_P$ and $I_V$ required by $Q_1$ may be changed by using larger or smaller resistance for $R_4$ and $R_5$, as is noted in the data sheet (Fig. 5-24). Time delay circuits using the PUT are of similar configurations. The main difference is that the components are valued so that $I_A > I_{V,\text{max}}$ after $Q_1$ conducts. Device $Q_1$ then stays in conduction, preventing oscillation, until the applied voltage is removed.

**solution**   Select the 2N6027 device (Fig. 5-24). Selection of a PUT is simplified by the fact that it is a recent device with relatively few types available. The 2N6027 is selected for this circuit because its values of peak and valley currents will cause oscillation. Other types would probably also work satisfactorily, although they might require adjustment of the programming resistors $R_4$ and $R_5$ to give the needed $I_P$ and $I_V$ values.

Calculations are needed to confirm proper operation. Voltage $V_s$ is determined first by Eq. (5-45):

$$V_s = \frac{20\text{ V} \times 33\text{ k}\Omega}{22\text{ k}\Omega + 33\text{ k}\Omega} = 12.0\text{ V}$$

The value of $I_A$ available to bring $Q_1$ into conduction can now be determined. Resistor $R_1$ must be set at its maximum value of 500 kΩ for worst case. Then, by Eq. (5-44):

$$I_A = \frac{20\text{ V} - 12\text{ V} - 0.7\text{ V}}{500\text{ k}\Omega + 470\text{ k}\Omega} = 7.53\ \mu\text{A}$$

The gate impedance $R_G$ is 22 kΩ × 33 kΩ/(22 kΩ + 33 kΩ) = 13.2 kΩ. The minimum (worst case) temperature is 20°C, and $V_s$ is 12 V. These values are close to the conditions in the specification for $I_{P,\text{max}} = 5\ \mu\text{A}$, thus ensuring that 7.53 μA is adequate.

For the transistor to oscillate, $Q_1$ must also turn off after discharging $C_1$. Anode current $I_A$ must decrease below the minimum valley current $I_V$, given as 70 μA for the same circuit conditions. Worst case, however, is at the high ambient temperature of 50°C. The curve of $I_V$ vs. ambient temperature in Fig. 5-24 shows a 20% decrease in $I_V$ at this temperature, reducing minimum $I_V$ to 70 μA less 20% = 56 μA. Additional temperature rise from power dissipated in $Q_1$ is negligible, since the average current is very small.

---

The General Electric PUT is a three-terminal planar passivated PNPN device in the standard plastic low cost TO-98 package. The terminals are designated as anode, anode gate and cathode.

The 2N6027 and 2N6028 have been characterized as Programmable Unijunction Transistors (PUT), offering many advantages over conventional unijunction transistors. The designer can select $R_1$ and $R_2$ to program unijunction characteristics such as $\eta$, $R_{BB}$, $I_P$ and $I_V$ to meet his particular needs.

The 2N6028 is specifically characterized for long interval timers and other applications requiring low leakage and low peak point current. The 2N6027 has been characterized for general use where the low peak point current of the 2N6028 is not essential. Applications of the 2N6027 include timers, high gain phase control circuits and relaxation oscillators.

**Fig. 5-24**   Excerpts of specification, 2N6027. (*Courtesy General Electric Company.*)

Absolute Maximum Ratings (25°C)

Voltage
  *Gate-cathode forward voltage                 +40 V
  *Gate-cathode reverse voltage              −5 V
  *Gate-anode reverse voltage               +40 V
  *Anode-cathode voltage                 ±40 V
Current
  *DC anode current                       150 mA
  Peak anode, recurrent forward
    (100 μs pulse width, 1% duty cycle)     1 A
  *(20 μs pulse width, 1% duty cycle)      2 A
  Peak anode, non-recurrent forward
    (10 μs)                       5 A
  *Gate current                     ±20 mA
Capacitive discharge energy           250 μJ
Power:
  *Total average power              300 mW
Temperature:
  *Operating ambient
   Temperature range          −50 to +100°C
Derate currents and powers 1%/°C above 25°C
$E = \frac{1}{2} CV^2$ capacitor discharge energy with no current limiting

$$R_G = \frac{R_1 R_2}{R_1 + R_2}$$

$$V_s = \frac{R_1 V}{R_1 + R_2}$$

$$V_T = V_p - V_s$$

Electrical Characteristics (25°C) (unless otherwise specified)

| | | Fig. No. | 2N6027 (D13T1) Min. | 2N6027 (D13T1) Max. | 2N6028 (D13T2) Min. | 2N6028 (D13T2) Max. | |
|---|---|---|---|---|---|---|---|
| Peak current ($V_s = 10$ V) | $I_P$ | 3 | | | | | |
|   ($R_G = 1$ MΩ) | | | | 2 | | .15 | μA |
|   ($R_G = 10$kΩ) | | | | 5 | | 1.0 | μA |
| Offset voltage ($V_s = 10$ V) | $V_T$ | 3 | | | | | |
|   ($R_G = 1$ MΩ) | | | 0.2 | 1.6 | .2 | 0.6 | V |
|   ($R_G = 10$kΩ) | | | 0.2 | 0.6 | .2 | 0.6 | V |
| Valley current ($V_s = 10$ V) | $I_V$ | 3 | | | | | |
|   ($R_G = 1$ MΩ) | | | | 50 | | 25 | μA |
| | | | 70 | | 25 | | μA |
|   ($R_G = 200$ Ω) | | | 1.5 | | 1.0 | | mA |
| Anode gate-anode leakage current | | | | | | | |
|   ($V_s = 40$ V, T = 25°C) | $I_{GAO}$ | 4 | | 10 | | 10 | nA |
|   (T = 75°C) | | | | 100 | | 100 | nA |
| Gate-to-cathode leakage current | | | | | | | |
|   ($V_s = 40$ V, anode-cathode short) | $I_{GKS}$ | 5 | | 100 | | 100 | nA |
| Forward voltage ($I_F = 50$ mA) | $V_F$ | | | 1.5 | | 1.5 | V |
| Pulse output voltage | $V_o$ | 6 | 6 | | 6 | | V |
| Pulse voltage rate of rise | $t_r$ | 6 | | 80 | | 80 | ns |

*JEDEC registered data

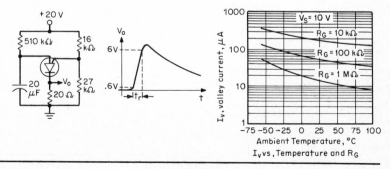

Fig. 5-24 (Cont.)

The actual maximum $I_A$ after discharge of $C_1$ is given by Eq. (5-46). Resistance $R_1$ is now $0\,\Omega$ for worst case, so that:

$$I_A = \frac{20\,V - 1\,V}{0\,\Omega + 470\,k\Omega} = 40.4\,\mu A$$

This is below the minimum $I_V$, and conditions for oscillation are met.

Maximum ratings of $Q_1$ also must be checked. Most of the voltage and current ratings are obviously well above those of the circuit. If the SCR gate is directly connected to the output without a series-limiting resistor, peak anode current could possibly be exceeded. This is not harmful if the capacitive discharge energy rating of $250\,\mu J$ is not exceeded. This is checked with the formula given in the specification: $E = \tfrac{1}{2}\,CV^2 = \tfrac{1}{2} \cdot 1.0\,\mu F \cdot (12\,V)^2 = 72\,\mu J$. Thus, current limiting of the output is not needed.

## 5.16  BIBLIOGRAPHY

Alvarez, E. C., and Fleckles, D. E.: *Introduction to Electron Devices*, 2d ed., McGraw-Hill, New York, 1974.

Bell, D. A.: *Fundamentals of Electronic Devices*, Reston Publishing Co., Reston, Va., 1975.

Cleary, J. F., (ed): *Transistor Manual*, 7th ed., General Electric Company, Syracuse, N.Y., 1964.

Cooper, W. D.: *Solid State Devices: Analysis and Applications*, Reston Publishing Company, Reston, Va., 1974.

Coughlin, R. F., and Driscoll, F. F.: *Semiconductor Fundamentals*, Prentice-Hall, Englewood Cliffs, N.J., 1976.

Grafham, D. R., and Hey, J. C., (eds.), *SCR Manual*, General Electric Company, Syracuse, N.Y., 1972.

Kaufman, M., and Seidman, A. H.: *Handbook for Electronics Engineering Technicians*, McGraw-Hill, New York, 1976, Chap. 8.

Kennedy, George: *Electronics Communications Systems*, 2d ed., McGraw-Hill, New York, 1977, Chap. 13.

Kiver, M. S.: *Transistor and Integrated Electronics*, 4th ed., McGraw-Hill, New York, 1972.

Mottershead, A.: *Electronic Devices and Circuits*, Goodyear Publishing Company, Pacific Palisades, Calif., 1973.

Sahm, W. H.: *General Electric Optoelectronics Manual*, Semiconductor Products Department, General Electric Company, Syracuse, N.Y.

Seidman, A. H., and Waintraub, J. L.: *Electronics Devices, Discrete and Integrated Circuits*, Charles E. Merrill Books, Inc., Columbus, Ohio, 1977.

*Semiconductor Applications Notes:* D.A.T.A. Inc., Orange, N.J. (published twice each year).

Sevin, L. J.: *Field-Effect Transistors*, Texas Instruments Electronics Series, McGraw-Hill, New York, 1965.

Tepper, I.: *Solid State Devices, vol. I: Theory*, Addison-Wesley, Reading, Mass., 1972.

———: *Solid State Devices, vol. II: Applications*, Addison-Wesley, Reading, Mass., 1974.

Texas Instruments, Inc. [J. R. Miller et al. (eds.)]: *Solid State Communications*, McGraw-Hill, New York, 1966.

Thomas, H. E.: *Handbook of Transistors, Semiconductors, Instruments, and Microelectronics*, Prentice-Hall, Englewood Cliffs, N.J., 1968.

Thornton, R. D., et al.: *Characteristics and Limitations of Transistors*, Wiley, New York, 1966.

# Chapter 6

# Audio Amplifiers

## DWIGHT V. JONES
### Technical Manager, General Electric Company, Syracuse, N.Y.

## 6.1 INTRODUCTION

Both germanium and silicon transistors have been used in audio amplifier circuit design, but the use of silicon predominates in diodes, transistors, and field-effect transistors (FETs). Transistors offer greater flexibility in circuit design than their predecessor (tubes), because of the existence of complementary transistors.

This chapter will deal exclusively with junction types of semiconductors since they account for almost all the active circuit elements in audio amplifier design. The junction field-effect transistors (p-channel and n-channel) and monolithic integrated circuits (ICs) will be discussed since they all play an important role in audio amplifier circuits.

## 6.2 TERMINATED BIPOLAR TRANSISTOR STAGE[1]*

Figure 6-1 shows a network being supplied from a generator which has an internal voltage $V_g$ and source impedance $Z_g$. A load impedance $Z_l$ is connected to the output terminals. The properties of a terminated common-emitter stage can be determined by using the equivalent circuit of Fig. 6-2 as the network in Fig. 6-1. Table 6-1 gives the properties of the terminated common-emitter stage. The equations reflect the effect of the source and load impedance, and the values used for the $h$ parameters must be adjusted for the circuit bias and the temperature conditions that are being analyzed. The basic properties of a terminated common-collector (emitter follower) stage are shown in Table 6-2.

*Superscript numbers refer to References, Sec. 6.10.

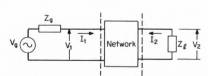

**Fig. 6-1** The terminated network.

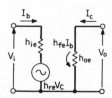

**Fig. 6-2** Bipolar transistor equivalent circuit with common-emitter $h$ parameters.

Figure 6-3 shows the input resistance of a common-emitter stage as a function of load resistance. This curve indicates that the input impedance approaches $h_{ie}$ as $Z_l$ approaches zero. A study of the equations in Table 6-1 will show that the current gain increases (approaches $h_{fe}$) as $Z_l$ approaches zero, while the voltage gain decreases.

**TABLE 6-1    Properties of the Terminated Common-Emitter Stage**

$$Z_i = h_{ie} - \frac{h_{fe}h_{re}Z_l}{1 + h_{oe}Z_l}$$

$$Z_o = \frac{h_{ie} + Z_g}{h_{oe}(Z_g + h_{ie}) - h_{fe}h_{re}}$$

$$A_v = \frac{-h_{fe}Z_1}{h_{ie}(1 + h_{oe}Z_l) - h_{fe}h_{re}Z_l}$$

$$A_i = \frac{h_{fe}}{1 + h_{oe}Z_l}$$

**TABLE 6-2    Properties of the Terminated Common-Collector Stage**

$$Z_i = \frac{h_{ie} + Z_l(1 + h_{fe})}{1 + h_{oe}Z_l}$$

$$Z_o = \frac{h_{ie} + Z_g}{1 + h_{fe} + h_{oe}Z_g}$$

$$A_v = \frac{Z_l(1 + h_{fe})}{h_{ie} + Z_l(1 + h_{fe})}$$

$$A_i = \frac{-(1 + h_{fe})}{1 + h_{oe}Z_l}$$

The common-collector connection has the highest input resistance and lowest output resistance. For this reason the so-called emitter follower finds wide use where either high-input impedance or low-output impedance is desired. The direct dependence of these impedances on either the generator or load impedances must be remembered. An emitter follower stage is often used to accomplish impedance transformation.

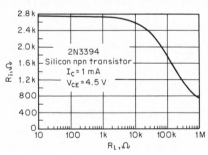

**Fig. 6-3**    Input resistance vs. load resistance for a common-emitter stage.

## 6.3    TERMINATED FIELD-EFFECT TRANSISTOR STAGE[2]

The FET equivalent circuit, Fig. 6-4, is used as the network in Fig. 6-1. Since $Z_l$ is usually small compared with the FET's output impedance, the input impedance is approximately

$$Z_i = \frac{1}{jw\,[C_{gs} + (1 + A_v)\,C_{gd}]} \tag{6-1}$$

and
$$A_v = Y_{fs}Z_l \tag{6-2}$$

Since $y_{fs}$ is transconductance, this is the same as the popular expression for voltage gain with vacuum tubes, $g_m R_l$. The FET responds only to input voltage variations since its input is a reverse-biased junction; thus, current amplification is not a factor.

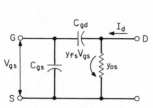

**Fig.  6-4** FET   common-source   equivalent circuit.

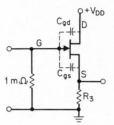

**Fig. 6-5**   FET source follower.

The common-drain (source follower) connection, shown in Fig. 6-5, is used to attain increased input impedance at higher frequencies where the Miller capacitance $[(1 + A_v) C_{gd}]$ is limiting the common-source connection.[2] In Fig. 6-5, $A_v = y_{fs}R_3/(1 + y_{fs}R_3)$, and since the voltage amplification is less than 1, there is no Miller capacitance. Also the effect of capacitance $C_{gs}$ is reduced since the signal at each end of $C_{gs}$ is in phase. The common-drain input impedance becomes

$$Z_i = \frac{1}{jw \ [C_{gs} \ (1 - A_v) + C_{gd}]} \tag{6-3}$$

## 6.4   BIASING[3,4]

### Forward- and Reverse-Biased Junctions

A diode is shown in Fig. 6-6 under both forward- and reverse-bias conditions. In Fig. 6-6a the electrons are attracted to a more positive voltage region just across the junction. Likewise the holes (+ charge) will flow across the junction toward the cathode. Thus there will be a large current flow across the junction for only a small applied voltage of about 0.1 V for germanium and about 0.6 V for silicon.

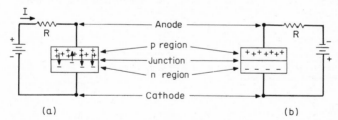

**Fig. 6-6**   PN junction diode or rectifier. (a) Forward bias. (b) Reverse bias.

In Fig. 6-6b the electrons in the n region see a more negative potential just across the junction which repels them away from the junction. Likewise the holes in the p region are repelled from the area near the junction owing to the more positively biased n region. The region where there are no carriers represents the dielectric of the associated junction capacitance; this capacitance decreases as the depletion layer widens with increased reverse-junction voltage.

### Basic Transistor Bias and Conduction

In transistor amplifier applications the collector base junction is always reverse-biased, and the emitter base junction is forward-biased in class A operation. The emitter base junction is operated with zero or reverse bias for half-cycle intervals in class B. Since the emitter base junction is forward-biased by $V_{BE}$ in Fig. 6-7, electrons will flow from the emitter region across the forward-biased junction into the base region. They diffuse through the narrow base region until they come under the influence of a voltage gradient as they approach the collector base junction. They flow across the collector base junction into the collector region. The emitter base junction functions in a manner similar to that of a forward-biased diode, and the collector base junction is similar to a reverse-biased diode. The arrows indicate the direction of electron flow in Fig. 6-7. The direction of conventional current flow (hole conduction) would be in the opposite direction. The magnitude of the collector current $(-I_C)$ through the load $(R_L)$ is controlled by the $V_{BE}$ potential, since increased $V_{BE}$ will increase the conductance across the emitter base junction. The ratio of collector current change to base emitter voltage change is the transconductance. The ratio of the collector current to base current is known as the dc current gain of the transistor ($h_{FE} = I_C/I_B$). The transistor input impedance at the base (Fig. 6-7) is low since we look into a forward-biased junction, whereas the output impedance at the collector is high owing to a reverse-biased collector junction.

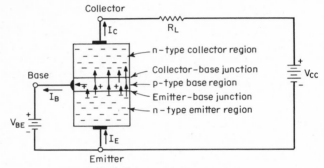

**Fig. 6-7**   Conduction in an npn junction transistor.

The operating principle of pnp transistors is similar to that of npn transistors except that holes (+) are the carriers from emitter to collector and the bias supplies will have reversed polarities from those used for npn transistors.

### Field-Effect Transistor (FET) Bias and Conduction

The junction FET, shown in Fig. 6-8, has a conducting channel between source and drain. The conductivity of the channel is modulated by an electric field operating at right angles to the channel, thus the term *field-effect transistor*. The electric field is created by applying a reverse bias to the junction or junctions in parallel as shown in Fig. 6-8. The width of the depletion region (area void of free carriers) increases as the gate-to-channel reverse junction voltage increases. This decreases the channel conductance, and when the reverse bias is large enough to cause the two depletion layers to meet, the channel is "pinched off" and the conductivity between source and drain drops to essentially zero.

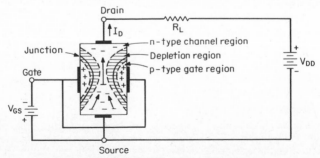

**Fig. 6-8**   Conduction in an FET (n-channel type).

The ratio of drain current change to gate source voltage change is the transconductance. The FET differs from the conventional junction transistor in that it has the high-input impedance of a reverse-biased junction, unlike the lower input impedance of the regular transistor with its forward-biased input junction. The p-channel FET requires the opposite polarity for its bias supplies.

## Large Signal Swing and Operating Point

When the ac-signal swing at the output of a transistor stage is large, the small signal parameters which were based on linear circuit characteristics no longer apply. This would be true for power output stages and the power driver stages. In these applications the peak-to-peak signal swing covers a major portion of the loadline as shown in Fig. 6-9. If the swing exceeds point $A$, severe clipping distortion will take place as the transistor is driven close to saturation at the knee of the characteristic. Also if the swing exceeds point $B$, severe clipping distortion will take place as the transistor current is cut off. The maximum signal output before clipping occurs, then, when the swing takes place about an operating point (OP) that is halfway between points $A$ and $B$. The operating bias current for this class A amplifier stage is just over 1 A, and the collector-to-emitter voltage is 12 V. In this example the supply voltage is 24 V, and the collector load line represents a collector load resistance of 12 $\Omega$. If no base bias were applied, the operating point would be at $B$ in Fig. 6-9, and conduction would occur for just 180° of a period, which is class B operation. The dc ratio of forward current transfer $h_{FE}$ approaches the large signal characteristics and thus is used in design of power stages in place of the small signal parameter $h_{fe}$.

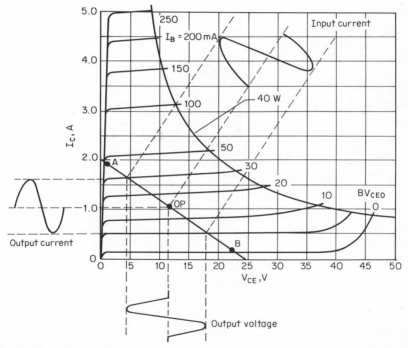

**Fig. 6-9**  Characteristics of a class A common-emitter power stage. (*Courtesy Motorola, Inc., Power Transistor Handbook, 1st ed.*)

## Bias Considerations with Variations in $h_{FE}$, $I_{CBO}$, $V_{BE}$, and Temperature

A basic problem in the design of transistor amplifiers is that of maintaining the proper dc bias current and voltage. The biasing problem is due primarily to the change of transistor parameters ($h_{FE}$, $I_{CBO}$, $V_{BE}$) with temperature and the variation of these parameters between transistors of the same type. This can readily be seen by referring to Fig. 6-10 where the transistor is operated in the common-emitter mode and is biased by a constant base current $I_B$. Figure 6-10 shows the common-emitter collector characteristics of two different transistors with the same collector load line superimposed on them. For the transistor characteristic shown with solid lines and a base current $I_{B2}$, the operating point is at A. On the other hand, if a transistor with a higher $h_{FE}$ is used, or the original transistor's $h_{FE}$ and $I_{CBO}$ are increased owing to an increase in temperature, the transistor characteristic shown with dashed lines could result. For the same base current ($I_{B2}$) the bias point is at B, and distortion would result since the transistor begins to saturate during the positive half-cycle of the signal base current.

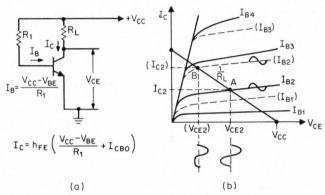

(a)                                     (b)

**Fig. 6-10**  Simple bias circuit.

The factors which must be considered in the design of transistor bias circuits are:

1. The maximum and minimum values of $h_{FE}$ at the operating point for the type of transistor used, and over the desired temperature range of operation
2. The magnitude and variation of $I_{CBO}$ with temperature
3. The variation of $V_{BE}$ with temperature
4. The tolerance of the supply voltages and also bias resistors

The variation of $h_{FE}$ with temperature is shown for a silicon transistor in Fig. 6-11.

$I_{CBO}$ is the collector-to-base current flowing, with the emitter open, and it increases exponentially with temperature. A silicon transistor has a very low $I_{CBO}$ value at room temperature (25°C) (see Fig. 6-12). For this reason $I_{CBO}$ often is not a factor in bias considerations with silicon transistors.

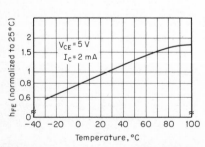

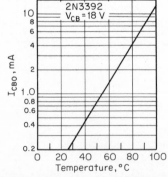

**Fig. 6-11**  $h_{FE}$ vs. temperature, 2N3392 transistor.

**Fig. 6-12**  Variation in $I_{CBO}$ vs. temperature for a 2N3392 planar silicon transistor. (*Courtesy of General Electric Co.*)

The variation of $V_{BE}$ with temperature is shown in Fig. 6-13 for a silicon transistor and indicates a change of about 1.3 mV/°C. This variation must be considered in the design of bias networks, especially dc-coupled stages and power-dissipating stages.

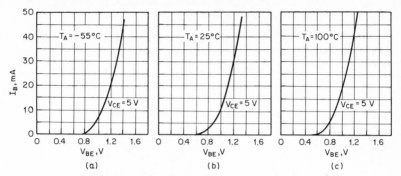

**Fig. 6-13**   Input characteristics for a silicon transistor.

## Basic Transistor Bias Circuit

The bias circuit in Fig. 6-10a can be used only where a range of collector voltage can be tolerated that is as great as the specified range of $h_{FE}$, and where the maximum $h_{FE} \times I_{CBO}$ is less than the desired bias current at the highest operating temperature. The bias circuit in Fig. 6-14 offers greater stability in collector current with changes in $h_{FE}$ or $I_{CBO}$, because of collector voltage feedback. However, this circuit also has ac feedback through the bias network ($R_1$) which reduces the gain and input impedance slightly. This feedback can be reduced by using two series resistors in place of $R_1$ and connecting a capacitor between their common point and ground.

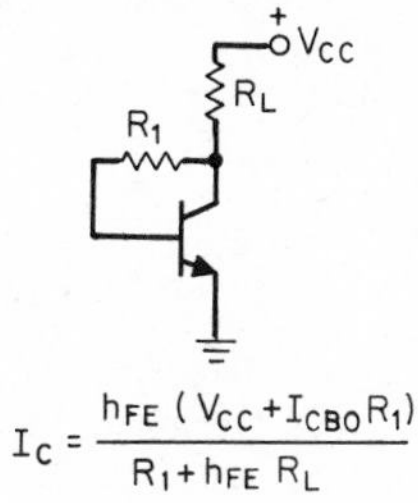

$$I_C = \frac{h_{FE}(V_{CC} + I_{CBO}R_1)}{R_1 + h_{FE} R_L}$$

**Fig. 6-14**   Voltage feedback bias stabilization.

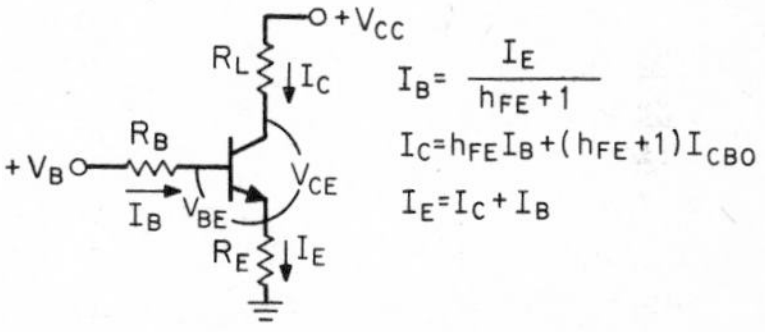

$$I_B = \frac{I_E}{h_{FE} + 1}$$

$$I_C = h_{FE} I_B + (h_{FE} + 1) I_{CBO}$$

$$I_E = I_C + I_B$$

**Fig. 6-15**   Basic transistor bias circuit.

Often more stability is required than is provided by the circuits of Figs. 6-10 or 6-14. Additional bias stabilization may be achieved by using current feedback with a resistor in series with the emitter of the transistor as shown in Fig. 6-15. There are several variations of this circuit, all of which may be obtained by the general design procedure outlined below. For the circuit of Fig. 6-15, the following equations apply:

$$I_E = (h_{FE} + 1)(I_B + I_{CBO}) \tag{6-4}$$

$$V_B = \left(\frac{R_B}{h_{FE} + 1} + R_E\right) I_E + V_{BE} - I_{CBO}R_B \tag{6-5}$$

With silicon transistors the $I_{CBO}$ has such a low value that it can be equated to zero for practical purposes, and therefore Eqs. (6-4) and (6-5) simplify to:

$$I_E = (h_{FE} + 1) I_B \tag{6-6}$$

$$V_B = \left(\frac{R_B}{h_{FE} + 1} + R_E\right) I_E + V_{BE} \tag{6-7}$$

If Eq. (6.7) is written for the worst bias conditions that would occur at the lowest operating temperature, the bias current $I_E$ will be at its minimum value, and the transistor's $h_{FE} = h_{FE}\text{min}$ and $V_{BE} = V_{BE}\text{max}$; therefore, at lowest temperature:

$$V_B = \left(\frac{R_B}{h_{FE}\text{min} + 1} + R_E\right) I_E\text{min} + V_{BE}\text{max} \qquad (6\text{-}8)$$

At the highest temperature of operation $I_E$ will have its maximum value, and the most severe bias conditions would occur for $h_{FE} = h_{FE}\text{max}$ and $V_{BE} = V_{BE}\text{min}$; therefore, at the highest temperature:

$$V_B = \left(\frac{R_B}{h_{FE}\text{max} + 1} + R_E\right) I_E\text{max} + V_{BE}\text{min} \qquad (6\text{-}9)$$

From these two equations the value of $R_B$ can be calculated:

$$R_B = \frac{(I_E\text{max} - I_E\text{min}) R_E + V_{BE}\text{min} - V_{BE}\text{max}}{(I_E\text{min}/h_{FE}\text{min} + 1) - (I_E\text{max}/h_{FE}\text{max} + 1)} \qquad (6\text{-}10)$$

### 6.4a    CALCULATING THE BIAS CIRCUIT DESIGN FOR A TYPICAL NPN TRANSISTOR

**problem 6-1**    Calculate the resistor values for a typical npn common-emitter transistor stage shown in Fig. 6-16. Design the circuit to operate over an ambient temperature range of 0 to 50°C with the emitter current stabilized to ± 10% maximum variation.

2N3392 Specification
$h_{FE} = 150 - 300$
( $V_{CE} = 5$ V, $I_C = 2$ mA, $T_A = 25\,°C$ )

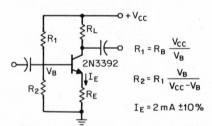

$$R_1 = R_B \frac{V_{CC}}{V_B}$$

$$R_2 = R_1 \frac{V_B}{V_{CC} - V_B}$$

$$I_E = 2\,\text{mA} \pm 10\%$$

**Fig. 6-16**    Voltage-divider bias circuit.

**theory**    (See previous two sections.) The equations for $R_1$ and $R_2$ in Fig. 6-16 relate this circuit to the basic bias circuit given in Fig. 6-15. The operating bias current must be high enough so that $I_{CBO}$ content is a negligible part. Also, the operating point is chosen so that with the extremes of $h_{FE}$ and temperature, the transistor does not cut off ($I_E = 0$) or saturate under conditions of maximum input signal.

**solution**    Given, $I_E\text{max} = 2.2$ mA; $I_E\text{min} = 1.8$ mA.
   1. Determine the values of $h_{FE}\text{min}$ and $h_{FE}\text{max}$: From the 2N3392 specification, $h_{FE}$ ranges from 150 to 300 at 25°C, when $I_c = 2$ mA and $V_{CE} = 5$ V. Then $h_{FE}\text{min} = 0.8 \times 150 = 120$ at 0°C and $h_{FE}\text{max} = 1.2 \times 300 = 360$ at 50°C (temperature multipliers from Fig. 6-11).
   2. Determine the maximum $V_{BE}$ change with a 50°C temperature variation ($V_{BE}\text{max} - V_{BE}\text{min}$): 50°C × 1.3 mV/°C = 0.065 V where 1.3 mV/°C is the temperature coefficient (see Fig. 6-13).
   3. Calculate the value of $R_B$ using Eq. (6-10):

$$R_B = \frac{(0.4 \times 10^{-3}) R_E + 0.065}{14.9 \times 10^{-6} - 6 \times 10^{-6}}$$

$$= \frac{0.4 R_E + 65}{8.9 \times 10^{-3}} = 45 R_E + 7300$$

   4. Using the equation from step 3 (above), choose suitable values of $R_B$ and $R_E$. If $R_E$ is chosen to be 470 Ω, then $R_B = 28.5$ KΩ.

Also a minimum collector-to-emitter voltage can be checked at this point for compatibility with the selection of $R_L$ and $V_{CC}$. If $R_L = 10$ kΩ and $V_{CC} = 28$ V, then the minimum $V_{CE} = 28$ V − (10,470 Ω × 2.2 × $10^{-3}$ A) = 5 V. The lowest minimum $V_{CE}$ would be 3.6 V if the supply voltage has a ±5% tolerance.

5. Calculate the $V_B$ using Eq. (6-8): If $V_{BE} = 0.6$ V when $I_E = 2$ mA, $T_A = 25°$C, then at 0°C $V_{BE}$ = 0.6 + (1.3 × $10^{-3}$) 25 = 0.63.

$$V_B = \left(\frac{28.5 \times 10^3}{121} + 470\right) 1.8 \times 10^{-3} + 0.63 = 1.9 \text{ V}$$

6. Now calculate $R_1$ and $R_2$ in Fig. 6-16.

$$R_1 = 28.5 \times 10^3 \frac{28}{1.9} = 420 \text{ k}\Omega$$

$$R_2 = 420 \text{ k}\Omega \frac{1.9}{28 - 1.9} = 30 \text{ k}\Omega$$

Choose the nearest standard value for $R_1$ and $R_2$.

### 6.4b   CALCULATING THE BIAS CIRCUIT DESIGN WITH FEEDBACK

**problem 6.2**   Calculate the resistor values when feedback is employed as in Fig. 6-17. Design the circuit to operate from 0 to 50°C and $I_E = 2$ mA ±10% maximum variation.

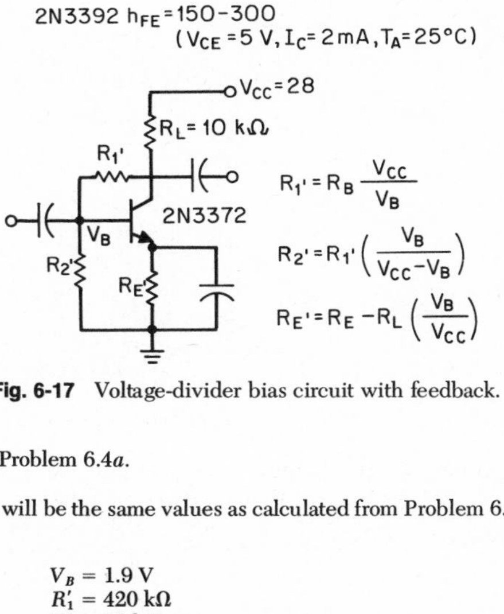

**Fig. 6-17**   Voltage-divider bias circuit with feedback.

**theory**   Same as for Problem 6.4a.

**solution**   $R_B$ and $V_B$ will be the same values as calculated from Problem 6.4a; thus $R_B = 45 R_E$ + 7300.

$$V_B = 1.9 \text{ V}$$
$$R_1' = 420 \text{ k}\Omega$$
$$R_2' = 30 \text{ k}\Omega$$
$$R_E' = 470 - 10 \text{ k}\Omega \frac{1.9}{28} = 470 - 678$$

This indicates that $R_E'$ can be zero in Fig. 6-17 and provide bias stability equal to Fig. 6-16 at somewhat less voltage amplification, because of the negative feedback from the collector back to base. If $R_E'$ is zero, then its shunt capacitor is also not required.

### 6.4c   CALCULATING THE INPUT IMPEDANCE OF A BOOTSTRAPPED BIAS NETWORK STAGE

**problem 6.3**   The bias circuits discussed previously have a relatively low input impedance when they are designed for good bias stability. When a higher input impedance is needed, a bootstrapped base bias network as shown in Fig. 6-18 is often used. Calculate the input impedance and see how much it is increased compared with $R_2$ of Fig. 6-16.

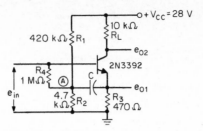

**Fig. 6-18** Bootstrapped bias network. *(Sergio Bernstein-Bervery, "Designing High Input Impedance Amplifiers," Electronic Equipment Engineering, August 1961.)*

**theory**   The higher input impedance is obtained in Fig. 6-18 by coupling the ac voltage swing at the emitter to point A in the bias network. This positive feedback from the emitter is a stable condition since the voltage amplification at the emitter is always less than unity. The signal at each end of $R_4$ is in phase, which has the effect of increasing the resistance of $R_4$ to $R_4' = R_4/(1 - A_V)$, where $A_V = e_o/e_{in}$.

The approximate input impedance for Fig. 6-18 is given in the following equation:

$$Z_{in} \approx \frac{R_4' R_e (h_{FE} + 1)}{R_4' + R_e (h_{FE} + 1)} \tag{6-11}$$

where $R_e$ is the effective parallel resistance combination in the emitter circuit. Therefore,

$$R_e = \frac{1}{1/R_3 + 1/R_2 + 1/R_1}$$

Equation (6-11) expresses the circuit input impedance as the parallel combination of $R_4'$ and the transistor's input resistance looking into its base. This equation neglects the shunting effect of the collector base impedance of the transistor, $l/h_{ob}$.

**solution**   The voltage gain for an emitter follower is slightly less than 1; therefore, assume $A_V = 0.95$.

$$R_4' = \frac{1 \text{ m}\Omega}{1 - 0.95} = 20 \text{ m}\Omega$$

$$R_e = \frac{1}{1/470 + 1/4700 + 1/420,000} = 427 \ \Omega$$

$$Z_{in} \approx \frac{20 \times 10^6 \times 427(201)}{20 \times 10^6 + 427(201)} = 85.5 \text{ k}\Omega$$

This input impedance is about three times that of Fig. 6-16 where $R_2 = 30$ k$\Omega$.

### Temperature-Compensated Bias Network

Diodes are used as temperature-sensitive elements since their temperature characteristic is similar to that of a transistor's emitter base junction. Thus the diode compensates for most of the $V_{BE}$ change with temperature and also stabilizes the bias current and voltages. A forward-biased diode or diodes can be used in place of the lower resistor of the base bias voltage divider (as in Fig. 6-19). One diode has a voltage drop approximately equal to $V_{BE}$ of the transistor, and the second diode will provide an equal fixed voltage across $R_3$ which maintains a constant current in $R_3$. Therefore the desired stabilized bias current is established by the value of $R_3$. Integrated circuits often use diodes or transistors as the bias-stabilizing element. Forward-biased diodes are quite often used to bias push-pull driver and output stages near their conduction threshold.

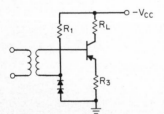

**Fig. 6-19** Forward-biased diode a temperature-sensitive element.

### Thermal Runaway

Thermal runaway can occur if the operating bias voltage and current are not properly stabilized. This may cause destruction of the transistor in power stages, or in low-level stages it can cause severe distortion due to the bias shift as in Fig. 6-10. Junction temperature $T_J$ is determined by the total power dissipation in the transistor $P_D$, the ambient temperature $T_A$, and the thermal resistance $\theta$.

$$T_J = T_A + P_D\theta_{JA} \qquad (6\text{-}12)$$

If the ambient temperature is increased, the junction temperature will increase an equal amount provided that the power dissipation is constant. Thermal runaway will occur when the rate of increase of junction temperature with respect to the power dissipation is greater than the thermal resistance $(\Delta T_J/\Delta P_D > \theta_{JA})$.

The thermal resistance term can be subdivided into three parts in Fig. 6-20 where

$\theta_{JA}$ = total thermal resistance (junction to ambient)
$\theta_{JC}$ = transistor thermal resistance (junction to case)
$\theta_{CS}$ = insulator thermal resistance (case to heat sink)
$\theta_{SA}$ = heat sink thermal resistance (heat sink to ambient)

Then,

$$\theta_{JA} = \theta_{JC} + \theta_{CS} + \theta_{SA} \qquad (6\text{-}13)$$
and
$$T_J = T_A + P_D(\theta_{JC} + \theta_{CS} + \theta_{SA}) \qquad (6\text{-}14)$$

Transistor specification sheets will state either $\theta_{JA}$, $\theta_{JC}$, or both and also maximum ratings for $T_J$ and $P_D$. The power dissipation in a transistor may be approximated as $P_D = V_{CE}I_C$, which is the hyperbolic (40-W) curve in Fig. 6-9. In practice the thermal resistance of the heat sink and insulator must be selected so that worst-case conditions of $T_A$, $P_D$, and $\theta_{JC}$ for the transistor as given in Eq. (6-14) do not exceed the maximum specified $T_J$.

### Second Breakdown

Second breakdown is the phenomenon at the point on the transistor characteristic of $I_C$ vs. $V_{CE}$ where there is an abrupt transition to a significantly lower voltage ($<15$ V) with an increase in $I_C$ dependent on the limiting series impedance in the collector load (see Fig. 6-21). This point is reached after the first avalanche or sustaining breakdown region and is, therefore, called second breakdown. The more critical applications are those involving a transistor

**Fig. 6-20**   Transistor thermal diagram.

conducting current (point A, Fig. 6-22) with an inductive collector load, and the transistor is driven toward cutoff (point B, Fig. 6-22). The energy stored in the inductive load continues to drive current through the transistor as $V_{CE}$ increases from A to B. The second breakdown current level can be increased if the time interval in going from A to B is decreased, and also if the reverse-base drive current is decreased. Unless the second breakdown current capability is equal to or greater than the upper dashed breakdown characteristics shown in Fig. 6-22, the transistor would not operate satisfactorily with this inductive load. Many power transistor specification sheets will show the safe area of operation on the $I_C$ vs. $V_{CE}$ characteristics with boundary limits.

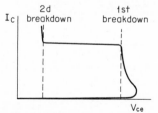

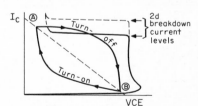

**Fig. 6-21**   Transistor breakdown characteristic.    **Fig. 6-22**   Inductive load operation.

### Biasing the Field-Effect Transistor

Since FETs are voltage-sensitive devices, the bias considerations involve the channel voltage $V_{DS}$ and the gate bias $V_{GS}$. One of the simpler bias circuits for FETs is shown in Fig. 6-23. Here $V_{GS}$ is operated at zero bias where $g_m$ is highest and represents a good point for low-noise operation of an FET. The input signal swing must be less than the conduction threshold of the gate diode (silicon = 0.5 V).

A large value of drain load resistance increases the voltage amplification ($A_V = g_m R_L$) but also increases the change of $V_{DS}$ with drain current ($I_D$) variation for a given FET type. The bias stability is improved by feedback through a self-biasing resistor $R_3$, as in Fig. 6-24. $R_3$ is selected so that the voltage drop induced by the FET channel current generates the proper gate bias. The variation in channel current is the FET parameter that requires

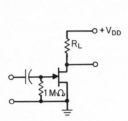

**Fig. 6-23**  Simple bias circuit for FET.

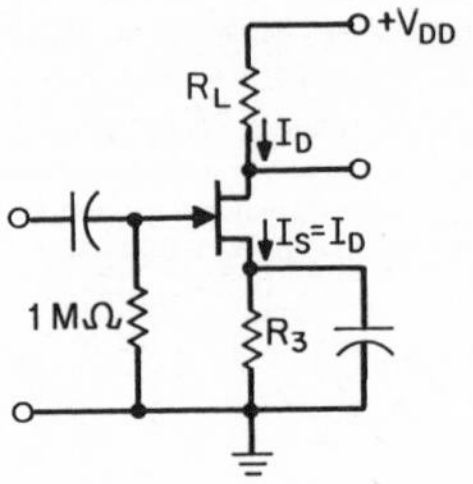

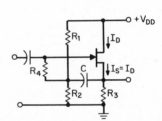

**Fig. 6-24**  Source-biasing FET circuit.

the most consideration in the design of the bias circuit. The stabilization of channel current with $R_3$ can be assessed by drawing the slope of the source bias resistor on the transfer characteristic as in Fig. 6-25. Since the channel current $I_D$ (drain) = $I_S$ (source), then $I_D R_3 = V_{GS}$ for the Fig. 6-24 bias circuit. If $R_3$ is 1 kΩ, then the variation in $I_D$ will be 3:1 compared with 5:1 if $R_3$ is zero for the three FET transfer characteristics shown in Fig. 6-25. $R_3$ is usually bypassed to improve the amplification.

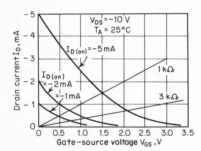

**Fig. 6-25**  FET common-source transfer characteristic and source-biasing resistor.

**Fig. 6-26**  Bootstrapped source follower.

A bootstrapped bias circuit like that used with transistors can also be used with FETs to increase the input impedance while maintaining good bias stabilization. The bootstrapped source follower shown in Fig. 6-26 would have a gate bias given by

$$V_{GS} = \frac{R_2 V_{DD}}{R_1 + R_2} - I_D R_3 \qquad (6\text{-}15)$$

The last term must be the largest for a reverse bias on the gate junction.

## 6.5    COUPLED STAGES[5]

Cascading FET stages normally present no special problems, except at high frequencies, since they are essentially unilateral devices. The transistor, on the other hand, is not unilateral, and, therefore, each stage in a cascade produces a loading effect on both the preceding and the following stages. An additional factor which must be considered is the effect of emitter degeneration. An unbypassed resistor in the emitter lead acts as though it were inserted in series with the base, multiplied by $(1 + h_{fe})$. If the resistor is bypassed by a capacitor, the impedance introduced into the base circuit will be frequency-dependent. At low frequencies it will approach the resistance multiplied by $1 + h_{fe}$, whereas at high frequencies the effect will be minimum.

### Analyzing RC-Coupled Stages for Input Impedance and Gain

In simplifying the two-stage $RC$-coupled amplifier of Fig. 6-27$a$, the two bias resistors may be replaced by a single equivalent one that has a value equal to that of the two in parallel. The simpler version is shown in Fig. 6-27$b$, which also assumes that the battery has zero internal impedance. We can further modify the circuit to the form shown in Fig. 6-27$c$ where the emitter circuit impedance is reflected into the input circuit, amplified by the factor $1 + h_{fe}$. The steps to follow are these:

1. The input impedance and current amplification of the second stage may be obtained from equations in Table 6-1.

2. This input impedance is added to that of the reflected emitter impedance in the second transistor base.

3. This impedance is now shunted by the effective bias resistance $R_{b2}$. The result is the total impedance to the right of the coupling capacitor $C_2$.

4. The reactance of $C_2$ is now added, and then the whole is shunted by $R_{c1}$. The result is the effective load impedance seen by transistor $Q_1$.

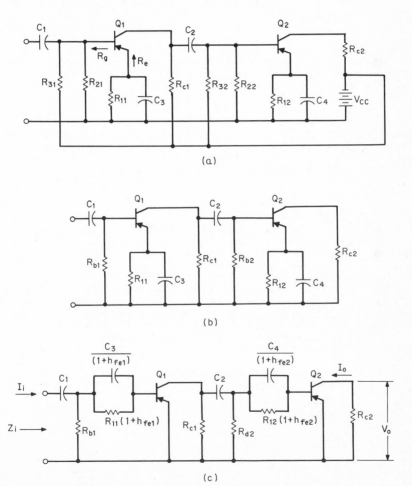

(a)

(b)

(c)

**Fig. 6-27**  ($a$) Two-stage $RC$-coupled amplifier; ($b$) Simplified version of ($a$). ($c$) Transfer of emitter impedance to base circuit.

5. If this impedance is used as the load for the first stage, its input impedance and current amplification may be calculated in the same manner as for the second stage in step 1.

6. Again, the base lead mesh is added to the input impedance, and the whole is shunted by $R_{b1}$ and added to the reactance of $C_1$ to give the input impedance of the overall amplifier.

7. The input current $I_i$ divides between $R_{b1}$ and the total impedance to the right of this point inversely as the impedances. Thus the portion of $I_i$ flowing into $Q_1$ is determined.

8. This is multiplied by the current amplification of the first stage as obtained in step 5.

9. This output current of $Q_1$ divides between $R_{c1}$ and the impedance to its right and again between $R_{b2}$ and the impedance to the right of that resistor. This gives the input current to transistor $Q_2$.

10. This current is multiplied by the current amplification obtained in step 1 to give the output current $I_o$. The output voltage will be $- R_{c2}I_o$.

11. The input voltage is $Z_iI_i$; hence the overall voltage amplification is $- R_{c2}I_o/Z_iI_i$.

The overall power gain will be the power into the load $R_{c2}(I_o)^2$ divided by the input power $R_i(I_i)^2$, where $R_i$ is the real part of the total input impedance $Z_i$. In all these steps the impedances become complex at low frequencies where the capacitor reactances are appreciable; thus the additions and multiplications would involve complex numbers.

An alternative to reflecting the external emitter impedance into the base leg, as shown in Fig. 6-27c, would be to reflect the base source resistance $R_g$ into the emitter leg as $R_e$. These resistances are indicated in Fig. 6-27a where

$$R_e \approx h_{ib} + \frac{R_g}{h_{fe} + 1} \tag{6-16}$$

If $h_{ib}$ is not given on a specification sheet, it may be approximated: $h_{ib} \approx 26/I_E$. The low-frequency voltage amplification of the first stage in Fig. 6-27a drops 3 dB when the reactance of $C_3$ equals the parallel impedance of $R_e$ and $R_{11}$. The coupling capacitors $C_1$ and $C_2$ also reduce the low-frequency response, but the impedances they couple are usually higher than those associated with the emitter, and therefore smaller capacitor values can be used.

### Direct-Coupled Stages

Direct coupling of stages for ac amplification offers several advantages. The reduced number of circuit components required for bias stabilization is shown in Fig. 6-28. Since the first stage serves as the base bias bleeder for the second stage, there is less power supply current drain than with RC-coupled stages having the same bias stability. In the same way, there is no ac signal loss similar to that which occurs in the bias network of RC-coupled stages. Not only are coupling capacitors eliminated, as indicated above, but the associated phase shift and attenuation of low frequencies are also eliminated. The fact that both npn and pnp transistors are available gives the circuit designer many possibilities for direct-coupled circuit configurations. Monolithic integrated circuit stages are direct-coupled since coupling capacitors are not readily available in monolithic format.

Coupling capacitors often are required between stages to block or isolate two different

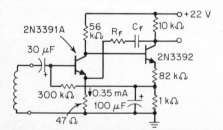

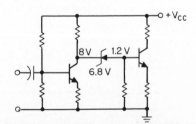

**Fig. 6-28** Two-stage direct-coupled feedback preamp for magnetic pickup.

**Fig. 6-29** Avalanche diode interstage coupling.

dc-voltage levels. This function can be performed with an avalanche diode (zener) having the correct voltage as shown in Fig. 6-29. This approach is also used in integrated (microelectronic) circuits.

### FET Coupled to Conventional Transistor

Several methods are used to couple from an FET to a bipolar transistor. The various connections offer different advantages. In Fig. 6-30 the transistor is connected to the FET in a close-coupled feedback arrangement. This feedback to the source of the FET neutralizes its gate-to-source capacitance and also decreases the normal voltage amplification loss of the FET source follower configuration. In the feedback loop the $g_m$ of the FET is multiplied by the $H_{fe}$ of the transistor. This circuit has the bootstrapped bias network, as in Fig. 6-26, to maintain a high-input impedance.

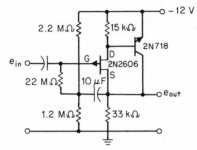

Fig. 6-30    FET, bipolar cascade.

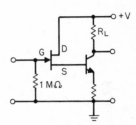

Fig. 6-31    Temperature-compensated amplifier.

The FET and bipolar transistor connection in Fig. 6-31 can be designed for a voltage amplification substantially independent of temperature because the transconductance of the FET has a negative temperature coefficient similar numerically to the positive $h_{fe}$ coefficient of the bipolar transistor. The collector voltage is also stabilized with temperature, since the base current is the FET channel current and it has a negative temperature coefficient of similar magnitude to the positive coefficient of the transistor's current transfer ratio.[6]

## 6.6  PREAMPLIFIERS[7]

### 6.6a  CALCULATING THE OPTIMUM SOURCE IMPEDANCE FOR A PREAMPLIFIER TO OPERATE AT LOWEST NOISE

**problem 6.4**    Calculate the optimum source impedance at 1 kHz for the preamplifier in Fig. 6-28 to operate at lowest noise. Assume that the noise characteristics in Fig. 6-32 apply also to the 2N3391A. The term $\sqrt{Hz}$ is the rate at which noise varies with the frequency band-width; for example, if the bandwidth increases by a factor of 2, the noise increases by 1.4.

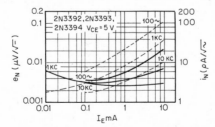

**Fig. 6-32**    Equivalent input noise voltage and current vs. bias current for a typical silicon transistor. *(Courtesy of General Electric Co.)*

**theory**    The noise generated in an amplifier stage can be considered as originating from a constant current noise generator in parallel with the input and a constant voltage noise generator in series with the input, as in Fig. 6-33.[8] When the input terminals of the amplifier are shorted, $e_n$

is responsible for the entire noise output; thus, this output divided by the voltage amplification for the stage gives the value of $e_n$. Likewise, when the input terminals of the amplifier are open, $i_n$ is responsible for the entire noise output. With optimum source resistance, both noise generators will contribute equally to the noise output.

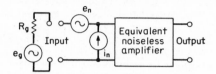

**Fig. 6-33**   Equivalent noise generation network.

Both $e_n$ and $i_n$ must be measured under the same operating conditions, but they are independent of circuit feedback and input impedance. Since $e_n$ and $i_n$ are based solely on the active amplifying device, they are convenient characteristics to use for selecting a transistor or FET with favorable noise characteristics for a particular application.

$$\text{Optimum } R_g = \frac{e_n}{i_n} \tag{6-17}$$

**solution**   From Fig. 6-32:

$$e_n \approx 0.00\ 3\ \mu\text{V}/\sqrt{\text{Hz}},$$

$$i_n \approx 4.5\ \text{pA}/\sqrt{\text{Hz}}$$

Therefore

$$\text{Optimum } R_g = \frac{0.003 \times 10^{-6}}{4.5 \times 10^{-12}} = 666\ \Omega$$

### Noise Characteristics of Transistors and FETs

The noise in transistors and FETs varies with bias current, temperature, frequency, and source impedance. The variation with frequency is shown in Fig. 6-34. The noise in area A results from random diffusion and recombination of carriers in the base region, and also from the thermal noise generated in the base resistance $r_b'$. Thus for low noise a transistor should have low $r_b'$ and a narrow base region (that is, high $h_{fe}$). For most transistors used in audio preamps region B is not a problem, since it is above the audio spectrum. The third area is excess noise, also called $1/f$ noise, since it varies inversely with frequency. The frequency spectrum of noise for an FET is similar to that of Fig. 6-34 except the $1/f$ noise corner occurs at a lower frequency—sometimes as low as 5 Hz.

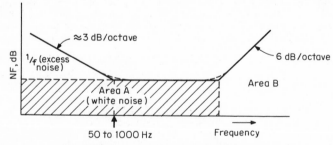

**Fig. 6-34**   Frequency spectrum of transistor noise.

A semiconductor device specification sheet may give the noise in terms of spot noise at given frequencies, or noise for a specified bandwidth that often includes parts of both the $1/f$ and white noise areas. The latter is more meaningful for audio preamplifiers since it usually covers most of the audible spectrum, while spot noise covers only a very narrow bandwidth at a given frequency. The third method of noise specification for semiconductors involves the current and voltage noise generators ($e_n$ and $i_n$) as described in Fig. 6-33.

The junction FET generally has a lower noise current $(i_n)$ characteristic than transistors; thus the FET has superior noise performance when the source impedance is above 20 to 50 k$\Omega$.

The base bias resistors of a transistor amplifier are in parallel with the input, and in order for their noise contribution to be negligible, this shunting resistance must be large with respect to $e_n/i_n$. The bias resistors should be low-noise types, such as deposited-carbon or metal-film types. Any impedance in series with the input (such as the coupling capacitor) should be small with respect to $e_n/i_n$ so as not to affect the noise factor. This requires a large coupling capacitor in order to have a reactance that is small compared with $e_n/i_n$ at 100 Hz; for example, in Fig. 6-32 for 100 Hz and bias current of 300 $\mu$A we see that $e_n i_n \approx 380$ $\Omega$. This requires a coupling capacitor of about 50 $\mu$F to have a reactance which is negligible compared with 380 $\Omega$.

### Frequency Considerations in Preamp Design

The output capacitance of a common-emitter stage is approximately $(h_{fe} + 1)C_{ob}$. The collector base capacitance $C_{ob}$ may cause a decline in the high-frequency response due to negative feedback to the input. This Miller effect increases with the voltage amplification $(A_v)$, that is, when the collector load impedance is increased in a common-emitter stage. The current flowing through $C_{ob}$ is $(A_v + 1)$ times the current which would flow through the same capacitance when connected from base to ground.[9]

### Preamp Design Influenced by the Impedance of the Transducer

The capacitor microphone and crystal phonograph cartridge are capacitive and thus have a high impedance that decreases with increasing frequency. Insufficient preamplifier input impedance will cause a drop in the low-frequency response. The required high-input impedance may be achieved by the use of (1) an FET, (2) the emitter follower transistor connection, (3) resistor in series with the input of a common-emitter configuration, or (4) an unbypassed emitter resistor in a common-emitter stage.

### Equalization Networks in Audio Circuits

Frequency-selective amplification is called *equalization*. Preequalization or preemphasis is used by the recording industry and also in fm broadcasting, in an attempt to optimize dynamic range and the signal-to-noise ratio. This has led to preemphasis on the upper end of the audio spectrum to achieve equal probability of overload at all frequencies. The frequency distribution of peak energy in music programming shows a decline in the spectrum above 1 to 4 kHz. The receiving end of fm broadcasting and playback of recordings then must incorporate equalization networks that have a complementary frequency characteristic if the output signal is to be like the original program material. The standard Record Industry Association of America (RIAA) playback equalization characteristic for phonograph records is shown in Fig. 6-35. This consists of bass boost and treble cut, assuming that a velocity (i.e., magnetic) pickup is used.

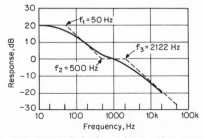

**Fig. 6-35**   RIAA playback equalization characteristic.

The equalization is usually accomplished with resistor-capacitor $(RC)$ networks. The two basic $RC$ equalization networks and their frequency responses are shown in Fig. 6-36.[10] The frequency at which the response has fallen or increased 3 dB from the flat-response range is termed the *turnover frequency*, the *3-dB point*, or sometimes the *inflection point.* A single $RC$-equalization time constant will have a maximum rate of change of 6 dB/octave, which is 20 dB/decade of frequency. Starting from the turnover frequency, an $RC$ curve requires two octaves in either direction to become essentially a straight line (either flat or sloping 6 dB/octave). A straight-line representation (shown dashed in Figs. 6-35 and 6-36) clearly depicts the turnover or inflection points and is within 3 dB of the actual curve.

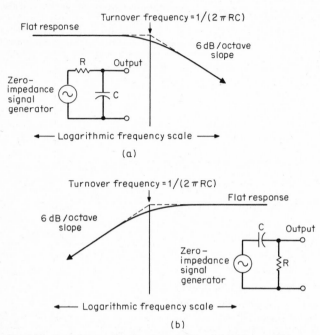

**Fig. 6-36**   Basic $RC$ equalization characteristics. $(a)$ treble cut; $(b)$ bass cut.

Bass boost is actually achieved by attenuating the treble frequencies, and this curve would be the complement of Fig. 6-36$b$. Also, treble boost is achieved by attenuating the base frequencies, and this would be the complement of Fig. 6-36$a$.

The fm deemphasis (Fig. 6-37) consists of treble cut beginning at the same turnover frequency as used for RIAA. This treble cut is the same as the basic curve of Fig. 6-36$a$ and is used for the sound portion of television sets and in fm receivers. The turnover is often stated in terms of an $RC$ time constant equal to 75 $\mu$s.

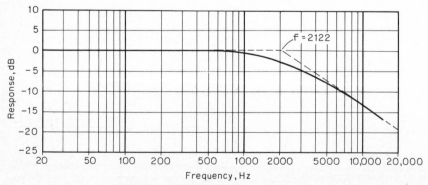

**Fig. 6-37**   Standard fm receiver deemphasis.

The National Association of Broadcasters (NAB) playback equalization characteristic for tape at 15 and 7.5 in/s consists entirely of bass boost (36 dB), beginning at 3180 Hz and terminating at 50 Hz, as in Fig. 6-38. Also shown is the playback equalization for 3¾ and 1⅞ in/s tape speed. Here, the higher turnover frequency is 1770 Hz instead of 3180 Hz, and the amount of bass boost is 31 dB.

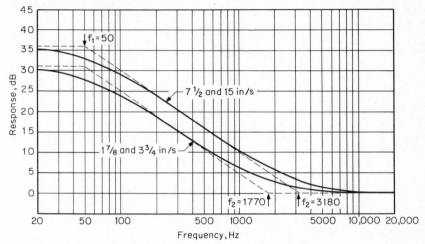

**Fig. 6-38**   NAB tape playback equalization characteristics.

The output of a magnetic phonograph cartridge is proportional to the needle velocity, and the output of a piezoelectric cartridge is proportional to needle displacement (i.e., groove amplitude). This variation of groove amplitude with frequency for RIAA record equalization is shown in Fig. 6-39a. The complement, in Fig. 6-39b, represents the playback amplifier equalization required for a flat response from a constant-amplitude cartridge. Thus the playback amplifier must impart bass cut and treble boost, which is just the opposite to what is required for a magnetic cartridge (see Fig. 6-35). The equivalent circuit for a piezoelectric phonograph cartridge is a voltage generator with the cartridge capacitance in series with the load resistance. Thus the base cut can be achieved by using the proper time constant of load resistance and cartridge capacitance. This $RC$ time constant should usually be between 1000 and 3000 $\mu$s to accomplish the low-frequency equalization.

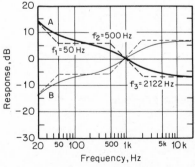

**Fig. 6-39**   Curve A: Response of piezoelectric cartridge to RIAA recording. Curve B: Playback equalization for amplitude response cartridge.

## Tone Control

A tone control can be defined as a variable audio filter used to vary the frequency response of an amplifier. Treble attenuation can reduce the amplitude of harmonics compared with that of the fundamental and thus reduce harmonic distortion, provided the predominant harmonics are in the attenuated frequency range. This is one of the reasons it is desirable to decrease the high-frequency response of the lower-quality audio systems. Bass boost can give a reduction in distortion if the fundamental is amplified more than the harmonics, and, conversely, bass cut (attenuation) can cause an increase in distortion. Thus bass boost and treble cut are the tone controls that are most useful in maintaining low distortion.

### Volume and Loudness Control

If the volume control is located too early in an amplifier, the overall signal-to-noise factor may suffer if the noise power generated in the following stages becomes comparable with the signal level at those stages. If the control is located too late in the amplifier chain, there may be overloading in a prior stage in which the signal levels become excessive, or there may be cross-modulation products due to nonlinearity. Consideration of these factors will normally indicate the proper location for the volume control.

A loudness control is desirable for boosting the level of the lower portion of the audio spectrum as the overall sound level is decreased. This is to compensate for the nonlinear response of the human ear, which is shown in Fig. 6-40. As the level is reduced, a greater intensity is required at the lower frequencies for equal loudness.

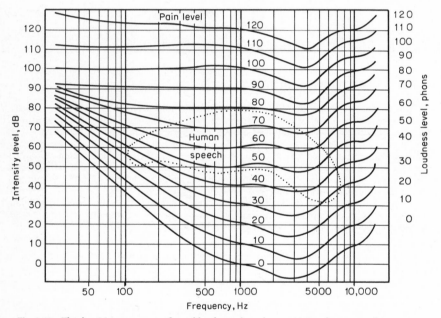

**Fig. 6-40**   Fletcher-Munson curves of equal loudness show the sensitivity of an average human ear.

### Preamp Design for Magnetic Cartridge

The type of circuit in Figs. 6-28 and 6-41 is often used as a preamplifier for magnetic pickups since negative feedback to the emitter increases the input impedance and is effective with an inductive source. The input impedance increases with frequency because of the frequency-selective negative-feedback network. The impedance of a magnetic pickup will increase with frequency but is below that of the preamplifier input, which can easily be made to accommodate most magnetic pickup impedances.

In Fig. 6-41, the feedback network is frequency-sensitive to accomplish RIAA equalization. The 500-Hz breakpoint is the frequency where the reactance of $C_3$ is equal to $R_4$, and $R_5$ in parallel with the total circuit shunting resistance is equal to the reactance of $C_3$ at 50 Hz. The reactance of $C_2$ is then made equal to $R_4$ at 2122 Hz to accomplish the treble cut.

Partial loading of the magnetic phonograph cartridge with an 82-kΩ resistor gives a flat amplifier output response. The output emitter follower stage isolates the feedback network from preamplifier output loading. The output should not be loaded with less than 3 kΩ and preferably with about 10 kΩ or more to maintain the dynamic range. The S/N ratio is optimized by adding $R_7$ and thereby transferring voltage amplification from the second stage to the input stage, since $R_7$ decreases the loading on the collector of $Q_1$. The input stage is operated at a bias current of 350 μA for low-noise operation, since the 2N2925 planar silicon transistor has noise characteristics similar to those shown in Fig. 6-32.

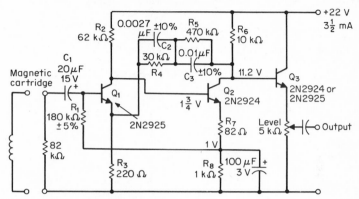

**Fig. 6-41** Phonograph preamplifier for magnetic cartridges. (*From D. V. Jones and R. F. Shea, Transistor Audio Amplifiers, Wiley, New York, 1968, p. 128.*)

## 6.7 CLASS A POWER AMPLIFIERS[11]

### Operating Characteristics and Load Lines

Class A operation of a transistor amplifier indicates that the collector current is not cut off for any portion of the cycle. For power stages the operating point is set at some midlocation on the static characteristics, and the signal excursion drives the output current uniformly above and below this point. We can see from Fig. 6-9 that during the portion of the cycle when the transistor is driven to point A (on the load line near saturation), most of the supply voltage appears across the load resistance, while at point B most of the supply voltage is across the transistor. The device dissipation is least when the amplifier is driven to maximum output, and greatest dissipation occurs with no signal input. With the operating point selected at the midpoint of the load line, it would be possible to deliver half of the dc input power to the load, and half would be dissipated by the active device. Thus the maximum efficiency of a transformer coupled class A stage (see Fig. 6-42) occurs at maximum output power and cannot exceed 50%.

$$P_{o,max} = \frac{V_{cc}^2}{2R_l'} \qquad (6\text{-}18)$$

Actually, this power output and 50% efficiency cannot be achieved in practice because of transistor saturation voltage at peak current and because of the dc voltage drop across the primary of the transformer and the ac coupling loss between the primary and secondary windings. Since the power supplied from the battery ($P_{dc}$) equals the power dissipated in the transistor ($P_T$) plus the power output ($P_o$), then

$$P_T = P_{dc} - P_o \qquad (6\text{-}19)$$

The power dissipated by the transistor with no signal input is the product of the quiescent voltage and current, and this is twice the amount that would occur at full signal with maximum efficiency of 50%. The efficiency is defined as $P_o/P_{dc}$. The dc resistance of the transformer primary is usually quite small, which makes the dc load line almost vertical, as in Fig. 6-42. The intersection of the ac and dc load lines is the quiescent operating point ($V_{CQ}$, $I_{CQ}$). The power output is

$$P_o = V_{rms} \times I_{rms} = \frac{V_{cm}I_{cm}}{2}$$

$$= \frac{I_{cm}^2 R_l'}{2} = \frac{V_{cm}^2}{2R_l'} \qquad (6\text{-}20)$$

where $V_{cm}$ and $I_{cm}$ are peak values of voltage and current sine wave forms occurring in both the load $R_l'$ and in the transistor.

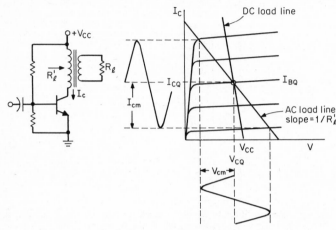

**Fig. 6-42**   Transformer-coupled load, common-emitter stage.

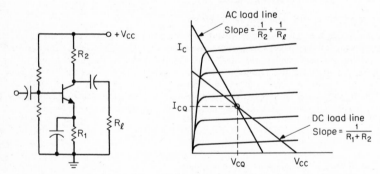

**Fig. 6-43**   *RC*-coupled load, common-emitter stage.

When the transistor is *RC*-coupled to the load as in Fig. 6-43, both ac and dc load lines are involved, as with transformer coupling. The power output is

$$P_o = \frac{V_{cm}I_{cm}}{2} \times \frac{R_2}{R_2 + R_l} \qquad (6\text{-}21)$$

Figure 6-44 shows a class A transistor amplifier directly coupled to the load. There is only one load line since the ac and dc resistance of the collector load are the same. The power output is

$$P_o = \frac{V_{cm}I_{cm}}{2} \qquad (6\text{-}22)$$

The maximum efficiency (25%) with an ideal transistor is one-half of what is possible for transformer coupling since the voltage across the dc load is equal to that across the transistor. Line *A* of Fig. 6-44 represents the saturation resistance of a transistor, and at a current $I_{C1}$ we have the saturation voltage $V_{CE,sat}$. The $V_{CE,sat}$ usually given on transistor specification sheets represents a transistor driven hard into saturation near line $A$. This is the point of interest when the transistor is used as a switch, but for linear amplification the signal swing must be limited so that the excursions on the load line do not exceed line $B$ if serious distortion is to be avoided. Depending on the transistor type, line $B$ represents a $V_{CE}$ that is two or three times the $V_{CE,sat}$ of the specification sheet. The low-current, high-voltage end of the load line is limited by the transistor voltage rating.

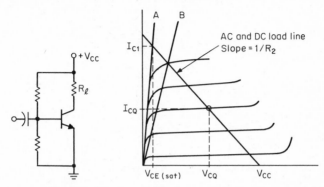

**Fig. 6-44**   Direct-coupled load, common-emitter stage.

The load line is an ellipse when the load has a reactive component. The load line in Fig. 6-42 assumes an ideal transformer, but in actual practice there would be some leakage inductance making the load line slightly elliptical. A dangerous condition may exist for the transistor if the load is removed from the secondary of the transformer. The impedance presented by the primary changes from nearly a resistive characteristic to that corresponding to a high inductance. In Fig. 6-45 we see that the signal swing, with the secondary of the transformer loaded, is from $A$ to $B$ on the resistive load line. If the load is removed from the transformer secondary, the transistor collector, which represents a current source, forces current through the high inductance of the primary. The load line is now an ellipse (distorted by saturation at $C$) with the slope of the major axis determined by the primary impedance of the transformer when the secondary is open. The operating region now extends outside the safe operating region for the transistor. Transistor failure may occur owing to excessive power dissipation or second breakdown.

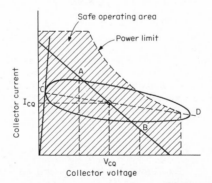

**Fig. 6-45**   Resistive vs. inductive ac load-line operation of common-emitter stage.

### 6.7a   CALCULATING THE POWER GAIN OF A CLASS A AUDIO OUTPUT STAGE

**problem 6.5**   Calculate the power gain (PG) of the TV audio output stage in Fig. 6-46. The driver stage is a dc-coupled emitter follower to provide both a stiff bias and stiff signal source for the output stage. It also provides impedance transformation, and so the input impedance will be high enough not to load the ratio detector. A 100-mV signal from the detector will drive this amplifier to full output of ¾ W. Assume $h_{FE} = 40$ for the D40N3.

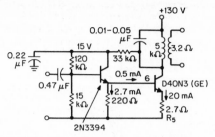

**Fig. 6-46**   TV audio amplifier, ¾ W output.

**theory**   The capacitor across the primary of the transformer is selected to give a high-frequency rolloff to balance the low-frequency rolloff due to the output transformer. The circuit biasing takes advantage of the dc-voltage drop in the transformer primary winding. Thus both stages utilize feedback bias stabilization as shown in Fig. 6-17. In a class A stage above approximately 20 mW the load resistance is very small compared with the transistor output impedance. Therefore the large signal current amplification is essentially equal to the dc ratio $h_{FE}$; it is used in the design of power stages in place of the small signal parameter $h_{fe}$. The power gain for a class A stage is

$$PG \approx h_{FE}^2 \times \frac{R_l}{R_{in}} \tag{6-23}$$

and the input resistance for the stage is

$$R_{in} = \left(\frac{26}{I_{E(mA)}} + R_s\right) h_{FE}$$

**solution**

$$R_{in} = \left(\frac{26}{20} + 2.7\right) 40 = 160 \ \Omega$$

$$PG = (40)^2 \times \frac{5000}{160} = 50,000$$

Therefore PG ≈ 47 dB

## 6.8   CLASS B AND AB PUSH-PULL AMPLIFIERS[12]

### Basic Class B and Class AB Operation

Class B operation of a transistor amplifier indicates that the collector current (Fig. 6-47) is cut off for one-half cycle (180°). Two transistors operating class B push-pull alternate their conduction and cutoff periods. During periods of low or zero signal input the power supply drain and transistor dissipation are low. In the class B operation the transistor is biased to cutoff, and the input signal drives the conducting transistor collector voltage to a low value approaching the saturation voltage. When this transistor is cut off, the voltage rises to approximately twice the quiescent voltage ($2V_{CQ}$).

The average current shown in Fig. 6-47 is

$$I_{C,av} = \frac{I_{cm}}{\pi} \tag{6-24}$$

The power supplied to each transistor is this current mutiplied by the supply $V_{CC}$. Thus, for a push-pull pair of transistors

$$P_{dc} = 2V_{CC} \times \frac{I_{cm}}{\pi} \tag{6-25}$$

The peak voltage swing (neglecting saturation voltage) is equal to the supply voltage $V_{CC}$, and thus the output power is

$$P_o = \frac{1}{2}V_{CC}I_{cm} \tag{6-26}$$

The maximum efficiency is, therefore, the total power output, as given by Eq. (6-26), divided by the total power supplied, as given by Eq. (6-25). This gives the efficiency at full swing as $\pi/4 = 78.5\%$. The transistor power dissipation peaks when the sine wave output signal swing is 40% of maximum, and the operating efficiency at this point is 50%.[13]

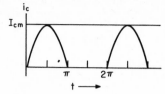

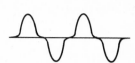

**Fig. 6-47** Class B output current.  **Fig. 6-48** Crossover distortion.

Transistors are not usually operated in the true class B mode because of crossover distortion (see Fig. 6-48) at low power output levels. This type of distortion results from nonconduction and nonlinearities in the transistor transfer characteristics at low (<0.6 V) base-to-emitter drive voltage. Therefore most push-pull stages operate with a $V_{BE}$ bias near the conduction threshold, which would be between 0.6 and 0.7 V. Thus the operation is near class B but with a small forward bias that results in class A operation for peak signal swings that are less than the bias voltage. This mode of operation is termed class AB, since the operating bias is between class A and class B. The quiescent current level can still be low enough for good efficiency.

### Linearity Considerations in Class B and AB Operation

Curves of output current vs. input current and also driving voltage have been constructed in Fig. 6-49. The nonalignment of the upper and lower halves of the composite curve is moderate for the $I_C$ vs. $I_B$ curve but severe for the $I_C$ vs. $V_G$ curve. The voltage drive has been more widely used because of its circuit simplicity. Since crossover distortion is produced by nonalignment of the two halves of the composite push-pull characteristic, the cure should be some means of realigning these two halves. This may be accomplished by applying the correct amount of forward bias so that the linear portions of the opposite halves are extensions of each other. Figure 6-50 illustrates this approach, and these curves are the same as the dashed curves of Fig. 6-49 displaced by 0.7 V. Two common methods employed to obtain this forward bias are illustrated in Fig. 6-51. In Fig. 6-51a a portion of the battery voltage is applied to the bases through the divider $R_1$, $R_2$. The voltage drop

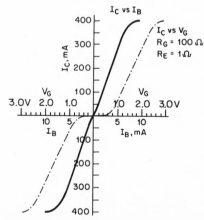

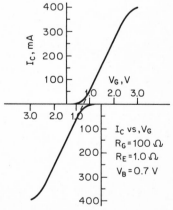

**Fig. 6-49** Output current vs. input current and voltage for a silicon transistor in push-pull class B operation. (*From D. V. Jones and R. F. Shea, Transistor Audio Amplifiers, Wiley, New York, 1968, p. 159.*)

**Fig. 6-50** Application of forward bias to eliminate crossover distortion, class AB. (*From D. V. Jones and R. F. Shea, Transistor Audio Amplifiers, Wiley, New York, 1968, p. 160.*)

across $R_1$ is just enough to eliminate the crossover misalignment. The variation of $V_{BE}$ with temperature is between 1.3 and 2.3 mV/°C; hence, it can produce quite a shift if a wide temperature range is to be accommodated. Figure 6-51$b$ illustrates a method which decreases the forward bias on the transistors with increasing temperature to compensate for the shift in emitter base voltage. Here the forward drop of the diode $D_1$, which replaces $R_1$, has essentially the same temperature variation as the emitter base voltage of the transistors. The distortion in push-pull class AB amplifiers is a function of the symmetry between the halves of the circuit, of base emitter bias, of power output level, and also of source and output impedances.

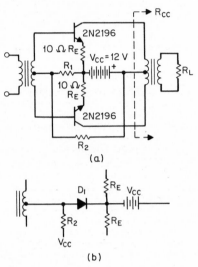

**Fig. 6-51**    Class AB biasing arrangements: ($a$) resistor biasing; ($b$) diode biasing.

### 6.8a    CALCULATING IMPEDANCE AND POWER GAIN FOR CLASS AB OUTPUT STAGE

**problem 6.6**    Calculate the collector-to-collector impedance and the power gain for the trans-former-coupled, push-pull ½-W output stage in Fig. 6-51. Assume the output transformer is 80% efficient, $V_{cm} = 10.5$ V, and $h_{FE} = 50$ for the 2N2196 transistor.

**theory**    The power output for a push-pull stage can be expressed as $P_o = \frac{1}{2}V_{cm}I_{cm}$. Then the collector load resistance is $R_l = V_{cm}/I_{cm}$, and the collector-to-collector impedance is four times the load resistance per collector, with a transformer-coupled load as in Fig. 6-51$a$. The load resistance multiplier results from transformer action where the impedance ratio varies as the square of the turns ratio. The power output for a push-pull stage with a transformer-coupled load is

$$P_o = \frac{2V_{cm}^2}{R_{c-c}} \tag{6-27}$$

where $V_{cm}$ is the peak collector voltage swing, and $R_{c-c}$ is the collector-to-collector load resistance. Thus for a specified output power and collector voltage the collector-to-collector load resistance can be determined. The collector voltage must exceed $V_{cm}$ by at least twice the saturation voltage plus any voltage drop across the stabilizing resistance in series with the emitter. The power gain (PG) is given by

$$PG = \frac{P_{out}}{P_{in}} = \frac{I_o^2 R_l}{I_{in}^2 R_{in}}$$

Since $I_o/I_{in}$ is approximately equal to $h_{FE}$ for a small load resistance, the power gain can be written as

$$PG = h_{FE}^2 \frac{R_{c-c}}{R_{b-b}} \tag{6-28}$$

where $R_{b-b}$ is the base-to-base input resistance and is four times the input to each transistor when a driver transformer is used.

**solution**    With 80% transformer efficiency, the transformer primary power level = 0.5 W/0.8 = 0.625 W.

$$R_{c-c} = \frac{2(V_{cm})^2}{P_o} = \frac{2(10.5)^2}{0.625} = 353 \ \Omega$$

Since the input to each transistor = $[(26/I_E) + R_e]h_{FE}$,

$$R_{b-b} = 4\left(\frac{26}{I_E} + R_E\right)h_{FE} = 4\left(\frac{26}{10} + 10\right)50 = 2520 \ \Omega$$

(Assuming $I_E$ averages ≈ 10 mA.)

$$PG = (50)^2\frac{353}{2520} = 350$$

The power gain to the load at the transformer secondary would be 350 × 0.8 = 280 or about 24½ dB.

## Single-Ended Output, Push-Pull Amplifier

The low-output impedance of transistor push-pull amplifiers permits direct or capacitive coupling to the speaker load without using an output transformer for impedance transformation. The elimination of the output transformer improves both the low frequency and the transient response. Low-output impedance also gives good speaker damping which is desirable for most speaker systems.

The two output coupling methods most commonly used are shown in Fig. 6-52. Both methods require two large electrolytic capacitors and have been referred to as series-connected output and also as single-ended output. The series-connected transistors will have equal collector currents, but they are dependent on base emitter biasing to give equal collector emitter voltages. This is the inverse of push-pull stages with transformer-coupled loads where the two collector voltages are equal, and the bias must be adjusted for the two dc collector currents to be equal.

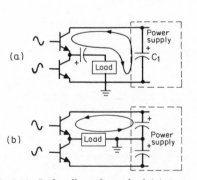

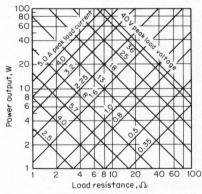

**Fig. 6-52**    Push-pull coupling to load: (a) capacitive coupling; (b) direct coupling.

**Fig. 6-53**    Peak load current and voltage vs. power output and load resistance.

The relationship of the peak load current and voltage to the load power has been expressed in Eq. (6-20). These relationships are quite useful in transformerless designs such as Fig. 6-52, since the peak load current is the peak transistor current, and the transistor voltage rating must be at least twice the peak load voltage. The relationship between peak load currents and voltage, power output, and load resistance is given in the design chart of Fig. 6-53. The required supply voltage $V_{cc}$ for a given maximum power output is given by

$$V_{cc} = 2V_{cm} + 6V_{CE,\text{sat}} + 2I_{cm}R_E \qquad (6\text{-}29)$$

where $V_{cm}$ is the peak load voltage at the given power output and $V_{CE,\text{sat}}$ is the saturation voltage at the peak load current $I_{cm}$. This expression allows for a collector emitter voltage at the peak current interval of the three times $V_{CE,\text{sat}}$ to avoid serious distortion. The last term in Eq. (6-29) represents voltage drop in the external emitter circuit.

### Transformerless Power Amplifiers

A transformerless circuit is shown in Fig. 6-54. This basic approach of complementary push-pull drive and capacitive coupling to the load was first presented by H. C. Lin[14] and has been optimized and promoted by General Electric Company[15] and others. When complementary (npn and pnp) driver transistors are used, they provide the necessary phase inversion to the series npn silicon power transistors with single-ended output. This approach is sometimes referred to as a *quasicomplementary amplifier.*

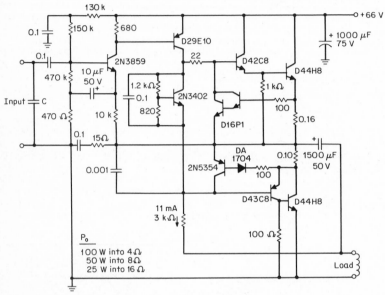

**Fig. 6-54** Quasicomplementary power amplifier. *(From Application Note 90.100, General Electric Company, Semiconductor Products Dept., Syracuse, N.Y.)*

A complementary push-pull pair of transistors may be driven with the same signal phase since the inherent characteristics of an npn transistor will cause it to conduct when its base is driven positive while the pnp transistor will be nonconductive. Then, when the signal phase reverses, the negative drive turns on the pnp transistor and turns off the npn transistor.

Figure 6-54 is a direct-coupled power amplifier with excellent low-frequency response, which also has the advantage of dc feedback for bias stabilization of all stages. This feedback system stabilizes the voltage division across the class AB push-pull output transistors. The complementary push-pull drive transistors also operate class AB and increase the current amplification from the collector of the low-power class A driver transistor (D29E10). The input stage (predriver) has a bootstrapped bias network (similar to Fig. 6-18) which gives an input impedance of 125 k$\Omega$. This impedance decreases at high frequencies with $C_1$ selected in conjunction with the drive source resistance to give an RC time constant of approximately $4 \times 10^{-6}$ s. The amplifier requires 1-V input signal for rated output with about 43-dB negative feedback. The closed-loop gain of the amplifier is determined by the resistor ratio, 10 k$\Omega$/470 = 21, or approximately 27 dB. Therefore the open loop gain is 70 dB. The amplifier is designed to limit the peak load current to about 8 A with the Darlington (D16P1) and the 2N5354 transistors sensing positive and negative excursions of load current above this level and shunting the drive from the quasicomplementary output. Peak output current required for 100 W into a 4-$\Omega$ load is 7 A.

The aluminum strap on the 2N3402 can be attached to the heat sink used for one of the output transistors (D44H8). This provides thermal feedback for bias stabilization of the output transistors and sets the quiescent current for the amplifier at about 20 mA. The frequency response with 12.5 W into an 8-$\Omega$ load is $-3$ dB at 30 Hz and 55 kHz.

## 6.9  MONOLITHIC INTEGRATED AUDIO AMPLIFIERS[16]

### Basic Monolithic Components

A monolithic integrated circuit (IC) is contained in one structure (silicon pellet) and thus differs from other forms of integrated circuits made up of discrete components on a common substrate (i.e., thick film, thin film, or hybrids). The components available to the designer of monolithic audio amplifiers are bipolar transistors, field-effect transistors, diodes, avalanche diodes, resistors, and capacitors. In addition to voltage and current feedback, an IC audio amplifier design may use thermal feedback to stabilize the bias. The very small physical size of an IC inherently provides close thermal coupling between circuit elements that cannot be matched using discrete transistors.

The all-diffused IC that is most commonly used at this time consists of a p-type silicon substrate into which all the circuit elements are diffused with processes that are related to transistor fabrication. Interconnections between elements are made with an aluminum metalization pattern on the top surface of the IC. Where isolation is required between circuit elements, the high impedance of a reverse-biased p-n junction (which completely surrounds each element or group of elements) provides an electrical barrier. Figure 6-55 shows the monolithic components that can be made with the standard all-diffused process. The structure on the left is an npn silicon transistor in which the emitter current would flow vertically down across the narrow base region and be collcted at the n region directly below as in a discrete transistor, for this would be the bottom of the pellet. Since all connections on ICs must be made on the top surface, the collector current flows laterally through the low-impedance n buried layer until it reaches a point of shortest path in the vertical direction up to the collector contact. This means that the current has to traverse the higher resistivity collector n region twice compared with once in a discrete transistor. A result is higher saturation resistance which must be considered in the IC design of power output stages. The n buried region not only minimizes the saturation resistance of the IC transistor but also isolates the p substrate to prevent vertical pnp action to the substrate. In fact, a substrate pnp has no buried layer (see Fig. 6-55). The substrate pnp has vertical current flow like a conventional transistor. It has lower $h_{FE}$ and higher base resistance than the npn. The main limitation of the substrate pnp is that it can be used only in a circuit where the collector is connected to the most negative common point. The p substrate must always be connected to the most negative potential in order to provide reverse bias for the isolating junctions.

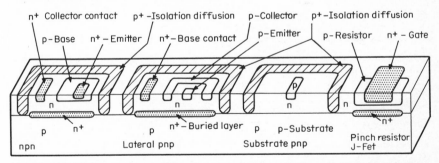

**Fig. 6-55**  Monolithic components diffused in a single structure. (*From R. K. Field, "The Tiny, Exploding World of Linear Micro-circuits," Electronics Design, **15** (15):49–66, July 19, 1967.*)

The lateral pnp shown in Fig. 6-55 gets its name from the lateral flow of transistor current from emitter to collector. The $h_{FE}$ is very low because of the wider base region which is determined by the physical layout of the collector and emitter areas. A lateral pnp is often used in conjunction with an npn transistor to increase the current amplification with a connection as shown in Fig. 6-56. This composite pnp monolithic transistor connection can provide the same circuit function and performance as a discrete pnp silicon transistor.

Most monolithic resistors are formed during the p-type (transistor base) diffusion cycle with a resistance value dependent on the length and width. Practical resistor values range from 30 to about 30,000 Ω. Although the resistors' absolute values may have considerable spread, the spread in resistor ratio can be held to ±5% or better. Thus monolithic circuits are designed to be more dependent on resistor ratios than on the absolute value for uniformity of circuit performance. The junction FET on the right in Fig. 6-55 can also be used to obtain a monolithic resistance up to 300 kΩ by pinching off the p-channel resistance.

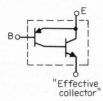

Fig. 6-56  Composite pnp transistor.

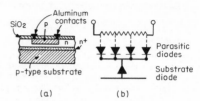

Fig. 6-57  Monolithic resistor: (a) p-type diffused resistor; (b) equivalent circuit.

Large-value capacitors are not practical in a monolithic structure because of the large area of silicon required. The collector base junction capacitor may be used where a low-value capacitor is needed for high-frequency stabilization. Regular diodes may be formed in a monolithic structure, but the more usual practice is to use a collector base shorted transistor connection which has a very low dynamic resistance for the area occupied. The reverse breakdown characteristic of a transistor emitter base junction is used as an avalanche (zener) diode for power supply regulation and also for coupling between stages to provide a shift in dc level.

Monolithic circuit components do have parasitic elements associated with them, but they usually have little effect on circuit performance at audio frequencies. The p-type resistor is diffused into n-type silicon and thus has parasitic diodes associated with it as shown in Fig. 6-57. These distributed parasitic diodes are blocked from conduction by the reverse-biased substrate isolation diode. This shows why it is important for the p-type substrate to be connected to the most negative potential.

## 6.10  REFERENCES

1. D. V. Jones and R. F. Shea, *Transistor Audio Amplifiers*, Wiley, New York, 1968, pp. 59, 61, 63, 64, 66, and 67.
2. Texas Instruments, Inc., FET Fact File.
3. Jones and Shea, op. cit., pp. 4–9 and 38–56.
4. General Electric Company, *Transistor Manual*, 4th ed., Syracuse, N.Y., pp. 7–9, 37–40, and 44–46.
5. Jones and Shea, op. cit., pp. 71–76.
6. Crystalonics, Inc., *Silicon Field-Effect Transistors*, Cambridge, Mass., p. 6.
7. Jones and Shea, op. cit., pp. 93–96, 100–103, 111–113, 122, 123, 127, and 128.
8. A. E. Sanderson and R. G. Fulks, "A Simplified Noise Theory and Its Application to the Design of Low-Noise Amplifiers," *IRE Trans Audio*, 9:4, July–August, 1961.
9. F. Langford-Smith, *Radiotron Designers' Handbook*, 4th ed., RCA Corp., Sydney, Australia, 1960, pp. 493, 637, 775, and 826.
10. H. Burstein, "RC Equalization Curves," *Electronics World*, **67**, April 1962.
11. Jones and Shea, op. cit., pp. 137–143.
12. Ibid., pp. 155–157, 163–170, and 176.
13. Motorola Semiconductor Products Division, "Motorola Power," Phoenix, Ariz.
14. H. C. Lin, "Quasi-Complementary Transistor Amplifier," *Electronics*, September, 1956.
15. D. V. Jones, "All-Transistor Stereo Tape System," *Electronics World*, July 1959.
16. Jones and Shea, op. cit., pp. 235–238.

Chapter **7**

# Tuned Amplifiers

## JOHN BAKUM

**Associate Professor, Electrical Engineering Technology Department,
Middlesex County College, Edison, N.J.**

## 7.1  INTRODUCTION

In many electronic applications it is necessary to select and amplify a relatively narrow band of frequencies. For example, narrow-band amplifiers are used extensively in communication systems, where information is contained in sideband frequencies on one or both sides of a center or carrier frequency. In order not to lose or distort information, and to prevent interference from adjacent communication channels, precise requirements are imposed on the frequency response characteristics of the system's amplifiers.

In general, narrow-band amplifiers are composed of two networks: (1) an amplifying element, such as a transistor with its associated circuit components; and, (2) a network that determines the proper frequency response. Although resistor-capacitor (*RC*) networks may be used to select the center frequency and shape the amplifier's frequency response characteristics, the majority of narrow-band amplifiers use *RLC* tuned circuits to perform this function. Narrow-band amplifiers are frequently termed *tuned amplifiers*.

The important performance characteristics of tuned amplifiers include: the bandwidth, the center frequency, the gain, and the noise figure. In this chapter a number of representative tuned amplifier configurations are analyzed after a review of the important characteristics of *RLC* tuned circuits.

## 7.2  CIRCUIT Q, TUNED CIRCUITS, AND BANDWIDTH

In general terms, the *quality factor Q* of a circuit (or circuit element) is a measure of the circuit's ability to store energy during a cycle as opposed to the energy dissipated over the cycle. Equation (7-1) defines $Q$.

$$Q = \frac{2\pi \text{ (peak energy stored during one-cycle)}}{\text{energy dissipated over one cycle}} \qquad (7\text{-}1)$$

As the definition of Eq. (7-1) implies, $Q$ has significance only in terms of alternating voltages and currents (usually sinusoidal).

Tuned circuits may be divided into two broad classifications: (1) series resonant circuits, and (2) parallel resonant circuits. Figure 7-1 illustrates a simple series resonant circuit. The resistor $R_s$ shown in Fig. 7-1 represents the losses associated with the reactive elements at the frequency of interest, and the inductance and capacitance may then, for analytical purposes, be considered ideal or lossless. The voltage source (also considered ideal) is sinusoidal, and the indicated voltage and current variables are phasor quantities.

Equation (7-2) gives the resonant frequency:

$$f_o = \frac{1}{2\pi\sqrt{LC}} \tag{7-2}$$

where $L$ = inductance, H
$C$ = capacitance, F
$f_o$ = resonant frequency, Hz

At the resonant frequency the peak energy stored in the circuit of Fig. 7-1 is given by Eq. (7-3):

$$W_p = \frac{L(I_{max})^2}{2} \tag{7-3}$$

where $I_{max}$ is the peak value of the sinusoidal current in amperes and $W_p$ the peak energy stored in joules.

The losses associated with the reactive elements are represented by the resistor $R_s$, and Eq. (7-4) gives the energy dissipated over one cycle,

$$W_{dis} = \frac{(I_{max})^2 R_s}{2f_o} \tag{7-4}$$

where $W_{dis}$ is in joules.

The $Q$ of the series resonant circuit may be obtained by substituting Eqs. (7-3) and (7-4) into Eq. (7-1), which yields Eq. (7-5).

$$Q = \frac{2\pi f_o L}{R_s} \tag{7-5}$$

Figure 7-2a illustrates a simple parallel resonant $RLC$ circuit. Once again, $R_s$ represents the losses associated with the reactive elements.

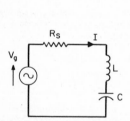

**Fig. 7-1**   Series resonant circuit.

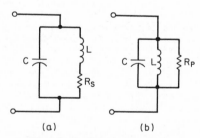

(a)          (b)

**Fig. 7-2**   Parallel resonant circuits.

Note that as far as the circuit elements are concerned, the circuit of Fig. 7-2a is essentially the same as that of Fig. 7-1; consequently, Eq. 7-5 still applies. It is generally more convenient, however, to represent the circuit losses at resonance by an equivalent parallel resistor $R_p$, as indicated in Fig. 7-2b. An expression for the $Q$ of the circuit at resonance in terms of $R_p$ may be derived by equating the terminal impedance expressions $Z(j\omega)$ at resonance for both networks of Fig. 7-2. Equation (7-6) results.

$$Q = \frac{R_p}{2\pi f_o L} \tag{7-6}$$

Equations (7-5) and (7-6) indicated that the $Q$ of a lossless circuit would be infinite; that is, $R_s = 0$ or $R_p = \infty$.

Perhaps the most important characteristic of tuned circuits is their ability to select a relatively narrow band of frequencies, either for amplification or in some cases rejection. Depending upon the particular application and the nature of the amplifying element, tuned amplifiers may have a tuned circuit at the input, output, or both. In order to illustrate the frequency-response characteristics, and the relationship between the $Q$ of the tuned circuit and the bandwidth of the amplifier, consider the simplified representation of a tuned amplifier of Fig. 7-3a. In Fig. 7-3a the load of the amplifier is a parallel resonant circuit; the tuned-circuit losses at resonance, as well as any external resistance, are represented by the parallel resistor $R_p$. Figure 7-3b is a plot of the amplifier's frequency response.

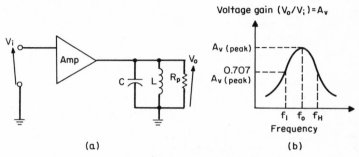

(a)    (b)

**Fig. 7-3**    Simplified tuned amplifier circuit.

Generally the voltage gain of an amplifier is proportional to the magnitude of the load impedance. At resonance the impedance of the parallel tuned circuit of Fig. 7-3a is at a maximum and equal to $R_p$. On either side of the resonant frequency the magnitude of the impedance falls off owing to the effect of one or the other reactive element. Thus, as shown in Fig. 7-3b, maximum voltage gain occurs at the resonant frequency of the tuned circuit.

The bandwidth, in hertz, of the amplifier of Fig. 7-3 is defined by $f_h$ and $f_l$ in accordance with Eq. (7-7).

$$\text{Bandwidth (BW)} = f_h - f_l \qquad (7\text{-}7)$$

As shown in Fig. 7-3b, $f_h$ and $f_l$ are the high and low frequencies, respectively, at which the voltage gain of the amplifier is 0.707 of the peak voltage gain. If the voltage gain is expressed in decibels, $f_h$ and $f_l$ are termed the 3-dB frequencies, i.e., those at which the gain is down 3 dB from the peak value. Strictly speaking, it is incorrect to refer to the 3-dB frequencies of the voltage gain, since decibels are defined in terms of the power gain. Only in the special case where the input and load resistances, and the input and output power factors, are equal is the power gain in decibels equal to the voltage (current) gain in decibels. However, the common practice is to describe the voltage (current) gain and the corresponding bandwidth in terms of decibels regardless of the resistances and power factors.

The bandwidth of the tuned circuit is a function of the resonant frequency $f_o$ and the $Q$ of the circuit. Equation 7-8 describes the relationship.

$$\text{Bandwidth (BW)} = \frac{f_o}{Q} \qquad (7\text{-}8)$$

Note that the higher the $Q$, the narrower the bandwidth, or in other words the greater the selectivity. Selectivity is a measure of a circuit's (or amplifier's) ability to choose a specific band of frequencies while rejecting all other frequencies.

Although Eqs. (7-5) and (7-6) were derived in terms of the resistance and the inductance of a tuned circuit, equivalent expressions may be derived in terms of the resistance and capacitance. Furthermore, the $Q$ of an individual circuit element at a specified frequency may be determined. In practice the capacitor in a tuned circuit is usually considered lossless compared with the inductor (i.e., capacitor $Q$'s are very large), and the $Q$ of the tuned circuit is approximately equal to that of the inductor's, which range from 30 to 200. The $Q$ of an inductor may be obtained from the manufacturer's specifications, or a $Q$-

meter may be used to measure the $Q$ at a particular frequency. Table 7-1 summarizes some of the important formulas involving $Q$.

TABLE 7-1    Basic Formulas Involving Q

| Series resonant | Parallel resonant |
|---|---|
| $f = \dfrac{1}{2\pi\sqrt{LC}}$ | $f_o = \dfrac{1}{2\pi\sqrt{LC}}$ |
| $Q = \dfrac{2\pi f_o L}{R_s} = \dfrac{1}{2\pi f_o C R_s}$ | $Q = \dfrac{R_p}{2\pi f_o L} = 2\pi f_o C R_p$ |
| $Q = \dfrac{1}{R_s}\sqrt{\dfrac{L}{C}}$ | $Q = R_p\sqrt{\dfrac{C}{L}}$ |
| $BW = \dfrac{f_o}{Q}$ | $BW = \dfrac{f_o}{Q}$ |

## 7.3  CALCULATING THE CAPACITANCE NECESSARY FOR A SPECIFIC RESONANT FREQUENCY AND DETERMINING THE BANDWIDTH

**problem 7.1**    A 640-$\mu$H inductor is to be used in a parallel tuned-circuit application where the center frequency is to be 455 kHz. The $Q$ of the inductor, measured at 455 kHz, is 70. Determine the capacitance of the required parallel capacitor and the resulting bandwidth.

**theory**    Equation (7-2) may be solved for $C$:

$$f_o = \frac{1}{2\pi\sqrt{LC}} \tag{7-2}$$

Therefore

$$C = \frac{1}{4\pi^2 f_o^2 L}$$

Equation (7-8) yields the resulting bandwidth:

$$\text{Bandwidth (BW)} = \frac{f_o}{Q} \tag{7-8}$$

**solution**    Substituting into the expression for $C$ yields:

$$C = \frac{1}{\begin{array}{c}4 \times (3.14)^2 \times (4.55 \times 10^5)^2 \times 6.40 \times 10^{-4}\end{array}}$$
$$= 191 \text{ pF}$$

And the bandwidth is:

$$BW = \frac{455 \text{ kHz}}{70} = 6.5 \text{ kHz}$$

## 7.4 CALCULATING THE BANDWIDTH OF A TUNED CIRCUIT WITH AN EXTERNAL LOAD RESISTOR

**problem 7.2**   A 240-k$\Omega$ load resistor is connected in parallel with a 640-$\mu$H inductor and a 191-pF capacitor. The resonant frequency of the tuned circuit is approximately 455 kHz. Assume that the capacitor is lossless and that the $Q$ of the inductor at 455 kHz is 70. Determine the bandwidth of the tuned circuit.

**theory**   The $Q$ of a tuned circuit without an external load resistor is termed the *unloaded Q, $Q_U$.* In this problem, since the capacitor is assumed lossless, the unloaded $Q$ of the parallel tuned circuit at 455 kHz is 70. Equation (7-6) may be solved for $R_p$ in terms of $Q_U$:

$$Q_U = \frac{R_p}{2\pi f_o L} \tag{7-6}$$

and

$$R_p = Q_U 2\pi f_o L$$

$R_p$ represents the losses of the inductor at the resonant frequency. $R_L$, the external load resistor, is in parallel with $R_p$. The equivalent resistance of $R_L$ in parallel with $R_p$ may be used to calculate the *loaded Q, $Q_L$,* of the tuned circuit in accordance with Eq. (7-9).

$$Q_L = \frac{R_p \| R_L}{2\pi f_o L} \tag{7-9}$$

Once the loaded $Q$ has been calculated, Eq. (7-8) yields the bandwidth of the tuned circuit with an external load resistor.

$$\text{BW} = \frac{f_o}{Q_L} \tag{7-8}$$

**solution**   Calculating $R_p$ yields:

$$R_p = 70 \times 2 \times 3.14 \times 4.55 \times 10^5 \times 640 \times 10^{-4}$$
$$= 128 \text{ k}\Omega$$

The equivalent resistance of $R_L$ in parallel with $R_p$ is:

$$R_{eq} = \frac{128 \text{ k}\Omega \times 240 \text{ k}\Omega}{128 \text{ k}\Omega + 240 \text{ k}\Omega} = 83.5 \text{ k}\Omega$$

Calculating the loaded $Q$ yields:

$$Q_L = \frac{83.5 \text{ k}\Omega}{2 \times 3.14 \times 4.55 \times 10^5 \times 6.40 \times 10^{-4}}$$
$$= 45.6$$

And the resulting bandwidth is:

$$\text{BW} = \frac{455 \text{ kHz}}{45.6} = 9.98 \text{ kHz}$$

## 7.5 ANALYSIS OF THE FREQUENCY RESPONSE OF A SINGLE-TUNED TRANSISTOR AMPLIFIER

**problem 7.3**   Figure 7-4 illustrates a single-tuned transistor amplifier stage. The output resistance and capacitance of the transistor are: $R_o = 20$ k$\Omega$ and $C_o = 20$ pF. The tuned circuit values are: $(N_1 + N_2)/N_1 = 4$, $(N_1 + N_2)/N_3 = 12$, $C_T = 300$ pF, $L_p = 1.2$ mH, and $Q$ (transformer primary) $= 100$. For the succeeding stage: $C_{12} = 10$ pF and $R_{12} = 625$ $\Omega$. Determine the amplifier's frequency-response characteristics ($f_o$ and bandwidth).

**theory**   Tuned amplifiers frequently employ transformers to couple the signal from one stage to the next. The inductance of the transformer winding and a tuning capacitor form the tuned circuit required for frequency selection. If only one side of the transformer is tuned, the amplifier is referred to as *single-tuned.* If both the primary and secondary of the transformer are tuned by means of tuning capacitors, the amplifier is termed *doubled-tuned.*

Bipolar junction transistor (BJT) tuned amplifiers are usually single-tuned with the tuned circuit at the output of the transistor. The reason for this is that the input resistance of a BJT is relatively low and a tuned circuit at the input would necessarily have a poor loaded $Q$ and correspondingly poor selectivity. The output resistance of a BJT, on the other hand, is fairly large, and a tuned circuit at the output may attain sufficient loaded $Q$ to achieve the desired selectivity.

Note in the circuit of Fig. 7-4 that there are *two* transformers in the output circuit of $Q_1$: a coupling transformer with $N_1 + N_2$ as the number of primary turns and $N_3$ the number of secondary turns, and an autotransformer with $N_1 + N_2$ primary turns and $N_1$ secondary turns. The autotransformer, formed by tapping the primary winding ($N_1 + N_2$), reduces the effect of the output resistance and capacitance of $Q_1$ upon the tuned circuit, since the reflected values of $R_o$ and $C_o$ in parallel with the tuned circuit are the actual values modified by the square of the autotransformer turns ratio. Equations (7-10) and (7-11) express the turns ratios of the coupling transformer and the autotransformer, respectively.

$$a_1 = \frac{N_1 + N_2}{N_3} \tag{7-10}$$

$$a_2 = \frac{N_1 + N_2}{N_1} \tag{7-11}$$

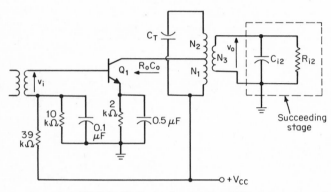

**Fig. 7-4**   Single-tuned transistor amplifier.

**solution**   Figure 7-5 is the equivalent circuit of the tuned circuit; the prime element values are the component values modified by the appropriate transformer turns ratio. It is usually more convenient for analysis purposes to refer all circuit element values to the primary winding of the coupling transformer as shown in Fig. 7-5.

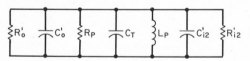

**Fig. 7-5**   Equivalent circuit of the tuned circuit referred to the primary side of the coupling transformer.

The input capacitance of the next stage referred to the primary, $C'_{i2}$, may be calculated using $a_1$, the coupling transformer turns ratio:

$$C'_{i2} = \left(\frac{1}{a_1}\right)^2 C_{i2} = \frac{1}{12^2} \times 10 \text{ pF} = 0.069 \text{ pF}$$

Similarly, using the autotransformer turns ratio $a_2$:

$$C'_o = \left(\frac{1}{a_2}\right)^2 C_o = \frac{1}{4^2} \times 20 \text{ pF} = 1.25 \text{ pF}$$

For $R'_{i2}$ and $R'_o$:

$$R'_{i2} = a_1^2 R_{i2} = 12^2 \times 625 \text{ } \Omega = 90.0 \text{ k}\Omega$$
$$R'_o = a_2^2 R_0 = 4^2 \times 20 \text{ } \Omega = 320 \text{ k}\Omega$$

The total capacitance in parallel with $L_p$ is:

$$C_{\text{total}} = C'_0 + C_T + C'_{i2} \tag{7-12}$$

For this example:

$$C_{total} = 1.25 \, pF + 300 \, pF + 0.069 \, pF = 301 \, pF$$

Note that the transformers have minimized the effect of $C_0$ and $C_{i2}$ on the tuning of the stage. The center frequency, $f_o$, may be calculated using Eq. (7-2).

$$f_o = \frac{1}{2\pi\sqrt{LC}}$$

$$= \frac{1}{2\pi\sqrt{1.2 \times 10^{-3} \times 3.01 \times 10^{-10}}} = 265 \, kHz \qquad (7\text{-}2)$$

The value of $R_p$, which represents the losses associated with $L_p$, may be calculated from Eq. (7-6):

$$Q_U = \frac{R_p}{2\pi f_o L}$$

$$R_p = Q_U 2\pi f_o L = 100 \times 2 \times 3.14 \times 2.65 \times 10^5 \times 1.2 \times 10^{-3} \qquad (7\text{-}6)$$
$$= 200 \, k\Omega$$

The total resistance in parallel with $L_p$ is

$$R_{total} = R_p \| R'_{i2} \| R'_0 = 200 \, k\Omega \| 90 \, k\Omega \| 320 \, k\Omega$$
$$= 52.0 \, k\Omega$$

Thus the loaded $Q_L$ is

$$Q_L = \frac{R_{total}}{2\pi f_o L}$$

$$= \frac{52.0 \, k\Omega}{2 \times 3.14 \times 2.65 \times 10^5 \times 1.2 \times 10^{-3}} = 26.0$$

And the bandwidth from Eq. (7-8):

$$BW = \frac{f_o}{Q_L}$$

$$= \frac{265 \, kHz}{26.0} = 10.2 \, kHz \qquad (7\text{-}8)$$

## 7.6   CALCULATING THE VOLTAGE GAIN OF A SINGLE-TUNED TRANSISTOR AMPLIFIER

**problem 7.4**   The transistor of the amplifier of Fig. 7-4 has the following parameters: $h_{fe} = 50$, $h_{ie} = 1 \, k\Omega$, $h_{oe} = 50 \, \mu S$, and $f_T = 30 \, MHz$. The transformer turns ratios are: $(N_1 + N_2)/N_1 = 4$ and $(N_1 + N_2)/N_3 = 12$. From the problem of Sec. 7.5 the total resistance in parallel with the primary of the coupling transformer, $R_{total}$, is 52.0 k$\Omega$ at the resonant frequency of 265 kHz. Calculate the voltage gain of the amplifier at $f_o(A_v = v_o/v_i)$.

**theory**   At frequencies below the beta cutoff frequency of the transistor, the voltage gain of a common emitter amplifier is given, to a good approximation, by Eq. (7-13), where $v_{ce}$ and $v_{be}$ are the ac collector-to-emitter and base-to-emitter voltages, respectively.

$$A_v = \frac{v_{ce}}{v_{be}} = \frac{-h_{fe}R_L}{h_{ie}} \qquad (7\text{-}13)$$

Note in Fig. 7-4 that while $v_i = v_{be}$ (ac quantities), $v_o$ does not equal $v_{ce}$, and the effects of the transformer turns ratios must be accounted for in computing $A_v = v_o/v_i$.

The beta cutoff frequency $f_\beta$ is the frequency at which the magnitude of $h_{fe}$ is 0.707 of its low-frequency value:

$$f_\beta = \frac{f_T}{h_{fe}} \qquad (7\text{-}14)$$

where $f_T$ is the frequency at which the magnitude of $h_{fe}$ is unity.

**solution**   The beta cutoff frequency of $Q_1$ is

$$f_\beta = \frac{30 \, MHz}{50} = 600 \, kHz$$

Since $f_o$ is less than $f_\beta$, Eq. (7-13) may be used to calculate the voltage gain of the amplifier. $R_L$ in Eq. (7-13) is the load resistance of the transistor from the collector to ac ground; thus the ac resistance between the inductor tap and ac ground must be computed. Since $R_{\text{total}}$ is known from the previous problem, the autotransformer turns ratio may be used to calculate $R_L$, reversing the steps used to transform $R_o$ to $R_o'$.

$$R_L = \left(\frac{1}{a_2}\right)^2 R_{\text{total}} = \frac{1}{4^2} \times 52.0 \text{ k}\Omega = 3.25 \text{ k}\Omega$$

Then, from Eq. (7-13):

$$\begin{aligned} A_v &= \frac{v_{ce}}{v_{be}} = \frac{-h_{fe}R_L}{h_{ie}} \\ &= \frac{-50 \times 3.25 \text{ k}\Omega}{1 \text{ k}\Omega} = -162.5 \end{aligned} \tag{7-13}$$

From the autotransformer turns ratio:

$$\frac{N_1 + N_2}{N_1} = \frac{v_{\text{primary}}}{v_{ce}} = 4$$

And the coupling transformer turns ratio yields:

$$\frac{N_1 + N_2}{N_3} = \frac{v_{\text{primary}}}{v_o} = 12$$

Putting the pieces together yields:

$$\begin{aligned} A_v &= \frac{v_o}{v_i} = \frac{v_{ce}}{v_{be}} \times \frac{v_{\text{primary}}}{v_{ce}} \times \frac{v_o}{v_{\text{primary}}} \\ &= (-162.5) \times 4 \times \frac{1}{12} = -54.2 \end{aligned} \tag{7-15}$$

## 7.7 ANALYZING THE FREQUENCY RESPONSE OF A DOUBLED-TUNED CIRCUIT

### A Critically Coupled, Double-Tuned Circuit

**problem 7.5**  Figure 7-6 illustrates the double-tuned interstage coupling network of a field-effect transistor (FET) amplifier. If $L_p = L_s = 220 \,\mu\text{H}$, $C_1 = C_2 = 560 \text{ pF}$, $M = 2.20 \,\mu\text{H}$, $Q_p = Q_s = 100$ ($Q$ values include loading effects), determine the frequency-response characteristics of the coupling network.

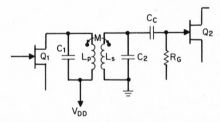

**Fig. 7-6**  Double-tuned FET amplifier (partial circuit).

**theory**  Since the input and output resistances of FETs are fairly high, both the primary and secondary sides of the coupling transformer may be tuned to the center frequency and relatively high $Q$'s obtained. Double-tuned circuits, such as that of Fig. 7-6, can result in superior performance with a flatter frequency response and sharper cutoffs at the edge of the passband than can be obtained with a single-tuned circuit.

The frequency response of a double-tuned circuit depends upon the mutual coupling between the two inductors as well as the frequency characteristics of each tuned circuit. Equation (7-16) defines the coefficient of coupling $k$:

$$k = \frac{M}{\sqrt{L_p L_s}} \tag{7-16}$$

where $M$ = mutual inductance between the two inductors, H
$L_p$ = inductance of the primary winding, H
$L_s$ = secondary inductance, H

The coefficient of coupling ($0 \le k \le 1$) describes the amount of coupling between the inductors.

When $k = 0$, the inductors are isolated from one another (no coupling); when $k = 1$, all lines of flux link the primary and secondary windings. Figure 7-7 illustrates the frequency response of a double-tuned circuit for various values of $k$.

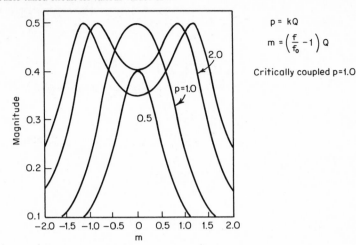

$$p = kQ$$

$$m = \left(\frac{f}{f_0} - 1\right)Q$$

Critically coupled p=1.0

**Fig. 7-7**   Frequency response of a double-tuned circuit.

The *critical coefficient of coupling* $k_c$ is given by Eq. (7-17),

$$k_c = \frac{1}{\sqrt{Q_p Q_s}} \qquad (7\text{-}17)$$

where $Q_p$ and $Q_s$ are the primary and secondary circuit $Q$'s (which include the loading effects), respectively. When $k < k_c$, the circuit is *undercoupled*, and the frequency response is similar to that of a single-tuned circuit. When $k > k_c$, the circuit is *overcoupled*, and the response has two separate peaks. The greater the overcoupling (large values of $k$), the greater the separation between the peaks and the deeper the dip in the response between the peaks (see Fig. 7-7). When $k = k_c$, the circuit is critically coupled and the response is *maximally flat*. The bandwidth of the maximally flat response is given by

$$\text{BW (maximally flat)} = \frac{\sqrt{2}\, f_o}{Q_A} \qquad (7\text{-}18)$$

where $Q_A$ is the average of $Q_p$ and $Q_s$, a useful approximation.

$$Q_A = \frac{Q_p + Q_s}{2} \qquad (7\text{-}19)$$

In the overcoupled case the center frequency is, of course, the resonant frequency of the tuned circuits. The upper and lower peak frequencies (see Fig. 7-8) are given by:

$$f_2 = f_o\left(1 + \frac{1}{2Q_A}\sqrt{k^2 Q_A^2 - 1}\right) \qquad (7\text{-}20)$$

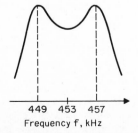

449   453   457

Frequency f, kHz

**Fig. 7-8**   Double-tuned response curve.

$$f_1 = f_o \left(1 - \frac{1}{2Q_A} \sqrt{k^2 Q_A^2 - 1}\right) \qquad (7\text{-}21)$$

**solution**   Compute the center frequency:

$$f_o = \frac{1}{2\pi\sqrt{LC}}$$

$$= \frac{1}{2\pi\sqrt{2.20 \times 10^{-4} \times 5.60 \times 10^{-10}}} = 453 \text{ kHz} \qquad (7\text{-}2)$$

To determine whether the circuit is overcoupled or critically coupled, calculate $k$ and $k_c$:

$$k = \frac{M}{\sqrt{L_p L_s}} \qquad (7\text{-}16)$$

$$= \frac{2.20 \ \mu\text{H}}{220 \ \mu\text{H}} = 0.01$$

$$k_c = \frac{1}{\sqrt{Q_p Q_s}} = \frac{1}{Q_A} \qquad (7\text{-}17)$$

$$= \frac{1}{100} = 0.01$$

Thus, the circuit is critically coupled and

$$\text{BW} = \frac{\sqrt{2} \times f_o}{Q_A}$$

$$= \frac{\sqrt{2} \times 453 \text{ kHz}}{100} = 6.41 \text{ kHz} \qquad (7\text{-}18)$$

### Analysis of an Overcoupled, Double-Tuned Circuit

**problem 7.6**   The circuit of Fig. 7-6 the primary and secondary windings are moved closer together so that the mutual inductance becomes 4.40 $\mu$H. All other circuit parameters remain as listed in Problem 7.5. Determine the frequency characteristics of the coupling network.

**theory**   Refer to Problem 7.5.

**solution**   Since $L$ and $C$ are unchanged, the center frequency remains 453 kHz. The coefficient of coupling is:

$$k = \frac{M}{\sqrt{L_p L_s}} \qquad (7\text{-}16)$$

$$= \frac{4.40 \ \mu\text{H}}{220 \ \mu\text{H}} = 0.020$$

Since $Q_A$ is unchanged, $k_c$ is 0.010, and the circuit is overcoupled. Calculating the peak frequencies:

$$f_2 = f_o \left(1 + \frac{1}{2Q_A} \sqrt{k^2 Q_A^2 - 1}\right) \qquad (7\text{-}20)$$

$$= 453 \text{ kHz} \left[1 + \frac{1}{2 \times 100} \sqrt{(0.02 \times 100)^2 - 1}\right]$$

$$= 457 \text{ kHz}$$

$$f_1 = f_o \left(1 - \frac{1}{2Q_A} \sqrt{k^2 Q_A^2 - 1}\right) \qquad (7\text{-}21)$$

$$= 453 \text{ kHz} \left[1 - \frac{1}{2 \times 100} \sqrt{(0.02 \times 100)^2 - 1}\right]$$

$$= 449 \text{ kHz}$$

With the aid of Fig. 7-7 the remainder of the frequency response may be sketched in. The form of the response is that of Fig. 7-8.

## 7.8 MULTISTAGE TUNED AMPLIFIERS

If the gain of a single stage of amplification is insufficient for a particular application, the obvious solution is to add additional amplifier stages. When tuned amplifiers are connected in cascade, the overall frequency response depends upon the tuning of the individual stages. Cascaded tuned amplifiers may be synchronously or stagger tuned. In synchronously tuned amplifiers the stages are all tuned to the same frequency; in stagger-tuned amplifiers the stages are tuned to slightly different center frequencies. Stagger tuning can yield flatter response with sharper (steeper) passband sides and correspondingly better selectivity than is possible with synchronous tuning. Aligning (tuning) stagger tuned amplifiers is, however, somewhat more difficult than aligning synchronously tuned amplifiers.

### Frequency-Response Characteristics of a Synchronously Tuned Multistage Amplifier

**problem 7.7**   Figure 7-9 is a block diagram of a two-stage synchronously tuned amplifier. If the overall bandwidth is to be 240 kHz with a center frequency of 10.7 MHz, find the $Q_L$ (loaded $Q$) of each stage. Assume $Q_{L1} = Q_{L2}$.

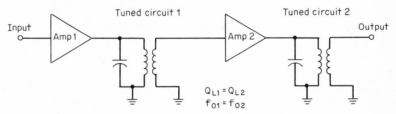

**Fig. 7-9**   Two-stage tuned amplifier.

**theory**   The overall bandwidth of $n$ identical single-tuned amplifier stages in cascade is:

$$\text{BW}_{\text{overall}} = \text{BW}_1 \sqrt{2^{1/n} - 1} \qquad (7\text{-}22)$$

where $n$ is the number of stages, and $\text{BW}_1$ is the bandwidth of a single stage.

**solution**   The required bandwidth of each stage may be determined by solving Eq. (7-22) for $\text{BW}_1$:

$$\text{BW}_1 = \frac{\text{BW}_{\text{overall}}}{\sqrt{2^{1/n} - 1}}$$

$$= \frac{240 \text{ kHz}}{\sqrt{2^{1/2} - 1}} = 373 \text{ kHz}$$

The $Q_L$ of each stage is given by:

$$Q_L = \frac{f_o}{\text{BW}} \qquad (7\text{-}8)$$

$$= \frac{10.7 \text{ MHz}}{373 \text{ kHz}} = 28.7$$

### Analysis of a Stagger-Tuned Amplifier

**problem 7.8**   The amplifier of Fig. 7-9 is to be stagger tuned for a maximally flat overall response with a center frequency of 10.7 MHz and a bandwidth of 240 kHz. Calculate the required center frequency, bandwidth, and $Q_L$ for each stage.

**theory**   In a two-stage tuned amplifier, if the center frequencies are displaced in accordance with Eqs. (7-23) and (7-24), and the individual stage bandwidths are in accordance with Eq. (7-25), a maximally flat frequency response occurs. (See Sec. 7.7.)

$$f_{o1} = f_o - 0.35 \, \text{BW}_{\text{overall}} \qquad (7\text{-}23)$$
$$f_{o2} = f_o + 0.35 \, \text{BW}_{\text{overall}} \qquad (7\text{-}24)$$
$$\text{BW}_1 = \text{BW}_2 = 0.707 \, \text{BW}_{\text{overall}} \qquad (7\text{-}25)$$

**solution**    Calculating the required center frequencies for the stages:

$$f_{o1} = f_o - 0.35\ BW_{overall}$$
$$= 10.7\ \text{MHz} - 0.35 \times 240\ \text{kHz} = 10.62\ \text{MHz} \tag{7-23}$$

And

$$f_{o2} = f_o + 0.35\ BW_{overall}$$
$$= 10.7\ \text{MHz} + 0.35 \times 240\ \text{kHz} = 10.78\ \text{MHz} \tag{7-24}$$

The bandwidth of the individual stages:

$$BW_1 = 0.707\ BW_{overall}$$
$$= 0.707 \times 240\ \text{kHz} = 170\ \text{kHz} \tag{7-25}$$

Finally the required $Q$'s:

$$Q_L = \frac{f_o}{BW}$$

$$Q_{L1} = \frac{f_{o1}}{BW_1} = \frac{10.62\ \text{MHz}}{170\ \text{kHz}} = 62.5 \tag{7-8}$$

$$Q_{L2} = \frac{f_{o2}}{BW_2} = \frac{10.78\ \text{MHz}}{170\ \text{kHz}} = 63.4$$

As a check, the expression for the bandwidth of a maximally flat response is:

$$BW = \frac{\sqrt{2} \times f_o}{Q_A} \tag{7-18}$$

For this example:

$$Q_A = \frac{Q_1 + Q_2}{2}$$
$$= \frac{62.7 + 63.4}{2} = 63.1 \tag{7-19}$$

$$BW_{overall} = \frac{\sqrt{2} \times 10.7\ \text{MHz}}{63.1} = 240\ \text{kHz}$$

## 7.9  VARIABLE-FREQUENCY TUNED CIRCUITS

Many tuned amplifier applications require that the center frequency of the amplifier be variable over some range of frequencies. Perhaps the most familiar example is an am broadcast band receiver. Since commercial am radio stations are assigned carrier frequencies by the Federal Communications Commission (FCC) between 540 kHz and 1.60 MHz, the tuned circuit at the input of the first stage of the receiver must also be variable over that range of frequencies. In addition, the local oscillator of an am receiver must maintain an output frequency 455 kHz above the incoming carrier frequency, and thus the oscillator frequency must be tunable from 995 kHz to 2.055 MHz.

Superheterodyne receivers employ a mixer stage that produces a constant frequency output, termed an *intermediate frequency* (i-f), that is the difference between the incoming carrier frequency and the local oscillator frequency. Thus the center frequency of i-f amplifiers is constant.

Since the resonant frequency of a tuned circuit is a function of the inductance and capacitance, either may be varied to change $f_o$. In practice, however, large frequency ratios are easier to obtain with variable capacitors than with variable inductors. Consequently variable air-gap capacitors (a series of movable parallel plates that dovetail with one another) are frequently used in tuned circuits whenarelatively large frequency range (am receivers, for example) is required. Equation 7-26 defines the frequency ratio,

$$FR = \frac{f_{high}}{f_{low}} \tag{7-26}$$

where $f_{high}$ and $f_{low}$ are the highest and lowest tuned frequencies, respectively.

Variable inductors generally employ a slug of powdered iron, or some other magnetic material, that moves through the axis of the inductor changing the magnetic properties and therefore the inductance.

In recent years, electronic tuning techniques, principally voltage-variable capacitors, are finding increasing application.

## Analysis of a Voltage-Variable Capacitor Tuning Circuit

**problem**   The tuned circuit of Fig. 7-10 employs a voltage-variable capacitor, or varactor, to adjust the resonant frequency of the circuit in response to changes in the level of the automatic frequency control (AFC) voltage. Such circuits are used in AFC applications, where it is necessary to correct an oscillator's output frequency in response to drift or other unwanted changes. While the circuit of Fig. 7-10 is simplified for illustrative purposes, the principle of operation is the same as that of more complex circuitry in communications equipment.

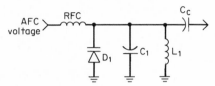

**Fig. 7-10**   Simplified varactor diode AFC circuit.

If $L_1 = 0.110 \ \mu$H with a $Q$ of 100 at $f = 65$ MHz, $C_1 = 47$ pF and the parameters of the varactor are: $C_T = 10$ pF at $V_R = 1$ V, tuning ratio (TR) = 2.5 for a maximum $V_R$ of 12 V, $R_s = 2 \ \Omega$, and $R_p = 10 \times 10^9 \ \Omega$, determine the range of resonant frequencies of the tuned circuit and the corresponding circuit $Q$'s. $V_R$ is the reverse bias on the diode. Assume the AFC voltage may vary between 1 and 12 V.

**theory**   Voltage-variable capacitors, or varactors, are silicon devices that use the depletion region of a reverse-biased pn junction as a voltage-controlled capacitor. The width of the depletion zone is proportional to the magnitude of the reverse-bias voltage; consequently, the junction capacitance is inversely proportional to the bias voltage. The exact relationship between the bias voltage and the capacitance depends on the geometry of the device. Varactors are finding increasing applications in tuning circuits, since they offer the possibility of remotely controlled tuning as well as the advantages of smaller size and weight over mechanically variable capacitors.

$C_T$ is the capacitance of the varactor at a particular value of reverse-bias voltage. Equation (7-27) defines the tuning ratio:

$$\text{TR} = \frac{C_T(V_{\min})}{C_T(V_{\max})} \tag{7-27}$$

Since the capacitance is inversely proportional to the magnitude of the bias voltage, the TR is the ratio of maximum to minimum $C_T$.

Varactors are manufactured with either *abrupt* or *hyperabrupt* junctions. The tuning ratios of abrupt-junction varactors typically range from 1.6 to 3.5, while tuning ratios in excess of 15 are possible with hyperabrupt junctions. Manufacturers' data sheets specify the *frequency ratio* of a diode; it is the square root of the tuning ratio. Applications which require a large frequency ratio, am radios, for example, employ hyperabrupt junction devices.

Unlike mechanically variable capacitors with typical $Q$'s of several thousand, the $Q$ of a varactor may be comparable with that of an inductor and thus may be of consequence in a particular application. Equations 7.28 and 7.29 describe varactor $Q$ at high and low frequencies, respectively:

$$Q_D = \frac{1}{2\pi f C_T R_s} \qquad f > 10^6 \text{ Hz} \tag{7-28}$$

$$Q_D = 2\pi f C_T R_p \qquad f < 10^4 \text{ Hz} \tag{7-29}$$

where $f$ = frequency, H
$\quad C_T$ = varactor capacitance, F
$\quad R_s$ = resistance of the bulk silicon, $\Omega$
$\quad R_p$ = reverse resistance of the diode, $\Omega$

Note that according to Eq. (7-29) the $Q$ of the varactor increases with frequency up to $10^4$ Hz, while according to Eq. (7-28), $Q$ decreases with frequency beyond $10^6$ Hz. For frequencies between $10^4$ and $10^6$ Hz, the $Q$ of a varactor is generally large enough to be ignored in comparison with that of the inductor, or may be obtained from the manufacturer's specifications.

**solution**   The range of $C_T$ may be determined from Eq. (7-27) and the given information:

$$TR = \frac{C_T(V_{min})}{C_T(V_{max})}$$

(7-27)

$$C_T(V_{max}) = \frac{C_T(V_{min})}{TR} = \frac{10 \text{ pF}}{2.5} = 4 \text{ pF}$$

The total parallel capacitance is the sum of the individual capacitors:

$$C_{total} = C_T + C_1$$
$$= 10 \text{ pF} + 47 \text{ pF} = 57 \text{ pF}$$
or
$$= 4 \text{ pF} + 47 \text{ pF} = 51 \text{ pF}$$

The range of frequencies is:

$$f_o = \frac{1}{2\rho\sqrt{LC}}$$

$$f_{o,low} = \frac{1}{2\pi\sqrt{0.11 \times 10^{-6} \times 57 \times 10^{-12}}} = 63.6 \text{ MHz}$$

(7-2)

$$f_{o,high} = \frac{1}{2\pi\sqrt{0.11 \times 10^{-6} \times 51 \times 10^{-12}}} = 67.2 \text{ MHz}$$

Since $f_o > 10^6$ Hz, the $Q$ of the varactor at the frequency and capacitance limits may be calculated as:

$$Q_D = \frac{1}{2\pi f \times C_T \times R_S}$$

(7-28)

$$= \frac{1}{2\pi \times 63.6 \times 10^6 \times 10 \times 10^{-12} \times 2} = 125$$

or

$$Q_D = \frac{1}{2\pi \times 67.2 \times 10^6 \times 4 \times 10^{-12} \times 2} = 296$$

The overall $Q$ of the tuned circuit, accounting for the $Q$'s of the varactor and inductor, $Q_D$ and $Q_1$, respectively, can be calculated as:

$$Q_{total} = \frac{Q_D Q_1}{Q_D + Q_1}$$

$$= \frac{125 \times 100}{125 + 100} = 55.6 \qquad \text{at } f = 63.6 \text{ MHz}$$

(7-30)

or

$$= \frac{296 \times 100}{296 + 100} = 74.8 \qquad \text{at } f = 67.2 \text{ MHz}$$

## 7.10  THE STABILIZATION OF TUNED AMPLIFIERS

At high frequencies the internal feedback signal of a transistor, through the collector capacitance $C_{ob}$, may lead to undesired oscillations. This is particularly true of high-gain amplifiers with loads that are inductive at the potentially unstable frequencies. Since the phase of the feedback signal is a function of frequency, it is difficult to design stabilization networks for high-gain, wide-band, amplifiers: passive elements that eliminate oscillation at one frequency may only accentuate the problem at another frequency. Fortunately the narrow-band characteristics of tuned amplifiers permit, in most cases, the design of relatively simple, yet reliable, stabilization circuits.

The most commonly used technique to stabilize a tuned amplifier is to provide an additional feedback signal of equal amplitude but opposite in phase to the internal feedback signal. If the added signal completely cancels the internal feedback signal, the amplifier is *unilateralized*. If the added signal stabilizes the amplifier, but does not entirely cancel the internal feedback signal, the amplifier is *neutralized*. In most cases the distinction is not an important one, and the terms are frequently used interchangeably.

Another approach to the design of stable tuned amplifiers is to use a cascade amplifier, a connection that minimizes the effect of $C_{ob}$. Finally, the simplest approach to stabilization, if the system requirements allow, is to reduce the gain of the amplifier, usually by intentionally mismatching at the load or the input, below the minimum gain that will support oscillation.

### Neutralizing a Tuned Amplifier

**problem 7.10**   In the tuned amplifier circuit of Fig. 7-11 $C_N$ is the neutralizing capacitor. If $C_{ob}$ of the transistor is 10 pF, determine the necessary value of $C_N$ to neutralize the amplifier. Assume $(N_1 + N_2)/N_1 = 10$ and $(N_1 + N_2)/N_3 = 6$.

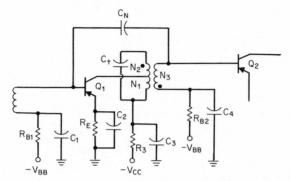

**Fig. 7-11**   Neutralized-tuned amplifier.

**theory**   The phase inversion property of a transformer provides a simple connection for a neutralizing circuit. In Fig. 7-11 the ac voltage from the base of $Q_2$ to ac ground is 180° out of phase with the ac voltage at the collector of $Q_1$ (note the transformer dot convention). To neutralize the amplifier, $C_N$ must be selected such that the signal coupled to the base of $Q_1$ by $C_N$ cancels the internal feedback through $C_{ob}$ from the collector to the base of $Q_1$. Accounting for the transformer turns ration, $C_N$ is given by:

$$C_N = \frac{N_1}{N_3} C_{ob} \tag{7-31}$$

**solution**   Calculating the required turns ratio yields:

$$\frac{N_1 + N_2}{N_3} \times \frac{N_1}{N_1 + N_2} = \frac{6}{10} = 0.6$$

Substituting into Eq. 7-31, we have

$$C_N = 0.6 \times 10\,\text{pF} = 6\,\text{pF}$$

## 7.11   AN INTEGRATED CIRCUIT TUNED AMPLIFIER

In recent years linear integrated circuits have been finding ever-increasing acceptance throughout the electronics industry. Several manufacturers now market ICs intended specifically for rf–i–f amplifier applications. The advantages of these specialized mono-lithic circuits include: low cost, high gain, and frequently an internal active circuit that provides a highly efficient automatic gain control (AGC) function.

**problem**   Figure 7-12 is the diagram of a single-stage tuned amplifier with a linear integrated circuit as the gain element. Determine the frequency response characteristics of the amplifier.

**theory**   The integrated circuit of Fig. 7-12 is typical of a number of ICs currently on the market intended for tuned amplifier applications. Such ICs usually consist of one or more differential amplifiers, with a separate connection for the AGC voltage which controls the gain of the amplifier by controlling the conduction of transistors that are part of the monolithic circuit.

The AGC voltage is generally derived from the detected (rectified) output signal, and it reduces the gain of the amplifier when the output signal is large in order to avoid distortion. Correspond-ingly the AGC voltage increases the gain of the amplifier when the strength is low.

The output of $IC_1$ in Fig. 7-12 is taken differentially in order to increase the voltage gain by 6 db and to achieve the maximum output signal swing. Single-ended operation, with a reduction in gain, may be used when stability is a problem.

Inductors $L_1$ and $L_2$ are wound on cylindrical core forms. Equation (7-32) yields the approximate self-inductance, in microhenries, of a single-layer air-core inductor:

$$L(\mu H) = \frac{(RN)^2}{9R + 10L} \tag{7-32}$$

where $L$ = length of the inductor, in
     $N$ = number of turns
     $R$ = radius of the inductor, in
The maximum efficiency for cylindrical inductors (maximum inductance for minimum physical size) occurs when $R/L = 1.25$.

$$L(\mu H) = \frac{RN^2}{17} \qquad R/L = 1.25 \tag{7-33}$$

For the amplifier of Fig. 7-12 the input tuned circuit consists of $C_2$ and $L_1$ while $C_6$ and $L_2$ form the output tuned circuit. $C_3$ and $C_5$ are decoupling capacitors that prevent undesired oscillations. $C_1$ and $C_7$ are used to adjust the effective source and load impedances presented to the amplifier. The smaller the values of $C_1$ and $C_7$, the larger the impedances presented to the amplifier.

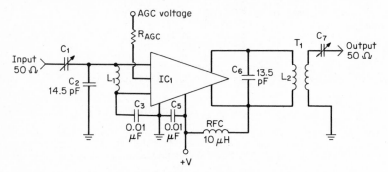

$L_1$– 10 turns, No. 22 AWG wire, 1/4 in. I.D., length = 1/2 in.

$T_1$ : Primary – 12 turns, No. 22 AWG wire, 1/4 in. I.D., length = 3/4 in.

Secondary – 2 turns, No. 22 AWG wire, 1/4 in. I.D., length = 1/4 in.

**Fig. 7-12**   Single-stage IC tuned amplifier.

**solution**   From the given information, the inductance of $L_1$ may be computed using Eq. (7-32).

$$L(\mu H) = \frac{(RN)^2}{9R + 10L}$$

$$L_1 = \frac{(\frac{1}{4} + 12)^2}{9 \times \frac{1}{4} + 10 \times \frac{1}{2}} = 0.862 \ \mu H \tag{7-32}$$

Similarly,

$$L_2 = \frac{(\frac{1}{4} \times 12)^2}{9 \times \frac{1}{4} + 10 \times \frac{3}{4}} = 0.923 \ \mu H$$

Equation (7-2) may be used to calculate the resonant frequencies of the input and output tuned circuits.

$$f_o = \frac{1}{2\pi\sqrt{LC}}$$

$$f_{o(input)} = \frac{1}{2\pi\sqrt{0.862 \times 10^{-6} \times 14.5 \times 10^{-12}}} = 45.0 \text{ MHz} \tag{7-2}$$

$$f_{o(output)} = \frac{1}{2\pi\sqrt{0.923 \times 10^{-6} \times 13.5 \times 10^{-12}}} = 45.1 \text{ MHz}$$

Thus the center frequency of the amplifier is approximately 45 MHz.
   The bandwidth of the amplifier depends on the impedances in parallel with the tuned circuits and the unloaded $Q$'s of the inductors. The input and output impedances of the integrated circuit may be obtained from the manufacturer's specification sheet. The $Q$'s of the inductors are

generally measured by a $Q$ meter. bt The effects of varying $C_1$ and $C_7$ are difficult to predict with exactness; fine adjustmennts are made on the completed circuit to arrive at the desired bandwidth and gain characteristics.

The bandwidth of each tuned circuit may be determined, once the total resistance in parallel with the tuned circuit is known, by means of Eq. (7-34):

$$BW = \frac{1}{2\pi R_T C} \tag{7-34}$$

where $R_T$ is the total parallel resistance in ohms and $C$ is the total parallel capacitance in farads.

## 7.12  BIBLIOGRAPHY

Ryder, J. D.: *Electronic Fundamentals and Applications*, 4th ed., Prentice-Hall, Englewood Cliffs, N.J., 1970.

Cirovic, Michael: *Basic Electronics, Devices, Circuits and Systems*, Reston, Va., 1974.

Pierce, J. F., and Paulus, T. J.: *Applied Electronics*, Charles E. Merrill Books, Inc., Columbus, Ohio, 1972.

Eimbinder, Jerry (ed.): *Designing with Linear Integrated Circuits*, Wiley, New York, 1969.

RCA Corp.: *RCA Designers Handbook, Solid State Power Circuits*, 1971.

Texas Instruments, Inc.: *Transistor Circuit Design*, McGraw-Hill, New York, 1963.

Johnson, D., and Hejhall, R.: "Tuning Diode Design Techniques," Motorola Application Note, AN-551, 1973.

Trout, B.: "A High Gain Integrated Circuit RF-IF Amplifier with Wide Range AGC," Motorola Application Note, AN-513, 1975.

Linvill, J., and Gibbons, J.: *Transistors and Active Circuits*, McGraw-Hill, New York, 1961.

# Chapter 8

# Feedback

## ROBERT A. BARTKOWIAK
### Professor of Engineering, The Pennsylvania State University

Feedback is a condition for a physical system in which a portion of the output is returned to the input. When feedback is used with a physical system, such as an electronic amplifier, the operating characteristics of that system are modified.

In the majority of feedback applications an amplifier circuit is arranged so that a signal is fed back to the input and subtracted from the impressed signal. The feedback in that case is called *negative* or *degenerative*. Conversely, when an amplifier circuit is arranged so the feedback signal adds to the impressed signal, the feedback is termed positive or regenerative.

As negative feedback has a stabilizing effect on amplifier gain, increases the bandwidth, and decreases the noise and distortion, it is used more often than positive feedback. Positive feedback, in contrast, increases the gain but decreases the bandwidth and stability. Although positive feedback is occasionally used in multiloop feedback amplifiers, it is used primarily for oscillators.

The application of feedback also affects the input and output impedances of an amplifier. However, the manner in which the impedance levels change depends on the type of feedback arrangement used.

The purpose of this chapter is to relate the changes in impedance, gain, and stability for typical feedback amplifiers.

## 8.1 CALCULATING THE OVERALL GAIN WHEN A GIVEN FEEDBACK TRANSFER RATIO IS APPLIED TO AN AMPLIFIER

**problem 8.1** Illustrated in Fig. 8-1 is an amplifier with an open-loop gain of 200. Determine the closed-loop gain if 15% of the output is fed back to the input so that it is 180° out of phase with the input.

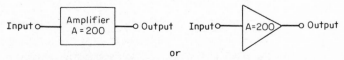

or

**Fig. 8-1** Amplifier symbols (with no feedback).

**theory**   An amplifier (voltage or current) normally has an open-loop gain $A$. A subscript, $v$, is often added to denote a voltage gain, and a subscript, $i$, is often added to denote a current gain. Using subscripts $o$ for output and $i$ for input, we define the open-loop voltage gain $A_v$ as:

$$A_v = \frac{v_o}{v_i} \tag{8-1}$$

Similarly, the open-loop current gain $A_i$ is:

$$A_i = \frac{i_o}{i_i} \tag{8-2}$$

With a fractional part $B$ of the output fed back to the input as in Fig. 8-2, a closed loop is formed, and the overall gain, with feedback, is called the *closed-loop gain*. The term $B$ is also called the *feedback transfer ratio*. The closed-loop voltage gain $A_{vf}$ and the closed-loop current gain $A_{if}$ are, respectively:

$$A_{vf} = \frac{A_v}{1 - A_v B} \tag{8-3}$$

$$A_{if} = \frac{A_i}{1 - A_i B} \tag{8-4}$$

Since each of the terms $A_i$, $A_v$, and $B$ are frequency-dependent, the overall gains are also frequency-dependent. Negative feedback results when:

$$|1 - AB| > 1 \tag{8-5}$$

Positive feedback results when:

$$|1 - AB| < 1 \tag{8-6}$$

Positive feedback may lead to undesirable oscillations.

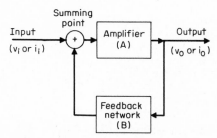

**Fig. 8-2**   Block diagram of an amplifier with feedback.

**solution**   Here no reference is made to the type of amplifier (voltage or current). Because the signal fed back is 180° out of phase with the input, the term $B$ is negative. Substituting into either expression for closed-loop gain, we obtain:

$$A_f = \frac{200}{1 - (-0.15)(200)} = \frac{200}{31} = 6.45$$

Hence, the application of negative feedback decreases the midband gain of an amplifier as shown in Fig. 8-3.

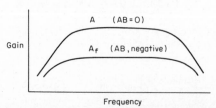

**Fig. 8-3**   The effect of negative feedback on midband gain.

## 8.2 CALCULATING THE VOLTAGE GAIN OF THE BJT AMPLIFIER WITH SERIES FEEDBACK

**problem 8.2**   Given the common-emitter (CE) amplifier shown in Fig. 8-4, calculate the voltage gain. Here $R_1 = 60$ kΩ, $R_2 = 30$ kΩ, $R_L = 5$ kΩ, $R_E = 100$ Ω, and the transistor parameters are $\beta = h_{fe} = 100$, $h_{ie} = 300$ Ω, and $h_{oe} = 1 \times 10^{-6}$ S.

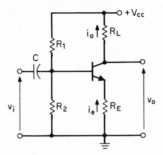

**Fig. 8-4**   The CE amplifier with series feedback.

**theory**   Both the amplifier and feedback networks of Figs. 8-1 and 8-2 are four-terminal networks where the "missing" terminals are understood to be connected to ground. Now we wish to explicitly show the four types of feedback system connections and must show each of the terminals. The four types of feedback system are shown in Fig. 8-5.

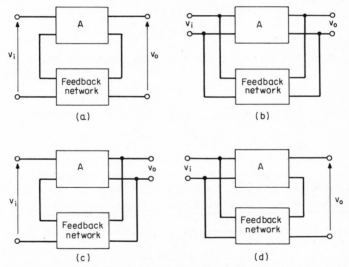

**Fig. 8-5**   The four types of feedback systems: (*a*) series input, series output; (*b*) parallel input, parallel output; (*c*) series input, parallel output; (*d*) parallel input, series output.

The emitter resistor $R_E$ of Fig. 8-4 is a common series element to the input and output so as to form a series-input/series-output feedback system. In short, this is called *series feedback*. Notice that the actual signal fed back is developed across $R_E$, and since the emitter current is approximately equal to the output current, $B$ by voltage division is:

$$B \approx \frac{R_E}{R_E + R_L} \tag{8-7}$$

Equation 8-7 is useful if $\beta R_E > h_{ie}$ and if the bias resistors are large. Otherwise we must return to the equivalent circuit shown in Fig. 8-6.

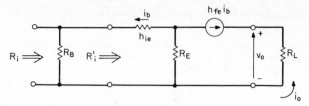

**Fig. 8-6**  The equivalent circuit of Fig. 8-4.

Then the input resistance to the transistor $R_i'$ is:

$$R_i' \approx h_{ie} + \beta R_E \qquad (8\text{-}8)$$

The equivalent resistance $R_B$ of the voltage divider $R_1$ and $R_2$ is:

$$R_B = R_1 \| R_2 = \frac{R_1 R_2}{R_1 + R_2} \qquad (8\text{-}9)$$

The input resistance to the complete circuit $R_i$ is:

$$R_i = R_B \| R_i' = \frac{R_B R_i'}{R_B + R_i'} \qquad (8\text{-}10)$$

and the current gain $A_{if}$ is:

$$A_{if} \approx -\beta \left( \frac{R_B}{R_B + h_{ie} + \beta R_E} \right) \qquad (8\text{-}11)$$

Finally, the voltage gain $A_{vf}$ is:

$$A_{vf} \approx A_{if} \left( \frac{R_L}{R_i} \right) \qquad (8\text{-}12)$$

**solution**   Without feedback, that is, with $R_E = 0$, $R_i \approx 300\ \Omega$, $A_i \approx -100$, and $A_v \approx -1670$. With $R_E = 100\ \Omega$, the feedback term $B$ is, from Eq. (8-7),

$$B = \frac{100}{100 + 5000} = \frac{100}{5100} \approx 0.0196$$

Substituting $A_v$ and $B$ into Eq. (8-3), we get for $A_{vf}$:

$$A_{vf} = \frac{-1670}{1 + (1670)(.0196)} = \frac{-1670}{33.7} \approx -49.5$$

## 8.3   CALCULATING THE CURRENT GAIN AND INPUT IMPEDANCE OF THE BJT AMPLIFIER WITH SERIES FEEDBACK

**problem 8.3**   Calculate the current gain and input impedance for the series feedback case of the preceding problem (Sec. 8.2). Using these values, recalculate $A_{vf}$.

**solution**   Referring to the previous problem and Figs. 8-4 and 8-6, we calculate $R_B$ by Eq. (8-9) and $A_{if}$ by Eq. (8-11). Thus,

$$R_B = R_1 \| R_2 = \frac{(30\ \text{k}\Omega)(60\ \text{k}\Omega)}{30\ \text{k}\Omega + 60\ \text{k}\Omega} = 20\ \text{k}\Omega$$

so that

$$A_{if} = -100 \left( \frac{20\ \text{k}\Omega}{20\ \text{k}\Omega + 300 + 10\ \text{k}\Omega} \right) \approx -100 \left( \frac{20\ \text{k}\Omega}{30\ \text{k}\Omega} \right)$$
$$= -67$$

Substituting into Eqs. (8-8) and (8-10), we get,

$$R_i' \approx h_{ie} + \beta R_E = 300 + (100)(100) = 10.3\ \text{k}\Omega$$

so that

$$R_i = R_B \| R_i' = \frac{(20\ \text{k}\Omega)(10.3\ \text{k}\Omega)}{20\ \text{k}\Omega + 10.3\ \text{k}\Omega} = 6.8\ \text{k}\Omega$$

Substituting into Eq. (8-12), we get for the voltage gain:

$$A_{vf} \approx A_{if}\left(\frac{R_L}{R_i}\right) = -67\left(\frac{5\ k\Omega}{6.8\ k\Omega}\right) \approx -49$$

The effects of series feedback, also called current feedback, on gain and impedance are summarized in Table 8-1.

**TABLE 8-1   Effects of Series and Shunt Feedback**

| Feedback type | Voltage gain $A_v$ | Current gain $A_i$ | Input resistance $R_i$ | Output resistance $R_o$ |
|---|---|---|---|---|
| Series | Decreases | Unaffected* | Increases | Increases |
| Shunt | Unaffected* | Decreases | Decreases | Decreases |

*A decrease may result, depending on the circuit values.

## 8.4   CALCULATING THE CURRENT GAIN AND THE INPUT IMPEDANCE OF THE BJT AMPLIFIER WITH SHUNT FEEDBACK

**problem 8.4**   Given the common-emitter amplifier shown in Fig. 8-7, calculate the current gain. Here $R_F = 100\ k\Omega$, $R_L = 5\ k\Omega$, and the transistor parameters are $\beta = h_{fe} = 100$, $h_{ie} = 300\ \Omega$, and $h_{oe} = 1 \times 10^{-6}$ S.

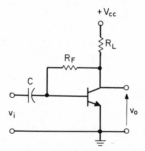

**Fig. 8-7**   The CE amplifier with shunt feedback.

**theory**   The feedback resistor $R_F$ creates a parallel-input/parallel-output system as in Fig. 8-5b. Hence, this type of feedback is called *shunt feedback*.

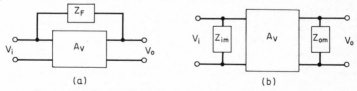

**Fig. 8-8**   (a) Shunt feedback. (b) The equivalent Miller impedances.

By the Miller effect, any impedance $Z_F$ shunting an amplifier with a voltage gain $A_v$, as in Fig. 8-8a, can be replaced by two equivalent Miller impedances at the input and output as shown in Fig. 8-8b. The equivalent impedance at the input $Z_{im}$ is

$$Z_{im} = \frac{Z_F}{1 - A_v} \tag{8-13}$$

and the equivalent impedance at the output $Z_{om}$ is

$$Z_{om} = \frac{A_v}{A_v - 1}Z_F \tag{8-14}$$

With an inversion amplifier $A_v$ is minus, and $Z_{im}$ and $Z_{om}$ are given by

$$Z_{im} = \frac{Z_F}{1 + A_v} \tag{8-15}$$

and

$$Z_{om} = \frac{A_v}{A_v + 1} Z_F \tag{8-16}$$

These impedances are combined in parallel with the input and output impedances of the amplifier. With a purely resistive feedback element $R_F$, $Z_{im}$ and $Z_{om}$ become resistive and are represented by $R_{im}$ and $R_{om}$ respectively.

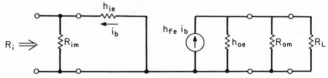

**Fig. 8-9**  The equivalent circuit of Fig. 8-7.

The equivalent circuit of Fig. 8-7 is shown in Fig. 8-9. Now, the input resistance of the complete circuit, $R_b$ is

$$R_i = R_{im} \| h_{ie} \tag{8-17}$$

and the current gain $A_{if}$ for $R_F \gg h_{ie}$ is

$$A_{if} \approx \frac{-\beta}{1 + \beta(R_L/R_F)} \tag{8-18}$$

**solution**   Without feedback, that is, with $R_F = \infty$, $R_i \approx 300 \ \Omega$, $A_i \approx -100$, and $A_v \approx -1670$. With $Z_F$ purely resistive and equal to $R_F$, we get, from Eq. (8-15),

$$R_{im} = \frac{R_F}{1 + A_v} \approx \frac{100 \ k\Omega}{1670} \approx 60 \ \Omega$$

Then the input resistance from Eq. (8-17) is

$$R_i = R_{im} \| h_{ie} = \frac{(60)(300)}{60 + 300} = 50 \ \Omega$$

Substituting into Eq. (8-18), we get for the current gain

$$A_{if} \approx \frac{-100}{1 + 100(5 \ k\Omega/100 \ k\Omega)} = \frac{-100}{1 + 5} \approx -16.7$$

## 8.5   CALCULATING THE VOLTAGE GAIN OF THE BJT AMPLIFIER WITH SHUNT FEEDBACK

**problem 8.5**   Calculate the voltage gain for the previous problem (Sec. 8.4).

**theory**   Having solved for $A_{if}$ and $R_i$ for a feedback amplifier, we may return to Eq. (8-12) which is valid for the shunt feedback case also.

**solution**   From Sec. 8.4, $R_i = 50 \ \Omega$ and $A_{if} \approx -16.7$. Substituting these values into Eq. (8-12), we obtain

$$A_{vf} \approx A_{if}\left(\frac{R_L}{R_i}\right) = -16.7 \left(\frac{5 \ k\Omega}{50 \ \Omega}\right) \approx -1670$$

The effects of shunt feedback, also called *voltage feedback*, on gain and impedance are summarized in Table 8-1.

## 8.6   CALCULATING THE OUTPUT IMPEDANCE FOR THE BJT SERIES FEEDBACK AMPLIFIER

**problem 8.6**   Calculate the output impedance for the circuit of Fig. 8-10a. Here a series feedback amplifier has a source voltage $v_s$ with source resistance $R_s$ connected to it. Consider that $R_s = 50 \ \Omega$, $\beta = h_{fe} = 100$, $h_{ie} = 300 \ \Omega$, and $h_{oe} = 1 \times 10^{-6}$ S.

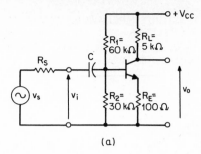

(a)

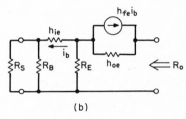

(b)

**Fig. 8-10**  (*a*) The series feedback amplifier with connected source. (*b*) The equivalent circuit for calculating the output impedance.

**theory**   Without feedback, the output impedance at midband is purely resistive. This resistance $R_o$ is given by

$$R_o \approx \frac{1}{h_{oe}} \tag{8-19}$$

With feedback, the output resistance is obtained by a Thévenin calculation. With the source removed and $R_s$ left in the circuit, the equivalent circuit of Fig. 8-10*b* results. Using network analysis, we find $R_o$ to be

$$R_o \approx \frac{h_{fe}}{h_{oe}} \left( \frac{R_E}{h_{ie} + R_E + R'_s} \right) \tag{8-20}$$

where $R_B$ is the parallel equivalent of the bias resistors [Eq. (8-9)] and

$$R'_s = R_s + R_B = \frac{R_s R_B}{R_s + R_B} \tag{8-21}$$

**solution**   Substituting into Eqs. (8-9) and (8-21), we get

$$R_B = R_1 \| R_2 = \frac{(60 \text{ k}\Omega)(30 \text{ k}\Omega)}{90 \text{ k}\Omega} = 20 \text{ k}\Omega$$

and

$$R'_s = R_s \| R_B = 50 \ \Omega \| 20 \text{ k}\Omega \approx 50 \ \Omega$$

Substituting into Eq. (8-20), we obtain

$$R_o \approx \frac{h_{fe}}{h_{oe}} \left( \frac{R_E}{h_{ie} + R_E + R'_s} \right) = \frac{100}{10^{-6}} \left( \frac{100}{300 + 100 + 50} \right)$$

$$= 10^8 \left( \frac{100}{450} \right) = 22 \text{ M}\Omega$$

Without feedback, from Eq. (8-19), we get

$$R_o \approx \frac{1}{h_{oe}} = \frac{1}{10^{-6}} = 1 \text{ M}\Omega$$

Note that $R_o$ is increased by series feedback.

## 8.7   CALCULATING THE OUTPUT IMPEDANCE FOR THE BJT SHUNT FEEDBACK AMPLIFIER

**problem 8.7**   Calculate the output impedance for the circuit of Fig. 8-11a when $R_s = 50\ \Omega$, $\beta = h_{fe} = 100$, $h_{ie} = 300\ \Omega$, and $h_{oe} = 1 \times 10^{-6}$ S.

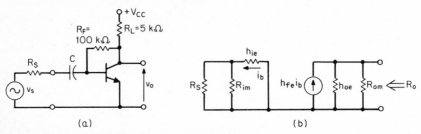

**Fig. 8-11**   (a) The shunt feedback amplifier with connected source. (b) The equivalent circuit for calculating the output impedance.

**theory**   As pointed out in Sec. 8.4, a shunting impedance between output and input is replaced by two equivalent impedances owing to the Miller effect. As in the series feedback case, the output resistance is obtained by a Thévenin calculation. With the source removed, $R_s$ left in the circuit, and the Miller effect resistances $R_{im}$ and $R_{om}$ connected, the equivalent circuit of Fig. 8-11b results. Now the output resistance $R_o$ is

$$R_o = \frac{1}{h_{oe}} \left\| R_{om} = \frac{R_{om}}{h_{oe}R_{om} + 1} \right.$$   (8-22)

**solution**   The gain $A_v$ without feedback is $A_v \approx -1670$. Then the Miller effect resistance on the output side $R_{om}$ is found by Eq. (8-14) to be

$$R_{om} = \frac{A_v}{A_v - 1}\, R_F = \frac{-1670}{-1671}(100\ \mathrm{k\Omega}) \approx 100\ \mathrm{k\Omega}$$

Substituting into Eq. (8-22), we obtain for $R_o$

$$R_o = \frac{R_{om}}{h_{oe}R_{om} + 1} = \frac{100\ \mathrm{k\Omega}}{(10^{-6})(10^5) + 1} = \frac{100\ \mathrm{k\Omega}}{1.1} \approx 91\ \mathrm{k\Omega}$$

Without feedback, $R_o \approx 1$ M$\Omega$. Note that $R_o$ is decreased by shunt feedback.

Table 8.1 summarizes the effects that series and shunt feedback have on gain and impedance.

## 8.8   CALCULATING THE GAIN AND OUTPUT IMPEDANCE OF AN FET FEEDBACK AMPLIFIER

**problem 8.8**   Given the FET circuit of Fig. 8-12a with a series feedback resistor $R_s$, calculate the voltage gain and output impedance. Here a metal oxide semiconductor field-effect transistor (MOSFET) is used with $R_D = 20$ k$\Omega$, $R_S = 400\ \Omega$, $g_m = 2000\ \mu$S, and $r_{ds} = 40$ k$\Omega$.

**theory**   The FET is characterized by a high-input impedance and a high-output impedance. Because of the high-input impedance the input current is negligible, and so the current gain is very large ($A_i \approx \infty$). For this reason $A_i$ is not meaningful here. As it is usually desirable to maintain high input and output impedances, series feedback is usually employed with FET amplifiers. In Fig. 8-12a, $R_S$ represents an unbypassed source resistance.

The ac equivalent for the circuit of Fig. 8-12a is shown in Fig. 8-12b. Note that the voltage-divider network forms the equivalent resistance $R_B$:

$$R_B = R_1 \| R_2 = \frac{R_1 R_2}{R_1 + R_2}$$   (8-23)

The voltage gain with feedback $A_{vf}$ is determined by writing circuit equations.

$$A_{vf} = \frac{-g_m r_{ds} R_D}{r_{ds} + R_D + R_S(1 + g_m r_{ds})}$$   (8-24)

The input impedance at midband is a resistance $R_{if}$, that is

$$Z_{if} = R_{if}$$   (8-25)

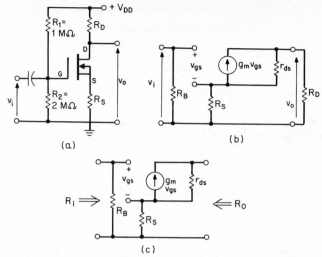

**Fig. 8-12** (*a*) The FET amplifier with feedback. (*b*) The equivalent circuit for $A_v$ calculation. (*c*) The equivalent circuit for impedance calculations.

The output impedance at midband is a resistance $R_{of}$ which is found by disconnecting the load resistance, in this case $R_D$, and again writing circuit equations. See Fig. 8-12*c*. Then,

$$R_{of} = r_{ds} + R_S(1 + g_m r_{ds}) \qquad (8\text{-}26)$$

Similarly, the input impedance at midband is a resistance $R_{if}$ which is found to be approximately equal to $R_B$.

$$R_{if} \approx R_B \qquad (8\text{-}27)$$

Notice that if there is no series feedback ($R_S = 0$), the equations reduce to

$$A_v = -g_m \frac{r_{ds} R_D}{r_{ds} + R_D} \qquad (8\text{-}28)$$

$$R_o = r_{ds} \qquad (8\text{-}29)$$
$$R_i = R_B \qquad (8\text{-}30)$$

**solution**    The equivalent resistance of the voltage divider is calculated by substitution into Eq. 8-23).

$$R_B = \frac{R_1 R_2}{R_1 + R_2} = \frac{(1\ \text{M}\Omega)(2\ \text{M}\Omega)}{3\ \text{M}\Omega} = 667\ \text{k}\Omega$$

Then
$$R_{if} \approx R_B = 667\ \text{k}\Omega$$

For the values given, from Eqs. (8-24) and (8-26), the voltage gain and output impedance are, respectively,

$$A_{vf} = \frac{-(2 \times 10^{-3})(40 \times 10^3)(20 \times 10^3)}{40 \times 10^3 + 20 \times 10^3 + 400(1 + 2 \times 10^{-3} \times 40 \times 10^3)}$$

$$= \frac{-(2)(40)(20 \times 10^3)}{60 \times 10^3 + 32.4 \times 10^3} = -17.3$$

and

$$R_{of} = 40 \times 10^3 + 400(1 + 2 \times 10^{-3} \times 40 \times 10^3)$$
$$= 40 \times 10^3 + 400(81) = 72.4\ \text{k}\Omega$$

Without feedback, the gain and impedances are, by Eqs. (8-28) to (8-30),

$$A_v = -2 \times 10^{-3} \frac{(40 \times 10^3)(20 \times 10^3)}{40 \times 10^3 + 20 \times 10^3} = -26.7$$

$$R_o = 40\ \text{k}\Omega$$
$$R_i = 667\ \text{k}\Omega$$

The effects of feedback on the FET amplifier are summarized in Table 8-2.

**TABLE 8-2    Effect of Series Feedback on FET Amplifier Performance**

| Voltage gain $A_v$ | Input resistance $R_i$ | Output resistance $R_o$ |
|---|---|---|
| Decreased | Unaffected* | Increased |

*Ideally ∞.

## 8.9    FEEDBACK AND THE OPERATIONAL AMPLIFIER

The operational amplifier (op amp) is a high-gain dc amplifier with a differential input. The equivalent circuit for the op amp is shown in Fig. 8-13. The following parameters define the op amp equivalent circuit:

$$v_d = v_2 - v_1 = \text{differential input voltage}$$
$$Z_{\text{in}} = \text{input impedance}$$
$$A_v = \text{open-loop voltage gain (positive value)}$$
$$Z_{\text{out}} = \text{output impedance}$$

Note that with an input voltage $v_2 = 0$ and $v_d = v_1$ the output is inverted. That is, there is a 180° phase shift between the output and input voltage at low and midband frequencies. With $v_1 = 0$ and $v_d = v_2$, the output is noninverted. That is, there is 0° phase shift between the output and input voltage at low and midband frequencies.

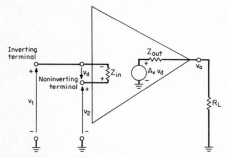

**Fig. 8-13**  The op amp equivalent circuit.

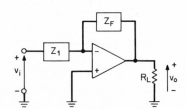

**Fig. 8-14**  The op amp connected as an inverting amplifier with feedback.

Historically, operational amplifiers were designed and primarily used to perform the mathematical operations of addition, multiplication, integration, etc. These mathematical operations were and are realized by using feedback elements such as resistors and capacitors.

Figure 8-14 shows how feedback is used with an inverting amplifier. For convenience $Z_{in}$ and the Thévenin equivalent of the amplifier have been omitted. If circuit equations are written, the voltage gain with feedback $A_{vf}$ is found to be

$$A_{vf} = \frac{v_o}{v_i} \approx \frac{-A_v Z_F}{Z_F + Z_1(1 + A_v)} \tag{8-31}$$

With a very high $A_v$ (ideally $A_v \approx \infty$) the voltage gain equation reduces to the ideal or classic equation:

$$A_{vf} \approx \frac{-Z_F}{Z_1} \tag{8-32}$$

Normally $Z_1$ is much smaller than $Z_{in}$ (ideally $Z_{in} \approx \infty$), and so the input impedance with feedback $Z_{if}$ is

$$Z_{if} \approx Z_1 + \frac{Z_F}{A_v} \tag{8-33}$$

Again considering a large $A_v$, the classic expression for $Z_{if}$ is obtained:

$$Z_{if} \approx Z_1 \qquad (8\text{-}34)$$

Notice that the feedback network portrayed by $Z_F$ in Fig. 8-14 typifies the parallel input, parallel output case of Fig. 8-5b. For such a case the output impedance is decreased by the Miller effect from its already low value $Z_{\text{out}}$. The output impedance with feedback $Z_{of}$ is approximated by

$$Z_{of} \approx \frac{Z_{\text{out}}}{1 + A_v B} \approx \frac{Z_{\text{out}}}{A_v B} \qquad (8\text{-}35)$$

where

$$B = \frac{Z_1}{Z_1 + Z_F} \qquad (8\text{-}36)$$

Note that $B$, the feedback transfer ratio, represents the voltage gain ($B < 1$) of the feedback network.

A special case arises when $Z_F = Z_1$ for the inverting amplifier case. Notice, from Eq. (8-32), that for $Z_F = Z_1$, $A_{vf} = -1$. The feedback amplifier now functions as a sign changer.

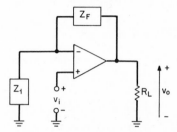

**Fig. 8-15**  The op amp connected as a noninverting amplifier with feedback.

With the op amp connected as a noninverting amplifier, the input is applied directly to the noninverting terminal as in Fig. 8-15. Then the voltage gain with feedback $A_{vf}$ is found by writing circuit equations and using the ideal properties:

$$A_{vf} \approx \frac{Z_1 + Z_F}{Z_1} = \frac{1}{B} \qquad (8\text{-}37)$$

In turn, the input and output impedances are calculated as

$$Z_{if} \approx \left( \frac{A_v Z_1}{Z_1 + Z_F} \right) Z_{\text{in}} = A_v B Z_{\text{in}} \qquad (8\text{-}38)$$

$$Z_{of} \approx \left( \frac{Z_1 + Z_F}{A_v Z_1} \right) Z_{\text{out}} = \frac{Z_{\text{out}}}{A_v B} \qquad (8\text{-}39)$$

Notice that the classic or ideal expression for $Z_{of}$ is the same for both the inverted and noninverted op amp cases. The noninverting feedback amplifier typifies the series-input, parallel-output case of Fig. 8-5c. Table 8-3 summarizes the classic gain and impedance equations for use with op amps.

**TABLE 8-3   Gain and Impedance Approximations for Operational Amplifier Feedback Circuits***

| Type | Voltage gain $A_{vf}$ | Input impedance $Z_{if}$ | Output impedance $Z_{of}$ |
|------|------------------------|---------------------------|----------------------------|
| Inverting | $-\dfrac{Z_F}{Z_1}$ | $Z_1$ | $\dfrac{Z_{\text{out}}}{A_v B}$ |
| Noninverting | $\dfrac{1}{B}$ | $A_v B Z_{\text{in}}$ | $\dfrac{Z_{\text{out}}}{A_v B}$ |

*Where $B = \dfrac{Z_1}{Z_1 + Z_F}$

## 8.10   CALCULATING THE GAIN AND IMPEDANCES OF THE FEEDBACK OP AMP CIRCUIT

**problem 8.9**   Calculate the gain, input impedance, and output impedance of the circuit of Fig. 8-16. The op amp parameters are $A_v = 10^5$, $Z_{\text{in}} = 1\ \text{M}\Omega$, $Z_{\text{out}} = 200\ \Omega$.

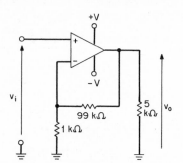

**Fig. 8-16**   A feedback op amp circuit.

**theory**   Referring to Sec. 8.9, we recognize this circuit as a noninverting feedback op amp. The appropriate formulas from Table 8-3 are used.

**solution**   Notice that in this problem $Z_F = 99\ \text{k}\Omega$ and $Z_1 = 1\ \text{k}\Omega$, and so the feedback term $B$ is

$$B = \frac{Z_1}{Z_1 + Z_F} = \frac{1\ \text{k}\Omega}{100\ \text{k}\Omega} = 0.01$$

Then, the voltage gain $A_{vf}$ is

$$A_{vf} = \frac{1}{B} = \frac{1}{0.01} = 100$$

the input impedance $Z_{if}$ is

$$Z_{if} = A_v B Z_{\text{in}} = (10^5)(10^{-2})(10^6) = 1000\ \text{M}\Omega$$

and the output impedance $Z_{of}$ is

$$Z_{of} = \frac{Z_{\text{out}}}{A_v B} = \frac{200}{(10^5)(10^{-2})} = 0.2\ \Omega$$

## 8.11   CALCULATING THE BANDWIDTH WITH THE APPLICATION OF FEEDBACK

**problem 8.10**   An amplifier has an open-loop gain $A = -10^4$ and lower and upper cutoff frequencies of 100 Hz and 20 kHz, respectively. Calculate the bandwidth when a feedback transfer ratio of 0.01 is applied.

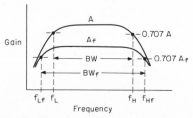

**Fig. 8-17**   Amplifier frequency response with and without negative feedback.

**theory**   Recall that the bandwith of an amplifier is defined as the range of frequencies over which the gain remains within 0.707 of its midband gain. The frequencies at which the gain is 0.707 of the midband gain are called cutoff, half-power, or 3-dB frequencies. With reference to Fig. 8-17,

$f_L$ equals the lower cutoff frequency,
$f_H$ equals the upper cutoff frequency.

and the bandwidth without feedback BW is

$$BW = f_H - f_L \qquad (8\text{-}40)$$

With negative feedback, the gain is $A_f (A_f < A)$, but the cutoff frequencies are extended to the values $f_{Lf}$ equals the lower cutoff frequency with feedback and $f_{Hf}$ equals the upper cutoff frequency with feedback. Then the bandwidth with feedback $BW_f$ is

$$BW_f = f_{Hf} - f_{Lf} \qquad (8\text{-}41)$$

where the cutoff frequencies with feedback are determined from

$$f_{Hf} = f_H(1 - AB) \qquad (8\text{-}42)$$

$$f_{Lf} = \frac{f_L}{1 - AB} \qquad (8\text{-}43)$$

The frequency response with feedback is often called the *closed-loop frequency response*.

**solution**    Without feedback, from Eq. (8-40),

$$BW = f_H - f_L = 20{,}000 - 100 = 19.9 \text{ kHz}$$

With feedback the cutoff frequencies are from Eqs. (8-42) and (8-43) are

$$f_{Hf} = 20 \times 10^3[1 - (-10^4)(10^{-2})] \approx 2 \text{ MHz}$$

$$f_{Lf} = \frac{100}{1 - (-10^4)(10^{-2})} \approx \frac{100}{100} = 1 \text{ Hz}$$

Then from Eq. (8-41)

$$BW_f = f_{Hf} - f_{Lf} \approx 2 \text{ MHz}$$

## 8.12    CALCULATING THE DESENSITIZING OF AMPLIFIER GAIN PARAMETERS WITH NEGATIVE FEEDBACK

**problem 8.11**    Variations in parameters of a certain transistor type produce a 20% change in amplifier voltage gain. It is desired to use these transistors in a feedback amplifier and obtain a voltage gain of $-150$ with only a 1% variation. Calculate the open-loop gain and feedback transfer ratio needed.

**theory**    Negative feedback results in a desensitizing of amplifier parameters. The fractional change in gain with feedback owing to a fractional change in open-loop gain is obtained by applying differential calculus to either Eq. (8-3) or (8-4). It is found that

$$\frac{dA_f}{A_f} = \frac{1}{1 - AB} \frac{dA}{A} \qquad (8\text{-}44)$$

where $dA_f$ equals the change in closed-loop gain and $dA$ equals the change in open-loop gain. Thus, when feedback is added, the fractional change in gain is divided by the desensitivity factor $D$.

$$D = 1 - AB \qquad (8\text{-}45)$$

**solution**    Substituting into Eq. (8-44), we obtain

$$0.01 = \frac{1}{1 - AB}(0.20)$$

or

$$1 - AB = \frac{0.20}{0.01} = 20$$

We then substitute $1 - AB = 20$ into Eq. (8-3) and obtain

$$-150 = \frac{A}{1 - AB} = \frac{A}{20}$$

or

$$A = (-150)(20) = -3000$$

Then from $1 - AB = 20$ we have

$$B = \frac{20 - 1}{-A} = \frac{19}{3000} = 0.0063$$

## 8.13   CALCULATING THE EFFECT OF FEEDBACK ON NOISE OR DISTORTION

**problem 8.12**   A two-stage amplifier has some noise introduced between stages. The gain of the first stage is 100, and the gain of the second stage is $-100$. If an overall feedback ratio of 0.01 is used, calculate the amplification experienced by an input signal and by the noise.

**theory**   Negative feedback is used to reduce noise or nonlinear distortion which may be inherent in an amplifier. If the noise or distortion, symbolized by the voltage $v_n$, is introduced with a desired input $v_i$ as in Fig. 8-18a, the output voltage is

$$v_o = v_i \left( \frac{A}{1 - AB} \right) + v_n \left( \frac{A}{1 - AB} \right) \tag{8-46}$$

Hence, the noise or distortion is reduced by the same factor as the open-loop gain. Even though it appears that the output has the same percentage of noise or distortion, the noise or distortion is generally of a different frequency so that with both $A$ and $B$ frequency-dependent, the noise or distortion is subjected to a different loop gain than the desired input.

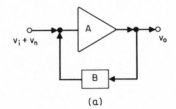

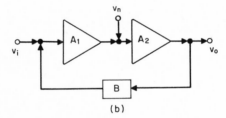

**Fig. 8-18**   (a) Noise or distortion at an amplifier input. (b) Noise or distortion within an amplifier.

If the noise or distortion is introduced within a cascaded amplifier as in Fig. 8-18b, the output voltage is given by

$$v_o = v_i \left( \frac{A_1 A_2}{1 - A_1 A_2 B} \right) + v_n \left( \frac{A_2}{1 - A_1 A_2 B} \right) \tag{8-47}$$

In this latter case, $A_1$ can be made large, thereby increasing the amplification of $v_i$ and decreasing the amplification of $v_n$. To achieve a high gain for $A_1$, an inner positive feedback loop may be utilized. Multiloop feedback is illustrated in Sec. 8-14.

**solution**   The amplification to the input $A_{vi}$ and to the noise $A_{vn}$ are from Eq. (8-47):

$$v_i^A = \frac{A_1 A_2}{1 - A_1 A_2 B} = \frac{(100)(-100)}{1 - (100)(-100)(0.01)} = -99$$

$$v_n^A = \frac{A_2}{1 - A_1 A_2 B} = \frac{-100}{1 - (100)(-100)(0.01)} = -0.99$$

## 8.14   MULTISTAGE AND MULTILOOP FEEDBACK AMPLIFIERS

A multistage feedback amplifier is a cascaded amplifier to which an overall feedback is applied. On the other hand a multiloop feedback amplifier is an amplifier having two or more feedback loops around different parts of the amplifier. Examples of both these configurations are shown in Fig. 8-19.

A great number of negative-feedback multistage amplifiers are possible. In each case for negative feedback, the signal fed back to the input must be 180° out of phase with the input signal. Four multistage transistor feedback amplifiers which represent the four types of configurations of Fig. 8-5 are shown in Fig. 8-20.

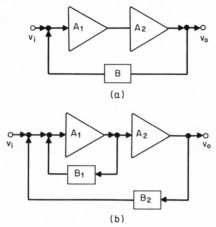

(a)

(b)

**Fig. 8-19**  (a) Multistage and (b) multiloop feedback amplifiers.

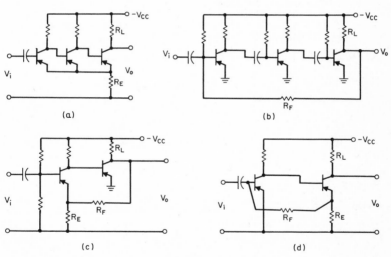

(a)    (b)

(c)    (d)

**Fig. 8-20**  Types of multistage negative feedback amplifiers (with feedback elements identified). (a) Series input, series output; (b) parallel input, parallel output; (c) series input, parallel output; (d) parallel input, series output.

If the total amplifier gain and the feedback networks are lumped, we get the approximate gain equations listed in Table 8.4. Note the similarities with the equations for the single-stage amplifiers.

Similarly, numerous multiloop configurations are possible. The circuit of Fig. 8-19b could represent a preamplifier with gain $A_1$ (stage 1) followed by an amplifier of gain $A_2$ (stage 2). If some noise or distortion occurs in the second-stage amplifier, it would be beneficial to provide positive feedback around stage 1, thereby increasing that gain. To prevent instability, though, negative feedback would be applied to the overall amplifier. See Sec. 8-13.

**TABLE 8-4     Gain Relationships for the Multistage Feedback Amplifiers of Fig. 8-20 (approx.)**

| Configuration | Gain with feedback $A_{vf}$ or $A_{if}$ | Feedback factor $B$ |
|---|---|---|
| Series input, series output (series feedback) | $A_{vf} \approx \dfrac{A_v}{1 - BA_v}$ | $\dfrac{R_E}{R_L}$ |
| Parallel input, parallel output (shunt feedback) | $A_{if} \approx \dfrac{A_i}{1 - BA_i}$ | $\dfrac{R_L}{R_F}$ |
| Series input, parallel output (voltage feedback pair) | $A_{vf} \approx \dfrac{A_v}{1 - BA_v}$ | $\dfrac{-R_E}{R_E + R_F}$ |
| Parallel input, series output (current feedback pair) | $A_{if} \approx \dfrac{A_i}{1 - BA_i}$ | $\dfrac{-R_E}{R_E + R_F}$ |

## 8.15   CALCULATING THE GAIN OF THE MULTISTAGE FEEDBACK AMPLIFIER

**problem 8.13**   The multistage feedback amplifier of Fig. 8-21 has an $A_v$ (without feedback) of 3000. Calculate the voltage gain with feedback applied as shown.

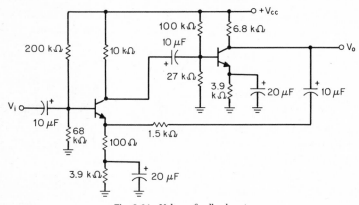

**Fig. 8-21**   Voltage feedback pair.

**solution**   If the midband equivalent circuit is drawn, we readily recognize this circuit as the voltage feedback pair of Fig. 8-20c. Here $R_E = 100\ \Omega$ and $R_F = 1.5\ \text{k}\Omega$. Then substituting into the appropriate formulas from Table 8.4, we obtain

$$B \approx \frac{-R_E}{R_E + R_F} = \frac{-100}{100 + 1500} = -0.0625$$

Then

$$A_{vf} \approx \frac{A_v}{1 - BA_v} = \frac{3000}{1 - (-0.0625)(3000)} \approx 16$$

## 8.16   CALCULATING THE IMPEDANCES OF THE MULTISTAGE FEEDBACK AMPLIFIER

**problem 8.14**   Calculate the input and output impedances of the feedback amplifier of Fig. 8-21 if the input and output impedances without feedback are $R_i = 2\ k\Omega$ and $R_0 = 3\ k\Omega$.

**theory**   As noted in earlier sections dealing with single-stage amplifiers, the input and output impedances with feedback differ from those without feedback. For multistage feedback amplifiers, these effects are summarized in Table 8.5. The factor $(1 - BA)$ is used in approximating the resistances with feedback.

**TABLE 8-5    Effects of Feedback on Input and Output Impedances of Multistage Feedback Amplifiers**

| Type of configuration | | | |
|---|---|---|---|
| Series input, series output | Parallel input, parallel output | Series input, parallel output | Parallel input, series output |
| $R_{if} \gg R_i$ <br> $R_{of} > R_o$ | $R_{if} < R_i$ <br> $R_{of} < R_o$ | $R_{if} \gg R_i$ <br> $R_{of} < R_o$ | $R_{if} < R_i$ <br> $R_{of} > R_o$ |

**solution**    This problem uses the same circuit as does the problem of Sec. 8.15. From that problem, $A_v = 3000$ and $B = -0.0625$ so that $(1 - BA_v) = 188.5$. For the series input, parallel output configuration $R_{if} \gg R_i$ and $R_{of} < R_o$ (Table 8-5). Thus

$$R_{if} \approx R_i(1 - BA_v) = 2000(188.5) = 377\ \text{k}\Omega$$
$$R_{of} \approx \frac{R_o}{1 - BA_v} = \frac{3000}{188.5} = 15.9\ \Omega$$

## 8.17   DETERMINING GAIN AND PHASE MARGIN

**problem 8.15**    A negative-feedback amplifier has a Bode plot as in Fig. 8-22$a$. Determine the gain and phase margin for the amplifier.

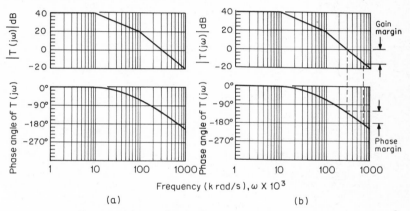

**Fig. 8-22**    ($a$) Typical Bode plot. ($b$) Gain and phase margin.

**theory**    The Bode plot or diagram is a graphical plot which shows the amplitude and phase of a transfer function as a function of frequency. The frequency is plotted on a logarithmic scale so that many decades of frequency may be observed.

The denominators of Eqs. (8-3) and (8-4) are frequency-dependent since both the $A$ and $B$ terms are frequency-dependent. In accordance, the desensitivity factor, sometimes called the *return difference*, is from Eq. (8-45).

$$D(j\omega) = 1 - A(j\omega)B(j\omega) \qquad (8\text{-}48)$$

In turn, the return ratio or loop transmission $T(j\omega)$ is defined as

$$T(j\omega) = -A(j\omega)B(j\omega) \qquad (8\text{-}49)$$

so that

$$D(j\omega) = 1 + T(j\omega) \qquad (8\text{-}50)$$

Note that if $D(j\omega)$ approaches zero, the gain with feedback approaches infinity, and the amplifier approaches the state of oscillation. Thus, a Bode plot of $T(j\omega)$ gives some insight into the stability of an amplifier. For oscillation,

$$T(j\omega) = -1 \qquad (8\text{-}51)$$

In order that a feedback amplifier remain stable, the magnitude $|T(j\omega)|$ should decrease to 1 (0 dB) before the phase angle of $T(j\omega)$ decreases from midband to $-180°$. Gain margin is defined as the negative decibel value of $|T(j\omega)|$ at which the phase angle is $-180°$ from midband. It is the additional gain that just makes the system unstable. Phase margin is defined as the additional phase lead which would cause the phase to be $-180°$ at $|T(j\omega)| = 1$.

**solution**    If the definitions of gain margin and phase margin are applied to the Bode plot as in Fig. 8-22b, it is seen that the amplifier is stable. Then, from the graph

$$\text{Gain margin} \approx 15 \text{ dB}$$

and

$$\text{Phase margin} \approx 60°$$

## 8.18    DETERMINING THE STABILITY OF A FEEDBACK AMPLIFIER

**problem 8.16**    The return ratio of a negative-feedback amplifier $T(j\omega)$ has a polar plot as in Fig. 8-23. Determine whether the amplifier is stable or unstable.

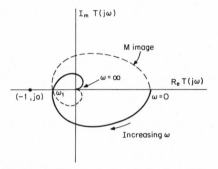

**Fig. 8-23**    A Nyquist plot for a feedback amplifier.

**theory**    The stability of a feedback amplifier may be determined by a number of graphical methods. One of these methods utilizes a Bode plot in which stability is determined by the presence of a gain and phase margin (see Sec. 8.17). Another method, called the Nyquist test or criterion, involves the inspection of a polar plot of $T(j\omega)$, called the Nyquist diagram or plot. For the negative feedback amplifiers commonly encountered, the test simplifies to a determination of whether the $T(j\omega)$ locus encircles the $(-1, j0)$ point. In this method, a Nyquist plot is drawn of $T(j\omega)$ from $\omega = 0^+$ to $\omega = \infty^+$. The mirror-image negative frequency plot from $\omega = \infty^-$ to $\omega = 0^-$ is drawn, and the figure is closed. If there is no encirclement of the point $(-1, j0)$, the amplifier is stable.

A third graphical technique of determining stability involves plotting the roots of the characteristic equation,

$$D(s) = 1 + T(s) = 0 \tag{8-52}$$

where $s = \sigma + j\omega$.

The roots as determined from Eq. (8-52) are the poles of the system, i.e., the values of complex frequency for which the closed-loop gain goes to infinity. Usually a gain parameter $K$ is added as a coefficient of $T(s)$, and the locus of solutions of the characteristic equation as $K$ is varied from 0 to $\infty$ is called the *root locus*. For the system to be stable, the root locus, for any gain parameter, must lie within the left-hand plane of the complex plane. Note that the Bode, Nyquist, and root locus methods provide the same basic information about amplifier stability.

**solution**    In this problem, the polar or Nyquist plot is given. As we traverse the $T(j\omega)$ loop from $\omega = 0^+$ to $\omega = \infty^+$ and back along the mirror image part, we do not enclose the critical point $(-1, j0)$. Thus the amplifier is stable.

## 8.19    CALCULATING A COMPENSATING NETWORK

**problem 8.17**    Calculate the values for a compensating network which, when connected to a feedback amplifier, will provide an additional 20 dB of gain reduction at a frequency of 100 kHz. The Bode plot of gain for the uncompensated and compensated cases is shown in Fig. 8-24.

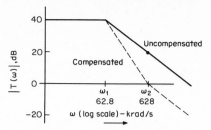

**Fig. 8-24** Compensation effect.

**theory** In Secs. 8.17 and 8.18, the stability of the network is shown to depend upon the phase and gain of the feedback amplifier. To be stable, a feedback amplifier must have sufficient gain and phase margin. Often the margin is provided by the use of $RC$ networks called *compensating* networks. Many compensating networks are available. Two such networks and their Bode plots are shown in Figs. 8-25 and 8-26.

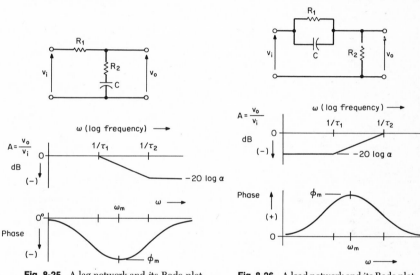

**Fig. 8-25** A lag network and its Bode plot.    **Fig. 8-26** A lead network and its Bode plot.

A network which introduces phase lag into the system is called a *lag* network, whereas a network which introduces phase lead into the system is called a *lead* network. Notice that the magnitude curves are approximated by straight lines which change slope at the break frequencies, $\omega_1 = 1/\tau_1$ and $\omega_2 = 1/\tau_2$. The maximum change in phase angle $\theta_m$ occurs at a frequency $\omega_m$. The design equations for each case are as follows:
Lag network:

$$\tau_1 = (R_1 + R_2)C \tag{8-53}$$

$$\tau_2 = R_2C \tag{8-54}$$

$$\alpha = \frac{R_2}{R_1 + R_2} \tag{8-55}$$

$$\theta_m = -\sin^{-1}\left(\frac{1-\alpha}{1+\alpha}\right) \tag{8-56}$$

$$\omega_m = \frac{\sqrt{\alpha}}{\tau_2} \tag{8-57}$$

Lead network:

$$\tau_1 = R_1 C \tag{8-58}$$

$$\tau_2 = \frac{R_1 R_2 C}{R_1 + R_2} \tag{8-59}$$

$$\alpha = \frac{R_2}{R_1 + R_2} \tag{8-60}$$

$$\theta_m = \sin^{-1}\left(\frac{1 - \alpha}{1 + \alpha}\right) \tag{8-61}$$

$$\omega_m = \frac{1}{\tau_1 \sqrt{\alpha}} \tag{8-62}$$

**solution** Notice that at the upper break frequency $\omega_2$ (corresponding to $f = 100$ kHz) the desired gain is 20 dB down from the uncompensated case, whereas a decade lower in frequency the gain is unchanged. A lag network is needed.

Assuming that the compensating network is inserted into the amplifier loop without changing the loading, the magnitude and phase response of the network may be added directly to those of the original loop to obtain the combined response. The three equations, Eqs. (8-53) to (8-55), are not independent, and so one of the parameters $R_1$, $R_2$, or $C$ must be chosen and the other two then found by use of the design equations. If $R_2$ is selected for loading considerations to be 1 kΩ, then from Eq. (8-54) $C$ is calculated:

$$C = \frac{\tau_2}{R_2} = \frac{1}{(628 \times 10^3)(1 \times 10^3)} \approx 0.00159 \, \mu\text{F}$$

From $-20 \log \alpha = -20$, $\alpha$ is found to be 0.1 so that from Eq. (8-55) $R_1$ is calculated as

$$0.1(R_1 + R_2) = R_2$$

or

$$R_1 = \frac{0.9}{0.1} R_2 = 9R_2$$

Thus, $R_1 = 9$ kΩ.

Note that from Eqs. (8-56) and (8-57) the maximum phase lag angle and the frequency at which it occurs are

$$\theta_m = -\sin^{-1}\left(\frac{1 - 0.1}{1 + 0.1}\right) = -\sin^{-1}\left(\frac{0.9}{1.1}\right) = -54.9°$$

$$\omega_m = \frac{\sqrt{0.1}}{\tau_2} = \sqrt{0.1} \, (628 \times 10^3) \approx 198 \times 10^3 \, \text{rad/s}$$

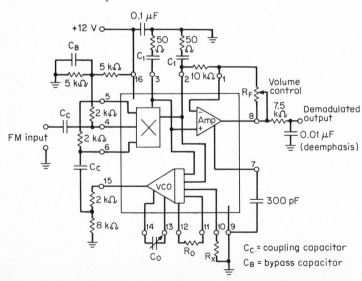

**Fig. 8-27** An fm demodulation circuit using the XR-215 PLL. (*Courtesy of Exar Integrated Systems, Inc.*)

## 8.20   CALCULATING THE CENTER FREQUENCY AND LOCK RANGE OF A PHASE-LOCKED LOOP

**problem 8.18**   Shown in Fig. 8-27 is an fm demodulation circuit using the XR-215 phase-locked loop (PLL). Calculate the value of timing capacitance $C_o$ and gain control resistance $R_o$ for an fm carrier frequency of 1 MHz and a lock range of $\pm 1\%$.

**theory**   The phase-locked loop (PLL) is a frequency feedback system composed of a phase comparator, a low-pass filter, an error amplifier, and a voltage-controlled oscillator (VCO). The block diagram of the PLL system is shown in Fig. 8-28.

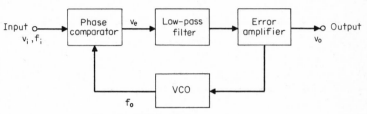

**Fig. 8-28**   Block diagram of a PLL system.

In operation, the VCO is a free-running oscillator which operates at a frequency $f_o$. When an input signal is applied to the system, the phase comparator compares the input-signal frequency $f_i$ to the VCO frequency $f_o$, and generates an error voltage $v_e$, which is related to the phase and frequency difference. The error voltage is filtered, amplified, and applied to the VCO. This forces the VCO frequency to change in a direction which reduces the frequency difference between $f_o$ and $f_i$.

The PLL essentially has three states: the free-running, capture, and lock states. The capture state during which the VCO is changing frequency is difficult to describe mathematically. In the lock state the PLL is synchronized or locked with the incoming signal, and the VCO frequency differs from the input only by a finite phase difference. In the lock state the PLL is approximated as a linear control system, and the following terms are used:

$K_d$ = phase comparator conversion gain, V/rad
$K_o$ = VCO conversion gain, rad/V·s
$\Delta \omega_L$ = lock range, the range of frequencies near $f_o$ over which lock can be maintained

The Exar XR-215 is one of the many commercially available monolithic PLLs. It is a general-purpose PLL which can be used for FM or FSK demodulation, frequency synthesis, and tracking filter applications. The block diagram is shown in Fig. 8-29.

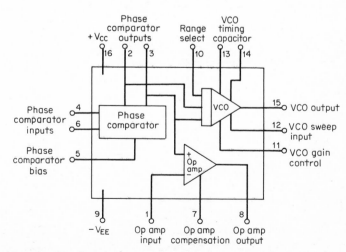

**Fig. 8-29**   Functional block diagram of the XR-215 PLL. (*Courtesy of Exar Integrated Systems, Inc.*)

Comparing Figs. 8-27 and 8-29, we note that as with most PLL monolithic circuits, external components are used to complete the circuit and to select the operating characteristics. These components are generally selected by data supplied by the manufacturer.

**solution**   The following design equations are applicable to the XR-215 PLL:

$$f_o = \frac{200}{C_o}\left(1 + \frac{0.6}{R_x}\right) \quad \text{Hz} \tag{8-63}$$

$$K_d \approx 0.08 \text{ V/rad} \quad \text{for} \approx 1 \text{ mV input} \tag{8-64}$$

$$K_o \approx \frac{700}{C_o R_o} \quad \text{rad/V} \cdot \text{s} \tag{8-65}$$

$$\Delta\omega_L = 2\pi \, \Delta f_L = K_d K_o \tag{8-66}$$

In Eqs. (8-63) to (8-66), the capacitances are in microfarads and the resistances are in kilohms. Resistor $R_X$ is a range extension resistor and is infinite (open circuited) for $f_o < 5$ MHz. From Eq. 8-63, the timing capacitor is calculated as (if $R_X = \infty$):

$$C_o = \frac{200}{f_o} \, \mu\text{F} = \frac{200}{1 \times 10^6} (10^{-6}\text{F}) = 200 \text{ pF}$$

Making a ratio of Eqs. (8-66) and (8-63), we have

$$\frac{\Delta f_L}{f_o} = \frac{K_d C_o}{2\pi(200)} K_o$$

and substituting the expression for $K_o$ (Eq. 8-65), we obtain,

$$\frac{\Delta f_L}{f_o} = \frac{K_d(700)}{2\pi(200)R_o}$$

Solving for $R_o$ and substituting values, we have

$$R_o = \frac{700 \, K_d}{200(2\pi)}\left(\frac{f_o}{\Delta f_L}\right) = \frac{700(0.08)}{200(2\pi)}\left(\frac{1}{0.01}\right) = 4.46 \text{ k}\Omega$$

Note that in the demodulator circuit the combination of $C_1$, the 50-$\Omega$ resistance, and the internal resistance of the comparator forms a low-pass $RC$ filter which determines the capture characteristic.

## 8.21  BIBLIOGRAPHY

Sentz, R. E., and Bartkowiak, R. A. *Feedback Amplifiers and Oscillators,* Holt, New York, 1968.
"Solid State Data Book—Linear Integrated Circuits and MOS Devices," RCA, Harrison, N.J., 1972.
XR-215 "Monolithic Phase-Locked Loop," Application Data, Exar Integrated Systems, Inc., Sunnyvale, Calif., October 1973.

# Chapter 9

# Oscillators

## JOHN T. MATTHEWS

**Professor and Department Head, Vocational Technical Education
Department, The University of Tennessee, Knoxville, Tenn.**

## 9.1 INTRODUCTION TO OSCILLATOR PRINCIPLES

Feedback, as discussed in Chap. 8, may be either negative or positive. When negative, an amplifier becomes more stable, with a resulting loss in gain. However, when positive feedback is applied to an amplifier, under most conditions it will become unstable and begin to oscillate.

The conditions under which a stable amplifier can become an oscillator are rather easy to predict. As will be shown, the feedback of the amplifier must be positive, the gain of the amplifier must be sufficient to overcome the losses due to feedback, and the feedback signal must, at some frequency, be in phase with the input signal. Infinitesimal noise-feedback voltage will normally have some component at the frequency having 360° phase shift. This tiny existing noise voltage is sufficient to start the amplifier to oscillate.

Depending on the net gain of the oscillating amplifier, the output oscillating voltage can be a nearly pure sine wave. When driven to near cutoff or saturation, the oscillator will develop distortion and have overtones, or harmonics. Generally the output is the sine wave that is desired, and the feedback network is selective to one frequency. If, however, one wishes to develop a signal at some harmonic of the fundamental oscillator frequency, the oscillator output may be deliberately distorted to ensure the presence of that harmonic.

The stability of an oscillator is a measure of how well it produces one frequency without drift. Drift in oscillator frequency is a function of the components' value stability. Physical and electrical dimensions of frequency-determining components are subject to change with changes in temperature, pressure, humidity, supply voltage, and loading. The most stable oscillators are constructed of precision resistors, capacitors, inductors, crystals, and active devices powered by regulated power supplies.

### The Barkhausen Criterion

**problem 9.1** The amplifier of Fig. 9-1 has the following values: $|A_v| = 20$, $|\beta| = 0.05$. Determine the closed-loop gain of the amplifier.

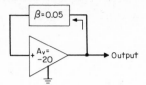

**Fig. 9-1**  Feedback amplifier connected as an oscillator.

**theory**   From feedback theory, it should be recalled that an amplifier with feedback, as in Fig. 9-1, displays a closed-loop gain of

$$A_f = \frac{A_v}{1 - \beta A_v} \tag{9-1}$$

where $A_v$ = open-loop gain
  $A_f$ = closed-loop gain
  $\beta$ = feedback factor

It can be seen that if the feedback factor $\beta$ is negative, the closed-loop gain $A_f$ will be less than the open-loop gain $A_v$. However, if the feedback factor is positive, the closed-loop gain may approach infinity. For example, if the feedback factor $\beta$ is the reciprocal of the open-loop gain $A_v$, the product $\beta A$ in Eq. (9-1) approaches 1: $\beta A_v = 1$. The denominator approaches zero, and the closed-loop gain approaches infinity.

**solution**

$$A_f = \frac{A_v}{1 - \beta A_v} = \frac{20}{1 - (0.05 \times 20)} = \frac{20}{1 - 1} = \frac{20}{0} = \infty \tag{9-1}$$

**discussion**   The condition in which oscillation will occur, and sustain itself, is thus described as an amplifier whose open-loop gain and positive-feedback factor meet the following criteria:

$$1 - \beta A_v = 0 \tag{9-2}$$

Stated another way,

$$\beta A_v = 1 \tag{9-3}$$

This criterion is generally called the *Barkhausen criterion* for oscillation. Actually, the positive feedback must not only satisfy this criterion, but it must satisfy it at the frequency where the feedback circuit produces a net phase shift of zero degrees. Since both $A_v$ and $\beta$ are complex numbers involving reactance factors in the circuit, Eq. (9-3) may be more accurately written as

$$\beta A_v = 1 \; \underline{/0°} = 1 + j0 \tag{9-4}$$

The zero-degrees phase shift condition for oscillation is, in reality, 360°. The oscillator/amplifier typically inverts its signal, producing a 180° phase shift. The frequency-selective positive-feedback network, connected back to an input point, must produce the additional 180° phase shift at the desired frequency.

Typically, the Barkhausen criterion should not be exceeded by a wide margin if a near-pure sine-wave output is desired. For greatest frequency stability, the oscillator feedback system should produce a rapidly changing phase shift for the narrow range of frequencies near the frequency of desired oscillation. This feedback condition will produce a more peaked feedback response at the selected frequency, and the oscillator will tend to drift in frequency less with temperature changes.

## Amplifier Stability

**problem 9.2**   Consider the circuit of Fig. 9-2. An amplifier must produce a stable gain of $A_v >$ −29. Assuming the output loading to be minimal, compute the value of unbypassed emitter resistance needed to ensure a circuit voltage gain greater than −29. Also compute the dc voltage that will result at the emitter and collector.

**theory**   The stability of an oscillator depends, to a great extent, on the initial stability of the amplifier from which the oscillator is derived. Biasing schemes to establish the thermal operating stability of the amplifier must then be the first design considerations. Sufficient amplifier gain must be present to overcome feedback losses; however, in solid-state circuitry, this presents no problem. As will be shown in the case of the phase-shift oscillator, a minimum gain of −29 is required just to overcome the feedback losses. Most modern silicon transistors can produce stable circuit voltage gains far exceeding this value.

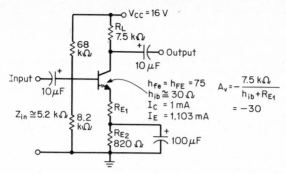

**Fig. 9-2** A stable voltage-divider biased amplifier that may be used as the basic circuit for an oscillator.

Voltage-divider biasing generally produces very stable amplifiers. The operating characteristics are relatively free of variations in the transistor's $h_{fe}$. The dc collector voltage is also quite stable. With voltage stability, the internal capacitances and resistances will remain essentially constant, thereby assisting in holding the operating frequency constant. Of course, other biasing designs will also provide acceptable results.

When an amplifier must act as a variable-frequency oscillator, the dc and ac designs must consider gain changes at the higher frequencies. Gain may change owing to the shunt capacitances of the transistors themselves. Selection of high-quality and high-frequency devices generally reduces these complications.

**solution**  Dc biasing for a collector current of 1 mA and a collector voltage approximately one-half the supply voltage will bias the amplifier for greatest output signal size and simplicity of design. The value of the emitter-base junction resistance $h_{ib}$ is about 30 Ω at 1 mA, and it is quite linear with respect to emitter current. The voltage-divider bias current sets the base voltage and swamps out any changes in base current. The emitter-base voltage at $I_E = 1$ mA is about 0.65 V.

Since the circuit voltage gain is a function of its current gain and resistance gain,

$$A_v = A_i A_r \tag{9-5}$$

For example, since

$$h_{ie} = (1 + h_{fe})(h_{ib}) \quad \text{and} \quad R'_{E_1} = (1 + h_{fe})R_{E_1} \tag{9-6}$$

$$A_r = \frac{R_L}{h_{ie} + R'_{E_1}} = \frac{R_L}{(1 + h_{fe})(h_{ib} + R_{E_1})} \tag{9-7}$$

Since the circuit current gain is essentially $h_{fe}$ ($A_i \cong h_{fe}$), Eq. (9-5) becomes

$$A_v = \frac{-h_{fe}R_L}{(1 + h_{fe})(h_{ib} + R_{E_1})} \tag{9-5}$$

Since $h_{fe} = 75$, then $h_{fe} \cong 1 + h_{fe}$ and therefore

$$A_v = \frac{-\cancel{h_{fe}}R_L}{(\cancel{1 + h_{fe}})(h_{ib} + R_{E_1})} \cong \frac{-R_L}{h_{ib} + R_{E_1}} \tag{9-8}$$

Since $h_{ib}$ is essentially inversely proportional to the emitter current and is known to be about 30 Ω at 1 mA for silicon, $h_{ib}$ is independent of $h_{fe}$. Circuit gain, then, for voltage-divider biasing is essentially independent of current gain. To calculate $R_E$ simplifies to

$$A_v = \frac{-R_L}{h_{ib} + R_{E_1}} = -29 = \frac{-7.5 \text{ k}\Omega}{30 \text{ }\Omega + R_{E_1}}$$

$$30 \text{ }\Omega + R_{E_1} = \frac{-7.5 \text{ k}\Omega}{-29} = 259 \text{ }\Omega$$

$$R_{E_1} = 259 \text{ }\Omega - 30 \text{ }\Omega = 229 \text{ }\Omega$$

The nearest standard resistor is 220 Ω for a gain of 30, which is greater than the required 29. The emitter dc voltage will be

$$V_E = (R_{E_1} + R_{E_2})I_E$$
$$V_E = (220 \text{ }\Omega + 820 \text{ }\Omega)(1 \text{ mA}) = 1.04 \text{ V} \tag{9-9}$$

The collector voltage will be

$$V_C = V_{CC} - V_{R_L} = V_{CC} - I_C R_L \tag{9-10}$$

Since $I_C \cong I_E$,

$$V_C = 16 - (1\ \text{mA} \times 7.5\ \text{k}\Omega) = 8.5\ \text{V}$$

## Frequency-Determining Components

Feedback elements all produce some phase shift since real resistors all contain dimensions of inductance and capacitance. The net reactance, capacitive or inductive, will determine the amount of phase shift. The most obvious reactive elements are, then, capacitors, inductors, and quartz crystals. However, by their structure, diodes also display capacitance which is a function of the width of the depletion area near the junction under reverse bias. Hence, the junction diode can be considered a voltage-controlled capacitor. Of course, all conductors display stray inductive and/or capacitive effects. These stray reactances often trigger unwanted oscillation.

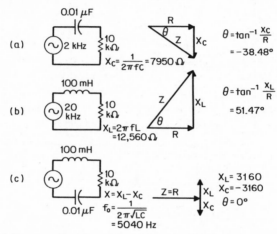

**Fig. 9-3**  Phase-shifting networks. (a) RC voltage lagging, (b) RL voltage leading, and (c) RLC series resonant.

A typical $RC$, $RL$, or $RLC$ network can produce a phase shift approaching 90°. When a circuit such as Fig-9.3a is used, it is obviously capacitive, and the voltage will lag the current. The amount of lagging of the phase angle is a function of frequency since the capacitive reactance is a function of frequency. As the value of the capacitance or the frequency is decreased, the phase angle will increase as indicated by Eqs. 9-11 and 9-12. The limit will be −90°.

$$X_C = \frac{1}{2\pi f C} \tag{9-11}$$

$$\theta = \tan^{-1}\frac{X_C}{R} \tag{9-12}$$

For the inductive circuit of Fig-9.3b, the inductive reactance increases with frequency or inductance. Therefore, the phase angle of back voltage leading the current will also increase with frequency. The limit, of course, will be +90°.

$$X_L = 2\pi f L \tag{9-13}$$

$$\theta = \tan^{-1}\frac{X_L}{R} \tag{9-14}$$

An $RLC$ circuit presents a circumstance in which at some frequency, known as resonance, $X_L = X_C$ in magnitude. Since they are 180° out of phase, the net effect is cancellation, with only the circuit resistance limiting the current. This situation is indicated in Fig-9-3c. This type of circuit is frequently used in feedback loops for narrow-band rejection if negative or for oscillation when positive. The frequency at which resonance occurs is, for all practical purposes,

$$f_r = \frac{1}{2\pi\sqrt{LC}} \qquad (9\text{-}15)$$

An *RLC* circuit such as Fig. 9-4a is sometimes called an *antiresonant circuit* and also a *tank circuit*. Equation (9-1) applies to this circuit as well as Fig. 9-3c; however, the net effect is opposite. At resonance the capacitor alternately is charged through the coil, then discharged the opposite direction through the coil. If there were no circuit resistance, the system would oscillate forever. However, such is not the case. At the antiresonant frequency, the tank circuit presents a very high impedance to the circuit current, even though the internal tank current is high. Circuit power is dissipated only in the resistance.

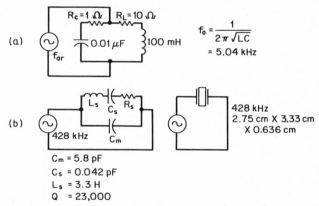

$$f_o = \frac{1}{2\pi\sqrt{LC}}$$
$$= 5.04 \text{ kHz}$$

**Fig. 9-4** *RLC* and quartz crystal tank circuits. (*a*) *RLC* antiresonant tank circuit, and (*b*) a quartz crystal equivalent circuit with physical and electrical dimensions.

The tank circuit is the basis for the Hartley, Colpitts, and other oscillators since it can be tuned to a range of frequencies by varying either $L$ or $C$. It is generally easier to make $C$ a variable capacitor.

Temperature changes cause expansion or contraction of the materials used in the manufacture of the frequency-determining components. Such changes affect both capacitance and inductances. The direction of change may be positive for some materials and negative for others. The net effect of temperature changes on an oscillator will be frequency drift. Precision components are constructed to keep these changes to a predictable minimum.

Quartz crystals are used to simulate the antiresonant tank circuit, but they have complex properties as shown in Fig. 9-4b. As can be seen, this equivalent circuit can act as a complex tank circuit or as a complex series-resonant circuit. The series-resonant frequency results from Eq. (9.15):

$$f_{sr} = \frac{1}{2\pi\sqrt{L_s C_s}} \qquad (9\text{-}16)$$

At a frequency slightly higher than that of series resonance, the net reactance of $X_{ls} - X_{Cs} = X_{Co}$. At this frequency antiresonance will occur.

$$f_{ar} = \frac{1}{2\pi\sqrt{L_s C_t}} \qquad (9\text{-}17)$$

where

$$C_t = \frac{C_s C_m}{C_s + C_m} \qquad (9\text{-}18)$$

The crystal's reactance varies with frequency as shown in Fig. 9-4. The resistance is very small; hence the $Q$ of a crystal is very high, in the order of 20,000 to 100,000.

The stability of the crystal is, again, a function of its dimensions. Crystals are very fragile and are generally kept in secure, temperature-controlled environments. Temperature or pressure changes can alter the crystal's dimensions and hence its operating characteristics.

### Sinusoidal and Nonsinusoidal Outputs

As suggested previously, an oscillator output is basically that of a positive-feedback amplifier. Normally, the feedback produces the zero-degree phase shift at only one frequency; hence the output should be a sine wave. However, if the phase shift changes slowly with a change in frequency, and if the feedback is large, the output waveform will have multifrequency components. Of course, if driven into saturation and cutoff, the output will resemble a square or rectangular waveform.

Most low-frequency applications require rather pure sine waves, and, therefore, most oscillators are devised for this purpose. Recently, however, function generators have come into use which produce square, triangular, and ramp waveforms as well as pulses. Nonsinusoidal waveforms are rich in harmonics of various orders. Amplifier testing with such waveforms is simplified since change of the complex waveform indicates frequency-response variations in the amplifier.

For VHF and UHF applications, it is difficult to design a stable oscillator to function at those frequencies. Instead, a well-controlled oscillator at a lower frequency is designed to produce a distorted waveform rich in second, third, fourth, or even fifth harmonics. The output of the oscillator, or some subsequent amplifier, is then tuned to respond to the higher harmonic desired. The selected harmonic then more closely resembles a sine wave.

While most sine waves are generated by feedback oscillators, nonsinusoidal waveforms are generally the result of switching circuitry. An astable circuit can be switched from saturation to cutoff quite easily at some desired frequency. The rectangular waveform may then be filtered to remove unwanted frequency components, finally yielding even a sine wave.

## 9.2  RC OSCILLATORS

Because inductors are cumbersome and difficult to handle, many circuits, especially at low frequencies, have been designed without them. As shown previously under Frequency-Determining Components, an $RC$ network can produce a phase shift approaching 90°. By using series networks of progressive phase shift, 180° is easily accomplished with three or more sections.

Other designs utilize bridging networks for more sharply defined "notch" filtering or feedback. Among the most common bridges used in oscillators are the Wien-bridge, the twin-T, and the bridged-T circuits. With the advent of IC op amps, active filters and oscillators using bridge networks have accomplished very high $Q$ circuits comparable with high-quality $LC$ circuits.

### The Phase-Shift Oscillator

Figure 9-5 shows the block diagram of a phase-shift oscillator while Fig. 9-6 shows an actual circuit. Note that the amplifier produces 180° phase shift. The feedback network must then produce an additional phase shift of 180°. The $RC$ feedback network also produces a significant loss of gain due to the nature of $RC$ networks. The voltage output of the three-stage feedback network is only 1/29 the size of the input. The amplifier must produce sufficient gain to overcome this loss at the frequency of operation.

Analyzing the three-section phase-shift network of Fig. 9-5 shows that $R_1$ passes $I_1$ in one direction and $I_2$ in the opposite direction. Current $I_2$ passes through both $R_1$ and $R_2$, and current $I_3$ passes through both $R_2$ and $R_3$.

From the above, it can be seen that the phase shift through $R_1$ and $C_1$, expressed as a voltage current function, can be given by

$$V_1 = I_1(R_1 - jX_C) - I_2R_1 \qquad (9-19)$$

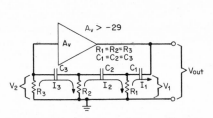

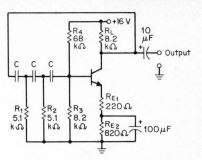

**Fig. 9-5**  A block diagram of an amplifier connected as a phase-shift oscillator.

**Fig. 9-6**  A practical phase-shift oscillator.

By similar reasoning, the phase shift through the second $RC$ section may be shown to be a function of

$$-I_1R_1 + I_2(R_1 + R_2) - jX_{C_2} - I_3R_2 = 0 \qquad (9\text{-}20)$$

Similarly, the last stage of the network sees

$$-I_2R_2 + I_3(R_2 + R_3) - jX_{C_3} = 0 \qquad (9\text{-}21)$$

If we assume $R_1 = R_2 = R_3$ and $C_1 = C_2 = C_3$, these three complex expressions can be solved to determine $I_3$ and then $V_2$. Once $V_2$ has been determined, the input-output voltage ratio, $V_1/V_2$, can be determined. The minimum amplifier gain needed to offset this feedback circuit loss results as well as the oscillation frequency.

$$\frac{V_1}{V_2} = \left(1 - j\frac{X_C}{R}\right)\left(2 - j\frac{X_C}{R}\right)^2 - 3 + j\frac{2X_C}{R} \qquad (9\text{-}22)$$

Since

$$X_C = -j\frac{1}{\omega C}$$

Eq. (9.22) expands to

$$\frac{V_1}{V_2} = -\left(\frac{5}{\omega^2 R^2 C^2} - 1\right) - j\left(\frac{6}{\omega RC} - \frac{1}{\omega^3 R^3 C^3}\right)$$

Since the second term, the reactive $j$ term, must be zero at the frequency of oscillation $f_o$, the gain expression $V_1/V_2$ at 180° becomes

$$\frac{V_1}{V_2} = -\left(\frac{5}{\omega^2 R^2 C^2} - 1\right) \qquad (9\text{-}23)$$

The second term in Eq. (9.22) then can be used to solve for $\omega_o$ at $f_o$

$$\frac{6}{\omega_o RC} - \frac{1}{\omega_o^3 R^3 C^3} = 0 \qquad (9\text{-}24)$$

$$\frac{6}{\omega_o RC} = \frac{1}{\omega_o^3 R^3 C^3}$$

$$\omega_o^2 = \frac{1}{6R^2 C^2}$$

$$\omega_o = \frac{1}{\sqrt{6}RC} \qquad (9\text{-}25)$$

Therefore,

$$f_o = \frac{1}{2\pi\sqrt{6}RC} \qquad (9\text{-}26)$$

By substituting Eq. (9.25) into Eq. (9.23), evaluating the signal loss in the frequency-selective feedback phase-shifting network is possible.

$$\frac{V_1}{V_2} = -\left(\frac{5}{\omega^2 R^2 C^2} - 1\right) = -\left[\frac{5}{(1/\sqrt{6}RC)^2 R^2 C^2} - 1\right]$$

$$= -\left[\frac{5(6R^2C^2)}{R^2C^2} - 1\right] = -(30 - 1) = -29 \tag{9-23}$$

By use of Eq. (9.26) and the results of Eq. (9.23), it is possible to determine circuit component values and the operating frequency of a phase-shift oscillator such as in Fig. 9-6. This oscillator uses the basic amplifier of Fig. 9-2 and the phase shift network of Fig. 9-5.

Since the effective input impedance of the amplifier of Fig. 9-2 is about 5.2 kΩ, it is best to use this value for $R$ of the phase shift network. In this way the input impedance of the amplifier can function as $R_3$ in the phase-shift network. Actually, the nearest standard resistor value for $R$ is 5.1 kΩ, which is close enough. It must be recalled that $Z_{in}$ changes somewhat with any changes in $h_{fe}$, so selecting a value of 5.1 kΩ is appropriate.

Note that $R_L$ in Fig. 9-6 has been increased to 8.2 kΩ to ensure that the voltage gain will exceed $-29$, the predicted loss of the phase-shifting network. To ensure oscillation, a higher gain may be used, but some distortion may become apparent.

**problem 9.3**   By use of the circuit of Fig. 9-6, what size phase-shifting capacitors should be used to produce an oscillator frequency of 3 kHz? Assume that $R = Z_{in} = 5.1$ kΩ.

**solution**

$$f_o = \frac{1}{2\pi RC\sqrt{6}} \tag{9-26}$$

$$C = \frac{1}{2\pi f_o R\sqrt{6}} \tag{9.27}$$

$$= \frac{1}{6.28 \times 3 \times 10^3 \times 5.1 \times 10^3 \times 2.45}$$

$$= 4{,}248 \text{ pF} = 0.004248 \text{ } \mu\text{F}$$

**problem 9.4**   If a three-section ganged variable capacitor with a range from 500 to 1500 pF were used in the phase-shift oscillator of Fig. 9-6, what would be the range of frequencies covered?

**solution**

$$f_1 = \frac{1}{6.28 \times 5.1 \times 10^3 \times 500 \times 10^{-12} \times 2.45} = 25{,}488 \text{ Hz} \cong 25.5 \text{ kHz}$$

$$f_2 = \frac{1}{6.28 \times 5.1 \times 10^3 \times 1500 \times 10^{-12} \times 2.45} = 8496 \text{ Hz} \cong 8.5 \text{ kHz}$$

$$\text{Range} = 8496 \text{ to } 25{,}488 \text{ Hz}$$

### The Wien-Bridge Oscillator

One of the most reliable $RC$-type oscillators uses the Wien bridge of Fig. 9-7. As with the phase-shift oscillator of Fig. 9-6, an $RC$ circuit is used as the frequency-selective network; however, the $RC$ Wien-bridge operation is quite different.

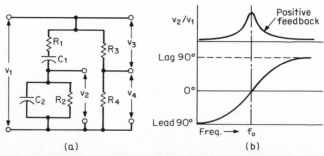

**Fig. 9-7**   (*a*) The Wien-bridge circuit showing feedback input voltage, $v_1$, feedback output voltages, $v_2$, positive, and $v_4$ negative, and (*b*) a comparison of the positive-feedback voltage and phase shift as a function of oscillating frequency.

The Wein-bridge oscillator uses either a two-stage amplifier or an operational amplifier to produce a full 360° phase shift. At the frequency of operation $f_o$, the reactive branch of the bridge produces a phase lead and lag voltage cancellation. If the net voltages $v_2$ and $v_4$ are equal and in phase at the desired frequency of operation, the positive feedback will cancel the negative feedback, and the circuit will oscillate. At any other frequency, the positive-feedback voltage $v_2$ will be too small to cancel the negative feedback $v_4$ and the circuit will not oscillate. Figure 9-7b illustrates the phase and feedback relationship. Notice that the positive-feedback voltage peaks at $f_o$, the frequency at which the phase shift is zero. There is no phase shift through the resistances $R_3$ and $R_4$, and so $v_4$ is always in phase with the input voltage $v_1$.

In practice, the Wien bridge is used with an operational amplifier or a two-stage amplifier. The resistive side of the Wien bridge acts as an in-phase voltage reference in a negative-feedback arrangement. The reference voltage $v_4$ is then connected to the inverting input as shown in Fig. 9-8. The positive feedback is through the phase-shift network, and so voltage $v_2$ is connected to the noninverting input.

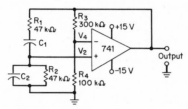

**Fig. 9-8**  A practical Wien-bridge oscillator using an op amp.

The closed-loop gain of the amplifier $A_f$ must be fixed at some value slightly greater than 3 since the positive-feedback factor $\beta$ at $f_o$ is to be one-third. That is, the value of $R_3$, the negative-feedback resistor, must be three times the value of $R_4$, connected as in Fig. 9-8.

The attenuation of the positive-feedback signal results from the bridge-circuit design where $R_1 = R_2$ and $C_1 = C_2$. The reactive branch of the bridge constitutes an ac voltage divider, the solution of which follows much the same procedure as with dc resistive values. For example:

$$\frac{V_4}{V_1} = \frac{R_4}{R_3 + R_4} \tag{9-28}$$

The evaluation is more complex since the reactances are out of phase with the resistances, and complex $j$ operators are generally used. Since at $f_o$ $v_4 = v_2$

$$\frac{v_4}{v_1} = \frac{v_2}{v_1} = \frac{R_4}{R_3 + R_4} = \frac{R(-j1/\omega C)/(R - j1/\omega C)}{(R - j1/\omega C) + [R(-j1/\omega C)/(R - j1/\omega C)]} \tag{9-29}$$

This simplifies somewhat to

$$\frac{v_2}{v_1} = \frac{-j(R/\omega C)}{[R - j(1/\omega C)]^2 - j(R/\omega C)}$$

which further simplifies to a form allowing the $j$ term to be extracted as zero at the desired frequency $f_o$ when $\omega = \omega_o$.

$$\frac{v_2}{v_1} = \frac{R/\omega C}{3R/\omega C + j(R^2 - 1/\omega^2 C^2)}$$

which at $f_o$ becomes

$$\frac{v_2}{v_1} = \frac{R/\omega_o C}{3R/\omega_o C} = \frac{1}{3} \tag{9-30}$$

Solving for $f_o$ using the reactive $j$ term yields

$$R^2 - \frac{1}{\omega_o^2 C^2} = 0 \qquad \text{therefore} \qquad \omega_o^2 = \frac{1}{R^2 C^2} \tag{9-31}$$

$$\omega_o = \frac{1}{RC} \tag{9-32}$$

$$f_o = \frac{1}{2\pi RC} \tag{9-33}$$

If the Wien-bridge oscillator is not loaded in such a way as to distort the calculated values, Eq. (9.33) will yield the frequency of oscillation.

**problem 9.5**  A Wien-bridge oscillator is to be designed using standard value resistors and capacitors that are variable two-ganged units. The oscillator should tune from 1 to 3 kHz. The amplifier gain should be 3. What value should be selected for $R_3$ if $R_4 = 100$ kΩ? Assume $R_1 = R_2 = 100$ kΩ. What range of values should be used for the two-ganged variable capacitor?

**solution**

$$A_v = 3 = \frac{R_3}{R_4} = \frac{R_3}{100 \text{ k}\Omega} \tag{9-34}$$

$$R_3 = 3 \times 100 \text{ k}\Omega = 300 \text{ k}\Omega$$

$$C_{f_1} = \frac{1}{2\pi f_1 R} = \frac{1}{6.28 \times 1 \times 10^3 \text{ Hz} \times 1 \times 10^5 \text{ }\Omega} \tag{9-35}$$

$$= \frac{1}{6.28 \times 10^8} = 1.59 \times 10^{-9} = 1590 \text{ pF}$$

$$C_{f_2} = \frac{1}{2\pi f_2 R} = \frac{1}{6.28 \times 3 \times 10^3 \text{ Hz} \times 1 \times 10^5 \text{ }\Omega} \tag{9-35}$$

$$= \frac{1}{1.88 \times 10^9} = 5.31 \times 10^{-10} = 531 \text{ pF}$$

Range of $C = 531$ to $1590$ pF.

### The Twin-T Oscillator

With the advent of the IC op amp, very high $Q$ circuits using $RC$ networks became economical. Active filters became commonplace. A group of "notch" filters took on new significance since they could easily reject a narrow band of frequencies. Notable in this group are the twin-T and the bridged-T circuits. The attenuation at frequencies some distance from $f_o$ is essentially zero, while at $f_o$ the attenuation is extremely high.

Figure 9-9 shows examples of the twin-T and bridged-T networks. The similarity is notable. The notch filters are actually combined high-pass and low-pass filters. By matching components, the roll-off of the filters causes an intersection at $f_o$ that may be 60 dB below the unattenuated frequencies as shown in Fig. 9-10.

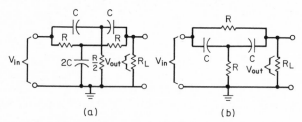

**Fig. 9-9**  (a) Twin-T filter network, and (b) bridged-T filter network.

When the twin-T filter is used in the negative-feedback loop of an operational amplifier, the amplifier has low gain at all but the notch frequency. Since the negative feedback is so low due to high feedback impedance at the notch frequency, the gain of the amplifier is extremely high. This causes instability and oscillation. The gain of the twin-T oscillator in Fig. 9-11 is set by the value of $R_4$, a variable resistance, since the op amp gain with negative feedback is

$$A_f = \frac{-Z_f}{Z_{in}} \tag{9-36}$$

Since $Z_f = Z_{\text{twin T}}$ and $Z_{\text{in}} = R_4$,

$$A_f = \frac{-Z_{\text{twin T}}}{R_4} \qquad (9\text{-}37)$$

Adjustment of $R_4$ is necessary to provide sufficient gain to start oscillation. Once started, $R_4$ acts as an amplitude control. For stable operating frequency, matched precision components should be used.

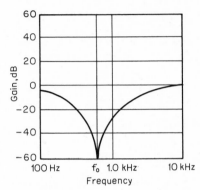

**Fig. 9-10** A graph of the typical frequency response of a notch filter.

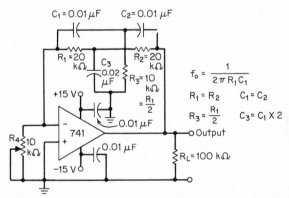

**Fig. 9-11** A practical twin-T oscillator using the 741 op amp. Note that capacitors $C_5$ and $C_6$ are used to bypass the power supply and do not affect the frequency of operation.

For the twin-T oscillator, the component values in the two legs must be observed. Notice that

$$R_3 = \frac{R_1}{2} \qquad (9\text{-}38)$$
$$C_3 = 2C_1 \qquad (9\text{-}39)$$

Selection of standard components that have values which are multiples of 2 may be difficult. Frequently, variable resistances are required to select a particular frequency.

The frequency of operation of the twin-T oscillator is computed from the expression

$$f_o = \frac{1}{2\pi R_1 C_1} \qquad (9\text{-}40)$$

The expression is the same as that used for the Wien-bridge oscillator and is also used for a bridged-T oscillator. It should be pointed out, however, that sometimes more than one filter section is used to provide a deeper notch. In that case the expression is different for $n$ sections. For example, for $n = 2$, for a two-section notch filter using the twin-T arrangement, the expression for operating frequency becomes

$$f_o = \frac{1}{2\pi n R_1 C_1} = \frac{1}{4\pi R_1 C_1} \qquad \text{for } n = 2 \tag{9-41}$$

For the bridged-T two-section filter, the expression becomes

$$f_o = \frac{n}{2\pi RC} = \frac{2}{2\pi RC} = \frac{1}{\pi RC} \qquad \text{for } n = 2 \tag{9-42}$$

It is important that the twin-T or bridged-T feedback circuit not be loaded appreciably. The open input impedance of the 741 op amp is about 2 M$\Omega$, while its output impedance is of the order of 75 $\Omega$.

**problem 9.6**   Select values of resistance and capacitance to produce an output frequency between 1.0 and 1.3 kHz from the twin-T oscillator using standard value components. What would be the expected deviation of frequency if the resistances averaged a change of $\pm 2\%$ with normal temperature changes? Assume no change in capacitance.

**solution**   First select capacitor values that are standard, since $C_3 = 2C_1$. The best selection is one which has a value half of another standard value, such as 0.01, 0.02, 0.04 $\mu$F. Selection of resistance follows from Eqs. (9.38) to (9.40).

$$C_3 = 2C_1 = 2 \times 0.01 \ \mu\text{F} = 0.02 \ \mu\text{F}$$
$$R_1 = \frac{1}{2\pi f_o C_1} = \frac{1}{6.28 \times 1 \text{ kHz} \times 0.01 \ \mu\text{F}} \tag{9-43}$$
$$= 15,900 \ \Omega$$

The nearest standard resistor is 15 k$\Omega$. There is also a standard resistor half this value; therefore, $R_3 = 7.5$ k$\Omega$, $\pm 5\%$. The resulting value of $f_o$ will be

$$f_o = \frac{1}{2\pi R_1 C_1} = \frac{1}{6.28 \times 15 \text{ k}\Omega \times 0.01 \ \mu\text{F}}$$
$$= 1.060 \text{ kHz} \tag{9-40}$$

If the value of the resistances change $\pm 2\%$

$$15,000 \times 0.02 = 300$$
$$15,000 - 300 = 14,700$$
$$15,000 + 300 = 15,300$$

$$f_1 = \frac{1}{6.28 \times 15,300 \ \Omega \times 0.01 \ \mu\text{F}} = 1039 \text{ Hz} \tag{9-40}$$

$$f_2 = \frac{1}{6.28 \times 14,700 \ \Omega \times 0.01 \ \mu\text{F}} = 1081 \text{ Hz} \tag{9-40}$$

Frequency deviation = 1.060 kHz $\pm 21$ Hz

## 9.3   THE *LC* FEEDBACK OSCILLATORS

*RC* oscillators are generally used for low-frequency applications up to about 1 MHz. It is the *LC* oscillator, however, that is most frequently used for radio-frequency applications. Controlling the phase shift of the *RC* oscillator amplifier can become somewhat of a problem at radio frequencies. Values of resistance and capacitance may also become difficult to control. On the other hand, small inductances can be used in conjunction with capacitors to tune *LC* feedback oscillators up to about 500 MHz.

As with *RC* oscillators, feedback must produce a net phase shift of zero degrees at the desired operating frequency. Since the signal at the emitter and collector are in phase with each other, feedback from collector to emitter may be easily employed. Of course, most arrangements take advantage of the 180° phase shift between the base and collector, allowing the feedback path to produce the other 180° shift, as in the Hartley and Armstrong oscillators.

The *LC* tank circuit forms the basis of many *LC* oscillators. The tank is generally in the collector circuit, and at the resonant frequency it appears resistive. At all other frequencies, the tank circuit appears either capacitively or inductively reactive. Figure 9-12 illustrates variations of the tank circuits to be discussed: (a) the Colpitts, (b) the Clapp, (c) the Hartley, and (d) the Pierce.

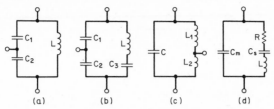

**Fig. 9-12**  Typical tank circuit configurations. (a) Colpitts, (b) Clapp, (c) Hartley, and (d) Pierce (crystal).

## The Colpitts Oscillator

**problem 9.7**    Figure 9-13 illustrates a practical Colpitts oscillator. Compute the component values required to start oscillation and operate at 2.36 MHz. $L = 10\ \mu H$, with a resistance of $2\Omega$. Use standard-value components where possible. Assume a feedback ratio of 1:10. Determine $Q$, $r_{tank}$, $r_L$, and $A_v$.

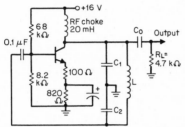

**Fig. 9-13**  A practical shunt-fed Colpitts oscillator.

**theory**    For a Colpitts oscillator to start up and oscillate at the proper frequency, the voltage gain of the circuit must meet or exceed the Barkhausen criterion. That is to say, to start,

$$A_v\beta \geq 1 \tag{9-3}$$

or

$$A_v \geq \frac{1}{\beta} \tag{9-44}$$

The tank circuit for the Colpitts oscillator is shown in Fig. 9-12a, as well as in Fig. 9-13. Notice that the output voltage $v_{out}$, collector to ground, appears across $C_1$, while the feedback voltage $v_{fb}$ base to ground, appears across $C_2$. The ratio of the voltages $v_f/v_{out}$, then, is the feedback ratio and is equal to the ratio of

$$\beta = \frac{X_{C_2}}{X_{C_1}} = \frac{1/\omega_o C_2}{1/\omega_o C_1} = \frac{C_1}{C_2} \tag{9-45}$$

or

$$\frac{1}{\beta} = \frac{C_2}{C_1} \tag{9-46}$$

Stated another way, for oscillator startup,

$$A_v \geq \frac{C_2}{C_1} \tag{9-47}$$

The antiresonant frequency for this Colpitts tank circuit occurs when

$$X_L = X_C \qquad \text{where } C = \frac{C_1 C_2}{C_1 + C_2} \tag{9-48}$$

$$2\pi fL = 1/2\pi fC \tag{9-49}$$

$$f_o^2 = \frac{1}{4\pi^2 LC} \tag{9-50}$$

$$f_o = \frac{1}{2\pi \sqrt{LC}} \tag{9-15}$$

**Solution** Since $L = 10\ \mu$H, $C$ can be determined from Eq. (9-50).

$$C = \frac{1}{f_o^2 4\pi^2 L}$$

$$= \frac{1}{(2.36 \times 10^6) \times 4(3.14)^2 \times 10 \times 10^{-6}}$$

$$= \frac{1}{5.57 \times 10^{12} \times 4 \times 9.87 \times 10 \times 10^{-6}} = \frac{1}{2199 \times 10^6}$$

$$= 0.000455\ \mu\text{F} = 455\ \text{pF}$$

(9-51)

If $\beta = 1/10$, then $1/\beta = 10 = C_2/C_1$ and $C_2 = 10C_1$.

Since

$$C = \frac{C_1 C_2}{C_1 + C_2} = \frac{C_1 \times 10C_1}{C_1 + 10C_1} = \frac{10C_1^2}{11C_1} = \frac{10C_1}{11}$$

$$C_1 = \frac{11C}{10} = \frac{11 \times 455\ \text{pF}}{10} = 500\ \text{pF}$$

(9-48)

$$C_2 = 10C_1 = 5000\ \text{pF}$$

A tank circuit with a high $Q$ is necessary to sustain oscillation and produce a near-sinusoidal waveform when driven beyond class A. The $Q$ of the tank circuit is a function of the inductive reactance of the coil compared with the resistance of the coil. That is,

$$Q = \frac{X_L}{R_s}$$

(9-52)

The larger the series resistance, the faster the energy within the tank is dissipated, and the sooner oscillation will cease. Notice that the coil's resistance is stated as $2\ \Omega$, and that the inductance is $10\ \mu$H. At the resonant frequency,

$$X_L = 2\pi fL = 6.28 \times 2.36 \times 10^6 \times 10 \times 10^{-6}$$
$$= 148\ \Omega$$

(9-13)

Therefore

$$Q = \frac{148\ \Omega}{2\ \Omega} = 74$$

(9-52)

The above relationship is important when calculating the ac load $r_L$ seen by the collector of the transistor. First of all, the collector sees the RFC as essentially an open circuit owing to its high $X_L$, which is in parallel with the output load of $C_c$ and $R_L$.

$$r_L = X_{L,\text{RFC}} \| r_\text{tank} \| R_L$$

(9-53)

$$r_\text{tank} = QX_L$$
$$= 74 \times 6.28 \times 2.36\ \text{MHz} \times 10\ \mu\text{H}$$
$$= 10{,}967\ \Omega$$

(9-54)

Since $X_{L\text{RFC}}$ is about 30 times larger than $r_\text{tank}$, it can be neglected in the calculation; however, $R_L$ is only about 4.7 k$\Omega$ and will present a significant loading effect. Therefore,

$$r_L = r_\text{tank} \| R_L$$

$$= \frac{10{,}967 \times 4700}{10{,}967 + 4700} = 3290\ \Omega = 3.29\ \text{k}\Omega$$

(9-55)

The signal gain $A_v$ of the circuit must exceed $1/\beta = 10$.

The circuit of Fig. 9-13 has an emitter current of about 1.2 mA; therefore,

$$h_{ib} \cong \frac{30}{I_E} \cong \frac{30}{1.2} \cong 25$$

$$A_v = \frac{r_L}{h_{ib} + R_{E_1}} = \frac{3.29\ \text{k}\Omega}{25\ \Omega + 100\ \Omega} = 26.3$$

(9-56)

This signal gain certainly exceeds the minimum of 10 required and should ensure oscillation start up. Once in operation, the circuit parameters will adjust to meet the criterion $A\beta = 1$. The ratio of $C_2/C_1$ may be increased in order to increase the output voltage. It should be remembered, however, that too much feedback will produce distortion. Remember, too, that when the ratio is changed, the change in capacitance will also change the frequency of operation.

The equation for resonant frequency is the result of a complex number derivation. It is accurate only when the tank circuit $Q$ is much greater than 10. For $Q$ less than 10, the equation to be used should be

$$f_o = \frac{1}{2\pi \sqrt{LC}} \times \sqrt{\frac{Q^2}{1 + Q^2}} \tag{9-57}$$

For practical considerations, the reactance of the RFC should be much greater than the reactance of the coupling capacitors.

**problem 9.8**   What would be the operating frequency of the Colpitts oscillator of Fig. 9-13 if the value of $C_1$ were changed to 330 pF? If $C_1$ were to remain 500 pF, but $C_2$ were to change to 6200 pF, what will be the new operating frequency? Which has the greater effect on frequency change, $C_1$ or $C_2$? Which has the greater change on $\beta$?

**solution**   For $C_1 = 330$ pF and $C_2 = 5000$ pF,

$$C = \frac{330 \times 5000}{330 + 5000} = 310 \text{ pF}$$

Change in $C = 34\%$ and the change in $\beta = 0.10$ to $0.066 = 34\%$. For $C_1 = 500$ pF and $C_2 = 6200$ pF,

$$C = \frac{500 \times 6200}{500 + 6200} = 463 \text{ pF}$$

Change in $C = 34\%$, and the change in $\beta = 0.10$ to $0.08 = 20\%$.

$$f_{o_1} = \frac{1}{2\pi \sqrt{LC}} = \frac{1}{6.28 \sqrt{10 \times 10^{-6} \times 311 \times 10^{-12}}}$$
$$= 2.85 \text{ MHz } (+21\% \text{ change}) \tag{9-15}$$
$$f_{o_2} = \frac{1}{6.28 \sqrt{10 \times 10^{-6} \times 463 \times 10^{-12}}} = 2.34 \text{ MHz } (-0.85\% \text{ change})$$

A change in $C_1$, the smaller capacitor, produces the greater frequency deviation since the summation of $C_1$ and $C_2$ will always be less than the smaller capacitor. For the same percentage change, $C_2$ has the greater effect on the change in $\beta$.

## The Clapp Oscillator

**problem 9.9**   Consider the Clapp oscillator of Fig. 9-14. Determine the $Q$ of the tank, the value of $\beta$, the value of $r_L$, the effective value of $C$, and the resonant frequency. Determine the gain $A_v$ of the circuit and whether it will start oscillation on its own.

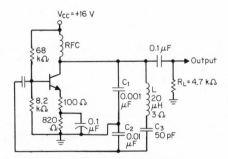

**Fig. 9-14**   A practical shunt-fed Clapp oscillator.

**theory**   The Clapp oscillator is a variation of the Colpitts oscillator in that the feedback voltage divider is capacitive. From Fig. 9-14, it can be seen that $C_1$ and $C_2$ control the feedback ratio $\beta$ as in the Colpitts. However, the presence of $C_3$, a very small capacitor, by comparison with $C_1$ or $C_2$, places it in control of the frequency.

The internal tank current of the Clapp antiresonant circuit passes through all three capacitors and the inductance. All components, therefore, must be considered as affecting the antiresonant frequency. However, since $C_3$ is so small, by comparison with $C_1$ and $C_2$, it is the determining capacitance. The values of $C_1$ and $C_2$ can change up to 20% with little effect on the effective capacitance, which will be slightly less than $C_3$.

**solution**    Using the equation to determine the effective capacitance for a series circuit yields:

$$C_{eff} = \frac{1}{1/C_1 + 1/C_2 + 1/C_3} \cong C_3 \qquad (9\text{-}58)$$

$$= \frac{1}{1000 + 100 + 20,000} = \frac{1}{21,100} = 47.4 \text{ pF}$$

$$f_o = \frac{1}{2\pi \sqrt{LC_{eff}}} = \frac{1}{6.28 \sqrt{20 \times 10^{-6} \times 47.4 \times 10^{-12}}} \qquad (9\text{-}15)$$

$$= \frac{1}{6.28 \sqrt{948 \times 10^{-18}}} = \frac{1}{6.28 \times 30.79 \times 10^{-9}}$$
$$= 5.172 \text{ MHz}$$

With the circuit's operating frequency known, the reactance of the inductors and the $Q$ of the circuit may be evaluated.

$$X_L = 2\pi fL = 6.28 \times 5.172 \times 10^6 \times 20 \times 10^{-6} \qquad (9\text{-}13)$$
$$= 650 \text{ }\Omega$$

$$Q = \frac{X_L}{R_s} = \frac{650 \text{ }\Omega}{3 \text{ }\Omega} = 217 \qquad (9\text{-}52)$$

The effective resistance of the tank-circuit load, as seen by the collector, may be determined from Eq. (9-54).

$$r_{tank} = QX_L$$
$$= 217 \times 650 \text{ }\Omega = 141 \text{ k}\Omega \qquad (9\text{-}54)$$

Using Eq. (9.55), we may evaluate the net load of the circuit.

$$r_L = r_{tank} \| R_L$$
$$= \frac{141 \text{ k}\Omega \times 4.7 \text{ k}\Omega}{141 \text{ k}\Omega + 4.7 \text{ k}\Omega} = 4.55 \text{ k}\Omega \qquad (9\text{-}55)$$

The ac voltage gain of the amplifier $A_v$, according to Eq. (9.56), is

$$A_v = \frac{r_L}{h_{ib} + R_{E_1}} = \frac{4.55 \text{ k}\Omega}{25 \text{ }\Omega + 100 \text{ }\Omega} = 36.4 \qquad (9\text{-}56)$$

The feedback ratio of the Clapp circuit is set by $C_1$ and $C_2$.

$$\beta = \frac{C_1}{C_2}$$

$$\beta = \frac{0.001 \text{ }\mu\text{F}}{0.01 \text{ }\mu\text{F}} = \frac{1}{10} \qquad (9\text{-}45)$$

Since the voltage gain of the circuit exceeds the value of $1/\beta$, the circuit should start to oscillate with no difficulty.

It can be concluded that if the $\beta$ of the circuit, fixed by $C_1$ and $C_2$, uses values, the smallest of which is at least 20 times larger than $C_3$, the frequency will be stable and primarily a function of the quality of $C_3$. If $C_3$ is made variable, it can be used to tune the oscillator, reducing the complexity of tuning as in the Colpitts oscillator.

## The Hartley Oscillator

**problem 9.9**    Figure 9-15b is a Hartley oscillator using a transistor amplifier. What is the range of frequencies through which it will tune? At the lowest frequency, compute the appropriate values for $C_c$ and $C_L$. Compute $\beta$, the gain $A_v$ required, and an appropriate value for $R_{E1}$, $C_E$, and RFC.

**theory**    Many similarities exist between the Hartley and the Colpitts oscillators. The major difference is in the feedback system. The Colpitts uses a capacitor voltage divider, whereas the Hartley uses an inductive voltage divider to determine the feedback ratio.

While the Hartley and Colpitts oscillators have many forms, depending on whether the active device is a transistor, an FET, or other device, the basic relationships are essentially the same. Figure 9-15 shows two forms of the Hartley, one using a FET and the other using a transistor.

The frequency of operation is determined, to a close approximation, from the constants of the tank circuit. The tuning capacitor $C_T$ allows the Hartley oscillator to be tuned over a wide range of frequencies. The lowest frequency is determined by the maximum capacitance of $C_T$.

Since the inductance is tapped, to control the feedback ratio, in effect $L$ is an autotransformer, where $L_1$ is the primary and $L_2$ is the secondary. Feedback via $L_2$ is coupled to the base through

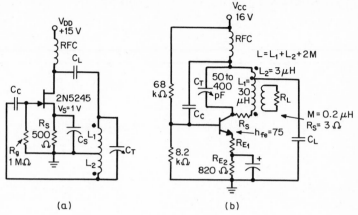

**(a)**                                    **(b)**

**Fig. 9-15** Practical Hartley oscillators: (*a*) FET shunt-fed Hartley and (*b*) transistor series-fed Hartley.

coupling capacitor $C_C$. The reactance of $C_C$ then should be very small compared with $Z_{in}$ at the lowest frequency of operation.

**solution**   The computation of the net inductance $L$ of the tank circuit must include the mutual inductance of the autotransformer if it is more than about 15% of $L_2$. For the tank circuit of Fig. 9-15*b*,

$$L = L_1 + L_2 + 2\,M$$
$$= 30\ \mu\text{H} + 3\ \mu\text{H} + 2(0.2\ \mu\text{H}) = 33.4\ \mu\text{H} \tag{9-59}$$

Use of the fully closed (largest) capacitance then allows computation of the lowest frequency that may be generated. Since $C_T$ is variable from 50 to 400 pF,

$$f_L = \frac{1}{2\pi\sqrt{LC}} = \frac{1}{6.28\sqrt{33.4\times10^{-6}\times400\times10^{-12}}} \tag{9-15}$$

$$= \frac{1}{6.28\sqrt{13360\times10^{-18}}} = \frac{1}{6.28\times115.59\times10^{-9}} \tag{9-15}$$
$$= 1{,}375{,}605\ \text{Hz} \cong 1376\ \text{kHz}$$

$$f_H = \frac{1}{6.28\sqrt{33.4\times10^{-6}\times50\times10^{-12}}}$$

$$= \frac{1}{6.28\sqrt{1670\times10^{-18}}} = \frac{1}{6.28\times40.87\times10^{-9}}$$

$$= \frac{1}{256.66\times10^{-9}} = 3{,}896{,}150\ \text{Hz} \cong 3896\ \text{kHz}$$

From the value of $f_L$, 1,375,605 Hz, the inductive reactance of $L_2$, 3 $\mu$H, may be computed.

$$X_{L_2} = 2\pi f_{L_2} = 6.28\times1{,}375{,}605\ \text{Hz}\times3\times10^{-6}\ \text{H}$$
$$= 25.92\ \Omega \tag{9-13}$$

The capacitive reactance $X_{C_C}$ should be much less than 10% of $Z_{in}$ at the lowest frequency of operation, that is, less than 200 at 1.38 MHz, since $Z_{in}$ is about 2 k$\Omega$.

$$C_C = \frac{1}{2\pi f\times0.1\ Z_{in}}$$
$$= \frac{1}{6.28\times1.38\times10^6\times200\ \Omega} = 0.0006\ \mu\text{F} \tag{9-60}$$

To ensure proper coupling, $C_C$ should be much larger than 0.0006 $\mu$F. A 0.005- to 0.01-$\mu$F ceramic disk or mica capacitor should be employed since either would be less inductive and more convenient to use than many other types.

The feedback ratio $\beta$ is established by proper ac grounding of the tap between $L_1$ and $L_2$. This point must not be directly connected to ground or the collector will become grounded. The value of $C_L$ is established in the same manner as for $C_C$. In this case, $X_{C_L} \ll X_{L_2}$.

$$C_L \geq \frac{1}{2\pi f \times 0.1\, X_{L_2}}$$

$$\geq \frac{1}{6.28 \times 1.38 \times 10^6 \times 2.6} \geq 0.045\ \mu\text{F} \tag{9-61}$$

As with selecting $C_C$, a convenient value larger than $0.045\ \mu\text{F}$ should be chosen. Nothing smaller than $0.05\ \mu\text{F}$ should be used, preferably $0.1\ \mu\text{F}$ to ensure proper *ac* ground.

Assuming $C_L$ is sufficiently large, the feedback ratio is essentially equal to the ratio of $L_2$ to $L$.

$$\frac{L_2}{L} = \frac{3\ \mu\text{H}}{33.4\ \mu\text{H}} = 0.09 \tag{9-62}$$

In order to ensure starting, the gain of the amplifier must be greater than $1/\beta$.

$$\frac{1}{\beta} = \frac{L}{L_2} = \frac{33.4\ \mu\text{H}}{3\ \mu\text{H}} = 11.13 \tag{9-63}$$

Therefore, $A_v > 11.13$.

Computing the actual load seen by the collector is quite a problem because of several unknown factors. Assuming the signal is coupled from the tank circuit via the transformer coupling loaded by $R_L = 5\ \text{k}\Omega$, a complex parallel load exists.

$$r_L \cong R_L \| QX_L \| X_{(\text{RFC} + L_2)} \| r_{\text{coll}} \tag{9-64}$$

However, for all practical purposes, the reactance of the RFC and the collector resistance of the transistor are sufficiently large as to be unimportant in the calculations. This leaves the coupled load and the tank impedance $QX_{L_1}$.

$$X_{L_1} = 6.28 \times 1.38\ \text{MHz} \times 30\ \mu\text{H} = 260\ \Omega \tag{9-13}$$

$$Q = \frac{X_L}{R_S} = \frac{260\ \Omega}{3\ \Omega} = 87 \tag{9-52}$$

$$r_{\text{tank}} = QX_{L_1} = 87 \times 260\ \Omega = 22.6\ \text{k}\Omega \tag{9-54}$$

$$r_L = 5\ \text{k}\Omega \| 22.6\ \text{k}\Omega = 4\ \text{k}\Omega \tag{9-55}$$

To start oscillation and sustain it, $A_v > 11.13$. Since

$$A_v \geq \frac{r_L}{h_{ib} + R_{E_1}} \geq 11.13 \tag{9-56}$$

$$R_{E_1} = \frac{r_L}{A_v} - h_{ib} \tag{9-65}$$

If $I_E$ is $1.2\ \text{mA}$, $h_{ib} = 25\ \Omega$, then

$$R_{E_1} = \frac{4\ \text{k}\Omega}{11.13} - 25\ \Omega = 334\ \Omega \tag{9-65}$$

In other words, if $R_{E_1} = 334\ \Omega$, the gain of the amplifer would cause the circuit to sustain oscillation. To ensure starting, the gain must exceed 11.13; therefore, a smaller resistance, between $10\ \Omega$ and $334\ \Omega$, should be chosen. Standard resistances of 100, 150, 200, or $220\ \Omega$ will work well. With $100\ \Omega$, the gain will be about 32, while with $220\ \Omega$, the gain will be about 16.

The value of the bypass capacitor $C_E$ is determined from the values of $R_{E_2}$ and the operating frequency.

$$C_E = \frac{1}{2\pi f X_{C_E}} = \frac{1}{2\pi f \times 0.1 R_{E_2}}$$

$$= \frac{1}{6.28 \times 1.38 \times 10^6 \times 82} = 0.0014\ \mu\text{F} \tag{9-66}$$

To ensure good bypassing action, a ceramic capacitor with a value of $0.005\ \mu\text{F}$ or greater should be used.

The value of RFC is chosen such that its reactance is much greater than that of $X_{C_C}$. Since $X_{C_C}$ was selected to be about $200\ \Omega$ or less, $X_{\text{RFC}}$ should be at least $10\, X_{C_C}$.

$$X_{\text{RFC}} = 2\pi f L_{\text{RFC}} \tag{9-67}$$

$$L_{\text{RFC}} = \frac{X_{\text{RFC}}}{2\pi f} \geq \frac{10\, X_{C_C}}{2\pi f} \geq \frac{2000}{6.28 \times 1.38 \times 10^6} = 230\ \mu\text{H} \tag{9-68}$$

However, $X_{\text{RFC}}$ should also offer no loading to the collector; that is, it should appear as an open circuit. Therefore,

$$X_{\text{RFC}} > 10 Q X_{L_1} \geq 226\ \text{k}\Omega \tag{9-69}$$

Again, by use of Eq. (9.68),

$$L_{\text{RFC}} \geq \frac{X_{\text{RFC}}}{2\pi f} \geq \frac{226 \times 10^3}{6.28 \times 1.38 \times 10^6} = 26.08 \text{ mH} \tag{9-68}$$

In practice, the design for borderline operation is rarely used. Component cost and availability of physical space may determine how close to the borderline the circuit is actually built. Remember, too, that the calculated values are based on real values, not the color-coded values of the components. For example, the mutual inductance of a Hartley oscillator coil may not be marked or coded. Only by use of an impedance bridge or other instrument may the actual mutual inductance be determined.

## The Pierce and Miller Oscillators

**problem 9.10**  The Pierce oscillator of Fig. 9-16$b$ has a crystal with the following characteristics: $L$ = 2.3 H, $C_s$ = 0.04 pF, and $R_s$ = 2200 $\Omega$. $C_m$ = 8.5 pF. What is the series-resonant frequency of operation? What is the approximate $Q$ of the crystal? Since $C_c$ = 0.001 $\mu$F, what should be the minimum inductance of the RFC?

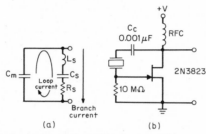

**Fig. 9-16**  ($a$) A typical crystal-equivalent *RLC* circuit and ($b$) a practical Pierce oscillator.

**theory**  The Pierce oscillator is a simple crystal oscillator employing a crystal as a series-resonant feedback element. A typical crystal has an equivalent series branch consisting of $R_s$, $C_s$, and $L_s$, which is effectively in parallel with the mounting capacitance $C_m$ as shown in Fig. 9-16$a$. At some frequency $L_s$ will resonate with $C_s$, effectively canceling the reactive effects of each other, resulting in a maximum branch current. It is this series-resonant effect that is used by the Pierce oscillator circuit. Maximum positive feedback from the drain to gate will occur only at the series-resonant frequency.

**solution**  The series-resonant frequency of the crystal may be computed using Eq. (9.15).

$$f_s = \frac{1}{2\pi\sqrt{LC_s}}$$

$$= \frac{1}{6.28\sqrt{2.3 \text{ H} \times 0.04 \times 10^{-12}}} = \frac{1}{6.28\sqrt{0.092 \times 10^{-12}}}$$

$$= \frac{1}{6.28 \times 0.3033 \times 10^{-6}} = 524,985 \text{ Hz}$$

Once the frequency has been computed, the effective inductive reactance of the crystal may be determined.

$$X_{L_s} = 2\pi f L_s = 6.28 \times 524,985 \times 2.3 = 7.58 \text{ M}\Omega \tag{9-13}$$

From the value of the inductive reactance, the effective $Q$ of the crystal may be computed.

$$Q = \frac{X_{L_s}}{R_s} = \frac{7.58 \times 10^6}{2200} = 3445 \tag{9-52}$$

The capacitive reactance of the coupling capacitor should be very small compared with the reactance of the RFC, typically less than $0.1 X_{\text{RFC}}$. From this relationship, it is possible to determine the minimum inductance of the RFC.

$$X_{C_C} = \frac{1}{2\pi f C_c} = \frac{1}{6.28 \times 524,985 \times 0.001 \times 10^{-6}} \tag{9-11}$$
$$= 303 \Omega$$
$$X_{\text{RFC}} = 10 X_{C_c} = 10 \times 303 \Omega = 3030 \Omega \tag{9-70}$$

Therefore,

$$L_{\text{RFC}} = \frac{X_{\text{RFC}}}{2\pi f} = \frac{3030}{6.28 \times 524{,}985} = 0.92 \text{ mH} \tag{9-68}$$

As a minimum value, 1 mH could be used, although a value of 10 mH will be more effective.

The effect of the crystal mount capacitance $C_m$ is to make a parallel (antiresonant) feedback arrangement possible. The Miller oscillator of Fig. 9-17 employs this arrangement. Notice that the maximum gate-source voltage signal will occur at the crystal's antiresonant frequency. The drain circuit is then tuned to this operating frequency, also.

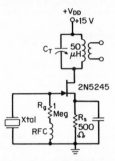

**Fig. 9-17**  A Miller oscillator tuned to the second harmonic.

**problem 9.11**  Using the crystal of the Pierce oscillator, compute the effective loop capacitance and the antiresonant frequency of operation of the Miller oscillator of Fig. 9-17.

**theory**  The loop current in the crystal circulates, in effect, much the same as in a tank circuit. Therefore, the capacitors seem to be in series. The effective loop capacitance of the capacitors in series is:

$$
\begin{aligned}
C_{\text{loop}} &= \frac{C_m C_s}{C_m + C_s} \\
&= \frac{8.5 \times 10^{-12} \times 0.04 \times 10^{-12}}{8.5 \times 10^{-12} + 0.04 \times 10^{-12}} = \frac{0.34 \times 10^{-24}}{8.54 \times 10^{-12}} \\
&= 0.0398 \text{ pF}
\end{aligned} \tag{9-71}
$$

As can be seen, the loop capacitance is only slightly less than the series capacitance. Therefore, the antiresonant frequency will be very near the resonant frequency.

$$
\begin{aligned}
f_p &= \frac{1}{2\pi \sqrt{LC_{\text{loop}}}} \\
&= \frac{1}{6.28 \sqrt{2.3 \times 0.0398 \times 10^{-12}}} \\
&= \frac{1}{6.28 \sqrt{0.0915}} = \frac{1}{6.28 \times 0.3026 \times 10^{-6}} \\
&= \frac{1}{1.90005 \times 10^{-6}} = 526{,}302 \text{ Hz}
\end{aligned} \tag{9-15}
$$

**problem 9.12**  It is desirable to tune the Miller oscillator to the second harmonic, or overtone, of the fundamental frequency of 526,302 Hz. What value of capacitance can be used to tune the oscillator output to this harmonic?

**theory**  Since the fundamental frequency is generated at the input to the amplifier, it is possible to tune the output tank circuit to the second or even third harmonic. The output will be greatly reduced, of course, depending upon the amount of harmonic content present. If the fundamental waveform is distorted, there is a substantial harmonic content.

Tuning the output tank circuit requires either a variable capacitor or inductance. While either may be used, variable capacitors are generally easier to obtain and use.

Using Eq. (9-15) again, we can derive the equation to determine the correct capacitance for a frequency of 1,052,604 Hz.

$$f_o = \frac{1}{2\pi\sqrt{LC}} \tag{9-15}$$

$$f_o^2 = \frac{1}{4\pi^2 LC}$$

$$C = \frac{1}{4\pi^2 f^2 L} \tag{9-72}$$

$$= \frac{1}{4 \times \pi^2 \times 1{,}052{,}604^2 \times 50 \times 10^{-6}}$$

$$= \frac{1}{4 \times 9.8696 \times 1.107975 \times 10^{12} \times 50 \times 10^{-6}}$$

$$= \frac{1}{2.18705 \times 10^9}$$

$$= 457.24 \, \text{pF}$$

## 9.4  NONSINUSOIDAL OSCILLATORS

The realm of linear electronics, until only recently, was dominated primarily by sinusoidal oscillators. However, various forms of relaxation oscillators have been developed for specialized uses. By nature, they produce nonsinusoidal waveforms. Common waveforms produced include the pulse, the square wave, the rectangular wave, the ramp wave, the triangle wave, and the trapezoidal waveform.

Most nonsinusoidal waveform generators, or oscillators, rely on the $RC$ charge time to determine frequency or timing intervals. Some use pulse generators to trigger switching circuits, as with a unijunction-triggered flipflop circuit, to generate square waves. Others utilize integrated circuitry of significant magnitude to produce voltage-controlled oscillators. Of course, for tight frequency control, various forms of crystal control may be used.

### The Unijunction Oscillator

The unijunction oscillator is actually a pulse generator with three types of outputs: (1) a positive-going pulse, (2) a negative-going pulse, and (3) an unlinear ramp waveform from the charge time of a capacitor.

**problem 9.13**  Figure 9-18 illustrates a typical basic unijunction oscillator. The unijunction device has an intrinsic standoff ratio $\eta = 0.58$. $R_T = 100 \text{ k}\Omega$ and $C_T = 0.05 \, \mu\text{F}$. What is the operating frequency? If a frequency of 400 Hz is desired, what new value of $R_T$ must be used?

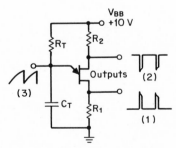

**Fig. 9-18**  A typical unijunction oscillator with expected waveforms at (1) base 1, (2) base 2, and (3) emitter.

**theory**  The unijunction transistor is actually a tiny bar of silicon to which a single junction has been alloyed. The aluminum wire is welded, in this case, 58 percent of the way from base 1 to the base 2 connection. In normal operation $R_1$ and $R_2$ have little, if any, effect on the frequency of operation. Instead, they simply provide a current limiting to the discharge path for $C_T$, the timing capacitor.

As shown in Fig. 9-19, capacitor $C_T$ charges through $R_T$ until the voltage at the emitter is sufficient to forward-bias the base 1–emitter diode. At this point, the base 1–emitter conductivity increases sharply, and current passes through $R_1$, base 1, and the emitter, suddenly discharging $C_T$. Of course, when the capacitor has discharged sufficiently to lower the emitter voltage well below forward bias, the charging cycle begins again. The result is a capacitor charge-discharge waveform at the emitter, a positive-going pulse across $R_1$, due to the turnon and turnoff of the discharge current, and a negative-going pulse at the base 2, due to the voltage drop across the changing interbase resistance $R_{BB}$ during the discharge time.

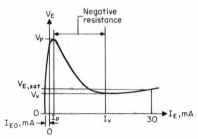

**Fig. 9-19** A sample characteristic curve for a unijunction transistor, showing peak voltage $V_p$, peak current $I_p$, valley voltage $V_V$, and valley current $I_V$.

**solution** Since the frequency of operation is an inverse function of the $RC$ charge-discharge time period, and since the voltage is an exponential function, the computation of frequency from the component values must resort to the use of natural logarithms. That is:

$$f_o = \frac{1}{R_T C_T \ln [1/(1 - \eta)]} \tag{9-73}$$

The solution is complex and approximate since the value of $\eta$ is not precisely known. For that matter, most of the time, neither is the firing potential where $V_E = V_P$. The two are related as follows.

$$\eta = \frac{R_{B1}}{R_{B1} + R_{B2}} \bigg|_{I_{E=0}} \tag{9-74}$$

The interbase resistance $R_{BB}$ is related in that

$$R_{BB} = R_{B1} + R_{B2} |_{I_{E=0}}$$

This is an easy way to discuss resistance ratios when the emitter current is zero.

The voltage across $R_{BB}$ is $V_{R_{BB}}$, and the voltage between base 1 and the junction point is generally indicated as $V_{R_{B1}}$.

Therefore

$$V_{R_{B1}} = \eta V_{R_{BB}} \bigg|_{I_{E=0}} \tag{9-75}$$

It follows, then, that the voltage $V_E = V_P$ must be the junction voltage more than $V_{R_{B1}}$, at the point the emitter fires.

$$V_E = V_P = V_{R_{B1}} + V_J = \eta V_{R_{BB}} + V_J \tag{9-76}$$

The emitter current before $V_P$ is reached is extremely small, in the microampere region. When the junction turns on, it increases the emitter current enormously, with a concurrent drop in the base-1 resistance $R_{B_1}$ and the voltage $V_{B_1}$. This is a negative-resistance phenomenon, excellent for oscillating conditions. The forward-diode voltage at which this surge of current begins is between 0.65 and 0.7 V normally and depends somewhat on the temperature.

Solution to the problem then follows easily by substitution of values into Eq. 9-73.

$$f_o = \frac{1}{100 \times 10^3 \times 0.05 \times 10^{-6} \ln [1/(1 - 0.58)]}$$

$$= \frac{1}{5 \times 10^{-3} \ln (2.38)} = \frac{1}{5 \times 10^{-3} \times 0.868}$$

$$= \frac{1}{4.338 \times 10^{-3}} = 231 \text{ Hz}$$

To increase the frequency of operation to 400 Hz, the resistance $R_T$ must be reduced to

$$R_T = \frac{1}{400 \text{ Hz} \times 0.05 \ \mu\text{F} \times \ln 2.38}$$
$$= 57,663 \ \Omega \cong 57.6 \text{ k}\Omega \tag{9-77}$$

For this reason, $R_T$ is generally two resistors, one fixed and the other a potentiometer.

**problem 9.14** Assuming $R_1$ and $R_2$ to be very small compared with $R_{BB}$, compute the firing voltage $V_P$ for the circuit of Fig. 9-18. Assume that $V_{BB} = R_{BB} = 10$ V.

**solution** Using Eq. (9-78) and substituting known values, the peak voltage may be computed.

$$V_P = \eta V_{R_{BB}} + V_J$$
$$= 0.58 \times 10 \text{ V} + 0.65 \text{ V} = 6.45 \text{ V} \tag{9-78}$$

**problem 9.15** What are the maximum and minimum values of $R_T$ that can be used in Fig. 9-18 if $I_P = 5 \ \mu\text{A}$ and $I_\text{valley} = 10$ mA?

**theory** It can be seen that the voltage across $R_T$ is always the difference between $V_{BB}$ and $V_P$. $R_T$ then must never be so large that the peak current cannot be reached.

**solution** At the point of firing,

$$R_T < \frac{V_{BB} - V_P}{I_P}$$
$$< \frac{10 \text{ V} - 6.45 \text{ V}}{5 \times 10^{-6} \text{ A}} = \frac{3.55 \text{ V}}{5 \times 10^{-6}} = 710,000 \ \Omega = 710 \text{ k}\Omega \tag{9-79}$$

Obviously the value of $V_{BB}$ can be increased or the value of $C_T$ increased to give the latitude in $RC$ values that may be needed.

$R_T$ cannot be so small that the device will not turn off where $V_E = V_\text{valley}$.

$$R_T > \frac{V_{BB} - V_\text{valley}}{I_\text{valley}}$$
$$> \frac{10 \text{ V} - 0.8 \text{ V}}{10 \times 10^{-3} \text{ A}} = 920 \ \Omega \tag{9-80}$$

## The Astable Multivibrator

The astable multivibrator is actually a cross-coupled two-stage amplifier as shown in Fig. 9-20. Output waveforms approximating a square wave may be taken from either collector. The two waveforms are equal, but opposite, because when a transistor is on the other is off. Typical waveforms are shown in Fig. 9-21.

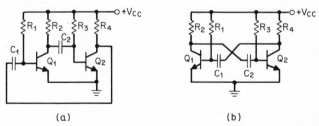

(a)             (b)

**Fig. 9-20** The transistor astable multivibrator (a) drawn as a two-stage amplifier for simplicity and (b) shown in its more conventional cross-coupled presentation.

**problem 9.16** Assume that the astable multivibrator of Fig. 9-20 is balanced with the following values: $R_1 = 47$ k$\Omega$ and $R_3 = 47$ k$\Omega$, $R_2 = 1$ k$\Omega$ and $R_4 = 1$ k$\Omega$, $C_1 = 0.05 \ \mu\text{F}$ and $C_2 = 0.05 \ \mu\text{F}$. What are the time periods $T_1$ and $T_2$? What is the frequency of operation? What values of $R_1$ and $R_2$ are required for 400 Hz operation?

**theory** The multivibrator is a variation of a saturated switch. The switch circuit is an inverting amplifier, which, when in an off condition (high resistance), can be turned on and driven into saturation quite suddenly with sufficient base current drive. The switched inverter will remain on and saturated as long as the base current is beyond that required for saturation.

The multivibrator uses a charging capacitor to provide the interval of time between on and off conditions. As shown in Figs. 9-20 and 9-21, when $Q_1$ is off, its collector voltage is equal to $V_{CC}$. Since $Q_2$ is on, $V_{C2}$ is at saturation, about 0.1 V. $C_1$ then is essentially in series with $R_1$ between ground and $V_{CC}$. $C_1$ thus discharges through $R_1$, holding $Q_1$ in cutoff until $IR_1$ decreases sufficiently to allow $V_{BE_1}$ to reach the turn-on voltage, about 0.6 V. When $Q_1$ begins to turn on, $V_{CE_1}$ begins to drop toward saturation, and the sudden change in voltage is coupled to the base of $Q_2$ by the charging action of $C_2$. $Q_2$ is driven rapidly into cutoff, as shown in Fig. 9-21. The positive feedback causes the cycle to repeat itself. Since at the instant of switching the capacitor is charged to $V_{CC}$ but is now effectively switched to the opposite polarity in the circuit, the instantaneous voltage across the circuit is $2V_{CC}$.

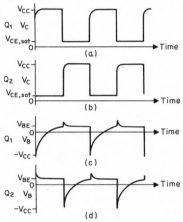

**Fig. 9-21** Output waveforms from a transistor astable multivibrator: (a) at collector of $Q_1$, (b) at collector of $Q_2$, (c) base of $Q_1$, and (d) at the base of $Q_2$.

**solution**    The charging of a capacitor through a resistor instantaneosly follows the function $(1 - e^{-T/RC})$. Since we wish to determine the time of each period of on time, in order to determine the frequency of operation, a relationship can be stated for the point when $Q_1$ begins to conduct.

$$T_1 = R_1 C_1 \ln \frac{2V_{CC} - V_{CE,\text{sat}} - V_{BE}}{V_{CC} - V_{BE}} \tag{9-81}$$

When the derivation is understood, the equation can be simplified somewhat to give a close approximation.

$$V_{CC} - V_{BE}(1 - e^{-T/RC}) = V_{CC} - V_{CE,\text{sat}} \tag{9-82}$$

$$1 - e^{-T/RC} = \frac{V_{CC} - V_{CE,\text{sat}}}{V_{CC} - V_{BE}}$$

$$-e^{-T/RC} = \frac{V_{CC} - V_{CE,\text{sat}}}{V_{CC} - V_{BE}} - 1$$

$$e^{-T/RC} = -\frac{V_{CC} - V_{CE,\text{sat}}}{V_{CC} - V_{BE}} + \frac{V_{CC} - V_{BE}}{V_{CC} - V_{BE}}$$

$$= -\frac{2V_{CC} - V_{CE,\text{sat}} - V_{BE} -}{V_{CC} - V_{BE}}$$

$$\frac{T}{RC} = \ln \frac{2V_{CC} - V_{CE,\text{sat}} - V_{BE}}{V_{CC} - V_{BE}}$$

$$T = RC \ln \frac{2V_{CC} - V_{CE,\text{sat}} - V_{BE}}{V_{CC} - V_{BE}} \tag{9-81}$$

If we evaluate the equation for real typical values, we find that the voltages virtually cancel, leaving:

$$T = RC \ln \frac{2(10 \text{ V} - 0.1 \text{ V} - 0.7 \text{ V})}{10 \text{ V} - 0.7 \text{ V}}$$

$$= RC \ln \frac{2(9.2)}{9.3} = RC \ln 1.98 \qquad (9\text{-}83)$$

$$\cong 0.68 \, RC$$

If the multivibrator is balanced, $T_1 = T_2$, then the frequency of operation becomes:

$$f_o = \frac{1}{T_1 + T_2} \cong \frac{1}{0.68 \times 2RC} \qquad (9\text{-}84)$$

Since, within reason, the supply voltage has little effect on the frequency, it can generally be neglected in the computation. The period of the astable multivibrator of the problem becomes

$$\begin{aligned} T_1 &= 0.68 \times R_1 C_1 \\ &= 0.68 \times 47 \text{ k}\Omega \times 0.05 \text{ }\mu\text{F} \qquad (9\text{-}85) \\ &= 1.598 \times 10^{-3} \text{ s} \end{aligned}$$

The frequency of operation follows from Eq. (9-83).

$$f_o = \frac{1}{0.68 \times 2RC}$$

$$= \frac{1}{2 \times 1.598 \times 10^{-3} \text{ s/cycle}} = 313 \text{ Hz} \qquad (9\text{-}86)$$

To produce 400 Hz, the values of $R_1$ and $R_3$ must be changed to:

$$R = \frac{1}{0.68 \times 2 \, fC}$$

$$= \frac{1}{0.68 \times 2 \times 400 \times 0.05 \times 10^{-6}} = \frac{1}{27.2 \times 10^{-6}} \qquad (9\text{-}87)$$

$$= 36{,}765 \text{ }\Omega \cong 36.8 \text{ k}\Omega$$

**problem 9.17**    What size capacitors are required in the IC multivibrator of Fig. 9-22 to produce a frequency of 10,000 Hz?

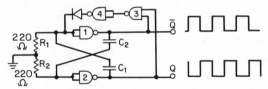

**Fig. 9-22**    An astable multivibrator made from a TTL quad two-input NAND gate IC, 7400 or equivalent, operating from a +5-V power supply.

**theory**    The IC multivibrator is essentially a set of inverting amplifiers cross-coupled to operate much like the multivibrator of Fig. 9-20. Resistors $R_1$ and $R_2$ bias the inverters into operating range. The values should be between 200 and 220 $\Omega$. To change the frequency, the coupling capacitors must be varied. Gates 3 and 4 are used to ensure that both outputs will never go on or off at the same time.

**solution**    The frequency of operation is approximately equal to the reciprocal of the period $T$.

$$f_o = \frac{1}{T} \qquad (9\text{-}88)$$

$$T \cong R_1 C_1 = R_2 C_2 \qquad (9\text{-}89)$$

$$f_o \cong \frac{1}{R_1 C_1} \qquad (9\text{-}90)$$

$$C_1 \cong \frac{1}{R_1 f_o} \qquad (9\text{-}91)$$

$$\cong \frac{1}{220 \times 10{,}000 \text{ Hz}}$$

$$\cong 0.45 \text{ }\mu\text{F}$$

### The Monostable Multivibrator

The monostable multivibrator, unlike other oscillators, puts out only one pulse when triggered. It then relaxes until it again receives a command trigger. Its pulse is a constant width, generally used as a form of time delay. Although the discrete-device monostable of Fig. 9-23a is quite functional, it is rarely used in new designs. Instead, the IC monostable of Fig. 9-23b is preferred owing to its simplicity, cost, and reliability.

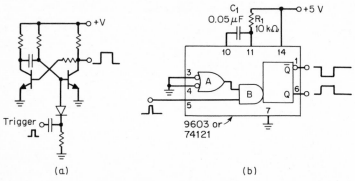

(a)                                            (b)

**Fig. 9-23**  The monostable multivibrator. (a) Discrete device circuit, and (b) a TTL IC circuit.

**problem 9.18**  Given the circuit of Fig. 9-23b, what pulse width will be generated when the input receives a positive trigger applied to input B? What value of resistance would be required to increase the pulse width to 1 ms?

**theory**  The 9603 or the 74121 monostable IC multivibrators are TTL devices that are triggerable from either a positive-going trigger pulse or a negative-going pulse edge, as shown in Fig. 9-24. The negative-going edge triggering is actually input to an inverter-driven NOR gate. The positive-going trigger must be input to the Schmitt trigger input B, which can accept a slow rise-time trigger of up to 1 V/s. The IC circuit is a sophisticated version of the discrete circuit, and thus the pulse width time $T_p$ has a similar derivation as in Eqs. (9-82) and (9-83).

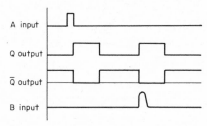

**Fig. 9-24**  Trigger and output pulses for the 9603/74121 monostable multivibrator. Note the trailing edge triggering for the A input, while the B input will trigger on the leading edge.

**solution**  The pulse width of the monostable of Fig. 9-23b is

$$T_p \cong R_1 C_1 \ln 2$$
$$\cong 10{,}000 \times 0.05 \, \mu\text{F} \times 0.693 \qquad (9\text{-}83)$$
$$= 0.000347 = 347 \, \mu\text{s}$$

To increase the pulse width to 1 ms, the resistance or capacitance may be increased. However, it is much easier to control changes in resistance by means of a potentiometer.

$$R_1 = \frac{T_p}{C_1 \ln 2}$$

$$= \frac{0.001}{0.05 \times 10^{-6} \times 0.693} \qquad (9\text{-}92)$$

$$= 28{,}860 \, \Omega$$

### The Voltage-Controlled Oscillator

While there have been a number of schemes to design voltage-to-frequency converters from discrete devices, the IC voltage-controlled oscillator (VCO) is the easiest and cheapest to use. A number of VCO devices are on the market, but one of the earliest, the Signetics NE566 Function Generator will be discussed here. It has both square-wave and triangle-wave outputs as shown in Fig. 9-25.

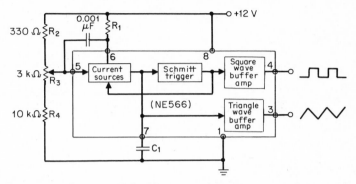

**Fig. 9-25**   A voltage-controlled oscillator using the NE566 Function Generator IC.

**problem 9.19**   What values of $R_1$ and $C_1$, and control voltage $V_c$, might be used to produce an output frequency of 10 kHz using the circuit of Fig. 9-25? What will the new frequency be if the control voltage is increased by 0.5 V? What is the value of resistance between pin 5 and ground for each frequency generated in this problem?

**theory**   The NE566 Function Generator IC is actually a circuit containing a constant-current source, a Schmitt trigger, and two buffer amplifiers, one for the square-wave and one for the triangle-wave output. The output frequency can be varied over a 10:1 range by means of the control voltage applied to the current-source circuit.

The control voltage for the NE566 is supplied by means of the voltage divider $R_2R_3R_4$. If the supply voltage is +12 V and the control voltage $V_c$ is 10.5 V at a frequency of 5 kHz, the frequency can be decreased to 3.5 kHz by increasing the control voltage to 11.0 V. By the same reasoning, the frequency could be increased to about 7.5 kHz by reducing $V_c$ to about 10.0 V.

**solution**   The approximate frequency, as a function of control voltage, control resistance, and timing capacitance, is expressed by Eq. (9.93).

$$f_o = \frac{2(V_{CC} - V_C)}{R_1 C_1 V_{CC}} \tag{9-93}$$

This expression holds true over the linear range of the device and within the temperature limits of the device for:

$$V_{CC} \geq V_C \geq 0.75\, V_{CC} \tag{9-94}$$
$$2\ \text{k}\Omega < R_1 < 20\ \text{k}\Omega \tag{9-95}$$

That is to say, the control voltage cannot be less than 3/4 of $V_{CC}$ nor as large as $V_{CC}$. The control resistance $R_1$ should be between 2 and 20 k$\Omega$.

For a starting point, it is a good practice to compute the range of control voltage limited by the voltage divider.

$$
\begin{aligned}
V_{C,\text{low}} &= \frac{10\ \text{k}\Omega \times 12\ \text{V}}{10\ \text{k}\Omega + 3\ \text{k}\Omega + 330\ \Omega} \\
&= 9\ \text{V} \\
V_{C,\text{high}} &= \frac{13\ \text{k}\Omega \times 12\ \text{V}}{10\ \text{k}\Omega + 3\ \text{k}\Omega + 330\ \Omega} \\
&= 11.7\ \text{V} \\
V_{C,\text{mid}} &= \frac{11.5\ \text{k}\Omega \times 12\ \text{V}}{13.33\ \text{k}\Omega} \\
&= 10.35\ \text{V}
\end{aligned}
\tag{9-96}
$$

If the voltage-control divider network is set so that the resistance between pin 5 and ground is 11.5 kΩ, the control voltage will be 10.35 V. This will allow the oscillator to be set to yield the widest variation in frequency output with changes in voltage input.

If a small common-size capacitor is selected, a resistance value for $R_1$ may be computed and supplied by a potentiometer. A convenient size value for $C_1$ might be 0.005 μF. By use of Eq. (9-93),

$$f_o = \frac{2(V_{CC} - V_C)}{R_1 C_1 V_{CC}} \tag{9-93}$$

$$R_1 = \frac{2(V_{CC} - V_C)}{f_o C_1 V_{CC}} \tag{9-97}$$

Substituting values yields

$$R_1 = \frac{2(12 \text{ V} - 10.35 \text{ V})}{10 \text{ kHz} \times 0.005 \text{ μF} \times 12 \text{ V}}$$
$$= 5500 \Omega$$

To determine the new frequency when the control voltage is increased to 10.85 V, the new values are simply substituted into Eq. (9.93).

$$f_o = \frac{2(12 \text{ V} - 10.85 \text{ V})}{5.5 \text{ kΩ} \times 0.005 \text{ μF} \times 12 \text{ V}}$$
$$= 6970 \text{ Hz} \tag{9-93}$$

The resistance between pin 5 and ground for $V_c$ = 10.85 V is:

$$R = \frac{10.85 \text{ V} \times 13{,}330 \text{ Ω}}{12 \text{ V}} = 12{,}053 \text{ Ω} \tag{9-98}$$

## The Phase-Locked Loop

The phase-locked loop is an extension of the development of an IC voltage-controlled oscillator. As its name implies, the VCO can be made to operate at the same frequency as some other signal, such as an fm carrier frequency. Once locked onto the fm carrier frequency signal, the output of the VCO will track the frequency variations of the carrier, within its tracking range of frequency, about ±60%.

**problem 9.20**    Consider the phase-locked loop SCA fm demodulator circuit of Fig. 9-26. What is the VCO free-running frequency? What is the frequency-lock range $f_L$? What is the capture range $f_c$?

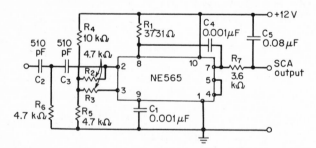

**Fig. 9-26**   A simplified phase-locked loop circuit to demodulate an fm SCA carrier signal.

**theory**    The block diagram of Fig. 9-27 shows the PLL to be composed of a differential amplifier phase detector/comparator, a voltage-controlled oscillator, and an amplifier to drive the low-pass filter.

The input signal to the demodulator passes through a two-stage high-pass filter to remove any lower-channel signals. The input terminals are dc-biased to the same voltage by $R_2$, $R_3$, $R_4$, and $R_5$ voltage dividers for single-supply operation. The output (demodulated) circuit, composed of $R_7 C_5$, is a low-pass filter to remove noise and provide deemphasis. It is actually a combination of resistors and capacitors that have been reduced to single components for simplicity of explanation.

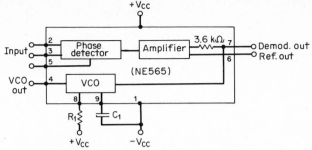

**Fig. 9-27** The block diagram of the phase-locked loop showing the timing elements of the VCO.

**solution** The components controlling frequency, lock range, and capture range are $R_1, C_1$, and $V_{CC}$. The free-running VCO frequency is approximately:

$$f_o \cong \frac{1}{4R_1C_1}$$

$$\cong \frac{1}{4 \times 3731 \times 0.001 \ \mu F} \tag{9-99}$$

$$\cong 67 \text{ kHz}$$

The job of the phase-locked loop is to capture a carrier-frequency signal and lock onto it. It must then be able to vary its VCO frequency within the range of the lock, in this case the fm deviation. This yields an output signal which is the fm, or sound, component of the carrier frequency. The NE562 PLL is a more sophisticated device and has a wider frequency response, whereas the NE565, owing to its simplicity, is easier to understand.

The lock range, once a signal is captured, is about ±60% for the NE565 and is a function of the center frequency and the supply voltage. The lock range $f_L$ may be determined from the empirical expression:

$$f_L \cong \pm \frac{8f_o}{V_{CC}} \text{ Hz}$$

$$\cong \pm \frac{8 \times 67 \text{ kHz}}{12 \text{ V}} \tag{9-100}$$

$$\cong \pm 44,667 \text{ Hz}$$

This means that the total lock range is $2 \times 44,667 = 89,334$ Hz.

The capture range is generally much narrower than the lock range and is a function of the output resistor and filter capacitance.

$$f_C \cong \pm \frac{1}{2\pi} \sqrt{\frac{2\pi f_L}{RC}}$$

$$\cong \pm \frac{1}{2\pi} \sqrt{\frac{2\pi \, (\pm 8f_o)}{RC \, V_{CC}}} \tag{9-101}$$

By simplifying the lock-range expression,

$$f_C \cong \pm \frac{1}{2\pi} \sqrt{\frac{2\pi 16 f_o}{R_7 C_5 V_{CC}}}$$

$$\cong \pm \frac{1}{2\pi} \sqrt{\frac{32 \times 3.1416 \times 67,000}{3.6 \times 10^3 \times 0.08 \times 10^{-6} \times 12}}$$

$$\cong \pm \frac{1}{6.283} \sqrt{\frac{6,735,575}{0.00346}} = \pm \frac{44,147}{6.283}$$

$$\cong 7026 \text{ Hz} \cong 7 \text{ kHz}$$

**problem 9.21** How can a phase-locked loop be used as a frequency multiplier? What values of $R_1$ and $C_1$ could be used to produce 30 kHz from a good 10-kHz signal source?

**theory** Since the PLL is primarily an interaction between the VCO and the phase detector/ comparator, and since they are not connected internally, each part can function independently. If a shift-register counter is set to divide by 3, and it is connected to the output of the VCO, as shown in Fig. 9-28, it will cause the phase dectector/comparator to lock the counter onto the signal instead of the VCO. The VCO can then be set up for 30 kHz, which will be divided down by the counter to 10 kHz, which, in turn, will be locked onto the incoming 10 kHz signal. If the VCO frequency is within the capture range, the VCO will lock onto the 10 kHz signal and operate at that frequency even though its free-running frequency is not exactly 30 kHz.

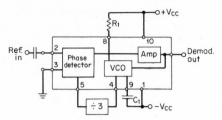

**Fig. 9-28**  A phase-locked loop connected as a frequency multiplier used as a multiply-by-3 circuit.

**solution**    Since the input signal is 10 kHz and the VCO is to operate at 30 kHz, values of $R_1 C_1$ are to be computed for 30 kHz. The optimum value of $R_1$ for the NE565 is about 4 kΩ. The nearest standard-value resistor is 4.3 kΩ and may be used for a trial calculation.

$$f_o = \frac{1}{4R_1 C_1} \tag{9-99}$$

Therefore, substitution and rearrangement will yield the value of $C_1$.

$$C_1 = \frac{1}{4 \times 30{,}000 \times 4300} = 0.00194 \ \mu F \tag{9-102}$$

If $C_1$ is changed to the nearest standard value of 0.002 $\mu F$, the VCO will run at slightly less than 30,000 Hz.

$$f_o = \frac{1}{4 \times 4.3 \times 10^3 \times 2 \times 10^{-9}}$$
$$= 29{,}070 \ \text{Hz}$$

The divide-by-3 counter output frequency will be 29,070 Hz/3 = 9690 Hz. This is certainly within the capture range of the phase detector/comparator. The phase-detector output will correct the error-voltage input to the VCO to shift its operating frequency up to 10 kHz × 3 = 30 kHz.

## 9.5  BIBLIOGRAPHY

Boylestad, Robert, and Nashelsky, Louis: *Electronic Devices and Circuit Theory,* Prentice-Hall, Englewood Cliffs, N.J., 1972.
Fairchild Semiconductor Division, Fairchild Camera and Instrument Corp.: *Databook for TTL,* Mountain View, Calif., 1971.
Lurch, E. N.: *Fundamentals of Electronics,* Wiley, New York, 1971.
Malvino, A. P.: *Electronic Principles,* McGraw-Hill, New York, 1973.
Mitchell, B. B.: *Semiconductor Pulse Circuits,* Holt, New York, 1970.
Natek, F. R.: *Applications of Linear Integrated Circuits,* Wiley, New York, 1975.
RCA Corporation: *Solid State Databook, Linear Integrated Circuits and MOS Devices,* Somerville, N.J., 1975.
Seeley, Samuel: *Electronic Circuits,* Holt, New York, 1968.
Signetics Corp., subsidiary of Corning Glass Works: *Linear Integrated Circuits,* vol. I, Sunnyvale, Calif., 1971.
Tepper, Irving: *Solid-State Devices,* vol. II, *Applications,* Addison-Wesley, Reading, Mass., 1973.
Texas Instruments, Inc.: *TTL Integrated Circuits Catalog,* Dallas, Texas, 1977.

# Chapter 10

# Power Supplies

## STEPHEN FERNANDES

**Product Development Manager, Lambda Electronics Corp.**

## 10.1 INTRODUCTION

This chapter presents several of the most commonly encountered power supply configurations. The discussion is limited to power supplies which transform ac input power into dc output power. While the objective of a regulated power supply may be to keep either the current or voltage constant at its output, only constant-voltage types will be considered here, because they constitute the vast majority of power supplies in existence.

Since the function of the regulated power supply as described above is to maintain a constant voltage at its output terminals, its performance can be judged in terms of variations of this parameter. Output voltage usually varies mainly with line voltage, load current, and degree of filtering. Parameters which describe performance are as follows:

1. *Ripple:* An ac component of the output voltage which is superimposed on the dc component.

2. *Line regulation:* The change in the steady-state value of output voltage which accompanies a change in the line voltage, with all other conditions constant.

3. *Load regulation:* The change in the steady-state value of output voltage which accompanies a change in the load current, with all other conditions constant.

The material in this chapter focuses on four areas:

1. Unregulated power supplies
2. Regulated power supplies
3. Switching power supplies
4. Thermal considerations

## 10.2 UNREGULATED POWER SUPPLIES

Unregulated power supplies are used where variations in the output voltage are not critical. They are found in many radio and TV circuits, and also in applications such as model trains and battery chargers. Simplicity and low cost are their trademarks. The types to be discussed are:

1. Half-wave rectifier power supplies
2. Full-wave rectifier power supplies

3. Full-wave bridge rectifier power supplies
4. Voltage-multiplier power supplies

A generalized block diagram of an unregulated power supply is shown in Fig. 10-1.

To facilitate comparison, the same transformer, rectifiers, and filter capacitor will be assumed in each of the examples. This will lead to slightly different output voltages in each case, but a lower output voltage merely indicates that a higher transformer secondary voltage is required with a particular type of power supply if a higher output voltage is necessary.

**Fig. 10-1**   Generalized block diagram of an unregulated power supply.

The specifications for all the four types mentioned above are as follows:
1. Load current varies from 0 to 2 A
2. Input voltage varies from 110 to 120 V ac rms, 60 Hz
3. Output voltage to be approximately 15 V dc*
4. Filter capacitor is 2000 $\mu$F, 50 V
5. Input diode drop is negligible
6. Transformer secondary voltage is shown in Fig. 10-2

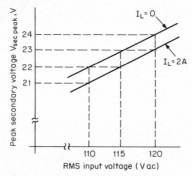

**Fig. 10-2**   Transformer secondary voltage vs. input voltage. *Note:* This information is not generally published. Measurements of secondary voltage under load can be taken by using an unregulated circuit (as in Fig. 10-5*a*) to keep the load relatively constant.

Note that the secondary voltage is given as a peak rather than an rms value to avoid the need for conversion. For a sinusoidal waveform as assumed, the ratio is

$$V_{\text{peak}} = V_{\text{rms}} \times \sqrt{2} \tag{10-1}$$

The difference in voltage as load current changes is due to the voltage drop in the transformer and contributes to load regulation.

## Half-Wave Rectifier Power Supplies

**problem 10.1**   Solve for the ripple, output voltage waveform, load regulation, and line regulation of the circuit shown in Fig. 10-3*a*.

**theory**   For a 60-Hz waveform, the total period of each cycle is 16.6 milliseconds (ms). When $V_{\text{sec}}$ is larger than $V_c$ and in the correct polarity, $D_1$ will conduct. While $D_1$ is conducting, the transformer provides both the load current and the charging current for $C_1$. In a half-wave rectifier power supply, $D_1$ typically conducts for approximately 3.6 ms/cycle at full load. During the remaining 13 ms of each cycle the capacitor provides the load current, and its voltage $V_c$ decreases. This is shown in Fig. 10-3*b*. The extent to which $V_c$ decreases is a function of capacitance and load current.

*Except for voltage-multiplier power supplies.

$$V_{pp} = \frac{I_L \times t}{C} \qquad (10\text{-}2)$$

where  $V_{pp}$ = peak-to-peak variation in $V_c$
   $I_L$ = load current, A
   $t$ = discharge time of the capacitor, s
   $C$ = capacitance, F

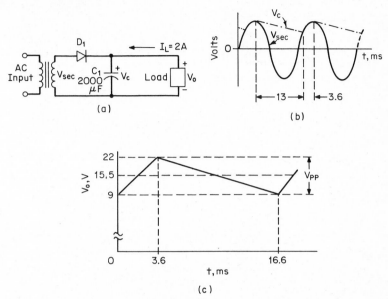

(a)

(b)

(c)

**Fig. 10-3**  (a) Half-wave rectifier power supply; (b) $V_c$ and $V_{sec}$ vs. time for half-wave rectifier power supply; (c) output voltage waveform for half-wave rectifier power supply.

**solution**   The output ripple voltage is found by substituting values into Eq. (10-2). $I_L$ and $C$ are given in Fig. 10-3a as 2 A and 2000 $\mu$F, respectively.

$$\text{Ripple} = V_{pp} = \frac{2\ \text{A} \times 13 \times 10^{-3}\ \text{s}}{2000 \times 10^{-6}\ \text{F}} = 13\ \text{V}$$

The peak voltage to which $V_c$ rises is given in the curves of Fig. 10-2. At 115 V ac, with $I_L$ = 2 A, $V_{sec}$ peak is 22 V. The output-voltage waveform, therefore, rises to 22 V, then decreases by 13 V as shown in Fig. 10-3c. The dc voltage lies midway between the maximum and minimum values (15.5 V).
   The load regulation is computed at 115 V ac for a load current change from 0 to 2 A. At 115 V ac $V_c$ at $I_L$ = 0 is 23 V from Fig. 10-2. $V_c$ at $I_L$ = 2 A is 15.5 V from Fig. 10-3c. The load regulation is, therefore, 23 − 15.5 = 7.5 V.
   The line regulation is computed at $I_L$ = 0 for a line voltage change from 110 to 120 V ac. At $I_L$ = 0, $V_c$ at 110 V ac is 22 V from Fig. 10-2. $V_c$ at 120 V ac is 24 V from Fig. 10-2. The line regulation is, therefore, 24 − 22 = 2 V.

## Full-Wave Rectifier Power Supplies

**problem 10.2**   Solve for the ripple, output-voltage waveform, load regulation, and line regulation of the circuit shown in Fig. 10-4a.

**theory**   The $V_{sec}$ peak of each half of the center-tapped transformer is shown in Fig. 10-2. With two diodes, $C_1$ is recharged during every half-cycle of the 60-Hz waveform. This reduces the period of every charging cycle to 8.3 ms. In a full-wave rectifier power supply at full load, each diode typically conducts for approximately 3.3 ms during alternate cycles. $C_1$ therefore discharges into the load for 5 ms per cycle. This is shown in Fig. 10-4b.

**solution**   The output ripple is found by substituting values into Eq. (10-2). $I_L$ and $C$ are given in Fig. 10-4a as 2 A and 2000 $\mu$f, respectively, and $T$ is 5 ms, as previously discussed.

$$\text{Ripple} = V_{pp} = \frac{2\,\text{A} \times 5 \times 10^{-3}\,\text{s}}{2000 \times 10^{-6}\,\text{F}} = 5\,\text{V}$$

At 115 V, with $I_L = 2$ A, $V_{sec}$ peak is 22 V as shown in Fig. 10-2. The output voltage, therefore, rises to 22 V, then decreases by 5 V as shown in Fig. 10-4c. The dc voltage lies midway between the maximum and minimum values (19.5 V).

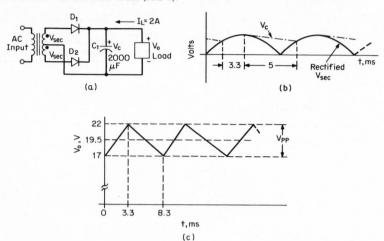

(a)                          (b)

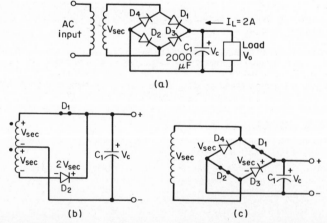

(c)

**Fig. 10-4**   (a) Full-wave rectifier power supply; (b) $V_c$ and rectified $V_{sec}$ for full-wave rectifier power supply; (c) output voltage waveform for full-wave rectifier power supply.

The load regulation is computed at 115 V ac for a load current from 0 to 2 A. At 115 V ac, $V_c$ at $I_L = 0$, $= 23$ V from Fig. 10-2. $V_c$ at $I_L = 2$ A, $= 19.5$ V from Fig. 10-4c. The load regulation is, therefore, $23 - 19.5 = 3.5$ V.

The line regulation is computed at $I_L = 0$ for a line voltage change from 110 to 120 V ac. At $I_L = 0$, $V_c$ at 110 V ac $= 22$ V from Fig. 10-2. $V_c$ at 120 V ac $= 24$ V from Fig. 10-2. The line regulation, is therefore, $24 - 22 = 2$ V.

### Full-Wave Bridge Rectifier Power Supplies

**problem 10.3**   Solve for the ripple, output-voltage waveform, load regulation, and line regulation of the circuit shown in Fig. 10-5a.

(a)

(b)                          (c)

**Fig. 10-5**   (a) Full-wave bridge rectifier power supply; (b) voltage across nonconducting diode in a full-wave rectifier power supply; (c) voltage across nonconducting diodes in a full-wave bridge rectifier power supply.

**theory**   Diodes $D_1$ and $D_2$ conduct simultaneously when $V_{sec}$ is greater than $V_c$ and in the correct polarity. Diodes $D_3$ and $D_4$ conduct during the next half-cycle. The discharge time of $C_1$ is approximately 5 ms, as in the full-wave rectifier example, but this circuit offers two advantages. One advantage is that the transformer has no center tap and is therefore simpler. The other advantage is that the diodes have to withstand less reverse voltage. To illustrate this, Figs. 10-5$b$ and 10-5$c$ show the reverse voltage across the nonconducting diodes in a full-wave and bridge circuit, respectively. The conducting diode is replaced by a short circuit in both cases. In Fig. 5$b$ the reverse voltage across $D_2$ is 2 $V_{sec}$. In Fig. 10-5$c$ the reverse voltage across either $D_3$ or $D_4$ is $V_{sec}$. The lower reverse voltage in the bridge configuration allows the use of diodes with lower reverse-voltage capability, which are generally less expensive than their higher-voltage counterparts. The cost saving with the less-expensive diodes and simpler transformer must be evaluated against the cost of two extra diodes.

**solution**   The output ripple, voltage waveform, load regulation, and line regulation are identical with those computed for the full-wave problem.

## Voltage-Multiplier Power Supplies*

**problem 10.4**   Solve for the ripple, output-voltage waveform, load regulation, and line regulation of the circuit shown in Fig. 10-6$a$.

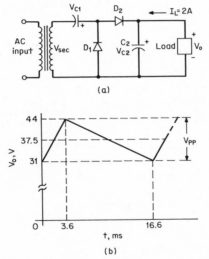

(a)

(b)

**Fig. 10-6**   (a) Half-wave doubler power supply; (b) output voltage waveform for half-wave doubler power supply.

**theory**   Figure 10-6$a$ shows a half-wave doubler circuit, which operates as follows: During one half-cycle, $D_1$ conducts, charging $C_1$ to $V_{sec}$ peak. During the next half-cycle, $D_2$ conducts, charging $C_2$ to $V_{sec}$ peak plus $V_{c1}$, which equals 2 $V_{sec}$ peak.

**solution**   Capacitor $C_2$ discharges for greater than one half-cycle, as in the half-wave rectifier example, and the output ripple is also 13 V.
At 115 V ac, with $I_L = 2$ A, $V_{sec}$ peak is 22 V. Therefore $C_2$ charges up to $2 \times 22 = 44$ V. This is shown in Fig. 10-6$b$. The dc value lies midway between the maximum and minimum values (37.5 V).
The load regulation is computed at 115 V ac for a load current change from 0 to 2 A. At 115 V ac, $V_{c2}$ at $I_L = 0$ is $2 \times 23 = 46$ V from Fig. 10-2. $V_{c2}$ at $I_L = 2$ A is 37.5 V from Fig. 10-6$b$. The load regulation is, therefore, $46 - 37.5 = 8.5$ V.
The line regulation is computed at $I_L = 0$ for a line voltage change from 110 to 120 V ac. At $I_L = 0$, $V_{c2}$ at 110 V ac is $2 \times 22 = 44$ V from Fig. 10-2. $V_{c2}$ at 120 V ac is $2 \times 24 = 48$ V from Fig. 10.2. The line regulation is, therefore, 4 V.

---

*The numbers given for output voltage under full-load conditions in this section are approximations. Because of the complex charge exchanges between capacitors, this type of circuit requires extensive analysis. Typical applications require much less load than used in these illustrative examples.

**problem 10.5**   Solve for the ripple, output voltage waveform, load regulation, and line regulation of the circuit shown in Fig. 10-7a.

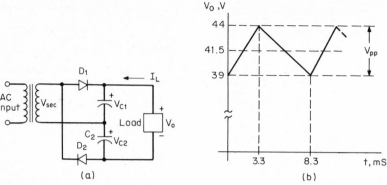

(a)                                                                        (b)

**Fig. 10-7**   (a) Full-wave doubler power supply; (b) output voltage waveform for full-wave doubler power supply.

**theory**   Figure 10-7a shows a full-wave doubler circuit, which operates as follows: During one half-cycle, $D_1$ conducts, charging $C_1$, and providing load current through $C_2$. On the next half-cycle, $D_2$ conducts, charging $C_2$, and providing load current through $C_1$. The total voltage appearing across the load is either $V_{sec}$ plus $V_{c1}$, or $V_{sec}$ plus $V_{c2}$, which equals 2 $V_{sec}$.

**solution**   Since $C_1$ and $C_2$ are charged on alternate half-cycles, the output ripple is 5 V as in the full-wave rectifier example.

At 115 V ac, with $I_L = 2$ A, $V_{sec}$ peak is 22 V. Therefore the output voltage rises to $2 \times 22 = 44$ V. This is shown in Fig. 10-7b. The dc value lies midway between the maximum and minimum values (41.5 V).

The load regulation is computed at 115 V ac for a load current change from 0 to 2 A. At 115 V ac, $V_o$ at $I_L = 0$ is $2 \times 23 = 46$ V from Fig. 10-2. $V_o$ at $I_L = 2$ A is 41.5 V from Fig. 10-7b. The load regulation is, therefore, $46 - 41.5 = 4.5$ V.

The line regulation is computed at $I_L = 0$, for a line voltage change from 110 to 120 V ac. At $I_L = 0$, $V_o$ at 110 V ac is $2 \times 22 = 44$ V from Fig. 10-2. $V_o$ at 120 V ac is $2 \times 24 = 48$ V from Fig. 10-2. The line regulation is, therefore, $48 - 44 = 4$ V.

*Note:* Figure 10-8 illustrates the extension of the half-wave doubler circuit to provide a five-times multipler. This particular circuit is used extensively in television circuits which may use 5000 V for the focus anode and 25,000 V for the accelerating anode of the picture tube. High-voltage diodes are used, and the entire assembly is usually encapsulated in an insulating material.

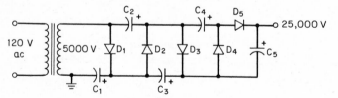

**Fig. 10-8**   Five-times multiplier power supply.

The performance results of these unregulated power supplies are summarized in Table 10-1.

## 10.3   REGULATED POWER SUPPLIES

In many applications, the changes in power supply voltage obtained with unregulated power supplies can not be tolerated. Ripple and regulation specifications of less than 100 mV are quite common. To achieve this, some form of voltage regulation must be used. The

**TABLE 10-1 Performance of Some Basic Power Supplies (Numerical values in volts)**

|  | Half-wave | Full-wave | Bridge | Half-wave doubler | Full-wave doubler |
|---|---|---|---|---|---|
| $V_o$ | 15.5 | 19.5 | 19.5 | 37.5 | 41.5 |
| Ripple | 13 | 5 | 5 | 13 | 5 |
| Line regulation | 2 | 2 | 2 | 4 | 4 |
| Load regulation | 7.5 | 3.5 | 3.5 | 8.5 | 4.5 |
| Features | Simple. inexpensive. higher ripple. | Efficient. Cost of transformer center-tap. Low ripple. More costly. | Efficient. Cost of extra diodes. Low ripple. | Higher voltage. Cost of extra parts. More costly diodes. | Higher voltage. Cost of extra parts. More costly diodes. Low ripple. |

two main types of regulators are series and shunt regulators, with the series type being more commonly used. To illustrate these types, the following circuits will be discussed.

1. Shunt-regulated power supplies
2. Series-pass-regulated power supplies with zener reference
3. Series-pass-regulated power supplies with feedback

To illustrate these types, full-wave rectification is shown at the input to the regulator.

### Shunt-Regulated Power Supplies

**problem 10.6** Solve for the load regulation, line regulation, and ripple attenuation of the circuit shown in Fig. 10-9. Assume: $V_c$ varies from 18 to 25 V, $V_{z,\text{nom}}$ is 15 V, $I_{zt}$ is 75 mA, $R_z$ is 2 Ω, $R_s$ is 15 Ω, $I_L$ is required to be 100 mA.

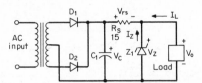

**Fig. 10-9** Zener shunt regulator.

**theory** This circuit is used in many low-current applications in which the input voltage is fairly constant. The zener diode provides essentially constant voltage, provided its current exceeds a certain value, $I_{zt}$, defined in the specifications published by the manufacturer. However, its voltage does vary with current in the zener region ($I > I_{zt}$) depending on its impedance. The equations applicable to this circuit are:

$$V_o = V_z = V_c - V_{rs} = V_c - I_s R_s \qquad (10\text{-}3)$$
$$I_s = I_z + I_L \qquad (10\text{-}4)$$
$$V_z = V_{z,\text{nom}} + (I_z - I_{zt})R_z \qquad (10\text{-}5)$$

where $V_{z,\text{nom}}$ denotes nominal zener voltage and $R_z$ denotes zener impedance.

**solution** At $V_c = 18$ V and $I_L = 0$:

$$I_z = I_s = \frac{(V_c - V_z)}{R_s} \text{ from Eqs. (10-3) and (10-4)}$$

$$I_z = \frac{(18 - 15)}{15} = 200 \text{ mA*}$$
$$V_z = 15 + 2(0.2 - 0.075) = 15.25 \text{ V from Eq. (10-5)}$$

At $V_c = 18$ V and $I_L = 100$ mA:

$$I_z = (0.2 - 0.1) \text{ A} = 100 \text{ mA from Eq. (10-4)}$$
$$V_z = 15 + 2(0.1 - 0.075) = 15.05 \text{ V from Eq. (10-5)}$$

Load regulation is, therefore, $15.25 - 15.05 = 0.2$ V.

*Note that this is based on an approximation of $V_z$.

At $V_c = 25$ V and $I_L = 0$:

$$I_z = \frac{(V_c - V_z)}{R_s} = \frac{(25 - 15)}{15} = 666 \text{ mA from Eq. (10-3)}$$

and (10-4)

$$V_z = 15 + 2(.666 - .075) = 16.18 \text{ V from Eq. (10-5)}$$

This $V_z$ can then be used to solve for $I_z$ again, which in turn yields a new $V_z$. The result can be approximated as follows:

$$I_z = 590 \text{ mA}, V_z = 16.1 \text{ V}$$

Line regulation is, therefore, $16.1 - 15.25 = 0.85$ V.

Since a change in line voltage of 7 V produces a change in output of only 0.85 V, any ripple at the input is likewise attenuated by the factor 7/0.85 or 8.23.

While this circuit provides better regulation than its unregulated counterpart, note that the zener diode must carry a large current at no load. This puts a great deal of power into the diode. In the example, $P_{z\text{max}} = 16.1 \times 0.590 = 9.5$ W. If $V_c$ does not vary much, and the load current remains essentially constant, then $R_s$ can be designed to lessen the power in the diode.

## Series-Regulated Power Supplies with Zener Reference

**problem 10.7**   Solve for the load regulation, line regulation, and ripple rejection of the circuit shown in Fig. 10-10. Assume: $V_c$ varies from 18 to 25 V, $V_{z\text{nom}}$ is 15 V, $I_{zt}$ is 75 mA, $R_z$ is 2 $\Omega$, $R_s$ is 35 $\Omega$, $H_{FE}$ is 100, $V_{be}$ varies from 0.7 V at 100 mA to 0.6 V at no load, $I_L$ is required to be 100 mA.

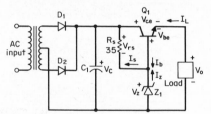

**Fig. 10-10**   Series-regulated power supply.

**theory**   This circuit uses a transistor to absorb the difference between the input and output voltages. It is more efficient and capable of providing more load current than the previous circuit. The equations applicable to this circuit are as follows:

$$V_o = V_c - V_{ce} = V_z - V_{be} \tag{10-6}$$

$$I_s = I_z + I_b = \frac{V_c - V_z}{R_s} \tag{10-7}$$

$$I_b = \frac{I_L}{H_{FE}} \tag{10-8}$$

In this circuit at no load the zener diode absorbs the base current, which is considerably less than the load current. Thus there is less power in the zener diode than in problem 10.6. As $V_c$ increases, $V_{ce}$ will increase to keep $V_o$ constant. Since $I_s$ also increases with $V_c$, $V_z$ and $V_o$ will also increase slightly. This affects line regulation. As $I_L$ increases, $V_{be}$ increases owing to the transconductance ($G_m$) of the transistor. Since $I_b$ also increases, $I_z$ decreases and the resultant decrease in $V_z$ adds to the increase in $V_{be}$, affecting load regulation.

**solution**

At $V_c = 18$ V and $I_L = 0$:

$$I_z = \frac{18 - 15}{35} = 85.7 \text{ mA from Eqs. (10-7) and (10-8)}$$

$$V_z = 15 + 2(0.0857 - 0.075) = 15.02 \text{ V from Eq. (10-5)}$$
$$V_o = 15.02 - 0.6 = 14.42 \text{ V from Eq. (10-6)}$$

At $V_c = 18$ V and $I_L = 100$ mA:

$$I_b = 100 \text{ mA}/100 = 1 \text{ mA from Eq. (10-8)}$$
$$I_z = (85.7 - 1)\text{mA} = 84.7 \text{ mA from Eq. (10.7)}$$
$$V_z = 15 + 2(0.0847 - 0.075) = 15.0194 \text{ V from Eq. (10.5)}$$
$$V_o = 15.0194 - 0.7 = 14.3194 \text{ V from Eq. (10-6)}$$

Load regulation is, therefore, $14.42 - 14.3194 = 0.1$ V.
At $V_c = 25$ V and $I_L = 0$:

$$I_z = \frac{25 - 15}{35} = 286 \text{ mA from Eqs. (10-7) and (10-8)}$$

$$V_z = 15 + 2(0.286 - 0.075) = 15.4 \text{ V from Eq. (10-5)}$$

$$V_o = 15.4 - 0.6 = 14.8 \text{ V from Eq. (10-6)}$$

Line regulation is, therefore, $14.8 - 14.42 = 0.38$ V; ripple is rejected by a factor $7/0.38 = 18.42$.

This circuit, therefore, regulates better than the previous one. Maximum zener power is now $15.4 \times 0.286 = 4.4$ W. Maximum transistor power is approximately $(25 - 14.8)0.1 = 1.02$ W. This circuit is, therefore, capable of providing much more load current than the previous one.

### Series-Regulated Power Supplies with Feedback

The regulated power supplies already discussed have two drawbacks—regulation is relatively poor, and the output voltage cannot be adjusted. To improve regulation, a high-gain feedback loop is used. A potentiometer is inserted in the feedback loop to vary the output voltage.

A generalized block diagram of such a power supply is shown in Fig. 10-11. The feedback amplifier compares $V_x$ with a reference voltage $V_R$ and adjusts the output voltage to make them equal.

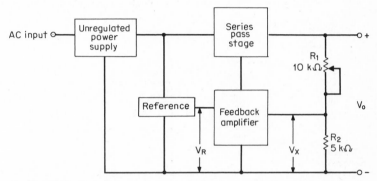

**Fig. 10-11**  Basic diagram of a series-regulated power supply with feedback.

**problem 10.8**  Assume $V_R = 6$ V, $R_1 = 10$ kΩ, $R_2 = 5$ kΩ. Solve for $V_o$ in Fig. 10-11.

**theory**  If the feedback amplifier has high gain, it draws negligible input current. Thus $R_1$ and $R_2$ form a voltage divider, and:

$$V_x = \frac{V_o R_2}{R_1 + R_2} \qquad (10-9)$$

**solution**  At equilibrium, $V_R = V_x = 6$ V.

$$V_o = V_x \frac{(R_1 + R_2)}{R_2} \qquad \text{from Eq. (10-9)}$$

$$V_o = \frac{6(15 \text{ k}\Omega)}{5 \text{ k}\Omega} = 18 \text{ V}$$

**problem 10.9**  Solve for the quiescent voltages and currents in the circuit of Fig. 10-12. Solve for the value of $R_A$. Assume $V_{be} = 0.6$ V, $h_{FE}$ of $Q_1 = 100$, $I_{c1} = I_{c2}$, load current is 100 mA.

**theory**  $Q_1$ and $Q_2$ form a differential amplifier which compares $V_{REF}$ to $V_{b1}$. If $V_{b1} > V_{REF}$, $Q_1$ conducts more, decreasing the base voltage of $Q_3$. This turns $Q_3$ and $Q_5$ off and causes $V_o$ to decrease until $V_{b1} = V_{REF}$. The circuit responds in the opposite way if $V_{b1} < V_{REF}$.

Thus the circuit constantly regulates the output voltage. As the loop gain is made higher, the circuit will detect and correct for increasingly smaller changes in $V_o$. Thus load and line regulation, as well as ripple, are functions of the gain of the amplifier.

Note that the reference has a direct effect on the output. Any changes in $V_{REF}$ are multiplied by $(R_A + R_B)/R_B$.

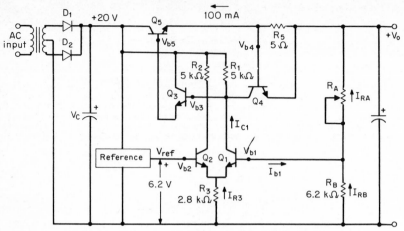

**Fig. 10-12**   Series-regulated power supply with feedback.

**solution**   At equilibrium:

$$V_{b1} = V_{b2} = V_{REF} = 6.2 \text{ V}$$
$$V_{R3} = V_{b1} - 0.6 = 5.6 \text{ V}$$
$$I_{R3} = 5.6/2.8 \text{ k}\Omega = 2 \text{ mA}$$
$$I_{c1} = I_{R3/2} = 1 \text{ mA}$$
$$V_{b3} = V_c - I_{c1}R_1 = 20 - (1 \text{ mA} \times 5 \text{ k}\Omega) = 15 \text{ V}$$
$$V_{b5} = V_{b3} - 0.6 = 14.4 \text{ V}$$
$$V_{b4} = V_{b5} - 0.6 = 13.8 \text{ V}$$
$$V_o = V_{b4} - (100 \text{ mA} \times 5) = 13.3 \text{ V}$$
$$IR_B = 6.2/6.2 \text{ k}\Omega = 1 \text{ mA}$$
$$I_{b1} = I_{c1}/h_{FE} = 1 \text{ mA}/100 = 0.01 \text{ mA}$$

Therefore $IR_B \gg I_{b1}$ and $IR_A = IR_B = 1$ mA. Since $V_o = V_{b1} + (IR_A \times R_A)$ then $R_A = (13.3 - 6.2)/1$ mA $= 7.1$ k$\Omega$.

Note that a 1-k$\Omega$ change in $R_A$ causes a 1-V change in $V_o$. This is known as 1000 ohM/volt programming.

**problem 10.10**   Solve for the maximum current available from the power supply in Fig. 10-12.

**theory**   Transistor $Q_4$ limits the output current of the power supply. When the voltage across $R_5$ reaches 0.6 V, $Q_4$ conducts, turning $Q_3$ and $Q_5$ off. This causes $V_o$ to decrease, and the power supply becomes current-limited.

**solution**

$$VR_5 = I_L \times 5 = 0.6 \text{ V}$$
$$I_L = 0.6/5 = 120 \text{ mA}$$

Thus the power supply will provide 120 mA before its output voltage decreases.

## Additional Comments on Regulated Power Supplies

The design of an amplifier like the one shown in Fig. 10-12 involves much attention to details such as stability, component tolerance, component stress, temperature effects, and transient response which are beyond the scope of this chapter. Many circuits have been developed to stabilize the reference voltage, improve current limit operation, and improve transient response, and these will be found in commercially available power supplies. With the advent of integrated circuit and hybrid technology, many of these circuits were incorporated into small packages. As a result, many power supplies today utilize small integrated circuits to regulate the output voltage to within 5 mv.

Most power supplies have an output capacitor which lowers the output impedance, improves transient response, and helps to stabilize the amplifier.

## 10.4  SWITCHING REGULATOR POWER SUPPLIES

The regulated power supplies discussed in Sec. 10.3 are called *linear power supplies*. As input voltage increases, input power increases. The equation for power dissipation within the power supply is:

$$P_{\text{diss}} = P_{\text{input}} - P_{\text{output}}$$

Since output power is constant, the power supply must dissipate more power at high input voltage, and this limits its efficiency. A 100-W output linear power supply may dissipate 200 W internally.

A switching regulator achieves high efficiency by minimizing internal power losses and holding input power relatively constant as input voltage increases. It does this at the expense of added complexity.

**problem 10.11**   In the circuit of Fig. 10-13, $V_{c1}$ is 48 V, $t_{\text{on}}$ is 20 $\mu$s, $t_{\text{off}}$ is 20 $\mu$s. Solve for the output voltage.

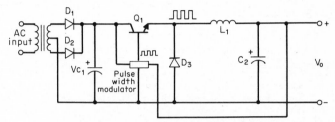

**Fig. 10-13**  Basic switching regulator power supply.

**theory**   Figure 10-13 shows a basic switching regulator power supply. The input voltage is rectified and filtered in the conventional manner and applied to $Q_1$. $Q_1$ is driven on and off by a series of pulses and delivers pulses of voltage to $L_1$. $L_1$ and $C_2$ constitute a low-pass filter which removes most of the ac component from the voltage pulses, so that the output voltage is dc. If $Q_1$ is on for $t_{\text{on}}$ and off for $t_{\text{off}}$, the output voltage is:

$$V_o = \frac{V_{c1}\, t_{\text{on}}}{t_{\text{on}} + t_{\text{off}}} \tag{10-10}$$

If the period ($t_{\text{on}} + t_{\text{off}}$) of the drive pulses remains constant, regulation can be achieved by varying $t_{\text{on}}$.

**solution**   From Eq. (10-10)

$$V_o = \frac{48 \text{ V} \times 20\ \mu\text{s}}{40\ \mu\text{s}} = 24 \text{ V}$$

**problem 10.12**   Assume that $V_{c1}$ in the previous problem increases to 72 V as the input voltage increases. Solve for $t_{\text{on}}$ such that $V_o$ remains constant ($t_{\text{on}} + t_{\text{off}}$ is held constant).

**solution**

$$V_o = 24 \text{ V} = \frac{72 \text{ V} \times t_{\text{on}}}{40\ \mu\text{s}}$$

$$t_{\text{on}} = \frac{24 \text{ V} \times 40\ \mu\text{s}}{72 \text{ V}} = 13.33\ \mu\text{s}$$

**Discussion**   Problems 10.11 and 10.12 show that line regulation can be achieved by making the duty cycle $t_{\text{on}}/(t_{\text{on}} + t_{\text{off}})$ inversely proportional to line voltage. As load is applied to the output, the duty cycle increases to compensate for added voltage drops in the circuit.

This type of power supply is more efficient than its linear counterpart but still requires an input transformer unless it is used with a dc input voltage. To reduce the size of the power supply, an off-line switching regulator may be used. This circuit is shown in Fig. 10-14.

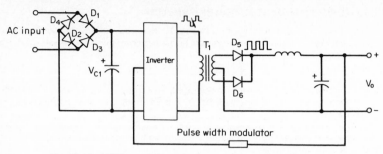

**Fig. 10-14**   Off-line switching regulator power supply with inverter.

The inverter uses high-voltage transistors to convert the dc voltage ($V_{c1}$) into ac pulses (usually at 20 kHz). A high-frequency transformer ($T_1$)—which is much smaller than an equivalent 60-Hz transformer—then steps up the ac voltage to the desired level. Operation of the $LC$ filter and the effects of varying the duty cycle are as in the previous example.

Small magnetic components and higher efficiency have resulted in switching power supplies which are substantially smaller than their linear counterparts with the same output power. For example, a 500-W switching supply is typically available in a 640-in$^3$ package, compared with 1520 in$^3$ for a linear supply. In terms of watts/cubic inch, a switching supply can achieve 0.9 W/in$^3$ while a linear supply usually provides 0.4 W/in$^3$. Switching power supplies have been used where smaller weight and size are important, and their high efficiency has appealed to those interested in energy conservation.

Circuitry used in switching regulator supplies, being complex and nonstandardized, is beyond the scope of this chapter.

## 10.5  THERMAL CONSIDERATIONS

One of the biggest problems encountered in power supplies is heat dissipation. A 33% efficient, 100-W output power supply dissipates 200 W internally. This power loss manifests itself as heat and can destroy components if the design does not allow for adequate cooling. One common method of cooling a component is to attach it to a heat sink. Thermal performance is usually described in terms of thermal impedance.

The thermal impedance ($\theta$) between two points is defined as the temperature differential ($T$) which occurs between the points when a specific amount of power ($P$) is applied to the system.

$$\theta = \frac{T}{P} \qquad °C/W$$

A lower $\theta$ results in a lower temperature rise for a given power.

The thermal impedances of interest are defined below (refer to Fig. 10-15).

$\theta_{jc}$: Thermal impedance from junction to case
$\theta_{ch}$: Thermal impedance from case to heat sink
$\theta_{ha}$: Thermal impedance from heat sink to ambient

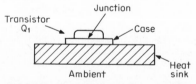

**Fig. 10-15**   Example of transistor mounting.

**problem 10.13**   The transistor $Q_1$ in Fig. 10-15 is dissipating 10 W (P). If $\theta_{jc}$ is 1°C/W, $\theta_{ch}$ is 0.5°C/W, $\theta_{ha}$ is 3.5°C/W, and ambient temperature $T_a$ is 25°C, solve for the junction temperature $T_j$.

**theory**    The temperature differential from junction to case ($\Delta T_{jc}$) is:

$$\Delta T_{jc} = \theta_{jc} \times P \tag{10-11}$$

Similarly,

$$\Delta T_{ch} = \theta_{ch} \times P \tag{10-12}$$

and

$$\Delta T_{ha} = \theta_{ha} \times P \tag{10-13}$$

Therefore,

$$T_j = T_a + (\theta_{jc} + \theta_{ch} + \theta_{ha})P \tag{10-14}$$

**solution**    From Eq. (10-14):

$$T_j = 25 + (1.0 + 0.5 + 3.5)10 = 75°C$$

**problem 10.14**    Another transistor ($Q_2$) is added to the heat sink in the previous problem. It is dissipating 20 W, and $\theta_{jc}$ is 1°C/W. Solve for the junction temperature of each transistor ($T_{j1}$ and $T_{j2}$).

**theory**    The total power on the heat sink has increased. This affects only the temperature differential across the heat sink.

$$T_{ha} = \theta_{ha}(P_1 + P_2) \tag{10-15}$$

**solution**    from Eqs. (10-11), (10-12), and (10-15):

$$
\begin{aligned}
T_{j1} &= T_a + (P_1 + P_2)\theta_{ha} + P_1(\theta_{jc} + \theta_{ch}) \\
&= 25 + (30 \times 3.5) + 10(1.5) \\
&= 145°C \\
T_{j2} &= T_a + (P_1 + P_2)\theta_{ha} + P_2(\theta_{jc} + \theta_{ch}) \\
&= 25 + (30 \times 3.5) + 20(1.5) \\
&= 160°C
\end{aligned}
$$

(Note that for most heat sinks as the total power increases $\theta_{ha}$ decreases.)

Commercial heat sinks are available in many configurations for specific applications. Some of them utilize the advantage of moving air or water cooling, which lowers the effective thermal impedance.

One other note of interest here is that transistors are often insulated from their heat sinks. The insulator increases the thermal impedance by allowing the case temperature to exceed the heat-sink temperature.

$$T_j = T_a + P(\theta_{ha}) + P(\theta_{jc}) + P(\theta_{insulator})$$

## 10.6    BIBLIOGRAPHY

Millman, J., and Halkias, C.: *Integrated Electronics: Analog and Digital Circuits and Systems*, McGraw-Hill, New York, 1972, secs. 18-9 to 18-11.

Wurzburg, H.: *Motorola Voltage Regulator Handbook*, Motorola, Inc., Phoenix, 1976.

Wurzburg, H.: *Motorola Silicon Rectifier Handbook*, Motorola, Inc., Phoenix, 1973.

Chapter **11**

# Battery Uses and Special Cells

## DR. HOWARD J. STRAUSS

**Vice President, ESB Ray-O-Vac Management Corp.**

## 11.1 INTRODUCTION

The large variety of devices now powered by batteries reflects the versatility and convenience of such power sources. To a large degree, this stems from the commercial production of a great many kinds, sizes, and shapes of batteries, the continual introduction of new types, and the ready availability of specialized forms. For many years, the principal uses for batteries have been in devices requiring relatively low power levels, such as flashlights, lanterns, portable radios, clocks, and toys. Such applications have been largely satisfied by Leclanche cells, also referred to as carbon-zinc cells, or simply as dry cells. In recent years, the general popularity of portable radios, cassette recorders, dictating equipment, electrically driven motion-picture cameras, portable instrumentation, and many other battery-powered devices, has resulted in the introduction of improved ("super") Leclanche cells, as well as other kinds of dry cells, such as mercury, silver, and alkaline cells, all capable of higher power and energy levels than were available from conventional Leclanche cells.

Other important applications for batteries can be classified as low-energy, high-power applications, as for engine starting; high-energy, moderate-power applications, as for powering electric vehicles, mine equipment, and industrial trucks; and high-energy, moderate-power applications, as for emergency lighting and general standby power. These applications have generally been satisfied by secondary or rechargeable batteries, principally of the lead-acid type, but to a small degree also by nickel-iron (Edison) and nickel-cadmium batteries.

In recent years, entirely new demands have been placed on batteries, and these have resulted in the introduction of many new battery systems, as well as in new designs for conventional systems. Typical of some of the newer uses are such low-power devices as hand-held calculators, electric watches (particularly digital display types), and implantable pulse generators (i.e., pacemakers), and such high-power devices as portable power tools, television receivers, and print-out calculators. In addition, entirely new areas of application have emerged, as in high-reliability memory devices used in remote-reading instrumentation and in noise-free devices such as geological survey and communication equipment.

## Types of Batteries

Most recently, the general need to efficiently generate and use energy has triggered interest in developing very large electrochemical energy storage systems for load-leveling and peak-shaving applications. Thus, the sizes of batteries that are now, or will shortly be, available range from cells that weigh a few grams to those that weigh several tons, and with power capabilities that range from a few microwatts to hundreds of megawatts. In addition, systems have been developed that may have particular merit in terms of special properties, such as good shelf life and good mechanical integrity. The result is a broad spectrum of readily available battery power sources that are in an advanced state of technical or commercial development.

Fundamentally, batteries are classified as *primary* if they cannot be recharged or are intended for a single discharge and as *secondary* if they can be recharged or are intended to be charged and discharged, i.e., cycled, a number of times. There is a third category generally referred to as *reserve* batteries. Such batteries are intended to be chemically stable during extended storage and to be activated immediately prior to use, usually by heating or the addition of water or electrolyte. The most common primary batteries of interest to electronic engineers are carbon-zinc, alkaline, mercury, magnesium, silver-zinc, and solid-state. The principal secondary batteries used in the electronic field are nickel-cadmium, lead-acid, silver-zinc, and silver-cadmium. The most important reserve batteries are the cuprous chloride-magnesium, silver chloride-magnesium, and thermal cells. Calculations pertinent to these types will be discussed later in this chapter.

## Battery Terminology

BATTERY. An arrangement of two or more cells connected in series and/or parallel so as to supply a given voltage and current.

CAPACITY. The amount of electrical energy a battery is capable of delivering under specified discharge conditions. Capacity is usually expressed in ampere-hours or milliampere-hours.

CELL. The basic unit capable of converting chemical energy into electrical energy. It consists of two electrodes immersed in a common electrolyte, one electrode being capable of receiving electrons during the chemical reaction, the other being capable of releasing electrons as a result of the chemical reaction.

CHARGING. The process of supplying electric energy to a cell for the purpose of converting the electric energy into stored chemical energy.

CUTOFF VOLTAGE. The minimum voltage at which the battery can deliver useful energy under the specific discharge conditions.

CYCLE. One sequence of charge and discharge of a cell.

DISCHARGE. The withdrawal of electrical energy from a cell through an external circuit. A deep discharge is one in which the energy delivered by the cell is practically all the energy which the cell is capable of delivering. A shallow discharge is one in which only a small fraction of the ultimate capacity of the cell is withdrawn.

DRAIN. The current at which a battery is discharged; also termed drain rate.

ENERGY DENSITY. The ratio of the energy delivered to its weight or volume. Energy density is usually expressed in watt-hours per pound or watt-hours per cubic inch.

ENERGY OUTPUT. The delivered capacity times the average voltage during the discharge of the battery. It is expressed in watt-hours or milliwatt-hours.

FLOAT CHARGING. A method of maintaining a rechargeable battery at full charge by continuously maintaining a constant voltage across the battery of such magnitude that the charging current just compensates for the various losses that occur in the battery. See also Trickle Charging.

IMPEDANCE. The first stable current that a battery supplies at the start of a discharge. (The definition of *stable current* is not fixed, and initial drain rates are, therefore, subject to individual practices.)

INTERNAL RESISTANCE. The opposition to the flow of a current within the cell. It is measured in ohms.

NOMINAL VOLTAGE. The nominal voltage of a battery is subject to an individual definition but is generally understood to be the voltage of a fully charged battery when delivering energy at very low discharge rates.

OPEN-CIRCUIT VOLTAGE. Voltage when the battery is delivering no energy to an external circuit.

POLARIZATION. The difference between the open-circuit voltage of an electrode and that which it actually delivers at any point in a discharge.

VOLTAGE REGULATION. The uniformity of the voltage with which the battery delivers its energy over the entire discharge period.

SHELF LIFE. That period of time (measured from the date of manufacture) for which a cell will retain a specified percentage of its original capacity. Usually the storage temperature is figured at 70°F and the period required for the capacity to fall to 90% of its original value is taken as the shelf life. The term *shelf life* is usually applied to primary batteries, while the equivalent term *charge retention* is applied to secondary batteries.

TRICKLE CHARGING.  A method in which a constant current is used to bring a battery to full charge and then maintain it in a fully charged condition. Trickle charging differs from float charging in that the latter is usually a constant-voltage process.

## Fundamentals of Battery Operation

Essentially all batteries consist of two half-cells in a common electrolyte. A half-cell, in turn, consists of a solid electrode immersed in an electrolyte. The electrolyte, which is usually a liquid but may be a solid, contains at least one chemical species capable of reacting with the electrode to either release or absorb electrons. When one half-cell absorbs electrons, the other must be capable of releasing electrons in order to complete the cell. Normally, the gain or release of electrons ceases very quickly owing to the fact that the electrodes become highly polarized and electron transfer is impeded. However, if

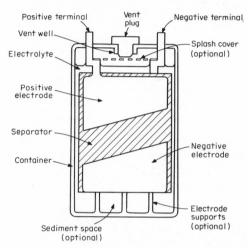

**Fig. 11-1**  Essential components of a battery. In one form or another, all batteries, both primary and secondary, have the components indicated above.

the two electrodes are immersed in a common electrolyte and are connected by a wire through an external circuit, the charge on each electrode is neutralized and the chemical reaction can proceed. The electronic current flowing through the external circuit is capable of doing work and represents the useful energy released by the battery. The electric circuit is completed inside the battery by the passage of charged particles, called *ions*, through the electrolyte, as illustrated in Fig. 11-1. Since the operation of a battery involves two chemical reactions, each of which involves the transference of a charged particle to or from the electrolyte to a reaction site on an electrode, the rate at which this transfer can occur is governed by (1) the potential generated between each electrode and the charged particle and (2) the ability of the reacting particle to move to the electrode under the influence of the electrostatic field set up between the opposing electrodes and the difference in concentration of the reacting species at the surface of the electrode and in the bulk of the electrolyte.

The potential of a cell is the algebraic sum of the potentials of the individual electrodes. Those cells having a high potential are generally capable of higher rates of discharge than those of low potential, simply because of the fact that the electrode potentials exert a larger force on the reacting ions in the electrolyte, permitting them to move more rapidly through the electrolyte. Anything that impedes the ability of these ions to move will tend to reduce the rate at which the battery can be discharged. Some factors that influence the rate of movement are temperature, viscosity of the electrolyte, the surface area across which the diffusing material can move, and the linear distance it has to move.

Decreasing temperatures tend to reduce ionic movement rates, but since most batteries generate heat during charge or discharge, the manner in which a battery performs under different conditions of temperature can be expected to be very complex. In addition, both the actual and apparent areas between the electrode and the electrolyte are of prime importance. Thus, batteries having solid electrodes (i.e., nonporous) can be expected to behave quite differently from those having porous electrodes in which there is a high interfacial area between the electrode and the electrolytes. As can be expected, the spacing between the electrodes, since it affects the length of the diffusion path, is also a major factor, as is anything that is placed between the electrodes, such as separators. These latter also interfere with diffusion and therefore greatly influence the rate at which the battery can be discharged. With these factors in mind, it is now possible to understand some of the characteristics of actual batteries. The characteristics of the commercially important battery systems are summarized in Table 11-1.

## 11.2   PRIMARY BATTERIES

### Background Information

Except for special-purpose batteries, the principal commercial primaries are packaged in unitized containers. In addition, they are so designed as to have the required electrolyte absorbed in the active materials and separators so that no free liquid is present. For this reason they are also referred to as *dry* cells. While this term is generally used to describe

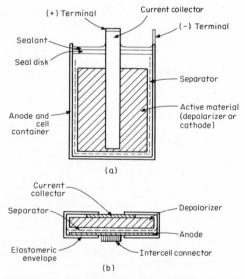

**Fig. 11-2**   Design of (*a*) cylindrical and (*b*) flat dry cells.

all cell types having no free electrolyte, the common dry cell refers to the carbon-zinc system, or Leclanche cells. Carbon-zinc cells are very widely distributed and are, by far, the most important commercial primary battery.

   The carbon-zinc system uses a carbon current collector in contact with a cathode of manganese dioxide, an electrolyte of ammonium chloride, and a zinc anode. In the familiar cylindrical carbon-zinc cell, the carbon is in the form of a centrally located rod surrounded by the manganese dioxide mix. The zinc anode is in the form of a can which also serves to contain the cell. Carbon-zinc cells are also available in flat, rectangular form in which case the same physical components are used but in a different geometric configuration. The essential designs of these forms are illustrated in Fig. 11-2. The nominal voltage of a carbon-zinc cell is 1.5 V, and through series and parallel combinations, batteries are available in voltages from 1.5 to 510 V. Usually batteries in the higher voltage ranges are made up of flat cells because of the greater convenience in series stacking.

**TABLE 11-1  Characteristics of Commercially Important Dry Battery Systems**

| | Carbon-zinc (Leclanche) | Alkaline-manganese | Mercury-zinc | Silver-zinc | Zinc-air |
|---|---|---|---|---|---|
| Open-circuit voltage | 1.5 | 1.5 | 1.35 | 1.86 | 1.35 |
| Voltage range under load | 0.8–1.4 | 0.8–1.2 | 1.25 | 1.4–1.6 | 1.0–1.2 |
| Energy density, Wh/lb | 30 | 20 | 50 | 65 | 85 |
| Energy storage, Wh/in | 1.8 | 4 | 4.6 | 7.5 | 8 |
| Low-temperature performance 0°C capacity as a % of 25°C capacity | 25 | 50 | 35 | 35 | 30 |
| Shelf life at 25°C (months to 80% capacity)* | 18 | 30 | 36 | 30 | 18 |
| Shelf life at 45°C (months to 80% capacity)* | 6 | 12 | 15 | 12 | 6 |
| Typical applications | Flashlights, lanterns, photoflash, radios, toys, hearing aids | Electronics, lighting, toys, tape recorders, transceivers, movie cameras | Hearing aids, pacemakers, tape recorders, watches, paging systems, instrumentation | Hearing aids, camera drives, electric watches, electronics, instrumentation | Hearing aids, barricade lights, train signals, tape recorders, toys |

*Values shown are general averages for normal usage.

Because of their long shelf life and higher power capability, alkaline cells are now an important and fast-growing commercial primary type. Like carbon-zinc cells, the alkaline system uses a manganese dioxide cathode and a zinc anode. In this case, the electrolyte is a strongly alkaline solution, usually about 30% KOH, and the zinc is in a finely divided form. Because gassing can be minimized, even to the point where the gassing is virtually negligible, alkaline cells can be hermetically sealed, making them particularly useful in devices, such as cameras, where positive protection against even accidental electrolyte leakage is desirable. The voltage characteristics of alkaline cells are generally similar to those of carbon-zinc cells, being about 0.1 V less commensurate circumstances, making the two cell systems generally interchangeable.

Mercury cells are similar to alkaline cells but with mercuric oxide (HgO) substituted for the $MnO_2$ depolarizer. Mercury cells have outstanding shelf life and very high capacities for a given volume. The voltage characteristics are similar to alkaline cells, being about 0.15 V less under commensurate conditions. Silver cells are similar to mercury cells but with silver oxides ($Ag_2O$ and $AgO$) used for the depolarizer. Silver cells have higher voltages than the other types, being about 0.2 V higher than carbon-zinc cells under comparable conditions.

## Commercial Sizes

Dry cells are commercially available in a very large variety of sizes and physical configurations, as shown in Table 11-2, and can consist of single cells or of multiple cells in series and parallel connections to achieve practically any desired current capacity or other characteristics. Since dry cells are used in a great many devices in which they have to be repeatedly replaced, it has been necessary to carefully standardize the various cell sizes.

**TABLE 11-2    Overall Dimensions of Commercial Dry Cells***

| Cell size designation | Diameter, in | Height, in |
|---|---|---|
| Cylindrical cells | | |
| N | 0.44 | 1.17 |
| AAA | 0.38 | 1.74 |
| AA | 0.56 | 1.96 |
| C | 1.02 | 1.96 |
| D | 1.31 | 2.40 |
| F | 1.31 | 3.60 |
| #6 | 1.53 | 6.56 |
| Button cells | | |
| M5 | 0.31 | 0.14 |
| M8 | 0.46 | 0.13 |
| M15 | 0.46 | 0.21 |
| M20 | 0.62 | 0.24 |
| M30 | 0.630 | 0.44 |
| M40 | 0.63 | 0.66 |

| Cell size designation | Length, in | Width, in | Thickness, in |
|---|---|---|---|
| Flat cells | | | |
| F15 | 0.56 | 0.55 | 0.12 |
| F20 | 0.94 | 0.55 | 0.12 |
| F25 | 0.89 | 0.89 | 0.23 |
| F30 | 1.25 | 0.84 | 0.13 |
| F40 | 1.25 | 0.84 | 0.21 |

*Dimensions may vary slightly among producers or for special label and terminal arrangements.

## Characteristics

The closed-circuit voltage, or working voltage, of dry cells falls continuously during the course of a discharge from about its open-circuit voltage to some predetermined voltage. The duration of a discharge can be increased if the cutoff voltage can be decreased. Typical cutoff voltages range from 0.8 to 1.1 V per cell, depending upon the application. To extract the greatest possible amount of energy, the cutoff voltage should thus be as low as possible.

The service capacity of primary batteries is not a fixed number of ampere-hours because the battery functions at efficiencies dependent on the current drain, the cutoff voltage, and the discharge program itself. The efficiency of a primary battery improves as the current drain decreases, as illustrated in Fig. 11-3 which gives a generalized voltage-time curve

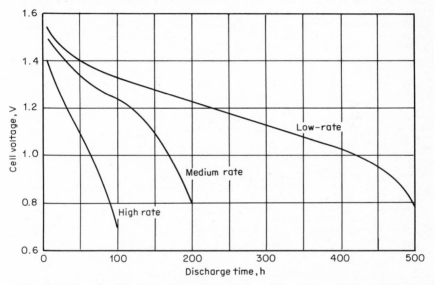

**Fig. 11-3**  Generalized discharge characteristics of a dry cell (intermittent discharge, 4 h/day).

for a carbon-zinc D-size cell discharged for 4 h/day. As can be seen from Fig. 11-3, the service life is greatly dependent on the current drain. As an example, the service life of most carbon-zinc cells under any set of circumstances can be tripled by halving the current drain. Other types of dry cells behave similarly, but perhaps not to that degree.

## Applications

In general, dry cells are most effectively used in those applications involving low current drains and intermittent use, particularly when the periods of nonuse are prolonged. Consequently, such cells have found wide use in radios, barricade flashers, telephones, toys, lighting systems, signaling circuits, flashlights, photoflash equipment, paging devices, and hearing aids, to mention only a few typical applications.

As a general industry practice, the capacity of virtually all types of dry cells is usually stated for discharges through a fixed resistor to a given end voltage. The capacities are reported in terms of service life, usually in hours; as a function of the initial drain, usually in milliamperes. The information so developed is presented either in graphical or tabular form. To obtain the initial drain, the nominal voltage of the battery is divided by the resistance of the load. Manufacturers' service life tables or graphs cite values for different cutoff voltages. In using graphical data, the vertical line corresponding to the calculated initial drain is followed vertically until it intersects the curve for the desired cutoff voltage. The ordinate at this point is the service life. In using tabular data, the calculated initial drain rate is found at the left and then followed horizontally across to the column corresponding to the desired cutoff voltage.

It is very important to use the graph or chart supplied by the manufacturer for any particular primary battery. Not only are there differences among manufacturers, but each manufacturer may use several formulations for each size, so as to develop such characteristics as best satisfy the intended use. For example, a battery intended for flashlight use is designed to accommodate a discharge of 0.5 A for 5 min at a time through a resistance of about 2.25 Ω. On the other hand, a battery designed for photoflash applications is designed to deliver high currents for very short durations through a very low resistive load. Each battery will not perform well if used in applications for which it was not designed. Consequently, batteries are specified in accordance with standard tests for which stipulated performances are published by the manufacturers.

It will be noted that the capacity for an intermittent discharge is considerably higher than that for a continuous discharge. This relates to the fact that in an intermittent discharge diffusion of the soluble reactants in the electrolyte is given enough time to equalize concentrations throughout the electrolyte, thereby replenishing that which is needed in the immediate vicinity of the electrodes.

The following problems are based on the common D-size carbon-zinc cell, this being the most widely used type and size. Other primary cell types are handled in precisely the same manner. However, care must be taken to use the tables and graphs that refer to the appropriate cell type and use conditions.

**problem 11.1**    How long can a D-size carbon-zinc cell be continuously discharged through a 7.5-Ω resistance until its voltage drops to 0.9 V?

**theory**    The discharge capability of each type and size of dry cell is provided by the manufacturer usually in the form of a graph, such as Fig. 11-4, or a table such as Table 11-3. For a discharge through a fixed resistance, the discharge current falls as the cell voltage decreases while the discharge proceeds. This involves a significant complication since the discharge characteristics of a Leclanche cell are very sensitive to the rate of discharge. To standardize the discharge conditions, the data are presented in terms of the *initial discharge current*. When the circuit is closed, the battery voltage will fall very rapidly from its open-circuit value to some lower, but stable, voltage, at which point the current, by convention, is taken as the initial discharge current. Although the cell voltage, and therefore current, may continue to fall, the initial discharge current is sufficiently stable as to be easily read. However, in the absence of any other information, the initial discharge current can be calculated by using the nominal voltage of the cell, 1.5 V, as the initial discharge voltage.

**solution**    Initial drain = $E_{nom}/R$ = 1.5/7.5 = 0.200 A = 200 mA. At the 200-mA abscissa on the $x$ axis on Fig. 11-4, move vertically until the 0.9-V cutoff voltage line is intersected. The service life can then be read as 10.0 h.

In the case of Table 11-3, data for 24 h/day are used since the discharge is continuous. For an initial drain of 200 mA, read the service life as 11.0 h in the 0.9-V cutoff voltage column. (Use an 8-Ω load to approximate 7.5 Ω.)

The difference between the values obtained from Fig. 21-4 and Table 11-3 is due to the fact that the cell mix formulations are different. Dry-cell manufacturers tailor their mixes to suit various general applications. For example, a cell designed for photoflash use will have a high pulse discharge rate capability which is generally obtained at the expense of

**TABLE 11-3    Continuous Discharge of a General-Purpose Carbon-Zinc D Cell (70°F)\***

| Load, Ω | Operating time, h, at a cutoff voltage of: | | | | |
|---|---|---|---|---|---|
| | 0.8 V | 0.9 V | 1.0 V | 1.1 V | 1.2 V |
| 2.25 | 1.5 | 0.9 | | | |
| 4 | 6.3 | 4.4 | 2.3 | | |
| 8 | 17 | 11 | 7.2 | 6.0 | 3.2 |
| 25 | 75 | 51 | 43 | 38 | 28 |
| 100 | 430 | 365 | 320 | 290 | 240 |

\*Values shown are general averages and may vary among producers.

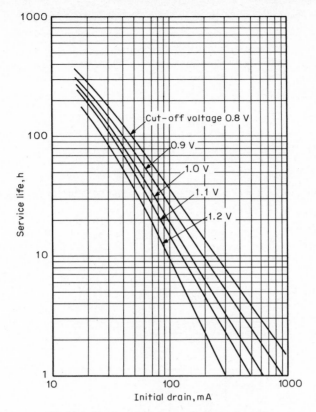

**Fig. 11-4** Continuous discharge of a D-size dry cell.

capacity. When such a cell is used in a flashlight, the light will be much brighter, but for a shorter time, than when a general-purpose cell is used. Consequently, it is necessary to use each manufacturer's data for each specific cell model in order to be able to calculate discharge information accurately.

**problem 11.2** How long would the cell of Problem 11.1 last if the discharge were for only 4 h/day?

**theory** Carbon-zinc cells "recover" if allowed to "rest" during the course of a discharge. This is because localized differences in electrolyte composition, caused by the discharge, are allowed to equalize by diffusion of water from the moister parts of the cell to the drier parts. Consequently, the service life of the cell tends to be significantly longer.

**solution** The methods are the same as for Problem 11.1. However, the figure or chart used must be for the discharge conditions cited. For an intermittent discharge of 4 h/day, Fig. 11-5 is used, but still at an initial drain of 200 mA. For these conditions the service life will be 15.3 h. In a manner similar to that of Problem 11.1, Table 11-4 yields a service life of about 16 h.

It should be noted that the rest periods helped the cell of Table 11-3 more than they did the cell of Fig. 11-4, again emphasizing that the characteristics of apparently similar cells can be quite different and that it is necessary to consult the actual manufacturer's data and carefully match cell characteristics to the intended use.

**problem 11.3** Using Fig. 11-4, construct a discharge curve for a D-size carbon-zinc cell discharged continuously through a 7.5 Ω resistance, and determine its capacity to a cutoff of 0.8 V.

**TABLE 11-4   Intermittent Discharge of a General-Purpose Carbon-Zinc D Cell (4 h per day at 70°F)***

| Load, Ω | Operating time, h, at a cutoff voltage of: | | | | |
|---|---|---|---|---|---|
| | 0.8 V | 0.9 V | 1.0 V | 1.1 V | 1.2 V |
| 2.25 | 1.5 | 0.9 | | | |
| 4 | 6.5 | 5.0 | 2.8 | 1.8 | |
| 8 | 20 | 17 | 15 | 9.2 | 5.1 |
| 25 | 98 | 89 | 81 | 70 | 60 |
| 100 | 430 | 380 | 360 | 345 | 310 |

*Values shown are general averages and may vary among producers. By comparison with Table 11-3, note that the additional service for intermittent discharges is most pronounced for the higher cutoff voltages.

**solution**   For an initial drain of 200 mA, the following service periods are read directly from Fig. 11-4:

| Cutoff voltage, V | Running time, h |
|---|---|
| 1.2 | 2.4 |
| 1.1 | 4.7 |
| 1.0 | 6.6 |
| 0.9 | 9.8 |
| 0.8 | 15.0 |

Figure 11-6 can then be drawn from the above data.

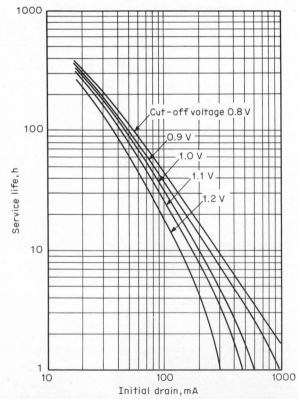

**Fig. 11-5**   Intermittent discharge of a D-size dry cell (4 h/day).

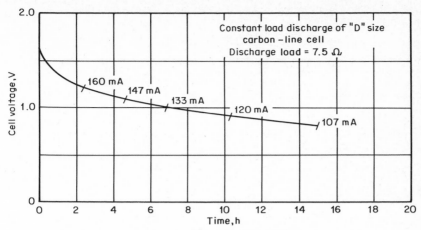

**Fig. 11-6**  Solution of Problem 11.3.

To determine the capacity, it is first necessary to know the average voltage. This can be obtained with sufficient accuracy by noting the voltage at the midpoint of each 2-h period and averaging.

| Time, h | Voltage, V |
|---------|------------|
| 1 | 1.34 |
| 3 | 1.17 |
| 5 | 1.07 |
| 7 | 1.00 |
| 9 | 0.93 |
| 11 | 0.88 |
| 13 | 0.84 |
| 15 | 0.80 |

Average = 1.004 V. $I_{av}$ = 1.004/7.5 = 0.134 A = 134 mA. Capacity = 134 mA × 15.0 h = 2.010 mAh.

**problem 11.4**  What is the service life for the conditions described in Problem 11.2 if the ambient temperature is 20°F?

**theory**  Since the mobility of the ions in the electrolyte and the electrode reaction rates are reduced as the temperature is lowered, most aqueous batteries are unable to deliver as much energy at low temperatures as they can at normal temperatures. It is frequently also necessary to go to a lower end voltage when operating at reduced temperatures. If the temperature is lowered enough to freeze the electrolyte, ionic mobility may be reduced to the point where the battery cannot deliver any useful energy. It is difficult to account for all the effects of low ambient temperatures since batteries have a poor thermal conductivity and are usually surrounded by insulating materials. This permits a battery to be exposed to a low ambient temperature for relatively long periods of time without actually reaching the temperature of its surroundings. Conversely, if a battery is discharged at low temperatures, the ionic mobility manifests itself as an increased internal resistance, and this, in turn, generates additional heat during the discharge. Since the thermal conductivity is poor, this heat tends to raise the temperature of the battery and, to some extent, helps compensate for the low ambient temperature. Consequently, the way a battery reacts at reduced temperatures depends, to a large extent, on its specific design and the specific discharge conditions. The manufacturer's data should be consulted where such information is required to a high degree of precision. Nevertheless, most dry cells react to a reduction in ambient temperature in a manner shown in Fig. 11-7.

**solution**  For dry-cell work, any discharge which is completed in less than 24 h (for continuous-drain conditions) can be classified as a high-rate discharge. Thus, from Fig. 11-7, the service life for the battery in question at 20°F will be 58% of the 70°F value. The service life of the battery is, therefore, 13.5(0.58) = 7.83 h.

While low temperatures reduce the electrical yield, warming a battery tends to increase the yield, as illustrated in Fig. 11-7. However, elevated temperatures bring about some deleterious effects as a result of accelerated, unwanted reactions within the cell, i.e., the chemical corrosion of the zinc anode by the water in the electrolyte,

$$Zn + H_2O \rightarrow ZnO + H_2$$

and drying out of the electrolyte. These factors can be very pronounced when the exposure to high temperatures occur over a significant period of time, as would be the case during storage. For extended storage, dry cells should be stored in a cool, or even refrigerated, location.

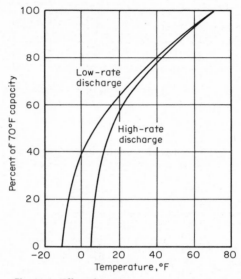

**Fig. 11-7**   Effect of temperature on capacity of carbon-zinc cells.

**problem 11.5**   What is the electrical efficiency of a D cell operated under the conditions described in Problem 11.1?

**theory**   The ultimate capacity of a battery is that which it will deliver at normal temperatures and at very low rates of discharge. At low rates, the polarization and internal loss effects are at a minimum, and the battery can deliver the maximum energy of which its active materials are capable. As the rate of discharge increases, the internal losses and resistance effects reduce the amount of energy which the battery can deliver, and, consequently, the stored chemical energy is not completely converted to electrical energy.

**solution**   The ultimate capacity of a D-size dry cell can be taken as the service life it will yield at very low rates of discharge, say, the 100- to 300-h rate, and to a low cutoff voltage, say, 0.8 V. Thus, at an initial drain of 25 mA and a cutoff voltage of 0.8 V, the service life, from Fig. 11-4, will be 265 h.

$$\text{Average voltage} = 1.40 \text{ V (load} = 60 \text{ } \Omega)$$
$$\text{Average current} = 0.023 \text{ A}$$
$$\text{Capacity} = 265(0.023) = 6.1 \text{ Ah}$$

For the conditions of Problem 11.1:

$$\text{Average voltage} = 1.25 \text{ V (load} = 7.5 \text{ } \Omega)$$
$$\text{Average current} = 0.17 \text{ A}$$
$$\text{Capacity delivered} = 18.2(0.17) = 3.1 \text{ Ah}$$
$$\text{Use efficiency} = \frac{3.1}{6.1} \times 100 = 51\%$$

## Internal Resistance

Since a battery is always in series with its load, the same current that passes through the load also passes through the battery, and there is a voltage drop within the cell equal to the product of the current and the resistance of the cell components themselves. The terminal voltage of the cell is actually the voltage generated by the positive and negative electrode systems less the voltage drop associated with the passage of current through the cell. The internal resistance can be of considerable importance since it is part of the electric circuit, as shown in Fig. 11-8, and, in many cases, as in electronic circuits, must be considered in the design of the circuit.

In other applications, the internal resistance of a battery will limit its ability to deliver high currents which are generally needed for pulse applications, as for photoflash, radio signaling, sonabuoy, and similar applications. As a matter of fact, one of the ways of measuring internal resistance is to note the initial stable current with a dead short across the battery. In this case, as can be seen from Fig. 11-8, the total resistance of the circuit is the internal resistance of the cell. Under these conditions, the initial stable current is called the *flash current*. Table 11-5 cites typical flash currents for different cell types and

**TABLE 11-5  Flash Currents and Internal Resistances for Dry Cells***

| Cell size designation | Flash current, A | Internal resistance, $\Omega$ | Flash current, A | Internal resistance, $\Omega$ |
|---|---|---|---|---|
| AAA | 4.3 | 0.34 | 5.5 | 0.25 |
| AA (general purpose | 5.3 | 0.27 | 9.6 | |
| AA (photoflash) | 7.8 | 0.19 | | |
| C (general purpose) | 5.7 | 0.25 | 12 | 0.11 |
| C (photoflash) | 6.9 | 0.21 | | |
| D (general purpose) | 7.2 | 0.20 | 20 | 0.07 |
| D (photoflash) | 1.16 | 0.13 | | |
| F | 9.6 | 0.15 | 30 | 0.05 |
| No. 6 | 32 | 0.05 | | |

*Values may vary somewhat among producers.

sizes. Since the resistance of the ammeter itself is important in such measurements, the industry has standardized on a total resistance of 0.01 $\Omega$, including the ammeter, when measuring flash currents.

A major part of the internal resistance of a cell is made up of the resistance of the active electrode materials and the electrolyte; these change with the age of the cell, the rate of discharge, and the amount of discharge that has already taken place. Nor does flash current always go hand in hand with capacity. As a matter of fact, a cell may have a very good shelf life and still suffer a considerable loss of flash current. Flash current and internal resistance are thus elusive quantities to measure. One of the best ways of measuring these quantities is to observe the transient cell voltage or current vs. time relationships immediately after the circuit is closed or opened. The instantaneous voltage or current changes, i.e., before enough chemical changes have taken place to alter the internal resistance of the cell, as illustrated in Fig. 11-9. Incidentally, the voltage drop

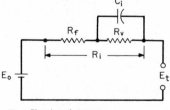

$R_f$ = fixed resistance
$R_v$ = variable resistance
$R_i$ = equivalent internal resistance
$C_i$ = equivalent internal capacitance
$E_o$ = voltage of cell at no load
$E_t$ = terminal voltage of cell

**Fig. 11-8** Equivalent circuit of a cell.

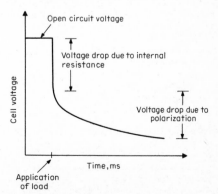

**Fig. 11-9** Characteristic oscilloscope trace for measurement of internal resistance.

caused by changes in the chemical compositions of the cell components is referred to as *polarization.*

Another, more classic, way of measuring internal resistance is to measure the *IR* characteristics of the cell using a 1000-Hz (or higher) alternating current. Since most (but certainly not all) electrode reactions are reversible, practically no net chemical reaction occurs in a cell subject to such high-frequency alternating current, and the impedance determined by simple *E-I* measurements can be taken as the internal resistance. This also turns out to be a convenient means for following the internal resistance as a cell is being discharged since the ac measurements can be superimposed on the dc measurements.

**problem 11.6**   An eight-cell battery utilizes two groups of D-size cells in parallel, each group consisting of four cells in series. What is the service life for a continuous discharge in which the initial drain is 120 mA and the battery cut-off voltage is 3.6 V?

**solution**   Since there are two groups in parallel, each group will have an initial drain of 60 mA, and since there are four cells in series in each group, each cell will have a cutoff voltage of 0.9 V. Under these conditions, Fig. 11-4 indicates that the running time will be 60 h. It should be noted that a 120-mA drain for only a single string of four cells in series would yield a service life of only 21.5 h, again demonstrating that carbon-zinc cells perform best at low current drains. A common practice, therefore, is to use parallel groups of cells to reduce the current discharge.

**problem 11.7**   The shelf life of a given D-size dry cell is given as 18 months. What would the shelf life be if the cells were stored at 106°F?

**theory**   The shelf life of a dry cell, by definition, is the period of time required for its capacity to drop to 90% of its original capacity at 70°F. In the absence of more precise data, the reactions resulting in the loss of capacity will follow the normal chemical reaction rate temperature dependence of doubling for each 10°C rise in temperature.

**solution**   Since $106 - 70 = 36°F \equiv 20°C$, the losses at 106°F will be about four times as fast as at 70°F, and the shelf life will be 18/4 = 4½ months.

**problem 11.8**   After being stored for 1 month at 106°F, what is the service life of a D-size carbon-zinc cell if it is discharged at an initial drain of 60 mA for 4 h/day to a cutoff of 1 V while still at 106°F?

**theory**   The losses during storage can be considered to be linear with time, particularly before the cell has reached its allowable shelf life.

**solution**   From Fig. 11-5, the cell will have a service life of 79 h at 70°F when the cell is fresh. By extrapolation from Fig. 11-7, the capacity of the cell will be increased to 150% of its normal capacity as a result of the high ambient temperature, and estimating the capacity loss due to shelf storage at 3% per month from closely related data given in Table 11-1 gives a shelf loss of $2 \times 3 = 6\%$. Two months of storage life are used up as a result of the cell having been stored for 1 month. The discharge, at 4 h/day, will take about 1 month to complete, during which time the shelf losses will continue while the battery is on open circuit. Thus, the service life will be:

$$79(1.50)(0.94) = 111 \text{ h}$$

Note that at 4 h/day, the period will be a little over 28 days, which checks with the time allowed for open-circuit losses during the discharge period. If it had not been checked, the calculation would have to be repeated with a different allowance for open-circuit time during the discharge period.

**problem 11.9**   Calculate the internal resistance of D-size carbon-zinc cells from the observation that four cells in series discharging through a 10-Ω resistive load yield a current of 5 A, while 10 cells in series, discharging through the same load, yield a current of 11 A.

**theory**   For a battery discharging through a total resistance equal to the sum of the external and internal resistances, $R = R_e + R_i$,

$$I = \frac{NE}{R_e + NR_i}$$

where $N$ = number of cells in series. Thus, for two different numbers of cells in series, $N_1$ and $N_2$, discharging through the same external resistance $R_e$, the following relations are applicable:

$$I_1 = \frac{N_1 E_1}{R_e N_1 R_i} \quad \text{and} \quad I_2 = \frac{N_2 E_2}{R_e + N_2 R_i}$$

Since the voltages generated by each cell are the same, $E_1 = E_2$, and dividing $I_2$ by $I_1$ yields:

$$\frac{I_2}{I_1} = \frac{N_2}{N_1} \times \frac{R_e + N_1 R_i}{R_e + N_2 R_i}$$

**solution**   Substituting the given values into the above equation yields:

$$\frac{11}{5} = \frac{10}{4} \times \frac{10 + 4R_i}{10 + 10R_i}$$

$$0.88 = \frac{10 + 4R_i}{10 + 10R_i}$$

$$8.8 + 8.8R_i = 10 + 4R_i$$

$$4.8R_i = 1.2$$

$$R_i = 0.25 \ \Omega$$

## 11.3   SECONDARY BATTERIES

### Background Information

In terms of their discharge characteristics, secondary batteries are essentially no different from primary batteries. The principal difference is that the discharge reactions in a secondary battery can be reversed by putting electric energy into the battery, in which case the active materials are returned to their original state of charge. The majority of electrode reactions are entirely reversible, including those used in carbon-zinc, alkaline, and other types of primary batteries. The principal difference lies in the fact that the electrode reactions in a secondary battery are readily reversible, and, in terms of chemical composition, their electrodes can be essentially restored to a state of full charge by the application of an electric current.

While the electrochemical differences between primary and secondary batteries are fairly small, the technologies have developed along entirely different lines. For the most part, primary batteries have been developed for discharges of relatively short duration at low power levels, while secondary batteries have generally been developed for long-term, high power level discharges. This is related to the fact that secondary batteries can be designed with free liquid electrolytes and large plate areas, resulting in cells of very low impedances.

Except for a few of the smaller sizes and automotive batteries, there is only general dimensional and capacity standardization of secondary batteries. Usually, a manufacturer will develop several electrode sizes and will then proceed to use an appropriate number of such electrodes to make up a variety of cell sizes. From such cells, a full battery of almost any capacity and voltage can be made by a suitable series and parallel arrangement of the cells. It is a general practice, therefore, to provide what are essentially custom batteries. Those batteries for which general standards have evolved are listed in Table 11-6.

Since rechargeability is the outstanding identifying characteristic of a secondary battery, the process of charging requires some discussion. Charging will occur whenever a dc potential greater than that of the battery is applied to the battery. In such a case, current will flow into the battery at a rate determined by the difference between the applied potential and the open-circuit (or reversible) potential of the battery. As can be seen from the charging portion of Fig. 11-10, which is representative of a typical charge-discharge curve for a conventional rechargeable battery, the potential of the battery rises as its state of charge increases. This process occurs until the potential is sufficiently high to initiate side reactions (such as the decomposition of water into hydrogen and oxygen), at which point at least part of the charging current is diverted into the spurious reactions. This is tantamount to saying that the efficiency with which the charging current is accepted is reduced and the inefficiencies are manifested by the evolution of hydrogen and/or oxygen. The efficiency with which a secondary battery is recharged generally increases as the charging rate decreases, particularly as the battery approaches its fully charged condition.

When the batteries become fully charged (i.e., the active materials are converted to their fully charged chemical state and accept no more electricity), the charging current goes entirely into the decomposition of water with an attendant copious evolution of gases. At this point, the potential of the battery remains constant since no net further chemical changes occur, other than a slight concentration of the electrolyte as the water is removed by electrolysis. That charge which is continued after the battery has been brought to a state of full charge is referred to as *overcharge* and is almost always characterized by the evolution of gases.

Overcharging is usually responsible for the maintenance which a secondary battery requires. Since water is removed during the overcharge period by electrolysis, it must be periodically restored to the battery. Frequently, where a very large number of cells are involved, this becomes a tedious task and is frequently poorly accomplished. A common problem is that the battery is frequently overfilled, in which case some of the electrolyte spills from the vent hole through which the water is being added. In addition, during the overcharge period, the gassing produces a very fine spray of electrolyte which escapes

**TABLE 11-6    Overall Dimensions of Rechargeable Batteries**

| Cell size designation | Diameter, in | Height, in |
|---|---|---|
| *Sealed cylindrical nickel-cadmium cells* | | |
| N | 0.44 | 1.17 |
| AAAA | 0.38 | 1.74 |
| 1/3AA | 0.56 | 0.66 |
| 1/2AA | 0.56 | 1.09 |
| AA | 0.56 | 1.96 |
| C | 1.02 | 1.96 |
| D | 1.31 | 2.40 |
| F | 1.31 | 3.60 |
| *Sealed nickel-cadmium button cells* | | |
| K20 | 1.00 | 0.25 |
| K28 | 1.00 | 0.35 |
| K32 | 1.36 | 0.21 |
| K45 | 1.36 | 0.39 |
| K65 | 2.00 | 0.41 |
| K75 | 2.00 | 0.60 |

| *3-cell prismatic sealed lead-acid batteries* | | | |
|---|---|---|---|
| Cell size designation | Length, in | Width, in | Height, in |
| PB2.6 | 1.38 | 5.30 | 2.45 |
| PB6.0 | 2.72 | 2.72 | 3.73 |
| PB9.0 | 3.71 | 2.21 | 5.67 |

| *6-cell automotive lead-acid batteries* | | | |
|---|---|---|---|
| Group number | Length, in | Width, in | Height, in |
| 22 | 9.50 | 6.88 | 8.63 |
| 23 | 10.75 | 6.88 | 8.44 |
| 24 | 10.25 | 6.81 | 8.88 |
| 27 | 12.06 | 6.81 | 8.88 |
| 29 | 13.13 | 6.81 | 9.13 |
| 32 | 14.25 | 5.50 | 8.94 |
| 41 | 11.53 | 6.88 | 6.88 |
| 42 | 9.56 | 6.81 | 6.81 |
| 53 | 13.00 | 4.69 | 8.25 |
| 60 | 13.06 | 6.63 | 8.88 |
| 72 | 9.06 | 7.25 | 8.25 |
| 74 | 10.25 | 7.25 | 8.75 |
| 77 | 12.06 | 7.25 | 8.75 |

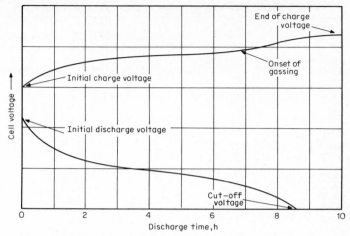

**Fig. 11-10**  Typical secondary battery charge-discharge curve.

with the gases. Both of these conditions result in a loss of electrolyte which is subsequently replaced with water; this process, after a period of time, results in low electrolyte concentration and consequently poor battery performance. Under special conditions, secondary batteries can be designed in which no water is decomposed and there is no escape of electrolyte. Such batteries need no periodic addition of water and consequently require little or no maintenance.

In secondary batteries a peculiar usage has evolved to describe the rate of charge or discharge in terms of the capacity of the battery. In this system, if the rated capacity $C$ (in ampere-hours) of a battery is withdrawn in a time $t$ (in hours), the discharge rate is said to be the $C/t$ rate. Similarly, if the battery is charged in $t'$ hours, the recharge rate is said to be the $C/t'$ rate. Thus, a 100-Ah battery that is discharged at the $C/5$ rate will be discharged in 5 h, meaning that the discharge current is 20 A. If this battery is recharged at the $C/10$ rate, the recharge current will be 10 A.

### Types of Batteries

A great many secondary systems have been used or proposed for future use. However, of these only five, listed below with the chemical reactions on which they are based, have achieved any significant commerical importance.

Lead-acid system:

$$Pb + 2H_2SO_4 + PbO_2 \rightarrow 2PbSO_4 + 2H_2O$$

Nickel-iron (Edison) system:

$$3Ni(OH)_3 + 3Fe \rightarrow Fe(OH)_2 + Fe(OH)_3 + 3Ni(OH)_2$$

Nickel-cadmium system:

$$2Ni(OH)_3 + Cd \rightarrow Cd(OH)_2 + 2Ni(OH)_2$$

Nickel-zinc system:

$$2Ni(OH)_3 + Zn \rightarrow Zn(OH)_2 + 2Ni(OH)_2$$

Silver-zinc system:

$$Ag_2O + Zn \rightarrow Ag + ZnO$$

The discharge reactions are shown with the arrow pointing to the right, while the charge reactions are those for which the arrows point to the left. The fundamental characteristics of each of these systems are shown in Table 11-7. Of these, the lead-acid system is unique in that one of the components of the electrolyte (sulfuric acid) is consumed during the

**TABLE 11-7    Characteristics of Commercially Important Secondary Battery Systems**

| | Lead-acid | Nickel-iron (Edison) | Nickel-cadmium | Nickel-zinc | Silver-zinc | Silver-cadmium |
|---|---|---|---|---|---|---|
| Open-circuit voltage | 2.15 | 1.4 | 1.35 | 1.8 | 1.86 | 1.4 |
| Voltage range under load | 1.75–1.9 | 1.1–1.3 | 1.1–1.3 | 1.5–1.7 | 1.3–1.5 | 1.1 |
| Energy density, Wh/lb | 12–14 | 16 | 18 | 40 | 60 | 30 |
| Energy storage, Wh/in³ | 1.5–2 | 1 | 2 | 3.6 | 3.6 | 2.8 |
| Power density, W/lb | 6 | 2 | 18 | 40 | 35 | 18 |
| Power storage, W/in³ | 0.8 | 0.35 | 1 | 3.2 | 2.6 | 1.3 |
| Cycle life (deep cycles) | 1000 | 3000 | 1500 | 300 | 50 | 200 |
| Low-temperature performance: 0°C capacity as a % of 25°C capacity | 60 | 35 | 65 | 40 | 35 | 50 |
| Charge retention at 25°C (months to 80% capacity) | 18 | 3 | 6 | 6 | 6 | 9 |
| Charge retention at 45°C (months to 80% capacity) | 6 | 1 | 1 | 1 | 1 | 2 |
| Typical applications | Engine starting, traction, standby, line filtering, telephone, telegraph | Traction, mine lighting, naval service, railroad service | Aerospace, transceivers, medical, photographic, power tools, appliances | Electric vehicles, lawn mowers, trolling motors, power tools | Aerospace, military | Aerospace, military |

discharge of the battery and regenerated during the charge. In all the other systems, the positive and negative electrode reactions involve the electrolyte, but in compensating ways so that there is no net change in the electrolyte composition as the battery is charged or discharged. The fact that there is such a change in the lead-acid system permits the course of the discharge to be followed by simply measuring the specific gravity of the electrolyte. In addition, the uniqueness continues in that both electrodes tend to approach the same chemical composition on full charge so that at full discharge the voltage of the cell tends to fall to zero.

In all cases, the negative electrodes are capable of reacting with oxygen, even while on charge, to produce water. Thus, if the positive electrode is brought to a state of full charge and begins to generate oxygen as the charge is continued, the oxygen will migrate to the negative electrode where it is reduced. This reduction is usually preferential to the charging of the negative electrode, and it is thus possible to design a cell in which the oxygen evolution on charge can be exactly compensated by other electrochemical reactions, thus establishing a state of equilibrium in which the battery can accept a continuous charging current without any further net reaction and without any loss of water. This is the principle of the sealed version of the nickel-cadmium and nickel-zinc systems.

This principle is also used, to some extent, to prepare sealed lead-acid batteries. However, it is more common to take advantage of another unique feature of the lead-acid system, namely, that the overvoltage necessary to evolve hydrogen and oxygen at the negative electrode and positive electrode, respectively, is very high. Thus, referring to Fig. 11-10, when a lead-acid battery is approaching full charge and gas begins to evolve from its electrodes, there is a sudden and very large rise in the voltage required to continue to pass a charging current through the cell. If the voltage driving the charging current is fixed at a value which is high enough to charge the electrodes but not high enough to discharge hydrogen and/or oxygen, the cell voltage will rise until it is equal to that of the charging source, at which point the charging current will fall to zero, and no further electrochemical reactions will occur; consequently, no water will be decomposed and there will be no gassing. Certain lead-acid batteries have been developed, through the use of very pure electrode materials, such that the overvoltages at which gases are evolved are very high; with these batteries it is, therefore, easy to set the charger voltage to a level which is not sufficient to evolve gases. Such a cell can, of course, be sealed and will require no periodic additions of water. This is the principle upon which the new maintenance-free lead-acid batteries are based.

**problem 11.10**    What is the running time of a D-size sealed nickel-cadmium cell, rated at 4.0 Ah, when discharged at 0.4, 4, and 40 A?

**theory**    Small nickel-cadmium cells are generally available in the same sizes as dry cells and are handled through essentially the same distribution channels. However, their capacity is considerably less sensitive to the discharge regime than are primary dry cells, and manufacturers need, therefore, not supply extensive discharge data. For example, the inclusion of rest periods into the discharge regime does not particularly improve the total capacity yielded by a nickel-cadmium cell. In addition, the capacity of a nickel-cadmium cell is insensitive to the discharge rate when the discharge rate is below the 10-h rate (that is, $C/10$). At higher rates, the capacity does decrease as the discharge rate increases, but not nearly as dramatically as for primary dry cells. Thus, the manufacturer need supply only a limited number of typical discharge curves and the rated capacity of the battery, as in Fig. 11-11, in order to provide sufficient information to cover almost any application. Intermediate discharge rates can, of course, be estimated by interpolation. In citing the rated capacity of the cell, a discharge rate is chosen (such as the $C/10$ rate) which is sufficiently low to develop the maximum capacity of the cell. The relationship between capacity and discharge rate is given in Fig. 11-12, which indicates that the capacity is essentially a constant for all discharge rates less than $C/5$.

The discharge rate can also be given in terms of the running time to be expected. Thus, if a particular battery runs for 5 h at a particular discharge current, the battery is said to be discharged at the 5-h rate.

**solution**    For a drain of 0.4 A and a capacity of 4.0 Ah, the cell will be discharging at the 0.4/4.0 or $C/10$ rate (also written as the 0.1-$C$ rate). As seen from Fig. 11-11, the cell will yield 100% of its rated capacity, or 4.0 Ah, and will therefore run for 4.0/0.4 = 10 h. Similarly, at 4 A the cell will yield 95% of its rated capacity, or 3.8 Ah, for a running time of 3.8/4 = 0.95 h. At 40 A, the data of Fig. 11-11 require considerable extrapolation and may therefore yield poor accuracy. For greater

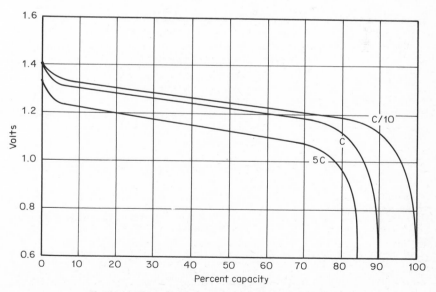

**Fig. 11-11**  Typical nickel-cadmium battery discharge curves.

accuracy at very high rates, the data of Fig. 11-12 should be used. At 40 A, the discharge rate is 40/4.0 or the 10-$C$ rate. From Fig. 11-12, at this rate the cell will yield 80% of its rated capacity, or 0.8(4.0) = 3.2 Ah for a running time of 3.2/40 = 0.08 h or 4.8 min. For the 0.95-h and 4.8-min discharges, the discharge rates can be described as the 1-h and 5-min rates, since it is customary to use whole numbers for the time-rates of discharge.

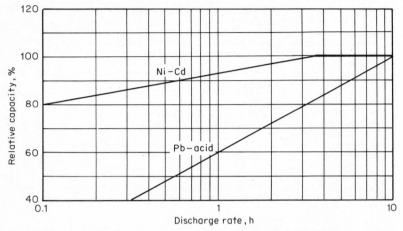

**Fig. 11-12**  Capacity of secondary batteries as a function of discharge rate.

**problem 11.11**   For the conditions given in Problem 11.10, what is the efficiency with which active materials are used at the three discharge rates?

**solution**   At the 10-h rate, the cell yields its maximum capacity, meaning that the active materials are essentially exhausted at the end of the discharge. However, for the 1-h and 5-min rates, the yields are 3.8 and 3.2 Ah, respectively, meaning that the efficiencies with which the active materials are used are 3.8/4.0 × 100 = 95% and 3.2/4.0 × 100 = 80%.

**problem 11.12**   If at the end of the 10$C$ discharge of the cell described in Problem 11.11 the discharge rate is reduced to the 0.1$C$ rate, how much longer will the cell run?

**theory**   At the high rate of discharge, the total voltage is reached before the electrochemical energy of the cell has been exhausted. The discharge is really terminated by virtue of the fact that the cell's internal resistance results in an $IR$ drop which reduces the voltage of the cell to below the cutoff point. However, a considerable amount of electrochemical energy is still left in the cell, and lowering the discharge current reduces the $IR$ drop within the cell, allowing the cell voltage to rise above the cutoff voltage; the discharge can then continue.

**solution**   At the 10$C$ rate only 80% of the total available capacity is used. Therefore, the remaining capacity, 0.20(4.0) = 0.8 Ah, can be exhausted at the $C$/10 rate, and the cell will run for an additional 0.8/0.4 = 2 h.

**problem 11.13**   If the cell of Problem 11.10 is first discharged at the 10$C$ rate and the current then cut back to the $C$ rate when the cutoff voltage is reached, then again cut back to the $C$/10 rate when the cutoff voltage is reached again, how long will the cell run at each rate?

**solution**   From Problem 11.10, the cell will run for 4.8 min at the 10$C$ rate and use 3.2 Ah. At the $C$ rate, the cell will yield 3.8 Ah, or 0.6 Ah more, which at a 4-A drain means that the cell will run 0.15 h, or 9 min longer. At the $C$/10 rate, the cell will yield 4.0 Ah, or an additional 0.2 Ah. At a drain of 0.4 A, the cell will run 0.5 h, or 30 min longer.

**problem 11.14**   What is the average power delivered by the battery of Problem 11.10 when discharged at the $C$/10, $C$, and 10$C$ rates?

**solution**   As can be seen from Fig. 11-11, the average discharge voltage for the $C$/10 rate is 1.23 V. (For higher accuracy, the area under the curve should be integrated to give total delivered watt-hours. This, divided by the delivered ampere-hours, will yield the true average voltage.) Since the discharge current is 0.4 A at the $C$/10 rate, the average power delivered will be 1.23 × 0.4 = 0.49

W. Similarly, for the $C$ rate, or a discharge current of 4 A, the average discharge voltage will be 1.20 V, and the average power delivered will be $1.20 \times 4 = 4.8$ W. For the $10C$ rate, for which the average discharge voltage will be 0.95 V, the average delivered power will be $0.95 \times 40 = 38$ W.

**problem 11.15**    The manufacturer states that the effective internal resistance of the cell described in Problem 11.10 is 0.005 $\Omega$. What is the maximum power that can be drawn from this cell, and what is its short-circuit current?

**theory**    When the battery is standing on open circuit, the delivered power is, of course, zero owing to the fact that the current is zero. At the other end of the scale, namely, when the battery is short-circuited, the delivered power is again zero since the voltage approaches zero even though the current may be very high. At intermediate points, the battery delivers useful voltage power, and since the average voltage is a function of the current drain, but not linearly proportional to it, the relative influence of voltage and current are not uniform. For example, at very low drain rates, the current is very low and the power levels are, therefore, consequently low. As the current drain is increased, the contribution to power of the current becomes more significant, but here the voltage begins to decay. It turns out that for maximum power the load resistance should be equal to the effective internal resistance of the battery. The manner in which power output varies with discharge rate is illustrated in Fig. 11-13.

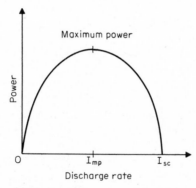

**Fig. 11-13**    Power as a function of discharge rate. $I_{mp}$ = current at maximum power. $I_{sc}$ = short-circuit current.

**solution**    The maximum power will occur when the load resistance and the internal resistance are equal. The total resistance in this case will be 0.010 $\Omega$, and the current flowing through the cell and load will be 1.35/0.010 = 135 A (since the open-circuit voltage of the cell is 1.35 V). The maximum power generated by the cell is, therefore, $1.35 \times 135 = 182$ W. The short-circuit current will be 1.35/0.005 = 270 A; that is, the internal resistance of the cell is the total circuit resistance against which the total voltage of the electrode system acts.

For the most part, the general performance characteristics of all secondary batteries will be similar to that of the nickel-cadmium battery used in our examples. Thus, there is very little difference in treating the calculations pertinent to vented Ni-Cd, Ag-Zn, Ag-Cd, Pb-acid, and other types of secondary batteries. Of course, the specific discharge characteristics and cell values vary from system to system, from cell type to cell type, and from manufacturer to manufacturer. Consequently, it is necessary to consult the manufacturer's performance data and ratings for calculations relating to any specific battery.

However, the lead-acid battery is a major exception. As can be seen from the chemical reactions discussed earlier, the electrolyte does not enter into the net discharge reaction for most battery types. However, in the lead-acid battery the sulfuric acid of the electrolyte does enter into the chemical reactions involved in the charge and discharge of the battery, and there is a net change in the composition of the electrolyte as the battery is cycled. For this system, the electrolyte becomes more dilute as the battery is discharged and more concentrated as the battery is charged, providing a very convenient means for checking the state of charge through measurements of the specific gravity of the electrolyte. This can be done with considerable accuracy, even in the field, by means of a simple hydrometer.

**problem 11.16**    A stationary Pb-acid battery with a rated capacity of 1200 Ah contains 72 liters of electrolyte having a specific gravity of 1.200 when fully charged. What is the specific gravity of the electrolyte at a 50% depth of discharge and when the battery is fully discharged?

**theory**    The equation for the charge or discharge of a lead-acid battery indicates that a certain amount of sulfuric acid is consumed, or generated, for the passage of a given amount of electricity and a given amount of chemical reaction taking place. It is, therefore, possible to relate the amount of sulfuric acid consumed (or generated) and the amount of electricity generated (or absorbed), a relationship defined by Faraday's law which states that for a given quantity of materials consumed (or generated) a fixed amount of electricity must pass through the system. The stoichiometry of the lead-acid battery equation is such that the reaction of 1.83 g of $H_2SO_4$ is required for the passage of 1 Ah through the battery.

Since the electrolyte of a lead-acid battery contains water as well as sulfuric acid, a certain concentration of sulfuric acid in a given volume of electrolyte will be equivalent to a fixed electrical equivalent, in ampere-hours. Table 11-8 gives the electrochemical equivalent for various concentrations of sulfuric acid solutions.

**TABLE 11-8    Electrochemical Equivalent of Sulfuric Acid Solutions**

| Concentration weight % | Specific gravity | Ah/liter |
|---|---|---|
| 4 | 1.020 | 8 |
| 6 | 1.040 | 26 |
| 9 | 1.060 | 35 |
| 12 | 1.080 | 44 |
| 15 | 1.100 | 53 |
| 18 | 1.120 | 62 |
| 20 | 1.140 | 71 |
| 21.5 | 1.150 | 77 |
| 23 | 1.160 | 81 |
| 26 | 1.180 | 90 |
| 28 | 1.200 | 100 |
| 31 | 1.220 | 110 |
| 33 | 1.240 | 120 |
| 36 | 1.260 | 130 |
| 38 | 1.280 | 141 |
| 40 | 1.300 | 151 |
| 42 | 1.320 | 162 |
| 45 | 1.340 | 173 |
| 47 | 1.360 | 184 |
| 49 | 1.380 | 195 |
| 51 | 1.400 | 206 |

There is usually considerably more $H_2SO_4$ in the electrolyte of a lead-acid battery than is required to complete discharge of the positive and negative electrodes, and, consequently, the acid concentration, as measured by the specific gravity, will fluctuate between two limits, the upper one for a fully charged battery, the lower for a fully discharged battery. As the total volume of electrolyte increases relative to the capacity of the electrodes, the variation in the acid gravity from full charge to full discharge will, of course, decrease, giving the manufacturers a convenient means of controlling the nature of the discharge according to the needs of the application. In solving Problem 11.16, it is important to note that the change in acid gravity is not proportional to the depth of discharge since the total amount of electrolyte is involved in the change in acid gravity as well as the total acid actually used in the discharge. Thus, while acid gravity is a very convenient and reliable means of determining the state of charge, the manufacturers' data for any given cell type must be consulted.

**solution**    From Table 11-8, sulfuric acid of 1.200 specific gravity has an acid equivalent of 100 Ah/liter. The total electrolyte, therefore, has an acid equivalent of $72 \times 100 = 7200$ Ah. At 50% depth of charge that battery has produced 600 Ah, meaning that the residual equivalent acid in the electrolyte is $7200 - 600 = 6600$ Ah, or $6600/72 = 92$ Ah/liter. From Table 11-8, by interpolation, this corresponds to an acid specific gravity of 1.184.

At 100% depth of discharge, the battery has delivered 1200 Ah, leaving an acid equivalent in the electrolyte of $7200 - 1200 = 6000$ Ah, or $6000/72 = 83$ Ah/liter. From Table 11-8, this corresponds to an acid specific gravity of 1.164.

**problem 11.17**    A cell from an automotive starting battery contains 0.60 liter of electrolyte and is rated at 40 Ah at the 5-h rate. In order that the electrolyte not freeze under any condition above 5°F, the specific gravity of the electrolyte must not be allowed to fall below 1.150 even when fully discharged. What is the minimum specific gravity at the full-charge condition?

**solution**    From Table 11-8, at a specific gravity of 1.150 the electrolyte has an electrical equivalent of 77 Ah/liter. For the cell containing 0.6 liter, the electrolyte electrical equivalent in the fully discharged state will therefore be $0.6 \times 77 = 46.2$ Ah. In the fully charged state, it will have an electrical equivalent of 40 Ah more, or 86.2 Ah. This adjusts to about 144 Ah/liter (since the cell had only 0.6 liter of electrolyte). From Table 11-8, this corresponds to an electrolyte specific gravity of about 1.285.

**problem 11.18**    What is the charge efficiency for a secondary battery that is recharged for 10 h at 2.5 A after a discharge of 5 h at 4 A?

**solution**    Electrical capacity withdrawn = $4 \times 5 = 20$ Ah. Electrical input on charge = $2.5 \times 10 = 25$ Ah. Charge (or amperes-hour) efficiency = $20/25 \times 100 = 80\%$

**problem 11.19**    For the cell of Problem 11.18, the average voltage on charge was 2.25 V, and 1.90 V on discharge. What was the cyclical energy efficiency?

**solution**    Energy withdrawn = $20 \times 1.90 = 38$ Wh. Energy expended on charge = $25 \times 2.25 = 56$ Wh. Energy efficiency = $38/56 \times 100 = 68\%$

**problem 11.20**    How much water does the cell of Problem 11.18 consume for the given cycle?

**theory**    During a charge, the electrical input which does not go into the conversion of the active materials to a higher state of charge goes into spurious or "side" reactions. While there can be many side reactions, depending on the nature of the battery and the conditions of the charge, essentially the principal reaction which uses that part of the electrical input not going into the active charging is the decomposition of water into hydrogen and oxygen. As before, by Faraday's law, the passage of a given amount of electricity will decompose a given weight of water. It turns out that exactly 2.98 Ah of electricity is required to decompose 1 g of water.

**solution**    The charge input (25 Ah) was 5 Ah more than the capacity withdrawn. Therefore, there was a 5-Ah surplus of electricity which will go into the decomposition of water. Amount of water decomposed = $5/2.98 = 1.68$ g.

## 11.4   REFERENCES

Ray-O-Vac Batteries Technical Information, Philadelphia (1972).
Nickel-Cadmium Battery Applications Engineering Handbook, 2d ed., General Electric Co., Syracuse, N.Y., 1975.
Everready Battery Applications, 2d ed., 1971.
Burgess Battery Engineering Handbook, 1969.

# Chapter 12

# Op Amp Applications

## JACK L. WAINTRAUB

**Assistant Professor, Electrical Engineering Technology
Department, Middlesex County College, Edison, N.J.**

## 12.1 INTRODUCTION

The operational amplifier (*op amp*) may be regarded as a universal amplifier because of
its versatility and the ease with which it can be utilized in a large variety of applications.
The advent of integrated circuits made op amps available in small size, with excellent
characteristics, and at low cost. Operational amplifiers generally contain a dozen or more
transistors and diodes and a multitude of resistors. The op amp is used in almost all
amplifier applications. It is also employed to perform mathematical operations, filtering,
wave generation and shaping, analog-to-digital (A/D) conversion and digital-to-analog (D/
A) conversion, and so on.

## 12.2 OP AMP CHARACTERISTICS

The basic symbol for the op amp is given in Fig. 12-1. There are, in general, two input
terminals and one output terminal. A signal connected to the *inverting* (−) terminal
results in an output which is 180° out of phase with the input. If a signal is connected to
the *noninverting* (+) input, the output and input signals are in phase. The op amp is
considered as a *linear-analog* device that is used in both linear and nonlinear
applications.

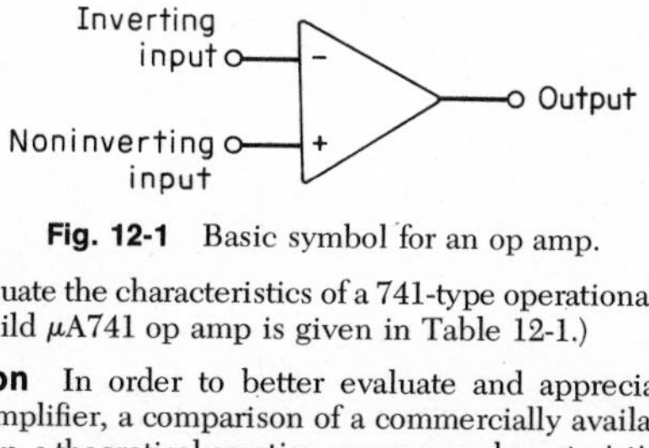

**Fig. 12-1** Basic symbol for an op amp.

**problem 12.1** Evaluate the characteristics of a 741-type operational amplifier. (A portion of the
data sheet for a Fairchild μA741 op amp is given in Table 12-1.)

**theory and solution** In order to better evaluate and appreciate the characteristics of a
practical operational amplifier, a comparison of a commercially available and an *ideal* op amp is
made. The ideal op amp, a theoretical creation, possesses characteristics that are ideally attributed
to an op amp.

On the data sheet (Table 12-1), under the heading *general description* are typical applications
for the device, the type of process used in its manufacture, and other general information.

### TABLE 12-1    Performance Specifications of a μA741 Operational Amplifier
(Fairchild Corp.)

- *No Frequency Compensation Required*
- *Short-Circuit Protection*
- *Offset Voltage Null Capability*
- *Large Common-Mode and Differential Voltage Ranges*
- *Low Power Consumption*
- *No Latch-UP*

**GENERAL DESCRIPTION**—The μA741 is a high performance monolithie operational amplifier constructed on a single silicon chip, using the Fairchild Planar epitaxial process. It is intended for a wide range of analog applications. High common-mode voltage range and absence of "latch-up" tendencies make the μA741 ideal for use as a voltage follower. The high gain and wide range of operating voltages provide superior performance in integrator, summing amplifier, and general feedback applications. The μA741 is short-circuit protected, has the same pin configuration as the popular μA709 operational amplifier, but requires no external components for frequency compensation. The internal 6 dB/octave roll-off insures stability in closed loop applications.

**ABSOLUTE MAXIMUM RATINGS**

| | |
|---|---|
| Supply voltage | $\pm 22$ V |
| Internal power dissipation | 500 mW |
| Differential input voltage | $\pm 30$ V |
| Input voltage | $\pm 15$ V |
| Storage temperature range | $-65$ to $+150°$C |
| Operating temperature range | $-55$ to $+125°$C |
| Lead temperature (soldering, 60 s) | $300°$C |
| Output short-circuit duration | Indefinite |

**PHYSICAL DIMENSIONS**
in accordance with
JEDEC (TO-99) outline

ORDER PART NO. U5B7741312

**ELECTRICAL CHARACTERISTICS** ($V_s = \pm 15$ V, $T_A = 25°$C unless otherwise specified)

| Parameter | Conditions | Min | Typical | Max | Units |
|---|---|---|---|---|---|
| Input offset voltage | $R_S \leqslant 10$ kΩ | | 1.0 | 5.0 | mV |
| Input offset current | | | 30 | 200 | nA |
| Input bias current | | | 200 | 500 | nA |
| Input-resistance | | 0.3 | 1.0 | | MΩ |
| Large-signal voltage gain | $R_L \geqslant 2$ kΩ, $V_{out} = \pm 10$ V | 50,000 | 200,000 | | |
| Output voltage swing | $R_L \geqslant 10$ kΩ | $\pm 12$ | $\pm 14$ | | V |
| | $R_L \geqslant 2$ kΩ | $\pm 10$ | $\pm 13$ | | V |
| Input voltage range | | $\pm 12$ | $\pm 13$ | | V |
| Common-mode rejection ratio | $R_s \leqslant 10$ kΩ | 70 | 90 | | dB |
| Supply-voltage rejection ratio | $R_s \leqslant 10$ kΩ | | 30 | 150 | μV/V |
| Power consumption | | | 50 | 85 | mW |
| Transient response (unity gain) | $V_{in} = 20$ mV, $R_L = 2$ kΩ, $C_L \leqslant 100$ pF | | | | |
| Rise time | | | 0.3 | | μs |
| Overshoot | | | 5.0 | | % |
| Slew rate (unity gain) | $R_L \geqslant 2$ kΩ | | 0.5 | | V/μs |
| The following specifications apply for $-55°$C $\leqslant T_A \leqslant +125°$C: | | | | | |
| Input offset voltage | $R_s \leqslant 10$ kΩ | | | 6.0 | mV |
| Input offset current | | | | 500 | nA |
| Input bias current | | | | 1.5 | μA |
| Large-signal voltage gain | $R_L \geqslant 2$ kΩ, $V_{out} = \pm 10$ V | 25,000 | | | |
| Output voltage swing | $R_L \geqslant 2$ kΩ | $\pm 10$ | | | V |

Under *absolute maximum ratings* one finds *limits* on parameters *that cannot be exceeded for safe operation.* The supply voltages are not to exceed $\pm 22$ V for this model. (The μA741 is designed for military use. For the μA 741C, which is for commercial applications, the maximum supply voltages are $\pm 18$ V.) The maximum allowable power dissipation varies with the type of package and mounting. Maximum allowable input and operating voltages, as well as operating temperature, also are given.

The *electrical characteristics* are listed in a separate section of the data sheet. Minimum, typical, and maximum values are given at supply voltages of ± 15 V and a temperature of 25°C. The manufacturer supplies a range of characteristic values and guarantees that a particular unit will have characteristics within the given range of values.

A comparison between the electrical characteristics of a commercially available op amp and an ideal op amp is provided in Table 12-2. The important features are:

1. High input resistance (nearly zero input current).
2. Low input capacitance.
3. Large open-loop gain ($A_{vo}$). Open-loop gain is the gain of an amplifier without feedback.
4. Low output resistance.
5. Fast rise time and low overshoot (wide bandwidth).
6. High slew rate. Slew rate is the maximum rate of change of the output voltage in its linear region per unit time. It is an indication of how fast the output of an op amp can change.
7. Low input offset voltage and current. Input offset voltage and current are defined as the differential voltage and difference in current required at the input terminals for zero output.
8. High common-mode rejection ratio (CMRR) (defined in Sec. 12.5). (The output should be a function of the difference between the input voltages only.)

**TABLE 12-2   Comparison between Electrical Characteristics of a Practical and an Ideal Op Amp**

|  | Min. | Typical | Max. | Ideal |
|---|---|---|---|---|
| 1. Input resistance | 0.3 MΩ | 2.0 MΩ |  | ∞ |
| 2. Input capacitance |  | 1.4 pF |  | 0 |
| 3. Large signal voltage gain |  |  |  |  |
| (open-loop gain), $R_L \geq 2$ kΩ | 50,000 | 200,000 |  | ∞ |
| 4. Output resistance |  | 75 Ω |  | 0 |
| 5. Transient response |  |  |  |  |
| (*a*) Rise time |  | 0.3 μs |  | 0 |
| (*b*) Overshoot |  | 5% |  | 0 |
| 6. Slew rate ($R_L \geq 2$ kΩ) |  | 0.5 V/μs |  | ∞ |
| 7. Input offset |  |  |  |  |
| (*a*) Voltage |  | 1 mV | 5 mV | 0 |
| (*b*) Current |  | 20 nA | 200 nA | 0 |
| 8. Common-mode rejection ratio |  |  |  |  |
| (CMRR) | 70 dB | 90 dB |  | ∞ |

## 12.3  FEEDBACK AMPLIFIERS

Op amps are connected in various feedback circuits to yield an amplifier with a controlled gain. Basic feedback amplifiers may be classified as inverting, noninverting, and the voltage follower.

### Inverting Amplifier

**problem 12.2**   Design an inverting amplifier to have a voltage gain of −50 and an input resistance of 100 kΩ.

**theory**   The inverting amplifier is an op amp with feedback that yields a controlled gain and an output signal inverted with respect to the input. The circuit of an inverting amplifier is provided in Fig. 12-2. Resistor $R_F$, the feedback resistor, is connected between the output and inverting terminals. The series resistor $R_1$ is connected between the inverting terminal and the input signal to be amplified. The closed-loop gain of the feedback amplifier is designated by $A_{rc}$

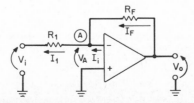

**Fig. 12-2**   An inverting amplifier (Problem 12.2).

In calculating the gain of the inverting amplifier, the op amp is assumed to be ideal. Therefore, the open-loop gain is infinite, and input current $I_i$ is zero. Because the output voltage $V_o$ is finite and the open-loop gain is infinite, the differential voltage $V_A$ is zero. Node A, therefore, is called a *virtual ground.*

Summing the currents at node A yields:

$$I_1 = I_F$$

and

$$\frac{V_i}{R_i} = \frac{-V_o}{R_F}$$

Solving for the voltage gain $A_{vc}$ yields:

$$A_{vc} = \frac{V_o}{V_i} = \frac{-R_F}{R_1} \tag{12-1}$$

Because node A is at ground potential, the input resistance $R_i$ is:

$$R_i \simeq R_1 \tag{12-2}$$

**solution**   For the circuit of Fig. 12-2, by Eq. (12-2) $R_i = R_1$. Let $R_1 = 100\ \text{k}\Omega$. Solving Eq. (12-1) for $R_F$, $R_F = |-A_{cc}R_1| = 50 \times 100 \times 10^3 = 5\ \text{M}\Omega$.

### Noninverting Amplifier

An op amp can be connected to deliver an output voltage that is in phase with the input signal. The resulting circuit is called a *noninverting amplifier* (Fig. 12-3). The input signal is applied to the noninverting terminal, and the inverting terminal is connected to the feedback resistor.

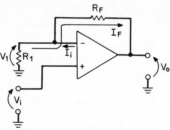

**Fig. 12-3**   A noninverting amplifier (Problem 12.3).

**problem 12.3**   Design a noninverting amplifier that yields an output voltage of 1 V for an input signal of 10 mV.

**theory**   To determine the closed-loop gain of a noninverting amplifier (Fig. 12-3), again assume that the currents through resistors $R_1$ and $R_F$ are equal ($I_i \simeq 0$). Hence,

$$\frac{V_o - V_i}{R_F} = \frac{V_1}{R_1}$$

Using the concept of a virtual ground, $V_1 = V_i$; hence,

$$A_{vc} = \frac{V_o}{V_i} = \frac{R_F + R_1}{R_1}$$
$$= 1 + \frac{R_F}{R_1} \tag{12-3}$$

If $R_F >> R_1$, then

$$A_{vc} \simeq \frac{R_F}{R_1} \tag{12-4}$$

**solution**   The desired closed-loop gain is $A_{vc} = 1/0.001 = 100$. By Eq. (12-3), $100 = 1 + R_F/R_1$ or, solving for $R_F/R_1$, $R_F/R_1 = 99$. Let $R_1 = 2.2\ \text{k}\Omega$; then, $R_F = 99 \times 2.2\ \text{k}\Omega = 217.8\ \text{k}\Omega$.

### Voltage Follower

**problem 12.4**   Design an impedance matching circuit which employs an op amp.

**theory and solution**   An impedance matching circuit is used to couple a low-impedance load to a source having a high-internal impedance. This type of circuit is referred to as a *buffer*. The voltage follower (employing an op amp) of Fig. 12-4 may be used as a buffer. The voltage follower has unity gain, low-output and high-input impedance, and an output voltage in phase with the input signal.

Summing the voltages around the amplifier yields $V_i + V_A = V_o$. Because $V_A = 0$, $V_i = V_o$ and

$$A_{vc} = 1 \tag{12-5}$$

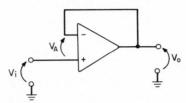

**Fig. 12-4**    A voltage follower (Problem 12.4).

## 12.4   PRACTICAL CONSIDERATIONS

Although it was shown that the physical op amp is a good approximation to the ideal amplifier, it has a number of limitations that must be considered. Physical op amps have a finite gain, a finite input resistance, a limited bandwidth, inherent offset currents and voltages, and are prone to parasitic oscillations.

### Frequency Compensation

The open-loop gain $A_{vo}$ is generally in the order of 100 dB and falls off to unity gain (0 dB) at a frequency called the *crossover frequency* $f_c$. The open-loop frequency response of a typical op amp is illustrated in Fig. 12-5. It is seen that the open-loop gain at very low frequencies is 100 dB. The low-frequency break point $f_L$ is at approximately 100 Hz, and the crossover frequency $f_c$ is 10 MHz.

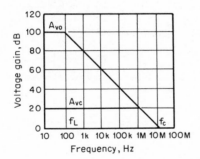

**Fig. 12-5**    Open- and closed-loop responses of a typical op amp.

The op amp is normally not used in its open-loop mode but rather in a negative-feedback circuit. The response of an op amp with feedback is superimposed on the graph of Fig. 12-5. The closed-loop gain $A_{vc}$ is 20 dB. Note that with feedback, the response is flat for better than 1 MHz. An improvement in bandwidth is, therefore, obtained with negative feedback.

Amplifiers that exhibit an open-loop response of $-60$ dB/decade (or $-18$ dB/octave) at the crossover point are unstable. Instability in amplifiers results in the amplifier oscillating. In order to prevent oscillations, frequency compensation is required. Commercial op amps have provision for external compensation, and some units are internally compensated.

Frequency compensation is usually achieved by connecting an $RC$ network or a capacitor between the compensation terminals of the op amp. The values of $R$ and $C$ are specified by the manufacturer. In particular, the value of the compensation capacitor influences the shape of the response curve.

Frequency-response characteristics of an op amp are shown in Fig. 12-6. Plot A is a straight-line approximation of the actual open-loop response, called a *Bode plot*. A Bode

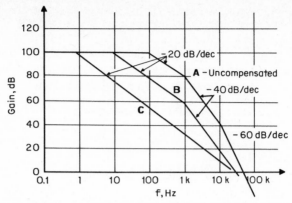

**Fig. 12-6**   Frequency-response curves of an amplifier. Curve $A$ is for an uncompensated amplifier; curves $B$ and $C$ are for a compensated amplifier.

plot is a simple way of describing frequency-response characteristics. It is easy to construct, easy to read, and contains the pertinent frequency-response information.

Plots $B$ and $C$ illustrate the response of the same amplifier with different levels of compensation. Typical op amp connections for frequency compensation and families of response curves are provided by the manufacturer (Fig. 12-7).

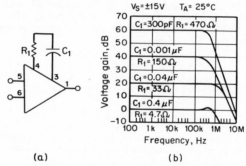

(a)                                 (b)

**Fig. 12-7**   Frequency compensation of an op amp. (a) Typical connections. (b) Frequency response curves for different values of $R_1$ and $C_1$. (*Courtesy Fairchild Corporation.*)

## Offset Compensation

Ideally, with no input voltage applied to an op amp, the output voltage is zero. In a practical amplifier connected in the inverting or noninverting mode, however, the output

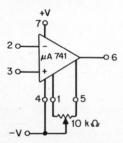

**Fig. 12-8**   An example of offset compensation.

is in the range of a few microvolts to millivolts for zero input signal. This voltage stems from internal component differences, and the effect is explained in terms of an *input offset voltage* $V_{io}$. Because of the input offset voltage, the output is said to contain an *error voltage*.

The *input offset current* $I_{io}$ is a quantity that also contributes to the output error voltage. Ideally, current does not flow into the two input terminals of an op amp. In practice there is a dc bias current at each of the input terminals. If the currents are unequal, their difference is the input offset current. Current $I_{io}$ is typically in the order of nanoamperes.

To compensate for the error voltage, manufacturers of op amps have developed various output voltage null circuits. One such example, recommended for the μA 741, is shown in Fig.

12-8. Pins 1 and 5 are marked offset null, and a 10-kΩ trimmer between those terminals acts to adjust the output to zero volts with no input signal. The wiper arm is connected to the negative (−V) power supply.

## 12.5   OP AMPS IN MATHEMATICAL APPLICATIONS

Operational amplifiers are capable of performing various mathematical operations. They can be connected to add, subtract, multiply, divide, and raise numbers to a power, as well as to extract roots. Op amps also can be used to integrate and differentiate mathematical functions.

### The Summing Amplifier

**problem 12.5**   Design a circuit (*summing amplifier*) that yields an output proportional to the sum of three input voltages. The input voltages are (*a*) $V_1 = 1$ V, $V_2 = 2$ V, $V_3 = 3$ V and (*b*) $V_1 = 10$ V, $V_2 = 100$ V, $V_3 = 200$ V. The amplifier saturates at ±10 V.

**theory**   The summing amplifier (Fig. 12-9) delivers an output voltage that is proportional to the sum of the input voltages. It is basically an inverting amplifier with multiple inputs connected through input resistors to summing node A. The output voltage is:

$$V_o = -\frac{R_F}{R_1}V_1 - \frac{R_F}{R_2}V_2 - \cdots - \frac{R_F}{R_N}V_N \qquad (12\text{-}6)$$

If $R_1 = R_2 = \cdots = R_N = R$, then Eq. (12-6) reduces to:

$$V_o = -(V_1 + V_2 + \cdots + V_N)\frac{R_F}{R} \qquad (12\text{-}7)$$

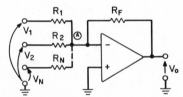

**Fig. 12-9**   A summing amplifier (Problem 12.5).

**solution**   (*a*) Let $R_F = R_1 = R_2 = R_3 = 100$ kΩ. By Eq. (12-7), for three inputs, $V_o = -(V_1 + V_2 + V_3) = -(1 + 2 + 3) = -6$ V.

(*b*) The same values of resistors used in (*a*) cannot be used here. The input voltages are much higher and will cause the op amp to saturate. The highest input level is 200 V. If a *scale factor* of 1/100 is used, then the input of 200 V will contribute only 200/100 = 2 V to the output.

The scale factor is realized by adjusting the ratio of the feedback resistor to the input resistors. In this problem,

$$\frac{1}{100} = \frac{R_F}{R_1} = \frac{R_F}{R_2} = \frac{R_F}{R_3}$$

Let $R_1 = R_2 = R_3 = 1$ MΩ; then, $R_F = 10^6/100 = 10$ kΩ. By Eq. (12-7), $V_o = -(10 + 100 + 200)(10^4/10^6) = -3.1$ V.

### The Difference Amplifier

**problem 12.6**   The circuit of Fig. 12-10 is a *difference amplifier* (or *subtractor*). If the input voltages are $V_1 = 200$ mV and $V_2 = 800$ mV, show that the output voltage is equal to their difference.

**theory**   The difference (or differential) amplifier yields an output voltage that is proportional to the difference in the inverting and noninverting input signals. By applying superposition, the individual effect of each input on the output can be calculated.

The output voltage $V_{o1}$ due to $V_1$ with $V_2$ set to zero is:

$$V_{o1} = -\frac{R_F V_1}{R_1}$$

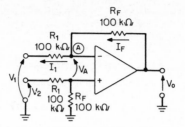

**Fig. 12-10**    A difference amplifier (Problem 12.6).

The output due to $V_2$ with $V_1$ set to zero is:

$$V_{o2} = \frac{R_F V_2}{R_1}$$

The true output $V_o$ is the sum of the two individual outputs:

$$V_o = V_{o1} + V_{o2}$$

$$= (V_2 - V_1)\frac{R_F}{R_1}$$

(12-8)

**solution**    Let $R_1 = R_F = 100$ kΩ. By Eq. (12-8), $V_o = (800 - 200) \times 100/100 = 600$ mV.

### Common-Mode and Difference-Mode Signals

The input signals to a difference amplifier, in general, contain two components: a *common-mode* and a *difference-mode* signal. The common-mode signal voltage $V_c$ is the average of the two input signals, $V_1$ and $V_2$: $V_c = (V_1 + V_2)/2$. The difference-mode signal $V_d$ is the difference between the two inputs: $V_d = V_1 - V_2$.

Ideally, an amplifier affects the difference-mode signals only. However, the common-mode signal is also amplified to some degree. The common-mode rejection ratio (CMRR), which is defined as the ratio of the difference signal voltage gain to the common-mode signal voltage gain, is a figure of merit for op amps. The greater the value of CMRR, the better the performance of an op amp.

### The Integrator

**problem 12.7**    Design a circuit that delivers a voltage ramp (Fig. 12-11) when the input is a 1.5-V step function.

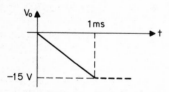

**Fig. 12-11**    A voltage ramp.

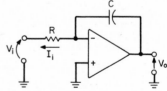

**Fig. 12-12**    An integrator.

**theory**    The *integrator* of Fig. 12-12 is an op amp circuit that is capable of performing the mathematical operation of integration. It is similar to the inverting amplifier (Fig. 12-2) with a capacitor replacing feedback resistor $R_F$.

The mathematical expression defining the operation of the circuit is:

$$V_o = \frac{-1}{RC}\int_0^t v_i\, dt$$

(12-9)

It also can be written as:

$$\Delta v_o = \frac{-v_i\,\Delta t}{RC}$$

(12-10a)

which states that the change in output voltage with respect to a change in time is equivalent to the negative of the input voltage divided by the time constant $RC$ of the circuit.

The input current $I_i$ is $I_i = v_i/R$; hence, Eq. (12-10a) may be rewritten as:

$$\Delta v_o = \frac{-I_i \Delta t}{C} \qquad (12\text{-}10b)$$

Equation (12-10b) states that changes in output voltage are proportional to the input current.

In its application as a mathematical element, an electrical signal applied to the input results in an output that assumes the form of an integrated function. For example, consider a perfect sinusoidal input. The output will be a cosine function. Recall that the difference between the sine and cosine functions is not their shape but a 90° phase shift. Hence, the integrator may be used as a 90°–phase shifter *(quadrature shifter)*.

**solution**    Equation (12-10a) suggests that if the input voltage $v_i$ is constant, then the output voltage changes *linearly* with time. The slope of the resulting ramp is obtained by rearranging Eq. (12.10a):

$$\text{Slope} = \frac{\Delta v_o}{\Delta t} = \frac{-v_i}{RC} \qquad (12\text{-}11)$$

From Fig. 12-11, the slope is equal to $(-15 \text{ V})/(1 \text{ ms}) = -15 \times 10^3 \text{ V/s}$. The integrator of Fig. 12-13, with the battery and switch generating a step function of 1.5 V, produces the desired ramp.

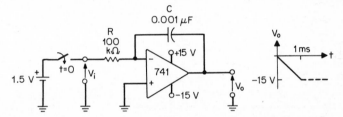

**Fig. 12-13**    An integrator used to generate a voltage ramp output (Problem 12.7).

Solving Eq. (12-11) for $RC$ yields: $RC = 1.5/(15 \times 10^3) = 10^{-4}$ s. Let $C = 0.001 \ \mu\text{F}$; then $R = 10^{-4}/C = 100 \text{ k}\Omega$. These are reasonable values for $R$ and $C$.

The output (Fig. 12-13) is a ramp that continues to climb in a negative direction with time. In practice, the ramp is limited to the saturation voltage of the op amp. In this problem, it is approximately equal to $-15$ V.

## The Differentiator

**problem 12.8**    A circuit is to deliver very narrow positive-going pulses *(spikes)* at 1-ms intervals. Design such a circuit and determine the necessary drive for it.

**theory**    The differentiator is the complement of the integrator. It produces an output that is proportional to the rate of change of the input. The circuit of a differentiator is illustrated in Fig. 12-14a. Note that the op amp is capacitor-coupled and resistor $R$ is in the feedback path. The defining equation for the differentiator is:

$$v_o = -RC \frac{dv_i}{dt} \qquad (12\text{-}12)$$

which can be approximated by:

$$v_o = -RC \frac{\Delta v_i}{\Delta t} \qquad (12\text{-}13)$$

**solution**    Since a differentiator produces an output that is proportional to the rate of change of the input, then a square-wave input applied to a differentiator results in an output during transitions only. (Input and output waveforms are provided in Fig. 12-14b and 12-14c, respectively.)

Let us apply Eq. (12-13) to verify the output waveform of Fig. 12-14c. At $t = 0$, a transition from 0 to 1 V occurs at the input. Assuming a perfect square wave, then the transition occurs in zero time, and $\Delta v_i/\Delta t = \infty$. This means that the output level is infinitely high and lasts for zero time only.

To be practical, assume that the input wave has a rise and fall time of 0.1 μs. Hence,

$$\frac{\Delta v_i}{\Delta t} = \frac{1 \text{ V}}{0.1 \ \mu\text{s}} = 10^7 \text{ V/s}$$

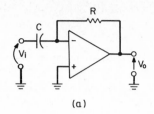

(a)

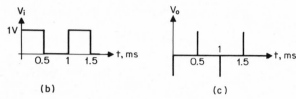

(b)                              (c)

**Fig. 12-14**  Basic differentiator. (a) Circuit. (b) Input waveform. (c) Output waveform. (Problem 12.8.)

At a transition, by Eq. (12-13), $v_o = \pm RC(10^7)$. Letting $C = 1\,\mu\text{F}$ and $R = 10\,\text{k}\Omega$, $v_o = \pm 10^{-6} \times 10^4$ $\times 10^7 = \pm 10^5$ V.

Obviously, the op amp cannot deliver a voltage of $10^5$ V. The actual output, therefore, is limited to the saturation voltage, or voltage of the power supplies.

During the constant (flat) portion of the input wave, the capacitor blocks the input from reaching the op amp. The output, therefore, remains at zero. Because only positive pulses are required, the *pulse repetition rate (PRR)* is 1/(1 ms) = 1000 pulses/s.

### The Logarithmic Amplifier

The logarithm is a tool that facilitates the mathematical operations of multiplication, division, exponentiation, and the extraction of roots. Multiplication, for example, is accomplished by obtaining logarithms of the numbers to be multiplied. The logarithms are then added, and the antilogarithm of the sum is obtained. The antilogarithm is the product of the original numbers. In order to construct an electronic circuit capable of performing such an operation, the basic logarithmic amplifier (converter) is considered first.

**problem 12.9**  Calculate the output voltage $V_o$ of the basic logarithmic amplifier of Fig. 12-15 for an input of 3 V. Resistor $R = 10\,\text{k}\Omega$, and $I_S$ for the diode is 10 nA.

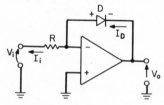

**Fig. 12-15**  Logarithmic amplifier with diode (Problem 12.9).

**theory**  The principal element in the logarithmic amplifier of Fig. 12-15 is diode $D$ in the feedback loop. A diode has an inherent exponential and, therefore, logarithmic characteristic. Diode current $I_D$ is:

$$I_D = I_S(\epsilon^{qV_D/kT} - 1) \tag{12-14}$$

where $I_S$ = reverse saturation current
$\epsilon$ = base of natural logarithms, 2.718
$V_D$ = voltage across the diode
$q$ = charge of electron, $1.6 \times 10^{-19}$ C
$k$ = Boltzmann's constant, $1.38 \times 10^{-23}$ J/K
$T$ = temperature, K

At room temperature, Eq. (12-14) reduces to:

$$I_D \simeq I_S \epsilon^{39 V_D} \tag{12-15}$$

To obtain an output-input relationship for the logarithmic amplifier of Fig. 12-15, the currents at the inverting terminal are summed:

$$I_i = I_D$$

or

$$\frac{V_i}{R} = I_S \epsilon^{39 V_D}$$

Taking the natural logarithm of both sides yields:

$$\ln \frac{V_i}{R} = \ln I_S + 39 V_D \tag{12-16}$$

Because of the virtual ground at the input terminals, the output voltage $V_o$ is the negative of the voltage across the diode: $V_o = -V_D$. Solving Eq. (12-16) for $V_o$ results in:

$$V_o = (-26 \text{ mV}) \left( \ln \frac{V_i}{R} - \ln I_S \right) \tag{12-17}$$

Equation (12-17) shows that the output voltage is indeed proportional to the log of the input. The effect of term $\ln I_S$ can be reduced by replacing the diode with a transistor in the common-base configuration (Fig. 12-16). The transistor exhibits a logarithmic relationship between collector current $I_C$ and the base-emitter voltage $V_{BE}$. The base-emitter voltage of a silicon transistor at room temperature may be expressed by:

$$V_{BE} = (26 \text{ mV}) \left( \ln \frac{V_i}{R} - \ln I_{ES} \right) \tag{12-18}$$

where $I_{ES}$ is the base-emitter reverse saturation current.

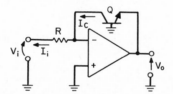

**Fig. 12-16**   Logarithmic amplifier with the diode replaced by a transistor in the feedback loop.

Output voltage $V_o$ is:

$$V_o = (-26 \text{ mV}) \left( \ln \frac{V_i}{R} - \ln I_{ES} \right) \tag{12-19}$$

In terms of common logarithms,

$$V_o = (-60 \text{ mV}) \left( \log \frac{V_i}{R} - \log I_{ES} \right) \tag{12-20}$$

**solution**   By Eq. (12-17), $V_o = (-26 \text{ mV}) [\ln (3/10^4) - \ln 10^{-9}] = -749.6 \text{ mV}$.

In the preceding discussion of log amplifiers, operation at room temperature was assumed. In practice, the output voltage is temperature-dependent. Compensation for

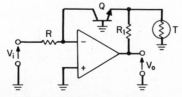

**Fig. 12-17**   Logarithmic amplifier employing temperature compensation.

temperature effects may be accomplished by adding a resistor and thermistor to the basic logarithmic amplifier (Fig. 12-17). A thermistor with a temperature coefficient of 0.3%/°C is used. The value of resistor $R_1$ is chosen to be 15 times the cold resistance value of the thermistor.

A practical logarithmic amplifier, that exhibits good thermal characteristics and a wide (100 dB) dynamic operating range, is given in Fig. 12-18. The output is proportional to the logarithm of the input for input currents from 10 nA to 1 mA. The circuit employs two op amps, two matched transistors to eliminate the effect of $I_{ES}$, and a thermistor for temperature compensation. The output voltage is:

$$V_o = -\left(\ln\left|\frac{V_i}{R_{in}}\right| + 5\right) V \tag{12-21}$$

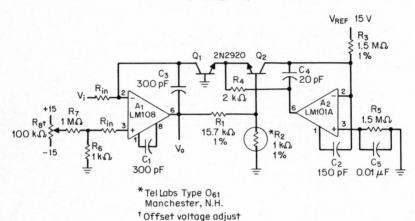

*Tel Labs Type O$_{61}$
  Manchester, N.H.
†Offset voltage adjust

**Fig. 12-18**  Logarithmic converter with temperature compensation and 100 dB dynamic range. *(Courtesy National Semiconductor Corporation.)*

### The Antilog Amplifier

In order to perform mathematical operations using logarithms, one must be able to perform the antilog operation. This operation converts a logarithm to the number it represents.

**problem 12.10**  Develop an antilog amplifier circuit.

**theory**  To obtain the antilog of the logarithm of a number, the exponential of the logarithm is taken:

$$\epsilon^{\ln z} = z \tag{12-22a}$$

Hence, the antilog function is basically the exponential function. A device with exponential characteristics, such as a diode or transistor, is used as the principal element.

**solution**  A circuit similar to the logarithmic amplifier with the diode (or transistor) moved from the feedback path to the input generates the antilog function. The circuit of a basic antilog amplifier using a transistor ($Q$) in the common-base configuration is illustrated in Fig. 12-19. The output voltage at room temperature is:

$$V_o = -I_{ES}R_F\epsilon^{39V_i} \tag{12-22b}$$

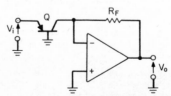

**Fig. 12-19**  Antilog amplifier (Problem 12.10).

If the input signal represents a logarithm of a number, $V_i = \ln z$. Substituting for $V_i$ in Eq. (12-22b) yields:

$$V_o = -I_{ES}R_F \epsilon^{39(\ln z)}$$
$$= -I_{ES}R_F \epsilon^{39} \epsilon^{\ln z}$$

If $I_{ES}R_F \epsilon^{39} = k$ (a constant), then,

$$V_o = -kz \tag{12-23}$$

A practical antilog amplifier with temperature compensation is shown in Fig. 12-20. Note the similarity of this circuit with the logarithmic amplifier of Fig. 12-18. With the component values shown, the output is:

$$V_o = 10^{-V_i} \quad \text{V} \tag{12.24}$$

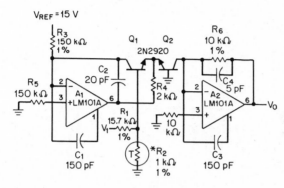

* Tel Labs type O₆₁
Manchester, N.H.

**Fig. 12-20**   Compensated antilog amplifier. (*Courtesy National Semiconductor Corporation.*)

## Voltage Multipliers and Dividers

**problem 12.11**   Using logarithmic and antilogarithmic amplifiers, develop a block diagram of a voltage multiplier and divider.

**theory**   Multiplication of two numbers is equivalent to the addition of their logarithms. Taking the antilog of the result yields the product. The antilog of the difference of the logarithms of two numbers is their quotient.
To multiply $X$ and $Y = Z$:
1. Values of $\ln X$ and $\ln Y$ are found.
2. $\ln X + \ln Y = \ln Z$.
3. $\ln^{-1}(\ln X + \ln Y) = \ln^{-1}(\ln Z)$, which is $XY = Z$.
To divide $X$ by $Y = Z$:
1. Values of $\ln X$ and $\ln Y$ are found.
2. $\ln X - \ln Y = \ln Z$.
3. $\ln^{-1}(\ln X - \ln Y) = \ln^{-1}(\ln Z)$, or $X/Y = Z$.

**solution**   The block diagram of a multiplier/divider based on the steps outlined in the preceding discussion is illustrated in Fig. 12-21.

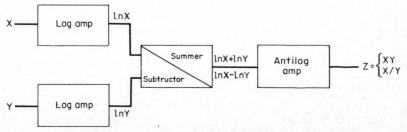

**Fig. 12-21**   Block diagram of a multiplier/divider (Problem 12.11).

**problem 12.12**  Develop a circuit diagram of a multiplier/divider.

**solution**  A practical multiplier/divider is shown in Fig. 12-22. This circuit is capable of multiplying two numbers and dividing by a third. The circuit uses the logarithmic and antilogarithmic amplifiers of Figs. 12-18 and 12-20, respectively. With the component values shown, the output voltage is:

$$V_o = \frac{V_1 V_3}{10 V_2} \quad \text{V} \tag{12-25}$$

It should be apparent from Eq. (12-25) that the circuit may be used to multiply two numbers, to divide one number by another, and to divide the product of two numbers by a third.

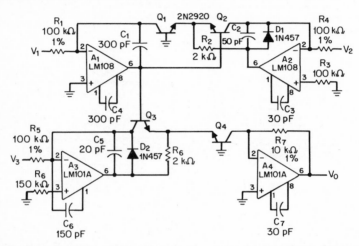

**Fig. 12-22**  Practical multiplier/divider circuit (Problem 12.12). *(Courtesy National Semiconductor Corporation.)*

### Extraction of Roots and Exponentiation

Extracting roots of numbers and raising numbers to a power are fairly simple operations with logarithms. To raise a number to a power, one multiplies the logarithm of the number by the value of the exponent and then takes the antilogarithm:

$$x^n = \ln^{-1}(n \ln x) \tag{12-26}$$

To extract the root of a number, one multiplies the logarithm of the number by the reciprocal of the root value and takes its antilogarithm:

$$\sqrt[n]{x} = \ln^{-1}\left(\frac{1}{n} \ln x\right) \tag{12-27}$$

The multiplier/divider of Fig. 12-22 can be modified to perform root extractions and exponentiation.

### 12.6  WAVE GENERATION AND SHAPING

The operational amplifier is used extensively in the generation of sinusoidal, square, triangular, sawtooth, and other nonsinusoidal waveforms. Op amps lend themselves to wave-shaping applications as well. (In Problem 12.8, a differentiator was used to convert a square waveform into narrow pulses.)

### Square-Wave Generator

**problem 12.13**  Design an astable (free running) multivibrator (MV) to deliver a square waveform having a period of 100 μs.

**theory**  Several conditions are necessary to produce oscillations:
1. A high-gain amplifier.
2. Feedback.
3. Frequency-sensitive components.

All three conditions are necessary in an astable MV (Fig. 12-23). The op amp is a National Semiconductor Co. LM 139 device. Capacitor $C_1$ is the frequency-sensitive component and feedback is provided by resistors $R_2$, $R_3$, and $R_4$. If one lets $R_1 = R_2 = R_3$, the period of oscillation $T$ in seconds is:

$$T = 2(0.694)R_4C_1 \tag{12-28}$$

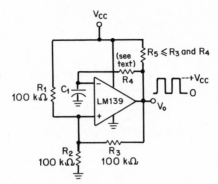

**Fig. 12-23**   Square-wave generator (Problem 12.13). (*Courtesy National Semiconductor Corporation.*)

**solution:**   By Eq. (12-28),

$$R_4C_1 = \frac{T}{2 \times 0.694}$$
$$= (100 \times 10^{-6})/(2 \times 0.694) = 72 \ \mu s$$

By choosing $C_1 = 0.01 \ \mu F$, then $R_4 = (72 \times 10^{-6})/10^{-8} = 7.2 \ k\Omega$. All other component values are as shown in Fig. 12-23.

An alternate circuit may be obtained with the crystal-controlled oscillator of Fig. 12-24. The frequency of the crystal is equal to 1 over the period of oscillation.

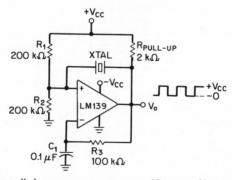

**Fig. 12-24**   Crystal-controlled square-wave generator. (*Courtesy National Semiconductor Corporation.*)

## Pulse Generator

**problem 12.14**   Calculate the values of $R_4$ and $R_5$ for the pulse generator of Fig. 12-25. The output pulse width is 0.2 $\mu s$ and the PRF (*pulse repetition frequency*) is 1 MHz (T = 1 $\mu s$).

**theory**   The variable-duty cycle pulse generator of Fig. 12-25 is essentially the same as the astable MV of Fig. 12-23. In Fig. 12-25, however, two diodes and two variable resistors constitute the feedback circuit. The ON time $t_1$ of the pulse is varied by $R_4$, and the OFF time $t_2$ is varied by $R_5$.

The approximate ON time $t_1$ is:

$$t_1 \simeq 0.694R_4C_1 \tag{12-29a}$$

The approximate OFF time $t_2$ is:

$$t_2 \simeq 0.694R_5C_1 \tag{12-29b}$$

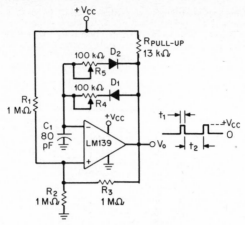

**Fig. 12-25** Variable-duty cycle pulse generator (Problem 12.14). (*Courtesy National Semiconductor Corporation.*)

**solution** For a pulse width of 0.2 $\mu$s and $C_1 = 80$ pF (given), $t_1 = 0.2 \times 10^{-6} = 0.694R_4(80 \times 10^{-12})$. By solving for $R_4$, $R_4 = 3.6$ k$\Omega$. The OFF time is $t_2 = T - t_1 = 1 - 0.2 = 0.8$ $\mu$s. By Eq. (12-29b), $R_2 = (0.8 \times 10^{-6})/(0.694 \times 80 \times 10^{-12}) = 14.4$ k$\Omega$.

### Sinusoidal Oscillator

The *Wien bridge oscillator* (Fig. 12-26) is a popular op amp circuit for generating a sine wave. An oscillator which generates a sine wave by filtering a square wave is also discussed in this section.

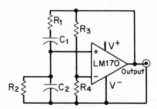

**Fig. 12-26** Wien bridge oscillator (Problem 12.15). (*Courtesy National Semiconductor Corporation.*)

**problem 12.15** Calculate the component values for the Wien bridge oscillator of Fig. 12-26 for a frequency of oscillation equal to 1.6 kHz.

**theory** Positive feedback is provided by the $R_1C_1$ and $R_2C_2$ networks. The gain of the amplifier is controlled by resistors $R_3$ and $R_4$. The condition of sustained oscillations is:

$$\frac{R_3}{R_4} = \frac{R_1}{R_2} + \frac{C_2}{C_1} \tag{12-30}$$

The frequency of oscillation $f_o$ is:

$$f_o = 0.159/\sqrt{R_1R_2C_1C_2} \qquad \text{Hz} \tag{12-31}$$

Letting $R_1 = R_2 = R$ and $C_1 = C_2 = C$, Eq. (12-31) reduces to:

$$f_o = \frac{0.159}{RC} \qquad \text{Hz} \tag{12-32}$$

**solution** To satisfy the frequency requirement, from Eq. (12-32),

$$RC = 0.159/f_o$$
$$= 0.159/(1.6 \times 10^3) = 10^{-4} \text{ s}.$$

Selecting a value for $C = 0.1$ $\mu$F, $R = 10^{-4}/C = 10^{-4}/10^{-7} = 1$ k$\Omega$. Hence, $R_1 = R_2 = 1$ k$\Omega$, and $C_1 = C_2 = 0.1$ $\mu$F.

By Eq. (12-30), $R_3/R_4 = R_1/R_2 + C_2/C_1 = 2$; then, $R_3 = 2R_4$. By letting $R_4 = 470\ \Omega$, $R_3 = 2 \times 470$ $= 940\ \Omega$.

To ensure good frequency stability, a thermistor may be used in place of $R_3$. Another approach is to use a positive-temperature coefficient device, such as a light bulb, in place of $R_4$.

**problem 12.16**   Design an oscillator which generates both sinusoidal and square waves at a frequency of 10 kHz.

**theory** .   A square-wave generator operating in conjunction with a filter (Fig. 12-27) is used to generate sine waves. A sine wave fed into the LM 111 op amp results in a square wave at its output. The square wave, in turn, is fed to a filter (tuned circuit) which causes the LM 101A to oscillate. The frequency of oscillation is adjustable by $R_3$.

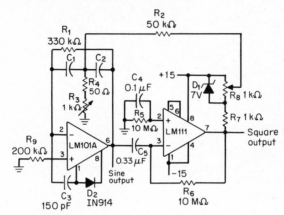

**Fig. 12-27** Tunable sine-wave oscillator (Problem 12.16). *(Courtesy National Semiconductor Corporation.)*

If $C_1 = C_2$, the frequency of oscillation is:

$$f_o = 0.159/(C_1\sqrt{R_1 R_3}) \qquad (12\text{-}33)$$

The circuit is fairly stable with frequency. Amplitude-level control is available by adjusting $R_8$; amplitude variations do not affect frequency.

**solution:**
For a frequency, $f_o = 10$ kHz, from Eq. (12-33),

$C_1 = 0.159/(f_o\sqrt{R_1 R_3})$. By letting $R_3 = 500\Omega$, $C_1 = 0.159/(10^4 \times \sqrt{330 \times 10^3 \times 500}) = 1240$ pF.

## Function Generator

A function generator supplies a variety of time-dependent functions, most commonly square, triangular, and sinusoidal waveforms. Triangular waves are obtained by integrating a square wave; a sine wave may be obtained by filtering a square wave.

The function generator of Fig. 12-28 generates variable-frequency and variable-amplitude square, sinusoidal, and triangular waveforms. An integrator was added to the circuit of Fig. 12-27 to obtain triangular waves.

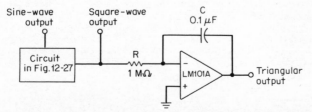

**Fig. 12-28**   Example of a function generator.

## Voltage Comparator

The op amp may be used as a voltage comparator. A voltage comparator compares the voltage at one input terminal with respect to a reference voltage at the other terminal. A comparator is also used in wave-shaping applications. The comparator accepts waveforms of various shapes and produces a well-defined rectangular output.

**problem 12.17**    A voltage comparator circuit, used for wave shaping, is illustrated in Fig. 12-29$a$. Draw the output waveform for a triangular input (Fig. 12-29$b$) if $+V = 10$ V and $V_{ref} = 1$ V.

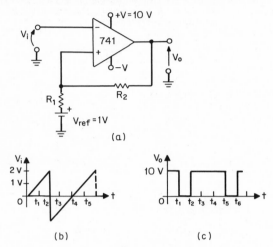

**Fig. 12-29**    Voltage comparator. ($a$) Circuit. ($b$) Input waveform. ($c$) Output waveform. (Problem 12.17.)

**theory and solution**    For $0 < t < t_1$, the input rises from 0 to 1 V. Since the reference is constant at 1 V, the input is at a lower level than the reference. The op amp, because of its very high gain, provides an output of $+V$ (10 V), which is the saturation voltage of the circuit.

For $t_1 < t < t_2$, the inverting input is more positive than the reference; the output, therefore, is 0 V if $-V = 0$ V. For $t_2 < t < t_5$, the output is again 10 V. Output $V_o$ is thus a rectangular waveform, as shown in Fig. 12-29$c$.

## 12.7   ACTIVE FILTERS

A comprehensive discussion of active filters is provided in Sec. 19.4.

## 12.8   OP AMPS IN INSTRUMENTATION

### Instrumentation Amplifier

The op amp is used extensively in instrumentation as the building block for the instrumentation amplifier. The essential characteristics of an instrumentation amplifier are high gain, high input resistance, low offset, and high common-mode rejection ratio.

The high gain is necessary because the amplifier has to be sensitive to very low amplitude signals. Minimum loading on the signal source requires a high-input resistance. The low offset is necessary for accuracy in the measurement being made. A high value of CMRR is required to ensure that only the differential input is amplified and the common-mode signal is greatly attenuated. The difference input signal is generally derived from a transducer, such as a strain gage or thermistor, and applied through a bridge circuit to the instrumentation amplifier.

An example of an instrumentation amplifier is provided in Fig. 12-30. The differential input signal is applied to two op amps which are connected as voltage followers. This connection results in an input resistance of $10^{12}$ $\Omega$ and a CMRR that is independent of the input stage.

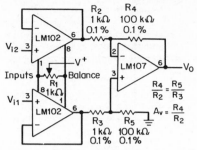

**Fig. 12-30**   Instrumentation amplifier. *(Courtesy National Semiconductor Corporation.)*

The LM 107 op amp provides a high differential gain. A high CMRR is obtained when $R_4/R_2 = R_5/R_3$. Note that $\pm0.1\%$ resistors are used. To ensure a CMRR of 100 or more, one of the resistors (usually $R_5$) is made variable to realize the necessary resistor ratio.

The output voltage of the instrumentation amplifier is:

$$V_o = (V_{i1} - V_{i2})\frac{R_4}{R_2} \tag{12-34}$$

The voltage gain is:

$$A_{vc} = \frac{R_4}{R_2} \tag{12-35}$$

For the component values in the circuit of Fig. 12-30, $A_{vc} = 100/1 = 100$, or 40 dB. Resistor $R_1$ is used to adjust the input offset voltage.

A variable-gain instrumentation amplifier is illustrated in Fig. 12-31. This circuit is basically the same as that of Fig. 12-30 with the addition of an active attenuator (LM 101A). An active attenuator is used to vary the gain of the circuit while preserving its high value of CMRR.

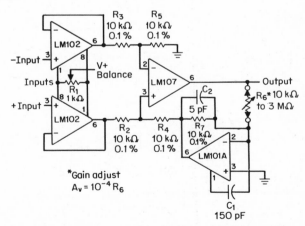

**Fig. 12-31**   Variable-gain instrumentation amplifier (Problem 12.18). *(Courtesy National Semiconductor Corporation.)*

**problem 12.18**   Calculate the output voltage range of the instrumentation amplifier of Fig. 12-31 for a differential input of 10 mV.

**theory and solution**   If the resistance ratio $R_4/R_2 = R_5/R_3$, then the output voltage is:

$$V_o = (V_{i1} - V_{i2})\frac{R_6}{R_4} \tag{12-36}$$

If $R_6 = 10$ k$\Omega$, by Eq. (12-36), $V_o = (10$ mV$) \times 10/10 = 10$ mV. For $R_6 = 3$ M$\Omega$, $V_o = (10$ mV$) \times 3000/10 = 3$ V. The output-voltage range, therefore, is from 10 mV to 3 V.

## Bridge Circuit

A bridge circuit converts small variations in resistance to a voltage. The voltage from the bridge is fed to an instrumentation amplifier. The output of the amplifier is then accurately displayed. Many instruments employ a bridge circuit, such as thermometers, pressure indicators, and light intensity meters. In the case of the thermometer, a thermistor is used as the variable-resistance element to sense changes in temperature.

An example of a bridge circuit is given in Fig. 12-32. The variable-resistance device has a resistance of $R$ Ω at rest. The potentiometer is set to the value of $R$ to balance the bridge at rest. When the bridge is balanced, $V_1 = V_2 = V/2$. If a condition occurs to unbalance the bridge owing to a change in the resistance of the device, then

$$V_2 = \frac{\Delta RV}{2R + \Delta R} \qquad (12\text{-}37)$$

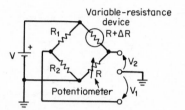

**Fig. 12-32**   An example of a bridge circuit.

But at rest, $V_1 = V/2$. The bridge output, $V_1 - V_2$, is, therefore, calculated by:

$$V_1 - V_2 = \frac{V}{2} - \frac{\Delta RV}{2R + \Delta R}$$

For $2R >> \Delta R$,

$$V_1 - V_2 = \frac{\Delta RV}{4R} \qquad (12\text{-}38)$$

**problem 12.19**   An electronic thermometer consisting of a thermistor in a bridge circuit, an instrumentation amplifier, and a voltmeter is illustrated in Fig. 12-33. The voltage gain of the instrumentation amplifier is 100. If the resistance of the thermistor at 21°C is 10 kΩ and at 20°C it is 10.2 kΩ, determine the output voltage at 20°C.

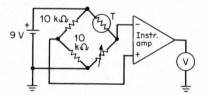

**Fig. 12-33**   An electronic thermometer (Problem 12.19).

**theory and solution**   The output voltage as a function of change in resistance is obtained by combining Eqs. (12-34), (12-35), and (12-38) to yield:

$$V_o = \frac{V}{4}\frac{\Delta R}{R}A_{vc} \qquad (12\text{-}39)$$

Substituting the given values in Eq. (12-39) yields:

$$V_o = \frac{9}{4} \times \frac{0.2}{10} \times 100 = 4.5 \text{ V}$$

## 12.9  BIBLIOGRAPHY

Coughlin, R., and Driscoll, E.: *Operational Amplifiers and Linear Integrated Circuits*, Prentice-Hall, Englewood Cliffs, N.J., 1977.

Deboo, G. J., and Burrous, C.: *Integrated Circuits and Semiconductor Devices: Theory and Application*, 2d ed., McGraw-Hill, New York, 1977.

Dobkin, R.: "Logarithmic Converters," National Semiconductor Application Note AN30, National Semiconductor Corp., Santa Clara, Calif., November 1969.

———: "Op Amp Circuit Collection," National Semiconductor Application Note AN31, National Semiconductor Corp., Santa Clara, Calif., February 1970.

Faulkenberry, I.: *An Introduction to Operational Amplifiers*, Wiley, New York, 1977.

Graeme, J., Tobey, G., and Huelsman, L. (eds.): *Operational Amplifiers: Design and Applications*, McGraw-Hill, New York, 1971.

Jung, W.: *IC Op Amp Cookbook*, Sams, Indianapolis, Ind., 1974.

Kaufman, M., and Seidman, A. H.: *Handbook for Electronic Engineering Technicians*, McGraw-Hill, New York, 1976.

*Linear Integrated Circuits*, vol. 6., Motorola Semiconductor Products Inc., Phoenix, Ariz., 1975.

*Linear Integrated Circuits*, National Semiconductor Corp., Santa Clara, Calif., February 1975.

Roberge, J.: *Operational Amplifiers*, New York, Wiley, 1975.

Seidman, A. H., and Waintraub, J. L.: *Electronics: Devices, Discrete and Integrated Circuits*, Charles E. Merrill Books, Inc., Columbus, Ohio, 1977.

Smathers, R., et al.: "LM139/LM239/LM339 A Quad of Independently Functioning Comparators," National Semiconductor Application Note AN74, National Semiconductor Corp., Santa Clara, Calif., January 1973.

Vander Kooi, M., and Cleveland, G.: "Micropower Circuits Using the LM4250 Programmable Op Amp," National Semiconductor Application Note AN71, National Semiconductor Corp., Santa Clara, Calif. July 1972.

Wait, J. V., Huelsman, L. P., and Korn, G.: *Introduction to Operational and Amplifier Theory and Applications*, McGraw-Hill, New York, 1975.

# Chapter 13

# Digital Logic

## JOEL H. LEVITT

**Research Associate, Queens College (CUNY); Visiting Associate
Professor of Electrical Engineering, Pratt Institute**

## 13.1 INTRODUCTION

A knowledge of digital logic is essential for an understanding of digital circuits and their operation. The purpose of this chapter is to show that digital logic is not only "logical" but straightforward and easy to understand. Numerous problems illustrate design methods and principles.

## 13.2 TRUTH TABLES AND BOOLEAN VARIABLES

Consider the circuit of Fig. 13-1. The lamp labeled $L$ lights if, and only if, the switches labeled $A$ and $B$ are closed. The situation can be described by a table listing all possibilities, called a *truth table*, given in Fig. 13-2. If we let symbol 0 represent OFF and symbol 1 represent ON, then $A = 1$ could be used to denote switch $A$ ON and $L = 1$ could be used to denote lamp $L$ ON.

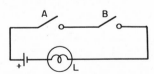

**Fig. 13-1**  A simple AND circuit.

| Switch A | Switch B | Lamp L |
|----------|----------|--------|
| off | off | off |
| off | on | off |
| on | off | off |
| on | on | on |

**Fig. 13-2**  Truth table for the AND circuit of Fig. 13-1.

This type of symbolic notation is the basis of Boolean algebra, named after mathematician George Boole (1815–1864). Symbols $A$, $B$, and $L$ are called *Boolean (logic) variables;* all such variables are two-valued, 0 or 1. In most digital circuits, device operation is constrained to two states. High reliability is, therefore, realized. The AND truth table for two variables, $A$ and $B$, can be rewritten more compactly as in Fig. 13-3.

Consider the circuit of Fig. 13-4. Lamp $L$ is ON when switch $A$ or switch $B$ is ON. The OR truth table is given in Fig. 13-5.

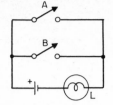

| A | B | L |
|---|---|---|
| 0 | 0 | 0 |
| 0 | 1 | 0 |
| 1 | 0 | 0 |
| 1 | 1 | 1 |

| A | B | L |
|---|---|---|
| 0 | 0 | 0 |
| 0 | 1 | 1 |
| 1 | 0 | 1 |
| 1 | 1 | 1 |

**Fig. 13-3** The AND truth table written more compactly.  **Fig. 13-4** A simple OR circuit.  **Fig. 13-5** The OR truth table.

Symbol · is often used to represent AND while symbol + is often used to represent OR. With this notation, the information in the OR truth table can be stated as $L = A + B$, and the information in the AND truth table can be stated as $L = A \cdot B$, or $L = AB$ with the · symbol understood. Individual entries in the OR table can be written as $0 + 0 = 0$, $0 + 1 = 1$, $1 + 0 = 1$, $1 + 1 = 1$. The last entry is a reminder that the 1's are logic symbols without numerical value and that we are using the symbol + to represent OR, not addition. Similarly, the four entries in the AND table could be written $0 \cdot 0 = 0$, $0 \cdot 1 = 0$, $1 \cdot 0 = 0$, $1 \cdot 1 = 1$. The results are summarized in Fig. 13-6.

Other basic Boolean relations are listed in Fig. 13-7. The truth of any suspicious Boolean equation can readily be verified using truth tables.

| AND | OR |
|---|---|
| $0 \cdot 0 = 0$ | $0 + 0 = 0$ |
| $0 \cdot 1 = 0$ | $0 + 1 = 1$ |
| $1 \cdot 0 = 0$ | $1 + 0 = 1$ |
| $1 \cdot 1 = 1$ | $1 + 1 = 1$ |

**Fig. 13-6** The AND and OR logic relationships.

| | | |
|---|---|---|
| $0 \cdot A = 0$ | $0 + A = A$ | $A + AB = A$ |
| $1 \cdot A = A$ | $1 + A = 1$ | $A(A+B) = A$ |
| $A \cdot A = A$ | $A + A = A$ | |

**Fig. 13-7** Other basic logic relationships.

### 13.3 DESCRIPTION OF SIMPLE ALARM CIRCUIT WITH BOOLEAN ALGEBRA

**problem 13.1**  A burglar alarm bell $B$ rings if a window switch $W$ or a door switch $D$ is activated. A key-operated switch $K$, however, silences the bell. Describe the situation with a logic equation and sketch a circuit that satisfies the conditions of the problem.

**solution**  Bell $B$ is ON when $K$ is ON and $W$ or $D$ is ON: $B = K(W + D)$. The circuit is sketched in Fig. 13-8.

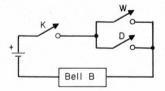

**Fig. 13-8**  A burglar alarm circuit. (Problem 13.1.)

### 13.4 COMPLEMENTS

A logic variable with a horizontal bar above it denotes the complement, or NOT, or negation, of the variable. For example, $\overline{A}$ is the complement of $A$. Whatever two conditions or states logic symbols 0 and 1 are allowed to represent, if $A = 0$, then by definition $\overline{A} = 1$, and vice versa. It follows that $A + \overline{A} = 1$, since either $A$ or its complement must be in the 1 state. Similarly, $A \cdot \overline{A} = 0$ since either $A$ or its complement must be in the 0 state. The bar notation may also be used to indicate the complement of an expression. For example, $\overline{A + B}$ is the complement of $(A + B)$. Some logic identities involving complemented variables are listed in Fig. 13-9.

$$A + \overline{A} = 1$$
$$A \cdot \overline{A} = 0$$
$$A + \overline{A} B = A + B$$
$$\overline{(\overline{A})} = A$$
$$\overline{A + B} = \overline{A} \cdot \overline{B}$$
$$\overline{A \cdot B} = \overline{A} + \overline{B}$$
$$\overline{A} \cdot B + A \cdot B = B$$

**Fig. 13-9**  Boolean identities involving complemented variables.

## 13.5  DE MORGAN'S LAW FOR COMPLEMENTING LOGIC EXPRESSIONS

The complement (negation) of a logic expression may be obtained by replacing all + symbols with ·, all · symbols with +, and all variables ($A$, $B$, etc.) with their complements ($\overline{A}$, $\overline{B}$, etc.). For example:

$$\overline{A\overline{B}C} = \overline{A} + B + \overline{C} \tag{13-1}$$
$$\overline{ABC} = \overline{A} + \overline{B} + \overline{C} \tag{13-2}$$

$$\overline{\overline{A} + \overline{B} + \overline{C}} = A \ B \ C \tag{13-3}$$
$$\overline{A + \overline{B}C} = \overline{A} + \overline{B} + \overline{C} = \overline{A}BC \tag{13-4}$$

## 13.6  CONVERTING A LOGIC EXPRESSION INTO A TRUTH TABLE

**problem 13.2**  Construct a truth table for the expression $(A + B) \cdot (A + C)$.

**solution**  Refer to Fig. 13-10. To construct a truth table, we must tabulate all possibilities. With $A$, $B$, $C$ each 0 or 1, there are 8 cases to consider. These are tabulated in the three leftmost columns. Then $A + B$ and $A + C$ are tabulated. Finally, the answer column is filled in, utilizing the $A + B$ and $A + C$ columns just constructed.

| A | B | C | A+B | A+C | (A+B) · (A+C) |
|---|---|---|-----|-----|---------------|
| 0 | 0 | 0 | 0 | 0 | 0 |
| 0 | 0 | 1 | 0 | 1 | 0 |
| 0 | 1 | 0 | 1 | 0 | 0 |
| 0 | 1 | 1 | 1 | 1 | 1 |
| 1 | 0 | 0 | 1 | 1 | 1 |
| 1 | 0 | 1 | 1 | 1 | 1 |
| 1 | 1 | 0 | 1 | 1 | 1 |
| 1 | 1 | 1 | 1 | 1 | 1 |

**Fig. 13-10**  Construction of a truth table for a given logic expression. (Problem 13.2.)

## 13.7  CONVERTING A TRUTH TABLE INTO A LOGIC EXPRESSION

**problem 13.3**  Given the truth tables of Fig. 13-11, find expressions for $F$ and $G$.

| A | B | F |   | A | B | G |
|---|---|---|---|---|---|---|
| 0 | 0 | 0 |   | 0 | 0 | 1 |
| 0 | 1 | 1 |   | 0 | 1 | 1 |
| 1 | 0 | 0 |   | 1 | 0 | 1 |
| 1 | 1 | 1 |   | 1 | 1 | 0 |

(a)                         (b)

**Fig. 13-11**  Converting a truth table into a logic expression. (Problem 13.3.)

**solution**    Reading down the column for $F$ of Fig. 13-11$a$, we find that $F$ is 1 in either of two cases. The first case is $A = 0$ and $B = 1$, which can be written $\overline{A} \cdot B = 1$. The second case is $A$ and $B$ both 1, that is, $AB = 1$. Thus $F = 1$ when $\overline{A} \cdot B$ OR $AB = 1$, which we write as $F = \overline{A} \cdot B + AB$. This can be simplified to $F = B$ using the last identity in Fig. 13-9. The result should have been obvious; $F$ is 1 whenever $B$ is 1 in the given truth table.

In Fig. 13-11$b$ we see that $G$ is 1 in three cases. We can write $G = \overline{A}\overline{B} + \overline{A}B + A\overline{B}$ which can be simplified to $G = \overline{A} + \overline{B}$. (A general method of simplification is covered in Secs. 13.17 to 13.19.)

For Fig. 13-11$b$, there is an easier approach in expressing $G$. Since $G = 0$ in only one case, it is easiest to list this case. When $G = 0$, $\overline{G} = 1$; therefore, we can write immediately $\overline{G} = AB$. The expression for $G$ can be found by taking the complement of both sides of the equation. De Morgan's law readily gives the complement of the expression on the right, yielding $G = \overline{A} + \overline{B}$.

## 13.8   VERIFICATION OF LOGIC EQUATIONS USING TRUTH TABLES

**problem 13.4**    Verify that $A + BC = (A + B)(A + C)$.

**solution**    A truth table for the expression is constructed in Fig. 13-12. Examining the rightmost column (the result column), we find agreement for each of the eight cases, thus verifying the given equation.

| A | B | C | (A + B)(A + C) | A + BC |
|---|---|---|:---:|:---:|
| 0 | 0 | 0 | 0 | 0 |
| 0 | 0 | 1 | 0 | 0 |
| 0 | 1 | 0 | 0 | 0 |
| 0 | 1 | 1 | 1 | 1 |
| 1 | 0 | 0 | 1 | 1 |
| 1 | 0 | 1 | 1 | 1 |
| 1 | 1 | 0 | 1 | 1 |
| 1 | 1 | 1 | 1 | 1 |

**Fig. 13-12**   Truth table used to verify a logic equation. (Problem 13.4.)

## 13.9   LOGIC GATES

Suppose we have a black box with an input and output terminal. The output can assume one of two possible voltages. Call the lower voltage the low output and the higher voltage the high output. Assume that when the input is low, the output is high, and vice versa. Such a device is called a NOT gate or *inverter*. It generates the complement of the input.

The symbol for a NOT gate and the truth table that describes its behavior are given in Fig. 13-13. The truth table is identical to the one that defines complements. Usually 0 is associated with the low level and 1 is associated with the high level. This association is called *positive logic*. We shall use positive logic throughout this chapter, except where explicitly noted.

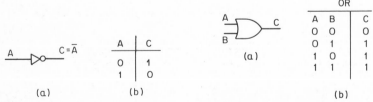

(a)                (b)                                                    (b)

**Fig. 13-13**   Symbol ($a$) and truth table ($b$) for an inverter.

**Fig. 13-14**   Symbol ($a$) and truth table ($b$) for an OR gate.

Suppose we want to realize a device with an output $C$ which is high when inputs $A$ *or* $B$, or both, are high. Such a device is called an OR gate (Fig. 13-14). If we want a gate which gives a high output only when $A$ *and* $B$ are high, we need an AND gate. Symbols for this and several other popular gates with corresponding truth tables are provided in Fig. 13-15. The NAND gate is equivalent to an AND followed by a NOT. The NOR gate is

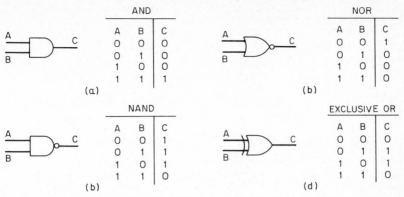

| AND | | |
|---|---|---|
| A | B | C |
| 0 | 0 | 0 |
| 0 | 1 | 0 |
| 1 | 0 | 0 |
| 1 | 1 | 1 |

(a)

| NOR | | |
|---|---|---|
| A | B | C |
| 0 | 0 | 1 |
| 0 | 1 | 0 |
| 1 | 0 | 0 |
| 1 | 1 | 0 |

(b)

| NAND | | |
|---|---|---|
| A | B | C |
| 0 | 0 | 1 |
| 0 | 1 | 1 |
| 1 | 0 | 1 |
| 1 | 1 | 0 |

(b)

| EXCLUSIVE OR | | |
|---|---|---|
| A | B | C |
| 0 | 0 | 0 |
| 0 | 1 | 1 |
| 1 | 0 | 1 |
| 1 | 1 | 0 |

(d)

**Fig. 13-15** Symbols and corresponding truth tables for (a) AND, (b) NAND, (c) NOR, and (d) EXCLUSIVE-OR gates.

equivalent to an OR followed by a NOT. The EXCLUSIVE OR behaves as an OR except for the 1,1 input case which is excluded (does not give a 1 output).

## 13.10  MULTI-INPUT GATES

Gates may have many inputs. For example, a 5-input AND gate will have a high output if, and only if, all five inputs are high. A 5-input OR gate will have a high output if one or more inputs are high. The truth tables for 4-input NAND and NOR gates are combined in Fig. 13-16. The four inputs are labeled $A$, $B$, $C$, $D$. The NAND's output is labeled $Y$ and the NOR's output is labeled $Z$.

| A | B | C | D | Y | Z |
|---|---|---|---|---|---|
| 0 | 0 | 0 | 0 | 1 | 1 |
| 0 | 0 | 0 | 1 | 1 | 0 |
| 0 | 0 | 1 | 0 | 1 | 0 |
| 0 | 0 | 1 | 1 | 1 | 0 |
| 0 | 1 | 0 | 0 | 1 | 0 |
| 0 | 1 | 0 | 1 | 1 | 0 |
| 0 | 1 | 1 | 0 | 1 | 0 |
| 0 | 1 | 1 | 1 | 1 | 0 |
| 1 | 0 | 0 | 0 | 1 | 0 |
| 1 | 0 | 0 | 1 | 1 | 0 |
| 1 | 0 | 1 | 0 | 1 | 0 |
| 1 | 0 | 1 | 1 | 1 | 0 |
| 1 | 1 | 0 | 0 | 1 | 0 |
| 1 | 1 | 0 | 1 | 1 | 0 |
| 1 | 1 | 1 | 0 | 1 | 0 |
| 1 | 1 | 1 | 1 | 0 | 0 |

**Fig. 13-16**  Truth table for 4-input NAND (output $Y$) and NOR (output $Z$) gates.

## 13.11  INHIBIT/ENABLE AND ANTICOINCIDENCE

A gate may be placed in a logic line in such a way that it either *inhibits* or *enables* the flow of digital data. The output of an AND gate with inputs $A$ and $B$ will follow input $A$ if $B = 1$ but will be 0 if $B = 0$. The $B$ input line can be thought of as inhibiting or enabling the gate. Any gate with two or more inputs can be used in a similar manner, as tabulated in Fig. 13-17. If it is desired to have a multi-input AND gate inhibited by a 1 rather than a 0, an inverter can be added in series with the control line. The standard *anticoincidence* or

| Inhibiting input | | Gate | Output |
|---|---|---|---|
| one | 0 | AND | Forced to 0 |
| one | 0 | NAND | Forced to 1 |
| one | 1 | OR | Forced to 1 |
| one | 1 | NOR | Forced to 0 |

**Fig. 13-17**  Inhibiting and enabling of standard gates.

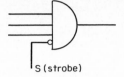

S (strobe)

**Fig. 13-18**  The INHIBIT gate.

*inhibit* gate is built this way. The symbol is given in Fig. 13-18. One line which controls a set of other lines is called a *strobe* line.

Another device, called a *transmission gate* or *bilateral switch,* permits a digital control signal to determine whether one terminal is effectively connected to another terminal. Signals of any level, therefore, (within a permitted range) are passed in either direction. Such a device can be used to inhibit or enable digital logic flow as an INHIBIT gate would, or in radically different applications. For example, a bilateral switch can be used to pass analog waveforms under digital control.

## 13.12  NUMBER SYSTEMS

### Decimal Number System

In the decimal number system there are 10 symbols: 0, 1, 2, 3, 4, 5, 6, 7, 8, 9. Beyond 9 we assign meaning to *position* of a symbol. The number 1264 represents 4 ones + 6 tens + 2 hundreds + 1 thousand. The first column on the right represents ones ($10^0$), the next column tens ($10^1$), the next hundreds ($10^2$), and the next thousands ($10^3$). The columns are weighted according to increasing powers of 10, which is called the *base* or *radix* of the number system. Note that the base is equal to the number of symbols.

To express numbers smaller than 1, negative powers of the base are used. Thus 0.32 represents $3 \times 10^{-1} + 2 \times 10^{-2}$. The decimal point is called, in general, a *base* or *radix point.*

### Binary Number System

Number systems, other than decimal, are structured in the same way using bases other than 10. The binary number system is a base-2 system. Two symbols are used, 0 and 1. Columns are weighted in powers of 2. Thus 1101, having column weightings of 8, 4, 2, 1, represents $8 + 4 + 0 + 1 = 13$ in the decimal system. Bases can be represented with subscripts, so that we can write $(1101)_2 = (13)_{10}$. The base is frequently implied, rather than explicitly written.

There is a close relationship between the binary number system and digital logic with two permitted states. Digital logic designers use the binary system extensively. Table 13-1 is a listing of powers of 2.

**TABLE 13-1    Powers of 2**

| $K$ | $2^K$ | $K$ | $2^K$ |
|---|---|---|---|
| −6 | 0.015625 | 7 | 128 |
| −5 | 0.03125 | 8 | 256 |
| −4 | 0.0625 | 9 | 512 |
| −3 | 0.125 | 10 | 1024 |
| −2 | 0.25 | 11 | 2048 |
| −1 | 0.5 | 12 | 4096 |
| 0 | 1 | 13 | 8192 |
| 1 | 2 | 14 | 16,384 |
| 2 | 4 | 15 | 32,768 |
| 3 | 8 | 16 | 65,536 |
| 4 | 16 | 17 | 131,072 |
| 5 | 32 | 18 | 262,144 |
| 6 | 64 | 19 | 524,288 |
| | | 20 | 1,048,576 |

## Octal Number System

The octal number system is a base-8 system. The eight symbols are 0 to 7; columns are weighted in powers of 8. Thus $(123)_8 = 1 \times 64 + 2 \times 8 + 3 \times 1 = (83)_{10}$.

## Hexadecimal Number System

The hexadecimal number system is a base-16 system. The 16 symbols are 0 to 9, A, B, C, D, E, F; columns are weighted in powers of 16. Thus $(123)_{16} = 1 \times 256 + 2 \times 16 + 3 \times 1 = (291)_{10}$ and $(C4)_{16} = 12 \times 16 + 4 \times 1 = (196)_{10}$. Table 13-2 provides the binary, octal, and

**TABLE 13-2   Decimal, Binary, Octal, and Hexadecimal Equivalents**

| Decimal Column value 10 1 | Binary 32 16 8 4 2 1 | Octal 64 8 1 | Hexadecimal 256 16 1 |
|---|---|---|---|
| 0 | 0 | 0 | 0 |
| 1 | 1 | 1 | 1 |
| 2 | 1 0 | 2 | 2 |
| 3 | 1 1 | 3 | 3 |
| 4 | 1 0 0 | 4 | 4 |
| 5 | 1 0 1 | 5 | 5 |
| 6 | 1 1 0 | 6 | 6 |
| 7 | 1 1 1 | 7 | 7 |
| 8 | 1 0 0 0 | 1 0 | 8 |
| 9 | 1 0 0 1 | 1 1 | 9 |
| 1 0 | 1 0 1 0 | 1 2 | A |
| 1 1 | 1 0 1 1 | 1 3 | B |
| 1 2 | 1 1 0 0 | 1 4 | C |
| 1 3 | 1 1 0 1 | 1 5 | D |
| 1 4 | 1 1 1 0 | 1 6 | E |
| 1 5 | 1 1 1 1 | 1 7 | F |
| 1 6 | 1 0 0 0 0 | 2 0 | 1 0 |
| 1 7 | 1 0 0 0 1 | 2 1 | 1 1 |
| 1 8 | 1 0 0 1 0 | 2 2 | 1 2 |
| 1 9 | 1 0 0 1 1 | 2 3 | 1 3 |
| 2 0 | 1 0 1 0 0 | 2 4 | 1 4 |
| 2 1 | 1 0 1 0 1 | 2 5 | 1 5 |
| 2 2 | 1 0 1 1 0 | 2 6 | 1 6 |
| 2 3 | 1 0 1 1 1 | 2 7 | 1 7 |
| 2 4 | 1 1 0 0 0 | 3 0 | 1 8 |
| 2 5 | 1 1 0 0 1 | 3 1 | 1 9 |
| 2 6 | 1 1 0 1 0 | 3 2 | 1 A |
| 2 7 | 1 1 0 1 1 | 3 3 | 1 B |
| 2 8 | 1 1 1 0 0 | 3 4 | 1 C |
| 2 9 | 1 1 1 0 1 | 3 5 | 1 D |
| 3 0 | 1 1 1 1 0 | 3 6 | 1 E |
| 3 1 | 1 1 1 1 1 | 3 7 | 1 F |
| 3 2 | 1 0 0 0 0 0 | 4 0 | 2 0 |
| 3 3 | 1 0 0 0 0 1 | 4 1 | 2 1 |
| 3 4 | 1 0 0 0 1 0 | 4 2 | 2 2 |
| 3 5 | 1 0 0 0 1 1 | 4 3 | 2 3 |
| 3 6 | 1 0 0 1 0 0 | 4 4 | 2 4 |
| 3 7 | 1 0 0 1 0 1 | 4 5 | 2 5 |
| 3 8 | 1 0 0 1 1 0 | 4 6 | 2 6 |
| 3 9 | 1 0 0 1 1 1 | 4 7 | 2 7 |
| 4 0 | 1 0 1 0 0 0 | 5 0 | 2 8 |
| 4 1 | 1 0 1 0 0 1 | 5 1 | 2 9 |
| 4 2 | 1 0 1 0 1 0 | 5 2 | 2 A |
| 4 3 | 1 0 1 0 1 1 | 5 3 | 2 B |
| 4 4 | 1 0 1 1 0 0 | 5 4 | 2 C |
| 4 5 | 1 0 1 1 0 1 | 5 5 | 2 D |
| 4 6 | 1 0 1 1 1 0 | 5 6 | 2 E |
| 4 7 | 1 0 1 1 1 1 | 5 7 | 2 F |
| 4 8 | 1 1 0 0 0 0 | 6 0 | 3 0 |
| 4 9 | 1 1 0 0 0 1 | 6 1 | 3 1 |
| 5 0 | 1 1 0 0 1 0 | 6 2 | 3 2 |
| 5 1 | 1 1 0 0 1 1 | 6 3 | 3 3 |

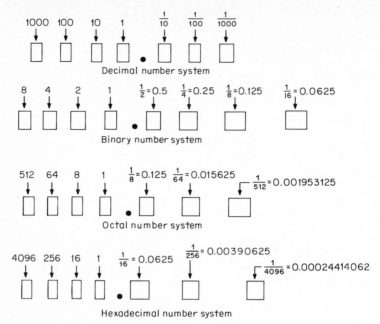

**Fig. 13-19**   Column weights in four number systems.

hexadecimal equivalents for some decimal numbers. Figure 13-19 shows column weightings, including fractional parts, in the number systems discussed.

### Binary-Octal Conversions

Numbers may readily be converted between the binary and octal systems because each group of three binary digits corresponds to one octal digit. The binary digits are grouped starting at the binary point. Thus $1111.11 = 001\ 111.110 = (17.6)_8$. Going the other way, $(17.6)_8 = 001\ 111.110 = 1111.11$.

### Binary-Hexadecimal Conversions

Numbers may readily be converted between the binary and hexadecimal (hex) systems because each group of four binary digits corresponds to one hex digit. The binary digits are grouped starting at the base point. Thus $1111.11 = 1111.1100 = (F.C)_{16}$. Note that 1111 corresponds to F as indicated in Table 13-2. Conversely, $(F.C)_{16} = 1111.1100 = (1111.11)_2$. Octal and hex systems are commonly used as a shorthand for binary.

### Conversions to Decimal

Numbers may be converted into base-10 numbers directly using the known column weightings.

> **problem 13.5**   Convert $(CA5.3C)_{16}$ to decimal.
>
> **solution**   $(CA5.3C)_{16} = 12 \times 256 + 10 \times 16 + 5 \times 1 + 3 \times (1/16) + 12 \times (1/256) = 3237.234375$

### Conversions from Decimal

Decimal numbers may be converted to any base. Repeated division of the integer portion and repeated multiplication of the fractional part, if any, by the base of the system being converted to, are used. The remainders and carries yield the answer.

> **problem 13.6**   Convert $(76.125)_{10}$ to binary.
>
> **solution**   See Fig. 13-20. Integer and fractional parts are treated separately. Repeated division of 76 by 2 yields remainders of 1001100. The first remainder obtained appears next to the base point in the answer. Note that odd integers in binary always end in 1; even numbers in binary always end in 0. The 0 or 1 is the remainder of the first division by 2.

Repeated multiplication of the fractional part in this example yields carries of 001. Again, the first carry must appear next to the base point in the final result; the answer is 1001100.001. Integer portions, of course, are written to the left of the base point, fractional portions to the right.

In general, the fractional portion may not lead to a finite string of 0s and 1s. In this case, an approximate answer is the best that can be obtained.

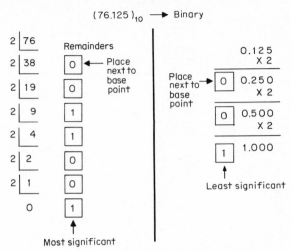

**Fig. 13-20**   Conversion from base 10 to binary. (Problem 13.6.)

## 13.13   NOTATION

1. Binary numbers are always written with the lowest weighted column on the right.

2. Logic expressions may appear in alphabetical order, for example, $ABCD$, or in reverse order as $D\overline{C}BA$.

3. It is usually convenient, when listing all possible combinations of 1s and 0s in a truth table, to proceed in numerical order using the binary number system. Heading may be ABCD or DCBA.

4. $BA = 1$ is shorthand for $B \cdot A = 1$.

5. $BA = 10$ is shorthand for $B = 1$, $A = 0$.

## 13.14   THE 16 POSSIBILITIES FOR TWO-INPUT LOGIC

Two inputs, $A$ and $B$, can be combined in no more than 16 ways, as shown in Table 13-3. This table was constructed using the binary number system as an aid in systematically

**TABLE 13-3   Listing of the 16 Possible Functions of Two Variables; Equivalent to 16 Truth Tables**

| AB | $F_0$ | $F_1$ | $F_2$ | $F_3$ | $F_4$ | $F_5$ | $F_6$ | $F_7$ | $F_8$ | $F_9$ | $F_{10}$ | $F_{11}$ | $F_{12}$ | $F_{13}$ | $F_{14}$ | $F_{15}$ | Weight |
|---|---|---|---|---|---|---|---|---|---|---|---|---|---|---|---|---|---|
| 00 | 0 | 1 | 0 | 1 | 0 | 1 | 0 | 1 | 0 | 1 | 0 | 1 | 0 | 1 | 0 | 1 | (1) |
| 01 | 0 | 0 | 1 | 1 | 0 | 0 | 1 | 1 | 0 | 0 | 1 | 1 | 0 | 0 | 1 | 1 | (2) |
| 10 | 0 | 0 | 0 | 0 | 1 | 1 | 1 | 1 | 0 | 0 | 0 | 0 | 1 | 1 | 1 | 1 | (4) |
| 11 | 0 | 0 | 0 | 0 | 0 | 0 | 0 | 0 | 1 | 1 | 1 | 1 | 1 | 1 | 1 | 1 | (8) |

listing all possibilities—not only for $A$ and $B$ but for the 16 possible functions listed as $F_0$ through $F_{15}$. For example, $F_3$ heads a column containing 0011, the binary representation for 3. (The most significant bit is at the left in 0011 and at the bottom of the $F_3$ column.)

Figure 13-21 illustrates the implementation for the 16 $F$'s. Small circles drawn at a gate's input are used as shorthand notation for an inverter (usually external). Small circles at the output refer to internal inversion and are part of the definition of the gate. For example, an AND gate symbol with a circle added at the output becomes a NAND gate symbol.

| Function | Implementation | Alternate |
|---|---|---|
| $F_0 = 0$ | Trivial   This would be a NEVER gate ! | |
| $F_1 = \overline{A+B} = \overline{A} \cdot \overline{B}$ | 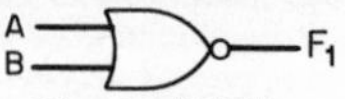 A B → $F_1$   Standard NOR | 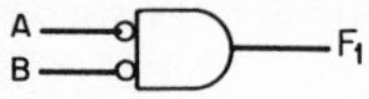 A B → $F_1$ |
| $F_2 = \overline{A} \cdot B$ | 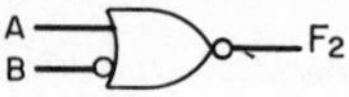 A B → $F_2$ | 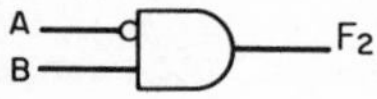 A B → $F_2$ |
| $F_3 = \overline{A}\,\overline{B} + \overline{A}B = \overline{A}$ | Trivial 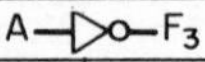   A →$F_3$ | |
| $F_4 = A \cdot \overline{B}$ | 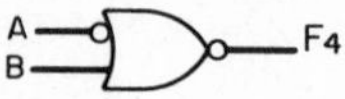 A B → $F_4$ | 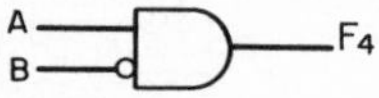 A B → $F_4$ |
| $F_5 = \overline{A} \cdot \overline{B} + A \cdot \overline{B} = \overline{B}$ | Trivial    B →$F_5$ | |
| $F_6 = \overline{A} \cdot B + A \cdot \overline{B}$ Exclusive – OR | 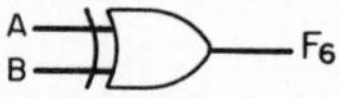 A B → $F_6$   Standard EXCLUSIVE–OR | 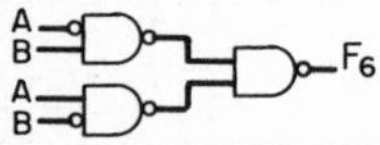 A B A B → $F_6$ |
| $F_7 = \overline{AB} = \overline{A} + \overline{B}$ | 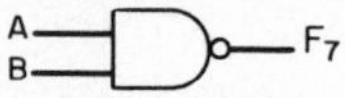 A B → $F_7$   Standard NAND | 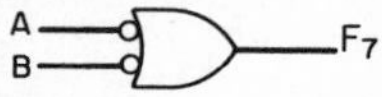 A B → $F_7$ |
| $F_8 = A \cdot B$ | 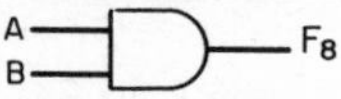 A B → $F_8$   Standard AND | 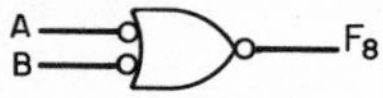 A B → $F_8$ |
| $F_9 = \overline{A} \cdot \overline{B} + A \cdot B$ | Complement the output of an EXCLUSIVE–OR gate | 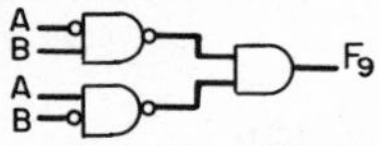 A B A B → $F_9$ |
| $F_{10} = \overline{A} \cdot B + A \cdot B = B$ | Trivial | |
| $F_{11} = \overline{A} + B$ | 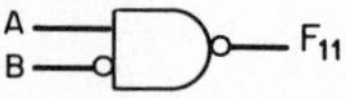 A B → $F_{11}$ | 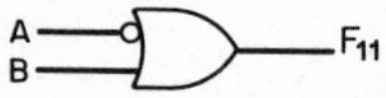 A B → $F_{11}$ |
| $F_{12} = A$ | Trivial | |
| $F_{13} = A + \overline{B}$ | 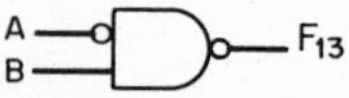 A B → $F_{13}$ |  A B → $F_{13}$ |
| $F_{14} = A + B$ | 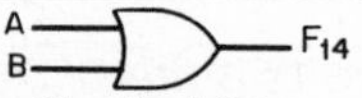 A B → $F_{14}$   Standard OR | 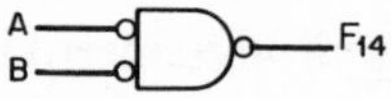 A B → $F_{14}$ |
| $F_{15} = 1$ | Trivial (ALWAYS gate) | |

**Fig. 13-21**   Implementation of the 16 logic functions of two variables.

## 13.15   SAVING INVERTERS

Two inversions in a row cancel out. Figure 13-22 shows how an output inversion (internal to gate—see the previous section) can cancel an input inversion requiring an external inverter. One inverter is saved.

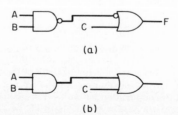

(a)

(b)

**Fig. 13-22**   (a) A 3-gate circuit (NAND, INVERTER, OR). (b) An equivalent 2-gate circuit (AND, OR).

## 13.16   COMBINATORIAL LOGIC DESIGN—SURVEY OF METHODS

Logic circuits discussed in previous sections, whose outputs depend only on present inputs, are called *combinatorial*. Combinatorial logic circuits may be designed using:

1. *Basic gates only.* Karnaugh mapping may be used to simplify logic expressions, as detailed in Secs. 13.17 to 13.19. This approach is the most economical solution for problems of limited complexity.

2. *Data selectors (multiplexers).* See Sec. 13.22.

3. *Field programmable logic arrays.* Integrated circuits are available containing large numbers of gates and programmable elements. (A programmable array is one which has internal connections that can be permanently made externally.) For example, a 24-pin Intersil IM5200 contains enough gates to implement expressions in "product term" form $(AB\overline{C}D + \overline{A}BCD + $ etc.) directly, up to 48 terms with 14 variables.

4. *Read-only memories (ROMs).* The ROM is most suitable for combinatorial logic when several variables (for example, 8) are functions of several other variables (for example, 8) involving a large truth table. It permits a one-package solution with no design work. The code is simply written once into the device (by field programming or at time of fabrication).

## 13.17   SIMPLIFICATION OF LOGIC EXPRESSIONS USING KARNAUGH MAPS

*Definition*   A Karnaugh map (K-map) is an array of squares labeled to represent a truth table. *Each square represents one line of a truth table.* A function plotted on a K-map appears as a geometric pattern, enabling one to take advantage of the human brain's ability to work with patterns when simplifying expressions.

Figure 13-23a shows a 2-variable truth table and an equivalent 2-variable K-map. Different labelings are indicated in Fig. 13.23c and d. Function $F = AB$, which appears in a vertical column as 0001 in the truth table, appears as just a 1 in the lower right-hand square of the K-map. Blank squares are understood to be 0s if labeled squares contain only

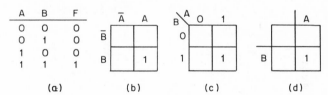

**Fig. 13-23**   (*a*) A 2-variable truth table. (*b*) An equivalent 2-variable K-map. (*c*) and (*d*) The same map labeled in other ways.

1s. Figure 13-24 shows five equivalent 3-variable maps. The term $ABC$, for example, would be entered as a 1 in the lower right corner of any of these maps. In Fig. 13-24e, variables $C, B, A$ are weighted 4, 2, 1, respectively, to determine an identifying number for each square. The term $\overline{A}BC$ would be written $CB\overline{A}$ (or read backwards) and associated with the binary number 110 = 6 for entry in square number 6.

Equivalent 4-variable maps are illustrated in Fig. 13-25. A 5-variable map can be constructed simply by duplicating a 4-variable map and placing it side by side with the original, as in Fig. 13-26. The original map represents $\overline{E}$ territory, the new map, $E$ territory.

A 6-variable map may be constructed by duplicating a 5-variable map and placing it underneath the original. The original 5-variable map represents $\overline{F}$ territory, the new map, $F$ territory, as shown in Fig. 13-27. Grouping of terms, to be discussed below, is also indicated.

*Logic Adjacency*   The labeling rule that we followed in constructing maps is that squares which are logically adjacent (i.e., squares that represent terms differing by one variable such as $ABC$ and $AB\overline{C}$) must be easily identifiable, as follows:

*2-Variable Maps*   Squares that touch edges (horizontally or vertically) are logically adjacent.

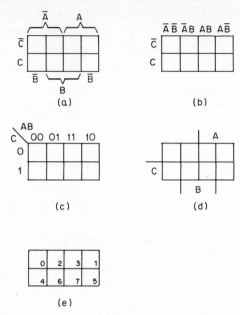

**Fig. 13-24** A 3-variable K-map labeled five different and equivalent ways.

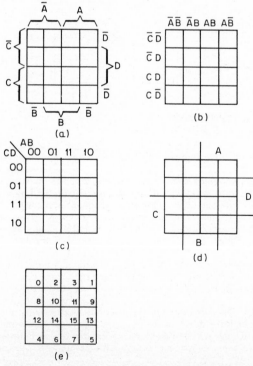

**Fig. 13-25** A 4-variable K-map labeled five different ways.

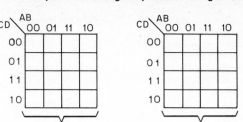

**Fig. 13-26** A 5-variable K-map obtained by placing two 4-variable maps side by side. Location of logically adjacent squares is described in text.

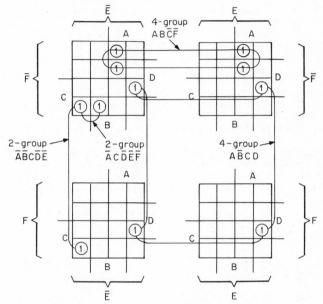

**Fig. 13-27** A 6-variable K-map consisting of two 5-variable maps (four 4-variable maps). Grouping of terms for function $F_8$ is illustrated. (Problem 13.7.)

*3- and 4-Variable Maps*   In addition to squares that touch as above, squares at extreme top and bottom of a column are logically adjacent, and squares at exteme left and right of a row are logically adjacent.

*5-Variable Maps*   In addition to logic adjacencies found as above, squares in corresponding positions (for example, 3d row, 2d column) in the two 4-variable maps used to construct these maps are logically adjacent.

*6-Variable Maps*   Same as 5-variable except that the two corresponding positions may be opposite each other horizontally or vertically in two of the four 4-variable maps used. With the numbered-squares method, the 64 squares would be numbered 0 to 63, and squares whose numbers differ by a power of 2 would be logically adjacent.

*Warning:* If the maps are constructed in a different way, then logically adjacent squares will appear in different positions. For example, assume that a 5-variable map is generated by folding the 4-variable map about its right-hand edge, rather than duplicating it. In this case, logically adjacent squares will appear in mirror-image positions about the folding axis, rather than in corresponding positions. This is also a popular procedure.

*Over Six Variables*   Maps of more than six variables are usually regarded as too tricky, and simplification using the computer can be considered.* Usually, if one deals with

*See Shiva and Nagel, "Bypass Multivariable Karnaugh Maps," *Electronic Design,* **22**:21, Oct. 11, 1974.

more than six variables, e.g., an 8-bit code, the simpler design techniques utilizing data selectors, logic arrays, or read-only memories are more appropriate.

*Simplification Rules—Grouping of Terms* Karnaugh-map simplification of Boolean functions is based on the theorem $ABC + AB\overline{C} = AB$. Terms that differ by one variable may be combined with the differing variable dropped. This leads to the following rules:

1. Two 1s that are logically adjacent (a group of two or 2-group) indicate that two terms may be combined with the one differing variable dropped.

2. Two groups of two that are logically adjacent to each other form a group of four (4-group) indicating that the four original terms may be combined into one term with two variables dropped.

3. Two groups of four logically adjacent indicate that the eight 1s represent eight terms that can be combined into one term with three variables dropped. (In general, if $2^N$ terms are logically adjacent to one another, then $N$ variables drop out.)

4. Each 1 must be used at least once when terms represented are written. If it cannot be combined into a group, the term it represents must be included in the logic expression.

5. A 1 may be used more than once if this yields a simpler result.

6. A term corresponding to a group of 1s which consists of 1s, each and every one of which is already part of a different group, is redundant and may be dropped.

7. Complement approach: Instead of finding $F$ by combining 1s, one can find $\overline{F}$ by combining 0s. This may give a simpler result, and this possibility should be checked. After $\overline{F}$ is obtained, $F$ can be generated by adding an inverter, or in some cases, by removing an inverter. For example, if you needed $F$ as the input $A$ for the alternate realization of $F_2$ in Fig. 13-21, you could use $\overline{F}$ and remove one inverter. (Tricks like this are sometimes regarded as use of negative logic, but there is no need to think of them this way.)

8. Don't cares: In some design problems a given combination of inputs either cannot occur or indicates a condition for which you do not care whether the logical output is a 0 or a 1. In these cases, you can enter on the K-map some symbol, such as $\Phi$, $d$, or $X$, which can later be chosen to be 0 or 1, whichever gives a simpler expression. (A don't care cannot be used as *both* a 0 and a 1.)

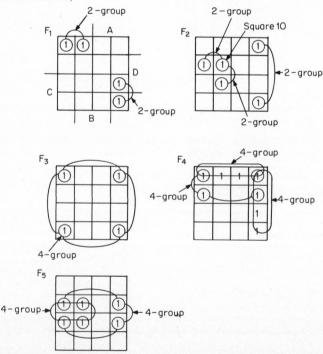

**Fig. 13-28**  Four-variable maps illustrating grouping of terms. (Problem 13.7.)

## 13.18  EXAMPLES OF SIMPLIFICATION USING KARNAUGH MAPS

**problem 13.7**    The following functions are to be simplified: (a) $F_1$ through $F_5$ as plotted on the maps of Fig. 13-28. (b) $F_6$ and $F_7$ as specified in truth table Fig. 13-29a. (c) $F_8$ as plotted on the 6-variable map of Fig. 13-27.

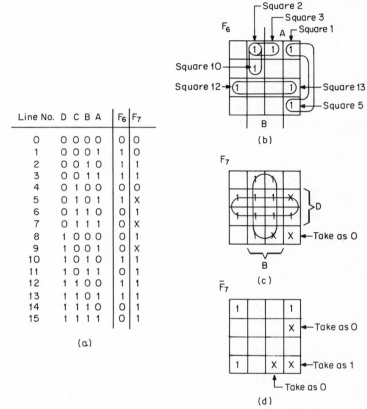

| Line No. | D C B A | $F_6$ | $F_7$ |
|----------|---------|-------|-------|
| 0 | 0 0 0 0 | 0 | 0 |
| 1 | 0 0 0 1 | 1 | 0 |
| 2 | 0 0 1 0 | 1 | 1 |
| 3 | 0 0 1 1 | 1 | 1 |
| 4 | 0 1 0 0 | 0 | 0 |
| 5 | 0 1 0 1 | 1 | X |
| 6 | 0 1 1 0 | 0 | 1 |
| 7 | 0 1 1 1 | 0 | X |
| 8 | 1 0 0 0 | 0 | 1 |
| 9 | 1 0 0 1 | 0 | X |
| 10 | 1 0 1 0 | 1 | 1 |
| 11 | 1 0 1 1 | 0 | 1 |
| 12 | 1 1 0 0 | 1 | 1 |
| 13 | 1 1 0 1 | 1 | 1 |
| 14 | 1 1 1 0 | 0 | 1 |
| 15 | 1 1 1 1 | 0 | 1 |

(a)

**Fig. 13-29**    (a) Truth table for functions $F_6$ and $F_7$ used to illustrate K-map simplification in Sec. 13.18. (b) and (c) Corresponding maps. (d) Map of complement of $F_7$. (Problem 13.7.)

**solution**    (a) For $F_1$ there are two groups of two. The ABCD labeling shown is convenient in going from a map to a function expressed in letters. The group of two 1s in the upper left corner spans B and $\overline{B}$. Therefore, B drops out leaving $\overline{A}\,\overline{C}D$ for this term. The group in the lower right corner spans D and $\overline{D}$; D drops out leaving $A\overline{B}C$. Thus $F_1 = \overline{A}\,\overline{C}D + A\overline{B}C$.

Similarly, $F_2 = A\overline{B}\,\overline{D} + \overline{A}CD + \overline{A}BD$. Note that the 1 in square no. 10 is used twice in order to form two groups of two. Function $F_3 = \overline{B}\,\overline{D}$ because the four 1s in the corners are logically adjacent to each other (ends of columns and ends of rows) and form a group of 4; A and C are the two variables that are spanned and drop out. Function $F_4$ contains three groups of 4; $F_4 = A\overline{B} + \overline{C}D + \overline{B}C$. Note how the 1 in the upper right corner is used three times to achieve simplicity. Similarly, $F_5 = \overline{A}D + BD$.

(b) Function $F_6$ given by the truth table in Fig. 13-29a may be plotted on a K-map in various ways. Term $\overline{D}\,\overline{C}\,\overline{B}A$ can be plotted by finding the square which is in A territory but not in D, C, B territory (Fig. 13-25d labeling). Or we can read 0001 from right to left and locate the 10 column and the 00 row (with Fig. 13-25c labeling). We also can use the prenumbered squares approach of Fig. 13-25e and immediately associate 0001 with square no. 1. The result, with any of these methods, is the map of Fig. 13-29b with 1s in squares 1, 2, 3, 5, 10, 12, and 13. *Once we have the map, we can use a different labeling scheme as an aid in writing the simplified expression.* (Figure 13-25d labeling is recommended.)

There are four groups of two 1s each as shown. The 1 in square no. 2 is used twice to permit two groups to be formed. Also, the 1s in squares 13 and 5 could form a group together. This term, however, would be redundant; each of its 1s has already been used. (See rule no. 6, in the previous section.) Reading of the terms using the $ABCD$ labeling yields $F_6 = D\overline{B}A + \overline{D}\overline{C}B + \overline{C}B\overline{A} + DC\overline{B}$.

If the don't cares are taken as 1s, the map for $F_7$ shows two groups of eight. Three variables, therefore, drop from each term yielding $F_7 = D + B$. The complementary map yields an even faster solution. Only the four corners must be expressed, and they combine yielding $\overline{F_7} = \overline{D}\overline{B}$. Taking the complement of both sides of the equation yields $F_7 = D + B$.

(c) Function $F_8$ is plotted on the 6-variable map of Fig. 13-27. This map may be regarded as two 5-variable maps or as four 4-variable maps. The groupings indicated are two 4-groups and two 2-groups. The upper 4-group spans $D$ and $\overline{D}$ as well as $E$ and $\overline{E}$. Therefore, $D$ and $E$ drop out of the term this group represents, leaving $AB\overline{C}F$. The lower 4-group spans $E$ and $\overline{E}$, $F$ and $\overline{F}$, leaving $A\overline{B}CD$. Including the two 2-groups yields $F_8 = AB\overline{C}\overline{F} + A\overline{B}CD + \overline{A}C\overline{D}EF + \overline{A}\overline{B}C\overline{D}E$.

## 13.19    RESTRICTIONS OF KARNAUGH MAPPING

It is sometimes stated that Karnaugh mapping gives a minimum-gate solution. This is true when you have one logic function $F_1$ and you examine both $F_1$ and $\overline{F}_1$. However, the following points must be remembered.

1. If you have two variables, say $F_1$ and $F_2$, to be implemented as functions of $A$, $B$, $C$, $D$, the minimum-gate solution may be achieved by introducing an intermediate variable $F$ containing terms common to $F_1$ and $F_2$. Alternatively, $F$ may be introduced if it contains terms in $F_1$ while $\overline{F}$ contains terms in $F_2$. Karnaugh mapping is a great aid in discovering common terms and complementary terms, but the results depend on the skill of the user.

2. A minimum-gate solution is not the same as a minimum-package solution. Using integrated circuits, a solution requiring far more gates than minimum may require fewer packages. For example, a data selector is a one-package solution to many problems.

3. A minimum-gate realization of an equation may not be an optimum solution to the real-world design problem in terms of parts, design, and assembly costs. One silicon controlled rectifier may yield a better solution than the gates, flip-flop, and power stage that may have been considered.

There is no mechanical procedure which can substitute for ingenuity. The good designer keeps Karnaugh mapping in its proper perspective as a useful design tool worth learning, but is aware of its limitations.

## 13.20    BINARY HALF-ADDER AND FULL-ADDER

Addition of binary numbers is illustrated in Fig. 13-30. Adding two zeros gives a zero; adding a zero and a 1 or a 1 and a zero gives a 1; adding two 1s gives a zero as the sum digit and generates a carry to the next column. A carry added to two zeros gives a 1; a carry plus a zero and a 1 gives a zero and generates a new carry; a carry plus two 1s gives a 1 as the sum digit and generates a carry. In the binary number system a carry always goes to a column weighted twice as heavily as the column which generated the carry. Whenever the answer is 2, which cannot be written with a single symbol in binary, the answer becomes 0 with a carry. Similarly, if the answer is 3, it becomes 1 with a carry.

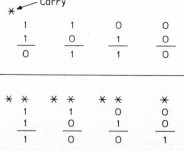

**Fig. 13-30**  Binary addition illustrated.

**problem 13.8**   Design a two-input logic circuit, called a *half-adder*, that adds two binary digits.

**solution**   Refer to Fig. 13-31*a*; the first step is to generate a truth table. Symbols $A$ and $B$ represent the digits to be added, $S$, the sum digit, and $C$, the carry bit ($C = 1$ indicates a carry). The table gives $S = \overline{A}B \oplus A\overline{B} = A + B$ (EXCLUSIVE-OR) and $C = AB$. A half-adder, therefore, can be implemented with just one EXCLUSIVE-OR gate and one AND gate, as shown in Fig. 13-31*b*. The symbol for a half-adder is provided in Fig. 13-31*c*.

| A | B | S | C |
|---|---|---|---|
| 0 | 0 | 0 | 0 |
| 0 | 1 | 1 | 0 |
| 1 | 0 | 1 | 0 |
| 1 | 1 | 0 | 1 |

(a)

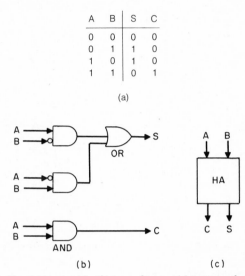

(b)                              (c)

**Fig. 13-31**   Half-adder. (*a*) Truth table. (*b*) Logic diagram. (*c*) Logic symbol. (Problem 13.8.)

**problem 13.9**   Design a three-input logic circuit, called a *full-adder*, that adds two binary digits and an input carry.

**solution 1**   Use two half-adders to handle the three inputs as shown in Fig. 13-32*a*. An OR gate produces the carry if either of the half-adders indicates that one is to be generated. The symbol for a full-adder is given in Fig. 13-32*b*.

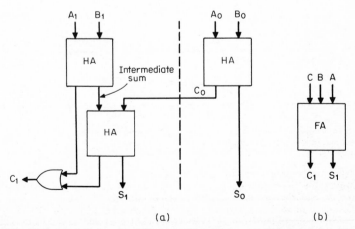

(a)                              (b)

**Fig. 13-32**   Full-adder built from two half-adders. (*a*) Logic diagram. (*b*) Logic symbol. (Problem 13.9.)

**solution 2**    Construct a truth table and K-maps as in Fig. 13-33. Sum $S = \overline{A}\,\overline{B}C + \overline{A}B\overline{C} + A\overline{B}\,\overline{C}$ $+ ABC$ and carry $D = AB + BC + CA$ are obtained. The sum can also be expressed in terms of $D$ as $S = (A + B + C)\overline{D} + ABC$.

| Line No. | C | B | A | S | D |
|----------|---|---|---|---|---|
| 0 | 0 | 0 | 0 | 0 | 0 |
| 1 | 0 | 0 | 1 | 1 | 0 |
| 2 | 0 | 1 | 0 | 1 | 0 |
| 3 | 0 | 1 | 1 | 0 | 1 |
| 4 | 1 | 0 | 0 | 1 | 0 |
| 5 | 1 | 0 | 1 | 0 | 1 |
| 6 | 1 | 1 | 0 | 0 | 1 |
| 7 | 1 | 1 | 1 | 1 | 1 |

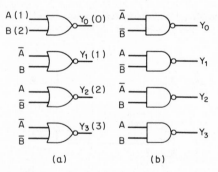

**Fig. 13-33**    Truth table and K-maps for sum $S$ and output carry $D$ for a full-adder with input carry $C$ and bits $B$ and $A$.

Binary adders and all types of arithmetic units are available as integrated circuits.

## 13.21    DECODERS/DEMULTIPLEXERS

A circuit which allows a binary number on control lines to select an individual output line (for example, $BA = 11$ selects output line no. 3), is called a *decoder*. There are two basic types as shown in Fig. 13-34: selected output line high and selected output line low.

A (1)
B (2)          $Y_0$ (0)          $\overline{A}$          $Y_0$
                                   $\overline{B}$

$\overline{A}$          $Y_1$ (1)          A          $Y_1$
B                                          $\overline{B}$

A          $Y_2$ (2)          $\overline{A}$          $Y_2$
$\overline{B}$                              B

$\overline{A}$          $Y_3$ (3)          A          $Y_3$
$\overline{B}$                              B

(a)                                        (b)

**Fig. 13-34**    Decoder circuits (2-line to 4-line). (*a*) Selected output line high, NOR gates used. (*b*) Selected output line low, NAND gates used.

Assume that output lines 10 to 15 are removed from a 4-line to 16-line decoder. Remaining, therefore, are output lines 0 to 9. The result is a 4-line to 10-line decoder, called a *decimal decoder* or *binary-to-decimal converter*.

A decoder can also be wired, with or without external gates (Fig. 13-35), so that the selected line is controlled by an input data line. That is, control lines select the output line to which one input data line will be gated. Used in this manner, the circuit is called a *demultiplexer*. Multiplexing is discussed in the next section. Many decoders/demultiplexers are available as integrated circuits.

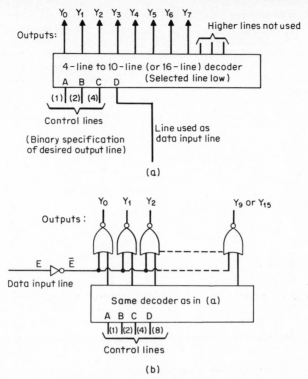

**Fig. 13-35** (*a*) Wiring of a decoder for "demultiplex" function to switch one input line to one of several possible output lines. (*b*) Addition of individual line gates to increase the number of usable output lines.

## 13.22   DATA-SELECTORS/MULTIPLEXERS

A data-selector (*multiplexer*) is a circuit that allows the user to select which of several input lines is gated to a single output line. Figure 13-36 shows a 4-line to 1-line multiplexer and its truth table. Lines $B$ and $A$ select which of the lines $D_0, D_1, D_2, D_3$ will be connected to output line $Y$. Circuits of this type are available in a one-package integrated circuit. They have two basic applications:

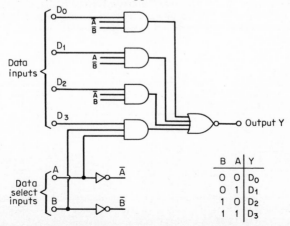

**Fig. 13-36** Internal wiring diagram of a data selector (4-line to 1-line) and its truth table.

## Logic Function Generation

In Fig. 13-36 $D_0, D_1, D_2, D_3$ can each be permanently wired to logic 0 or logic 1 levels, thus making $Y$ any desired function of $A$ and $B$. With 16-line to 1-line data selectors, the output can be made any function of control lines $ABCD$. This is a one-package solution requiring essentially no design time and is easy to modify. The selector size required (number of input lines) can be cut in half using a *folding* approach which usually requires an extra inverter.

**problem 13.10**   Wire a data selector using the folding technique to implement the function $F$ given in the truth table of Fig. 13-37a.

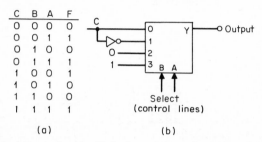

| C | B | A | F |
|---|---|---|---|
| 0 | 0 | 0 | 0 |
| 0 | 0 | 1 | 1 |
| 0 | 1 | 0 | 0 |
| 0 | 1 | 1 | 1 |
| 1 | 0 | 0 | 1 |
| 1 | 0 | 1 | 0 |
| 1 | 1 | 0 | 0 |
| 1 | 1 | 1 | 1 |

(a)                                    (b)

**Fig. 13-37**   (a) An 8-line truth table. (b) Implementation with a 4-input data selector and an inverter. (Problem 13.10.)

**solution**   Mentally set aside the highest weighted variable, $C$. Entries for $BA$ appear in pairs; for example, 00 in upper half of table $(C = 0)$ and 00 in lower half $(C = 1)$. With $BA = 00$, $F = 0$ when $C = 0$ and $F = 1$ when $C = 1$, that is, $F = C$ when $BA = 00$. Since $BA = 00$ selects input line no. 0, connecting $C$ to input line no. 0 takes care of two lines in the truth table (the $BA = 00$ lines). The same argument applies to $BA = 01$ except here $F = \overline{C}$. For $BA = 10$, $F = 0$ in both cases; for $BA = 11$, $F = 1$ in both cases. The simple wiring is shown in Fig. 13-37b.

## Multiplex Applications

In the most common multiplex application, input lines are sequentially scanned and gated to the output line. A binary counter driving the control lines can provide the sequential stepping. If each input line carries a digitized telephone conversation, several conversations can be communicated using a single line. On the other end of the line a demultiplexer (see previous section) sequentially switches the communication (data) line to individual output lines as shown in Fig. 13-38. Some type of synchronizer circuit must also be provided to ensure that the receiver's counter is periodically resynchronized with the transmitter's counter. The synchronizing event can be anything detectable, such as power line zero crossings or extra-wide pulses, on data line $Y$.

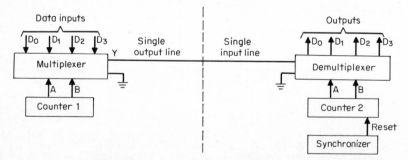

**Fig. 13-38**   A multiplex transmission system.

## 13.23 FLIPFLOPS: INTRODUCTION

Sequential logic circuits are those whose outputs depend on the sequence of inputs; that is, prior inputs influence present outputs. Such circuits must possess memory. The simplest sequential circuit is a flipflop (FF) (Fig. 13-39). It has two stable states, labeled the 0 and 1 states. It remains in a given state until it is made to change its state. Because it remembers which of the two states it is in, it stores one bit of information.

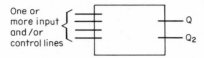

Fig. 13-39   A flipflop (FF), general case.

A main output line labeled $Q$ in Fig. 13-39 will be high when the FF is in the 1 state and low when the FF is in the 0 state (positive logic definition). Another output line labeled $Q_2$ is often present and yields the complement of $Q$.

*WARNING:* It is common practice to label the $Q_2$ line as $\overline{Q}$ even though $Q_2$ is not always the complement of $Q$. Many flipflops can be held indefinitely in a condition (not considered stable, as explained in the next section) where $Q_2 = Q$. This is sometimes called a *forbidden* condition, even though it may be exactly what is wanted in special cases.

There are two classes of flipflop; clocked (*synchronous*) and unclocked (*asynchronous*). With a clocked FF, one of the input/control lines is called a *clock* line. It provides synchronizing pulses that permit a change of state at certain times only.

## 13.24 UNCLOCKED R-S FLIPFLOPS (NOR GATE AND NAND GATE TYPES)

Figure 13-40 shows unclocked $R$-$S$ flipflops made from (*a*) NOR and (*b*) NAND gates. Symbol $R$ stands for *reset* ($Q$ goes to 0) and $S$ stands for *set* ($Q$ goes to 1). The basic truth table for an $R$-$S$ FF is given in Fig. 13-40c: $SR = 10$ sets the FF and $SR = 01$ resets it.

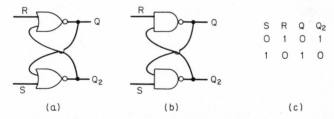

Fig. 13-40   *R-S* flipflops. (*a*) NOR implementation. (*b*) NAND implementation. (*c*) Basic truth table.

A detailed truth table for the NOR-gate $R$-$S$ FF is given in Fig. 13-41a. This type of FF remembers the previous state when $SR = 00$. Condition $SR = 11$ yields the special case $Q = Q_2 = 0$ (because a 1 input forces a 0 output from a NOR gate). If $SR = 00$ follows $SR = 11$, the result is either 01 or 10 (indeterminate). Some texts recommend ensuring that $S = R = 1$ never occurs with this type FF. If, however, $SR = 11$ is always followed by $SR = 01$ or 10 before attempting to remember the prior state with $SR = 00$, there will be no indeterminacy. Figure 13-41b gives the NAND-gate $R$-$S$ FF behavior. Here $SR = 11$ allows remembering the previous state. $SR = 00$ forces $Q = Q_2 = 1$. Note that the 1 state implies $Q = 1$. However, $Q = 1$ does not guarantee the 1 state if $Q = Q_2$ is allowed for this FF. For any FF, the $Q = Q_2$ condition is not considered a stable state because it cannot be remembered by the FF as can the 1 state ($QQ_2 = 10$) and the 0 state ($QQ_2 = 01$).

| S | R | Q | $Q_2$ |
|---|---|---|---|
| 0 | 0 | No change if previous $QQ_2 = 01$ or $10$ <br> indeterminate if previous $QQ_2 = 00$ (now 01 or 10) | |
| 0 | 1 | 0 | 1 |
| 1 | 0 | 1 | 0 |
| 1 | 1 | 0 | 0 |

(a)

| S | R | Q | $Q_2$ |
|---|---|---|---|
| 0 | 0 | 1 | 1 |
| 0 | 1 | 0 | 1 |
| 1 | 0 | 1 | 0 |
| 1 | 1 | no change if previous $QQ_2 = 01$ or $10$ <br> indeterminate if previous $QQ_2 = 11$ (now 01 or 10) | |

(b)

**Fig. 13-41**  Detailed truth table for $R$-$S$ flipflops made from ($a$) NOR and ($b$) NAND gates.

## 13.25  CLOCKED FLIPFLOPS: THE J-K FLIPFLOP

As defined in Sec. 13.23, a clocked FF utilizes pulses that permit a change of state at certain times only. The $R$ and $S$ (or equivalent) lines are examined prior to a change of state and can be regarded as *steering lines* that steer the FF into a desired state when the clock pulse permits a change of state. The most versatile clocked FF is the $J$-$K$ flipflop. Symbol $J$ corresponds to $S$ and $K$ to $R$, except that there are no indeterminate conditions and the device exists only in clocked form. Condition $JK = 00$ allows remembering of the prior state, while $JK = 11$ guarantees flipping into the state opposite the present one. All output changes occur only when permitted by a clock pulse.

Figure 13-42 shows a $J$-$K$ FF in block diagram form (with clock pulse line labeled $cp$), the truth table, and an alternative form of the truth table called an *excitation table*. In these tables, the subscript $n$ refers to the time before a clock pulse has allowed outputs to change, while the subscript $n + 1$ refers to the time after outputs have changed. To go from $Q_n = 0$ to $Q_{n+1} = 1$ requires $JK = 10$ (set) or $JK = 11$ (flip). The entry in the excitation table, therefore, is $1X$ where $X =$ don't care.

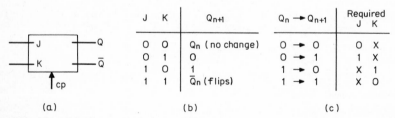

| J | K | $Q_{n+1}$ | $Q_n \rightarrow Q_{n+1}$ | | Required <br> J   K | |
|---|---|---|---|---|---|---|
| 0 | 0 | $Q_n$ ( no change) | 0 | → 0 | 0 | X |
| 0 | 1 | 0 | 0 | → 1 | 1 | X |
| 1 | 0 | 1 | 1 | → 0 | X | 1 |
| 1 | 1 | $\bar{Q}_n$ (flips) | 1 | → 1 | X | 0 |

(a)                              (b)                              (c)

**Fig. 13-42**  The $J$-$K$ flipflop. ($a$) Block diagram. ($b$) Truth table. ($c$) Excitation table (alternative form of truth table).

Clocking requirements for a $J$-$K$ FF vary with the design of the circuit. One manufacturer may offer four or more $J$-$K$ FF's within the same logic family which behave differently. The behavior depends on when $J$ and $K$ must be maintained stable and whether the outputs will change during, or just after, the leading or trailing pulse edge. How the FF behaves with inverted clock pulses (clock line normally high, going low for short pulse times) may also be a consideration.

## 13.26  SHIFT REGISTERS

A row of flipflops connected so that digital data can be shifted down the row is called a *shift register*. The most common type, using the $J$-$K$ FF, is illustrated in Fig. 13-43$a$. Assume that all the FF's are initially in the zero state. This can be achieved by pulsing a line connected to a direct reset terminal on each FF (if provided) or by shifting in a string of 0s.

Data entry is as follows. Suppose input line $W$ is initially $W_1$ (a 1 or a 0). Since $J = W$ and $K = \overline{W}$, after the first clock pulse, $A = W_1$. Now suppose the input becomes $W_2$. After the second clock pulse $A = W_2$ and $B = W_1$. The table of Fig. 13-43$b$ shows how the information is shifted down the line. Depending on hardware, the outputs will change during, or just after, the leading or trailing pulse edges.

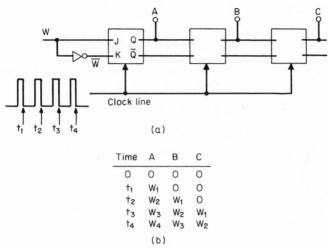

| Time | A | B | C |
|------|-----|-----|-----|
| 0 | 0 | 0 | 0 |
| $t_1$ | $W_1$ | 0 | 0 |
| $t_2$ | $W_2$ | $W_1$ | 0 |
| $t_3$ | $W_3$ | $W_2$ | $W_1$ |
| $t_4$ | $W_4$ | $W_3$ | $W_2$ |

(b)

**Fig. 13-43**    A shift register. ($a$) Wiring diagram and clocking waveform. ($b$) Tabulation of output versus time.

## 13.27    SHIFT REGISTER APPLICATIONS

### Serial-to-Parallel Conversion

If a 3-bit number, or code word, $W_3W_2W_1$ (each $W_i$ is 0 or 1) is presented to the input of a three-stage shift register and each input is entered via a clock pulse, then $W_3$, $W_2$, and $W_1$ will be simultaneously available in parallel at the flipflop output terminals.

### Parallel-to-Serial Conversion

If input gating is added to each FF so that information can be loaded in parallel, the information can then be shifted out serially.

### Multiplication and Division

Just as numbers in the decimal number system can be multiplied or divided by powers of 10 by shifting the radix point (or holding the decimal point position fixed and shifting the number), binary numbers can be multiplied or divided by powers of 2 by shifting. Additional gating may be used so that a register will shift right or left. Right/left IC shift registers are available.

### Time Delays and Data Buffering

Digital information can be intentionally delayed by passing it through a shift register. If parallel outputs are not needed, shift registers with hundreds of stages are easily fabricated with ICs. Long shift registers also can be used for data buffering. Information shifted in can be shifted out at a different time (after a waiting period) and at a different rate.

### Recirculating Memories

If the output of a shift register is wired back to the input, the data can be made to recirculate. At any output the data appears in serial form again and again.

### Self-Decoding Counters

If a cleared shift register is preloaded with a single 1 (stage no. 0), the single 1 will move one stage with each clock pulse. After three clock pulses, for example, only output no. 3 will be high. This is an example of a *self-decoding counter*. The disadvantage of this device, compared with counters discussed in the next section, is that far more FFs are required.

### 13.28   COUNTERS

### Introduction

If several *J-K* FFs are connected as in Fig. 13.44*a* or *b*, a binary counter is formed. With $CBA = 000$ initially, $CBA = 001$ after the first pulse, 010 after the second pulse, 011 after the third, and so on. With three FFs shown, $CBA = 111$ after the seventh pulse and 000 after the eighth. This is called a *modulo eight* counter.

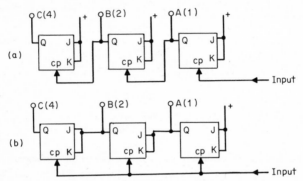

**Fig. 13-44**   (*a*) Asynchronous counter (also called a ripple counter). (*b*) Synchronous counter.

In general, a modulo $N$ counter has $N$ states, labeled 0 through $N$-1. It resets to 0 after the $N$th pulse. For a binary counter with $K$ FFs, $N = 2^K$. Table 13.1, which gives $K$ and $N = 2^K$, tells how many flipflops are required to count to a given number. For example, a three-FF counter resets on the eighth pulse; it can count up to 7 (binary 111). A four-FF counter resets on the fifteenth pulse; it can count up to 15 (binary 1111).

Signal flow in counter diagrams in this section has been drawn from right to left in order to make operation clearer. In the binary number system, the least significant digit is the rightmost digit; it changes with every count. It is 0 for all even numbers and 1 for all odd numbers. In Fig. 13-44 the rightmost flipflops, each with output line labeled $A$, change for every input pulse. Terminals $J$ and $K$ are returned to a positive voltage (logic 1).

### Ripple Counter Operation

The ripple counter of Fig. 13-44*a* yields binary counting if the *J-K* flipflops chosen are the type whose output changes are enabled by the trailing edge of clock pulses. All $J$'s and $K$'s are tied to logic 1. Therefore, each FF flips to the opposite state when its clock line falls from 1 to 0. Count 000 is followed by 001 followed by 010; the fall in $A$ causes the rise in $B$. Because 011 is followed by 100, the fall of $A$ causes the fall of $B$ which results in a rise of $C$.

The input pulse has to ripple through the full length of the counter. During the ripple time, outputs will be briefly incorrect, which is the main difficulty with this type of counter. For driving visual or mechanical loads, the brief "glitches" can be ignored. For circuits in which timing is critical, the outputs can be run through clocked gates so that one does not look at the counter output until all lines have settled. This is called *strobing*. If one is interested in decoding a particular state, a resynchronzer can be used. In this case, an earlier state is decoded in advance and the result is gated out at a particular time.

## Synchronous Counter Operation

In this counter (Fig. 13-44b), any type J-K FF may be used. All clock input lines are tied together, thereby allowing all flipflops to change synchronously (or nearly so). Each FF will either toggle (switch) or not toggle, depending on whether $J = K = 0$ or $J = K = 1$ before the outputs change. If A was 0, then the next pulse will not change B; if A was 1, then the next pulse will change B.

## Commercial Counter Features

Commercially available IC counters may exhibit the following features:

1. Internal reset, yielding a count other than $2^K$. For example, decade counters which count 0, 1, . . . , 9, 0 are readily available.

2. Up/down counters that have either a steering control (up/down) or two-pulse input lines (one for up pulses, one for down pulses) are available. (The counters examined here will count down if the $\overline{Q}$ outputs are used to drive adjacent FFs, instead of the Q outputs.)

3. Reset (to all 0s) and set (to all 1s, or to 1001 on decade counters) is usually provided. If a 2-input reset gate is available which resets on two 1s (as in the 7490 counter), the counter can be wired without additional gates to reset when any two output lines are high. For example, a decade counter may become a modulo-3 counter, if output lines $A(1)$ and $B(2)$ operate the reset gate. It becomes a modulo-6 counter if lines $B(2)$ and $C(4)$ are used and a modulo-9 counter if $A(1)$ and $D(8)$ are used. If the decade counter is a modulo-2 counter (1 FF) and a modulo-5 counter in series (as is the 7490), then the single package may be used as a 2-, 3-, 5-, 6-, 9-, or 10-base counter.

By adding FFs as needed, the design of Fig. 13-44b provides modulo-2, 4, 8, 16, 32, etc., counting without modification. With external gating operating a reset line, any count can be used to achieve reset. Two basic problems may be encountered: (a) Reset may be too brief with not all FFs reset. (b) With a ripple counter, temporarily incorrect outputs may cause a premature reset (partial or total).

## Divide-by-N Counters

Consider a modulo-3 binary counter initially in state 00. It counts 01, 10, 00. The return to 00 takes place after receipt of the third pulse. If the return to 00 is taken to be the output event, the circuit yields one output event for every three input pulses. This behavior is that of a divide-by-3 counter.

In general, a divide-by-N circuit need not be a weighted counter (counter with weighted output lines). The only requirement is that there be an output line which falls (or rises) after every N pulses.

Some commercial counters allow preloading of any number (e.g., from a register or thumbwheel switches) and then counting up, or down, to zero. (With some counters, the thumbwheel switch or register must load the complement of the number wanted.) Thus, N can be set to be any number within the range of the counter. If a divide-by-5 circuit is cascaded with a divide-by-30 circuit, a divide-by-150 circuit is obtained. Some counters, such as the MC4016, can feed carries and borrows to each other and reload the thumbwheel switch numbers in such a way that they become unit cascadable. For example, setting 5 on the units counter and 3 on the 10s counter yields division by 35, rather than 150.

## 13.29   A MODULO-3 BINARY COUNTER DESIGN

**problem 13.11**   Design a synchronously operated modulo-3 binary counter using J-K flipflops.

**solution**   The counter, when started in state 00, must count 01, 10, 00. In this design example we shall specify in advance that if the forbidden state 11 arises because of noise, it shall return to 00 upon receipt of the next pulse. Figure 13-45 provides the complete solution. The table of Fig. 13-45a lists the present state, the next desired state, and the required J's and K's, is called a *state transition table*. The Karnaugh maps for the J's and K's to aid in simplification are given in Fig. 13-45b, and the final wiring diagram in Fig. 13-45c.

The state transition table is constructed by referring to the J-K excitation table, Fig. 13-42c. Going from $A = 0$ to $A = 1$, for example, requires $JK = 11$ (for toggle) or $JK = 10$ (for set to 1). Therefore, the entry is $J = 1$, $K = X$ (don't care). While the map for $K_B$ yields $K_B = 1$ as the

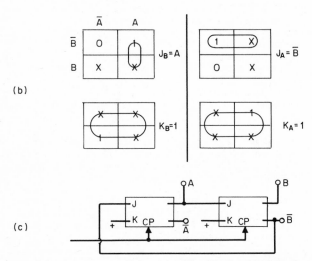

| State No. | Present state | | Next state | | Required J and K | | | |
|---|---|---|---|---|---|---|---|---|
| | B | A | B | A | $J_B$ | $K_B$ | $J_A$ | $K_A$ |
| 0 | 0 | 0 | 0 | 1 | 0 | X | 1 | X |
| 1 | 0 | 1 | 1 | 0 | 1 | X | X | 1 |
| 2 | 1 | 0 | 0 | 0 | X | 1 | 0 | X |
| 3 (forbidden) | 1 | 1 | 0 | 0 | X | 1 | X | 1 |

(a)

(b)

(c)

**Fig. 13-45** Modulo-3 binary counter design. (*a*) Tables. (*b*) Maps. (*c*) Wiring diagram. (Problem 13.11.)

simplest result, it shows that $K_B = \overline{A}$ is a less simple result. A wiring diagram with $K_B = \overline{A}$, however, is no more complicated than the one given. Since $\overline{A}$ is available, $K_B$ could be wired to $\overline{A}$ instead of to + (logic 1 level).

## 13.30 DESIGN OF COUNTER WITH ARBITRARY COUNT SEQUENCE

**problem 13.12** Deisgn a three-FF circuit to go through the list of states, or "count," in the following way: state 000, 001, 111, 011, back to 000. Expressed decimally, the count sequence is 0, 1, 7, 3 and then back to 0. (States 2, 4, 5, and 6 are skipped.)

**solution 1** Figure 13-46 provides the state transition table, K maps, and wiring diagram. The procedure is the same as in the previous section except that, for simplicity, no a priori choices are made as to transitions leading out of forbidden states. This is a common approach. What happens if the circuit initially or owing to noise goes into one of the forbidden states? Depending on the choice of 1s or 0s for the X's during simplification, the counter may loop through forbidden states and never return to a permitted state until externally reset.

If one wants to examine all possibilities after simplification, one can make a table listing all the forbidden states (as present states), the J's and K's as specified during simplification (all 1s and 0s, no X's), and the next state which would follow each present state. Then one can see if each forbidden state leads (directly or indirectly) to a permitted state, or to a loop, and make changes only where needed. In many practical applications, a manual reset button is all that is needed (or turning the power off and on; automatic reset when power comes on is often provided). In some critical applications, going into a forbidden state would cause an unacceptable error even for a brief entry. For these applications, all forbidden states could be detected and could be made to ring a bell, flash a warning light, halt the system, etc.

**solution 2** Another approach which cuts down the K mapping by a factor of 2, but usually leads to more complicated gating (six gates instead of two in this problem), is to use the toggle capability of J-K flipflops. Figure 13-47 gives the solution. Symbol $T_A$ represents the tied J and K inputs of

| State No. | Present state | | | Next state | | | Required J and K | | | | | |
|---|---|---|---|---|---|---|---|---|---|---|---|---|
| | C | B | A | C | B | A | $J_C$ | $K_C$ | $J_B$ | $K_B$ | $J_A$ | $K_A$ |
| 0 | 0 | 0 | 0 | 0 | 0 | 1 | 0 | X | 0 | X | 1 | X |
| 1 | 0 | 0 | 1 | 1 | 1 | 1 | 1 | X | 1 | X | X | 0 |
| 7 | 1 | 1 | 1 | 0 | 1 | 1 | X | 1 | X | 0 | X | 0 |
| 3 | 0 | 1 | 1 | 0 | 0 | 0 | 0 | X | X | 1 | X | 1 |

(a)

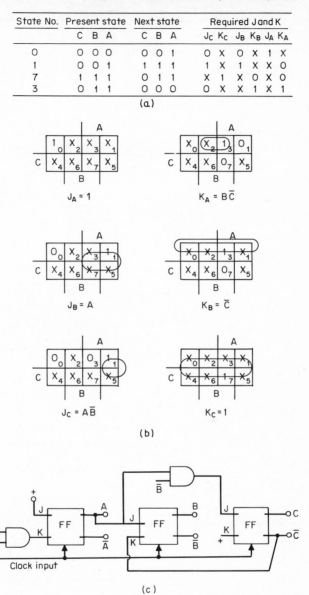

(b)

(c)

**Fig. 13-46**  (*a*) State transition table. (*b*) K-maps. (*c*) Wiring diagram. (Problem 13.12.)

flipflop A. When $T_A = 0$, it disables the separate clock input. When $T_A = 1$, the clock input is enabled, ensuring that each clock pulse toggles the FF. (A common notation, not being used here, represents the clock input of an always enabled toggle FF as $T$.) Each entry in the toggle function column is made by noting whether the corresponding variable must change when going from present state to next state. For example, to go from 000 to 001, only A must change. Hence, only $T_A$ must be 1 to achieve the change.

| State No. | Present state | | | Next state | | | Required toggle functions | | |
|---|---|---|---|---|---|---|---|---|---|
| | C | B | A | C | B | A | $T_C$ | $T_B$ | $T_A$ |
| 0 | 0 | 0 | 0 | 0 | 0 | 1 | 0 | 0 | 1 |
| 1 | 0 | 0 | 1 | 1 | 1 | 1 | 1 | 1 | 0 |
| 7 | 1 | -1 | 1 | 0 | 1 | 1 | 1 | 0 | 0 |
| 3 | 0 | 1 | 1 | 0 | 0 | 0 | 0 | 1 | 1 |

(a)

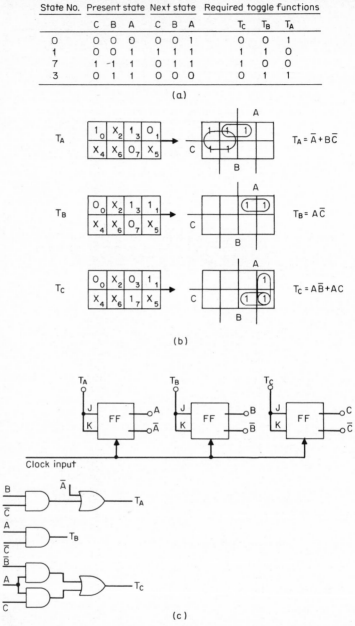

$T_A$     $T_A = \bar{A} + B\bar{C}$

$T_B$     $T_B = A\bar{C}$

$T_C$     $T_C = A\bar{B} + AC$

(b)

(c)

**Fig. 13-47** (*a*) State transition table using toggle function. (*b*) K-maps. (*c*) Wiring diagram. (Problem 13.12.)

## 13.31   WIRED-OR LOGIC

If a gate has output $C$ low under certain input conditions and another gate has output $E$ low under certain input conditions, can terminals $C$ and $E$ be wired together to go low when $C$ or $E$ is low? The answer is that it depends entirely on the hardware with which the gates are implemented. With most implementations, the equipment is destroyed if the logic asks one gate to pull high and the other to pull low. With some implementations (for example, TTL open collector logic), it is permitted and is called *wired-OR logic*. Two alternative ways of denoting gates that are wired-OR connected are shown in Fig. 13-48. This type of logic has been, to a great extent, superseded by tri-state logic.

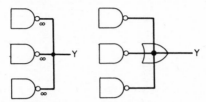

**Fig. 13-48**   Two alternative methods of denoting the wired-OR connection. Depending on hardware and application, a resistor connected from $Y$ to the power supply may also appear.

## 13.32   TRI-STATE LOGIC

The term *tri-state logic* refers to a hardware implementation where a gate can be high, low, or disconnected (i.e., floating at a high-impedance level). This allows gate outputs to be tied in parallel, with digital selection of the active gates. A data line between two digital devices can be bidirectional. In this case, either end can send or receive data if each end has input gates and tri-state output drivers tied to the line. To avoid a tug-of-war, care must be taken to ensure that two drivers on the same line are never activated simultaneously.

## 13.33   BIBLIOGRAPHY

Bartee, T. C.: *Digital Computer Fundamentals,* 4th ed., McGraw-Hill, New York, 1977.
Hill, F. J., and Peterson, G. R.: *Introduction to Switching Theory and Logical Design,* 2d ed., Wiley, New York, 1974.
Kline, R. M.: *Digital Computer Design,* Prentice-Hall, Englewood Cliffs, N.J., 1977.
Lancaster, D.: *TTL Cookbook,* Sams, Indianapolis, Ind., 1976.
O'Malley, J.: *Introduction to the Digital Computer,* Holt, New York, 1972.
Peatman, J. B.: *The Design of Digital Systems,* McGraw-Hill, New York, 1972.

Chapter **14**

# Computer-Aided Circuit Design

## H. JANET KELLY AND MICHAEL J. KELLY

IBM System Products Division, East Fishkill,
Hopewell Junction, N.Y.

## 14.1 INTRODUCTION

Ten years ago bench testing was the accepted modus operandi for the circuit designer, with an occasional use of computer-aided circuit design (CACD) programs to verify designs. The past decade, however, witnessed a profusion of CACD programs.[1] Today's programs permit the analysis of circuits with thousands of elements at speeds that make such analysis economically feasible. The circuit designer now accepts simulation as a way of life and CACD programs as a prerequisite for large-scale integration (LSI) design.[2]

The focus of this chapter is on analysis and use of CACD tools. The complexity, size of the circuits, and type of analysis to be performed will dictate the type of CACD program to be used.[3] CACD may be described as a means by which synthesis is realized through an iterative analytical process.

Many excellent CACD programs are in use in the electronics industry today. The literature[1,3,4] gives information on some of those more commonly used, including details on their availability, limitations, and analytical features. Only one program, ASTAP, is featured in this chapter. Proficiency in the use of the program, however, permits easy adaptation to almost any other user-oriented program.

## 14.2 ADVANCED STATISTICAL ANALYSIS PROGRAM*

The Advanced Statistical Analysis Program (ASTAP)[5] provides a CACD tool that incorporates the principal modes of analysis: direct current (dc), transient, and frequency (ac), into a single programming system capable of handling large nonlinear circuits. Additional major features include a general modeling capability and simplicity of use to permit its application without prior programming experience. There are no restrictions on the

*The authors wish to express appreciation to A. J. Jimenez, G. W. Mahoney, D. A. Mehta, H. Qassemzadeh, T. R. Scott, and W. R. Weeks for their contributions to ASTAP.

number or types of nonlinearities in the circuits to be analyzed, nor are there limitations on the size of networks apart from the limitations imposed by the host computer. ASTAP requires a minimum problem-program region of 200 kilobytes. The speed of execution and large problem-size capability of ASTAP are a consequence of the mathematical and programming techniques used in its development.[6-10]

Three basic parts common to all CACD programs are indicated in Fig. 14-1. The user first determines what is to be analyzed, the type of analysis desired, and the forms of the results to be attained. The input phase of the program is basically a discipline-oriented language for describing the circuit, specifying functional relationships, and identifying analysis options and the form of the output. The analysis phase incorporates the formulation techniques and mathematical content to support the analytical capabilities of the program. The output provides the tabulated and graphical results of the analysis, diagnostic messages, and other information supported by the program, such as the input listings and topological structure of the circuit. The user must then evaluate the results of the output.

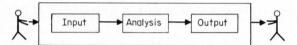

**Fig. 14-1**    Problem-solving through human and computer interaction.

The input to ASTAP is classified into discrete categories called *heading statements.* Models are named and described, execution and utility controls are identified, conditions for analysis are stipulated, and the overall organization of the input data is provided.

The complete list of ASTAP heading statements follows:

MODEL DESCRIPTION
MODEL
ELEMENTS
FUNCTIONS
FEATURES
EXECUTION CONTROLS
ANALYZE
RUN CONTROLS
INITIAL CONDITIONS
OPERATING POINT
OUTPUTS
RERUNS
UTILITY CONTROLS

## 14.3  MODELS

Every electrical schematic is a model of a circuit. It is intended to describe the physical reality of the circuit to be analyzed. An integrated circuit consists of many elements, including semiconductor devices. The semiconductor devices themselves are described as electrical models made up of basic electrical parameters. The circuit designer usually employs available electrical models of semiconductor devices and, in most instances, calls these models from a model library where they were previously stored. This is analogous to a programmer calling a subroutine from a computer library rather than actually writing the subroutine.

The accuracy of the analysis of an integrated circuit is directly proportional to the accuracy of the semiconductor device models. It is important, therefore, that the CACD program used have available an associated library of accurate device models, as well as a mechanism for easily changing stored models and introducing new models into the library.

## 14.4  ASTAP LANGUAGE CONVENTIONS

The input to ASTAP consists of language statements for describing the circuit. The heading statements MODEL DESCRIPTION, EXECUTION CONTROLS, and UTILITY CONTROLS divide the input data into three major sections. The MODEL

DESCRIPTION and EXECUTION CONTROLS sections contain lower-level heading statements. There is no program limitation on the number of statements that may appear under a particular heading, nor is there any restriction on the order in which the major headings appear in the input data. Any number of comment statements may appear following MODEL DESCRIPTION, MODEL, EXECUTION CONTROLS, and ANALYZE.

## MODEL DESCRIPTION Heading

Under the MODEL DESCRIPTION heading, all the models that constitute the circuit to be analyzed must be described, beginning with the main model designated for analysis. Within the main model, additional models may be used, and these, too, must be described under a MODEL heading unless the model identified is in the program library. Any model hierarchy may be used. That is, model $A$ may contain model $B$; model $B$ may contain models $C$ and $D$, and so on, to any level.

## ELEMENTS Heading

The statements that describe the elements and model references follow the ELEMENTS heading. They may be listed in any order and include the element, mutual inductance, parameter, and model reference statements. The general form is:

> element name, from node-to node = element value

The element name may be any unique set of alphanumeric characters, but the first character must be a key letter indicating the type of element, as follows:

| Key letter | Type |
|---|---|
| R | Resistance |
| L | Inductance |
| C | Capacitance |
| G | Conductance |
| M | Mutual inductance |
| E | Voltage source |
| J | Current source |

The FROM node and TO node are the assigned node names of the element. In the ASTAP program, voltage is assumed positive at the FROM node with respect to the TO node in all elements, except the voltage source. For the voltage source, the direction of increasing potential corresponds to the direction of assumed positive current flow. The reader should be cautious in the use of the conventions (Fig. 14-2), which are peculiar to ASTAP. Values of elements may be specified as constants, tables, equations, or expressions.

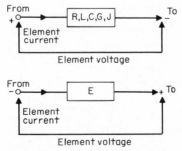

**Fig. 14-2**   ASTAP element conventions.

Under the ELEMENTS heading, the model reference statement must contain the reference name corresponding to the unique name assigned to a particular model at the time it is used, or at the time it was stored, in the model library. The general form is:

> reference name = MODEL   model name (node-node- . . . -node)

where reference name corresponds to a particular device, such as T1; model name identifies the model to be inserted in the calling circuit, such as 5D123; and the node list contains the names of the nodes to which the referenced model is connected.

## MODEL Heading

The function of the MODEL heading statement is to designate the start of a model description. It contains the three subheadings (ELEMENTS, FUNCTIONS, and FEATURES) to be used as required to describe the models. The first model identified by a unique name such as MODEL AMPLIFIER ( ) must contain the open and closed parentheses which identify this model as the circuit model to be analyzed. All other models described under the heading model are followed by a unique model name, the disposition (if required), and the external nodes of the model, as in the following general form:

> MODEL    model name (disposition) (node-node- . . . -node)

The model name must be unique and may be any combination of alphanumeric characters. That is, the model name may be TRANS, TR5213, or any other combination of alphanumerics. The disposition field, which includes the parentheses, is an optional field. When it is included, it controls the disposition of the model with respect to a model library via the use of one of the keywords, RETAIN, REPLACE, DELETE, or PRINT.

The disposition keyword RETAIN directs the program to save the model named in the model library. Once retained, the model may be referenced in any future model description without the model statements needing to be included in the input data. The keyword REPLACE directs the program to replace a model in the library with the new model of the same name. The keyword DELETE directs the program to remove the model from the library following its use in the specified analysis. PRINT causes the model to be printed as part of the input listing.

The names of the nodes are those that connect the model to another circuit and are listed in the external node list (node-node- . . . -node).

A parameter statement is used to define an auxiliary circuit quantity and may appear anywhere under the elements heading. A parameter is similar to an element except that it has *no nodal connections* and its meaning is defined by the user. Parameters must begin with one of the following letters: A, B, F, H, K, O, P, U, V, W, Y, or Z.

The use of element values, models, and parameters will be illustrated and further explained in the examples that follow. It should be noted that no units need be specifically identified with the element values. The units are implicit and are assumed to be a consistent set, as in Table 14-1.

**TABLE 14-1    Consistent Sets of Units**

| Quantity | Units (1) | Units (2) | Units (3) |
|---|---|---|---|
| Voltage | volt | volt | millivolt |
| Current | milliamp | milliamp | milliamp |
| Resistance | kilohm | kilohm | ohm |
| Conductance | millisiemens | millisiemens | siemens (mho) |
| Inductance | microhenry | henry | microhenry |
| Capacitance | picofarad | microfarad | microfarad |
| Time | nanosecond | millisecond | microsecond |
| Frequency | gigahertz | kilohertz | megahertz |

## FUNCTIONS Heading

The FUNCTIONS heading is used to supply definitions for certain items such as tables and equations, as illustrated in the example problems. The general form for the table description is:

> TABLE    table name, x,y, x,y, . . . x,y

where TABLE is the keyword (it may be abbreviated by the letter T). Table name is the unique name assigned to the table, and *x,y,* are the coordinate points with *x* as the independent variable and $y$ as the dependent variable. The coordinate points of the independent variable must always be given in algebraic-increasing order.

The equation statement is specified as a FORTRAN-like expression in the following general form:

> EQUATION    equation name (dummy argument list) = (expression)

EQUATION is the key word (it may be abbreviated by Q). Equation name is the unique name assigned to the equation. Dummy argument list is a list of dummy variables separated by commas. Expression is a FORTRAN-like expression using the variables in the dummy argument list.

The variables in the dummy argument list must begin with an alphabetic character other than I, J, K, L, M, or N and must not exceed six characters in length. The equation may also include ASTAP system-supplied functions, which will be illustrated later.

### FEATURES Heading

Statements under the FEATURES heading are used to provide additional specifications for the model description. For example, the user may explicitly designate any node in the circuit as the ground node through the statement:

> GROUND = (node name)

The user may also specify any two nodes of a model as a designated port through the use of

> PORTS = (node-node, node-node, . . . node-node)

Each port is specified by the pair of nodes identified. The specification of the ports of a model is required to calculate transfer functions in ac analysis.

### EXECUTION CONTROLS Heading

EXECUTION CONTROLS is the second major heading statement. Subheadings under this section describe the type of analysis, starting point, analysis controls, designated outputs, and analysis reruns. Each of the subheadings will now be explained.

### ANALYZE Heading

The ANALYZE heading is used to specify the model to be analyzed and the type of analysis to be performed. The general form of the statement is

> ANALYZE model name (mode of analysis)

where ANALYZE is a keyword, model name is the main model to be analyzed, and mode of analysis can be dc, transient, transient only, or ac.

A dc analysis on a model named AMPLIFIER is requested using the statement

> ANALYZE AMPLIFER (DC)

The dc analysis finds the equilibrium, or steady-state, operating point of a linear or nonlinear circuit. For a circuit that contains capacitors or inductors, which exhibits a transient response to time-varying driving functions, the dc solution corresponds to the steady-state condition that exists after all transients have decayed.

Typical outputs of a dc analysis are the steady-state levels of element voltages and currents associated with linear devices, node voltages, and user-defined quantities such as power dissipation.

A transient analysis on a model named AMPLIFIER may be requested in one of two ways:
1. ANALYZE AMPLIFIER (TRANSIENT)
2. ANALYZE AMPLIFIER (TRANSIENT ONLY)

A transient analysis produces a time-domain solution. It provides the transient response of a linear or nonlinear circuit to driving functions specified by the user. The driving functions may be time-varying or constant.

Typical outputs of a transient analysis are plots of element voltages or currents and node voltages as a function of time. The TRANSIENT option, in effect, requests a dc analysis followed by a transient analysis. A dc analysis provides the initial conditions from the beginning (time = 0) of the transient solution. The TRANSIENT ONLY option requests that only a transient analysis be performed, in which case a particular set of initial conditions desired by the user must be specified.

An ac analysis of a model named AMPLIFIER is requested by the statement

> ANALYZE AMPLIFER (AC)

An ac analysis produces a frequency-domain solution. It provides the small-signal frequency response of a linear circuit. Nonlinear circuits are linearized at a given operating point, and the frequency response of the linearized network is computed.

The operating point for a nonlinear network may be supplied by the user, or it may be obtained automatically from a previous dc or transient analysis. When a transient analysis is used to obtain the operating point, the user may specify a particular time at which the operating point is to be computed.

When the operating point is obtained automatically, the topological structure of the network for ac analysis must be the same as that used for dc or transient analysis. The user may manually input an operating point for ac analysis through the use of special language statements. With this approach there is no restriction on the topological structure of an ac problem.

### RUN CONTROLS Heading

The run control statements, which are placed under the RUN CONTROLS heading, are in the form of either a command or a statement equating the run control to a numeric value. ASTAP contains over 40 run controls that can be used to control such items as the duration of analysis. For example, STOP TIME = 200; the resolution of the output, i.e., PRINT INTERVAL = .5; and the accuracy of the results, i.e., MAXIMUM PASSES = 1000.

All controls affecting the accuracy of results have default values. For most problems they offer a reasonable trade-off between solution accuracy and the computation time required. The user normally sets only a few controls to specify the desired analysis duration and output resolution, and relies on the default settings for those controls affecting the accuracy of the solution.

### INITIAL CONDITIONS Heading

Under INITIAL CONDITIONS heading, the user may specify the initial state of a network at the start of an analysis performance in the transient, transient-only, or dc mode. The initial state of a network is specified by stipulating the voltages across capacitors and current through inductors. In addition, initial conditions for other current or voltage variables, element values, and parameter values can be specified in those applications in which they may be required.

Each capacitor voltage and inductor current used to specify the initial state is entered, as are individual initial condition statements. The general form of the initial condition statement is:

> V capacitor name = numeric constant
> I inductor name = numeric constant

where $V$ is a key letter that indicates the voltage across an element and $I$ is a key letter that indicates the current through an element.

The set of all capacitor voltages and all inductor currents is sufficient to specify the initial state. When not specified, they are assumed to be zero by default.

### OPERATING POINT Heading

The operating point of a nonlinear circuit must be specified prior to an ac analysis of the circuit. As indicated previously, two means of specifying the operating point, manual and automatic, are provided by the program. The operating point is specified manually by entering the values of the voltages or currents of nonlinear elements at the operating point of the circuit. These are specified by using operating point statements under the OPER-ATING POINT heading.

In the automatic procedure, the operating point is derived from a dc or transient analysis, and the OPERATING POINT heading is not used.

The general form of the operating point statement is

> V element name = numeric constant
> I element name = numeric constant

where $V$ and $I$ prefacing an element denote the voltage across, and current through, an element, respectively.

## OUTPUTS Heading

The request for displaying the results of analysis is made under the OUTPUTS heading. All output from the program, with the exception of the input statements listing, summaries of execution data, and diagnostic messages, must be requested by the user. Possible forms of program output are printed tabular listings and printer plots of the response curves. All element voltages and currents may be requested as outputs. Each output variable, however, must be requested individually.

Node voltages are additional response variables available as output. A node voltage is specified by prefixing the letter N to the node name. The node voltage obtained is that between the node specified and ground. This implies specification of a ground node in the main model.

Transfer function parameters are output variables available from an ac analysis. To specify the output of transfer function parameters, each parameter must be individually requested in an output statement.

An output statement is in the form of a request, which specifies the type of output, followed by a list of output variables. The general form of output statements is:

> PRINT variable list
> PLOT (plot options) variable list vs ind. variable

where PRINT and PLOT are the keywords. Variable list is a list of the individual output variables requested. Plot options are a series of options that pertain to the entire plot, and the independent variable is that specified by the user. If no independent variable is specified, the independent variable is assumed to be time for a transient analysis and frequency for an ac analysis.

## RERUNS Heading

A RERUNS heading is used when a reanalysis of the specified circuit is required following parameter modifications. The rerun is limited to reanalysis of the described circuit in which only parameters whose values are defined as numeric constants can be modified to other numeric constants. Any number of reruns of the basic circuit may be made with the outputs for the reruns appearing in sequence.

Elements themselves cannot be directly modified in a rerun; however, constant-valued elements can be modified by defining the element value as a parameter and modifying the parameter. This is illustrated in the first example problem.

The rerun statement consists of a parameter name followed by the values the parameter takes on subsequent reruns. The general form is:

> parameter name = (constant 1, constant 2, . . . )

where parameter name designates the parameter to be modifed and constant 1, constant 2, . . . , are numeric constants.

The first analysis is made with the basic circuit where the element value is equated to a parameter and the parameter value is, in turn, equated to a constant. On subsequent reruns, the parameter takes the value of constant 1, then constant 2, etc., until the number of constants is exhausted.

## UTILITY CONTROLS Heading

The UTILITY CONTROLS heading is the last of the major headings. Under it are listed special optional requests to be executed on any run. Included are requests for printing the status of the program, such as PRINT NOTES, and information about the model libraries, such as PRINT USER LIBRARY. The UTILITY CONTROLS is a little-used, but important, heading. Through it the user can keep fully informed on daily changes that may be made to either the program or the library.

## 14.5  LANGUAGE SYNTAX

The language rules and conventions previously described are summarized in Appendix C. The information in the Appendix exceeds what is required to understand the examples that follow. All relevant language information has been included, however, to permit the user a simple and comprehensive reference to the ASTAP language. It should be specifically pointed out that ASTAP has statistical features associated with each of the major modes of analysis. The Appendix includes the language features to support statistics (which is not treated in this chapter).

As the reader proceeds through each of the examples, the Appendix can be used as a reference to language conventions. When additional details are required, the user is referred to *ASTAP Program Reference Manual.*[5]

## 14.6  ANALYSIS OF CAPACITOR DISCHARGE

**problem 14.1**   Figure 14-3 shows an *RC* circuit. Assume that the capacitor has been charged to 10 V and that the switch is closed at $t = 0$. Determine the current and voltage waveforms. Solve the same problem when $C = 10$ $\mu$F and $R = 1$ k$\Omega$ and for $C = 20$ $\mu$F and $R = 1$ k$\Omega$.

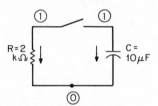

**Fig. 14-3**   Circuit for calculating the discharge of a capacitor (Problem 14.1).

**solution**   After the switch is closed, the voltages across the capacitor and resistor are equal. Likewise, only one current is involved. The current in the circuit is equal to the voltage across the capacitor divided by the resistance. If the solution for the voltage across the capacitor is obtained, the current can be obtained from Ohm's law.

The voltage across the capacitor, $V_C$ is:

$$V_C = V_0 e^{-t/RC}$$

or, in this case,

$$V_C = 10e^{-t/20}$$

where $V_0 = 10$ V, the initial voltage to which the capacitor has been charged, and $RC$, the time constant, is 20 ms. For this problem the second set of units from Table 14-1 will be used. The current is equal to $V_R/R$ and, since $V_C = V_R$,

$$I = V_C/R = 5e^{-t/20}$$

The solutions for all three sets of elements are presented in Table 14-2. For the first and third analyses, the $RC$ time constant is twice that for the second analysis. This means that the capacitor will discharge twice as fast for the second analysis.

**TABLE 14-2   Solutions for Three Sets of Elements (Problem 14.1).**

|  | First analysis | Second analysis | Third analysis |
|---|---|---|---|
| $R$ | 2 k$\Omega$ | 1 k$\Omega$ | 1 k$\Omega$ |
| $C$ | 10 $\mu$F | 10 $\mu$F | 20 $\mu$F |
| $V_C$ | $10e^{-t/20}$ | $10e^{-t/10}$ | $10e^{-t/20}$ |
| $I$ | $5e^{-t/20}$ | $10e^{-t/10}$ | $10e^{-t/20}$ |

Figure 14-4 shows the ASTAP input coding required to solve this problem. Only those lines with a card number are part of the input statements. The unnumbered lines, such as PART 1 OF THE INPUT DATA, are added by the program compiler. Lines 1, 2, 3, 8, 9, 10, 13, 15, 18, and 20 are the heading statements previously described.

**Fig. 14-4** Capacitor discharge input data.

In line 2 the name of the main model is stated as CAPACITOR DISCHARGE. The open/closed parentheses tell the program that the group of circuit elements to follow constitute the main model. In this example the main model is the only model. The first seven lines of input code are the MODEL DESCRIPTION portion, which describes the circuit topology to the program. Under the heading ELEMENTS on line 3 the elements of the circuit are described. Line 4 specifies to the program that resistor $R$ is connected between nodes 1 and 0. Node names are enclosed in circles in Fig. 14-3.

Letter R is a key letter, and any element beginning in R is taken to be a resistor. By specifying 1 as the FROM node and 0 as the TO node, the voltage and current are identified as positive from node 1 to node 0. Line 5 identifies an element called C, a capacitor, by the key letter C. Its FROM and TO nodes are the same as for the resistor.

The values of R and C have been specified through the use of parameters. Line 4 specifies the value of R to be P1, and line 6 specifies the value of P1 to be 2. The values of R and C have been specified by parameters so that the values of R and C can be changed to take advantage of the rerun feature.

The second part of the input coding, lines 8 to 19, describes the type of analysis to be performed on the circuit of part 1. Line 9 calls for a TRANSIENT ONLY analysis of the model CAPACITOR DISCHARGE. In the TRANSIENT ONLY mode, the analysis begins immediately from a specified initial state of the network, which requires that a complete specification be made under the INITIAL CONDITIONS heading. In this example the voltage across the capacitor is a sufficient specification of the initial state.

Lines 10 to 12 constitute the RUN CONTROLS portion. A START TIME of 0 and a STOP TIME OF 200 are specified. If no start time is specified, it defaults to zero. If no stop time is specified, the program will not execute.

Lines 13 and 14 specify the initial conditions. In this case line 14 specifies the voltage across capacitor $C$ to be 10 V. The V placed before the letter C identifies it as the voltage across capacitor $C$.

Lines 15 to 17 indicate that, in addition to the analysis performed on the circuit of part 1, two additional analyses are desired. During the first run, parameters P1 and P2 have the values assigned in lines 6 and 7. During the first rerun P1 = 1 and P2 = 10. During the second rerun the rerun list of P1 is exhausted, so it maintains its former value of P1 = 1, and P2 assumes the value 20.

Line 19 requests a printed list and plot of the voltage across capacitor $C$ and the current through resistor $R$. The letter I before the letter R specifies the current through element R. The code LABEL = (CAPACITOR DISCHARGE) requests that the label CAPACITOR DISCHARGE be printed on the output.

Line 20 is an optional end card. It indicates that there is no PART 3 OF THE INPUT DATA, i.e., no UTILITY CONTROLS.

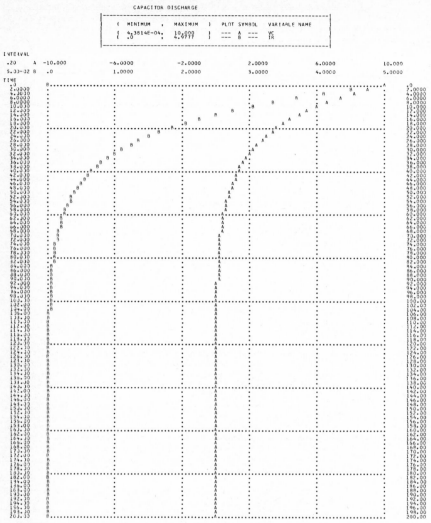

**Fig. 14-5** Output plots for the initial run of the capacitor discharge problem. $R = 2\text{ k}\Omega$, $C = 10\ \mu\text{F}$.

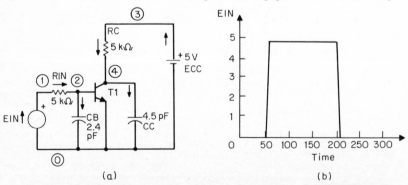

**Fig. 14-6** A logic inverter. (*a*) Circuit. (*b*) Input waveform EIN vs. time (Problem 14.2).

Although both a printed list and a plot were obtained, only the output plot is included. Lists will be shown in subsequent problems. Figure 14-5 shows the plots obtained for the first analyses. Plots for the two reruns are similar to those of Fig. 14-5.

## 14.7  ANALYSIS OF SATURATING INVERTER

**problem 14.2**   The circuit of Fig. 14-6a is similar to the logic inverter circuit found in many digital systems. It is characterized by a low input voltage and high output voltage or a high input and low output voltage. For the purposes of this example a voltage close to zero is defined as low, and a voltage of about five volts is considered high. Find the time responses of the output voltage to the input waveform given in Fig. 14-6b.

**solution**   The time response of the output voltage is obtained by performing a transient analysis. Figure 14-7 is the ASTAP input coding for this problem. The first line is the MODEL DESCRIPTION heading statement under which all models are described. Note that there are two models, lines 2 and 13, in this example. The first model, line 2, is the main model, or the circuit of Fig. 14-6a, which is to be analyzed. The program recognizes the model of line 2 as the main model by the empty pair of parentheses. This line also includes the name given to the main model, INV.

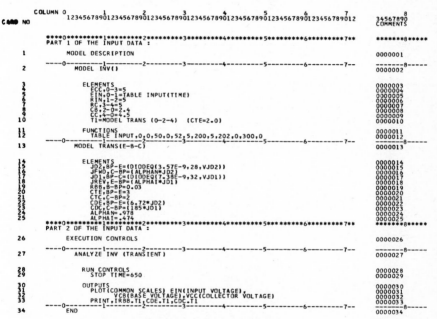

**Fig. 14-7**   Logic inverter input data.

Line 3 is the ELEMENTS heading for model INV. Here the circuit elements are described to the program. Voltage source ECC is identified on line 4. The key letter E identifies the element as a voltage source. From Fig. 14-6a it is evident that ECC is connected from node 0 to node 3 and that its value, 5 V, is entered on line 4. The arrows in the figure indicate the assumed direction of positive current flow. The unit for ECC is volt, where the choice of a consistent set of units is column I in Table 14-1.

Line 5 specifies that EIN is the voltage source connected from node 0 to node 1 and that its value is a table of voltage versus time identified as TABLE INPUT (TIME) where TIME is the argument. TABLE INPUT (TIME) will be further explained when lines 11 and 12 are discussed.

Lines 6 and 7 identify the resistor elements RIN and RC as well as their nodes and values. Since they are resistors, their names must begin with the key letter R.

Lines 8 and 9 identify capacitor elements CB and CC, their nodes and values. Since they are capacitors, their names must begin with the key letter C.

Line 10 identifies the transistor of Fig. 14-6a as T1 and equates it to a model named TRANS, which at this point is undefined. Line 10 further states that T1 is connected to nodes 0, 2, and 4. Code CTE = 2.0 requests that when model TRANS is used to model T1, the value of element

CTE within the model be changed to equal 2.0. Any value of any element in a model may be changed. The new value is used only for the specified analysis. If the model is in a library, the library value of the element remains unchanged.

Line 11 is the FUNCTION heading. Under this heading, the details of TABLE INPUT (TIME) are described on line 12. The numbers following TABLE INPUT are the pairs of values that describe the waveform of Fig. 14-6b. For values of time outside the range of the table, the values are determined by the program through extrapolation.

It is necessary to describe active device elements in terms of the basic circuit elements that constitute the device model. Lines 13 to 25 describe the model TRANS. Again, the heading MODEL is introduced on line 13 along with the model name TRANS and its nodes (E-B-C) for emitter, base, and collector. Line 14 is the ELEMENTS heading for this model, and lines 15 to 25 describe the elements of the model.

Figure 14-8 is the circuit configuration of the element list of lines 15 to 25 and is the commonly used Ebers-Moll model.[11] It could have been called from the model library but is included in the listing to make the example more complete. Elements JD2 and JD1 represent the diode currents of the transistor emitter and collector junctions, respectively. Element JFWD represents the forward collector current and JREV the reverse collector current.

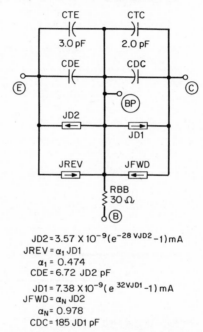

$$JD2 = 3.57 \times 10^{-9} (e^{-28 \, VJD2} - 1) \, mA$$
$$JREV = \alpha_1 \, JD1$$
$$\alpha_1 = 0.474$$
$$CDE = 6.72 \, JD2 \, pF$$

$$JD1 = 7.38 \times 10^{-9} (e^{32 \, VJD1} - 1) \, mA$$
$$JFWD = \alpha_N \, JD2$$
$$\alpha_N = 0.978$$
$$CDC = 185 \, JD1 \, pF$$

**Fig. 14-8**   Ebers-Moll transistor model equivalent circuit modeled by model TRANS.

Elements JD2, JFWD, JD1, JREV, CDE, and CDC are nonlinear elements whose values are defined through FORTRAN-like expressions. Letter J is the key letter for current. Expressions must be enclosed in parentheses. Terms ALPHAN and ALPHAI are parameters used in the expressions for JFWD and JREV. A full set of FORTRAN arithmetic operators is available for the expressions. The set is addition, subtraction, multiplication, division, and exponentiation denoted by +, −, *, /, and **, respectively.

Elements JD2 and JD1 are defined by use of the function DIODEQ, supplied by ASTAP.* Function DIODEQ is the equation for the current of a diode.

$$I = I_s (e^{qV/kT} - 1)$$

where $I_s$ is the saturation current, $q/kT$ can be considered a constant for a given temperature, and $V$ is the voltage across the diode.

If the function DIODEQ did not exist in the ASTAP system-supplied library, the user could have defined it by means of an equation. In this case lines 15 and 17 would then be replaced by:

JD2, BP-E = EQUATION DIODEQ (3.57 E-9, 28, VJD2)
JD1, BP-C = EQUATION DIODEQ (7.38 E-9, 32, VJD1)

*For a complete set of functions supplied by ASTAP, see Ref. 5.

and DIODEQ would be defined under an additional FUNCTIONS heading as an equation. The code would read:

FUNCTIONS
EQUATION DIODEQ (IS, C, VJ) = (IS *(EXP (C*VJ) − 1)).

Term IS represents the saturation current, C is a constant to represent $q/kT$, and VJ is the voltage across the current source that represents the diode.

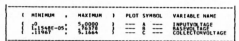

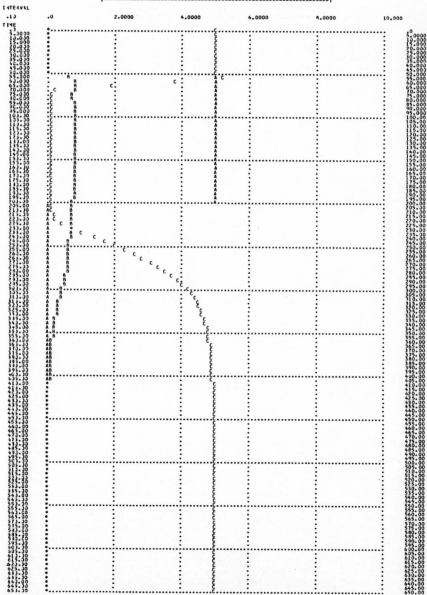

**Fig. 14-9**  Output plot from the analysis of the logic inverter.

In the preceding equation definition, IS, C, and VJ are dummy arguments used to define the variables. They are replaced by the corresponding constants or variables enclosed in parentheses in the equation reference. The equation is, in this case, the system-supplied equation.

Statement 26 marks the beginning of the second part of the input data, the EXECUTION CONTROLS. Statement 27 calls for the TRANSIENT mode of solution on the main model INV.

In this problem, the time response of the network to a time-dependent voltage source is desired. It is assumed that the network is biased at its dc quiescent point when the input waveform is applied.

The transient mode of analysis first performs a dc analysis to find the dc quiescent point before any time-dependent voltage or current sources are applied. In finding the quiescent point, the values of time-dependent sources are set equal to their values at the start time. By using the transient mode of analysis, the user does not have to specify, or even have any knowledge about, the initial state of the network. Line 28 is the RUN CONTROLS heading, and line 29 is a run control establishing the stop time as 650. The start time defaults to zero.

Line 30 is the OUTPUTS heading. Lines 31 and 32 request that variables EIN, VCB, and VCC be plotted; they further request plotting by use of the COMMON SCALES option. The COMMON SCALES option produces mutliple plots that use the same values along the $y$ axis. Line 31 also indicates that EIN is to be renamed INPUT VOLTAGE, VCB is to be renamed BASE VOLTAGE, and VCC is to be renamed COLLECTOR VOLTAGE, when they are referenced on the output sheets.

Figure 14-9 shows the resultant plot. Statement 33 requests a printed list of the current through resistor RBB in model T1 and the value of capacitor CDE and CDC in model T1. The tabulated output (one page of four shown) requested in line 33 is provided in Fig. 14-10.

ANALYZE INV (TRANSIENT)

| | INITIAL VALUES | | | | | | |
|---|---|---|---|---|---|---|---|
| TIME | .0 | | | | | | |
| STEP SIZE | .0 | | | | | | |
| PASS COUNTER | .0 | | | | | | |
| IRBB.T1 | .0 | | | | | | |
| CDE.T1 | .0 | | | | | | |
| CDC.T1 | .0 | | | | | | |

| TIME | .0 | 5.0000 | 10.000 | 15.000 | 20.000 | 25.000 | 30.000 |
|---|---|---|---|---|---|---|---|
| STEP SIZE | 130.00 | 4.0792 | 5.0000 | 5.0000 | 5.0000 | 5.0000 | 5.0000 |
| PASS COUNTER | .0 | 5.0000 | 6.0000 | 7.0000 | 8.0000 | 9.0000 | 10.000 |
| IRBB.T1 | -6.7394E-06 | -4.6978E-06 | -4.1576E-06 | -3.6825E-06 | -3.2643E-06 | -2.8957E-06 | -2.5704E-06 |
| CDE.T1 | -2.6958E-11 | 2.3603E-11 | 2.0786E-11 | -1.8326E-11 | -1.6174E-11 | -1.4290E-11 | -1.2637E-11 |
| CDC.T1 | -1.3653E-06 | -1.3653E-06 | -1.3653E-06 | -1.3653E-06 | -1.3653E-06 | -1.3653E-06 | -1.3653E-06 |

| TIME | 35.000 | 40.000 | 45.000 | 50.000 | 55.000 | 60.000 | 65.000 |
|---|---|---|---|---|---|---|---|
| STEP SIZE | 5.0000 | 5.0000 | 5.0000 | .28279 | .31211 | 1.1562 | 2.8708 |
| PASS COUNTER | 11.000 | 12.000 | 13.000 | 32.000 | 57.000 | 75.000 | 79.000 |
| IRBB.T1 | -2.2832E-06 | -2.0292E-06 | -1.8045E-06 | -1.3419E-02 | -.56296 | .84744 | .84591 |
| CDE.T1 | 1.1185E-11 | 9.9088E-12 | 8.7846E-12 | 5.4474E-10 | .67674 | 18.919 | 21.143 |
| CDC.T1 | -1.3653E-06 | -1.3653E-06 | -1.3653E-06 | -1.3653E-06 | -1.3653E-06 | -1.3653E-06 | -1.3653E-06 |

| TIME | 70.000 | 75.000 | 80.000 | 85.000 | 90.000 | 95.000 | 100.00 |
|---|---|---|---|---|---|---|---|
| STEP SIZE | .16908 | 1.4981 | 2.5143 | 3.5704 | 4.8625 | 5.0000 | 5.0000 |
| PASS COUNTER | 132.00 | 149.00 | 151.00 | 155.00 | 157.00 | 159.00 | 161.00 |
| IRBB.T1 | .88867 | .84905 | .84913 | .84923 | .84928 | .84929 | .84931 |
| CDE.T1 | 16.333 | 14.326 | 15.366 | 16.035 | 16.449 | 16.700 | 16.849 |
| CDC.T1 | 21.891 | 120.73 | 182.61 | 220.89 | 244.50 | 258.76 | 267.21 |

| TIME | 105.00 | 110.00 | 115.00 | 120.00 | 125.00 | 130.00 | 135.00 |
|---|---|---|---|---|---|---|---|
| STEP SIZE | 5.0000 | 5.0000 | 5.0000 | 5.0000 | 5.0000 | 5.0000 | 5.0000 |
| PASS COUNTER | 163.00 | 164.00 | 165.00 | 166.00 | 167.00 | 168.00 | 169.00 |
| IRBB.T1 | .84931 | .84931 | .84932 | .84932 | .84932 | .84932 | .84932 |
| CDE.T1 | 16.936 | 16.987 | 17.012 | 17.026 | 17.036 | 17.043 | 17.047 |
| CDC.T1 | 272.16 | 275.04 | 276.46 | 277.27 | 277.83 | 278.21 | 278.48 |

Fig. 14-10   Print output (one page shown) from the analysis of the logic inverter.

## 14.8  ANALYSIS OF SMALL-SIGNAL AMPLIFIER

**problem 14.3**   A schematic of a two-stage, common-emitter small-signal amplifier is given in Fig. 14-11a. Its input is excited with a sinusoidal signal of 0.1 V in magnitude and zero-degree phase angle. Identical transistors $Q_1$ and $Q_2$ are the same as the transistor used in Sec. 14.7. Determine the magnitude of the output voltage, expressed in decibels, and its phase angle in degrees over the frequency range of 10 Hz to 1 MHz.

**solution**   This problem can be solved by using the ac analysis feature of ASTAP, which produces a frequency-domain solution and provides the small-signal frequency response of a linear network. The small-signal amplifier of Fig. 14-11a is a nonlinear network because transistors $Q_1$ and $Q_2$ are nonlinear. For this problem the choice is made to first find the operating point from a dc analysis. The choice is made for two reasons: First, it enables the user to verify that the dc operating points of the transistors are in the linear region before ac analysis is performed. Second, because $Q_1$ and $Q_2$ are biased identically, the operating points of the transistors are identical and can be found from the simpler circuit of Fig. 14-11b.

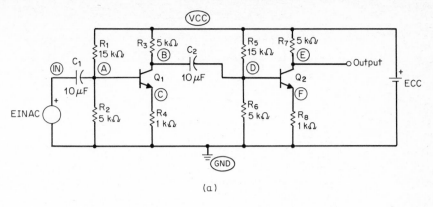

(a)

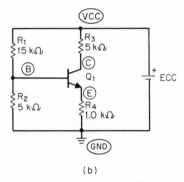

(b)

**Fig. 14-11**  A two-stage, common-emitter small-signal amplifier. (*a*) Schematic. (*b*) Single common-emitter stage used to find the operating point (Problem 14.3).

### Coding for the DC Solution

Input statements for the dc analysis are given in Fig. 14-12. The model used for transistors $Q_1$ and $Q_2$ is the same as in the previous problem, but its name has been changed. Although the model was retained in the program library, as instructed in Problem 14.2, the change of name requires that it be listed again. The system of units is given in column II of Table 14-1.

The code under MODEL DESCRIPTION describes the circuit topology and has been named COMMON EMITTER AMPLIFIER. Under the heading FEATURES, line 4 specifies the node GND as the reference, or ground, node. Under EXECUTION CONTROLS, line 26 requests a dc analysis of the COMMON EMITTER AMPLIFIER circuit. Under OUTPUTS, line 28 requests NB, NC, NE nodal voltages at nodes B, C, and E, respectively, with regard to the reference node specified in line 4. Element IRBB.Q1 is the conventional current through RBB.Q1 from the FROM to the TO node; VJD1.Q1 is the voltage across JD1.Q1 from the FROM to the TO node.

### Output for the DC Solution

Figure 14-13 shows the results of the dc analysis. Because $V_{CB}$ is positive, the transistor is in its linear region.

$$V_{CB} = NC-NB$$
$$= 2.3913 - 1.2060 = 1.1853$$

The value of $V_{CB}$ could also be calculated from VJD1 + IRBB * RBB. The negative sum must be taken since the voltage across RBB is defined from node B to node BP, and the voltage across the current source JD1 is defined from BP to C. The voltage of interest is from node C to node B:

$$V_{CB} = -(VJD1 + IRBB \times RBB)$$
$$= -(-1.1856\text{ V} + 1.1723 \times 10^{-2}\text{ mA} \times 0.03\text{ k}\Omega)$$
$$= 1.1853\text{ V}$$

Fig. 14-12 Common-emitter input data.

Fig. 14-13 Output from the dc analysis of common-emitter circuit.

### Coding for the AC Solution

The input listing for the ac analysis is given in Fig. 14-14. Line 7 specifies the input signal. The keyword COMPLEX indicates that EINAC is a complex ac source, and the keyword MAGPH, that the arguments are the magnitude and phase. The other statements under the MODEL DESCRIPTION heading describe the circuit topology, as has been discussed previously.

Under RUN CONTROLS, the start and stop frequency specifications, lines 36 and 37, are required. They are set equal to the frequency range requested in the problem statement. POINTS PER DECADE specifies that 10 points per decade are to be printed. It is not necessary to specify POINTS PER DECADE in this case because the default value is 10.

Under the OUTPUT heading, line 45 requests a printed list and a plot of the voltage at node E (the output node) with respect to ground. Since the voltage at E is complex, there are two components to it. No plot options were specified; therefore, the default plot generated gives the magnitude of the voltage at node E in decibels and the phase of the voltage at node E in degrees. Both quantities versus frequency are plotted on a logarithmic scale.

The second output request (line 46) results in five additional listings and 10 additional plots. A request for output from internal nodes of the circuit is usually made for diagnostic purposes. If the circuit behavior deviates from what the user expects, the user can trace the response through the circuit and pinpoint where the behavior deviated from what was expected.

Under the heading OPERATING POINT, it is necessary to specify all voltages and currents upon which other elements in the circuit depend. In this example, specifying VJD1 and VJD2 in

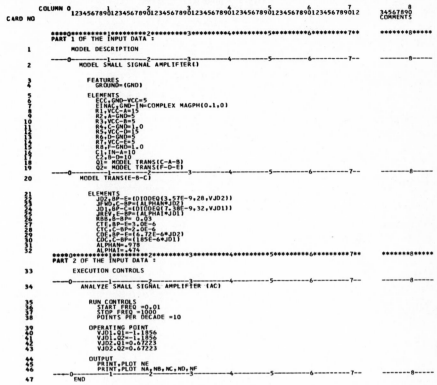

```
                    L I S T   O F   D A T A   C A R D S
                    ***********************************

         COLUMN 0         1         2         3         4         5         6         7          8
CARD NO  12345678901234567890123456789012345678901234567890123456789012    34567890
                                                                           COMMENTS

         ****0*********1*********2*********3*********4*********5*********6*********7**    *******8*****
         PART 1 OF THE INPUT DATA :
1        MODEL DESCRIPTION
         ---0---------1---------2---------3---------4---------5---------6---------7--    -------8-----
2        MODEL SMALL SIGNAL AMPLIFIER()

3        FEATURES
4          GROUND=(GND)

5        ELEMENTS
6          ECC,GND-VCC=5
7          EINAC,GND-IN=COMPLEX MAGPH(0.1,0)
8          R1,VCC-A=15
9          R2,A-GND=5
10         R3,VCC-B=5
11         R4,C-GND=1.0
12         R5,VCC-D=15
13         R6,D-GND=5
14         R7,VCC-E=5
15         R8,F-GND=1.0
16         C1,IN-A=10
17         C2,B-D=10
18         Q1= MODEL TRANS(C-A-B)
19         Q2= MODEL TRANS(F-D-E)
         ---0---------1---------2---------3---------4---------5---------6---------7--    -------8-----
20       MODEL TRANS(E-B-C)

21       ELEMENTS
22         JD2,BP-E=(DIODEQ(3.57E-9,28,VJD2))
23         JFWD,C-BP=(ALPHAN*JD2)
24         JD1,BP-C=(DIODEQ(7.38E-9,32,VJD1))
25         JREV,E-BP=(ALPHAI*JD1)
26         RBB,B-BP= 0.03
27         CTE,BP-E=3.0E-6
28         CTC,C-BP=2.0E-6
29         CDE,BP-E=(6.72E-6*JD2)
30         CDC,C-BP=(185E-6*JD1)
31         ALPHAN=.978
32         ALPHAI=.474
         ****0*********1*********2*********3*********4*********5*********6*********7**    *******8*****
         PART 2 OF THE INPUT DATA :
33       EXECUTION CONTROLS
         ---0---------1---------2---------3---------4---------5---------6---------7--    -------8-----
34       ANALYZE SMALL SIGNAL AMPLIFIER (AC)

35       RUN CONTROLS
36         START FREQ =0.01
37         STOP FREQ =1000
38         POINTS PER DECADE =10

39       OPERATING POINT
40         VJD1.Q1=-1.1856
41         VJD1.Q2=-1.1856
42         VJD2.Q1=0.67223
43         VJD2.Q2=0.67223

44       OUTPUT
45         PRINT,PLOT NE
46         PRINT,PLOT NA,NB,NC,ND,NF
         ---0---------1---------2---------3---------4---------5---------6---------7--    -------8-----
47       END
```

**Fig. 14-14**   Small-signal amplifier input data.

```
              FREQUENCY =    1.0010        ANGULAR FREQUENCY =    6.2895
         REAL          IMAGINARY       MAGNITUDE     LOG MAGNITUDE      PHASE
                                                         (DB)          (DEG)
NE     .86108        5.3115E-03       .86110         -1.2990         .35342
NA    9.9998E-02     4.5674E-04      9.9999E-02      -20.000         .26170
NB    -.18800       -3.1626E-04       .18800         -14.517        -179.90
NC    9.3666E-02     4.2773E-04      9.3667E-02      -20.568         .26164
ND    -.18799       -1.1749E-03       .18800         -14.517        -179.64
NF    -.17609       -1.1002E-03       .17609         -15.085        -179.64

              FREQUENCY =    2.0010        ANGULAR FREQUENCY =   12.573
         REAL          IMAGINARY       MAGNITUDE     LOG MAGNITUDE      PHASE
                                                         (DB)          (DEG)
NE     .86110        2.3037E-03       .86110         -1.2990         .15328
NA    9.9999E-02     2.2849E-04       .10000         -20.000         .13092
NB    -.18800       -1.0398E-04       .18800         -14.517        -179.97
NC    9.3667E-02     2.1384E-04      9.3668E-02      -20.568         .13080
ND    -.18800       -5.3354E-04       .18800         -14.517        -179.84
NF    -.17609       -4.9905E-04       .17609         -15.085        -179.84

              FREQUENCY =    3.0010        ANGULAR FREQUENCY =   18.856
         REAL          IMAGINARY       MAGNITUDE     LOG MAGNITUDE      PHASE
                                                         (DB)          (DEG)
NE     .86110        1.1434E-03       .86110         -1.2989        7.6077E-02
NA     .10000        1.5235E-04       .10000         -20.000        8.7292E-02
NB    -.18800       -9.0823E-06       .18800         -14.517        -180.00
NC    9.3668E-02     1.4263E-04      9.3668E-02      -20.568        8.7122E-02
ND    -.18800       -2.5550E-04       .18800         -14.517        -179.91
NF    -.17609       -2.7573E-04       .17609         -15.085        -179.91

              FREQUENCY =    4.0010        ANGULAR FREQUENCY =   25.139
         REAL          IMAGINARY       MAGNITUDE     LOG MAGNITUDE      PHASE
                                                         (DB)          (DEG)
NE     .86110        4.4531E-04       .86110         -1.2989        2.9630E-02
NA     .10000        1.1427E-04       .10000         -20.000        6.5474E-02
NB    -.18800        5.6448E-05       .18800         -14.517         179.98
NC    9.3668E-02     1.0667E-04      9.3668E-02      -20.568        6.9248E-02
ND    -.18800       -1.5839E-04       .18800         -14.517        -179.95
NF    -.17609       -1.4694E-04       .17610         -15.085        -179.95
```

**Fig. 14-15**   Print output from analysis of the small-signal amplifier.

the transistor model sets values for the other elements in the model. That is, from line 22, VJD2 determines JD2, and in line 23, JFWD is a function of JD2, etc. Under OPERATING POINT, VJD1 and VJD2 are specified for both transistors.

### Output for the AC Solution

Figure 14-15 is 1 page of 11 that results from the PRINT requests of lines 45 and 46. Since the output covers 5 decades with 10 points per decade, 50 points are printed. These 50 points provide the output voltage in decibels and its angle in degrees. For example, at 2.001 kHz the magnitude of the output, node E, is 0.86110 V. The log magnitude expressed in decibels is 20 log (0.86110) or −1.2989 dB, and the phase angle is 0.15328°. At node B, which is the output of the first stage, the phase of the voltage is at an angle of −179.97°. The phases of nodes B and E are reasonable. Since the first stage is known to be inverting, a near 180° phase difference is anticipated. The output, node E, is nearly in phase with the input, because the second stage causes another phase reversal.

Figure 14-16 illustrates the plot generated as a result of line 45. The log magnitude of the voltage at node E is designated by symbol A, and its phase in degrees by symbol B, versus frequency on a logarithmic scale. The plot that results from the plot request of line 46 is similar to that of Fig. 14-16.

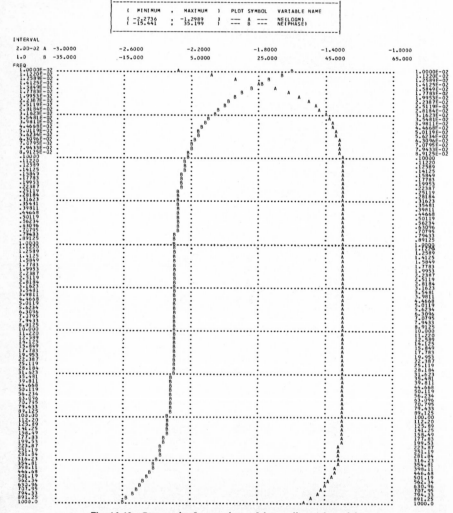

**Fig. 14-16** Output plot from analysis of the small-signal amplifier.

**Combined DC/AC Solution**

This problem can be solved by combining the dc and ac analyses in one run. For a combined analysis, however, the circuit topology for both analyses must be the same. Because Fig. 14-11a represents the total circuit, it is the one that must be used in the combined analysis.

Input cards required for the combined analysis are given in Fig. 14-17. The compiled listing used in the other examples is not used here; it is considerably different from the input cards. The main differences between the combined dc/ac input cards of Fig. 14-17 and the individual dc and ac listings of Figs. 14-12 and 14-14 are highlighted in Fig. 14-17 by an asterisk. They are:

1. The use of parameter POINT = (ACPUN (0)). Parameter POINT is a dummy parameter used to cause the system-supplied function ACPUN to be invoked during the dc analysis requested by the first ANALYSIS statement. Function ACPUN causes the operating point to be saved for the subsequent ac analysis. The argument of ACPUN (zero in this example) is required and is the time at which the operating point is desired. Zero is the only time which makes sense for a dc analysis. For a combined transient and ac analysis, however, a value other than zero has physical significance.

2. The use of two sets of EXECUTION CONTROLS.

3. The necessity for requesting the parameter POINT as output in both the dc and ac analyses.

**Fig. 14-17** Input cards for combined dc/ac analysis of the small-signal amplifier.

The dc output requests for nodes B, C, and E of Fig. 14-11b had to be changed to correspond to the nodes of Fig. 14-11a, that is, to nodes A, B, and C. Nodes D, E, and F could have been used as well because the dc conditions of the two transistors are identical. This second option is illustrated because of the added simplicity of permitting the program to calculate the operation point. Since the results from the combined dc/ac analysis are identical to those obtained by separate analyses, the output from the combined analysis is not shown.

## 14.9  ANALYSIS OF AN ACTIVE FILTER

**problem 14.4**  The circuit of Fig. 14-18a is a bandpass filter.[12] Determine its center frequency, gain, and bandwidth. (The model for a nonideal operational amplifier, that is, one having a finite input impedance (Fig. 14-18b), will be used.)

**solution**  The model of Fig. 14-18b has already been linearized. ac analysis of the circuit can, therefore, be performed directly. The system of units is that of column 2 in Table 14-1.

**input coding**  The input code for this problem is provided in Fig. 14-19. In this problem the ac driving source, EINAC, is expressed as a complex number in line 6. The keyword COMPLEX means that EINAC is an ac source and that the arguments of COMPLEX are its real and imaginary parts (1, 0). Line 12 establishes a parameter proportional to the frequency to be used in the OUTPUT statement.

In lines 25 and 26, the MAG and PHASE in parentheses after N4 show that the magnitude and phase of N4 are to be plotted versus PFREQ. REAL in parentheses tells that PFREQ is a real

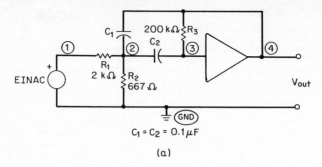

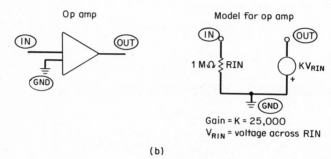

**Fig. 14-18**  Bandpass filter. (*a*) Schematic. (*b*) Model for operational amplifier (Problem 14.4).

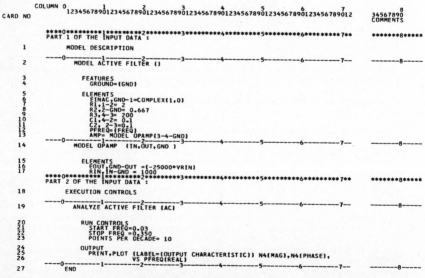

**Fig. 14-19**  Active filter input code.

ANALYZE ACTIVE FILTER (AC)

```
                     OPERATING POINT
PFREQ                    3.0000E-02
N4                        .0


                        FREQUENCY =      .30000D-01      ANGULAR FREQUENCY =    .18850
                     REAL            IMAGINARY        MAGNITUDE       LOG MAGNITUDE      PHASE
                                                                          (DB)         (DEG)
PFREQ                3.0000E-02        .0             3.0000E-02       -30.458            .0
N4                  -1.9258E-02      -.97715          .97734          -.19908          -91.129


                        FREQUENCY =      .40000D-01      ANGULAR FREQUENCY =    .25133
                     REAL            IMAGINARY        MAGNITUDE       LOG MAGNITUDE      PHASE
                                                                          (DB)         (DEG)
PFREQ                4.0000E-02        .0             4.0000E-02       -27.959            .0
N4                  -3.6274E-02      -1.3409          1.3413           2.5508          -91.550


                        FREQUENCY =      .50000D-01      ANGULAR FREQUENCY =    .31416
                     REAL            IMAGINARY        MAGNITUDE       LOG MAGNITUDE      PHASE
                                                                          (DB)         (DEG)
PFREQ                5.0000E-02        .0             5.0000E-02       -26.021            .0
N4                  -6.1205E-02      -1.7413          1.7424           4.8228          -92.013


                        FREQUENCY =      .60000D-01      ANGULAR FREQUENCY =    .37699
                     REAL            IMAGINARY        MAGNITUDE       LOG MAGNITUDE      PHASE
                                                                          (DB)         (DEG)
PFREQ                6.0000E-02        .0             6.0000E-02       -24.437            .0
N4                  -9.7218E-02      -2.1938          2.1959           6.8324          -92.538


                        FREQUENCY =      .70000D-01      ANGULAR FREQUENCY =    .43982
                     REAL            IMAGINARY        MAGNITUDE       LOG MAGNITUDE      PHASE
                                                                          (DB)         (DEG)
PFREQ                7.0000E-02        .0             7.0000E-02       -23.098            .0
N4                   -.14955         -2.7195          2.7236           8.7028          -93.148
```

**Fig. 14-20**   One page of the printed output from the active filter analysis.

number. The use of PFREQ instead of FREQ for the independent plot variable results in a linear scale of frequency, instead of a default logarithmic scale.

Figure 14-20 is one page of the printed output and Fig. 14-21 is a plot of the output. From Fig. 14-20, the maximum magnitude of N4 is seen to be 48.8 V at a frequency of 0.15744 kHz or 157.44 Hz. Thus the center frequency is 157.44 Hz, and the center frequency gain is 48.8. The bandwidth is defined as the half-power points (points at which the voltage gain drops to $1/\sqrt{2}$ of its peak). If one reads from the plot of Fig. 14-21, the bandwidth appears to be about 16 Hz.

## 14.10   ANALYSIS OF A NAND CIRCUIT

**problem 14.5**   Figure 14-22 shows the circuit of a two-input transistor-transistor logic (TTL) NAND gate.[13] If both of its inputs are high, the output is low; otherwise its output is high. For the purposes of this problem, high is described as a voltage above 2.5 V, low as any voltage below 0.5 V. Input B is held at a high voltage. Determine the output waveshape when the voltage of Fig. 14-22b is applied to input A.

**solution**   The voltage of Fig. 14-22b can be described by the ASTAP-supplied function SINSQ. SINSQ provides a train of pulses with a sine-squared rise and fall (Fig. 14-23). The general form of the sine-squared function SINSQ is SINSQ (TO, TR, TH, TF, TP, HO, H) where the arguments are those identified in Fig. 14-23.

Any interval can be zero, but only the arguments H and HO can have negative values. The pulse can be made to fall first and rise last by making H less than HO. If only one pulse is desired, argument TP must be made larger than the STOP TIME of the transient analysis. All seven arguments must appear when defining a voltage or current source in the input data.

The transistor model and units of column I, Table 14-1, will be used here. The model TRANS is described again because it was not retained formerly. A model for the Schottky diode is given in Fig. 14-24.

### Input Coding

Figure 14-25 shows the input listing for this problem. Line 8 specifies the time-varying driving function. Line 47, as in the first problem, calls for a transient analysis. This implies that a dc analysis will be performed to obtain the initial conditions of the circuit prior to the transient analysis. The dc analysis supplies the steady-state conditions that exist at the start of the transient analysis.

Line 49 asks for a TOPOLOGY, which lists all the nodes of the circuit, all elements connected to these nodes, and the other node to which each element is connected. The time-varying voltage sources are set to their value at the start time. In this problem there is only one time-varying voltage source, line 8, and its value at the start time, 0, is 3 V.

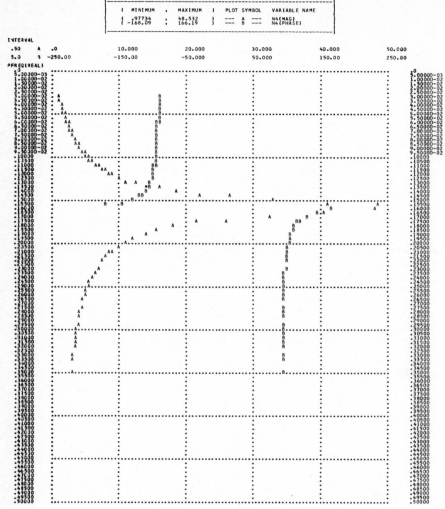

**Fig. 14-21**  Output plot from the active filter analysis.

### Output from the NAND Circuit

Figure 14-26 shows one page of TOPOLOGY CHECK that resulted from the TOPOLOGY request of line 49. Figure 14-27 shows one page of five from the PRINT output requested by lines 53 and 54. The plotted output requested by line 54 is shown in Fig. 14-28. The behavior of the circuit can be seen to obey NAND operation. The plot requested by line 54 is not shown because it would not add anything new.

## 14.11  CONCLUSION

The needs of the circuit designer continue to be addressed through the development of comprehensive simulators such as ASTAP that provide the required modes of analysis and the facilities of a user-oriented language. The capability now exists to handle the large networks peculiar to LSI at reasonable, if not fully acceptable, computational speeds. The circuit designer is becoming the chip designer, and the chip now contains many of the functions previously provided by a computer.

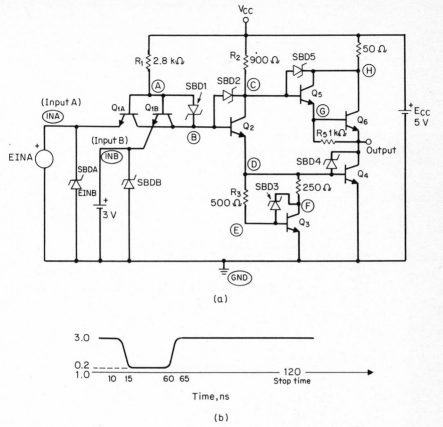

(a)

(b)

**Fig. 14-22**  A TTL two-way NAND. (*a*) Schematic. (*b*) Voltage waveform applied to node INA (Problem 14.5).

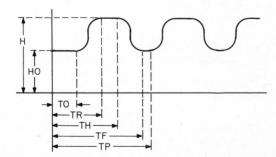

**Fig. 14-23**  Illustration of the sine-square function SINSQ (TO, TR, TH, TF, TP, HO, H).

The examples of this chapter illustrate the application of one of many CACD programs now in use. Many additional features of ASTAP were not explained, and other CACD programs have unique features of their own. The development of CACD tools is continuing unabated, with new features being incorporated on an almost continuous basis. Significant efforts are also in process to integrate CACD tools into comprehensive CAD systems.

In contrast to the batch type of runs illustrated in this chapter, many CACD users today

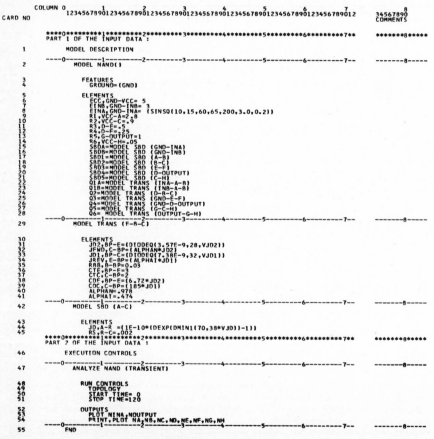

RS = .002 kΩ

(R)

JD = (1E-10∗(DEXP (DMAX1 (70,38VJD))1))

(a)      (b)

**Fig. 14-24**  Schottky barrier diode. (a) Schematic. (b) ASTAP model. DEXP and DMAX1 are FORTRAN functions.

```
                        L I S T   O F   D A T A   C A R D S
                        ****************************************
           COLUMN 0       1         2         3         4         5         6         7         8
  CARD NO      1234567890123456789012345678901234567890123456789012345678901234567890123456789012      34567890
                                                                                                      COMMENTS

            ****0*********1*********2*********3*********4*********5*********6*********7**      *******8*****
            PART 1 OF THE INPUT DATA :
    1          MODEL DESCRIPTION
            ----0---------1---------2---------3---------4---------5---------6---------7--      -------8-----
    2             MODEL NAND()

    3             FEATURES
    4                GROUND=(GND)

    5             ELEMENTS
    6                ECC,GND-VCC= 5
    7                EINB,GND-INB= 3
    8                EINA,GND-INA=  (SINSQ(10,15,60,65,200,3.0,0.2))
    9                R1,VCC-A=2.8
   10                R2,VCC-C=.9
   11                R3,D-E=.5
   12                R4,D-F=.25
   13                R5,G-OUTPUT=1
   14                R6,VCC-H=.05
   15                SBDA=MODEL SBD (GND-INA)
   16                SBDB=MODEL SBD (GND-INB)
   17                SBD1=MODEL SBD (A-B)
   18                SBD2=MODEL SBD (B-C)
   19                SBD3=MODEL SBD (E-F)
   20                SBD4=MODEL SBD (D-OUTPUT)
   21                SBD5=MODEL SBD (C-H)
   22                Q1A=MODEL TRANS (INA-A-B)
   23                Q1B=MODEL TRANS (INB-A-B)
   24                Q2=MODEL TRANS (D-B-C)
   25                Q3=MODEL TRANS (GND-E-F)
   26                Q4=MODEL TRANS (GND-D-OUTPUT)
   27                Q5=MODEL TRANS (G-C-H)
   28                Q6= MODEL TRANS (OUTPUT-G-H)
            ----0---------1---------2---------3---------4---------5---------6---------7--      -------8-----
   29             MODEL TRANS (F-B-C)

   30             ELEMENTS
   31                JD2,BP-E=(DIODEQ(3.57E-9,28,VJD2))
   32                JFWD,C-BP=(ALPHAN*JD2)
   33                JD1,BP-C=(DIODEQ(7.38E-9,32,VJD1))
   34                JREV,E-BP=(ALPHAI*JD1)
   35                RBB,B-BP=0.03
   36                CTE,BP-E=3
   37                CTC,C-BP=2
   38                CDE,BP-E=(6.72*JD2)
   39                CDC,C-BP=(185*JD1)
   40                ALPHAN=.978
   41                ALPHAI=.474
            ----0---------1---------2---------3---------4---------5---------6---------7--      -------8-----
   42             MODEL SBD (A-C)

   43             ELEMENTS
   44                JD,A-R =(1E-10*(DEXP(DMIN1(70,38*VJD))-1))
   45                RS,R-C=.002
            ****0*********1*********2*********3*********4*********5*********6*********7**      *******8*****
            PART 2 OF THE INPUT DATA :
   46          EXECUTION CONTROLS
            ----0---------1---------2---------3---------4---------5---------6---------7--      -------8-----
   47             ANALYZE NAND (TRANSIENT)

   48             RUN CONTROLS
   49                TOPOLOGY
   50                START TIME= 0
   51                STOP TIME=120

   52             OUTPUTS
   53                PLOT NINA,NOUTPUT
   54                PRINT,PLOT NA,NB,NC,ND,NE,NF,NG,NH
            ----0---------1---------2---------3---------4---------5---------6---------7--      -------8-----
   55          END
```

**Fig. 14-25**  NAND input code.

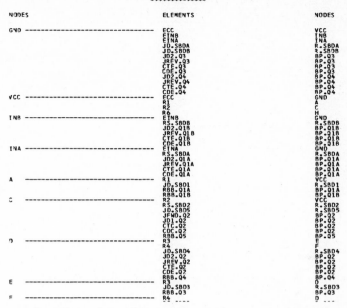

**Fig. 14-26**   One page of the topology output for the NAND circuit.

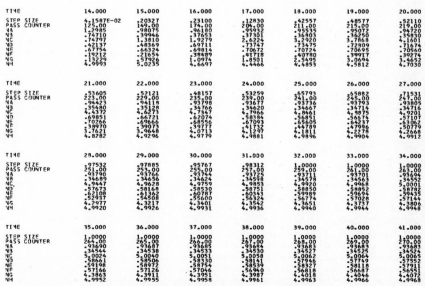

**Fig. 14-27**   Part of the print output for the NAND circuit.

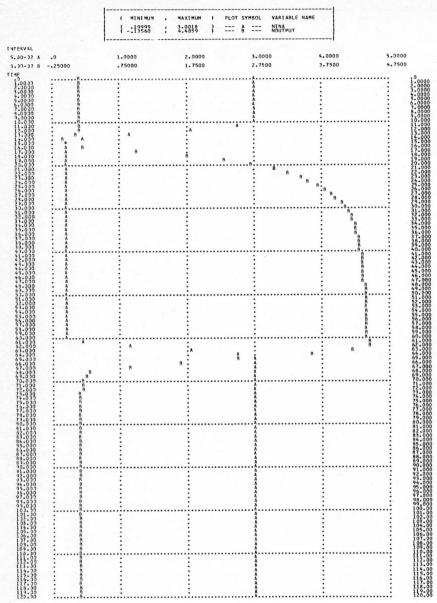

**Fig. 14-28**  Output plot for the NAND circuit.

are taking advantage of graphical terminals that permit the designer to interact directly with the computer in either a dedicated mode via microprocessors or in a time-sharing mode where the computer is a large processor. To be successful, the engineers and technicians engaged in integrated semiconductor design must appreciate the intrinsic value of CACD and use it extensively. They must also push the state of the art in the development of CACD tools by demanding programs that not only serve functional needs, but also enhance the man-machine communication.

## 14.12   REFERENCES

1. Jensen, R. W., and McNamee, L. P.: *Handbook of Circuit Analysis Languages and Techniques,* Prentice-Hall, Englewood Cliffs, N.J., 1976.
2. Kelly, M. J.: CAD Limits LSI, Proc. Eighth Asilomar Conference on Circuits, Systems, and Computers, December 1974, pp. 379–384.
3. Blattner, D. J.: Choosing the Right Programs for Computer-Aided Design, *Electronics,* vol. 49, no. 9, April 29, 1976.
4. A Survey of Computer-Aided Design and Analysis Programs, Technical Report AFAPL-TR-76-33, Air Force Aero Propulsion Laboratory, April 1976.
5. *ASTAP Program Reference Manual,* SH20-1118, IBM Corporation, Mechanicsburg, Pa., 1973.
6. Weeks, W. T., et al.: Algorithms for ASTAP: A Network Analysis Program, *IEEE Trans. Circuit Theory,* CT-20, No. 6, pp. 628–634, November 1973.
7. Gear, C. W.: The Automatic Integration of Stiff Ordinary Differential Equations, *Proc. 1968 IFIPS Congress,* pp. A81–A85.
8. Rose, D. J., and Willoughby, R. A. (eds.): *Sparse Matrices and Their Applications,* Plenum, New York, 1972.
9. Hachtel, G. D., Brayton, R. K., and Gustavson, F. G.: The Sparse Tableau Approach to Network Analysis and Design, *IEEE Trans. Circuit Theory,* CT-18, No. 1, pp. 101–113, January 1971.
10. Mahoney, G. W.: Program Called ASTAP Makes Fast Work of Analysing Large-Scale Circuits, *Electronics,* pp. 109–113, April 18, 1974.
11. Searle, C. L., Boothroyd, A. R., Angelo, E. J., Jr., Gray, P. E., and Peterson, D. O.: Elementary Circuit Properties and Transistors, Semiconductor Electronic Education Committee, vol. 3, Wiley, New York, 1964.
12. Millman, J., and Halkias, C. C.: *Integrated Electronics Analog and Digital Circuits and Systems,* McGraw-Hill, New York, 1972, p. 556.
13. Morris, R. L., and Miller, J. R.: *Designing with TTL Integrated Circuits,* McGraw-Hill, New York, 1971.

# Chapter 15

# Analog-Digital Conversion

## DANIEL H. SHEINGOLD

**Manager of Technical Marketing,
Analog Devices, Inc., Norwood, Mass.**

## 15.1 INTRODUCTION

Information in digital form can be easily processed, stored, communicated, and displayed, indestructibly and without error. The availability of low-cost devices to handle digital data affords many opportunities for the application of digital techniques to the measurement, manipulation, and control of real-world variables, such as voltages, velocities, pressures, flows, and temperatures. Such variables can be transduced to electrical form—voltage, current, or impedance. In order to communicate with the digital world, these variables must then be converted into digital form. In turn, data are often returned to analog form for display or control of real-world variables (Fig. 15-1).

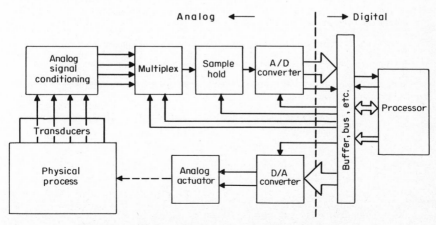

**Fig. 15-1** Application of converters in a digital processing system.

This chapter deals with problems encountered in the understanding, selection, and application of analog-to-digital (a/d) and digital-to-analog (d/a) converters. Because d/a converters (DACs) are simpler in concept and configuration than a/d converters (ADCs) and, in fact, are used as components of some ADCs, we shall first consider the DAC.

## DIGITAL-ANALOG CONVERSION

### 15.2   CODING AND THE BASIC CONVERSION RELATIONSHIP

**problem 15.1**   (*a*) What is the resolution of a 12-bit DAC, as a percentage of full scale? (*b*) How many bits are needed for a resolution of 5 mV out of +10 V full scale? (*c*) What output voltage might be expected from an 8-bit DAC (Fig. 15-2), having a fixed reference and a full-scale range of 0 to +10 V, when the parallel *binary* input code 1 0 0 1 1 0 0 1 is applied? (*d*) What is the output if the same set of bits represents a *binary-coded decimal* (BCD) input, and the DAC is configured appropriately? (*e*) What is the output if the same input word represents a *complementary binary* ("negative-true") input and the DAC has an appropriate conversion relationship? (*f*) What is the output if the same input word represents a *2's complement* bipolar input and the nominal output range is ±10 V (noninverted)? (*g*) What change in the input word is needed to adapt an *offset-binary* DAC for a 2's-complement input?

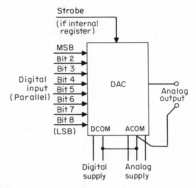

**Fig. 15-2**   An 8-bit digital-to-analog converter. (Problem 15.1.)

**theory**   A DAC accepts a digital input *code*. The input code usually appears in *parallel*, i.e., simultaneously, on a set of input lines. However, it may also appear in *serial*—as a train of levels or pulses on a single line. If a given code is applied to a DAC, one cannot know what the output of the DAC will be unless two pieces of information are known: the *quantitative meaning* of the code and the *conversion relationship* embodied in the converter. There are many kinds of codes, but the most popular is *binary*, in which the code simply represents a number in the binary numbering system.

For example, given the 4-*bit* code 1 0 1 1, we may interpret it as a binary number with a value of $(1 \times 2^3)$ plus $(0 \times 2^2)$ plus $(1 \times 2^1)$ plus $(1 \times 2^0) = 8 + 2 + 1 = 11$. The 1 at the extreme left is called the most-significant bit (MSB), and the 1 at the extreme right is called the least-significant bit (LSB). The maximum value of a 4-bit code in binary is 15 (all 1s); the minimum is zero. The value of the least-significant bit, relative to the total number of values (1:16) is the *resolution* of the binary number. There are no finer subdivisions.

In converter practice, it is helpful to use the *fractional equivalent* of binary numbers. In this representation, the binary number is divided by $2^n$, where $n$ is the number of bits. For example, the fractional value $N$ of 1011 is:

$$N = \frac{(1 \times 2^3) + (0 \times 2^2) + (1 \times 2^1) + (1 \times 2^0)}{2^4}$$
$$= (1 \times 2^{-1}) + (0 \times 2^{-2}) + (1 \times 2^{-3}) + (1 \times 2^{-4})$$
$$= 11/16$$

In this way, the maximum value is ¹⁵⁄₁₆ (or $1 - \frac{1}{16}$); the minimum value is 0 (see Table 15-1). In converter practice, the MSB is generally numbered bit 1, the next is bit 2, and so on to the LSB, which is bit $n$. The value of the $i$th bit is $2^{-i}$, and the value of "all-1s" is $(1 - 2^{-n})$. This notation has practical advantages.

**TABLE 15-1    A 4-Bit Example of Fractional-Binary Coding and Its Base-10 Equivalents**

| | | Code | | | |
|---|---|---|---|---|---|
| Decimal fraction | Binary fraction | MSB $(\times \frac{1}{2})$ | Bit 2 $(\times \frac{1}{4})$ | Bit 3 $(\times \frac{1}{8})$ | Bit 4 $(\times \frac{1}{16})$ |
| 0 | 0.0000 | 0 | 0 | 0 | 0 |
| $\frac{1}{16} = 2^{-4}$ (LSB) | 0.0001 | 0 | 0 | 0 | 1 |
| $\frac{2}{16} = \frac{1}{8}$ | 0.0010 | 0 | 0 | 1 | 0 |
| $\frac{3}{16} = \frac{1}{8} + \frac{1}{16}$ | 0.0011 | 0 | 0 | 1 | 1 |
| $\frac{4}{16} = \frac{1}{4}$ | 0.0100 | 0 | 1 | 0 | 0 |
| $\frac{5}{16} = \frac{1}{4} + \frac{1}{16}$ | 0.0101 | 0 | 1 | 0 | 1 |
| $\frac{6}{16} = \frac{1}{4} + \frac{1}{8}$ | 0.0110 | 0 | 1 | 1 | 0 |
| $\frac{7}{16} = \frac{1}{4} + \frac{1}{8} + \frac{1}{16}$ | 0.0111 | 0 | 1 | 1 | 1 |
| $\frac{8}{16} = \frac{1}{2}$ (MSB) | 0.1000 | 1 | 0 | 0 | 0 |
| $\frac{9}{16} = \frac{1}{2} + \frac{1}{16}$ | 0.1001 | 1 | 0 | 0 | 1 |
| $\frac{10}{16} = \frac{1}{2} + \frac{1}{8}$ | 0.1010 | 1 | 0 | 1 | 0 |
| $\frac{11}{16} = \frac{1}{2} + \frac{1}{8} + \frac{1}{16}$ | 0.1011 | 1 | 0 | 1 | 1 |
| $\frac{12}{16} = \frac{1}{2} + \frac{1}{4}$ | 0.1100 | 1 | 1 | 0 | 0 |
| $\frac{13}{16} = \frac{1}{2} + \frac{1}{4} + \frac{1}{16}$ | 0.1101 | 1 | 1 | 0 | 1 |
| $\frac{14}{16} = \frac{1}{2} + \frac{1}{4} + \frac{1}{8}$ | 0.1110 | 1 | 1 | 1 | 0 |
| $\frac{15}{16} = \frac{1}{2} + \frac{1}{4} + \frac{1}{8} + \frac{1}{16}$ | 0.1111 | 1 | 1 | 1 | 1 |

Irrespective of the number of bits, all values can be referred to a never-quite-achieved "full scale," with a normalized value of unity. In fact, each bit has a constant value (e.g., the MSB always is $\frac{1}{2}$), regardless of the number of bits.* Since these values are referred to full scale, they can be expressed in percent, parts per million (ppm) or even decibels (dB).† Table 15-2 lists bit weights, in fractional binary, for up to 20 bits.

**TABLE 15-2    Fractional-Binary Bit Weights**

| BIT | $2^{-n}$ | $\frac{1}{2}^n$ (Fraction) | "dB" | $\frac{1}{2}^n$ (Decimal) | % | ppm |
|---|---|---|---|---|---|---|
| FS | $2^0$ | 1 | 0 | 1.0 | 100 | 1,000,000 |
| MSB | $2^{-1}$ | 1/2 | −6 | 0.5 | 50. | 500,000 |
| 2 | $2^{-2}$ | 1/4 | −12 | 0.25 | 25 | 250,000 |
| 3 | $2^{-3}$ | 1/8 | −18.1 | 0.125 | 12.5 | 125,000 |
| 4 | $2^{-4}$ | 1/16 | −24.1 | 0.0625 | 6.2 | 62,500 |
| 5 | $2^{-5}$ | 1/32 | −30.1 | 0.03125 | 3.1 | 31,250 |
| 6 | $2^{-6}$ | 1/64 | −36.1 | 0.015625 | 1.6 | 15,625 |
| 7 | $2^{-7}$ | 1/128 | −42.1 | 0.007812 | 0.8 | 7,812 |
| 8 | $2^{-8}$ | 1/256 | −48.2 | 0.003906 | 0.4 | 3,906 |
| 9 | $2^{-9}$ | 1/512 | −54.2 | 0.001953 | 0.2 | 1,953 |
| 10 | $2^{-10}$ | 1/1024 | −60.2 | 0.0009766 | 0.1 | 977 |
| 11 | $2^{-11}$ | 1/2048 | −66.2 | 0.00048828 | 0.05 | 488 |
| 12 | $2^{-12}$ | 1/4096 | −72.2 | 0.00024414 | 0.024 | 244 |
| 13 | $2^{-13}$ | 1/8192 | −78.3 | 0.00012207 | 0.012 | 122 |
| 14 | $2^{-14}$ | 1/16,384 | −84.3 | 0.000061035 | 0.006 | 61 |
| 15 | $2^{-15}$ | 1/32,768 | −90.3 | 0.0000305176 | 0.003 | 31 |
| 16 | $2^{-16}$ | 1/65,536 | −96.3 | 0.0000152588 | 0.0015 | 15 |
| 17 | $2^{-17}$ | 1/131,072 | −102.3 | 0.00000762939 | 0.0008 | 7.6 |
| 18 | $2^{-18}$ | 1/262,144 | −108.4 | 0.000003814697 | 0.0004 | 3.8 |
| 19 | $2^{-19}$ | 1/524,288 | −114.4 | 0.000001907349 | 0.0002 | 1.9 |
| 20 | $2^{-20}$ | 1/1,048,576 | −120.4 | 0.0000009536743 | 0.0001 | 0.95 |

If $N$ is the fractional value of the binary number, the conversion relationship is $NV_{FS}$, where $V_{FS}$ is the nominal full-scale output voltage (Fig. 15-3). The actual maximum output magnitude is thus $V_{FS}(1 - 2^{-n})$, the MSB value is $V_{FS}/2$, and the LSB value is $2^{-n}V_{FS}$. Term $V_{FS}$ may be either positive or negative. The output may also be a positive or negative *current*, $NI_{FS}$, where $I_{FS}$ is the nominal full-scale current. Values of $V_{FS}$ or $I_{FS}$ depend on the magnitude and polarity of a

---

*In microprocessor practice, the bits are usually numbered in terms of positive exponents of 2, starting at the right: Bit 0 is the LSB, and bit $(n - 1)$ is the MSB.
†1 db $= 20 \log 2^{-i} = -6.02i$.

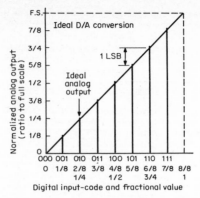

**Fig. 15-3**  Ideal conversion relationship of a 3-bit binary DAC with positive-true input.

*reference* (either internal or external) and the magnitude and polarity of the device's transfer "gain."

In interpretation of BCD coding, the input "word" is divided into groups of four bits *(quads)*, starting from the right. Each quad is permitted to have a maximum binary value of 1 0 0 1 (9). The quad's LSB thus has a fractional value of $\frac{1}{10}$ the weighting of the quad. The leftmost quad has a weighting of 1, and each quad to its right is successively weighted by an additional factor of $\frac{1}{10}$ (Table 15-3). Thus, a 12-bit (three-digit) BCD number, $M = 0\ 0\ 1\ 1\ 0\ 1\ 0\ 1\ 0\ 1\ 1\ 1$, would be (3 × 0.1) plus (5 × 0.01) plus (7 × 0.001), or 0.357. The maximum fractional BCD value is 0.999. In

**TABLE 15-3   Example of Binary-Coded Decimal (2 Digits)**

| | BCD Code | |
| --- | --- | --- |
| Decimal fraction | MSQ (× $\frac{1}{10}$) × 8 × 4 × 2 × 1 | 2d Quad (× $\frac{1}{100}$) × 8 × 4 × 2 × 1 |
| 0.00 = 0.00 + 0.00 | 0 0 0 0 | 0 0 0 0 |
| 0.01 = 0.00 + 0.01 | 0 0 0 0 | 0 0 0 1 |
| 0.02 = 0.00 + 0.02 | 0 0 0 0 | 0 0 1 0 |
| 0.03 = 0.00 + 0.03 | 0 0 0 0 | 0 0 1 1 |
| 0.04 = 0.00 + 0.04 | 0 0 0 0 | 0 1 0 0 |
| 0.05 = 0.00 + 0.05 | 0 0 0 0 | 0 1 0 1 |
| 0.06 = 0.00 + 0.06 | 0 0 0 0 | 0 1 1 0 |
| 0.07 = 0.00 + 0.07 | 0 0 0 0 | 0 1 1 1 |
| 0.08 = 0.00 + 0.08 | 0 0 0 0 | 1 0 0 0 |
| 0.09 = 0.00 + 0.09 | 0 0 0 0 | 1 0 0 1 |
| 0.10 = 0.10 + 0.00 | 0 0 0 1 | 0 0 0 0 |
| 0.11 = 0.10 + 0.01 | 0 0 0 1 | 0 0 0 1 |
| . | | |
| . | | |
| 0.20 = 0.20 + 0.00 | 0 0 1 0 | 0 0 0 0 |
| . | | |
| . | | |
| 0.30 = 0.30 + 0.00 | 0 0 1 1 | 0 0 0 0 |
| . | | |
| . | | |
| 0.90 = 0.90 + 0.00 | 1 0 0 1 | 0 0 0 0 |
| 0.91 = 0.90 + 0.01 | 1 0 0 1 | 0 0 0 1 |
| . | | |
| . | | |
| 0.98 = 0.90 + 0.08 | 1 0 0 1 | 1 0 0 0 |
| 0.99 = 0.90 + 0.09 | 1 0 0 1 | 1 0 0 1 |

**TABLE 15-4  Examples of Complementary (Negative-True) Codes**

| Decimal number | Natural binary | Complementary binary | BCD | Complementary BCD |
|---|---|---|---|---|
| 0  BIN  DEC | 0000 | 1111 | 00000 | 11111 |
| 1, 1/16, 1/10 | 0001 | 1110 | 00001 | 11110 |
| 2, 2/16, 2/10 | 0010 | 1101 | 00010 | 11101 |
| 3, 3/16, 3/10 | 0011 | 1100 | 00011 | 11100 |
| 4, 4/16, 4/10 | 0100 | 1011 | 00100 | 11011 |
| 5, 5/16, 5/10 | 0101 | 1010 | 00101 | 11010 |
| 6, 6/16, 6/10 | 0110 | 1001 | 00110 | 11001 |
| 7, 7/16, 7/10 | 0111 | 1000 | 00111 | 11000 |
| 8, 8/16, 8/10 | 1000 | 0111 | 01000 | 10111 |
| 9, 9/16, 9/10 | 1001 | 0110 | 01001 | 10110 |
| 10,10/16,10/10 | 1010 | 0101 | 10000· | 01111 |
| 11,11/16,11/10 | 1011 | 0100 | 10001 | 01110 |

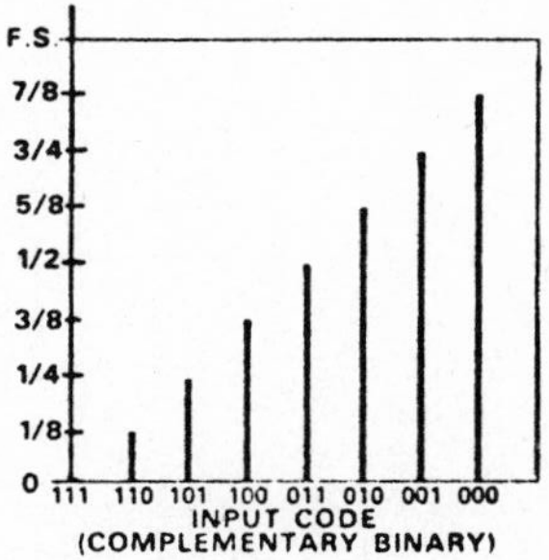

Conversion Relationship of
ideal 3-bit complementary
binary DAC

F.S  7/8  3/4  5/8  1/2  3/8  1/4  1/8  0

111 110 101 100 011 010 001 000
INPUT CODE
(COMPLEMENTARY BINARY)

some applications of BCD, there are additional "overrange" digits, which add whole-number significance. For example, 3½ digits has an additional digit with a weight of 1.000; 3¾ digits has two additional digits with a maximum value of 3.999.

The conversion relationship of a BCD d/a converter is $MV_{FS}$, where $M$ is the BCD fractional value. In addition to the magnitude bits, a BCD converter may accept an additional *polarity* bit to switch the polarity of the output.

*Complementary* codes, such as complementary binary, are simply codes for which all bits are complemented (negative-true codes). In other words, 1s and 0s are interchanged (Table 15-4). Reasons for DAC's requiring complementary input codes are the availability of high-accuracy switching components, limitations of space, and lower production cost (for some manufacturers).

For bipolar applications, the MSB becomes the sign bit, and the remaining bits represent the magnitude. A popular binary digital code is *2's complement*, which is formed by complementing it (1's complement), adding 1 LSB, and disregarding any overflow. An MSB of 0 indicates a positive number; 1 indicates a negative number. For example, the three-bit-plus-sign number, 0 1 0 1 (+5), is made negative by complementing all bits (1 0 1 0) and adding 1 LSB (1 0 1 1). When the result is checked, if the overflow carry is disregarded, 0 1 0 1 + 1 0 1 1 = 0 0 0 0.

Two's complement can be implemented with a binary DAC, if the MSB is complemented to an *offset binary* number (Table 15-5). The progression of values in offset binary is from all-0s at minus full-scale ($-FS$), to all 1s at (FS $-$ 1 LSB). Analog zero occurs at 1 0 0 0. Thus, a unipolar binary DAC's conversion relationship can be made 2's complement by adding a fixed negative offset at the output equal to half-scale, amplifying the output by a factor of 2, and complementing the MSB.

**TABLE 15-5    Examples of Codes for Positive and Negative Quantities**

| Number | Decimal Fraction Positive reference | Negative reference | Sign + magnitude | Two's complement | Offset binary | One's complement |
|---|---|---|---|---|---|---|
| +7 | +⅞ | −⅞ | 0 1 1 1 | 0 1 1 1 | 1 1 1 1 | 0 1 1 1 |
| +6 | +⁶⁄₈ | −⁶⁄₈ | 0 1 1 0 | 0 1 1 0 | 1 1 1 0 | 0 1 1 0 |
| +5 | +⅝ | −⅝ | 0 1 0 1 | 0 1 0 1 | 1 1 0 1 | 0 1 0 1 |
| +4 | +⁴⁄₈ | −⁴⁄₈ | 0 1 0 0 | 0 1 0 0 | 1 1 0 0 | 0 1 0 0 |
| +3 | +⅜ | −⅜ | 0 0 1 1 | 0 0 1 1 | 1 0 1 1 | 0 0 1 1 |
| +2 | +²⁄₈ | −²⁄₈ | 0 0 1 0 | 0 0 1 0 | 1 0 1 0 | 0 0 1 0 |
| +1 | +⅛ | −⅛ | 0 0 0 1 | 0 0 0 1 | 1 0 0 1 | 0 0 0 1 |
| 0 | 0+ | 0− | 0 0 0 0 | 0 0 0 0 | 1 0 0 0 | 0 0 0 0 |
| 0 | 0− | 0+ | 1 0 0 0 | (0 0 0 0) | (1 0 0 0) | 1 1 1 1 |
| −1 | −⅛ | +⅛ | 1 0 0 1 | 1 1 1 1 | 0 1 1 1 | 1 1 1 0 |
| −2 | −²⁄₈ | +²⁄₈ | 1 0 1 0 | 1 1 1 0 | 0 1 1 0 | 1 1 0 1 |
| −3 | −⅜ | +⅜ | 1 0 1 1 | 1 1 0 1 | 0 1 0 1 | 1 1 0 0 |
| −4 | −⁴⁄₈ | +⁴⁄₈ | 1 1 0 0 | 1 1 0 0 | 0 1 0 0 | 1 0 1 1 |
| −5 | −⅝ | +⅝ | 1 1 0 1 | 1 0 1 1 | 0 0 1 1 | 1 0 1 0 |
| −6 | −⁶⁄₈ | +⁶⁄₈ | 1 1 1 0 | 1 0 1 0 | 0 0 1 0 | 1 0 0 1 |
| −7 | −⅞ | +⅞ | 1 1 1 1 | 1 0 0 1 | 0 0 0 1 | 1 0 0 0 |
| −8 | −⁸⁄₈ | +⁸⁄₈ | | (1 0 0 0) | (0 0 0 0) | |

The offset-binary conversion relationship is displayed in Fig. 15-4. It is expressed by:

$$V_o = V_{FS}(2N - 1) \tag{15-1}$$

where $V_o$ is the output voltage and $N$ is the fractional value of the offset-binary input code.

**solution**    (a) From Table 15-2, 12-bit resolution ($2^{-12}$) is 0.024% of full scale.

(b) For a resolution of 5 mV/10 V, or 0.05%, 11 bits are required.

(c) For an input code of 1 0 0 1 1 0 0 1, and +10 V full scale, the output of a binary DAC is 10 V (½ + ¹⁄₁₆ + ¹⁄₃₂ + ¹⁄₂₅₆) = 5.98 V.

(d) For the same code and full-scale value, the BCD value is 10 V (0.9 + 0.09) = 9.9 V.

(e) For complementary-binary coding, the response is the same as that of a binary DAC having an input code 0 1 1 0 0 1 1 0. The output is 10 V (¼ + ⅛ + ¹⁄₆₄ + ¹⁄₁₂₈) = 3.98 V.

(f) The MSB is 1, which denotes a 2's-complement code for a negative number. Use the rules for 2's complement to obtain the magnitude (0 1 1 0 0 1 1 0 + 1 = 0 1 1 0 0 1 1 1). The output value is −10 V (½ + ¼ + ¹⁄₃₂ + ¹⁄₆₄ + ¹⁄₁₂₈) = −8.05 V. Since the leftmost bit represents the sign, the second bit becomes the MSB for magnitude.

(g) To convert the original code to offset binary, the MSB is complemented: 0 0 0 1 1 0 0 1. Checking the result of (f) by using Eq. (15-1) yields $V_o$ = 10 V (⅛ + ¹⁄₁₆ + ¹⁄₁₂₈ − 1) = −8.05 V.

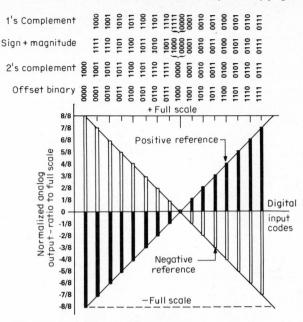

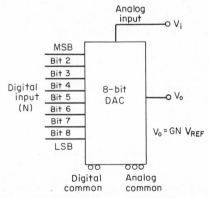

**Fig. 15-4**   Ideal conversion relationship of 4-bit bipolar DAC.

## 15.3   CONVERSION RELATIONSHIP OF A MULTIPLYING DAC

**problem 15.2**   Figure 15-5 is a block diagram of an 8-bit multiplying (variable-reference) DAC, with 4-quadrant response and a nominal gain of 1 at full scale. Compute the gains and sketch the output waveforms that result from the application of a ±10-V triangular wave and the following 2's-complement code: (a) 0 0 0 0 0 0 0 0, (b) 0 1 0 0 0 0 0 0, (c) 1 1 0 0 0 0 0 0, (d) 0 1 1 1 1 1 1 1, (e) 1 0 0 0 0 0 0 1, (f) 1 0 0 0 0 0 0 0. (g) Demonstrate how the DAC may be used as a divider.

**Fig. 15-5**   An 8-bit multiplying DAC. (Problem 15.2.)

**theory**   It was suggested in Sec. 15.2 that the full-scale output voltage of a DAC ($V_{FS}$) depends on the product of the reference voltage and the transfer gain (transconductance, in the case of current-output DACs). Thus, for a binary DAC, the output voltage $V_o$ is:

$$V_o = GNV_{REF} \qquad (15\text{-}2)$$

where $N$ = fractional value of the digital code
　　　$G$ = gain
　　　$V_{REF}$ = internal or external reference voltage

Since $V_o$ depends on the product of $N$ and $V_{REF}$, a DAC can be used as a multiplier of two signals—one digital, the other analog. A common application for this function is in *variable-gain amplification*, where the gain of an analog signal is controlled by a digital input.

DACs differ in their suitability for multiplication. A full *four-quadrant* DAC can multiply positive or negative digital codes by positive or negative analog signals. The output also obeys the rules of multiplication with respect to sign. Some DACs have only a limited multiplying capability, with provision for reference inputs of but a single polarity, and unipolar or bipolar digital input. They are, respectively, *one-* or *two-quadrant* multiplying DACs. Yet other DACs have internal references that are permanently engaged. They are *fixed-reference* DACs, with perhaps a limited range of gain-trim, but no variability.

The most frequently used codes are binary, for single-polarity digital inputs, and offset binary (and/or 2's complement), for bipolar digital inputs. Figure 15-6 illustrates the range of gains possible with a 4-bit multiplying DAC. Analog output vs. analog input, as a function of the input digital code is plotted. Note the asymmetry that exists because the number of available codes is even. The extra code is used for *minus* full scale, because the MSB indicates that it is of negative polarity.

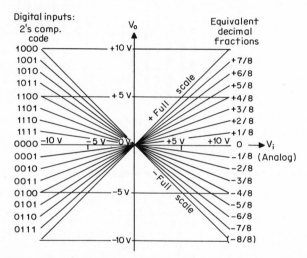

**Fig. 15-6**   Ideal multiplication relationship of 4-bit bipolar 4-quadrant multiplying DAC. Analog output vs. analog input as a function of the digital code.

As is the case with analog multipliers, multiplying DACs can be used for division through the use of negative feedback. If the overall transfer function is positive, and there is no access to the internal circuitry, an external inverting op amp may be used. (Care must be exercised to ensure system stability.) As reference to Fig. 15-7 shows, the feedback path around amplifier Al is closed through the multiplying DAC. Since Al is a high-gain amplifier, and must maintain its negative input terminal near zero, the following condition must hold:

$$V_i = - GNV_o \tag{15-3}$$

Therefore,

$$V_o = -\frac{1}{G}\frac{V_i}{N} \tag{15-4}$$

Term $N$ may have only positive values, but $V_i$ may be either positive or negative. As with all feedback dividers, errors and response time are inversely proportional to $N$ and become very large as $N$ approaches zero. The values of $V_i$ and $N$ must be such that $V_o$ is never greater than its full-scale value. Diodes from the amplifier summing point to ground may help prevent latchup. The circuit is simplified for one-quadrant operation if the multiplier already has an inverting transfer function and the feedback resistor of the output op amp is available.

**solution**   The waveshapes appear in Fig. 15-8, with the alphabetic designations matching the parts of the problem.

(*a*) 0 0 0 0 0 0 0 0 in 2's complement has a value of zero; hence, the output is zero also. Any leakage output that is observed in an actual device is known as feedthrough.

(*b*) For 0 1 0 0 0 0 0 0 in 2's complement, $N = 2^{-1} = \frac{1}{2}$. Hence, the gain will be $\frac{1}{2}$ ($G$ is defined as unity).

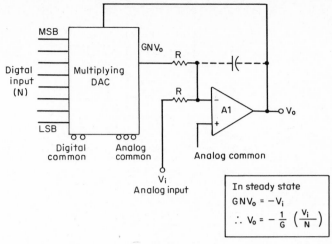

**Fig. 15-7** Using a multiplying DAC for division.

(c) 1 1 0 0 0 0 0 0 is a negative number with a magnitude of 0 0 1 1 1 1 1 1 + 1, or 0 1 0 0 0 0 0 0; hence, the gain is $-\frac{1}{2}$.

(d) 0 1 1 1 1 1 1 1 is a positive number of magnitude $(1 - 2^{-7})$, or full scale minus 1 LSB.

(e) 1 0 0 0 0 0 0 1 is a negative number with a magnitude of 0 1 1 1 1 1 1 0 + 1, or 0 1 1 1 1 1 1 1; hence, the gain is $-(1 - 2^{-7})$.

(f) 1 0 0 0 0 0 0 0 is a negative number with a magnitude of 0 1 1 1 1 1 1 1 + 1, equivalent to unity in fractional binary, or minus full-scale gain.

(g) Figure 15-7 shows a way of performing division, as discussed above.

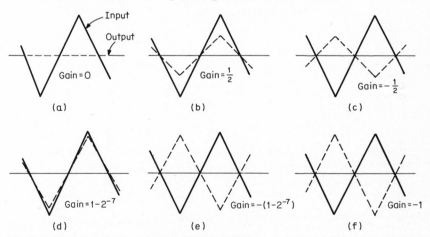

**Fig. 15-8** Response of a 4-quadrant multiplying DAC to various codes: (a) 0 0 0 0 0 0 0 0. (b) 0 1 0 0 0 0 0 0. (c) 1 1 0 0 0 0 0 0. (d) 0 1 1 1 1 1 1 1. (e) 1 0 0 0 0 0 0 1. (f) 1 0 0 0 0 0 0 0. (Problem 15.2.)

## 15.4  BASIC PARALLEL DAC WITH BINARY RESISTANCE VALUES

**problem 15.3**  The circuit of Fig. 15-9 demonstrates the concept of a basic binary DAC. (a) What is the output voltage for the code 1 0 0 0 0 0? (b) What is the full-scale output voltage? (c) Which switches are closed to embody the code for −7.03 V? (d) If resistors and switches are added to extend the circuit for 12-bit resolution, what is the resistance value of the LSB resistor? (e) Is this circuit practical for the manufacture of high-resolution DACs? Why?

**theory**  The basic elements of a complete DAC are the reference; a resistance network to provide a set of weighted voltages, currents, or gains; a set of switches to determine which "bits"

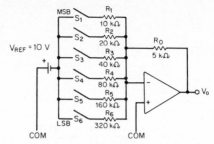

**Fig. 15-9** A 6-bit basic DAC circuit. (Problem 15.3.)

will contribute to the output; and a transducer to provide an output having the desired format (voltage or current), level, and impedance. In addition (not shown), the converter requires a switch drive and logic translation from the digital input format and levels.

In Fig. 15-9, because the summing point is at virtual ground, when a switch is closed, $-V_{REF}$ appears across the associated resistor. This causes a current of magnitude $V_{REF}/R_i$ to flow from the summing point through the resistance, switch, and reference, back to ground. Similarly, currents flow from the op-amp summing point through other circuits having closed switches. The only way the sum of these currents can be provided is from the output through feedback resistor $R_o$. Voltage $V_o$ must be just the right value to enforce the condition:

$$\frac{V_o}{R_o} = \frac{V_{REF}}{R_1} + \frac{V_{REF}}{R_2} + \frac{V_{REF}}{R_3} + \cdots + \frac{V_{REF}}{R_n} \tag{15-5}$$

Hence, the contribution of the $i$th bit to the output is

$$\Delta V_{oi} = \frac{R_o}{R_i} V_{REF} \tag{15-6}$$

Ratios $R_o/R_i$ are made identical to the bit contributions, $2^{-i}$.

While this circuit demonstrates the principle, it is little used in practice for medium- or high-resolution (8- or 12-bit and higher) converters. The wide range of required resistance is very difficult to implement efficiently. In addition, it is very demanding on the switches, requiring a very large ratio between leakage resistance when *off* and series resistance when *on*. The circuit also has poor dynamics, since it imposes a variable load on the op-amp summing point and on the reference. Switching time-constants are also variable, causing "glitches" (erroneous intermediate codes that appear as transient spikes) while switching between two *adjacent* codes that require a large number of switches to be opened and closed: e.g., from 0 1 1 1 1 1 1 1 to 1 0 0 0 0 0 0 0.

The circuits used in modern DACs, while rather varied in detail, almost universally involve resistive attenuation networks that do not require ratios in excess of 8:1 and, in most cases, only 2:1. To minimize dynamic errors, the accurately weighted bit-currents are constant, and the switches simply steer the currents from one destination (summing point) to another (ground). Because current flow is not interrupted, delays and transient disturbances are minimized.

**solution**    (a) Code 1 0 0 0 0 0 means that the MSB is only one; thus S1 is closed. The output is $-(5 \text{ k}\Omega/10 \text{ k}\Omega)(-10 \text{ V}) = +5$ V.

(b) Since the MSB is one-half full scale, the full-scale output is +10 V.

(c) To find which switches are closed, let us try a method of *successive approximations*. The switched legs contribute, successively, 5, 2.5, 1.25, 0.62, 0.31, and 0.16 V. We know that S1 must be closed, since 7.03 V > 5 V. If we close S2, the output is 7.5 V, which is too high; S2 must therefore be open. Closing S3 contributes 1.25 V, for a total of 6.25 V; closing S4 contributes 0.62 V, for a total of 6.87 V; and closing S5 contributes 0.31 V, for a total of 7.18 V. This is too high, so S5 must be open. Finally, closing S6 contributes 0.16 V, for a total of 7.03 V. Thus, switches 1, 3, 4, and 6 are closed and switches 2 and 5 are open. This is equivalent to code 1 0 1 1 0 1.

(d) The resistance value of the LSB resistor is $2^{12} \times 5 \text{ k}\Omega = 20.5 \text{ M}\Omega$.

(e) No, for reasons mentioned above.

## 15.5   D/A CONVERTER USING R-2R LADDER AND CMOS CURRENT-STEERING SWITCHES

**problem 15.4**    Figure 15-10 is a simplified circuit of a 4-bit DAC employing Complementary Metal-Oxide Semiconductor (CMOS) logic and switching, and an R-2R ladder network. (a) If the logic voltage applied to switch 1 ($V_{s1}$) is +10 V, what are the approximate voltages to ground at points A, B, C, D, E? (b) To which side of the switch will current $I_1$ be steered? (c) What are the magnitudes of voltages $V_2$, $V_3$, and $V_4$, and currents $I_1$, $I_2$, $I_3$ and $I_4$? (d) What is the output voltage for "all bits on" ($V_{s1}$, $V_{s2}$, $V_{s3}$, and $V_{s4}$ all equal to +10 V)? (*e*) If the analog reference voltage becomes +5 V, what is the output voltage for input code 1 1 1 1? (*f*) What is the output voltage if the analog reference is $-10$ V, for the same code? (*g*) If $V_{s1}$ is $-10$ V, what are the approximate voltages to ground at points *A, B, C, D, E*? (*h*) To which side of the switch will $I_1$ be steered? (*i*) Can the currents that were steered to the ground bus in this problem serve a useful purpose? (*j*) What is the purpose of diodes $D_1$ and $D_2$?

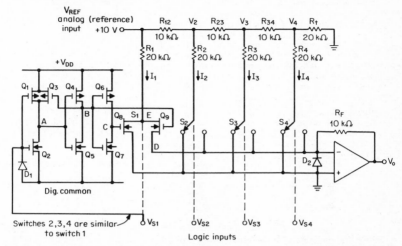

**Fig. 15-10** Simplified circuit of a 4-bit binary DAC employing an R-2R ladder network and CMOS switches in current-steering mode. (Problem 15.4.)

**theory** The R-2R ladder network has a set of series resistors, $R_{12}$, $R_{23}$, $R_{34}$, etc., of resistance $R$, and a set of shunt resistors, $R_1$, $R_2$, $R_3$, etc., of resistance $2R$. There is a final terminating resistor $R_t$ of value $2R$. The switches are in series with the shunt legs, and the current in each leg is steered either to the amplifier summing point (at virtual ground) or to a common line which is grounded.

The performance of the R-2R network can be most easily understood by performing the analysis backwards:

$$I_4 = \frac{V_4}{R_4} \tag{15-7}$$

Because $R_4$ and $R_t$ are in parallel and equal, their parallel combination equals $R_{34}$; hence,

$$V_3 = 2V_4 \tag{15-8}$$

Also, $R_3 = R_4$; the current through $R_3$ is:

$$I_3 = \frac{2V_4}{R_3} = \frac{2V_4}{R_4} = 2I_4 \tag{15-9}$$

Because $R_{34} + (R_4 \| R_4) = 2R_{34} = R_3$, the parallel combination of resistance to ground from $V_3$ is equal to $R_{23}$ and

$$V_2 = 2V_3 = 4V_4 \tag{15-10}$$

Continuing in this manner,

$$I_1 = 2I_2 = 4I_3 = 8I_4 = \frac{V_{REF}}{R_1} \tag{15-11}$$

and $$V_{REF} = 2V_2 = 4V_3 = 8V_4 \tag{15-12}$$

Hence, the currents through the shunt legs and the voltages at the nodes form a binary progression. It should be apparent that the nature of this progression is unaffected by the number of "cells," whether there be 4, 10, or 12.

Binary-weighted currents $I_1$, $I_2$, $I_3$, and $I_4$, steered by the switches, contribute either to the output voltage through the op amp or to the common current. If binary 1s steer the currents to the summing-point line, and 0s steer the currents to the common line, the current in the common line is, therefore, the complement of the current in the summing-point line. In bipolar applications, the current in the common line is used by connecting it to a second inverting op amp summing-point, instead of to ground.

In the CMOS switches, transistors $Q_{1-2}$, $Q_{4-5}$, and $Q_{6-7}$ are a set of logic inverters. Transistors $Q_{8-9}$ perform analog switching. Positive feedback is provided by $Q_3$ to speed up the switching action. Since the voltages at $B$ and $C$ are in opposite phase, the switching transistors are driven out of phase: when $Q_8$ is open, $Q_9$ is conducting.

In a practical design, it is necessary to consider the resistance of the switches $(R_{on})$, which appears in series with resistance $R_i$ of the shunt resistors. If $R_{on}$ is not negligible compared with $R_i$, the resulting DAC may be nonlinear. This may be corrected if the switch geometry causes the $R_{on}$ to increase, from leg to leg, in binary progression. Because the currents are decreasing in binary progression, there is an equal drop across all $R_{on}$, producing only a small scale-factor error.

The circuit shown can be made to operate from TTL or CMOS logic levels. Diodes $D_1$ and $D_2$ protect the CMOS switches from destruction when reverse logic or summing-point (overload) voltages are applied. Since the fraction of $V_{REF}$ that appears across the actual switches is quite small, the DAC can be used bilaterally, for both positive and negative analog inputs in multiplying-DAC applications. An advantage of CMOS is its low steady-state power drain.

**solution**    (a) Since $Q_{1-2}$, $Q_{4-5}$, and $Q_{6-7}$ invert when the input is high, point $A \approx 0$ V, $B \approx V_{DD}$, and $C \approx 0$ V. When point $C$ is low, $Q_8$ is open and when $B$ is high, $Q_9$ is closed. This permits $I_1$ to flow to the summing point $D$. The voltage at the summing point and at point $E$ (separated only by $R_{on}$), is also approximately 0 V.

(b) Current $I_1$ is steered through $Q_9$ to the amplifier summing point.

(c) $I_1 = 500$ μA, $I_2 = 250$ μA, $I_3 = 125$ μA, $I_4 = 62.5$ μA, $V_2 = 5$ V, $V_3 = 2.5$ V, and $V_4 = 1.25$ V.

(d) $V_o = -10$ kΩ $(500 + 250 + 125 + 62.5)$ μA $= -9.375$ V, same as all 1s: $-10$ V $(1 - 2^{-4}) = -9.375$ V. If accuracy and resolution are both to 4 bits, this is rounded off to $-9.4$ V.

(e) If $V_{REF} = +5$ V, the output is halved to $-4.7$ V.

(f) If the input is $-10$ V, the output polarity is inverted to $+9.4$ V.

(g) Point $A = +V_{DD}$, $B = 0$ V, $C = +V_{DD}$, $D = 0$ V, $E = 0$ V.

(h) Current flow is through $Q_8$ to the common bus.

(i) Yes, the currents steered to the ground bus represent the complementary output current. They can be supplied to an op amp instead of the "summing-point bus" (which is instead grounded), if the DAC is to respond to negative-true logic. In bipolar applications, the complementary currents are summed in an inverting amplifier, then summed with the positive-true currents and a 1 LSB offset for offset-binary conversion.

(j) The diodes provide protection against damage from excessive reverse gate voltages.

## 15.6   DAC ERRORS

**problem 15.5**   A 6-bit DAC with nominal + 10-V full-scale range has the following measured output values:

| | Code | Voltage, V |
|---|---|---|
| All 1s | 111111 | 10.04 |
| MSB | 100000 | 5.20 |
| Bit 2 | 010000 | 2.56 |
| Bit 3 | 001000 | 1.40 |
| Bit 4 | 000100 | 0.90 |
| Bit 5 | 000010 | 0.58 |
| Bit 6 (LSB) | 000001 | 0.40 |
| All 0s | 000000 | 0.20 |

(a) How much offset error is there? List corrected values. (b) What is the gain error? List corrected values. (c) List the theoretical output values for the listed inputs. (d) Calculate the errors. (e) Would you expect the response of this DAC to be monotonic? If not, which transitions would be nonmonotonic?

**theory**   Zero-offset error in a DAC is the amount by which its output differs from zero when the input code calls for zero. If the output is adjusted for the correct zero value, all other output codes are translated by the same value (Fig. 15-11a).

Gain error is the amount by which the output span (from the most-negative code to the most-positive code) differs from the theoretical value. It is caused by overall resistance-ratio errors and, in fixed-reference DACs, reference errors (Fig. 15-11b). The span is adjusted for a magnitude of $V_{FS}(1 - 2^{-n})$ in unipolar binary DACs and $2V_{FS}(1 - 2^{-n})$ in bipolar DACs. The "zero" offset of bipolar DACs is often adjusted at $-V_{FS}$. In some cases, a separate trim adjustment is provided for the MSB to ensure zero output at the midscale zero value.

Any deviation from a "best straight line" is termed a linearity error (Fig. 15-11c). It is not usual to find provisions for external trimming of linearity errors. Two kinds of linearity errors are found. One is differential nonlinearity, which is caused by differences in bit size. It leads to errors of differing magnitude when the bits are summed in various combinations.

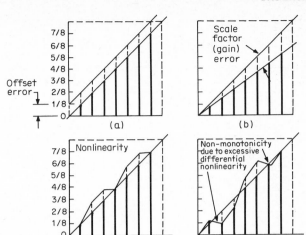

**Fig. 15-11** Errors in d/a conversion. (*a*) Zero offset. (*b*) Scale factor. (*c*) Nonlinearity. (*d*) Nonmonotonic behavior. (Problem 15.5.)

Theoretically, each step from one code to the next, in LSB increments, should be equal to 1 LSB. A deviation in the step size from this value is called a *differential-linearity* error. If, at certain transitions that involve simultaneous switching of large numbers of bits, the differential nonlinearity is greater than 1 LSB and of appropriate polarity, an increase in the digital input actually results in a *decrease* of the analog output. A departure from the expected monotonic response is known as *nonmonotonicity* (Fig. 15.11*d*). Besides being inherently inaccurate, nonmonotonic behavior is intolerable in many applications, such as displays, control systems, and a/d converters that employ DACs.

Nonlinearity of a more conventional type, *integral nonlinearity*, may also be introduced by nonlinear amplifiers and feedback resistors. Even if all the bit values are individually perfectly accurate, their sum may not be accurate. In some cases, the changing load presented by a current- or gain-output DAC's output impedance to the output op amp's offset voltage may also introduce apparent linearity errors.

**TABLE 15-6   Solution of Problem 15.5**

| Condition | Code | Voltage, V | Corrected for offset, V | Corrected for gain, V | Theoretical values, V | Bit errors, V |
|-----------|------|-----------|------------------------|----------------------|----------------------|---------------|
| All 1s | 1 1 1 1 1 1 | 10.04 | 9.84 | 9.84 | 9.84 | |
| MSB | 1 0 0 0 0 0 | 5.20 | 5.00 | 5.00 | 5.00 | 0.00 |
| Bit 2 | 0 1 0 0 0 0 | 2.56 | 2.36 | 2.36 | 2.50 | −0.14 |
| Bit 3 | 0 0 1 0 0 0 | 1.40 | 1.20 | 1.20 | 1.25 | −0.05 |
| Bit 4 | 0 0 0 1 0 0 | 0.90 | 0.70 | 0.70 | 0.62 | +0.08 |
| Bit 5 | 0 0 0 0 1 0 | 0.58 | 0.38 | 0.38 | 0.31 | +0.07 |
| Bit 6 | 0 0 0 0 0 1 | 0.40 | 0.20 | 0.20 | 0.16 | +0.04 |
| All 0s | 0 0 0 0 0 0 | 0.20 | 0.00 | 0.00 | 0.00 | |

**solution**   The required answers are summarized in Table 15-6. (*a*) The offset voltage is +0.20 V.

(*b*) There is no gain error. The correct all 1's value is $10 \text{ V}(1 - 2^{-6}) = 9.84$ V.

(*c*) The theoretical bit values (rounded off) are binary fractions of 10 V.

(*d*) The residual errors are the differences between the corrected values and the theoretical values.

(*e*) No, the DAC is nonmonotonic. The three least-significant bits have errors adding up to +0.19 V. Consider the transition from 0 0 0 1 1 1 to 0 0 1 0 0 0. The calibrated output corresponding to 0 0 0 1 1 1 is 1.28 V; the output corresponding to 0 0 1 0 0 0 is 1.20 V, a *decrease* in output for a 1-bit *increase* in digital input. This can be seen in terms of the differential linearity error. Since 0 0 0 1 1 1 is 0.19 V high and 0 0 1 0 0 0 is 0.05 V low, the differential-linearity error at that transition is −0.24 V, which exceeds the value of 1 LSB. This DAC is, in fact, nonmonotonic at *six* transitions

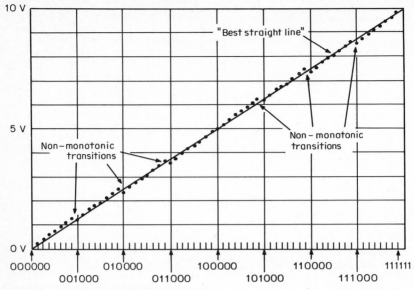

**Fig. 15-12**  Actual conversion relationship of DAC in Problem 15.5, showing nonmonotonic transitions and comparison with "best straight line."

(Fig. 15-12). Since the algebraic sum of the bit errors is zero, however, the only form of nonlinearity is differential nonlinearity.

## 15.7  TEMPERATURE EFFECTS IN THE DAC

**problem 15.6**  A 12-bit DAC has a differential-nonlinearity specification of $<\frac{1}{2}$ LSB at 25°C and a differential-nonlinearity temperature coefficient of 5 ppm/°C for the range 0 to 70°C. Is it reasonable to expect this converter to be monotonic at 70°C?

**theory**  Converter properties vary with temperature. A converter which has a total error within $\pm\frac{1}{2}$ LSB at 25°C, and hence has accuracy which is compatible with its resolution, may have gross errors at temperatures near the extremes of its rated range. Resistance values change with temperature; resistance *ratios* change (to a lesser degree) with temperature; reference voltages and op amp bias currents and offset-voltage drift with temperature; switch resistances and leakage drift with temperature. As a result, one can expect offset, gain, and relative bit-weights (hence nonlinearity) to change with temperature.

A temperature coefficient (tempco or TC) is specified to indicate the maximum change of a parameter that may be expected in a given temperature range. These coefficients, when verified by a manufacturer, are checked at two or three (sometimes more) points in the range. Unless a given temperature-sensitive parameter is known to be proportional to temperature, the presence of a tempco on a data sheet does not necessarily imply that it can be applied for small temperature changes at any arbitrary portion of the range.

Temperature coefficients are usually specified for offset, gain, and differential linearity errors. The importance of the drift depends on both its magnitude and the application. For example, if a 12-bit DAC is used for an oscilloscope display, a few bits of drift of offset (position) or gain (size) may be unimportant. Excessive differential nonlinearity or nonmonotonicity, however, may significantly distort the displayed pattern.

**solution**  The converter can be nonmonotonic if the differential nonlinearity is $> 1$ LSB. If a converter has a differential nonlinearity of $\frac{1}{2}$ LSB ($2^{-13}$, or 122 ppm, for a 12-bit converter) at 25°C, and the temperature variation adds $(70° - 25°)(5$ ppm/°C$) = 225$ ppm, the maximum differential nonlinearity could be as great as 347 ppm. This is substantially greater than the allowable 1 LSB. (At 12 bits, 1 LSB $= 2^{-12} = 244$ ppm.) It is, therefore, not reasonable to expect the converter to be monotonic at 70°C.

## 15.8  DAC SPECIFICATIONS

Specifications that are commonly used to characterize the performance of DACs are summarized in Table 15-7.

**TABLE 15-7    Digital-to-Analog Converter Specifications**

**Accuracy, Absolute**    The difference between the actual analog output and the output that is expected when a given digital code is applied to the converter. Sources of error include gain, zero (offset), linearity, and noise. Error is usually commensurate with resolution, e.g., less than $2^{-(n+1)}$, or $\frac{1}{2}$ LSB of full scale. Accuracy, however, may be much better than resolution in some applications. For example, a reference supply having only 16 discrete digitally chosen levels has a resolution of 4 bits ($\frac{1}{16}$). Its accuracy, however, may be within 0.01% of each ideal value. Absolute accuracy measurements should be made under a set of standard conditions with sources and meters traceable to an internationally recognized standard.

**Accuracy, Relative**    The deviation of the analog value at any code (relative to the full analog range of the device transfer characteristics) from its theoretical value (relative to the same range), after the full-scale range (FSR) has been calibrated. It is expressed in percent, parts per million (ppm), or fractions of 1 LSB. Since the discrete analog values corresponding to the digital input values ideally lie on a straight line, the relative-accuracy error can also be interpreted as a measure of nonlinearity. (See *Linearity*.)

**Compliance-Voltage Range**    For a current-output DAC, the maximum range of output-terminal voltage for which the device provides the specified current-output characteristics.

**Feedthrough**    Undesirable signal coupling around switches or other devices that are supposed to be turned off or provide isolation. Feedthrough is variously specified in percent, parts per million, fractions of 1 LSB, or fractions of 1 V, with a given set of inputs at a specified frequency.

**Four-Quadrant DAC**    A multiplying DAC, for which both the reference signal and the number represented by the digital input are variable and may be of either positive or negative polarity. A four-quadrant multiplier is expected to obey the rules of multiplication for algebraic sign.

**Gain**    The analog scale-factor setting that provides the normal conversion relationship of a converter.

**Gain Tempco**    *See entry following* Temperature Coefficient.

**Least-Significant Bit (LSB)**    The bit that carries the smallest value, or weight. Its analog weight, relative to full scale, is $2^{-n}$, where $n$ is the number of binary digits. It represents the smallest analog change that can be resolved by an $n$-bit converter.

**Linearity**    Linearity error of a converter (also *integral nonlinearity*), expressed in percent or parts per million of full-scale range or (sub)multiples of 1 LSB, is a deviation of the analog values, in a plot of the measured conversion relationship, from a straight line. The straight line may be a "best straight line," determined empirically by manipulation of the gain and/or offset to equalize maximum positive and negative deviations of the actual transfer characteristic from this straight line. It can also be a straight line passing through the endpoints of the transfer characteristic after they have been calibrated ("endpoint linearity"). The latter is both conservative and easier to measure, and it is similar to *relative-accuracy* error. The user should determine which definition is employed for a given commercial device.

For a multiplying DAC, the *analog* linearity error, at a specified digital code, is defined in the same way as for multipliers. It is the deviation from a "best straight line" through the plot of the analog output-input response.

Multiplying DACs that have a specified nonlinear function (e.g., variable-gain companding DACs) may, nevertheless, have specified linearity over portions of the response. "Step" nonlinearity consists of step-size deviations from the ideal within a *chord*, a group of linearly related steps in the transfer function.

**Linearity, Differential**    Any two adjacent digital codes should result in measured output values that are exactly 1 LSB apart ($2^{-n}$ of full scale for an $n$-bit converter). Any deviation of the measured "step" from the ideal difference is called *differential nonlinearity*, expressed in (sub)multiples of 1 LSB. Negative differential nonlinearity greater than 1 LSB can lead to nonmonotonic response in a DAC.

**Linearity Tempco**    *See entry following* Temperature Coefficient.

**Monotonic**    A DAC is said to be monotonic if the output either increases or remains constant as the digital input increases. This results in an output that is always a single-valued function of the input. The specification *monotonic* (over a given temperature range) is sometimes substituted for a specification of *differential nonlinearity*. A nonmonotonic DAC, used in an ADC, can lead to *missing codes*.

**Most-Significant Bit (MSB)**    The bit that carries the largest value or weight. Its analog weight, relative to a DAC's full-scale span, is $\frac{1}{2}$. In bipolar DACs, the MSB indicates the polarity of the number represented by the rest of the bits.

**Multiplying DAC**    A multiplying DAC differs from a fixed-reference DAC in being designed to operate with varying (or ac) reference signals. The output signal of such a DAC is proportional to the product of the analog input voltage and the fractional equivalent of the digital input number. (See *Four-Quadrant DAC*.)

**TABLE 15-7   Digital-to-Analog Converter Specifications** (*Continued*)

**Noise, Peak and RMS**   Peak noise is the magnitude of the largest noise deviation expected within the output bandwidth of the DAC. For gaussian noise, peaks greater than four times rms can be expected to occur only $63 \times 10^{-6}$ of the time. Large spikes coupled into the device from elsewhere in the system may have little effect on the rms value, yet they may be considerably greater in magnitude (and effect).

**Offset, Bipolar**   Most bipolar converters, instead of generating negative currents to correspond to negative numbers, use a unipolar DAC. Its output is offset by one-half of full-scale span (1 MSB). For best results, the offset voltage or current is derived from the same reference supply that determines the gain of the converter. This reduces the effect of the subtraction on drift of the converter's zero point.

**Offset Tempco**   *See entry following* Temperature Coefficient.

**Power-Supply Sensitivity**   The sensitivity of a converter to changes in the power supply voltages is normally expressed in terms of percent-of-full-scale change in analog output value (or fractions of 1 LSB) for a 1% dc change in the power supply, for example, $0.05\%/\%\Delta V_s$. Power supply sensitivity may also be expressed in relation to a specified dc shift of supply voltage. A converter may be considered "good" if the change in reading at full scale does not exceed $\pm\frac{1}{2}$ LSB for a 3% change in power supply. Even better specifications are necessary for converters designed for battery operation.

**Quantizing Uncertainty or Error**   The analog continuum is partitioned into $2^n$ discrete ranges for $n$-bit processing. All analog values within a given range of output (of a DAC) are represented by the same digital code, usually assigned to the nominal midrange value. For applications in which an analog continuum is to be restored, there is an inherent quantization uncertainty of $\pm\frac{1}{2}$ LSB (due to limited resolution), in addition to the actual conversion errors. For applications in which discrete output levels are desired (e.g., digitally controlled power supplies or digitally controlled gains), this consideration is not relevant.

**Resolution**   An $n$-bit binary converter should be able to provide $2^n$ distinct and different analog output values corresponding to the set of $n$-bit binary words. A converter that satisfies this criterion is said to have a *resolution* of $n$ bits. The smallest output change that can be resolved is $2^{-n}$ of the full-scale span.

**Settling Time**   The time required, following a prescribed digital data change, for the output of a DAC to reach and remain within a given fraction (usually $\pm\frac{1}{2}$ LSB) of the final value. Typical prescribed changes are full scale, 1 MSB, and 1 LSB at a *major carry* (for example, 0 1 1 1 1 1 1 1 to 1 0 0 0 0 0 0 0). Settling time of current-output DACs is quite fast. The major share of the settling time of a voltage-output DAC is usually contributed by the settling time of the output op amp.

**Slew Rate (or Slewing Rate)**   A limitation in the maximum rate of change imposed by a basic circuit constraint, such as the limited current available to charge a capacitor. The slewing rate of a voltage-output DAC is usually the same as the slewing rate of its output op amp.

**Switching Time**   The time required for a switch to change from one state to the other (delay time plus rise time from 10 to 90%). It does not include the settling time.

**Temperature Coefficient (Tempco, TC)**   The change in the parameter divided by the corresponding temperature change. In general, temperature instabilities are expressed in percent per degree Celsius, parts per million per degree Celsius, fractions of 1 LSB/°C, or as a change in a parameter over a specified temperature range. Measurements are usually made at room temperature and at the extremes of the specified range. Parameters of interest include *gain*, *linearity*, *offset* (bipolar), and *zero*. The last three are expressed in percent or parts per million of full-scale range (FSR) per degree Celsius.

**Gain Tempco**   This is affected by two factors. In fixed-reference converters, the reference *source* varies with temperature. This variation is added to the inherent temperature sensitivity, in all DACs, of the switches, reference circuitry, and span resistors.

**Linearity Tempco**   Sensitivity of linearity ("integral" and/or differential linearity) to temperature over the specified range. Monotonic behavior requires that negative differential nonlinearity be less than 1 LSB at any temperature in the range of interest. The *differential nonlinearity tempco* may be expressed as a ratio, as a maximum change over a temperature range, and/or as a statement that the device is monotonic over the specified temperature range.

**Offset Tempco**   The temperature coefficient of the all-DAC-switches-off (minus-full-scale) point of a bipolar converter depends on three major factors: the tempco of the reference source, the voltage zero-stability of the output amplifier, and the tracking capability of the bipolar-offset resistors and the span resistors.

**Unipolar Zero Tempco**   The temperature stability of a unipolar fixed-reference DAC is principally affected by current leakage (current-output DAC) and offset voltage and bias current of the output op amp (voltage-output DAC).

**Zero- and Gain-Adjustment Principles**   The output of a unipolar DAC is set to zero volts in the all-bits-off condition. The gain is set to FS $(1 - 2^{-n})$ with all bits on. The "zero" of an offset-binary (2's complement) bipolar DAC is set to $-$FS with all bits off. The gain is set for $+$FS $[1 - 2^{-(n-1)}]$ with all bits on.

## ANALOG-DIGITAL CONVERSION

### 15.9  SUCCESSIVE-APPROXIMATIONS ADC

**problem 15.7**  Figure 15-13 is a block diagram of a successive-approximations ADC. (*a*) With the aid of a timing diagram, describe the process for arriving at the output code 1 1 0 1 0 1 0 1. (*b*) If the internal clock of a 12-bit converter runs at 500 kHz, what is the minimum time required for performing a single conversion?

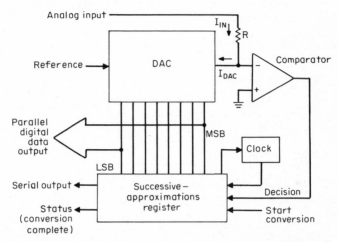

**Fig. 15-13**  Block diagram of a successive-approximation ADC. (Problem 15.7.)

**theory**  There are a large and diverse number of ways to perform analog-to-digital conversion. This chapter deals in depth with only the most popular converter architectures: *successive approximations* and *integration*.

The technique of successive approximations is popular because it can combine useful resolutions, up to and beyond 12 bits, with a fairly short conversion time (less than 12 μs for 12-bit conversion). A further advantage is that conversion time is fixed and independent of the input magnitude, permitting efficient interfacing with microprocessors. Its principal shortcomings are susceptibility to input changes during conversion (including noise) and higher cost per bit than integrating techniques.

Successive approximations is similar to weighing a mass on a precision balance, using accurately known weights whose values form a binary (or a BCD) progression. We have already, in effect, described the successive-approximations process in Sec. 15.4 in which a coding was derived from a voltage value. Each weight is added in its turn, starting with the heaviest; if it tips the scale, it is removed, otherwise it is allowed to remain. Then the next-heaviest is tried, then the next. When all the weights have been tried, the sum of those remaining on the scale is an accurate representation of the unknown weight. If the scale is biased by ½ LSB (that is, the first transition occurs at +½ LSB instead of at zero), the sum of the weights remaining on the scale is within ±½ LSB of the correct value.

A successive-approximations converter consists of a DAC, a voltage- or current-comparator, a clock, a shift register, control logic, and an output register (Fig. 15-13). The basic control input is a *start-conversion* line. Since the output data are invalid until the conversion is complete, a *status of conversion* [or end-of-conversion (EOC), conversion complete (CC), or BUSY] line signals that the converter is *busy* while the conversion is in progress.

**solution**  (*a*) When the conversion command is applied, the *status* line goes to the *busy* state, the register is cleared, except for a 1 in the MSB position, and the clock is gated on (Fig. 15-14). The 1 in the MSB position causes the MSB current (½ full scale) to appear at the DAC output (comparator input) with the analog input signal. If the DAC output is less than the signal input, the comparator output signals KEEP; if the DAC output is greater, the comparator signals REJECT.

On the next clock pulse, the MSB is latched at 1 or 0, depending on the decision. Bit 2 (¼ full scale) is added to the DAC output. If the DAC output (¼ + ½ or ¼ + 0) is less than the signal, the decision is to KEEP; otherwise it is REJECT. On the next clock pulse, bit 2 is latched, and bit 3 (⅛ full scale) is added to the DAC output. If the DAC output (⅛ + ¼ + ½, ⅛ + ½, ⅛ + ¼, or ⅛ + 0) is less than the signal, the decision is KEEP; otherwise it is REJECT. The process is repeated until the LSB has been tried and latched, after which the clock is turned off and the *status* line

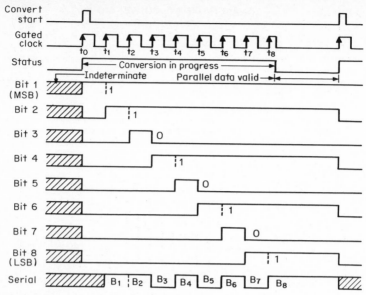

**Fig. 15-14**   Timing of an 8-bit successive-approximation conversion of data: 1 1 0 1 0 1 0 1.

*signals* CONVERSION COMPLETE. The parallel digital data output is now available, and the converter is ready for a new conversion.

Serial data are also available, since the bits have been cycled serially. Each serial bit (nonreturn to zero, NRZ) becomes valid with the leading edge of each clock pulse. The trailing edges of the clock pulses may be used to clock a receiving shift-register.

(*b*) The conversion time for an *n*-bit successive-approximations converter is at least equal to the time for *n* clock pulses. Some converters have delays at the start or end of conversion equal to one or more clock pulses. Thus, the minimum time for performing a single 12-bit conversion with a 500-kHz clock is about 25 $\mu s$.

## 15.10   CONVERSION TIME AND SAMPLING RATE

**problem 15.8**   (*a*) For a full-scale sinusoidal input, what is the maximum frequency that permits less than 1 LSB time-uncertainty? Assume a 12-bit converter with a 25-$\mu s$ conversion time. (*b*) If a sample-hold having an aperture uncertainty of 5 ns were used ahead of the converter, what is the maximum permissible frequency for 1 LSB error due to time uncertainty? (*c*) What minimum sampling rate is theoretically necessary to permit complete recovery of a signal having no components higher than this in frequency? (*d*) If the sample-hold has a 5-$\mu s$ acquisition time, what is the maximum throughput rate, using the above (12-bit) converter? (*e*) Is the criterion of (*c*) met?

**theory**   If the input signal to an ADC changes during the conversion process, there is an uncertainty as to the value of the analog input at a specific time—the beginning of conversion. In particular, if the input to a successive-approximation converter changes, the digital output may take on any value in the range through which the analog input changed. In order for the time uncertainty to be less than $\pm 1$ LSB, the maximum rate-of-change of signal must be less than 1 LSB per conversion interval.

The maximum rate-of-change of a sinusoid is $2\pi f V_m$, where $f$ is a given frequency and $V_m$ is the peak value of the sine wave (Fig. 15-15*a*). If $\tau_c$ is the conversion interval,

$$2\pi f V_m \leqslant \frac{2^{-n} \times 2V_m}{\tau_c} \tag{15-13}$$

Therefore

$$f \leqslant \frac{2^{-n}}{\pi \tau_c} \tag{15-14}$$

The solution to (*a*) will show that this is an intolerably low frequency. If the input voltage could be accurately sampled and retained at the precise instant desired, and if conversion were to take

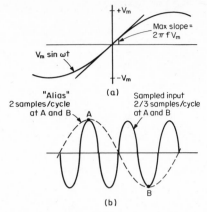

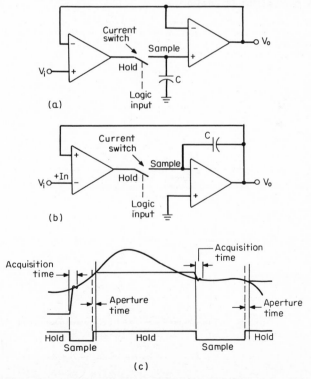

**Fig. 15-15** (*a*) Relationship between conversion time and sine-wave frequency for a given resolution. (*b*) Aliasing due to inadequate sampling rate.

place while that input value was in *hold*, there would be no time uncertainty, irrespective of the length of time required for conversion.

A sample-hold circuit (Fig. 15-16) contains a capacitor, a switch, an input buffer, and an output buffer. For speed and accuracy while tracking, it is configured as either an *RC* lag circuit or an integrator, connected in a feedback loop. During *sample* (or *track*), the loop seeks to follow the input signal, charging the capacitor as fast as it can. During *hold*, the switch is open, and the charge remains on the capacitor, except for leakage, which causes a slow "droop." Two critical

**Fig. 15-16**   Sample-hold circuits. (*a*) Follower. (*b*) Integrator. (*c*) Timing errors. (Problem 15.8).

dynamic parameters are *acquisition time*, when the switch closes, and *aperture time*, when the switch opens.

*Acquisition time* (see Fig. 15-16c) is the time required for the capacitor voltage to change—from the value that has been *held* to the latest signal value—to within a required fraction of full scale. Its application will be seen in (d). The *aperture time* is the interval between the application of the *hold* command and the actual opening of the switch. It consists of a delay (which depends on the logic and on the switching device, typically 50 ns), and an uncertainty, due to jitter, of typically 5 ns for general-purpose devices.

When a sample-hold is used in an application where timing is critical, the timing of the *hold* command can be advanced to compensate for the known component of aperture delay. The jitter, however, imposes the ultimate limitation on timing accuracy. When a sample-hold is used with an ADC, the timing uncertainty due to the conversion time is reduced by the ratio of the conversion time to the aperture jitter. In other words, the aperture uncertainty $\tau_a$ replaces the conversion delay $\tau_c$ in Eq. (15-14).

If a signal is sampled at too slow a rate and is later reconstructed, distortion may occur owing to *aliasing*. This is the creation of signals at other frequencies having the same sampling pattern (Fig. 15-15b). To avoid aliasing, the sampling rate must be greater than twice the highest frequency of interest. Further, all higher frequencies must be filtered out before sampling (and, at the receiving end, after reconstitution). This, in greatly simplified form, is in consequence of the *sampling theorem*.

**solution** (a) By Eq. (15-14), the maximum frequency is $2^{-12}/(\pi \cdot 25 \cdot 10^{-6}) = 3.11$ Hz!

(b) By Eq. (15-14), the maximum frequency is $2^{-12}/(\pi \cdot 5 \cdot 10^{-9}) = 15.5$ kHz.

(c) The theoretical minimum sampling rate, from the sampling theorem, is $2 \times 15.5 = 31$ kHz.

(d) The time for sampling and conversion is about $(5 + 25)$ $\mu$s. Therefore, the maximum throughput rate is $1/(30\ \mu\text{s}) = 33.3$ kHz.

(e) The criterion of (c) is met, but just barely. In practical cases, because of the effects of gaussian noise and the imperfectness of filter characteristics, the sampling rate should be *three* or more times the highest frequency.

## 15.11 INTEGRATING CONVERSION

**problem 15.9** Figure 15-17a shows a block diagram of a dual-slope (integrating) ADC. (a) Describe how it works. (b) What are its advantages and disadvantages relative to the successive-approximation ADC? (c) How are errors minimized? (d) How would the effects of 60-Hz pickup (and its harmonics) at the input be minimized?

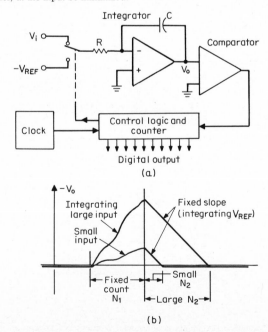

(a)

(b)

**Fig. 15-17** Dual-slope integrating ADC. (a) Circuit. (b) Integrator output. (Problem 15.9.)

**theory and solution**    (*a*) An integrating ADC measures the time required for an integrator output to traverse a voltage range, proportional to the average value of the input, at a constant (reference) rate. In Fig. 15-17*a*, at the start of conversion, the integrator is unclamped and starts to integrate the input signal $V_i$. At the same time, the counter starts to count clock pulses. When $N_1$ clock pulses have been counted after a period $t_1$, the counter rolls over and switches the input.

A reference ($-V_{REF}$) of opposite polarity to the signal is applied to the integrator, the integrator starts to integrate in the reverse direction at a constant rate, and the counter starts to count. When the output of the integrator reaches the starting value, the comparator is tripped, and the conversion is complete. The clock is stopped, and the integrator is clamped at its starting value. The number of counts $N_2$, indicating time $t_2$, can be shown to be proportional to the average value of the input $\overline{V}_i$. The integrator output is illustrated in Fig. 15-17*b*.

Since the integrator output after the first integration ($V_1$) is equal to the change in the integrator output after the second integration ($V_2$),

$$V_1 = V_2 = \frac{Q_1}{C} = \frac{Q_2}{C}$$

$$= \frac{\overline{I_1 t_1}}{C} = \frac{\overline{I_2 t_2}}{C}$$

$$= \frac{\overline{V_i} t_1}{RC} = \frac{V_{REF} t_2}{RC}$$

Hence

$$\frac{t_2}{t_1} = \frac{\overline{V}_i}{V_{REF}} \tag{15-15}$$

Since time is directly proportional to count,

$$N_2 = \frac{\overline{V}_i}{V_{REF}} N_1 \tag{15-16}$$

As long as the values of $R$, $C$, and the clock rate remain constant during a conversion interval, the accuracy is independent of them. It depends, rather, on amplifier offset, drift, and nonlinearity, and the dynamics of the switches, the integrator, and the comparator.

The counter's output may be binary or BCD. Because most integrating ADCs are used in digital panel meters (DPMs) or other display devices, BCD is the more usual approach.

(*b*) Integrating ADCs are much slower than successive-approximation types. However, they have considerably greater potential accuracy, no missing codes, better noise rejection, and, because they rely on fewer high-accuracy parts, they tend to be lower in cost. It is their simplicity, low cost, and compatibility with integrated-circuit technology that make them the preferred approach for DPMs.

(*c*) Errors in integrating converters are minimized in two principal ways: one analog, the other digital. Both, however, require an extra time interval for error correction. The best known is the *autozeroing* technique. After a conversion has been completed, the input is disconnected; the reference charges a capacitor, $C_1$ (Fig. 15-18), and applies its voltage, through a buffer, to the

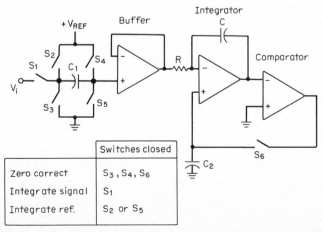

**Fig. 15-18**    Autozero circuit for an integrating converter (analog section). (Problem 15.9.)

integrator input. The comparator and integrator are connected in a local feedback loop having a closed-loop gain of unity. A second capacitor, $C_2$, is charged to the reference voltage plus the buffer and integrator offsets, and the integrating capacitor to the comparator offset.

When the conversion command is received, $C_1$ is disconnected from the reference and ground. The formerly grounded end is connected to the input signal, and $C_2$ is disconnected from the comparator output. The net voltage applied to the integrating resistor is equal to the input, and the integrator integrates the proportional current. The output of the integrator, starting from the comparator's zero-offset, integrates in the appropriate direction (depending on the input polarity). This pegs the comparator in the appropriate direction, giving a polarity indication.

When count $N_1$ occurs, and the integral of the signal plus the comparator offset appears at the integrator output, the buffer input is switched either to ground or to the reference. Depending on polarity, the integrator integrates back and triggers the comparator when it crosses the comparator offset voltage, to end conversion. All offsets have been stored and ignored.

A digital scheme, used in a 13-bit monolithic integrating ADC, is known as quad-slope. An initial dual-slope conversion cycle, using an offset reference with the input grounded, provides a digital measure of the zero and gain errors, in the form of a count. Then, with the signal applied, a second dual-slope cycle occurs. The error count is subtracted from the main count, providing excellent compensation.

(d) The integrating technique achieves excellent noise rejection, because any high-frequency noise is simply averaged out. In addition, there is (theoretically) complete rejection of the fundamental and all harmonics of the frequency whose period is equal to the signal-integrating period. Hence, in order to obtain high rejection of 60 Hz and its harmonics, the integrating period is set at 16.7 ms, or multiples thereof.

## 15.12   ADC CONVERSION RELATIONSHIP AND ERRORS

**problem 15.10**    Figure 15-19 shows the ideal conversion relationship of a 3-bit ADC. Sketch variations of this conversion relationship that show (a) zero-offset error, (b) scale-factor error, (c) linearity error. (d) Is Fig. 15-19 fully descriptive of the output under analog overrange conditions?

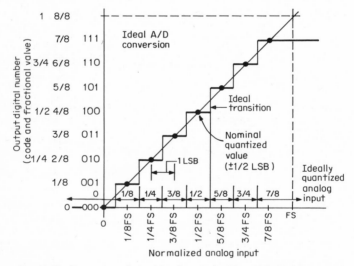

**Fig. 15-19**    Conversion relationship of a unipolar 3-bit ADC. (Problem 15.10.)

**theory**    Analogous to the DAC, two facts must be known in order to determine what the output code of an ADC means: the *code* employed and the *conversion* relationship. Generally, the same codes are encountered with ADCs and DACs. The *analog* input of an ADC, however, can have *any* value in the defined range of operation, and some values outside that range as well. Since it is desired to obtain $2^n$ codes that represent the analog range, it is necessary to *quantize* the analog continuum, i.e., to divide it into $2^n$ ranges, or *quanta*, each of which is assigned to represent a set of analog values. All analog values within a given range are represented by the same digital code, which generally corresponds to the magnitude of the nominal midrange value. These midrange values correspond to the bar heights of the DAC conversion relationship.

There is, therefore, in the decision to use an ADC an agreement to tolerate an inherent *quantization uncertainty* of $\pm\frac{1}{2}$ LSB. This is in addition to the errors encountered in implementing the conversion. The only sure way to reduce this quantization uncertainty is to increase the number of bits.

It is easier to determine the location of a transition from one code to the next than it is to determine a midrange value. Therefore, the errors and adjustments of an ADC are defined and measured in terms of the analog values at which transitions occur, with respect to the ideal transition values.

Like the DAC, an ADC has an *offset* error: the first transition may not occur at exactly ½ LSB. In addition, there are a *scale-factor (gain)* error [the difference between the values at which the first transition and the last transition occur is not equal to FS(1 − 2LSB)] and a *linearity* error (the differences between transition values are not all equal or uniformly changing). If the differential linearity is large enough, there is the possibility that one or more codes will be missed (the counterpart of nonmonotonic d/a conversion).

The differences between the various approaches to a/d conversion are more radical than the differences among DACs. We may, therefore, find qualitative departures in the conversion relationship among types of ADCs. For example, successive-approximation types and parallel-comparator types generally have no way to signal overrange. On the other hand, they do not roll over and provide what may be erroneous codes—if uninterpreted—in the overrange condition. Types employing counters, principally integrating converters, can roll over. They also can signal overrange by putting out a *carry*.

Integrating converters do not have a differential-linearity problem, since they do not use DACs or large numbers of precision elements. The differing relationships of these elements cause a differential nonlinearity. Hence, integrating converters, in general, do not have missing codes.

**solution**    (*a*) Figure 15-20*a* shows zero-offset error.

(*b*) Figure 15-20*b* shows gain error.

(*c*) Linearity error is illustrated in Fig. 15-20*c* and *d* shows missed codes.

(*d*) As discussed above, Fig. 15-19 shows analog overrange for a successive-approximation or parallel-comparator ADC, but not for an integrating type.

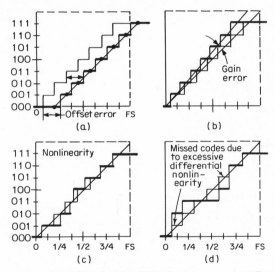

**Fig. 15-20** Conversion relationship of a 3-bit ADC, showing errors. (*a*) Offset error. (*b*) Gain error. (*c*) Nonlinearity. (*d*) Missing codes. (Problem 15.10.)

## 15.13 NONMONOTONIC DACs AND MISSING CODES IN ADCs

**problem 15.11**    A portion of the range of a successive-approximation ADC is provided in Fig. 15-21. The characteristic of the DAC used in it is shown horizontally. It is seen to be nonmonotonic. Which code, if any, will be missing at the ADC output? How?

**theory**    In the illustration, the horizontal bars represent the measured DAC output values corresponding to six adjacent codes. The DAC is nonlinear, in that the next-least-significant bit (0 1 0) is 1½ LSBs too large. Thus, the 5 quanta, or steps, are not all equal to 1 LSB; quantum 2 is 2½ LSB, and quantum 4 is −½ LSB. The differential-linearity error, the difference between the actual quantum width and the ideal 1 LSB, is +1½ LSB for quantum 2 and −1½ LSB for quantum 4.

In a successive-approximations converter, the bit combinations are tested in order, starting with

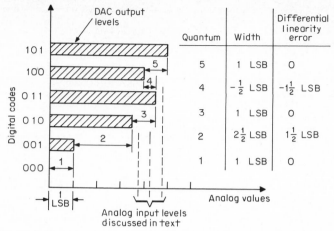

**Fig. 15-21** Role of a nonmonotonic DAC in producing missing codes in an ADC. (Problem 15.11.)

the MSB. In the present example, if the analog signal is greater than both 1 0 0 and 0 1 1, but less than 1 0 1, then 1 0 0 is tried. It will be accepted because it is smaller than the analog input. When the next bit, 0 1 0, is added to 1 0 0, the result will be too great; 0 1 0 will be rejected. Then the LSB, 0 0 1, is added, and 1 0 1 is tried; it is too large, so 0 0 1 will be rejected. The final answer is 1 0 0.

If the analog signal is between 1 0 0 and 0 1 1, 1 0 0 will be accepted, since it is less than the analog input. Again, the next bit to be tested will be 0 1 0, added to 1 0 0. The final answer turns out, as above, to be 1 0 0.

If the analog signal is between 1 0 0 and 0 1 0 in magnitude, 1 0 0 will be rejected when tested, because it is too large. When 0 1 0 is tested, it is accepted, because it is less than the analog signal. Then 0 0 1 is added to 0 1 0. Since the combination is greater than the analog signal, it will be rejected. The answer turns out to be 010.

**solution** The missing code is 011. As we have shown in the previous discussion, there are no conditions under which it will be chosen.

## 15.14 A/D CONVERTER SPECIFICATIONS

Specifications that are commonly used to characterize the performance of A/D converters are summarized in Table 15-8.

**TABLE 15-8 Analog-to-Digital Converter Specifications**

**Accuracy, Absolute** The error of an ADC at a given output code is the difference between the theoretical and the actual analog input voltages required to produce that code. Because the code can be produced by any analog voltage in a finite band (see *Quantizing Uncertainty*), the "input required to produce that code" is defined as the midpoint of the band of inputs that yields the code (Fig. 15-19). For example, if 5 V ($\pm$1.2 mV) theoretically produces a 12-bit half-scale code of 1 0 0 0 0 0 0 0 0 0 0 0, then a converter for which any voltage from 4.997 to 4.999 V produces that code has an absolute error of ($\frac{1}{2}$)(4.997 + 4.999) − 5 V = −2 mV.

Absolute error comprises gain error, zero error, and nonlinearity, together with noise. Absolute-accuracy measurements should be made under a set of standard conditions with sources and meters traceable to an internationally recognized standard.

**Accuracy, Relative** Relative accuracy error, expressed in percent, parts per million (ppm), or fractions of 1 LSB, is the deviation of the analog value of any code (relative to the full analog range of the device transfer characteristic) from its theoretical value (relative to the same range), after the full-scale range (FSR) has been calibrated. Since the discrete points of the theoretical transfer characteristic lie on a straight line, this deviation also can be interpreted as a meausre of nonlinearity (see *Linearity* in Table 15-7). The "discrete points" of an a/d transfer characteristic are the midpoints of the quantization bands at each code (see *Accuracy, Absolute*).

**Conversion Time** The time required for a complete measurement by an ADC. It is usually delineated by the interval from the application of the conversion command to the return of the *status* line to its "ready" state after all bits have been latched.

**TABLE 15-8   Analog-to-Digital Converter Specifications** (*Continued*)

**Dual-Slope Converter**   An integrating ADC in which an unknown signal is converted to a proportional time interval and then measured digitally. This is done by integrating the unknown for a predetermined time. Then a reference input of opposite polarity is switched to the integrator, which integrates "back" from the level determined by the unknown until the starting point is reached. The time for the second integration is proportional to the average of the unknown signal level over the predetermined integrating period.

**Feedthrough**   See Table 15-7.

**Gain**   See Table 15-7.

**Gain Tempco**   *See entry following* Temperature Coefficient below.

**Least-Significant Bit**   See Table 15-7.

**Linearity**   See Table 15-7.

**Linearity, Differential**   See Table 15-7. It is an important specification, because a differential nonlinearity error greater than 1 LSB can lead to nonmonotonic behavior of a DAC and missed codes in an ADC that employs such a DAC (Fig. 15-21).

**Linearity Tempco**   *See entry following* Temperature Coefficient below.

**Offset Tempco**   *See entry following* Temperature Coefficient below.

**Power Supply Sensitivity**   See Table 15-7. High-accuracy ADCs, intended for battery operation, require excellent rejection of large supply variations.

**Quad-Slope Converter**   An integrating ADC that goes through two cycles of *dual-slope* conversion, once with zero input, and once with the analog input being measured. The errors determined during the first cycle are subtracted digitally from the result in the second cycle. The scheme results in an extremely accurate converter. Typical temperature coefficients for a monolithic IC using the technique are 1 ppm/°C *max* for both gain and offset.

**Quantizing Uncertainty or Error**   See Table 15-7. Accuracy specifications for integrating ADC types often include the effects of quantizing uncertainty in the catchall statement "plus-or-minus one count."

**Ratiometric Converter**   The output of an ADC is a digital number proportional to the ratio of the input to a reference. Most requirements for conversions call for an absolute measurement against a fixed reference. In some cases, where the measurement is affected by a reference voltage subject to drift (e.g., the voltage applied to a bridge), to eliminate the effect of variations, it is advantageous to use that same reference as the reference for the conversion. Ratiometric conversion also serves as a substitute for analog signal division (where the denominator changes by less than 1 LSB during the conversion).

**Successive Approximations**   A high-speed method of comparing an unknown against a group of weighted references. (See text.)

**Temperature Coefficient**   See Table 15-7.

**Gain Tempco**   Two factors principally affect converter gain-instability with temperature. In fixed-reference converters, the reference *source* varies with temperature. This variation is added to the inherent temperature sensitivity of the basic ratiometric device.

**Linearity Tempco**   Sensitivity of linearity to temperature over the specified range. To avoid missed codes in successive-approximation devices, it is necessary that the differential nonlinearity be less than 1 LSB at any temperature in the range of interest. The *differential-nonlinearity temperature coefficient* may be expressed as a ratio, as a maximum change over a specified temperature range, and/or as a statement that there are no missed codes when operating within a specified temperature range.

**Offset Tempco**   The temperature coefficient of the all-DAC-switches-off (minus full-scale) point of a bipolar successive-approximations converter depends on the tempco of the reference source, the stability of the input buffer and the comparator, and the tracking capability of the bipolar-offset resistors and the gain resistors.

**Unipolar Zero Tempco**   The zero tempco of an ADC is dependent only on the zero stability of the integrator and/or the input buffer and the comparator. It may be expressed in microvolts per degree Celsius, or in percent or parts per million of full-scale per degree Celsius.

**Zero- and Gain-Adjustment Principles**   The zero adjustment of a unipolar ADC is set so that the transition from all-bits-off to LSB-on occurs at $+\frac{1}{2}$ LSB: $2^{-(n+1)}$ of full scale. (See Fig. 15-19.) The gain is set for the final transition to all-bits-on to occur at FS $[1 - (\frac{3}{2})2^{-n}]$. The "zero" of an offset-binary or 2's complement bipolar ADC is set so that the first transition occurs at $-$FS $(1 - 2^{-n})$ and the last transition at $+$FS $(1 - 3 \times 2^{-n})$.

## 15.15 BIBLIOGRAPHY

Many of the figures and tables in this section originally appeared in publications of Analog Devices, Inc., and they are reproduced here with the permission of Analog Devices. The principal source of these was *Analog-Digital Conversion Notes*, listed below. In addition to these publications, the catalogs and data books of conversion-product manufacturers provide much valuable and timely information.

"Applications Manual for Operational Amplifiers," Teledyne Philbrick, 1968.

Brodie, D.: Test Your Data-Conversion I.Q., *EDN Magazine*, May 20, 1975.

Coughlin, R. F., and Driscoll, F. F.: *Operational Amplifiers and Linear Integrated Circuits*, Prentice-Hall, Englewood Cliffs, N.J., 1977.

—— and ——: *Semiconductor Fundamentals*, Prentice-Hall, Englewood Cliffs, N.J., 1976.

*Data Conversion Handbook*, Hybrid Systems Corporation, 1974.

Freeman, W., and Ritmanich, W.: Cut a/d Conversion Costs by Using Software and d/a Converters, *Electronic Design*, 9, April 26, 1977.

Fullagar, D., Bradshaw, P., Evans, L., and O'Neill, B.: Interfacing Data Converters and Microprocessors, *Electronics*, Dec. 9, 1976.

Gordon, B. M., et al.: *The Analogic Data-Conversion Systems Digest*, Analogic Corp., Wakefield, Mass., 1977.

Gordon, B. M.: Digital Sampling and Recovery of Analog Signals, *EEE Magazine*, May 1970.

——: On What's Wrong with Converter Specs, *EEE Magazine*, February 1969.

Graeme, J.: *Designing with Operational Amplifiers*, McGraw-Hill, New York, 1977.

Hnatek, E.: *A User's Handbook of D/A and A/D Converters*, Wiley-Interscience, New York, 1970.

Hoeschele, D. F., Jr.: *Analog-to-Digital, Digital-to-Analog Conversion Techniques*, Wiley, New York, 1968.

Jung, W. G.: *I.C. Op-Amp Cookbook*, Sams, Indianapolis, Ind., 1974.

*I.C. Converter Cookbook*, Sams, Indianapolis, Ind., 1978.

Molinari, F., and Fishman, A.: Test Your Data-Acquisition System I.Q., *EDN Magazine*, Nov. 5, 1976.

Pouliot, F.: Have You Considered V/f Converters?, *Analog Dialogue* 9-3, 1975.

Roberge, J.: *Operational Amplifiers, Theory and Practice*, Wiley, New York, 1975.

Schmid, Hermann: *Electronic Analog/Digital Conversions*, Van Nostrand Reinhold, New York, 1970.

Sheingold, D. H. (ed.): *Analog-Digital Conversion Handbook*, Analog Devices, Inc., Norwood, Mass., 1972.

——: *Analog-Digital Conversion Notes*, Analog Devices, Inc., Norwood, Mass., 1977.

Sherwin, J.: Specifying D/A and A/D Converters, *Electronic Products Magazine* wallchart, (undated).

Smith, J. I.: *Modern Operational Circuit Design*, Wiley-Interscience, New York, 1971.

Spofford, W. R., Jr.: Putting D/A Converters to Work, *Electronics*, Oct. 26, 1970.

Wagner, R.: Laser-Trimming on the Wafer, *Analog Dialogue*, 9-3, 1975.

Whitmore, J.: Resistance-Ratio to Digital Conversion, *Analog Dialogue*, 9-3, 1975.

Wold, I.: CMOS Converters as I/O Devices, *Analog Dialogue*, 11-1, 1977.

Chapter **16**

# Video Amplifiers

## ROGER C. THIELKING

Chairperson and Associate Professor of Electrical Technology,
Onondaga Community College, Syracuse, N.Y.

## 16.1 INTRODUCTION

Since the development of television, video amplifiers have become very numerous. These circuits, by definition, provide amplification of the visual information, which usually has frequency components ranging from direct current up to several megahertz. Video amplifiers must have a flat response over most of this range in order to provide acceptable picture quality. The circuits and signal levels used in examples in this chapter are generally representative of video stages in commercial television receivers. They have been selected for general applicability of the calculation techniques to other types of video and broad-band amplifiers, as well.

Individual stages of video amplifiers are essentially similar to other amplifier stages. They usually include unbypassed resistors and have relatively low gain and low output impedances so as to maintain good high-frequency response. Multistage negative feedback, which is common in audio and some other amplifiers, is difficult to apply here because of oscillation problems at video frequencies.

Many of the calculations involving video amplifiers are the same as or very similar to those of audio and other amplifiers. Such calculations would include, for example, dc biasing and mid- and low-frequency gain and input/output impedances. Since these topics are thoroughly covered in other chapters, they are treated very briefly here. Emphasis is given, rather, to the high-frequency aspects of various circuits. Two low-frequency techniques, for frequency compensation and for dc restoration, are also included, as these occur more commonly in video amplifiers than they do in amplifiers of other types.

A note of caution is needed about the calculations in this chapter. Most of the high-frequency relationships involve small amounts of stray capacity, and other parameter values, which are difficult to measure or estimate accurately. There may also be unrecognized circuit effects which are not taken into account. The results of the calculations, then, may have considerable error. They should be used only as a rough guide, together with experimentation and measurements with the actual circuits.

## 16.2 TRANSISTORS FOR VIDEO AMPLIFIERS

General-purpose silicon bipolar junction transistors have been so improved in recent years that many of them are suitable for operation at video frequencies. Medium- and high-power transistors, on the other hand, are often poorer in high-frequency response so that the choice of devices is more limited. There are also a number of field-effect transistors that operate well into the megahertz range. They are normally avoided, however. Their high input impedance is not needed for the relatively low impedance video circuitry, and they offer no other advantages over the lower-cost junction devices.

### Equivalent Circuits

Transistors are useful for video amplification only at frequencies below their $f_T$, the frequency at which the common-emitter current gain reduces to unity. The widely used hybrid-pi ac equivalent circuit, Fig. 16-1$a$, is valid for frequencies up to $f_T$ and therefore is very adequate for video work.

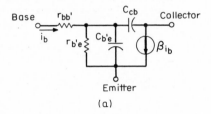

(a)

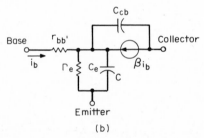

(b)

**Fig. 16-1**   Transistor Models: ($a$) hybrid-pi. ($b$) T-equivalent.

Another form of the equivalent circuit, which is functionally identical to the hydrid-pi (Fig. 16-1), is the T-equivalent model shown in Fig. 16-1$b$. This arrangement is easier to use in certain circuits.

Most of the parameters for these circuits are readily determined from available data, as described below.

### Transistor Parameters

Parameter values for the equivalent circuits are needed for many of the circuit calculations. These values sometimes change at high frequencies approaching $f_T$ and may be determined as follows:

$\beta$   At low frequencies the value of $\beta$ is the same as the $h_{fe}$ specified by the manufacturer, but at high frequencies it will reduce to unity at the $f_T$ of the transistor, as shown in Fig. 16-2. The following relationships give good approximation at all frequencies:

$$\text{for } f < \frac{f_T}{h_{fe}} \qquad \beta = h_{fe} \tag{16-1}$$

$$\text{for } \frac{f_T}{h_{fe}} < f < f_T \qquad \beta = \frac{f_T}{f} \tag{16-2}$$

Values of $h_{fe}$ and $f_T$ are given by the manufacturer. $f_T$ is also discussed below.

$r_e$   The emitter junction resistance varies with the dc operating point. Actual values come very close to the value given by junction theory:

$$r_e = \frac{26 \text{ mV}}{I_E} \tag{16-3}$$

where $I_E$ is the dc value of emitter current, calculated by conventional dc analysis of the particular circuit.

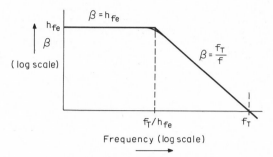

**Fig. 16-2**   Variation of $\beta$ with frequency.

$r_{b'e}$   The emitter resistance, as seen at the base, must be multiplied by the base-to-emitter current gain:

$$r'_{be} = (\beta + 1) \, r_e \tag{16-4}$$

then, use of Eqs. (16-1) to (16-3) yields

for $\qquad\qquad\qquad r'_{be} = (h_{fe} + 1)26 \text{ mV}/I_E \text{ for } f < f_T/h_{fe} \tag{16-5}$

or $\qquad\qquad\qquad r'_{be} \cong h_{fe} \times 26 \text{ mV}/I_E \tag{16-6}$

for $f_T/h_{fe} < f < f_T, \qquad r'_{be} = \left(\frac{f_T}{f} + 1\right)\frac{26 \text{ mV}}{I_E} \tag{16-7}$

$C'_{be}$ and $C_e$  The emitter junction diffusion capacitance is very nearly the same in both the hybrid-pi and T circuits and has a reactance equal to $r_e$ at $f_T$. Thus,

$$C'_{be} = C_e = \frac{1}{2\pi f_T r_e} \tag{16-8}$$

and, from Eq. (16-5),

$$C'_{be} = C_e = \frac{I_E}{2\pi f_T \times 26 \text{ mV}} \tag{16-9}$$

$r'_{bb}$   The base spreading resistance is rarely specified by the manufacturer and is very difficult to measure. Fortunately, it has a low value (usually between 10 and 50 $\Omega$) so that, in most circuits, it can be replaced with a short circuit without causing error. It can, however, be significant at times. Manufacturers sometimes specify the time constant $r'_{bb}C_{cb}$, so that an estimate of $r'_{bb}$ can be obtained by dividing $r'_{bb}C_{cb}$ by $C_{cb}$ (see below).

$C_{cb}$   Collector feedback capacitance is obtained from manufacturers' specifications. It varies with collector voltage. Data in graphical form, as shown in Fig. 16-3 for the 2N6223 transistor, are preferred when available. Then the capacitance may be determined for the particular dc operating point of the circuit. From Fig. 16-3, for example, if the transistor is used at $V_{CB} = 10$ V, then $C_{cb} = 2$ pF; but if $V_{CB}$ is reduced to 5 V, $C_{cb} = 4$ pF.

$f_T$   The gain-bandwidth product, while not appearing directly in the equivalent circuits, is needed to calculate several of the above parameters. Data are given by manufacturers in

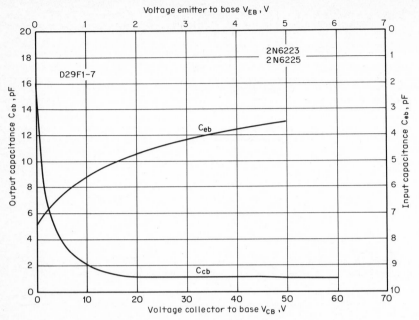

**Fig. 16-3**  Input and output capacitance vs. voltage.

various ways. As with $C_{cb}$, the value varies with operating point, and graphical data in the form of gain bandwidth contours may be used when available. These curves are shown in Fig. 16-4 for the 2N6223. In problem 16.1, this transistor is used with $V_{CE} = -12.3$ V and $I_c = 9.6$ mA. This operating point is located, in Fig. 16-4 between the 500- and 600-MHz

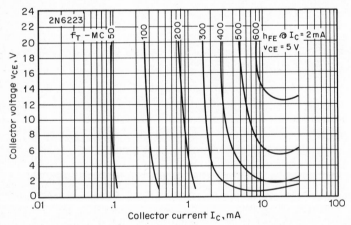

**Fig. 16-4**  Gain bandwidth product vs. collector current.

contours. An interpolated value may be determined, but taking the lower value (500 MHz) is easier and is a more conservative approach.

Formulas for transistor parameters are summarized in Table 16-1. The transistors selected for the problems which follow are good examples of devices now available but are not necessarily optimum for particular applications.

**TABLE 16-1  Transistor Parameter Formulas**

$$\text{For } f < \frac{f_T}{h_{fe}} \qquad \beta = h_{fe} \tag{16-1}$$

$$\text{For } \frac{f_T}{h_{fe}} < f < f_T \qquad \beta = \frac{f_T}{f} \tag{16-2}$$

$$r_e = \frac{26 \text{ mV}}{I_E} \tag{16-3}$$

$$r_{b'e} = (\beta + 1)r_e \tag{16-4}$$

$$\text{For } f < \frac{f_T}{h_{fe}} \qquad r_{b'e} = (h_{fe} + 1)\frac{26 \text{ mV}}{I_E} \tag{16-5}$$

$$\text{For } \frac{f_T}{h_{fe}} < f < f_T \qquad r_{b'e} = (f_T + 1)\frac{26 \text{ mV}}{I_E} \tag{16-7}$$

$$C_{b'e} = C_e = \frac{I_E}{2\pi f_T \times 26 \text{ mV}} \tag{16-9}$$

## 16.3  CALCULATING UPPER BANDWIDTH LIMIT AND INPUT IMPEDANCE OF A LOW-GAIN, LOW-LEVEL, UNCOMPENSATED VIDEO AMPLIFIER STAGE

**problem 16.1**  Figure 16-5 illustrates two stages of dc coupled video amplification driving a load with 25-pF capacitance. The gain of each stage is approximately 3. The input signal, at the base of $Q_1$, is at a dc level of +2.0 V. $C_{cb}$ data and gain-bandwidth contours for the transistor $Q_2$ are given in Figs. 16-3 and 16-4. The $h_{fe}$ of $Q_2$ is 140. Stray capacities are 5 pF, collector to ground, and 4 pF, base to ground, at $Q_2$. Calculate the upper bandwidth limit of the second stage and the input impedance into the base terminal of $Q_2$.

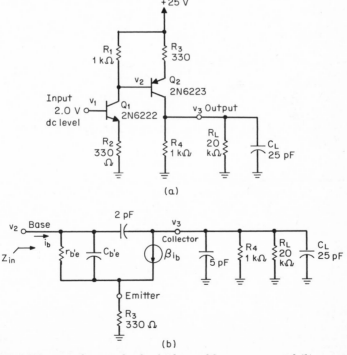

**Fig. 16-5**  (a) Two-stage, low-gain, low-level video amplifier, uncompensated. (b) ac equivalent circuit of the second stage.

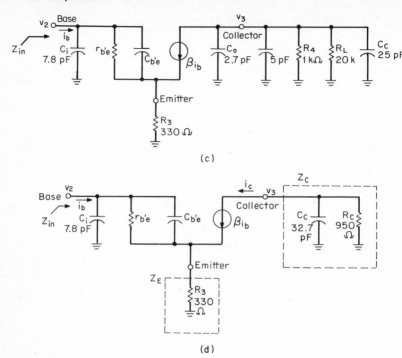

**Fig. 16-5 (cont.)**  $(c)$ ac equivalent circuit with Miller transformation of $C_{cb}$. $(d)$ Simplified ac equivalent circuit.

**theory**   The circuit configuration shown in this example is widely used in video amplifiers. Dc coupling is usually very desirable and is easily done with alternate stages of npn and pnp transistors, with the collector of one directly connected to the base of the next.

The dc operating point of $Q_2$ must first be determined. This, in turn, depends upon the collector current of $Q_1$. The input dc level of 2 V at its base is reduced by the emitter junction drop (0.7 V), to give 1.3 V at the emitter; thus $I_C$ of $Q_1 \cong 1.3$ V/330 $\Omega = 3.9$ mA. Collector resistance (1 k$\Omega$ shunted by $h_{FE} \times R_3$) is 980 $\Omega$. At $Q_2$, $V_B = 21.14$ V and $V_E = 21.84$ V, which results in the following values for the operating point of $Q_2$:

$$V_{CE} = -12.3 \text{ V} \tag{16-10}$$
$$I_E \cong I_C = 9.6 \text{ mA} \tag{16-11}$$

Now we may proceed with ac analysis. The ac equivalent circuit, shown in Fig. 16-5$b$, includes the hybrid-pi equivalent of the transistor (Fig. 16-1$a$) previously described. Note that all stray capacities are included.

Parameters for the transistor equivalent circuit are determined, as described in Sec. 16.2, as follows. From the operating point [Eqs. (16-10) and (16-11)], using Fig. 16-4, we have

$$f_T = 500 \text{ MHz} \qquad \text{(typical)} \tag{16-12}$$

From Eqs. (16-1) and (16-2),

$$f_T/h_{fe} = \frac{500 \text{ MHz}}{140} = 3.6 \text{ MHz} \tag{16-13}$$

For $f < 3.6$ MHz
$$\beta = 140 \tag{16-14}$$

For 3.6 MHz $< f <$ 500 MHz
$$\beta = \frac{500 \text{ MHz}}{f} \tag{16-15}$$

$C_{cb}$ is determined, from Fig. 16-3, to be $\approx 2$ pF. $r'_{bb}$ may be assumed to be small enough to neglect, and values of other parameters are not needed for the calculations.

$C_{cb}$ may be transformed into equivalent capacitors to ground, $C_i$ (base to ground) and $C_o$ (collector to ground), by means of the Miller effect. The formulas are:

$$C_o = C_{cb}\left(1 + \frac{1}{A}\right)$$    (16-16)

$$C_i = C_{cb}(1 + A)$$    (16-17)

where $A$ is the midfrequency base-to-collector stage gain. For this circuit,

$$A = \frac{R_4\|R_L}{R_3} = 2.89$$    (16-18)

then

$$C_o = 2 \text{ pF}\left(1 + \frac{1}{2.89}\right) = 2.7 \text{ pF}$$    (16-19)

$$C_i = 2 \text{ pF}(1 + 2.89) = 7.8 \text{ pF}$$    (16-20)

The transformed circuit is shown in Fig. 16-5c. It may be simplified by combining parallel components, as in Fig. 16-5d.

Analysis of this circuit indicates that there are three possible mechanisms for limiting the high-frequency response:

1. Impedance in the collector circuit. The stage gain $v_3/v_2$ is equal (approximately) to $Z_C/Z_E$. Therefore, the 32.7-pF collector capacitance will cause 3-dB reduction when its reactance is equal to the 950-Ω collector resistance.

2. Reduced transistor gain at high frequencies, as given in Eq. (16-15). This factor will cause a 3-dB reduction in collector current at approximately

$$f = \frac{f_T}{2}$$    (16-21)

3. Input impedance. This will not affect the voltage gain $v_3/v_2$ of the second stage, but it may affect the response of the first stage because the input impedance is a part of the collector load of that stage. Input impedance consists of the reactance of $C_i$ in parallel with the product of $\beta$ times the emitter circuit impedance (approximately).

Formulas for this problem are summarized in Table 16-2.

**TABLE 16-2    Formulas Used for Problem 16.1**

| | |
|---|---|
| $V_{CE} = -12.3 \text{ V}$ | (16-10) |
| $I_E \cong I_C = 9.6 \text{ mA}$ | (16-11) |
| $f_T = 500 \text{ MHz (typical)}$ | (16-12) |
| $\dfrac{f_T}{h_{fe}} = 500 \text{ MHz}/140 = 3.6 \text{ MHz}$ | |
| For $f < 3.6 \text{ MHz}$    $\beta = 140$ | (16-13) |
| | (16-14) |
| For $3.6 \text{ MHz} < f < 500 \text{ MHz}$    $\beta = \dfrac{500 \text{ MHz}}{f}$ | |
| | (16-15) |
| $C_o = C_{cb}\left(1 + \dfrac{1}{A}\right)$ | |
| | (16-16) |
| $C_i = C_{cb}(1 + A)$ | (16-17) |
| $A = \dfrac{R_4\|R_L}{R_3} = 2.89$ | |
| | (16-18) |
| $C_o = 2 \text{ pF}\left(1 + \dfrac{1}{2.89}\right) = 2.7 \text{ pF}$ | |
| | (16-19) |
| $C_i = 2 \text{ pF}(1 + 2.89) = 7.8 \text{ pF}$ | (16-20) |
| $f = \dfrac{f_T}{2}$ | |
| | (16-21) |

**solution**    The collector impedance $Z_C$ causes a 3-db gain reduction when $X_C = R_C$, or

$$\frac{1}{2\pi f \times 32.7 \text{ pF}} = 950 \text{ }\Omega$$

Solving, we have

$$f = \frac{1}{2\pi \times 32.7 \text{ pF} \times 950 \text{ }\Omega} = 5.1 \text{ MHz}$$

The effect of reduced transistor current gain must also be examined. From Eq. (16-21),

$$f = \frac{500 \text{ MHz}}{2} = 250 \text{ MHz}$$

Since this frequency is greater, the bandwidth limiting factor is the collector impedance, and the bandwidth of the stage is 5.1 MHz.

Input impedance is given by*

$$Z_{in} = \frac{1}{j\omega 7.8 \text{ pF}} \| \left( \beta \times 330 \text{ }\Omega \right)$$

At frequencies below 3.6 MHz, as given in Eq. (16-14), $\beta = 140$ and

$$Z_{in} = \frac{1}{j\omega 7.8 \text{ pF}} \| 46.2 \text{ k}\Omega$$

(It is better to leave the expression in this form than to combine the capacitive and resistive values, because other parallel components will be added when the collector circuit of $Q_1$ is analyzed.) At frequencies above 3.6 MHz, $\beta$ is given by Eq. (16-15) as 500 MHz/$f$ so that

$$Z_{in} = \frac{1}{j\omega 7.8 \text{ pF}} \| \left( \frac{165 \text{ MHz}}{f} \right) \text{k}\Omega$$

The input impedance becomes the load for the previous stage ($Q_1$) and may affect the frequency response of that stage. Calculations of frequency response and input impedance for the first stage would be similar to those shown for the second stage. Note that the second-stage calculations must be made first, in order to know the collector load for the first stage.

## 16.4    FREQUENCY COMPENSATION WITH AN EMITTER CAPACITOR

**problem 16.2**    Figure 16-6 shows the same amplifier as in Problem 16.1, but with $C_1$ added at the emitter for frequency compensation. Calculate the value of $C_1$ and the resulting upper bandwidth limit and input impedance of the second stage containing $Q_2$.

**theory**    The ac equivalent circuit, shown in Fig. 16-6b, is the same as in Problem 16.1 (Fig. 16-5d) with $C_1$ added.

The gain of the stage is:

$$A \cong \frac{Z_C}{Z_E} = \frac{950 \text{ }\Omega \| \dfrac{1}{j\omega \times 32.7 \text{ pF}}}{330 \text{ }\Omega \| \dfrac{1}{j\omega \times C_1}} \tag{16-22}$$

If high-frequency gain is to be the same value as for intermediate and low frequencies, the capacitive reactances in Eq. (16-22) must have the same ratio as the resistances:

$$\frac{1/(j\omega \times 32.7 \text{ pF})}{1/(j\omega \times C_1)} = \frac{950 \text{ }\Omega}{330 \text{ }\Omega} \tag{16-23}$$

which can be solved for $C_1$.

High-frequency response will not be limited by the capacity in the collector circuit, since it is now compensated for in the emitter circuit. Therefore, the only limit to high-frequency response is that of the transistor itself, given in Eq. (16-21).

Input impedance is also determined as in Problem 16.1.

**solution**    $C_1$ is calculated from Eq. (16-23). Solving for $C_1$ yields

$$C_1 = \frac{950 \text{ }\Omega}{330 \text{ }\Omega} \times 32.7 \text{ pF} = 94 \text{ pF}$$

The only limit to frequency response is now that of the transistor characteristic, from Eq. (16-21):

$$f \cong \frac{f_T}{2} = \frac{500 \text{ MHz}}{2} = 250 \text{ MHz}$$

Input impedance, determined as in Problem 16.1, is

$$Z_{in} = \frac{1}{j\omega 7.8 \text{ pF}} \| \left( 330 \text{ }\Omega \| \frac{1}{j\omega 94 \text{ pF}} \right) \beta$$

*The $j$ operator is defined mathematically as $\sqrt{-1}$ and indicates a 90° phase difference between the reactive and resistive portions of the impedance. The combined impedance is the vectorial sum, with a magnitude equal to the square root of the sum of the squares of the resistance and reactance magnitudes.

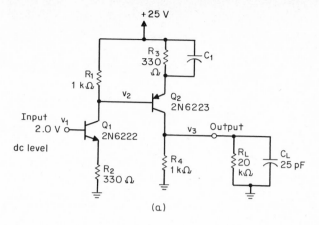

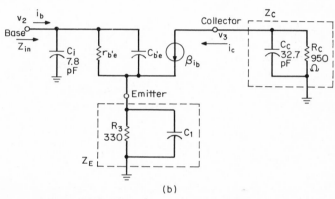

**Fig. 16-6** (*a*) Amplifier of Problem 16.1, with an emitter compensating capacitor $C_1$ added. (*b*) Simplified ac equivalent circuit.

For $f < 3.6$ MHz, $\beta = 140$, and the impedance becomes:

$$Z_{in} = \frac{1}{j\omega 8.5 \text{ pF}} \,\Big\|\, 46.2 \text{ k}\Omega$$

For 3.6 MHz $< f < f_T$, $\beta = f_T/f$, and the impedance becomes:

$$Z_{in} = \frac{1}{j\omega \left(7.8 + \dfrac{f \times 94 \text{ pF}}{500 \text{ MHz}}\right)\text{pF} \,\Big\|\, \left(\dfrac{165 \text{ MHz}}{f}\right) \text{k}\Omega}$$

The 8.5-pF reactive component of the input impedance, which is independent of frequency, may be compensated for in a similar manner with a capacitor in the emitter circuit of $Q_1$.

It is important to note that the above calculations are based on analysis which assumes linear operation of the transistor. In practice, under some signal conditions, the emitter capacitor may cause momentary transistor cutoff and a resulting transient response situation which is similar, in some respects, to a reduced high-frequency bandwidth limit. An example illustrating this effect is given in the next problem.

## 16.5    TRANSIENT RESPONSE WITH EMITTER COMPENSATION

**problem 16.3**    The second stage of the amplifier of Problem 16.2 has an idealized square-wave input signal applied, with levels of +19.5 to +22.5 V, as shown in Fig. 16-7*a*. Calculate the rise and fall times of the transitions in the output waveform at the collector.

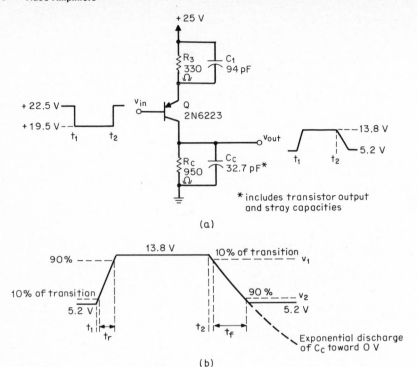

**Fig. 16-7** (*a*) Second stage, with idealized square-wave input. (*b*) $v_{out}$ waveform, showing details of $t_r$ and $t_f$ (not to scale).

**theory** The frequency response of this circuit, as calculated in the previous problem, was based on linear analysis which assumes that the transistor is continuously conducting. However, under certain conditions, an abrupt change in signal level can cause transient cutoff of a transistor and a resulting degradation of performance.

At time $t_1$, the transition causes increased current in the transistor. The analysis of Problem 16.2, which showed an upper bandwidth limit of 250 MHz, is valid. Transition time and upper bandwidth are then related by the formula:

$$ft = 0.35 \qquad (16\text{-}24)$$

where $f$ is the upper bandwidth limit and $t$ is the transition time (the time for the voltage to change from 10 to 90% of the final value), as shown in Fig. 16-7*b*.

At time $t_1$, the transition causes increased current in the transistor. The analysis of Problem 16.2, maintain their voltage, and $Q_2$ will be temporarily cut off as a result. The output waveform is then controlled by the exponential discharge of the capacitors until the voltage transition is completed, at which time $Q_2$ will again come into conduction. The waveform has the same relative shape at both capacitors because their circuits have the same time constant.

The collector capacitor voltages, during discharge, are determined by the exponential decay given by:

$$v_2 = v_1 e - t_f/R_c C_c$$

and, solving for the fall time $t_f$, yields*

$$t_f = R_c C_c \times \ln\left(\frac{v_1}{v_2}\right) \qquad (16\text{-}25)$$

where $R_c$ and $C_c$ are from Fig. 16-7*a* and $v_1$ and $v_2$ are shown in Fig. 16-7*b*.

The equivalent upper bandwidth limit which corresponds to the fall time can be determined with Eq. (16-24).

*ln is the natural logarithm.

**solution**  The rise time $t_r$ at the output (during which the transistor is conducting) is determined by Eq. (16-24). The upper bandwidth limit $f$ is 250 MHz (see Problem 16.2). Then,

$$250 \text{ MHz} \times t_r = 0.35, \text{ and solving yields } t_r = 1.4 \text{ ns.}$$

To find the fall time $t_f$ which occurs during cutoff of the transistor, the output signal voltage levels must be known. They are determined by conventional analysis. Levels at the emitter are shifted by +0.7 V from those at the base, to 20.2 and 23.2 V, and these become 13.8 and 5.2 V, respectively, at the collector. The 10 and 90% values, then, are:

$$v_1 = 12.94 \text{ V}$$
$$v_2 = 6.06 \text{ V}$$

Then, from Eq. (16-25), with values of $R_c$ and $C_c$ from Problem 16.2,

$$t_f = 950 \ \Omega \times 32.7 \text{ pF} \left( \ln \frac{12.94 \text{ V}}{6.06 \text{ V}} \right)$$
$$= 0.76 \times 31 \text{ ns} = 24 \text{ ns}$$

The equivalent bandwidth which corresponds to this fall time, from Eq. (16-24), is:

$$f \times 24 \text{ ns} = 0.35$$
$$f = 15 \text{ MHz}$$

It is important to note that the calculation results are strongly dependent on the signal amplitude and that a smaller signal will have better transient response. Also, the reduction of effective bandwidth affects only a portion of a particular type of signal and would not always be objectionable.

## 16.6  INDUCTIVE SHUNT COMPENSATION, WITH EMITTER COMPENSATION

**problem 16.4**  In Fig. 16-8, a video output stage drives a capacitive load. Frequency compensation is provided by (1) the inductor $L_1$ in shunt with the load, and (2) with an emitter capacitor $C_1$. Transistor $Q_2$ has a feedback capacity $C_{cb}$ of 3 pF, and stray capacities to ground at the collector are 6 pF. The gain-bandwidth product $f_T$ of $Q_2$ is 60 MHz or greater over the operating range of the circuit. Calculate approximate values of the $L_1$ and $C_1$ to maximize circuit bandwidth, and the resulting upper bandwidth limit.

**theory**  In output stages the signal level may be very large. As was seen in Problem 16.3, if emitter capacitor compensation only is used, transient response problems are likely and become worse at larger signal levels. Therefore, in these stages inductive compensation may be superior. A smaller size of emitter capacitor may also be helpful, as will be shown here.

The purpose of inductive compensation is to "tune out" the capacitance by providing a parallel resonant circuit at a high frequency. The ac equivalent circuit for the output is shown in Fig. 16-8b and is simplified in Fig. 16-8c. The output capacity of the transistor $C_o$ is approximately equal to the feedback capacity $C_{cb}$, as indicated by Eq. (16-16). This 3-pF value, added to the 6-pF stray and the 30-pF load, gives a total collector capacity $C_c = 39$ pF.

If $L_1$ is replaced with a short circuit, then the uncompensated upper bandwidth limit $f_1$ would be the frequency at which the resistive and reactive impedances are equal:

$$R_2 = \frac{1}{2\pi f_1 C_c}$$

or
$$f_1 = \frac{1}{2\pi R_2 C_c} \tag{16-26}$$

$L_1$ is added to resonate with $C_c$ somewhat above $f_1$ so as to increase the collector impedance and thus extend the frequency response. By careful choice of the value, a nearly flat, extended frequency response is obtained.

Reactances $x_L$ and $x_C$ are equal at a frequency above $f_1$. The ratio of these frequencies is $k$ and is given by:

$$k = R_2 \sqrt{\frac{C_c}{L_1}} \tag{16-27}$$

Analysis of the collector impedance vs. frequency for various values of $k$ shows that optimum improvement in frequency response is obtained when

$$1.4 < k < 1.6 \tag{16-28}$$

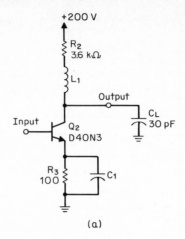

(a)

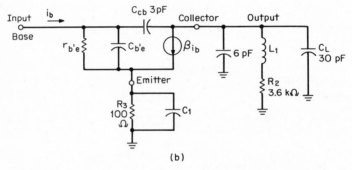

(b)

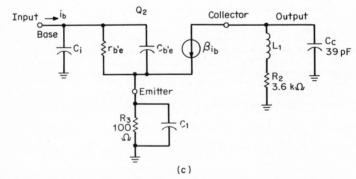

(c)

**Fig. 16-8** (a) Video output stage with inductive shunt and emitter compensation. (b) ac equivalent circuit. (c) Simplified ac equivalent circuit.

and that for values of $k$ within this range the upper bandwidth limit of the compensated circuit $f_2$ is:

$$f_2 \cong 1.7 f_1 \tag{16-29}$$

($f_1$ is the uncompensated bandwidth limit.)

The relative frequency response curves are shown in Fig. 16-9. $k$ is valued at 1.5 for these data, but for other values of $k$ within the range of Eq. (16-28), the curves would be nearly the same.

Since the response of the collector circuitry is decreased by 3 dB at $f_2$, the emitter capacitor should be valued to give compensating gain. This occurs when emitter resistance and capacitive reactance are equal:

$$R_3 = \frac{1}{2\pi f_2 C_1} \qquad \text{hence} \qquad C_1 = \frac{1}{2\pi f_2 R_3} \qquad (16\text{-}30)$$

With $C_1$ added, analysis of the complete circuit will show that the new upper bandwidth $f_3$ is:

$$f_3 \cong 2.4 f_2 = 4.1 f_1 \qquad (16\text{-}31)$$

These values for $L_1$ and $C_1$ produce a nearly flat frequency response as shown in Fig. 16-9 with about a 1-dB increase at frequencies near $f_1$. This accentuation of high frequencies is often

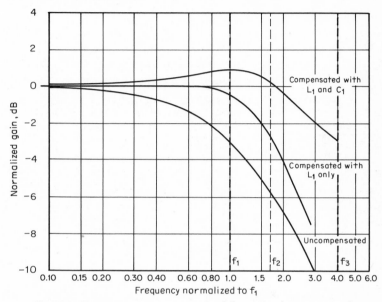

**Fig. 16-9**  Frequency response, compensated ($k = 1.5$) and uncompensated.

tolerable and is, indeed, usually desirable for television video signals where it helps to enhance details in the picture. If desired, the component values can be adjusted to give more nearly flat response, at the expense of some reduction in bandwidth.

Formulas for the calculations are summarized in Table 16-3.

**TABLE 16-3  Formulas Used for Problem 16.4**

| | |
|---|---|
| $f_1 = \dfrac{1}{2\pi R_2 C_c}$ | (16-26) |
| $f_2 \cong 1.7 \, f_1$ | (16-29) |
| $C_1 = \dfrac{1}{2\pi f_2 R_3}$ | (16-30) |
| $f_3 \cong 2.4 \, f_2 = 4.1 \, f_1$ | (16-31) |

**solution**  The uncompensated upper bandwidth limit $f_1$ from Eq. (16-26) is

$$f_1 = \frac{1}{2\pi \times 3.6k \times 39 \text{ pF}} = 1.13 \text{ MHz}$$

From Eq. (16-28) we shall select $k = 1.5$. Equation (16-27) may now be used to calculate $L_1$:

$$1.5 = 3.6 \, k \, \sqrt{\frac{39 \text{ pF}}{L_1}}$$

Solving for $L_1$ yields

$$\frac{39\ \text{pF}}{L_1} = \left(\frac{1.5}{3.6\ k}\right)^2 = 1.74 \times 10^{-7}/\Omega^2$$

$$L_1 = \frac{39\ \text{pF} \times \Omega^2}{1.74 \times 10^{-7}} = 225\ \mu\text{H}$$

The upper bandpass limit with inductive compensation only, $f_2$, is calculated from Eq. (16-29):

$$f_2 = 1.7 \times 1.13\ \text{MHz} = 1.92\ \text{MHz}$$

For additional compensation, the emitter capacitor $C_1$ is calculated from Eq. (16-30):

$$C_1 = \frac{1}{2\pi 1.92\ \text{MHz} \times 100\ \Omega} = 829\ \text{pF}$$

The frequency response for the entire circuit is determined from Eq. (16-31):

$$f_3 = 4.1 \times 1.13\ \text{MHz} = 4.65\ \text{MHz}$$

These data may be related to the response curve of Fig. 16-9 to determine approximate gain at any frequency.

The above calculations assume that the response is unaffected by frequency effects of the current source in the equivalent circuit of the transistor (see Fig. 16-8c). This may be checked by using Eq. (16-21), giving the bandwidth of the collector current source:

$$f \cong \frac{60\ \text{MHz}}{2} = 30\ \text{MHz}$$

and this is, indeed, enough above the circuit bandwidth of 4.65 MHz that it will have little effect.

*Note:* Transient response, in this type of circuit, is difficult to calculate for transient conditions causing transistor cutoff. It is dependent upon the damped resonance of $L_1$ and $C_c$ together with signal levels and is unlikely to cause signal degradation beyond that caused by the 4.65-MHz bandwidth. The response can, however, be measured using a square-wave generator and oscilloscope. A moderate amount of overshoot would be expected, due to the 1-dB "peak" in the frequency response near $f_1$.

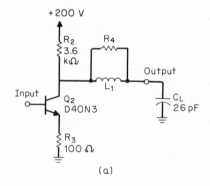

(a)

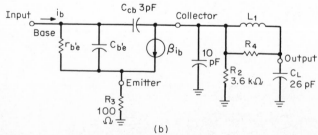

(b)

**Fig. 16-10**   (a) Series-compensated output stage. (b) ac equivalent circuit.

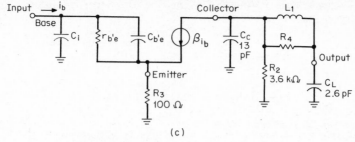

**Fig. 16-10 (cont.)** *(c)* Simplified ac circuit.

## 16.7 SERIES AND SERIES-SHUNT INDUCTIVE COMPENSATION

**problem 16.5** The output stages of Figs. 16-10 and 16-11 are similar to that of the preceding problem, 16.4. These circuits have, respectively, series and series-shunt inductive compensation. The load capacitance for both, $C_L$, is 26 pF, and the stray capacitance to ground is 10 pF. Other parameters are the same as in Problem 16.4. Calculate the values of the compensating components $L_1$, $L_2$, and $R_4$ (as required), and the resulting upper bandwidth limit.

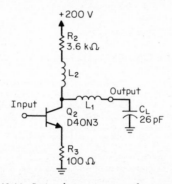

**Fig. 16-11** Series-shunt compensated output stage.

**theory** The ac equivalent circuits, Fig. 16-10b and c, are developed as in Sec. 16.6 for series compensation. (The equivalent circuits for series-shunt compensation would be the same, but with $L_2$ added and $R_4$ deleted.)

If the compensation is deleted by replacing the inductors with short circuits, both of these circuits become identical with the uncompensated version of Problem 16.4 and the same upper bandwidth limit $f_1$ applies.

With both of these compensation methods the circuit and load capacitances are separated from each other with the series inductor, all of which form a series resonant circuit at a sufficiently high frequency to extend the bandwidth. These circuits give optimum results when the load capacity is twice that of the circuit, as is the case in these problems.

Maximum bandwidth $f_2$ with nearly flat response is obtained with components having the following relationships:

Series compensation (Fig. 16-10)

$$R_4 \cong 7.8 R_2 \qquad (16\text{-}32)$$
$$L_1 \cong 0.82 (R_2)^2 C_L \qquad (16\text{-}33)$$
$$f_2 \cong 2.1 f_1 \qquad (16\text{-}34)$$

Series-shunt compensation (Fig. 16-11)

$$L_1 \cong 0.77 (R_2)^2 C_L \qquad (16\text{-}35)$$
$$L_2 \cong 0.21 (R_2)^2 C_L \qquad (16\text{-}36)$$
$$f_2 \cong 2.3 f_1 \qquad (16\text{-}37)$$

**solutions**    From Problem 16.4 the uncompensated bandwidth $f_1$ is 1.13 MHz.
*Series Compensation (Fig. 16-10):*

| | |
|---|---|
| From Eq. (16-32) | $R_4 \cong 7.8 \times 3.6 \text{ k}\Omega = 28 \text{ k}\Omega$ |
| From Eq. (16-33) | $L_1 \cong 0.82(3.6 \text{ k}\Omega)^2 \times 26 \text{ pF} = 276 \text{ }\mu\text{H}$ |
| From Eq. (16-34) | $f_2 = 2.1 \times 1.13 \text{ MHz} = 2.37 \text{ MHz}$ |

*Series-Shunt Compensation (Fig. 16-11):*

| | |
|---|---|
| From Eq. (16-35) | $L_1 \cong 0.77(3.6 \text{ k}\Omega)^2 \times 26 \text{ pF} = 259 \text{ }\mu\text{H}$ |
| From Eq. (16-36) | $L_2 \cong 0.21(3.6 \text{ k}\Omega)^2 \times 26 \text{ pF} = 71 \text{ }\mu\text{H}$ |
| From Eq. (16-37) | $f_2 = 2.3 \times 1.13 \text{ MHz} = 2.6 \text{ MHz}$ |

*It must be emphasized that these formulas and calculations can be applied only when the circuit capacity is equal to half of the load capacity.* For other ratios of capacity, less bandwidth improvement will result. Optimum component values may be determined by using computer-aided circuit analysis or by experimentation, using the above formulas only as a rough guide.

An emitter capacitor may be used for additional compensation as in Problem 16.4. However, it will give less improvement and probably no advantage over the simpler shunt compensation circuit.

## 16.8    INPUT IMPEDANCE OF AN EMITTER FOLLOWER–DRIVEN OUTPUT STAGE

**problem 16.6**    The video output stage of Section 16.6 is driven with an emitter follower $Q_1$ as shown in Fig. 16-12. The transistor parameters are:

$$Q_1: \quad \begin{aligned} h_{fe} &= 250 \\ f_T &= 200 \text{ MHz} \\ C_{cb} &= 5 \text{ pF} \end{aligned} \qquad Q_2: \quad \begin{aligned} h_{fe} &= 50 \\ f_T &= 60 \text{ MHz} \\ C_{cb} &= 3 \text{ pF} \end{aligned}$$

Stray capacities are 4 pF, emitter of $Q_1$, and 6 pF, collector of $Q_2$, to ground. Calculate the input impedance at the base of $Q_1$.

**theory**    The input impedance of transistor $Q_1$ is approximately equal to the product of its $\beta$ and its emitter impedance. Part of the emitter impedance is the input impedance of $Q_2$; therefore $Z_{\text{in}}$ ($Q_2$) must be calculated first. In this type of output stage, as will be seen in the calculations, the input impedance proves to have such a large amount of capacitance that it would be very difficult to drive it at high frequencies with a conventional common-emitter stage. One way around this problem is the use of an emitter follower, as illustrated by this problem. An alternate approach follows in Problem 16-7.

An ac equivalent circuit of the $Q_2$ stage, shown in Fig. 16-12b, follows from Fig. 16-8c. The Miller effect capacity, $C_i$ is valued as in previous sections [see Problem 16.1, Eqs. (16-16) and (16-17)]. In Fig. 16-12c, $C_{cb}$ of $Q_1$ is grounded at the collector so that there is no Miller effect multiplication.

**solution**    Input impedance at the base of $Q_2$:*

$$Z_{\text{in}}(Q_2) = \frac{1}{j\omega 111 \text{ pF}} \left\| \beta \left( 100 \text{ }\Omega \left\| \frac{1}{j\omega 820 \text{ pF}} \right. \right) \right.$$

From Eqs. (16-1) and (16-2),

$$\beta = 50 \quad \text{for } f < \frac{60 \text{ MHz}}{50} = 1.2 \text{ MHz}$$

$$\beta \cong \frac{60 \text{ MHz}}{f} \quad \text{for } f > 1.2 \text{ MHz}$$

Thus, at frequencies below 1.2 MHz,

$$Z_{\text{in}}(Q_2) \cong \frac{1}{j\omega 111 \text{ pF}} \left\| 50 \left( 100 \text{ }\Omega \left\| \frac{1}{j\omega 820 \text{ pF}} \right. \right) \right.$$

$$= \frac{1}{j\omega 111 \text{ pF}} \left\| 50 \times 100 \text{ }\Omega \left\| \frac{50}{j\omega 820 \text{ pF}} \right. \right.$$

$$= \frac{1}{j\omega 111 \text{ pF}} \left\| 5 \text{ k}\Omega \left\| \frac{1}{j\omega 16.4 \text{ pF}} \right. \right.$$

---

*The $r'_{be}$ and $c'_{be}$ of the transistors are also part of the input impedances, but their impedance values are comparatively small and so their effect is negligible.

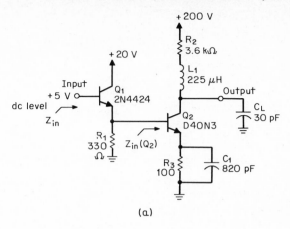

(a)

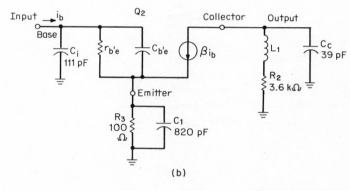

(b)

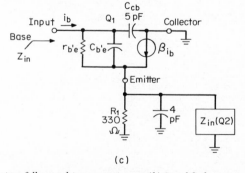

(c)

**Fig. 16-12** (a) Emitter follower–driven output stage. (b) Simplified ac equivalent circuit of the output stage (from Fig. 16-8c). (c) ac equivalent circuit $Q_1$.

Combining the 111 pF and 16.4 pF values yields

$$Z_{in}(Q_2) = \frac{1}{j\omega 127 \text{ pF}} \left\| 5 \text{ k}\Omega \right| \qquad \text{for } f < 1.2 \text{ MHz} \tag{16-38}$$

and at higher frequencies,

$$Z_{in}(Q_2) \cong \frac{1}{j\omega 111 \text{ pF}} \left\| \frac{60 \text{ MHz}}{f} \left( 100 \text{ } \Omega \left\| \frac{1}{j\omega 820 \text{ pF}} \right) \right. \qquad \text{for } f > 1.2 \text{ MHz} \tag{16-39}$$

Input impedance at the base of $Q_1$:[*]

$$Z_{in} = \frac{1}{j\omega 5 \text{ pF}} \left\| \beta \left( 300 \ \Omega \left\| \frac{1}{j\omega 4 \text{ pF}} \right\| Z_{in}(Q_2) \right) \right.$$  (16-40)

Again, from Eqs. (16-1) and (16-2),

$$\beta = 250 \quad \text{for } f < \frac{200 \text{ MHz}}{250} = 0.8 \text{ MHz}$$  (16-41)

$$\beta = \frac{200 \text{ MHz}}{f} \quad \text{for } f > 0.8 \text{ MHz}$$  (16-42)

Substituting Eqs. (16-38) and (16-41) into (16-40) yields:

$$Z_{in} \cong \frac{1}{j\omega 5 \text{ pF}} \left\| 250 \left( 330 \ \omega \left\| \frac{1}{j\omega 4 \text{ pF}} \right\| \frac{1}{j\omega 127 \text{ pF}} \right\| 5 \text{ k}\Omega \right) \right.$$

Combining the 330-$\Omega$ and 5-k$\Omega$ parallel resistances and the 4- and 127-pF capacitances yields:

$$Z_{in} \cong \frac{1}{j\omega 5 \text{ pF}} \left\| 250 \left( 310 \ \Omega \left\| \frac{1}{j\omega 131 \text{ pF}} \right) \right.$$

$$= \frac{1}{j\omega 5 \text{ pF}} \left\| 250 \times 310 \ \Omega \right\| \frac{250}{j\omega 131 \text{ pF}}$$

$$= \frac{1}{j\omega 5 \text{ pF}} \left\| 77 \text{ k}\Omega \right\| \frac{1}{j\omega 0.6 \text{ pF}}$$

Combining the capacitances yields:

$$Z_{in} \cong \frac{1}{j\omega 5.6 \text{ pF}} \left\| 77 \text{ k}\Omega \quad \text{for } f < 0.8 \text{ MHz} \right.$$

Substituting Eqs. (16-38) and (16-42) into (16-40) yields:

$$Z_{in} \cong \frac{1}{j\omega 5 \text{ pF}} \left\| \frac{200 \text{ MHz}}{f} \left( 310 \ \Omega \left\| \frac{1}{j\omega 131 \text{ pF}} \right) \quad 0.8 \text{ MHz} < f < 1.2 \text{ MHz} \right. \right.$$

Substituting Eqs. (16-39) and (16-42) into (16-40) gives:

$$Z_{in} \cong \frac{1}{j\omega 5 \text{ pF}} \left\| \frac{200 \text{ MHz}}{f} \left[ \frac{1}{j\omega 111 \text{ pF}} \left\| \frac{60 \text{ MHz}}{f} \left( 100 \ \Omega \left\| \frac{1}{j\omega 820 \text{ pF}} \right) \right] \quad \text{for } f > 1.2 \text{ MHz} \right. \right. \right.$$

The value of $Z_{in}$ may easily be calcalculated for any high frequency of interest. Since the bandwidth of the output stage is 4.65 MHz (Problem 16.4), this is likely to be the highest such frequency. For this frequency value, for example,

$$Z_{in} = \frac{1}{j\omega 5 \text{ pF}} \left\| \frac{200 \text{ MHz}}{4.65 \text{ MHz}} \left[ \frac{1}{j\omega 111 \text{ pF}} \left\| \frac{60 \text{ MHz}}{4.65 \text{ MHz}} \left( 100 \ \Omega \left\| \frac{1}{j\omega 820 \text{ pF}} \right) \right] \right. \right. \right.$$

$$= \frac{1}{j\omega 5 \text{ pF}} \left\| 43 \left[ \frac{1}{j\omega 111 \text{ pF}} \left\| 12.9 \left( 100 \ \Omega \left\| \frac{1}{j\omega 820 \text{ pF}} \right) \right] \right. \right.$$

$$= \frac{1}{j\omega 5 \text{ pF}} \left\| 43 \left( \frac{1}{j\omega 111 \text{ pF}} \left\| 12.9 \times 100 \ \Omega \right\| \frac{12.9}{j\omega 820 \text{ pF}} \right) \right.$$

$$= \frac{1}{j\omega 5 \text{ pF}} \left\| 43 \left( \frac{1}{j\omega 111 \text{ pF}} \left\| 1290 \ \Omega \right\| \frac{1}{j\omega 64 \text{ pF}} \right) \right.$$

$$= \frac{1}{j\omega 5 \text{ pF}} \left\| \frac{43}{j\omega 111 \text{ pF}} \right\| 43 \times 1290 \ \Omega \left\| \frac{43}{j\omega 64 \text{ pF}} \right.$$

$$= \frac{1}{j\omega 5 \text{ pF}} \left\| \frac{1}{j\omega 2.6 \text{ pF}} \right\| 55.5 \text{ k}\Omega \left\| \frac{1}{j\omega 1.5 \text{ pF}} \right.$$

Combining capacitances yields:

$$Z_{in} = \frac{1}{j\omega 9.1 \text{ pF}} \left\| 55.5 \text{ k}\Omega \quad \text{at } f = 4.65 \text{ MHz} \right.$$

Impedance at lower frequencies will, of course, be greater.

---

[*]The $r'_{be}$ and $c_{b'e}$ of the transistors are also part of the input impedances, but their impedance values are comparatively small and so their effect is negligible.

It must be noted, also, that the capacitive load driven by $Q_1$ can possibly cause transient response problems as detailed in Problem 16.3. In this example the $RC$ time constant in the emitter of $Q_1$ is smaller than $R_3C_1$ at $Q_2$ emitter. This ensures that $Q_1$ will not cause any further degradation of transient response.

## 16.9   INPUT IMPEDANCE OF A CASCODE-DRIVEN OUTPUT STAGE

**problem 16.7**   Figure 16-13 is another version of the output stage used in Sec. 16.6, modified so that it is cascode driven by $Q_1$. Transistor parameters are:

$$Q_1: \quad \begin{aligned} h_{fe} &= 250 \\ f_T &= 200 \text{ MHz} \\ C_{cb} &= 5 \text{ pF} \end{aligned} \qquad Q_2: \quad \begin{aligned} h_{fe} &= 50 \\ f_T &= 60 \text{ MHz} \\ C_{cb} &= 3 \text{ pF} \end{aligned}$$

Stray capacity, collector of $Q_2$ to ground, is 6 pF as before. Calculate the input impedance at the base of $Q_1$.

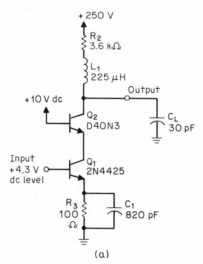

(a)

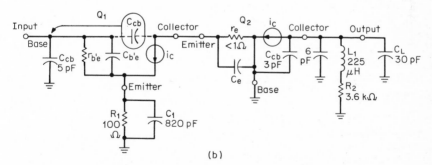

(b)

**Fig. 16-13**   (*a*) Cascode-drive output stage. (*b*) ac equivalent circuit.

**theory**   This circuit is often a good alternative to the emitter follower driver of Problem 16.6. By driving $Q_2$ in common-base mode, its $C_{cb}$ becomes ac grounded at the base, and there is no large Miller-effect capacitance. The frequency response and compensation characteristics are the same as calculated in Problem 16.4 because the collector current of $Q_1$ becomes, for practical purposes, the collector current of $Q_2$.

Figure 16-13*b* gives the ac equivalent circuit. The T-equivalent circuit (see Sec. 16.2), used for $Q_2$, is usually more appropriate for the common-base configuration. The input impedance at the emitter of $Q_2$ is so small* that $Q_1$ will have a negligibly small voltage gain (base to collector); thus

*See Sec. 16-2 for calculation of $r_e$.

there is no Miller-effect multiplication of its $C_{cb}$, and it may be considered to connect to ground, as shown.

The input impedance at the base of $Q_1$ is approximately $\beta$ times the emitter impedance, in parallel with $C_{cb}$.

**solution**   From Fig. 16-13$b$

$$Z_{in} = \frac{1}{j\omega 5 \text{ pF}} \left\| \beta \left( 100 \text{ }\Omega \middle\| \frac{1}{j\omega 820 \text{ pF}} \right) \right.$$

From Eqs. (16-1) and (16-2),

$$\beta = 250 \quad \text{for } f < \frac{200 \text{ MHz}}{250} = 0.8 \text{ MHz}$$

and

$$\beta \cong \frac{200 \text{ MHz}}{f} \quad \text{for } f > 0.8 \text{ MHz}$$

$$\text{Then } Z_{in} = \frac{1}{j\omega 5 \text{ pF}} \left\| 250 \left( 100 \text{ }\Omega \middle\| \frac{1}{j\omega 820 \text{ pF}} \right) \right.$$

$$= \frac{1}{j\omega 5 \text{ pF}} \left\| 250 \times 100 \text{ }\Omega \middle\| \frac{250}{j\omega 820 \text{ pF}} \right.$$

$$= \frac{1}{j\omega 5 \text{ pF}} \left\| 25 \text{ k}\Omega \middle\| \frac{1}{j\omega 3.3 \text{ pF}} \right.$$

Combining capacitances yields:

$$Z_{in} = \frac{1}{j\omega 8.3 \text{ pF}} \left\| 25 \text{ k}\Omega \quad \text{for } f < 0.8 \text{ MHz} \right.$$

and

$$Z_{in} = \frac{1}{j\omega 5 \text{ pF}} \left\| \frac{200 \text{ MHz}}{f} \left( 100 \text{ }\Omega \middle\| \frac{1}{j\omega 820 \text{ pF}} \right) \right. \quad \text{for } f > 0.8 \text{ MHz}$$

At the upper bandpass limit of the output, 4.65 MHz (see Problem 16.4),

$$Z_{in} = \frac{1}{j\omega 5 \text{ pF}} \left\| \frac{200 \text{ MHz}}{4.65 \text{ MHz}} \left( 100 \text{ }\Omega \middle\| \frac{1}{j\omega 820 \text{ pF}} \right) \right.$$

$$= \frac{1}{j\omega 5 \text{ pF}} \left\| 43 \left( 100 \text{ }\Omega \middle\| \frac{1}{j\omega 820 \text{ pF}} \right) \right.$$

$$= \frac{1}{j\omega 5 \text{ pF}} \left\| 43 \times 100 \text{ }\Omega \middle\| \frac{43}{j\omega 820 \text{ pF}} \right.$$

$$= \frac{1}{j\omega 5 \text{ pF}} \left\| 4.3 \text{ k}\Omega \middle\| \frac{1}{j\omega 19 \text{ pF}} \right.$$

Combining capacitances gives:

$$Z_{in} = \frac{1}{j\omega 24 \text{ pF}} \left\| 4.3 \text{ k}\Omega \quad \text{for } f = 4.65 \text{ MHz} \right.$$

## 16.10   LOW-FREQUENCY $RC$ COMPENSATION

**problem 16.8**   The small-signal video amplifier in Fig. 16-14 has a partially bypassed emitter resistor, which causes decreased gain at low frequencies. This is compensated for by adding $R_1$ and $C_1$ in the collector circuit. The current gain $h_{fe}$ of $Q_1$ is 150, and the source impedance of the driving signal is 1.0 k$\Omega$. Calculate the values of $C_1$ and $R_1$ to give optimum low-frequency compensation.

**theory**   Usually bypass capacitors are avoided in video amplifiers when very low frequency response is needed. However, such capacitors do permit increased gain with good dc bias stability, and the low-frequency degradation may sometimes be compensated for as in this example.

The ac equivalent circuit, using the T-equivalent circuit for $Q_1$, is given in Fig. 16-14$b$ with the signal source resistance added. Figure 16-14$c$ shows the mid and low-frequency equivalent circuit of the emitter, which determines emitter current and involves the time constants effecting low-frequency response. The effective source impedance is $r_s$ divided by $h_{fe}$ due to the current multiplication of $Q_1$. At high and mid frequencies $C_2$ is an ac short and gain is high; at very low

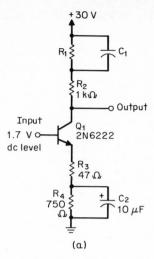

(a)

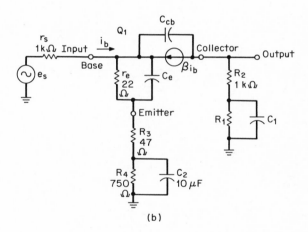

(b)

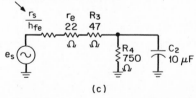

(c)

**Fig. 16-14** (*a*) Amplifier with low-frequency compensation. (*b*) ac equivalent circuit. (*c*) Circuit for emitter time constants.

frequencies $C_2$ is an open and gain is the same as at direct current. There is a transition between these gain levels defined by $f_1$ and $f_2$, the frequencies of 3-db deviation, as shown in Fig. 16-15. Analysis of the circuit in Fig. 16-14*c* will show that:

$$f_1 = \frac{1}{2\pi R_{\text{TH}} C_2} \qquad (16\text{-}43)$$

where $R_{TH}$ is the Thévenin equivalent resistance between $v_s$ and $C_2$, given by:

$$R_{TH} = \left(\frac{r_s}{h_{fe}} + r_e + R_3\right) \Big\| R_4 \tag{16-44}$$

and

$$f_2 = \frac{1}{2\pi R_4 C_2} \tag{16-45}$$

$r_e$ must be evaluated. Dc analysis indicates $I_E = 1.2$ mA. From Eq. (16-3),

$$r_e = \frac{26 \text{ mV}}{1.2 \text{ mA}} = 22 \ \Omega$$

In the collector circuit, components $R_1$ and $C_1$ are arranged to provide increased low-frequency gain, due to the increase in collector impedance as frequency decreases. This will offset the low-frequency attenuation caused by the emitter components. If values of $R_1$ and $C_1$ are used which will give the same inflection frequencies $f_1$ and $f_2$ as shown in Fig. 16-15, then the compensation is optimum, and the overall frequency response of the circuit is flat down to dc.

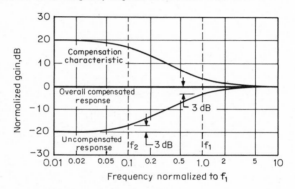

**Fig. 16-15**  Low-frequency response characteristics.

The inflection points in the collector circuit are:

$$f_1 = \frac{1}{2\pi(R_1\|R_2)C_1} \tag{16-46}$$

and

$$f_2 = \frac{1}{2\pi R_1 C_1} \tag{16-47}$$

Equations (16-46) and (16-47) may be solved simultaneously for $R_1$, giving:

$$R_1 = \frac{R_2 f_1}{f_2} - R_2 \tag{16-48}$$

If circuit constraints require $R_1$ to be smaller than in Eq. (16-48), improvement will still result with an extension of lower bandwidth limit and less decrease in dc gain than with the uncompensated circuit.

This type of circuit is practical only for relatively small signal levels. In this example, even though a 30-V supply is used, the output signal will be limited to about 2.4 V peak to peak. If the emitter is not bypassed, this circuit may be used to compensate for low-frequency attenuation which may occur in another stage.

The formulas used in Problem 16.8 are summarized in Table 16-4.

**solution**    $R_{TH}$ in the emitter circuit must be calculated first, from Eq. (16-44):

$$R_{TH} = \left(\frac{1 \text{ k}\Omega}{150} + 22 \ \Omega + 47 \ \Omega\right) \Big\| 750 \ \Omega = 69 \ \Omega$$

$f_1$ and $f_2$ may now be determined from Eqs. (16-43) and (16-45):

$$f_1 = \frac{1}{2\Pi \times 69 \ \Omega \times 10 \ \mu\text{F}} = 231 \text{ Hz} \tag{16-49}$$

$$f_2 = \frac{1}{2\pi \times 680 \ \Omega \times 10 \ \mu\text{F}} = 23.4 \text{ Hz} \tag{16-50}$$

**TABLE 16-4   Formulas Used for Problem 16.8.**

Emitter Circuit

$$f_1 = \frac{1}{2\pi R_{TH} C_2} \qquad (16\text{-}43)$$

$$R_{TH} = \left(\frac{r_s}{h_{fe}} + r_e + R_3\right)\bigg\| R_4 \qquad (16\text{-}44)$$

$$f_2 = \frac{1}{2\pi R_4 C_2} \qquad (16\text{-}45)$$

Collector Circuit

$$f_1 = \frac{1}{2\pi (R_1\| R_2) C_1} \qquad (16\text{-}46)$$

$$f_2 = \frac{1}{2\pi R_1 C_1} \qquad (16\text{-}47)$$

$$R_1 = \frac{R_2 f_1}{f_2} - R_2 \qquad (16\text{-}48)$$

Then, from Eqs. (16-48) to (16-50),

$$R_1 = 1\text{ k}\Omega \times \frac{231\text{ Hz}}{23.4\text{ Hz}} - 1\text{ k}\Omega = 8.87\text{ k}\Omega \qquad (16\text{-}51)$$

and from Eqs. (16-47), (16-50), and (16-51),

$$23.4\text{ Hz} = \frac{1}{2\pi \times 8.87\text{ k}\Omega \times C_1}$$

and solving for $C_1$ yields:

$$C_1 = \frac{1}{2\pi \times 23.4\text{ Hz} \times 8.87\text{ k}\Omega} = 0.77\ \mu\text{F}$$

This problem has assumed a negligible resistive load on the output. If a load is negligible with respect to $R_2$ but shunts $R_1$ significantly, the value of $R_1$ may be increased enough to compensate without changing other calculations.

High-frequency calculations may be made as in Problem 16.1. $C_1$ and $C_2$ would be effectively short circuits, and the calculations would have to include the effects of $r_e$ and $C_e$ on the emitter circuit impedance.

## 16.11   DC RESTORER COUPLING

**problem 16.9**   A television video signal, 4 V peak-to-peak amplitude and 63 $\mu$s between negative synchronization pulses (as shown in Fig. 16-17), is coupled between stages with the dc restorer illustrated in Fig. 16-16. The dc current gain $h_{FE}$ of $Q_2$ is 200. Calculate: (a) The dc level of the signal at the base of $Q_2$, (b) the percent droop caused by the circuitry, and (c) the maximum frequency of lower bandwidth limit needed in prior stages.

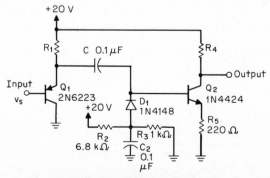

**Fig. 16-16**   DC restorer coupling circuit.

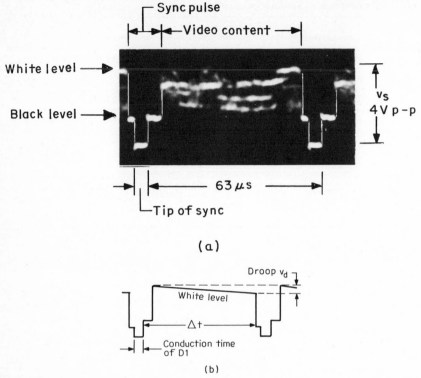

Fig. 16-17  (a) Typical television video signal. (b) Droop, exaggerated for illustration, with video content at white level.

**theory**  When full dc coupling cannot be maintained in a signal path, the dc information which would otherwise be lost may be restored provided the signal contains pulses that reference a dc level. Dc restoration is also useful when the dc level must be shifted between stages or when control of dc level, independent of the dc conditions in prior stages, is needed.

The restoration is accomplished through conduction of the diode $D_1$ at the negative extremes of the synchronization pulses. When $D_1$ conducts, $C_1$ is charged to the dc reference voltage determined by the voltage divider $R_2$ and $R_3$. Since $D_1$ may conduct for just a very few microseconds, the charging circuit must have a fast time constant which is comparable with (or less than) the synchronization tip width. This is provided by supplying the charging current through an emitter follower transistor $Q_1$, which has an output impedance of just a few ohms.

Between pulses, $D_1$ becomes reverse biased, and the base current $i_b$ of $Q_2$ is supplied by $C_1$. This causes it to become partially discharged, resulting in signal droop which is shown in exaggerated form in Fig. 16-17b. The amount of the droop voltage $v_d$ is:

$$v_d = \frac{i_b \, \Delta t}{C_1} \tag{16-52}$$

where $\Delta t$ is the time from the end of the sync tip to the beginning of the next sync pulse, about 60 μs. The percent droop is defined as:

$$\% \text{ droop} = \frac{100 v_d}{v_s} \tag{16-53}$$

where $v_s$ is the peak-to-peak signal level.

If one or more prior stages are ac coupled or have degraded dc response, this will also result in droop. Maximum droop, from this source, occurs when there is an abrupt shift in the video information from black level to white level. This represents, for this signal, a change of about 0.75 $v_s$. The amount of this droop is related to the RC time constant of the lower bandwidth limit by

$$v_d = 0.75 \, v_s \frac{\Delta t}{RC} \tag{16-54}$$

and the $RC$ time constant is related to the lower bandwidth limit $f$ by:

$$RC = \frac{1}{2\pi f} \tag{16-55}$$

Combining Eqs. (16-54) and (16-55) yields:

$$v_d = 1.5\pi v_s \, \Delta t f \tag{16-56}$$

Keeping $v_d$ of Eq. (16-56) comparable with or less than that of Eq. (16-52) will maintain the low-frequency transient response of the signal. Then, from these two equations:

$$1.5\pi v_s \, \Delta t f \leq \frac{i_b \, \Delta t}{C_1}$$

and solving for $f$ gives:

$$f \leq \frac{i_b}{1.5\pi v_s C_1} \tag{16-57}$$

If the earlier stage has poorer low-frequency response than this, much of the advantage of using the dc restorer becomes lost. Subsequent stages, if any, must be dc coupled.

Formulas used in the calculation are summarized in Table 16-5.

**TABLE 16-5    Formulas Used for Problem 16.9**

| | |
|---|---|
| $v_d = \dfrac{i_b \Delta t}{C_1}$ | (16-52) |
| % droop $= \dfrac{100 v_d}{v_s}$ | (16-53) |
| $f \leq \dfrac{i_b}{1.5\pi v_s C_1}$ | (16-57) |

**solutions**    (a) Dc level of the signal at the base of $Q_2$: The voltage established at $C_1$ by the voltage divider $R_2$ and $R_3$ is $20 \text{ V} \times 1 \text{ k}\Omega/7.8 \text{ k}\Omega = 2.56$ V. Conduction of $D_1$ during the sync tip clamps that point to 2.56 V, less 0.7-V diode drop. Therefore, the peak signal levels at the base of $Q_2$ are 1.86 V (tip of sync) and 5.86 V (white level).

The restorer must not load the voltage divider very much, since an unacceptable variability of the dc level would result.

(b) % droop: Maximum droop will occur with a "white level" signal of 5.86 V. Then,

$$i_b = \frac{5.86 \text{ V} - 0.7 \text{ V}}{\beta R_5} = \frac{5.16 \text{ V}}{200 \times 220 \ \Omega} = 0.12 \text{ mA} \tag{16-58}$$

Then, from Eq. (16-52),

$$v_d = \frac{0.12 \text{ mA} \times 60 \ \mu s}{0.1 \ \mu F} = 0.072 \text{ V}$$

and, by use of Eq. (16-53),

$$\% \text{ droop} = \frac{100 \times 0.072 \text{ V}}{4 \text{ V}} = 1.8\%$$

This amount of droop is well below the level that would be visually discernible in a television picture.

(c) Maximum frequency of lower bandwidth limit in prior stages: From Eqs. (16-57) and (16-58),

$$f \leq \frac{0.12 \text{ mA}}{1.5\pi \times 4 \text{ V} \times 0.1 \ \mu F} = 64 \text{ Hz}$$

If low-frequency response does not extend at least to this frequency, the effectiveness of the dc restorer circuit will be impaired.

## 16.12    BIBLIOGRAPHY

Blitzer, R.: *Basic Pulse Circuits*, McGraw-Hill, New York, 1964, chaps. 2 and 3.

Doyle, M. J.: *Pulse Fundamentals*, Prentice-Hall, Englewood Cliffs, N.J., 1963, chaps. 6 and 11.

Ghausi, M. S.: *Principles and Design of Linear Active Circuits*, McGraw-Hill, New York, 1965, chap. 16.

McWhorter, M. M., and Pettit, J. M.: *Electronic Amplifier Circuits*, McGraw-Hill, New York, 1961, chaps. 1 to 6.
Millman, J., and Taub, H.: *Pulse and Digital Circuits*, McGraw-Hill, New York, 1956, chaps. 2 to 4.
———and ———: *Pulse, Digital, and Switching Waveforms*, McGraw-Hill, New York, 1965, chaps. 2, 4, 5, and 8.
Shea, R. F. (ed.): *Amplifier Handbook*, McGraw-Hill, New York, 1966, chap. 25.
Thielking, R. C.: *RGB Video Amplifiers for Color TV Offer High Performance*, Application Note 90.88, General Electric Company, Syracuse, N.Y.
———: "Use RGB For Color TV," *Electronic Design*, Feb. 18, 1971.

Chapter **17**

# The Microprocessor

## MORTON HERMAN

**Senior Systems Engineer,**
**Standard Microsystems, Hauppauge, N.Y.;**
**Visiting Assistant Professor of Electrical Engineering,**
**Pratt Institute, Brooklyn, N.Y.**

## 17.1 INTRODUCTION

Figure 17-1 is a simplified block diagram of a general-purpose digital computer. Simply stated, a computer is a device that can automatically execute a sequence of instructions on given data. A digital computer processes data expressed in discrete (binary) form. The function of each section in Fig. 17-1 may be defined as follows:

MEMORY: In this section of a computer information (data) is stored and held for future use. In addition, the memory contains a sequence of instructions or commands, called a *program*, that is executed by the computer.

CENTRAL PROCESSING UNIT: (CPU): This section is the "brain" of the computer. The CPU is capable of reading the program from memory, interpreting each command, and executing the commands.

INPUT/OUTPUT (I/O): This section interfaces with the outside world. While the CPU is executing a program, printing or displaying the results may be necessary at some point. It may also be necessary for the CPU to read information entered by an operator to complete the task satisfactorily. All data transfers of this kind are handled through the I/O.

Traditionally, computers were very large and expensive machines. Today, however, integrated circuit technology has made it possible to fabricate circuits that contain

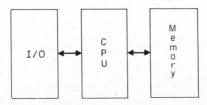

**Fig. 17-1** Simplified block diagram of a general-purpose digital computer.

thousands of transistors in a tiny volume of silicon. Integrated circuit technology has had its impact on computer technology and has made the microprocessor a reality. It is now possible to fabricate a CPU on a tiny silicon chip. Since the microprocessor is a CPU on a chip, the terms CPU and microprocessor are frequently used interchangeably.

In a similar manner, integrated circuit technology has been responsible for great advances in memory systems. In the past, computers used core memories which limited their speed and efficiency. A core memory consists of a number of doughnut-shaped disks of minute diameter that can store binary information (a logic 1 or logic 0). The storage of information is possible because the core is magnetized in one of two directions (either clockwise or counterclockwise). Today, although many computers employ core memories, semiconductor memories are used in increasing numbers because of their higher speed and packing density (number of bits per unit volume).

It is now possible to obtain a semiconductor memory chip that contains 65,536 binary digits *(bits)* of information, referred to as a 16-K memory. In addition, general purpose I/O interface chips have been fabricated that provide flexible interfacing between a microprocessor and an I/O device, such as a printer. By use of a relatively small number of inexpensive IC chips, a powerful computer, referred to as a *microcomputer*, is now possible.

## 17.2  PROGRAMMING THE MICROPROCESSOR

**problem 17.1**   Write a program that adds two numbers in memory and stores the result back into the memory.

**theory**   Instructions reside in memory as a string of bits. In the course of executing a program, the CPU reads the bits and interprets each binary pattern as a particular instruction. Since the machine understands only binary patterns, this form of representing instructions is referred to as *machine language.*

A programmer rarely writes programs directly in machine language. Instead, the programmer uses an *assembly language,* or a *higher-level language* such as BASIC. In assembly language, each microprocessor machine language instruction is represented in *mnemonic form* (a few key letters which communicate to the programmer the meaning of the instruction).

There is a one-to-one correspondence between the assembly and machine languages. In using an assembly language, there must be a mechanism that takes an assembly language program (called the *source program*) and operates on it to produce an executable machine language (called the *object program,* or *code*). The mechanism that performs this operation is an *assembler.* The assembler is itself a computer program that provides the object program as its output.

When a microprocessor program is assembled, common practice is to have the program written and run on a large computer. A large computer can provide the programmer with a paper tape output which, in turn, can be loaded into the microprocessor's memory and run. If an assembler program resides and is executed in a computer other than the one in which the object code will be run, the assembler is called a *cross assembler.*

In the use of higher-level languages, each higher-level instruction can represent a group of machine language instructions. The mechanism that converts the higher-level source program to the object code is called a *compiler.*

The format of machine language instructions residing in the memory is illustrated in Fig. 17-2. The first part of the instruction *(operation code)* tells the CPU what operation is to be performed; the second part of the instruction *(operand)* gives the particular location in memory associated with the operation code.

| Operation code | Operand |

**Fig. 17-2**   Computer instruction format.

A memory location is specified by a memory address. For example, assume that the CPU is to store the contents of one of its internal registers in location 287 in memory. The operation, *store register,* resides in the operation code part of the instruction, and the operand contains the number 287 in binary form.

Two popular microprocessors, the Intel 8080 and the Motorola 6800, have an 8-bit *(one byte)* data field and a 16-bit address field. The 16-bit address field permits the microprocessor to address $2^{16}$ or 65,536 memory locations. Since the data field is 8 bits in length, each memory location has at least 8 bits associated with it. A maximum of 16 bits is needed to specify the operand. Because of this requirement, the microprocessor must read instructions from memory, called an *instruction fetch*, in a number of 8-bit memory reads. Both the Intel and Motorola microprocessors use 8 bits to define the operation code. Instructions, therefore, may be 1, 2, or 3 bytes in length depending on the length of the operand.

It is often convenient, when writing in machine language, to use *hexadecimal* instead of binary notation. Hexadecimal notation is a way of representing 4 bits with a single hexadecimal digit. Conversion from binary to hexadecimal is provided in Table 17-1. For example, the hexadecimal representation of 1100 0011 is C3.

**TABLE 17-1   Binary-to-Hexadecimal Conversion**

| Decimal | Binary | Hexadecimal |
|---------|--------|-------------|
| 0  | 0000 | 0 |
| 1  | 0001 | 1 |
| 2  | 0010 | 2 |
| 3  | 0011 | 3 |
| 4  | 0100 | 4 |
| 5  | 0101 | 5 |
| 6  | 0110 | 6 |
| 7  | 0111 | 7 |
| 8  | 1000 | 8 |
| 9  | 1001 | 9 |
| 10 | 1010 | A |
| 11 | 1011 | B |
| 12 | 1100 | C |
| 13 | 1101 | D |
| 14 | 1110 | E |
| 15 | 1111 | F |

**solution**   The flowchart and program for adding two numbers and storing the result in memory are illustrated in Fig. 17-3a and b, respectively. A flowchart provides a visual sequence of the steps to be executed in the program.

In the column labeled FUNCTION in Fig. 17-3b, the arrow (←) means *gets replaced by* and the parentheses surrounding a name refer to the contents of the memory location. These names are converted to an absolute memory location during program assembly.

For example, (ACC) ← (LOC 1) reads "the contents of the accumulator are replaced by the contents of location 1." In this problem, LOC 1 becomes an absolute address, 000A. The accumulator, residing in the CPU, acts as a master buffer or holding register. Usually, data can be manipulated only when they are first loaded into the accumulator.

(a)

**Fig. 17-3**   Adding two numbers and storing sum in memory. (a) Flowchart. (b) Program. (Problem 17.1.)

| Flowchart step | Assembly language | Function | Memory location in hexadecimal | Machine language | Comments |
|---|---|---|---|---|---|
| READ IN NUMBER 1 | LDA, LOC1 | (ACC) ← (LOC1) | 0 0 0 0 | 2 3 | operation code, load accumulator |
| | | | 0 0 0 1 | 0 A | low byte of operand address |
| | | | 0 0 0 2 | 0 0 | high byte of operand address |
| READ IN NUMBER 2, ADD | ADD, LOC2 | (ACC) ← (ACC) + (LOC2) | 0 0 0 3 | 7 A | operation code, add to accumulator |
| | | | 0 0 0 4 | 0 B | low byte of operand address |
| | | | 0 0 0 5 | 0 0 | high byte of operand address |
| STORE ANSWER | STA, LOC3 | (LOC3) ← (ACC) | 0 0 0 6 | 6 5 | operation code, store accumulator |
| | | | 0 0 0 7 | 0 C | low byte of operand address |
| | | | 0 0 0 8 | 0 0 | high byte of operand address |
| HALT | HLT | HALT | 0 0 0 9 | 7 6 | operation code, halt |
| | | | 0 0 0 A | A 2 | LOC1, the addend |
| | | | 0 0 0 B | 3 F | LOC2, the augend |
| | | | 0 0 0 C | E 1 | LOC3, the sum |

(b)

**Fig. 17-3 (Cont.)**

The first three instructions in column 1 are 3 bytes each in length. The low part of the 16-bit operand resides in the second byte, and the high part, in the third byte of the instruction. This method of using the second and third bytes as the low and high parts of an address appears to be an accepted convention. The HALT instruction is 1 byte long and tells the CPU to stop. LOC 3 (absolute memory location 000C) stores the sum of the addition: A2 + 3F = E1.

**problem 17.2**  Determine the contents of the internal CPU registers during the execution of the program of Problem 17.1.

**theory**  A block diagram of a simplified CPU is illustrated in Fig. 17-4. The individual blocks are defined as follows:

*ACC*umulator (**ACC**): The accumulator is a register which stores data to be processed or accepts new data (the result).

*Arithmetic Logic Unit* (**ALU**): This block performs addition, subtraction, and logic operations.

*Instruction Register* (**IR**): The operation code is stored in the IR during an instruction fetch.

*Memory Address Register* (**MAR**): This register holds the address of the memory location to be read or written.

*Memory Data Register* (**MDR**): Information read from memory, with the exception of the operation code, is stored in the MDR.

*Program Counter* (**PC**): This register holds the address of the next sequential instruction in the program.

*Timing and control:* This block provides timing signals and controls the sequence of operations.

**solution**  Table 17.2 illustrates the contents of these registers as the program is executed. The notation MEM(MAR) reads "the memory location as specified by the contents of the MAR." Figure 17-5 is a flowchart indicating the internal CPU operations that are continually performed in the course of its execution. The diamond-shaped symbol is a *decision block*. It asks a question and, depending upon the answer, the operation of the CPU can take one of two paths.

**problem 17.3**  Write a program that multiplies 08 by 0A.

**theory**  It is easy enough to write a program that adds 08, 0A times. But what if one wanted to multiply 08 by 500? It becomes apparent that the program would be very large and quickly use up memory space. Memory space is a very important commodity in a computer, and programmers are forever trying to reduce the number of instructions in a program.

The problem can be solved using a minimum of memory space if one introduces another accumulator into the microprocessor and two new instructions. The new instructions are *jump* if the accumulator is not equal to zero (mnemonic JNZ) and *decrement* the accumulator (mnemonic DEC).

The JNZ instruction is capable of making a decision. If the accumulator is not equal to zero, then the next instruction is fetched from the address as specified by bytes 2 and 3 of the instruction (the operand associated with JNZ). Otherwise the next sequential instruction in memory is fetched. This jump instruction is referred to as a *conditional jump instruction.* Generally many conditional jump instructions are contained in a microprocessor instruction set, all jumping on a different condition.

An *unconditional jump instruction* (mnemonic JMP) also exists. This forces the computer to fetch the next instruction from the memory location as specified by bytes 2 and 3 of the instruction, regardless of the present condition of the computer. The DEC instruction subtracts 1 from the accumulator.

It is now possible to continuously execute the same ADD instruction with the new sum residing in one accumulator. The other accumulator acts as a counter and allows the addition to occur a specified number of times while being decremented after each addition.

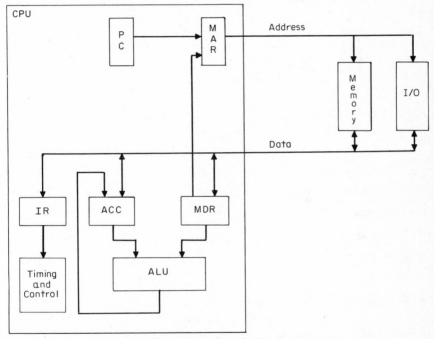

**Fig. 17-4**  Block diagram of a simplified CPU.

**solution**   The flowchart and assembly program for multiplying two numbers are provided in Fig. 17-6a and b, respectively. Numbers 08 and 0A are in memory locations labeled MULT1 and MULT2, respectively. It is important to note that each instruction which refers to an accumulator now must specify the accumulator involved (ACC1 or ACC2). REPEAT, AGAIN, MULT1, MULT2, and ANS are memory locations that are given symbolic names in the assembly program. These are converted to absolute memory addresses during the assembly program's conversion to machine language.

**problem 17.4**   Write a program that performs the mathematrical operation $(7 \times A + 4) \times 2$. Numbers 7, A, 4, and 2 reside in memory locations LOC1 through LOC4.

**theory**   The solution requires two multiplications. Although a program could be written to do this, much redundancy and wasted memory space would result. If a problem requires 300 multiplications, a straightforward method is out of the question.

**TABLE 17-2  Contents of Registers for Problem 17.2**

| PC | MAR | IR | MDR | ACC | MEM(000C) | Comments | Cycle |
|----|-----|----|-----|-----|-----------|----------|-------|
| 0000 | XXXX | XX | XXXX | XX | XX | COMPUTER IS RESET, (PC) ← 0000 | (PC) ← 0000 |
|  | 0000 |  |  |  |  | (MAR) ← (PC) | LDA FETCH |
| 0001 |  |  |  |  |  | (PC) ← (PC) + 1 |  |
|  |  | 23 |  |  |  | (IR) ← MEM(MAR) |  |
|  | 0001 |  |  |  |  | (MAR) ← (PC) |  |
| 0002 |  |  |  |  |  | (PC) ← PC + 1 |  |
|  |  |  | XX0A |  |  | (MDRLOW) ← MEM(MAR) |  |
|  | 0002 |  |  |  |  | (MAR) ← (PC) |  |
| 0003 |  |  |  |  |  | (PC) ← (PC) + 1 |  |
|  |  |  | 000A |  |  | (MDRHIGH) ← MEM(MAR) |  |
|  | 000A |  |  |  |  | (MAR) ← (MDR) |  |
|  |  |  |  | A2 |  | (ACC) ← MEM(MAR) | EXECUTION OF LDA |
|  | 0003 |  |  |  |  | (MAR) ← (PC) | ADD FETCH |
| 0004 |  |  |  |  |  | (PC) ← (PC) + 1 |  |
|  |  | 7A |  |  |  | (IR) ← MEM(MAR) |  |
|  | 0004 |  |  |  |  | (MAR) ← (PC) |  |
| 0005 |  |  |  |  |  | (PC) ← (PC) + 1 |  |
|  |  |  | 000B |  |  | (MDRLOW) ← MEM(MAR) |  |
|  | 0005 |  |  |  |  | (MAR) ← (PC) |  |
| 0006 |  |  |  |  |  | (PC) ← (PC) + 1 |  |
|  |  |  | 000B |  |  | (MDRHIGH) ← MEM(MAR) |  |
|  | 000B |  |  |  |  | (MAR) ← (MDR) |  |
|  |  |  | 003F |  |  | (MDRLOW) ← MEM(MAR) | EXECUTION OF ADD |
|  |  |  |  | E1 |  | (ACC) ← (ACC) + (MDRLOW) |  |
|  | 0006 |  |  |  |  | (MAR) ← (PC) | STA FETCH |
| 0007 |  |  |  |  |  | (PC) ← (PC) + 1 |  |
|  |  | 65 |  |  |  | (IR) ← MEM(MAR) |  |
|  | 0007 |  |  |  |  | (MAR) ← (PC) |  |
| 0008 |  |  |  |  |  | (PC) ← (PC) + 1 |  |
|  |  |  | 000C |  |  | (MDRLOW) ← MEM(MAR) |  |
|  | 0008 |  |  |  |  | (MAR) ← (PC) |  |
| 0009 |  |  |  |  |  | (PC) ← (PC) + 1 |  |
|  |  |  | 000C |  |  | (MDRHIGH) ← MEM(MAR) |  |
|  | 000C |  |  |  |  | (MAR) ← (MDR) |  |
|  |  |  |  |  | E1 | MEM(MAR) ← ACC | EXECUTION OF STA |
|  | 0009 |  |  |  |  | (MAR) ← (PC) | HALT FETCH |
| 000A |  |  |  |  |  | (PC) ← (PC) + 1 |  |
|  |  | 76 |  |  |  | (IR) ← MEM(MAR) |  |
| 000A | 0009 | 76 | 000C | E1 | E1 | HALT | EXECUTION OF HALT |

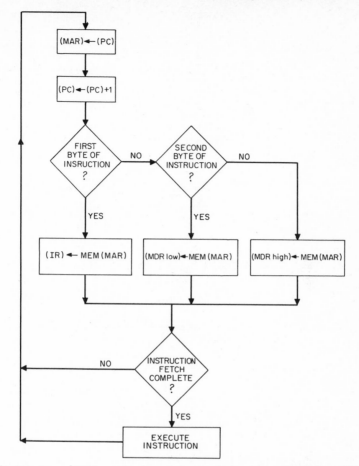

**Fig. 17-5** Internal CPU operation for an instruction fetch and execution (Problem 17.2).

In general, if one has to multiply many times in the course of a large and complex program, it is nice to have an instruction that saves the current contents in the PC register and jumps to an address in memory where a multiply program (called a *subroutine*) resides. After the subroutine is executed, another instruction returns the PC register to its original value prior to the jump and allows the main program to continue executing without any loss of continuity.

Microprocessors have the ability of doing this. To understand how this is accomplished, two additional instructions are defined. These are *call a subroutine* (mnemonic CALL) and *return from a subroutine* (mnemonic RET). Figure 17-7 illustrates the procedure of calling and returning from a subroutine at a random place in a main program.

During the execution of the CALL instruction, the current contents of the PC register, referred to as the *return address,* are stored in a *stack.* The stack is a special portion of memory reserved to save information related to subroutine jumps. The stack has associated with it a register internal to the microprocessor, called a *stack pointer.*

When the program encounters a CALL instruction, the return address is written into the memory location as specified by the stack pointer. After each byte is written, the stack pointer is decremented. The operation that writes information into the stack is referred to as *pushing the stack.* When the program encounters a RET instruction, it restores the return address to the PC register by reading this information from memory as specified by the stack pointer.

The operation that reads information from the stack is called *popping the stack.* After each byte is read, the stack pointer is incremented. The net result, therefore, of a CALL instruction followed sometime later by a RET instruction, leaves the stack pointer in its original position. The stack is often referred to as a *last-in, first-out* (LIFO) *memory.*

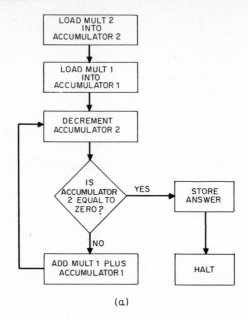

(a)

| | Assembly program | Function |
|---|---|---|
| | LDA2, MULT2 | (ACC2) ← (MULT2) |
| REPEAT: | DEC2 | (ACC2) ← (ACC2) − 1 |
| | LDA1, MULT1 | (ACC1) ← MULT1 |
| | JNZ2, AGAIN | IF ACC2 ≠ 0, (PC) ← AGAIN |
| | STA1, ANS | (ANS) ← (ACC1) |
| | HALT | HALT |
| AGAIN: | ADD1, MULT1 | (ACC1) ← (ACC1) + (MULT1) |
| | JMP, REPEAT | (PC) ← REPEAT |
| | | |
| MULT1: | 08 | |
| MULT2: | 0A | |
| ANS: | 50 | |

(b)

**Fig. 17-6**  Multiplying two numbers. (*a*) Flowchart. (*b*) Assembly program. (Problem 17.3.)

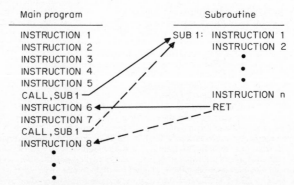

**Fig. 17-7** · Calling and returning from a subroutine.

Employing the stack to handle the subroutine allows one subroutine to call another. This is referred to as *subroutine nesting*. Having subroutines nested 15 deep is not uncommon. The depth of subroutine nesting is limited by the size of the stack.

Generally, the return address is not the only information that must be saved when jumping to a subroutine. If the contents of any register in the microprocessor are needed by the main program, these registers can be pushed on and popped off the stack, as required, with separate instructions.

**solution**   The flowchart of the main program is given in Fig. 17-8a. The multiply subroutine and the assembly program are provided in Fig. 17-8b and c, respectively. It is possible to view subroutine MULT as a mathematical function with variables MULT1 and MULT2. Once the main program specifies these two variables by loading them into their correct memory locations, subroutine MULT cranks out the answer. The main program is now able to use location PROD because it knows that the memory location contains the result.

**problem 17.5**   Monitor the status of an I/O device and perform subroutines that service the device when needed.

**theory**   In most computers, there are a number of tasks to be performed. A simple way of ensuring that the computer performs all the tasks is to write a main program that continuously monitors all I/O devices and performs the needed service when it recognizes that a particular device needs servicing. This procedure is called *polling*.

Each device has associated with it a *status register*. Further, each bit in the status register, when set by the device and read by the CPU, informs the CPU that a certain type of service is

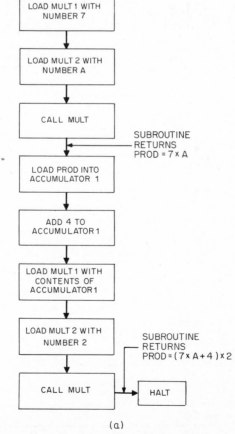

(a)

**Fig. 17-8**   Problem 17.4. (a) Flowchart of main program. (b) Subroutine flowchart. (c) Assembly program.

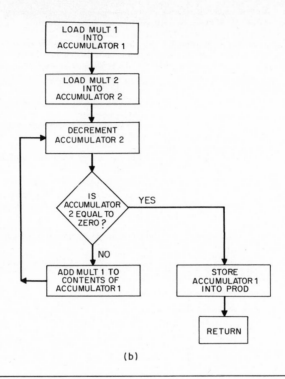

(b)

| Main Program | |
|---|---|
| Assembly program | Function |
| LDA1, LOC1 | (ACC1) ← (LOC1) |
| STA1, MULT1 | (MULT1) ← (ACC1) |
| LDA1, LOC2 | (ACC1) ← (LOC2) |
| STA2, MULT2 | (MULT2) ← (ACC1) |
| CALL, MULT | MEM (STACK POINTER) ← (PC), (PC) ← MULT |
| LDA1, PROD | (ACC1) ← (PROD) |
| ADD, LOC3 | (ACC1) ← (ACC1) + (LOC3) |
| STA1, MULT1 | (MULT1) ← (ACC1) |
| LDA1, LOC4 | (ACC1) ← (LOC4) |
| STA1, MULT2 | (MULT2) ← (ACC1) |
| CALL, MULT | MEM (STACK POINTER) ← (PC), (PC) ← MULT |
| HALT | HALT |

| Subroutine | | |
|---|---|---|
| | Assembly program | Function |
| MULT: | LDA2, MULT2 | (ACC2) ← MULT2 |
| REPEAT: | DEC2 | (ACC2) ← (ACC2) − 1 |
| | LDA1, MULT1 | (ACC1) ← (MULT1) |
| | JNZ2, AGAIN | IF ACC2 ≠ 0, (PC) ← AGAIN |
| | STA1, PROD | (PROD) ← ACC1 |
| | RET | (PC) ← MEM (STACK POINTER) |
| AGAIN: | ADD, MULT1 | (ACC1) ← (MULT1) + (ACC1) |
| | JMP, REPEAT | (PC) ← REPEAT |

(c)

**Fig. 17-8** (Cont.)

needed. The CPU then examines the bits in the status register and uses a conditional jump to call a subroutine that services the device correctly.

The flowchart in Fig. 17.9 illustrates the polling procedure. A simplified I/O interface is given in Fig. 17.10. The *command register* is written into by the CPU and tells the device exactly what to do. The *data register* handles data that are to be transferred across the interface. The *status register*, and all other registers associated with the I/O, can be treated as just another memory location (*memory mapped I/O*).

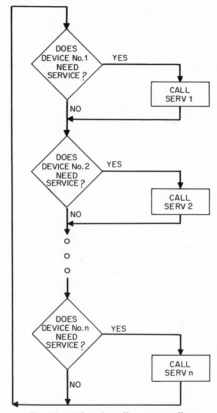

**Fig. 17-9**   Flowchart illustrating polling.

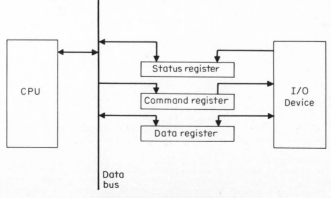

**Fig. 17-10**   Simplified I/O interface.

If the data word length of the microprocessor is 8 bits, one can define a status register that indicates service to a device in 8 different ways. In order for the microprocessor to check the status bits one at a time, it is necessary to introduce an instruction called *rotate accumulator* (mnemonic ROTLA or ROTRA), where L and R indicate the direction in which the accumulator is to be rotated.

Figure 17-11 illustrates the way an 8-bit accumulator can be rotated to the left through the carry bit C. The carry bit is a ninth bit associated with the accumulators. It derives its name from the fact that when two 8-bit numbers are added, a carry or *overflow* may occur.

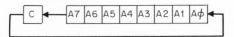

**Fig. 17-11**   Rotation of accumulator to the left through carry operation. $(A_{n+1}) \leftarrow (A_n)$, $(C) \leftarrow (A_7)$, $(A_0) \leftarrow (C)$.

Assume that the accumulator is first loaded with the status register contents and rotated eight times. The carry bit is checked after each rotation. It is possible to execute a conditional jump, called *jump if there is no carry* (mnemonic JNC), that decides whether to call the service subroutine. In this manner, each bit of the status register can be checked one at a time.

Because the status bits are set by the device and indicate what service is required, there must be a way to turn off the bits after service has been delivered. This is accomplished by loading all zeros into the status register after the entire register has been checked.

**solution**   The assembly program is provided in Fig. 17-12. Instruction CLRA puts all zeros into the accumulator. This is done prior to loading the accumulator contents back into the memory location STATUS, which is the status register.

| Assembly program | Function |
|---|---|
| LDA1, STATUS | $(ACC1) \leftarrow (STATUS)$ |
| ROTLC1 | $(C) \leftarrow (A_7)$, $(A_{n+1}) \leftarrow (A_n)$ |
| JNC, R1 | IF $C \neq 0$, $(PC) \leftarrow R1$ |
| CALL, SERV1 | MEM (STACK POINTER) $\leftarrow (PC)$, $(PC) \leftarrow SERV1$ |
| R1:    ROTLC1 | $(C) \leftarrow (A_7)$, $(A_{n+1}) \leftarrow (A_n)$ |
| JNC, R2 | IF $C \neq 0$, $(PC) \leftarrow R2$ |
| CALL, SERV2 | MEM (STACK POINTER) $\leftarrow (PC)$, $(PC) \leftarrow SERV2$ |
| R2:    ROTLC1 | $(C) \leftarrow (A_7)$, $(A_{n+1}) \leftarrow (A_n)$ |
| . | . |
| . | . |
| . | . |
| ROTLC1 | $(C) \leftarrow (A_7)$, $(A_{n+1}) \leftarrow (A_n)$ |
| JNC, R8 | IF $C \neq 0$, $(PC) \leftarrow R1$ |
| CALL, SERV8 | MEM (STACK POINTER) $\leftarrow (PC)$, $(PC) \leftarrow SERV1$ |
| R8:    CLRA1 | $(ACC1) \leftarrow 0$ |
| STA1, STATUS | $(STATUS) \leftarrow (ACC1)$ |

**Fig. 17-12**   Assembly program for Problem 17.5.

**problem 17.6**   Write a program that converts 8-4-2-1 BCD code to 7-segment code.

**theory**   This code conversion is often required in using displays. When data are displayed in decimal form, the binary data must be converted into a form that permits the display to operate. The classic method of conversion is to use a small-scale integrated (SSI) circuit, such as the 7447 TTL chip.

With a microprocessor, however, one can solve many of the classic hardware problems by using a program. Once the initial investment of a microprocessor has been made, solving a minor problem with additional hardware is wasteful if it can be solved by software. In fact, this feature makes the microprocessor such a powerful device.

A binary-coded decimal (BCD) code uses 4 bits to represent one decimal digit. For example, number 7 in decimal is represented as 0111 in BCD. Number 983 in decimal is 1001 1000 0011 in BCD.

A seven-segment code is a 7-bit code that can drive a standard LED decimal display. Figure 17.13a shows the structure of such a display. Each segment numbered S0 through S6 is an independent LED that lights when its corresponding input is a logic 1. Employing specific combinations of S0 through S6, one can display all digits from 0 to 9. For example, Fig. 17.13b shows digit 3 displayed and its code. The complete 7-segment code is summarized in Table 17-3.

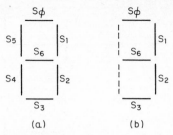

**Fig. 17-13** Standard LED display. (*a*) Seven-segment structure. (*b*) Digit 3 displayed. Segments S0, 1, 2, 3, and 6 are ON; S4 and 5 are OFF.

**TABLE 17-3  BCD to Seven-Segment Code Conversions**

| Decimal digit | BCD code | S0 | S1 | S2 | S3 | S4 | S5 | S6 |
|---|---|---|---|---|---|---|---|---|
| 0 | 0000 | 1 | 1 | 1 | 1 | 1 | 1 | 0 |
| 1 | 0001 | 0 | 1 | 1 | 0 | 0 | 0 | 0 |
| 2 | 0010 | 1 | 1 | 0 | 1 | 1 | 0 | 1 |
| 3 | 0011 | 1 | 1 | 1 | 1 | 0 | 0 | 1 |
| 4 | 0100 | 0 | 1 | 1 | 0 | 0 | 1 | 1 |
| 5 | 0101 | 1 | 0 | 1 | 1 | 0 | 1 | 1 |
| 6 | 0110 | 0 | 0 | 1 | 1 | 1 | 1 | 1 |
| 7 | 0111 | 1 | 1 | 1 | 0 | 0 | 0 | 0 |
| 8 | 1000 | 1 | 1 | 1 | 1 | 1 | 1 | 1 |
| 9 | 1001 | 1 | 1 | 1 | 0 | 0 | 1 | 1 |

To solve this problem, it is necessary to introduce a different method in which the microprocessor acquires the address of a memory location. Up to now, an instruction that involved any location outside the microprocessor contained a 16-bit operand. It is possible to have a register internal to the microprocessor that indicates the address of the operand. Such a register is called an *index register*.

An instruction using such a register consists of an operation code only, making it a 1-byte instruction. Under many conditions this kind of instruction is desirable because it saves memory space. Instructions of this kind are referred to as *move instructions* (mnemonic MOV). An 8-bit operation code specifies the source and destination of the data movement.

Assume an 8-bit microprocessor and 1 bit for each memory location. If the CPU reads in the BCD code to be converted, then four of those bits are undefined. It might be convenient to set these 4 bits to some known value; at this point the logic operation of the ALU comes into the picture. Assuming that the four least significant bits contain the BCD data, it is possible to perform a bit-by-bit AND operation of the accumulator with some predefined bit pattern. The results are then stored back into the accumulator. If the pattern 0000 1111 is used, the accumulator contains zeros in the four previously undefined bit positions and leaves the four least significant bits unchanged. This procedure is called *masking*.

**solution**  The flowchart and assembly program for the problem are provided in Fig. 17-14a and b, respectively. The base address is loaded into the index register which, after modification, is used to point to the correct seven-segment code. The method used to solve this problem is referred to as *table lookup*. The program requires a specified location in memory, pointed to by a base address which has stored in it the seven-segment code for digit zero. The next nine sequential memory locations have stored in them the seven-segment code for the next nine digits in ascending order.

The program adds the actual BCD data, after masking, to the index register. It then uses the MOV instruction to load the accumulator with the corresponding seven-segment code. The index register, since it specifies an address, must be loaded using two MOV instructions. The low byte resides in address BASELOW, and the high byte resides in the next sequential memory location called BASEHI. When the index register is added to the accumulator, only the low byte is added. Likewise, the instruction STA1 defines an internal data transfer from the accumulator to the index register (low byte).

Addresses A000 through A009 contain the seven-segment code. Because these locations are shown in hexadecimal, the most significant bit of the contents of each location is arbitrarily chosen to be a logic 0. Table 17-4 illustrates the sequence of contents of the index register, the accumulator, and the memory location labeled DIGIT, as the program proceeds.

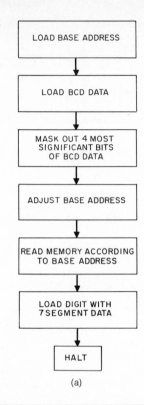

(a)

| Inst # | Assembly program | Function |
|---|---|---|
| 1 | MOV, INDEXLOW, BASELOW | (INDEXLOW) ← MEM (BASELOW) |
| 2 | MOV, INDEXHI, BASEHI | (INDEXHI) ← MEM (BASEHI) |
| 3 | LDA1, BCD | (ACC1) ← MEM (BCD) |
| 4 | AND, MASK | (ACC1) ← (ACC1) · MEM (MASK) |
| 5 | STI, INDEXLOW, ACC1 | (INDEXLOW) ← (ACC1) |
| 6 | MOV, ACC1, MEM | (ACC1) ← MEM (INDEXHI, INDEXLOW) |
| 7 | STA, DIGIT | MEM (DIGIT) ← (ACC1) |

BASELOW: 00    A000: 7E
 BASEHI: A0    1: 30
   BCD: X5    2: 6D
  MASK: 0F    3: 79
            4: 33
            5: 5B
            6: IF
            7: 70
            8: 7F
            9: 73          (b)

**Fig. 17-14**    Problem 17.6. (*a*) Flowchart. (*b*) Assembly program.

## 17.3  MICROPROCESSOR SUPPORT CIRCUITRY

**problem 17.7**    Design a two-phase clock for driving a microprocessor in accordance with the timing diagram of Fig. 17-15.

**theory**    Many microprocessors are fabricated using MOS technology. The unique nature of this technology dictates the use of a two-phase clock. One method for generating this clock is to employ an oscillator that yields a square-wave output. This is followed by a *twisted-ring counter*, commonly called a *Johnson counter*.

**TABLE 17-4   Contents of Key Registers in the Execution of Problem 17.6**

| After inst. no. | Index HI | Index LOW | ACCA | Digit |
|---|---|---|---|---|
| 1 | XX | 00 | XX | XX |
| 2 | A0 | 00 | XX | XX |
| 3 | A0 | 00 | X5 | XX |
| 4 | A0 | 00 | 05 | XX |
| 5 | A0 | 00 | 05 | XX |
| 6 | A0 | 05 | 05 | XX |
| 7 | A0 | 05 | 5B | XX |
| 8 | A0 | 05 | 5B | 5B |

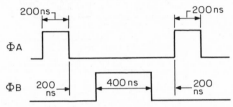

**Fig. 17-15**   Timing diagrams for a two-phase clock (Problem 17.7).

**solution**   A five-stage Johnson counter implemented with *S-R* flipflops is illustrated in Fig. 17-16a. The decoding of ΦA and ΦB is summarized in Fig. 17-16b. This counter has 10 states out of a possible 32 states which it would have if it were a normal binary counter. For this reason, the counter must be initialized to zero. If this is not done, it may cycle through erroneous states. For example, if the counter is initialized to 11001, it will cycle through 10 states. These states are different from the 10 states shown in the truth table of Fig. 17-16b.

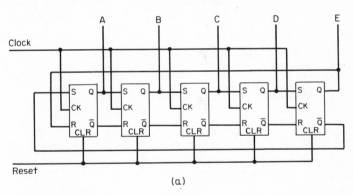

(a)

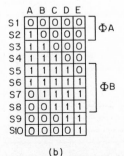

(b)

**Fig. 17-16**   A five-stage Johnson counter. (a) Circuit. (b) Truth table.

There are two attractive features of a Johnson counter. First, each state transition has a change in only one variable. For example, when S1 switches to S2, only A changes. This permits glitch-free decoding of states. If more than one variable changes during a stable transition, there is always the question of which variable changed first. This may result in momentarily decoding of a false state. Another feature of the Johnson counter is that any state, or combination of states, can be decoded with two input logic gates. A straight binary counter generally needs as many inputs as there are stages in the counter for decoding.

The timing, along with the decoding logic, for ΦA and ΦB is illustrated in Fig. 17-17. All state transitions occur during the rising edge of the clock pulses. Besides ΦA and ΦB, any other clock can be generated simply by using an additional logic gate. Because of this, the Johnson counter provides a flexible design for generating clock pulses.

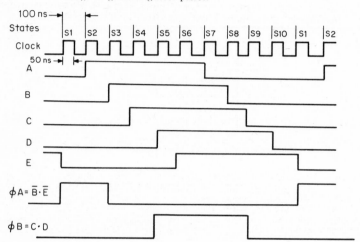

**Fig. 17-17**    Timing and decoding of a five-stage Johnson counter (Problem 17.7).

**problem 17.8**    Solve Problem 17.7 using a read-only memory (ROM).

**theory**    A ROM is a memory that has a particular bit pattern permanently stored in it and presents this pattern when addressed. If one sequentially accesses the ROM under the control of a clock, then it is possible to store in the ROM the bit pattern that generates ΦA and ΦB according to the timing diagram of Fig. 17-15.

Figure 17-18 shows a block diagram for accomplishing this. Inputs A0 through A3 are binary addresses that are presented to the ROM. These four address bits permit up to 16 locations to be accessed.

**solution**    A timing diagram showing the operation of the ROM as a clock generator and the programmed bits necessary to generate ΦA and ΦB are provided in Fig. 17-19. The positive

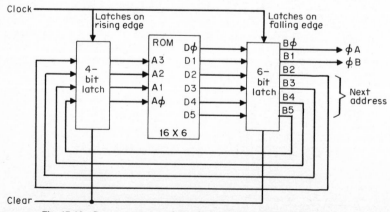

**Fig. 17-18**    Generating a two-phase clock using a ROM (Problem 17.8).

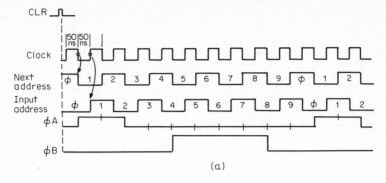

| Input address | Bφ | B1 | B2 | B3 | B4 | B5 |
|---|---|---|---|---|---|---|
| 0 0 0 0 | 1 | 0 | 0 | 0 | 0 | 1 |
| 0 0 0 1 | 1 | 0 | 0 | 0 | 1 | 0 |
| 0 0 1 0 | 0 | 0 | 0 | 0 | 1 | 1 |
| 0 0 1 1 | 0 | 0 | 0 | 1 | 0 | 0 |
| 0 1 0 0 | 0 | 1 | 0 | 1 | 0 | 1 |
| 0 1 0 1 | 0 | 1 | 0 | 1 | 1 | 0 |
| 0 1 1 0 | 0 | 1 | 0 | 1 | 1 | 1 |
| 0 1 1 1 | 0 | 1 | 1 | 0 | 0 | 0 |
| 1 0 0 0 | 0 | 0 | 1 | 0 | 0 | 1 |
| 1 0 0 1 | 0 | 0 | 0 | 0 | 0 | 0 |

(b)

φA   φB   Next address

**Fig. 17-19**  Problem 17.8. (*a*) Timing of the ROM. (*b*) Programmed bits.

transition of the clock latches up the next address and presents this address to the ROM. The negative transition presents the ROM output to the outside world by latching it up in the output latch. The output consists of the next address (4 bits) and 2 bits for ΦA and ΦB.

Once the memory address is presented to the ROM, a finite time is required before the output of the ROM is valid. This time is called the *access time*. Assuming that the delay through the input latch is negligible, the access time for the ROM must be less than 50 ns, which happens to be the time of the clock.

In practice, the input latch has a delay and the output latch has a minimum *setup time* (time the input data must be stable prior to the triggering edge of the clock). Figure 17-20 illustrates these times in detail. For this pattern, the following equations must be satisfied:

$$T_{ON} - T_{D,max} - T_{S,min} \geq T_{ACC} \tag{17-1}$$

where $T_{ON}$ = ON time of the clock
$T_D$ = delay through the input latch
$T_S$ = setup time of the output latch
$T_{ACC}$ = access time of the ROM

By using values $T_{D,max} = 10$ ns, $T_{S,min} = 5$ ns, and $T_{ON} = 50$ ns, the minimum access time is 35 ns.

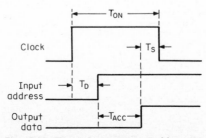

**Fig. 17-20**  Timing constraints in Problem 17.8.

**problem 17.9** Given 16 address lines from an MOS microprocessor, design a decoder that generates signals for 16 separate memory zones.

**theory** A total of 16 address lines is capable of addressing 65,536 ($2^{16}$) memory locations. For convenience, one refers to this memory size as 64 K to indicate the approximate amount of memory space. By taking the four most significant bits of the address and using a TTL 4-16 decoder, it is possible to generate 16 unique signals. The remaining 12 address lines will be common to the memory. This results in 16 zones, each 4096 ($2^{12}$) locations in size.

Because in a system it is often required to break up the memory into different sections as dictated by the function of the computer, the memory is divided. Each section of memory serves a specific function and has characteristics which are different from other memory sections.

In general, the microprocessor must interface to the outside world in some manner. The logic it interfaces with is usually TTL. Care must be exercised when interfacing MOS to TTL since MOS does not have the drive capability of a TTL gate.

Figure 17-21 shows an MOS-to-TTL interface. Current $I_{OL}$ is the current the MOS output can sink and maintain a valid logic 0 at the TTL input. Current $I_{IL}$ is the current that must be sunk from a single TTL load to maintain the input at a logic 0. For the interface to operate, the following expression must be satisfied:

$$I_{OL} \geq I_{IL} \tag{17-2}$$

Table 17-5 lists the values of $I_{IL}$ for the five families of TTL. Values of $I_{OL}$ are specified by the manufacturer of the microprocessor and typically range from 0.25 to 1.6 mA.

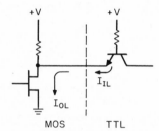

**Fig. 17-21**   MOS to TTL interface.

**TABLE 17-5   Values of $I_{IL}$ for TTL**

| TTL family | $I_{IL}$ |
|---|---|
| High performance | 2.0 mA |
| Standard | 1.6 mA |
| Schottky | 2.0 mA |
| Low-power Schottky | 0.36 mA |
| Low power | 0.18 mA |

**solution** Assume that all address lines emanating from the microprocessor have an $I_{OL}$ of 1 mA. It is necessary to use either low-power, or low-power Schottky, TTL gates at the interface. Since a low-power Schottky gate is faster than a low-power gate, the system's speed requirements dictate which of the two families is used.

Figure 17-22 shows the interface using a TTL 74LS154, 4-16 decoder and its truth table. (The LS in the number signifies low-power Schottky.) The 74LS154 has four inputs (A15 through A12) which are inputted directly from the microprocessor. Terminals A11 through A0 also come out of the microprocessor and form the 12-bit address line common to all memory. These 12 lines go to a number of memories, and there must be a TTL driver which resembles a low-power load at the input and is capable of driving a line having multiple loads.

This addressing scheme allows much flexibility in the system. A new memory can be added by choosing one of the zone selects and running the common 12-bit address to it. The outputs of the 74LS154 are defined in negative logic. In other words, when a specific output is active or specified by inputs A, B, C, D, it has a low voltage (0.2 V nominal) at the output. All other outputs have a high voltage (3.4 V nominal).

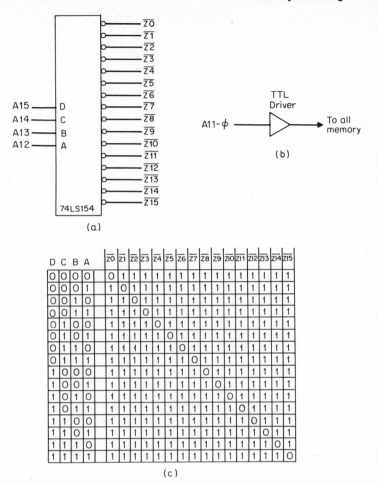

(a)

(b)

| D C B A | $\overline{Z0}$ | $\overline{Z1}$ | $\overline{Z2}$ | $\overline{Z3}$ | $\overline{Z4}$ | $\overline{Z5}$ | $\overline{Z6}$ | $\overline{Z7}$ | $\overline{Z8}$ | $\overline{Z9}$ | $\overline{Z10}$ | $\overline{Z11}$ | $\overline{Z12}$ | $\overline{Z13}$ | $\overline{Z14}$ | $\overline{Z15}$ |
|---|---|---|---|---|---|---|---|---|---|---|---|---|---|---|---|---|
| 0 0 0 0 | 0 | 1 | 1 | 1 | 1 | 1 | 1 | 1 | 1 | 1 | 1 | 1 | 1 | 1 | 1 | 1 |
| 0 0 0 1 | 1 | 0 | 1 | 1 | 1 | 1 | 1 | 1 | 1 | 1 | 1 | 1 | 1 | 1 | 1 | 1 |
| 0 0 1 0 | 1 | 1 | 0 | 1 | 1 | 1 | 1 | 1 | 1 | 1 | 1 | 1 | 1 | 1 | 1 | 1 |
| 0 0 1 1 | 1 | 1 | 1 | 0 | 1 | 1 | 1 | 1 | 1 | 1 | 1 | 1 | 1 | 1 | 1 | 1 |
| 0 1 0 0 | 1 | 1 | 1 | 1 | 0 | 1 | 1 | 1 | 1 | 1 | 1 | 1 | 1 | 1 | 1 | 1 |
| 0 1 0 1 | 1 | 1 | 1 | 1 | 1 | 0 | 1 | 1 | 1 | 1 | 1 | 1 | 1 | 1 | 1 | 1 |
| 0 1 1 0 | 1 | 1 | 1 | 1 | 1 | 1 | 0 | 1 | 1 | 1 | 1 | 1 | 1 | 1 | 1 | 1 |
| 0 1 1 1 | 1 | 1 | 1 | 1 | 1 | 1 | 1 | 0 | 1 | 1 | 1 | 1 | 1 | 1 | 1 | 1 |
| 1 0 0 0 | 1 | 1 | 1 | 1 | 1 | 1 | 1 | 1 | 0 | 1 | 1 | 1 | 1 | 1 | 1 | 1 |
| 1 0 0 1 | 1 | 1 | 1 | 1 | 1 | 1 | 1 | 1 | 1 | 0 | 1 | 1 | 1 | 1 | 1 | 1 |
| 1 0 1 0 | 1 | 1 | 1 | 1 | 1 | 1 | 1 | 1 | 1 | 1 | 0 | 1 | 1 | 1 | 1 | 1 |
| 1 0 1 1 | 1 | 1 | 1 | 1 | 1 | 1 | 1 | 1 | 1 | 1 | 1 | 0 | 1 | 1 | 1 | 1 |
| 1 1 0 0 | 1 | 1 | 1 | 1 | 1 | 1 | 1 | 1 | 1 | 1 | 1 | 1 | 0 | 1 | 1 | 1 |
| 1 1 0 1 | 1 | 1 | 1 | 1 | 1 | 1 | 1 | 1 | 1 | 1 | 1 | 1 | 1 | 0 | 1 | 1 |
| 1 1 1 0 | 1 | 1 | 1 | 1 | 1 | 1 | 1 | 1 | 1 | 1 | 1 | 1 | 1 | 1 | 0 | 1 |
| 1 1 1 1 | 1 | 1 | 1 | 1 | 1 | 1 | 1 | 1 | 1 | 1 | 1 | 1 | 1 | 1 | 1 | 0 |

(c)

**Fig. 17-22** Problem 17.9. (*a*) Interface using 74LS151 decoder. (*b*) TTL driver. (*c*) Truth table.

## 17.4 MEMORY INTERFACING

**problem 17.10** (*a*) Design a CPU-to-memory interface using a 1 K × 1 ROM. (*b*) Design the ROM to contain a 4 K × 8 configuration.

**theory** A general CPU-to-memory interface is illustrated in Fig. 17-23. This interface has three busses. A *bus* is a path through which data, addresses, or controls can be transmitted and received. Associated with each bus is a source and a destination.

For the address bus, the source is the microprocessor and the destination is the memory. The

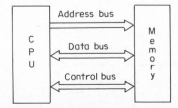

**Fig. 17-23** General CPU-to-memory interface.

address bus may go to many places, and there is thus the need of deciding which destination is the recipient of information. The address decoder of Problem 17.9 is one way this may be accomplished.

In the data bus in Fig. 17-23 the arrows point into the microprocessor and the memory, indicating that it is a *bidirectional bus*. Data can originate from the microprocessor and have their destination at the memory *(memory write operation)* or vice versa *(memory read operation)*.

For a ROM, the data bus is unidirectional, with the ROM being the source and the microprocessor the destination. For a read/write memory (RAM), the memory must be informed whether it is to be a source or a destination. In any computer, information of this type is derived from the CPU and is transmitted to all devices and memories through the control bus.

In the memory interface of Fig. 17-23, the control bus may consist of a single wire only. One logic level could indicate a memory read (memory is the source, microprocessor is the destination), and another logic level indicate a memory write (microprocessor is the source, memory is the destination). For a ROM, no control bus is needed because all operations are memory reads. For more complex memory interfacing, however, more than one wire is contained in the control bus.

One very important aspect in the description of the memory interface which is vital for successful operation is *timing*. For a memory read, the microprocessor presents its address and, some time later, expects data to be on the data bus. The memory also has a finite access time with respect to the memory read operation. Figures 17-24a and b illustrate typical read and memory read timings, respectively.

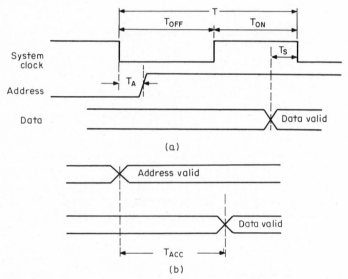

**Fig. 17-24** Typical timings. (*a*) Read cycle timing. (*b*) ROM or RAM read timing.

All signals and timings associated with the microprocessor are referenced to the system clock. Time $T_A$ is the time taken by the microprocessor to present its address after the falling edge of the clock pulse. Internal to the microprocessor, the falling edge of the clock may be the time when the new value in the PC register is loaded into the memory address register (MAR) and the internal propagation delay is equivalent to $T_A$.

Time $T_S$ is the time data must be stable prior to the next falling edge of the clock. Internal to the microprocessor, this falling edge could be the time when data are entered into a register, and $T_S$ then represents the register setup time. It is important to note that read time is a function of both the system clock period and the internal microprocessor characteristics. If $T_{ACC}$ is the access time of the memory, for reliable operation the read time $T_{READ}$ is:

$$T_{READ,max} = T - T_{A,max} - T_{S,min} \geq T_{ACC} \tag{17-3}$$

where $T$ is the period of the system clock.

**solution**  (*a*) The specifications of the microprocessor for the read cycle are summarized in Table 17-6. If one is using a 1 K × 1 ROM (1024, 1-bit memory locations) with $T_{ACC} = 700$ ns, it is clear that the minimum value of $T = 500$ ns cannot be used. A good choice for $T$ is 1000 ns. Because this yields a $T_{READ,max} = 850$ ns, it allows for 150 ns of slack. The cable delay through the busses can be significant. It is good practice to design with this in mind.

**TABLE 17-6   Timing Specifications for Read Operation (Problem 17.10)**

|         | Min     | Typical | Max      |
| ------- | ------- | ------- | -------- |
| $T$     | 500 ns  |         | 2000 ns  |
| $T_A$   | 0 ns    | 50 ns   | 100 ns   |
| $T_S$   | 50 ns   |         |          |

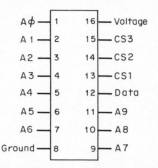

**Fig. 17-25**   Typical 1 K × 1 ROM and its pinout.

Figure 17-25 illustrates a typical 1 K × 1 ROM chip and its pinout. Terminals A9 through A0 are 10 addresses required to address the full 1-K memory capacity. Data leave the ROM via terminal 12. Terminals CS1, 2, and 3 are three-chip select inputs used when laying out large memory boards. The ROM delivers data only when CS3 = CS2 = CS1 = logic 1.

The address scheme for a 4-K memory board may be as follows:

A15–A12 selects the entire 4-K board.

A11, A10 select one of the four 1-K quadrants.

A9–A0 is common to all ROM chips.

(b) Figure 17-26 shows the 4 K × 8 design. The outputs of the ROMs can usually be connected

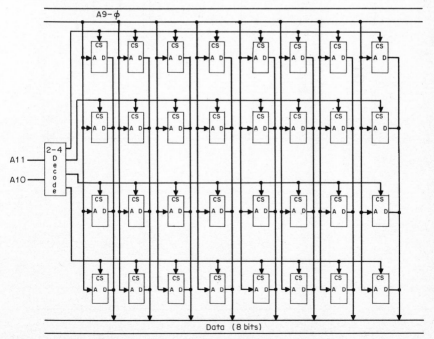

**Fig. 17-26**   A 4 K × 8 ROM memory board (Problem 17.10).

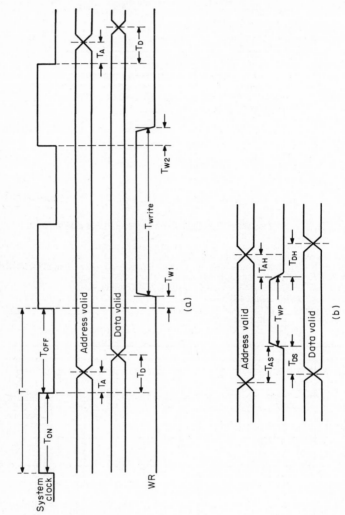

**Fig. 17-27** Problem 17.11. (a) Microprocessor write timings. (b) RAM write timings.

together since only one of the chips is selected at a time. The other outputs are inactive, and only the selected chip affects the data line. If other boards also are connected in the same manner, they can all feed a common data bus. The addressing scheme will decide which chip is to source the data bus.

**problem 17.11** (a) Design a microprocessor-to-memory interface using 1 K × 1 RAMs. (b) Design the RAM memory to have a 4 K × 8 configuration.

**theory** In this case, not only must the read cycle timing be satisfied, but the write cycle timing must be satisfied as well. Figures 17-27a and b illustrate the microprocessor and RAM-write timings, respectively.

In addition to the previously defined read timings, the microprocessor delivers a write/read signal which is designated by WR. Time $T_{W1}$ is the time between the rising edge of the system clock and the rising edge of the WR pulse. Time $T_{W2}$ is the time between the next rising edge of the system clock and the falling edge of the WR pulse. During a write cycle, data emanate from the microprocessor, and the delay of the data out of the microprocessor from the falling edge of the system clock is designated by $T_D$. Times $T_{ON}$ and $T_{OFF}$ are the ON and OFF times of the system clock, respectively.

With respect to the memory, $T_{AS}$ and $T_{DS}$ are the length of time the address and data must be stable prior to the write pulse. The length of time the address and data must be stable after the falling edge of the write pulse is designated by $T_{AH}$ and $T_{DH}$, respectively. Time $T_{WP}$ is the width of the write pulse.

The following relationships must be satisfied to write into a RAM memory:

$$T_{\text{WRITE, min}} = T - T_{W1,\max} + T_{W2,\min} \geq T_{WP} \tag{17-4}$$
$$T_{OFF} + T_{W1,\min} - T_{A,\max} \geq T_{AS} \tag{17-5}$$
$$T_{OFF} + T_{W1,\min} - T_{D,\max} \geq T_{DS} \tag{17-6}$$
$$T_{ON} + T_{A,\min} - T_{W2,\max} \geq T_{AH} \tag{17-7}$$
and
$$T_{ON} + T_{D,\min} - T_{W2,\max} \geq T_{DH} \tag{17-8}$$

In addition, Eq. (17-3) must be satisfied.

**TABLE 17-7a  Specifications for the Microprocessor for Problem 17.11**

|        | Min | Typical | Max  |
|--------|-----|---------|------|
| $T$    | 500 |         | 2000 |
| $T_A$  | 0   | 50      | 100  |
| $T_S$  | 50  |         |      |
| $T_D$  | 0   | 50      | 150  |
| $T_{W1}$ | 0 | 50      | 100  |
| $T_{W2}$ | 0 | 50      | 150  |
| $T_{ON}$ | 250 |       |      |
| $T_{OFF}$ | 250 |      |      |

**TABLE 17-7b  Specifications for the RAM for Problem 17.11**

|          | Minimum | Maximum |
|----------|---------|---------|
| $T_{WP}$ | 600     |         |
| $T_{AS}$ | 100     |         |
| $T_{DS}$ | 150     |         |
| $T_{AH}$ | 50      |         |
| $T_{DH}$ | 100     |         |
| $T_{ACC}$ | 0      | 700     |

**solution** (a) The specifications of the microprocessor and RAM are given in Tables 17-7a and 17-7b. Substituting these values in Eqs. (17-3) to (17-8) yields:

$$T - 100 \geq 700 \qquad T \geq 800 \text{ ns}$$
$$T - 100 - 0 \geq 600 \qquad T \geq 700 \text{ ns}$$
$$T_{OFF} - 100 \geq 100 \qquad T_{OFF} \geq 200 \text{ ns}$$
$$T_{OFF} - 150 \geq 150 \qquad T_{OFF} \geq 300 \text{ ns}$$
$$T_{ON} - 150 \geq 50 \qquad T_{ON} \geq 200 \text{ ns}$$
$$T_{ON} - 150 \geq 100 \qquad T_{ON} \geq 250 \text{ ns}$$

Therefore, the conditions on the microprocessor as dictated by the RAM are:

$$T \geq 800 \text{ ns}$$
$$T_{OFF} \geq 300 \text{ ns}$$
$$T_{ON} \geq 250 \text{ ns}$$

If one chooses $T = 1000$ ns, then if $T_{ON} = 500$ ns, all conditions are satisfied.

Figure 17-28 illustrates a typical 1 K × 1 RAM. The pinout of the RAM is the same as in Fig. 17-25 except that pin 13 accepts a control signal. This indicates to the memory whether the operation is to be a read or a write. Also, pin 12 is a bidirectional pin which means that the RAM becomes a destination when pin 13 is a logic 1 (write) and a source when pin 13 is a logic 0 (read). Some RAM chips have two separate data pins: one is used solely for read and the other for write operations.

(b) Figure 17-29 illustrates a 4 K × 8 RAM design.

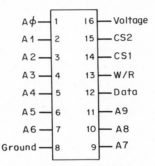

**Fig. 17-28**  Typical 1 K × 1 RAM and its pinout.

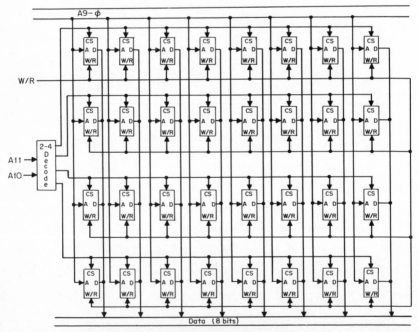

**Fig. 17-29**  A 4 K × 8 RAM memory board (Problem 17.11).

**problem 17.12** Design a memory interface that is capable of being used with a number of pin-compatible memory chips, but having different read and write timings.

**theory** Manufacturers often make an entire family of pin-compatible memory chips of various speeds. Since the timings will be different, a severe limitation is placed on the kinds of memory chips that may be used once the interface is designed. Some microprocessors overcome this limitation by allowing the memory to cause the microprocessor to enter a *wait state*. By entering a wait state, the $T_{READ}$ and $T_{WRITE}$ times can be extended in multiples of the clock period $T$, allowing the microprocessor to be compatible with an entire family of memory chips.

Figure 17-30 shows how the $T_{READ}$ time can be extended by one clock period. By designing the memory interface to include a wait time, which goes from the memory to the microprocessor, and by turning this signal on and off at the proper times, it is possible to lengthen the memory cycle and add to the flexibility of the interface. In general, if the wait time is for $N$ clock periods, then $T_{READ}$ and $T_{WRITE}$ are extended by a time equal to $T \times N$.

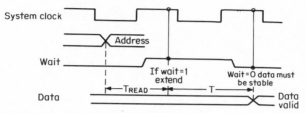

**Fig. 17-30** Extending the read time by using a wait state (Problem 17.12).

**solution** Suppose that a 1-K RAM exists in four different speed ranges as given in Table 17-8. Times $T_{READ}$ and $T_{WRITE}$ are now given by:

$$T_{READ} = T - T_{A,max} - T_{S,min} + N \times T \tag{17-9}$$

and

$$T_{WRITE} = T - T_{W1,max} - T_{W2,min} + N \times T \tag{17-10}$$

where $N$ is the number of wait cycles in the memory cycle. If $T = 1000$ ns, $T_{ON} = T_{OFF} = 500$ ns, and $T_A$, $T_S$, $T_{W1}$, and $T_{W2}$ have the values given previously. Table 17-9 lists the number of wait cycles needed for each RAM.

Equations (17-5) to (17-8) are satisfied, as can be verified. If this were not the case, an adjustment of $T$, $T_{ON}$, and $T_{OFF}$ would be necessary. RAM 4 has a $T_{ACC}$ of 2800 ns; using three wait cycles might be cutting things a bit close. An evaluation of all additional delays must be made before one can choose the proper number of wait states. Once the interface is designed, it is a simple matter to personalize the memory board during manufacture so it may be used with one of the four types of RAMs.

**TABLE 17-8   Write and Read Specifications for Four-Pin Compatible RAMs (Problem 17.12)**

|       | $T_{ACC,max}$ | $W_{P,min}$ | $T_{AS,min}$ | $T_{DS,min}$ | $T_{AH,min}$ | $T_{DH,min}$ |
|-------|------|------|-----|-----|-----|-----|
| RAM 1 | 700  | 600  | 100 | 150 | 50  | 100 |
| RAM 2 | 1400 | 1200 | 125 | 175 | 75  | 125 |
| RAM 3 | 2100 | 1800 | 150 | 200 | 100 | 150 |
| RAM 4 | 2800 | 2400 | 200 | 250 | 125 | 175 |

**TABLE 17-9   Number of Wait States for Each RAM (Problem 17.12)**

|       | No. of wait cycles | $T_{READ}$ | $T_{WRITE}$ |
|-------|-------|--------------|--------------|
| RAM 1 | 0     | 900          | 900          |
| RAM 2 | 1     | 1900         | 1900         |
| RAM 3 | 2     | 2900         | 2900         |
| RAM 4 | 2 or 3 | 2900 or 3900 | 2900 or 3900 |

## 17.5  SELECTING A MICROPROCESSOR

In selecting a microprocessor, it is essential to clearly define one's aims. Once the problem is defined, the choice of a microprocessor is much easier. The areas to be considered when choosing a microprocessor include:

1. *Instruction set.* Evaluation of the instruction set is by far the most important step. The microprocessor instruction set tells the potential user exactly what the microprocessor can and cannot do. With a good idea of one's needs, one can write programs and evaluate them for speed, length, and efficiency. These programs can be compared with the same task written with other microprocessor instruction sets.

Benchmark routines can be defined for a particular application. A *benchmark routine* is a frequently used routine, or program, for comparing microprocessors. Once these benchmark routines are defined, they are written in the assembly language of the particular microprocessor. The results are sometimes displayed graphically as points in the X-Y plane. Each point corresponds to the program length and the speed of program execution. The power of the benchmark method is that no hardware is necessary for an evaluation.

In addition to benchmark routines, the instruction set has to be used by humans. People writing programs must feel comfortable with the instruction set. If an instruction set resembles an already existing computer's instruction set, then this instruction set should be considered regardless of its benchmark results. The reason is that for established computers many debugged programs are available, and some of these might be used with little or no effort. It is also important to be aware of what each manufacturer supplies in the way of assemblers, programming manuals, development systems, etc.

2. *Microprocessor architecture.* Although all microprocessors are CPUs, their structures may differ considerably. Some microprocessors have two accumulators, internal RAMs, internal stacks, varying word length, etc. Each added or missing feature must be evaluated with respect to the user's specific application.

3. *Memory capacity.* As programs are written and hardware defined, one gets a feel for the total size of the memory. Since memory capacity varies greatly among microprocessors, care must be taken to choose a microprocessor that adequately covers the needed memory.

4. *Speed.* Certain computer operations must be completed within a specified amount of time. For example, if a computer is processing keys as they are depressed on a keyboard, then the computer must process the key and be finished by the time the next key is depressed. In addition, the application might call on the microprocessor to service several other devices. The potential user must make sure that all operations can be completed within the time constraints of the system.

5. *Interrupts.* Some devices require immediate servicing, regardless of what the microprocessor is doing. If the application calls for such a device, the microprocessor must be equipped with an *interrupt* feature.

6. *Direct memory access operation.* Direct memory access (DMA) operation is most useful when the application calls for large transfers of sequentially located blocks of data. For example, if a slow device (usually a tape device) must transfer 1024 bytes of data into sequential locations in memory, it is not good practice to tie up the microprocessor in handling this transfer.

With some external hardware and the DMA feature, loading a number of registers is possible under program control. Each register is told to start the transfer, the address where the first byte of data will be stored, and the number of bytes to be transferred. As each byte of data becomes ready for transfer to memory, the microprocessor is made to enter a hold state. The hardware can steal one cycle and act as the CPU. This practice is often referred to as *cycle stealing.*

Since the device is slow, the hold state is entered infrequently and the CPU's efficiency is greatly increased. Instead of handling a large amount of data transfers from the same device to memory, the CPU simply loads a few registers and is free to perform other tasks,

7. *Support Circuitry.* Some microprocessor manufacturers make a number of chips that are directly compatible with their microprocessors. Among compatible chips are DMA, clock generator, interrupt, and general I/O interface chips. In addition, manufacturers also supply equipment that can be used to test a system built around a microprocessor. An evaluation of the company's product line will aid greatly in the selection of a microprocessor.

## 17.6 BIBLIOGRAPHY

Barna, A., and Porat, D. I.: *Introduction to Microcomputers and Microprocessors*, Wiley, New York, 1976.

Hilburn, J. L., and Julich, P. N.: *Microcomputers/Microprocessors: Hardware, Software, and Applications*, Prentice-Hall, Englewood Cliffs, N.J., 1976.

*Intel 8080 Microcomputer Systems Users Manual*, Intel Corporation, Santa Clara, Calif., 1975.

Klingman, E. E.: *Microprocessor Systems Design*, Prentice-Hall, Englewood Cliffs, N.J., 1977.

Levine, M. E.: *Digital Theory and Practice Using Integrated Circuits*, Prentice-Hall, Englewood Cliffs, N.J., 1978.

McGlynn, D. R.: *Microprocessors*, Wiley, New York, 1976.

Peatman, J. B.: *The Design of Digital Systems*, McGraw-Hill, New York, 1972.

———: *Microcomputer-Based Design*, McGraw-Hill, New York, 1977.

*The TTL Data Book for Design Engineers*, Texas Instruments, Inc., Dallas, Tex., 1973.

Chapter **18**

# Transmission Lines

## LLOYD TEMES, P.E.

Associate Professor, Department of Electrical Technology,
College of Staten Island (CUNY)

### 18.1 INTRODUCTION

This chapter covers the characteristics of transmission lines used mainly below microwave frequencies. (Microwave transmission lines are covered in Chapter 21, Microwaves.)

Transmission lines serve to transfer electrical energy at various frequencies, with minimum loss, from point to point. They are used most often to transfer energy from a transmitter to an antenna, or from some form of transducer to a remote point where the waves will be processed for their practical use. In the latter case, an example would be their use in a closed-circuit television system.

Sections of a transmission line may be used as reactances, or to match different impedances, or to reduce standing waves on a line. They may also be used to provide a desired phase shift.

This chapter presents a series of practical problems covering the important characteristics of transmission lines. In addition, a number of tables of transmission line characteristics are also included.

### 18.2 CALCULATING THE TIME FOR AN ELECTRIC SIGNAL TO TRAVEL A GIVEN LENGTH OF CABLE

**problem 18.1** A pulse is launched on the input end of a transmission line. How long will it take the leading edge of the pulse to travel 30 m to the output end of the line? The transmission line has a velocity factor of 0.85. See Fig. 18-1.

**theory** The speed of electromagnetic waves in free space is $3 \times 10^8$ m/s. The velocity factor of a medium represents the ratio between the medium in question and the velocity in free space. Thus:

$$k = \frac{\text{velocity on transmission line}}{\text{velocity in free space}} \tag{18-1}$$

The formula for the relationship between time and distance is:

$$\text{Velocity} \times \text{time} = \text{distance} \tag{18-2}$$

This formula is as applicable to transmission lines as it is to all other cases of constant velocity. The velocity factors for some common types of transmission lines are shown in Table 18-1.

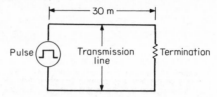

**Fig. 18-1**    A pulse travels 30 m on a transmission line.

**TABLE 18-1    Velocity Factors for Some Common Transmission Lines**

| Type | Velocity factor |
|---|---|
| Coaxial cable with polyethylene dielectric | 0.66 |
| 300-$\Omega$ twin lead | 0.82–0.84 |
| 150-$\Omega$ twin lead | 0.76–0.77 |
| 75-$\Omega$ twin lead | 0.68–0.71 |
| Two-wire line in air | 0.98 |

**solution**    By the definition of velocity factor:

$$k = \frac{\text{velocity on transmission lines}}{\text{velocity in free space}}$$

Velocity on transmission line $= k \times$ velocity in free space
Velocity on transmission line $= 0.85 \times 3 \times 10^8 = 2.55 \times 10^8$ m/s
Velocity $\times$ time $=$ distance

$$\text{Time} = \frac{\text{distance}}{\text{velocity}}$$
$$= \frac{30}{2.55 \times 10^8}$$
$$= 11.7 \times 10^{-8}\text{ s}$$
$$= 0.117\ \mu s \qquad \text{on transmission line}$$

## 18.3 DETERMINING WHETHER IT IS NECESSARY TO CONSIDER A SECTION OF CABLE AS A TRANSMISSION LINE

**problem 18.2**    A 60-kHz signal is to be carried on a cable which is 50 m long. Determine whether consideration must be given to this cable as a transmission line. See Fig. 18-2.

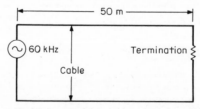

**Fig. 18-2**    A 60-kHz signal on a 50-m line.

**theory**    Transmission line considerations come into play when the wavelength of the signal being transmitted on the line is smaller than the physical length of the line or is on the same order of magnitude as the length of the line.

The first consideration must, therefore, be to determine the wavelength of the signal being transmitted. The wavelength can be determined from frequency by:

$$f\lambda = kC \qquad (18\text{-}3)$$

where $k$ is the velocity factor as described in Problem 18.1 and $C$ is the velocity of electromagnetic radiation in free space. However, since exact dimensions are not required in this calculation because it is only necessary to determine orders of magnitude, the velocity factor can be assumed to be 1.0. The actual velocity factor will be between 0.6 and 0.9. Thus Eq. (18-3) simplifies to:

$$f\lambda = C \qquad (18\text{-}4)$$

where $C$ will be taken to be $3 \times 10^8$ m/s.

**solution**    Determining wavelength by Eq. 18-4 yields:

$$f\lambda = C$$
$$\lambda = \frac{C}{f}$$
$$= \frac{3 \times 10^8}{60 \times 10^3}$$
$$= 5000 \text{ m}$$

The wavelength of the signal is 100 times as large as the cable length. Consideration of transmission line theory is not necessary.

## 18.4.   DETERMINING CHARACTERISTIC IMPEDANCE OF A TRANSMISSION LINE WHEN CAPACITANCE PER UNIT LENGTH AND INDUCTANCE PER UNIT LENGTH ARE KNOWN

**problem 18.3**    RG8A axial cable has a capacitance of 29.5 pF/ft and an inductance of 0.083 $\mu$H/ ft. Calculate the characteristic impedance of RG8/U cable.

**theory**    When two conductors are separated by a dielectric, capacitance exists. In the case of transmission lines, it is quite obvious that this situation exists. In order to provide data that would be useful for any length of transmission line, the information regarding the capacitance of a particular type of transmission line is described in terms of its capacitance per unit length.

Similarly an inductance is involved whenever a length of conductor is encountered. All transmission lines, therefore, have inductance. Again, in order to provide data for the inductance of a transmission line, which is valid for all transmission lines of the same type regardless of length, the inductance per unit length is recorded.

Another major reason for describing inductance and capacitance in terms of unit length quantities is that they are distributed reactances (along the line) rather than lumped reactances. See Fig. 18-3.

Table 18-2 presents capacitance and inductance per unit length for some commercially available transmission lines.

The characteristic impedance of a transmission line is dependent on the capacitance per unit length and inductance per unit length. The relationship between them is:

$$Z_o = \sqrt{L/C} \qquad (18\text{-}5)$$

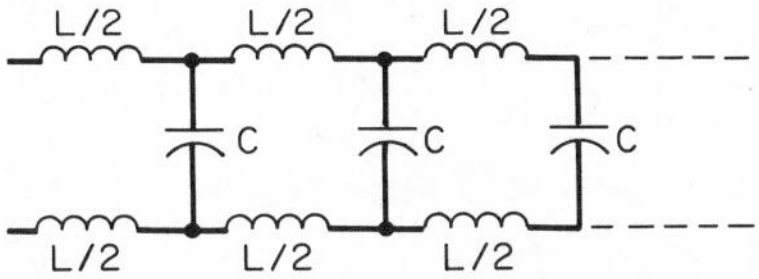

**Fig. 18-3**   Distributed capacitance and inductance of a transmission line.

**solution**    By Eq. (18-5):

$$Z_o = \sqrt{L/C}$$
$$= \sqrt{\frac{0.083 \times 10^{-6}}{29.5 \times 10^{-12}}}$$
$$= \sqrt{0.0028 \times 10^6}$$
$$= 0.053 \times 10^3$$
$$= 53 \ \Omega$$

**TABLE 18-2    Capacitance and Inductance per Foot for Some Common Transmission Line Types**

| Cable type | Capacitance per foot, pF | Inductance per foot, $\mu$H |
|---|---|---|
| RG-8A/U | 29.5 | 0.083 |
| RG-11A/U | 20.5 | 0.115 |
| RG-59A/U | 21.0 | 0.112 |
| 214-023 | 20.0 | 0.107 |
| 214-076 | 3.9 | 0.351 |

## 18.5    DETERMINING THE OPTIMUM TERMINATION OF A TRANSMISSION LINE

**problem 18.4**    A transmission line is required which is to connect a 300-$\Omega$ resistive load to a transmitter. Determine the required characteristic impedance of the transmission line to transfer power to the load without power being reflected back to the transmitter, regardless of the length of the transmission line. See Fig. 18-4.

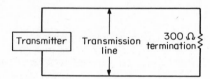

**Fig. 18-4**    A transmission line connecting a 300-$\Omega$ load to a transmitter.

**theory**    An infinitely long transmission line would provide a situation in which no power would be reflected back to the source of energy. Obviously such a situation, that of an infinitely long transmission line, cannot exist in reality. However, the operation of an infinite transmission line can be simulated by a finite transmission line having a termination equal to the characteristic impedance of the transmission line.

The distributed capacitance and inductance of an infinitely long transmission line would continuously draw power from the source since inductance and capacitance further and further down the line would be storing energy. Unlike lumped capacitance and inductance, an infinite line has an infinite number of inductances and capacitances storing energy. Thus, in order to provide a condition in which no power is reflected from a load on the finite transmission line, the characteristic impedance of the transmission line is chosen to be equal to the load impedance. See Fig. 18-5.

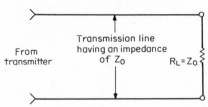

**Fig. 18-5**    The load should be resistive and equal to the characteristic impedance of the transmission line.

**solution**    Since the terminating load is given as a 300-$\Omega$ resistance, a transmission line having a characteristic impedance equal to this value would satisfy the conditions set forth in this problem.

$$Z_o = R_L$$
$$= 300 \; \Omega$$

## 18.6    DETERMINING CHARACTERISTIC IMPEDANCE FROM CABLE GEOMETRY

**problem 18.5**    What is the characteristic impedance of a coaxial cable transmission line whose cross section is shown in Fig. 18-6 and which has air as a dielectric?

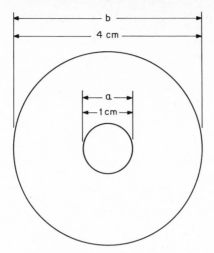

**Fig. 18-6**   Coaxial cable with air dielectric.

**theory**   Since characteristic impedance is dependent upon inductance and capacitance per unit length, it can be expected to be dependent on the geometry of the cable construction and upon the dielectric constant of the insulation which separates the conductors.

Table 18-3 presents the dielectric constants of some frequently encountered dielectric materials.

The characteristic impedance of a coaxial cable can be found using the following equation:

$$Z_o = \frac{138}{\sqrt{K}} \log \frac{b}{a}$$    (18-6)

where $K$ = dielectric constant of the insulating material (obtained from Table 18-3)
    $b$ = inside diameter of the outer conductor
    $a$ = outside diameter of the inner conductor

**TABLE 18-3   Dielectric Constant of Commonly Used Materials**

| Material | Dielectric constant |
|---|---|
| Air | 1.0 |
| Bakelite | 4.4–5.4 |
| Cellulose acetate | 3.3–3.9 |
| Formica | 4.6–4.9 |
| Window glass | 7.6–8.0 |
| Pyrex glass | 4.8 |
| Mica | 5.4 |
| Paper | 3.0 |
| Plexiglass | 2.8 |
| Polyethylene | 2.3 |
| Polystyrene | 2.6 |
| Porcelain | 5.1–5.9 |
| Quartz | 3.8 |
| Teflon | 2.1 |

**solution**   By Eq. (18-6):

$$Z_o = \frac{138}{\sqrt{K}} \log \frac{b}{a}$$

$$= \frac{138}{1} \log \frac{4}{1}$$

$$= 138 \log 4.0$$

$$= 138 \, (0.602)$$

$$= 83.08 \; \Omega$$

An alternative solution is to make use of available graphs such as that shown as Fig. 18-7.* Note, however, that if air is not the dielectric, the value obtained from the graph must be divided by $\sqrt{K}$:

$$\frac{b}{a} = \frac{4}{1} = 4$$

Entering the graph of Fig. 18-7 at 4 on the horizontal axis provides a characteristic impedance approximately equal to that calculated above:

$$83.08 \ \Omega$$

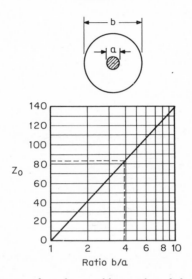

**Fig. 18-7**  Characteristic impedance for coaxial lines with air dielectric (simplified graph).

## 18.7  DETERMINING VELOCITY OF PROPAGATION AND VELOCITY FACTOR

**problem 18.6**   Calculate the velocity of electromagnetic waves on a transmission line having an inductance of 0.08 $\mu$H/ft and a capacitance of 29.5 pF/ft.

**theory**   The velocity with which an electromagnetic wave travels is dependent upon the medium it is traveling in. On a transmission line, the velocity of propagation of the electromagnetic wave is dependent on the inductance and capacitance per unit length. The relationship is:

$$v = \frac{1}{\sqrt{LC}} \tag{18-7}$$

Velocity factor of a medium is defined as the ratio of the velocity in the medium to that in free space. The velocity of electromagnetic waves in free space is $3 \times 10^8$ m/s, 186,000 mi/s, or $982 \times 10^6$ ft/s.

**solution**   By Eq. (18-7):

$$v = \frac{1}{\sqrt{LC}}$$

$$= \frac{1}{\sqrt{0.08 \times 10^{-6} \times 29.5 \times 10^{-12}}}$$

$$= \frac{1}{1.536 \times 10^{-9}} = 0.651 \times 10^9$$

*Complete graphs for coaxial and two-wire lines are found at the end of this section.

Since the inductance and capacitance per unit length were in terms of feet, the velocity above is in feet per second.

$$v = 651 \times 10^6 \text{ ft/s}$$

Converting to meters per second (1 m = 3.281 ft) yields,

$$v = \frac{651 \times 10^6}{3.281} = 198.4 \times 10^6 \text{ m/s}$$

$$= 198.4 \times 10^6 \text{ m/s}$$

Determining velocity factor $k$, we get

$$k = \frac{651 \times 10^6 \text{ ft/s}}{982 \times 10^6 \text{ ft/s}}$$

$$= 0.663$$

## 18.8    CALCULATING STANDING-WAVE RATIO FROM Z₀ AND LOAD RESISTANCE

**problem 18.7**    What would the standing-wave ratio be if a 75-Ω transmission line were used to feed a 300-Ω resistive load? See Fig. 18-8.

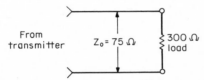

**Fig. 18-8**    A 75-Ω line feeding a 300-Ω resistive load.

**theory**    Standing waves are apparent stationary waves of voltage or current appearing on a transmission line (or antenna). They are stationary from the point of view that their maxima and minima always occur at the same physical points along the line (or antenna). Standing waves are created when a line is not terminated in its characteristic impedance. In this event, the incident waves from the generator are reflected to some degree at the end of the line. The reflected waves combine continuously with the incident waves causing *standing waves* to be formed along the line.

Standing-wave ratio (SWR) is the ratio of maximum current (or voltage) along a line to the minimum current (or voltage) along the line. The ratio is commonly expressed as a number larger than 1. (See Fig. 18-9.)

The voltage standing-wave ratio (VSWR) is equal to the current standing-wave ratio (ISWR).

$$\text{VSWR} = \frac{V_{\text{rms,max}}}{V_{\text{rms,min}}} \tag{18-8}$$

$$\text{ISWR} = \frac{I_{\text{rms,max}}}{I_{\text{rms,min}}} \tag{18-9}$$

$$\text{SWR} = \text{VSWR} = \text{ISWR} \tag{18-10}$$

Assuming that the standing waves on a particular transmission line are due totally to a mismatch between load impedance and characteristic impedance of the line, the SWR is related to $Z_o$ and $R_L$ according to:

$$\text{SWR} = \frac{Z_o}{R_L} \tag{18-11}$$

or

$$\text{SWR} = \frac{R_L}{Z_o} \tag{18-12}$$

whichever provides a quantity greater than unity.

The most sought after condition is that of an SWR of 1:1 since this indicates a perfect match and no reflected power, and thus all power incident in the load is absorbed by the load.

**solution**    By Eq. (18-12):

$$\text{SWR} = \frac{R_L}{Z_o}$$

$$= \frac{300}{75} = 4$$

This is described as SWR = 4:1.

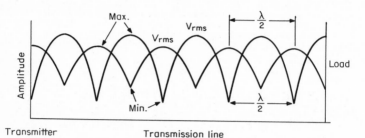

**Fig. 18-9**    Standing-wave amplitude values vary with distance along the line.

## 18.9   CALCULATING REFLECTION COEFFICIENT ON A LINE

**problem 18.8**    Determine the reflection coefficient resulting from mismatching a 50-Ω load and a 300-Ω transmission line. See Fig. 18-10.

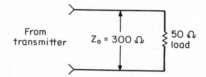

**Fig. 18-10**    A 50-Ω load being fed from a 300-Ω line.

**theory**    When a mismatch exists between transmission line characteristic impedance and load termination, some of the power transmitted down the line is reflected from the load and returns on the line. This results in outgoing and returning waves of voltage and current on the line. The ratio of reflected voltage to incident voltage (or alternately, reflected current to incident current) is defined as reflection coefficient $K_r$.

$$K_r = \frac{V_{\text{refl}}}{V_{\text{inc}}} \tag{18-13}$$

$$= \frac{I_{\text{refl}}}{I_{\text{inc}}} \tag{18-14}$$

Since the inequality between load resistance and line characteristic impedance is the cause of the reflections, a relationship exists between reflection coefficient, characteristic impedance, and load resistance.

$$K_r = \left| \frac{Z_o - R_L}{Z_o + R_L} \right| \tag{18-15}$$

**solution**    By Eq. (18-15):

$$K_r = \left| \frac{Z_o - R_L}{Z_o + R_L} \right|$$

$$= \frac{300 - 50}{300 + 50}$$

$$= \frac{250}{350}$$

$$= 0.7143$$

## 18.10  DETERMINING INCIDENT, REFLECTED, AND ABSORBED POWER

**problem 18.9**  A standing-wave ratio of 4:1 exists on a line. Determine reflection coefficient, percent incident power reflected from the load, and percent incident power absorbed by the load. See Fig. 18-11.

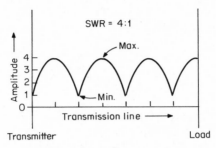

**Fig. 18-11**  Standing-wave ratio of 4:1.

**theory**  Since both standing-wave ratio and reflection coefficient are indications of mismatch between load and transmission line, a relationship should exist between them.

$$K_r = \frac{SWR - 1}{SWR + 1} \tag{18-16}$$

Power is proportional to the square of voltage, and reflection coefficient is the ratio of reflected to incident voltage; therefore, the percent of incident power that is reflected from the load is:

$$\% \text{ reflected power} = K_r^2 \tag{18-17}$$

**solution**  By Eq. (18-16):

$$K_r = \frac{SWR - 1}{SWR + 1}$$

$$= \frac{4 - 1}{4 + 1}$$

$$= \frac{3}{5}$$

$$= 0.6$$

And by Eq. (18-13):

$$\% \text{ reflected power} = K_r^2 \times 100\%$$
$$= 0.6^2$$
$$= 0.36 \times 100$$
$$= 36\%$$

Percent of incident power absorbed by the load = 100% minus percent reflected.

$$\% \text{ incident power absorbed} = 100 - 36$$
$$= 64\%$$

## 18.11  DETERMINING ATTENUATION LOSS ON A TRANSMISSION LINE WITH AN SWR OF 1:1

**problem 18.10**  Twin-lead cable designated as Amphenol 214-056 is in use to transfer power from a transmitter to a load. The frequency of the signal is 10 MHz. The SWR is measured and found to be 1:1 (no standing waves). Determine the signal attenuation on 200 ft of line. See Fig. 18-12.

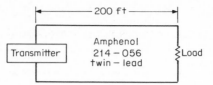

**Fig. 18-12**  Twin-lead cable connecting a load to a transmitter.

**theory**  Previously, we have considered transmission lines to be loss-less, i.e., no loss in power due to radiation from the line or loss due to heating of the conductor or dielectric. Losses due to radiation are difficult to estimate theoretically, and this is usually handled by direct instrument measurement. However, power losses can be estimated on the basis of information provided by the manufacturer. These data are usually presented by the manufacturers in the form of a graph as in Fig. 18-13. These data apply to situations in which SWR is 1:1. Additional information is needed for cases in which SWR is not 1:1. Such a case is discussed in Problem 18.11.

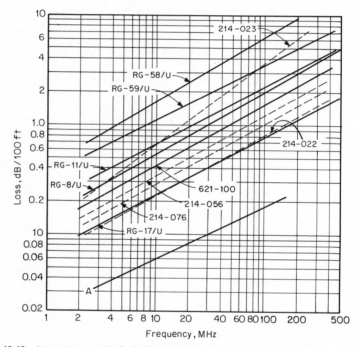

**Fig. 18-13**  Attenuation per 100 ft of cable for SWR = 1:1. Curve *A* is for 600-Ω open wire line using No. 12 conductors. (*Courtesy ARRL.*)

**solution**  Referring to Fig. 18-13, for 10 MHz we see that cable 214-056 has a loss of 0.35 dB/100 ft. Since we are concerned with 200 ft of cable, the dB loss is:

$$\text{dB loss for 200 ft} = 2 \times 0.35$$
$$= 0.7 \text{ dB}$$

## 18.12  DETERMINING LINE LOSS ON A LINE WITH SWR NOT EQUAL TO 1:1

**problem 18.11**  Three hundred feet of RG11/U cable is used as a transmission line for a 20-MHz signal. Calculate line loss of SWR = 3:1. See Fig. 18-14.

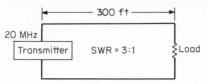

**Fig. 18-14**    Three hundred feet of cable operating with an SWR of 3:1.

**theory**    Data published by manufacturers will provide dB loss for the situation in which SWR = 1:1 as in Fig. 18-13. Additional loss due to SWR being other than 1:1 can be found from Fig. 18-15.

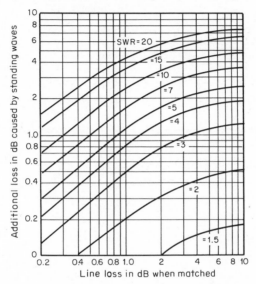

**Fig. 18-15**    Additional loss caused by various standing-wave ratios.

**solution**    From Fig. 18-13, we see that

$$\text{dB loss}/100 \text{ ft} = 0.9 \text{ dB}/100 \text{ ft}$$

For 300 ft:

$$\text{dB loss}/300 \text{ ft} = 3 \times 0.9 \text{ dB}/100 \text{ ft}$$
$$= 2.7 \text{ dB}/300 \text{ ft}$$

Figure 18-15 gives the additional loss due to SWR of 3:1 for a dB loss of 2.7 dB/300 ft.

$$\text{Additional dB loss} = 0.95$$
$$\text{Total dB loss} = 2.7 + 0.95$$
$$\text{dB loss} = 3.65 \text{ dB}/300 \text{ ft}$$

## 18.13    PROVIDING IMPEDANCE MATCHING WITH QUARTER-WAVE MATCHING SECTIONS

**problem 18.12**    A 300-$\Omega$ load is being fed by a 72-$\Omega$ transmission line. Design a quarter-wave matching section to be placed between the load and the transmission line so as to provide matched conditions for a 100-MHz signal. See Fig. 18-16.

**theory**    A segment of transmission line equal in length to one-quarter of the wavelength of the signal on the line and having been cut from a cable with a characteristic impedance equal to:

$$Z_o = \sqrt{Z_1 Z_2} \tag{18-16}$$

where $Z_1$ is the load resistance and $Z_2$ is the characteristic impedance of the transmission line, is called a quarter-wave transformer. See Fig. 18-17.

The effect of the quarter-wave section is to make it appear that the load and line are matched, thereby reducing the SWR to 1:1, and thus reducing line losses.

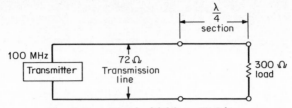

**Fig. 18-16**   A 300-$\Omega$ load fed from a 72-$\Omega$ line.

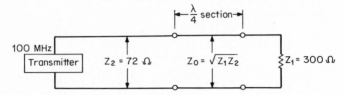

**Fig. 18-17**   A quarter-wave section used to match unequal impedances of Fig. 18-16.

**solution**   Determining wavelength of the 100-MHz signal, assuming velocity factor $= 1$,

$$f\lambda = C$$
$$(100 \times 10^6)\lambda = 3 \times 10^8$$
$$\lambda = \frac{C}{f}$$
$$= \frac{3 \times 10^8}{100 \times 10^6}$$
$$= 3 \text{ m}$$

A quarter-wavelength section of line would, therefore, be:

$$\frac{\lambda}{4} = \frac{3}{4}$$
$$= 0.75 \text{ m}$$

To calculate the required characteristic impedance using Eq. (18-16):

$$Z_o = \sqrt{Z_1 Z_2}$$
$$= \sqrt{300(72)}$$
$$= 147 \ \Omega$$

See Fig. 18-18.

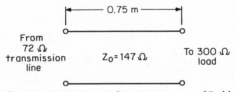

**Fig. 18-18**   Quarter-wave section satisfying requirements of Problem 18.12.

## 18.14   ELIMINATING STANDING WAVES BY USING STUBS

**problem 18.13**   A transmission line is operating with an SWR of 6:1. Design a short-circuited stub and determine its appropriate location in order to eliminate standing waves at 150 MHz. Assume a velocity factor of 1.0. See Fig. 18-19.

**theory**   A transmission line with a standing-wave ratio other than 1:1 presents an impedance that is not totally resistive along the line. The impedance varies from point to point on the line. One means of eliminating standing waves is to determine where on the line the resistive component is equal to the desired characteristic impedance and then shunt the line with an appropriate external reactance so as to tune out the reactive part of the line impedance at that point. (Inductance tunes out capacitance, and capacitance tunes out inductance.)

Short sections (less than λ/4) of transmission lines (stubs), when open ended or short-circuit terminated, possess the property of having an impedance that is reactive. By properly choosing a length of open-circuited or short-circuited transmission line and placing it in shunt with the original section of line at an appropriate position on the line, standing waves can be eliminated on the line from the input end of the line up to the stub. Stubs are made from the same transmission line as the original line and thus have the same characteristic impedance. Although the determination of the length and location of the stub usually requires use of rather complex mathematics, Fig. 18-20 provides a simplified approach to stub matching. In Fig. 18-20 the location of the stub is given in terms of a distance from $V_{\text{rms.max}}$. This distance is to be measured toward the load for open-circuited stubs and toward the transmitter for short-circuited stubs.

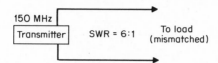

**Fig. 18-19**   Transmission line operating with an SWR of 6:1.

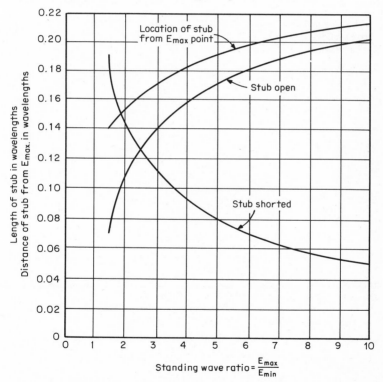

**Fig. 18-20**   Graph showing impedance matching with stubs.

**solution**   Entering the graph of Fig. 18-20 at SWR = 6, the length of the short-circuit stub is found to be 0.07λ, to be placed 0.198λ from the maximum voltage point.

The wavelength of the 150-MHz signal is found from:

$$f\lambda = C$$
$$150 \times 10^6\lambda = 3 \times 10^8$$
$$\lambda = \frac{3 \times 10^8}{150 \times 10^6} = \frac{3 \times 10^8}{1.5 \times 10^8}$$
$$= 2 \text{ m}$$

Thus the length of the stub and the distance from the voltage maximum are:

$$l_s = 0.07(2) = 0.14$$
$$d_s = 0.198(2) = 0.396 \text{ m toward the transmitter}$$
$$\text{(for short-circuited stub)}$$

See Fig. 18-21.

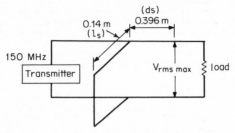

**Fig. 18-21**    Solution to Problem 18.13 (using stubs to eliminate standing waves).

## 18.15   DETERMINING THE PROPER DIFFERENCE IN LENGTH BETWEEN TWO TRANSMISSION LINE SEGMENTS TO PROVIDE A PARTICULAR PHASE SHIFT

**problem 18.14**   Two antennas are to be fed the same signal. However, in order to obtain certain radiation effects, the two antennas are to receive signals 180° out of phase with each other. The signal being transmitted has a frequency of 2 MHz. The transmission line has a velocity factor of 0.75. Calculate the difference in length ($\Delta L$) required for the two transmission lines feeding the antennas. See Fig. 18-22.

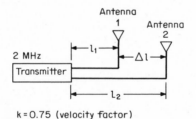

k = 0.75 (velocity factor)

**Fig. 18-22**   Signals fed to two antennas, 180° out of phase.

**theory**   Within each section of transmission line equal in length to one wavelength the transmitted signal undergoes a phase shift of 360°. In order to provide a phase shift proportionately less than 360°, a proportionately smaller section of transmission line is needed.

**solution**   Determining the wavelength of the 2-MHz signal:

$$f\lambda = C$$
$$2 \times 10^6 \lambda = 0.75(3 \times 10^8)$$
$$\lambda = \frac{0.75(3 \times 10^8)}{2 \times 10^6}$$
$$= 1.125 \times 10^2$$
$$= 112.5 \text{ m}$$

Since a phase difference of 180° is desired, a difference in length of the two cables is found from the proportions:

$$\frac{\Delta L}{\lambda} = \frac{\phi}{360}$$

$$\frac{\Delta L}{112.5} = \frac{180}{360}$$

$$\Delta L = \frac{180}{360} \times 112.5$$
$$= 56.25 \text{ meters}$$

or $\Delta L$ plus an integer multiple of $\lambda$.

## 18.16 DETERMINING WHETHER A PARTICULAR WAVEGUIDE GEOMETRY CAN BE USED TO CARRY A SIGNAL OF A PARTICULAR FREQUENCY

**problem 18.15**   Can the waveguide shown in Fig. 18-23 be used to carry a 5-GHz signal?

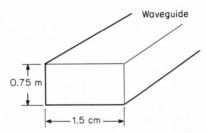

**Fig. 18-23**   Can this waveguide be used to carry a 5-GHz signal?

**theory**   As attempts are made to propagate higher and higher frequency signals on a standard traditional type transmission line, it is found that beyond 1.0 GHz efficient propagation can not be obtained with these lines. It becomes necessary to use waveguides, hollow metal tubes, for this function. Energy is radiated into and picked up from waveguides by small antennas located within them.

Waveguides have either a circular or rectangular cross section. A rectangular waveguide must have its larger cross-sectional dimension greater than one-half wavelength of the signal to be propagated, and it must have its smaller cross-sectional dimension less than one-half of the wavelength of the signal in order for it to be effective.

**solution**
$$f\lambda = 3 \times 10^8$$
$$5 \times 10^9 \lambda = 3 \times 10^8$$
$$\lambda = \frac{3 \times 10^8}{5 \times 10^9} = 0.6 \times 10^{-1} = 0.06 \text{ m}$$
$$= 6 \text{ cm}$$
$$\lambda/2 = 3 \text{ cm}$$

This waveguide (Fig. 18-23) cannot be used to carry a signal whose frequency is 5 GHz since its larger dimension is less than 3 cm. *Note:* See Chap. 21, Microwaves, for additional information on this topic.

## 18.17 BIBLIOGRAPHY

Adam, S. F.: *Microwave Theory and Applications*, Prentice-Hall, Englewood Cliffs, N.J., 1969.

Belden Corp. staff: *Electronic Cable Handbook*, Sams, Indianapolis, 1966.

DeMaw, Doug (ed.): *The Radio Amateur's Handbook*, American Radio Relay League, Hartford, Conn., 1976.

Department of the Air Force: *Antenna Systems* AF Manual 52-19, 1953.

Kaufman, Milton: *Radio Operator's License Q & A Manual*, Hayden, New York, 1975.

Langford and Smith: *Radiotron Designers Handbook*, Radio Corporation of America, Harrison, N.J., 1953.

McKenzie, Alexander: *Radiotelephone Examination Key and Answers*, McGraw-Hill, New York, 1972.

Schure, Alexander: *R-F Transmission Lines*, Rider, New York, 1956.

Shrader, R. L.: *Electronic Communication*, McGraw-Hill, New York, 1967.

Stewart, J. L.: *Circuit Analysis of Transmission Lines*, Wiley, New York, 1958.

Temes, Lloyd: *Communication Electronics for Technicians*, McGraw-Hill, New York, 1974.

Westman, H. P. (ed.): *Reference Data for Radio Engineers*, Sams, Indianapolis, Ind., 1970.

# Appendix to Chapter 18

On the following pages, a homograph, several graphs, and a table relating to transmission lines are presented without accompanying problems. These are provided for the readers' reference.

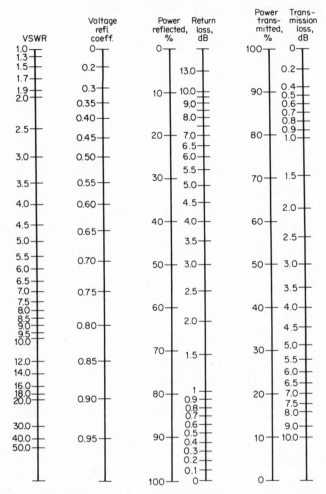

**Fig. 18A-1** Nomograph for transmission and reflection of power at high voltage standing wave ratios (VSWR). *(From Donald G. Fink, Electronics Engineers' Handbook, McGraw-Hill, 1975.)*

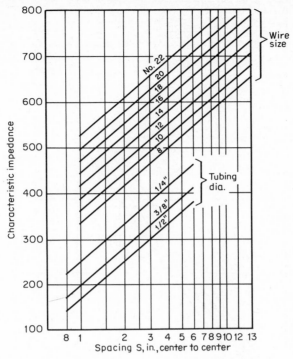

**Fig. 18A-2** Graph of characteristic impedance of two-wire parallel transmission lines with air dielectric. *(Courtesy A.R.R.L.)*

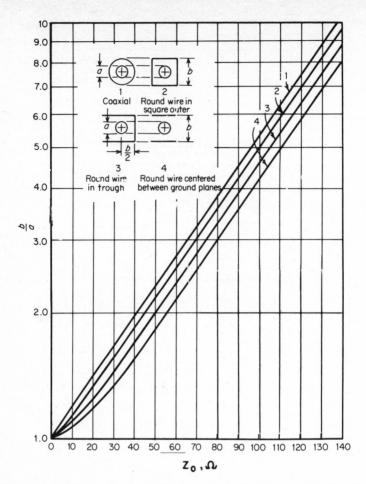

**Fig. 18A-3** Characteristic impedance of coaxial lines using solid dielectrics. *(From Donald G. Fink, Electronics Engineers' Handbook, McGraw-Hill, 1975.)*

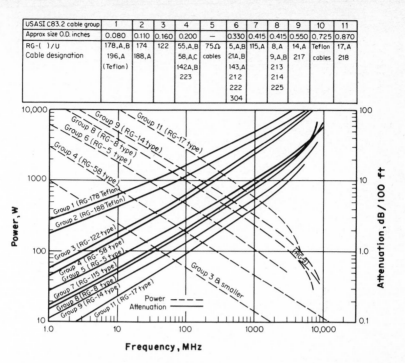

| USASI C83.2 cable group | 1 | 2 | 3 | 4 | 5 | | 6 | 7 | 8 | 9 | 10 | 11 |
|---|---|---|---|---|---|---|---|---|---|---|---|---|
| Approx size O.D. inches | 0.080 | 0.110 | 0.160 | 0.200 | — | | 0.330 | 0.415 | 0.415 | 0.550 | 0.725 | 0.870 |
| RG-( )/U Cable designation | 178,A,B 196,A (Teflon) | 174 188,A | 122 | 55,A,B 58,A,C 142A,B 223 | 75Ω cables | | 5,A,B 21A,B 143,A 212 222 304 | 115,A | 8,A 9,A,B 213 214 225 | 14,A 217 | Teflon cables | 17,A 218 |

Fig. 18A-4   Power rating and attenuation of flexible polyethylene coaxial cable. For Teflon cables multiply power ratings by 5. *(From Donald G. Fink, Electronics Engineers' Handbook, McGraw-Hill, 1975.)*

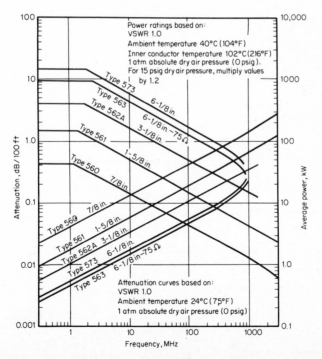

Fig. 18A-5   Power rating and attenuation for rigid coaxial cables. *(From Donald G. Fink, Electronics Engineers' Handbook, McGraw-Hill, 1975.)*

**TABLE 18A-1  Characteristics of Some Typical Coaxial and Twin-Lead Transmission Lines**

| RG No. | AWG & stranding material | Insulation | No. of shields & material | Jacket | Nom. O.D. in | Nom. imp. Ω | Nom. vel. of prop. | Nom. Cap., pF/ft | Nom. attenuation per 100 MHz | dB |
|---|---|---|---|---|---|---|---|---|---|---|
| 5A/U JAN-C-17A | 16 (Solid) silver coated copper | Polyethylene | 2 silver-coated copper | Gray noncontaminating vinyl | 0.328 | 50 | 66% | 30.8 | 100<br>200<br>400 | 2.40<br>3.50<br>5.25 |
| 6A/U Type | 21 (Solid) bare Copperweld | Polyethylene | 2 bare copper | Black polyethylene | 0.332 | 75 | 66% | 20.5 | 50<br>100<br>200<br>300<br>400 | 2.0<br>2.9<br>4.3<br>5.4<br>6.5 |
| 8/U JAN-C-17A | 13 (7 × 21) bare copper | Polyethylene | 1 bare copper | Black vinyl | 0.405 | 52 | 66% | 29.5 | 100<br>200<br>400 | 2.10<br>3.30<br>4.50 |
| 8/U Type | 11 (7 × 19) bare copper | Cellular Polyethylene | 1 bare copper | Black vinyl | 0.403 | 50 | 78% | 26 | 50<br>100<br>200<br>300<br>400 | 1.20<br>1.80<br>2.60<br>3.30<br>3.80 |
| 9/U JAN-C-17A | 13 (7 × 21) silver coated copper | Polyethylene | 2 inner-silver-coated outer-bare copper | Gray noncontaminating vinyl | 0.420 | 51 | 66% | 30.0 | 100<br>200<br>400 | 2.10<br>3.30<br>4.50 |
| 9B/U MIL-C-17D | 13 (7 × 21) silver coated copper | Polyethylene | 2 silver coated copper | Black noncontaminating vinyl | 0.420 | 50 | 66% | 30.8 | 100<br>200<br>400 | 2.10<br>3.30<br>4.50 |
| 11/U JAN-C-17A | 18 (7 × 26) tinned copper | Polyethylene | 1 bare copper | Black vinyl | 0.405 | 75 | 66% | 20.5 | 100<br>200<br>400 | 1.90<br>2.85<br>4.35 |

| Type | Conductor | Dielectric | Shield | Jacket | O.D. | Impedance | Velocity | Capacitance | Freq. | Atten. |
|---|---|---|---|---|---|---|---|---|---|---|
| 11/U Type | 14 (Solid) bare copper | Cellular Polyethylene | 1 bare copper | Black polyethylene | 0.405 | 75 | 78% | 17.3 | 50<br>100<br>200<br>300<br>400 | 1.0<br>1.5<br>2.3<br>2.9<br>3.4 |
| 11A/U MIL-C-17D | 18 (7 × 26) tinned copper | Polyethylene | 1 bare copper | Black non-contaminating vinyl | 0.405 | 75 | 66% | 20.5 | 100<br>200<br>400 | 1.90<br>2.85<br>4.35 |
| 54A/U JAN-C-17A | 18 (7 × .0152) bare copper | Polyethylene | 1 tinned copper | Clear polyethylene | 0.245 | 58 | 66% | 26.5 | 100<br>200<br>400 | 3.10<br>4.40<br>6.70 |
| 15/U JAN-C-17A | 20 (Solid) bare copper | Polyethylene | 2 tinned copper | Clear polyethylene | 0.201 | 53.5 | 66% | 28.5 | 100<br>200<br>400 | 4.10<br>6.20<br>9.50 |
| 53/U JAN-C-17A | 20 (Solid) bare copper | Polyethylene | 1 tinned copper | Black vinyl | 0.195 | 53.5 | 66% | 28.5 | 100<br>200<br>400 | 4.10<br>6.20<br>9.50 |
| 58/AU Type | 20 (19 × 32) tinned copper | Cellular Polyethylene | 1 tinned copper | Black vinyl | 0.195 | 50 | 78% | 26 | 50<br>100<br>200<br>300<br>400 | 3.6<br>4.8<br>7.6<br>8.5<br>10.1 |
| 58A/U JAN-C-17A | 20 (19 × .0071) tinned copper | Polyethylene | 1 tinned copper | Black vinyl | 0.195 | 50 | 66% | 30.8 | 100<br>200<br>400 | 5.30<br>8.20<br>12.60 |
| 58C/U MIL-C-17D | 20 (19 × .0071) tinned copper | Polyethylene | 1 tinned copper | Black non-contaminating vinyl | 0.195 | 50 | 66% | 30.8 | 100<br>200<br>400 | 5.30<br>8.20<br>12.60 |
| 59/U JAN-C-17A | 22 (Solid) bare Copperweld | Polyethylene | 1 bare copper | Black vinyl | 0.242 | 73 | 66% | 21.0 | 100<br>200<br>400 | 3.75<br>5.60<br>8.30 |

**TABLE 18A-1  Characteristics of Some Typical Coaxial and Twin-Lead Transmission Lines** (*Continued*)

| Type | Conductor | Dielectric | Shield | Jacket | OD | $Z_0$ | Velocity | Cap. | Freq. | Atten. |
|---|---|---|---|---|---|---|---|---|---|---|
| 59/U Type | 22 (Solid) bare Copperweld | Cellular Polyethylene | 1 bare copper | Black Polyethylene | 0.242 | 75 | 78% | 16.3 | 50<br>100<br>200<br>300<br>400 | 2.4<br>3.4<br>4.8<br>5.8<br>6.7 |
| 59/U Type | 20 (Solid) bare Copperweld | Cellular Polyethylene | 1 bare copper | Black Polyethylene | 0.242 | 75 | 78% | 17.3 | 50<br>100<br>200<br>300<br>400 | 2.3<br>3.2<br>4.5<br>5.5<br>6.3 |
| 59/U Type | 22 (Solid) bare Copperweld | Cellular Polyethylene | 1 bare copper | Gray, White, Black vinyl | 0.242 | 75 | 78% | 16.3 | 50<br>100<br>200<br>300<br>400 | 2.4<br>3.4<br>4.8<br>5.8<br>6.7 |
| 59B/U MIL-C-17D | .023 (Solid) bare Copperweld | Polyethylene | 1 bare copper | Black non-contaminating vinyl | 0.242 | 75 | 66% | 20.5 | 100<br>200<br>400 | 3.75<br>5.60<br>8.30 |
| 62/U JAN-C-17A | 22 (Solid) bare Copperweld | Semi-solid Polyethylene | 1 bare copper | Black vinyl | 0.242 | 93 | 84% | 13.5 | 100<br>200<br>400 | 3.10<br>4.40<br>6.30 |
| 62B/U MIL-C-17D | 24 (7 × 32) bare Copperweld | Semi-solid Polyethylene | 1 bare copper | Black non-contaminating vinyl | 0.242 | 93 | 84% | 13.5 | 100<br>200<br>400 | 3.10<br>4.40<br>6.30 |
| 71/U JAN-C-17A | 22 (Solid) bare Copperweld | Semi-solid Polyethylene | 2 tinned copper | Clear Polyethylene | 0.245 | 93 | 84% | 13.5 | 100<br>200<br>400 | 3.10<br>4.40<br>6.30 |
| 71B/U MIL-C-17D | 22 (Solid) bare Copperweld | Semi-solid Polyethylene | 2 tinned copper | Black polyethylene | 0.245 | 93 | 84% | 13.5 | 100<br>200<br>400 | 3.10<br>4.40<br>6.30 |
| 174/U MIL-C-17D | 26 (7 × 34) bare Copperweld | Polyethylene | 1 tinned copper | Black vinyl | 0.100 | 50 | 66% | 30.8 | 100<br>200<br>400 | 8.8<br>13.0<br>20.0 |

| Type | Conductor | Insulation | | Jacket | Nominal O.D. | Nominal impedance | Nominal velocity of propagation | Nominal capacitance (pF/ft) | Max. operating voltage | Nominal attenuation |
|---|---|---|---|---|---|---|---|---|---|---|
| 178B/U MIL-C-17D | 30 (7 × 38) Silver coated Copperweld | TFE TEFLON | 1 Silver coated copper | Brown FEP TEFLON | 0.072 | 50 | 69.5% | 29 | 400 | 29.0 Max. |
| 180B/U MIL-C-17D | 30 (7 × 38) Silver coated Copperweld | TFE TEFLON | 1 Silver coated copper | Brown FEP TEFLON | 0.142 | 95 | 69.5% | 15 | 400 | 17.0 Max. |
| 188A/U MIL-C-17D | 26 (7 × .0067) Silver coated Copperweld | TFE TEFLON | 1 Silver coated copper | White TFE TEFLON | 0.106 | 50 | 69.5% | 29 | 400 | 20.0 Max. |

**Shielded, Twin-Lead Line (TV Type)**

| AWG & stranding | Color | Nominal O. D. (inch) | Nominal velocity of propagation | Nominal capacitance (pF/ft) | Nominal attenuation per 100 | |
|---|---|---|---|---|---|---|
| | | | | | MHz | dB |
| 22 (7 × 30) | Brown | 0.305 × 0.515 | 69.8% | 7.8 | 57 | 1.7 |
| | | | | | 85 | 2.1 |
| | | | | | 177 | 3.2 |
| | | | | | 213 | 3.5 |
| | | | | | 473 | 5.4 |
| | | | | | 671 | 6.6 |
| | | | | | 887 | 7.7 |

Copperweld, 2 conductors, orange polyethylene insulation and web between conductors, cellular polyethylene oval insulation, Beldfoil shield, stranded tinned drain wire, polyethylene jacket.

**Unshielded, Twin-Lead Line (TV Type)**

| AWG & stranding | Color | Nominal O. D. (inch) | Nominal velocity of propagation | Nominal capacitance (pF/ft) | Nominal attenuation per 100 | |
|---|---|---|---|---|---|---|
| | | | | | MHz | dB |
| 22 (7 × 30) | Brown | 0.255 × 0.468 | 73.3% | 5.3 | 100 | 1.4 |
| | | | | | 300 | 2.8 |
| | | | | | 500 | 3.8 |
| | | | | | 700 | 4.8 |
| | | | | | 900 | 5.6 |

Copperweld, 2 conductors parallel, orange polyethylene insulation and web between conductors, cellular polyethylene oval jacket.

SOURCE: Belden Corp.

# Chapter 19

# Filters

## ARTHUR B. WILLIAMS

**Manager, Analog Development,
Coherent Communications Systems Corp.**

## 19.1 INTRODUCTION

A filter can be defined as a device which is placed between the terminals of an electrical circuit in order to modify the frequency components of a signal. Filters can be designed for operation from direct current to beyond 10,000 MHz. They are categorized by frequency-response shape, filter type (such as $LC$, crystal, and active), and the frequency range.

## 19.2 RESPONSE SHAPES

Filters can be classified into the following four basic categories on the basis of frequency response:

1. Low-pass filters pass low-frequency components (usually starting at direct current) up to a specified cutoff frequency and introduce high attenuation above this cutoff.

2. High-pass filters reject frequencies from direct current up to a cutoff frequency and pass frequency components above this cutoff.

3. Bandpass filters pass frequencies within a specified band and reject components outside this band.

4. Band-reject filters reject frequencies within a specified band and pass components outside this band.

These descriptions are somewhat ideal since, with actual filters, there is a transition region between the passband and the region of high attenuation (stopband). Figure 19-1 shows typical filter curves for each of the four categories.

## 19.3 DEFINITIONS OF BASIC PARAMETERS

In addition to the basic filter shapes, the technician or engineer should become familiar with the following terminology which is used when describing or designing filters.

### Cutoff Frequency ($F_c$)

The cutoff frequency defines the passband limit and usually corresponds to 3 dB of attenuation. Where low-pass and high-pass filters have only one cutoff frequency, band-pass and band-reject filters have two cutoff frequencies.

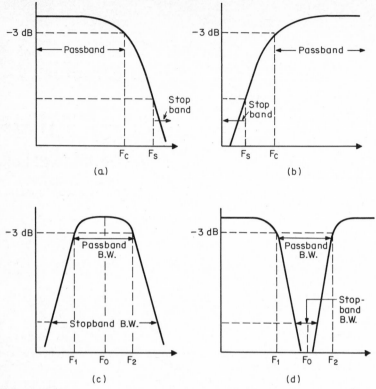

**Fig. 19-1**  Filter response shapes. (*a*) Low-pass; (*b*) high bandpass; (*c*) bandpass; (*d*) band-reject.

### Center Frequency ($F_0$)

Bandpass filters are geometrically symmetrical, i.e., symmetrical around a center frequency when plotted on linear-log graph paper with frequency on the logarithmic axis. The center frequency can be computed by:

$$F_0 = \sqrt{F_1 F_2} \tag{19-1}$$

where $F_1$ is the lower cutoff and $F_2$ is the upper cutoff frequency, as shown in Fig. 19-1.

For narrow filters, where the ratio of $F_2$ to $F_1$ is less than 1.1, the response shape approaches arithmetic symmetry. $F_o$ can then be computed by the average of the cutoff frequencies:

$$F_o = \frac{F_1 + F_2}{2} \tag{19-2}$$

### Stop-Band Cutoff Frequency ($F_s$)

The passband and stop band are separated by a transition region. $F_s$ is the frequency at which the minimum required attenuation is specified.

### Selectivity Factor (Q)

$Q_o$ is the ratio of the center frequency of a bandpass filter to the 3-dB bandwidth. If $F_1$ and $F_2$ correspond to the lower and upper 3-dB points, the selectivity factor can be expressed as:

$$Q_o = \frac{F_o}{F_2 - F_1} \tag{19-3}$$

An alternate method of expressing the selectivity of a filter is by the percentage of bandwidth (BW) which is defined by:

$$\%\text{BW} = \frac{F_2 - F_1}{F_o} \times 100 \tag{19-4}$$

### Shape Factor (SF)

Shape factor is the ratio of stop-band bandwidth to passband bandwidth of bandpass filters. The passband bandwidth usually is measured at the 3-dB points, and the stop-band bandwidth is measured at specified attenuation points such as 40 dB.

### Insertion Loss (IL)

Insertion loss is the reduction in level (measured in decibels) at the output of a filter compared with the level measured at the same terminals prior to insertion of the filter.

Additional filter terms will be introduced as they are required within the chapter.

## 19.4  SURVEY OF FILTER TYPES

Filters consist mainly of reactances. These reactances can take many physical forms such as coils, capacitors, crystals, and mechanical resonators. Amplifiers are combined with resistors and capacitors in active filters. The type of filter element chosen depends mainly on the parameters described above since each form of realization has its limitations. Figure 19-2 indicates the operating frequency ranges for the different filter types available.

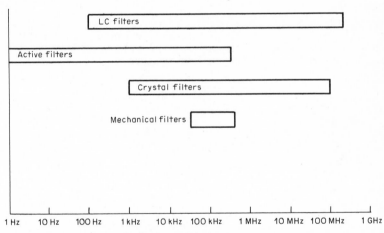

**Fig. 19-2**    Filter frequency ranges.

### LC Filters

Filters consisting of inductors and capacitors are used from near direct current up to a few hundred megahertz. UHF applications of *LC* filters become impractical because of the difficulties caused by parasitic capacities and inductances. Very low frequency filters use high inductance and capacitance values which require prohibitively large components. Therefore, *LC* filters are restricted to the range of approximately 100 Hz to 300 MHz.

Inductors are not purely reactive since resistive components due to winding and magnetic core losses are present. The ratio of the reactive to the series resistive component of inductors is the inductor quality factor $Q_c$.

$$Q_c = \frac{X_L}{R_{\text{ac}} + R_{\text{dc}}} \tag{19-5}$$

where $R_{\text{ac}}$ = ac core loss
$\quad\quad\ R_{\text{dc}}$ = dc resistance of the winding
$\quad\quad\ X_L$ = inductive reactance

Inductor $Q_c$ should be much larger than the filter selectivity factor $Q_0$ to construct satisfactory bandpass filters. In general, the more complex the filter, the higher the required inductor $Q_c$ for satisfactory performance. Coil $Q$'s as high as 600 are obtainable in the range of a few hundred kilohertz using ferrite cores. Outside this range, available $Q$ decreases.

### Active Filters

Active filters for use in the range of direct current to 500 kHz can be easily constructed using off-the-shelf operational amplifiers, resistors, and capacitors. They can be designed to offer performance comparable with that of *LC* filters. At very low frequencies, where *LC* filters are impossible, active filters can yield satisfactory results. Component size can be kept to a minimum by appropriate design techniques. Active filter circuits can be built using microelectronic technology where deposited *RC* networks are combined with operational amplifier chips. Laser trimming is then used if component adjustment is required.

Bandpass $Q$'s of a few hundred can be obtained at the lower range of the operating spectrum where amplifiers have very high open-loop gain. Reduction of open-loop gain restricts obtainable $Q$'s at the higher frequencies.

Design of active filters allows more flexibility than does design with *LC* filters. A desired input and output impedance can be provided which is independent of frequency. Voltage gain is available.

### Crystal Filters

Quartz crystal resonators have the equivalent electrical circuit of Fig. 19-3. The circuit $Q$ can be as high as 1,000,000. Because such enormous $Q$'s are obtainable, crystals are nearly perfect filter elements. Crystal filters also exhibit very high stability since the electrical parameters of the quartz remain essentially constant with time and temperature.

The circuit of Fig. 19-3 exhibits both series- and parallel-resonant frequencies extremely close to each other. The range of obtainable frequencies is limited by the mechanical restrictions in the manufacturing of the crystals. Below 1 kHz the quartz element becomes prohibitively large. Above 100 MHz the crystal becomes too small to control during manufacture.

Crystals are restricted to a limited range of $L$ and $C$ values of the equivalent circuit. Because of the circuit configuration, value limitations, and economic considerations, crystals are desirable as filter elements only when very high $Q$'s and high stability are required as for bandpass filters having very narrow percentage bandwidths.

**Fig. 19-3**   Crystal equivalent circuit.

### Mechanical Filters

A mechanical filter accepts an electrical signal, converts this signal into mechanical vibrations with a transducer, applies these vibrations to a series of interconnected disks, and reconverts the resultant vibrations to electrical output signals. By proper design of these metal disks, high $Q$ mechanical resonances can be obtained so that each disk is the mechanical equivalent of an electrical parallel resonant circuit. Since these disks are mechanically coupled together, the input signal is affected by the response of each disk in passing between the input and output transducers.

Mechanical filters are best suited for narrow bandwidth bandpass filters in the frequency range of 50 to 500 kHz. Bandpass $Q$'s of up to 1000 are obtainable with good frequency stability.

A serious penalty of mechanical filters is high insertion loss. This occurs mainly because of the inefficiency of the input and output transducers.

## 19.5 SELECTING THE RESPONSE FAMILY

All filters can be mathematically represented by an expression referred to as a *transfer function*. This expression is a ratio of two polynominals and can take the following general form:

$$T(S) = \frac{E_{\text{out}}}{E_{\text{in}}} = \frac{N_m S^m + N_{m-1} S^{m-1} + \cdots + N_1 S + N_0}{D_n S^n + D_{n-1} S^{n-1} + \cdots + D_1 S + D_0} \tag{19-6}$$

where the $N$'s are the numerator coefficients, the $D$'s are the denominator coefficients, and $S = J\omega$ ($J = \sqrt{-1}$, $\omega = 2\pi F$). The highest power of the denominator, $n$, is referred to as the order of the filter or the number of poles. The highest power of the numerator, $m$, is referred to as the number of zeros. The roots of the denominator are called the poles and the numerator roots are the zeros. However, not all filters have zeros.

The design techniques outlined in this chapter use tabulated element values so that application of the transfer function concept is not required for satisfactory filter designs. Nevertheless, the fact that all filter types can be represented by a transfer function is an important concept since these element values have all been derived directly from transfer functions.

### NORMALIZING FILTER REQUIREMENTS

Transfer functions can be selected in accordance with certain mathematical rules so that the corresponding low-pass filter curves all have a 3-dB point at 1 rad ($\omega = 1$). Each curve represents a set of element values for an $LC$ or active filter. This filter and its response are said to be "normalized" to 1 rad.

The general technique for designing filters is to first convert the filter requirement to a normalized low-pass requirement. We can then compare the resulting specifications with the normalized frequency-response curves and select a satisfactory low-pass filter. The corresponding low-pass element values are then denormalized to the required frequency range. If a high-pass, bandpass, or band-reject filter is desired, circuit transformations must be made as well.

### Normalization of Low-Pass Filters

**problem 19.1**   A low-pass filter having a 3-dB cutoff at 600 Hz and a minimum attenuation of 50 dB at 1800 Hz is required. Normalize these specifications to 1 rad. Then make a comparison to a family of normalized low-pass curves and select an appropriate filter.

**theory**   To normalize a low-pass filter requirement for 3 dB at 1 rad, first compute the filter steepness factor $A_s$, which is the ratio of the stop-band cutoff frequency $F_s$ to the 3-dB cutoff frequency $F_e$.

$$A_s = \frac{F_s}{F_c} \tag{19-7}$$

The normalized curves can now be entered, and a design selected that satisfies the minimum required stop-band attenuation at $A_s$ rad.

**solution**   Compute the low-pass filter steepness factor using Eq. 19-7:

$$A_s = \frac{F_s}{F_c} = \frac{1800 \text{ Hz}}{600 \text{ Hz}} = 3 \tag{19-7}$$

Using the sample family of curves of Fig. 19-4, determine which filter has a minimum attenuation of 50 dB at 3 rad. Clearly a normalized filter having a minimum complexity of $n = 5$ (fifth-order filter) satisfies this requirement.

### Normalization of High-Pass Filters

**problem 19.2**   A high-pass filter is required having a 3-dB cutoff of 900 Hz and a minimum attenuation of 50 dB at 300 Hz. Normalize this requirement to 1 rad and determine the minimum filter complexity required using the family of curves given in Fig. 19-4.

**theory**   Every normalized low-pass filter can be transformed into a normalized high-pass filter also having a 3-dB cutoff of 1 rad. Figure 19-5 shows the relationship between a normalized low-pass filter and the corresponding transformed high-pass filter. Both filters have identical attenuation at reciprocal frequencies; for example, the 12-dB points occur at 2 rad for the low-pass filter and 0.5 rad for the transformed high-pass filter.

Because of this relationship between normalized high-pass and low-pass filters, a high-pass steepness factor can be defined which is the reciprocal of the low-pass steepness factor. For a high-pass filter:

$$A_s = \frac{F_c}{F_s} \qquad (19\text{-}8)$$

The normalized low-pass curves can then be directly used to select a design having the required attenuation at $A_s$ rad.

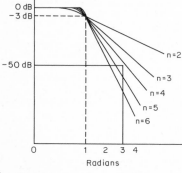

**Fig. 19-4**   Use of normalized curves.        **Fig. 19-5**   Low-pass to high-pass transformation.

**solution**    Compute the high-pass filter steepness factor using Eq. (19-8):

$$A_s = \frac{F_c}{F_s} = \frac{900\ \text{Hz}}{300\ \text{Hz}} = 3 \qquad (19\text{-}8)$$

Using the curves of Fig. 19-4, select the filter having a minimum attenuation of 50 dB at 3 rad. A fifth-order filter meets this requirement. The associated normalized low-pass filter must be transformed into a high-pass filter in the actual design procedure.

## Normalization of Bandpass Filters

**problem 19.3**    A bandpass filter is required having 3-dB points at 150 and 300 Hz and a minimum of 50-dB attenuation at 50 and 900 Hz. Normalize these requirements and select the appropriate filter from the curves of Fig. 19-4.

**theory**    Bandpass filters are generally classified into two categories, narrowband and wideband. The rule of thumb is that if the ratio of upper cutoff frequency to the lower cutoff frequency is over 1.5, the filter is a wide-band type.

A wide-band filter specification can be separated into individual low-pass and high-pass requirements. These requirements are then separately normalized, and a satisfactory low-pass and high-pass design is selected. The resulting filters must be cascaded to meet the overall specifications.

**solution**

$$\frac{\text{Upper cutoff frequency}}{\text{Lower cutoff frequency}} = \frac{300\ \text{Hz}}{150\ \text{Hz}} = 2 \qquad (19\text{-}9)$$

The filter is a wide-band type which can be separated into a low-pass and a high-pass requirement as follows:

$$
\begin{aligned}
\text{Low pass:} \quad & 3\ \text{dB at }300\ \text{Hz} \\
& 50\ \text{dB minimum at }900\ \text{Hz} \\
\text{High pass:} \quad & 3\ \text{dB at }150\ \text{Hz} \\
& 50\ \text{dB minimum at }50\ \text{Hz}
\end{aligned}
$$

Calculate the steepness factors:

$$\text{Low-pass steepness factor: } A_s = \frac{900\ \text{Hz}}{300\ \text{Hz}} = 3 \qquad (19\text{-}7)$$

$$\text{High-pass steepness factor: } A_s = \frac{150\ \text{Hz}}{50\ \text{Hz}} = 3 \qquad (19\text{-}8)$$

By use of the normalized curves of Fig. 19-4, the attenuation requirements for both the low-pass and high-pass filters can be met by an $n = 5$ design.

**problem 19.4**    A bandpass filter is specified having 3-dB points at 900 and 1100 Hz and a minimum attenuation of 50 dB at 700 and 1300 Hz. Normalize this filter to a low-pass requirement and select a satisfactory normalized filter from the curves of Fig. 19-4.

**theory**    In narrow-band bandpass filters the ratio of the upper cutoff frequency to the lower cutoff frequency is less than 1.5. These filters cannot be designed as separate low-pass and high-pass filters.

We have seen earlier how a low-pass filter can be transformed into a high-pass filter. This relationship allows us to design a high-pass filter by converting the high-pass requirement directly into a normalized low-pass specification, selecting a low-pass filter, and then transforming it into the desired high-pass filter.

A specific relationship also exists between low-pass and bandpass filters. The frequency response of the low-pass filter is transformed into the bandwidth of the bandpass filter with identical attenuation. Figure 19-6 shows how a typical bandpass filter response is related to a low-pass filter. Note that the 10- and 15-Hz bandwidth points of the bandpass filter have the same attenuation as the 10- and 15-Hz cutoff frequencies of the low-pass filter.

This relationship allows us to design narrow-band bandpass filters by converting the bandpass requirement to a low-pass specification and then using normalized low-pass curves. The bandpass to low-pass conversion proceeds as follows.

Calculate the geometric center frequency $F_0$ using Eq. (19-1) or (19-2).

At points of equal attenuation on both sides of $F_0$, the two frequencies must be geometrically related; that is, they must satisfy the following relationship:

$$F_a F_b = F_0^2 \qquad (19\text{-}9)$$

where $F_a$ and $F_b$ are, respectively, below and above $F_0$ and have equal attenuation.

Modify the bandpass filter specifications by calculating the corresponding geometric frequency for each stop-band frequency specified, using Eq. (19-9). For each pair of stop-band frequencies, two new pairs will result. Select the pair having the least separation which represents the more severe requirement. If the ratio of $F_b$ to $F_a$ is less than 1.1, the calculation involving Eq. 19-9 is not required if $F_a$ and $F_b$ both have the same separation from $F_0$ ($F_0 - F_a = F_b - F_0$).

Compute a bandpass steepness factor as follows:

$$A_s = \frac{\text{stop-band bandwidth}}{\text{3-dB bandwidth}} \qquad (19\text{-}10)$$

The stop-band bandwidth is the separation computed above.

Enter the normalized low-pass curves and select a filter having the required stop-band attenuation at $A_s$ rad.

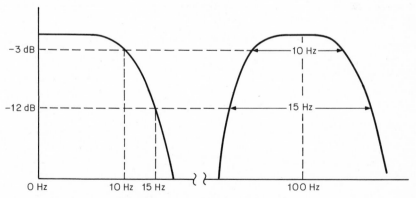

**Fig. 19-6**    Low-pass bandpass relationship.

**solution**

   *a.* Compute the geometric center frequency $F_0$:

$$F_0 = \sqrt{900 \times 1100} = 995 \text{ Hz} \qquad (19\text{-}1)$$

b. Compute two pairs of geometrically related stop-band frequencies:

$$\text{Using } F_a = 700 \text{ Hz} \qquad F_b = \frac{(995)^2}{700} = 1414 \text{ Hz} \tag{19-9}$$

$$F_b - F_a = 714 \text{ Hz}$$

$$\text{Using } F_b = 1300 \text{ Hz} \qquad F_a = \frac{(995)^2}{1300} = 762 \text{ Hz} \tag{19-9}$$

$$F_b - F_a = 538 \text{ Hz}$$

Clearly the second pair of frequencies represents the more severe requirement.

c. Calculate the bandpass steepness factor:

$$A_s = \frac{538 \text{ Hz}}{200 \text{ Hz}} = 2.69 \tag{19-10}$$

d. Select a normalized low-pass filter: If the normalized curves of Fig. 19-4 are used, an $n = 6$ design provides more than 50 dB of attenuation at 2.69 rad. The selected normalized low-pass filter must be transformed into the required bandpass filter.

## Normalization of Band-Reject Filters

**problem 19.5**    A band-reject filter is required having 3-dB points at 900 and 1100 Hz and a minimum attenuation of 50 dB at 970 and 1030 Hz. Normalize this filter to a low-pass requirement and select a normalized filter from the curves of Fig. 19-4.

**theory**    The previous section on narrow bandpass filters demonstrated how a bandpass filter design can be chosen by using normalized low-pass curves. A similar approach can be used to design band-reject filters.

Band-reject filters are directly related to high-pass filters. The frequency response of the high-pass filter is transformed into the bandwidth of the band-reject filter with identical attenuation. Figure 19-7 illustrates the high-pass to band-reject relationship for a typical filter. Observe how the 10- and 15-Hz bandwidth of the band-reject filter has the identical attenuation of the high-pass filter at the 10- and 15-Hz cutoff frequencies.

This relationship enables us to design a band-reject filter by first converting the band-reject requirement into a steepness factor and then using the normalized low-pass curves directly as we did for high-pass filters. The design method proceeds as follows.

1. Compute the geometric center frequency $F_0$ from the 3-dB points using Eq. (19-1) or (19-2).

2. The stop-band frequencies must be geometrically related in accordance with Eq. (19-9) as in the bandpass case. Modify the stop-band specification using this relationship and calculate two pairs of stop-band frequencies for each frequency specified. Select the pair with the *most* separation which represents the more severe requirement (steeper filter).

3. Compute a band-reject steepness factor:

$$A_s = \frac{\text{passband bandwidth}}{\text{stop-band bandwidth}} \tag{19-11}$$

4. Enter the normalized low-pass curves and select a filter having the required stop-band attenuation at $A_s$ rad.

5. In the actual design the normalized low-pass circuit must be transformed into a high-pass filter and then converted into the appropriate band-reject filter. This will be covered on pages 19-26 to 19-27.

## solution

a. Compute the geometric center frequency $F_0$:

$$F_0 = \sqrt{900 \times 1100} = 995 \text{ Hz} \tag{19-1}$$

b. Compute two pairs of geometrically related stop-band frequencies:

$$\text{Using } F_a = 970 \text{ Hz} \qquad F_b = \frac{995^2}{970} = 1021 \text{ Hz}$$

$$F_b - F_a = 51 \text{ Hz}$$

$$\text{Using } F_b = 1030 \text{ Hz} \qquad F_a = \frac{995^2}{1030} = 961 \text{ Hz} \tag{19-9}$$

$$F_b - F_a = 69 \text{ Hz}$$

The second pair of frequencies is the more severe requirement.

c. Calculate the band-reject steepness factor:

$$A_s = \frac{200 \text{ Hz}}{69 \text{ Hz}} = 2.9 \tag{19-11}$$

*d.* Select a normalized low-pass filter: If the normalized curves of Fig. 19-4 are used, an $n = 5$ design offers more than 50 dB of attenuation at 2.9 rad. The chosen filter must be transformed into a high-pass filter and then converted to a band-reject filter as will be demonstrated in Sec. 19.3.5.

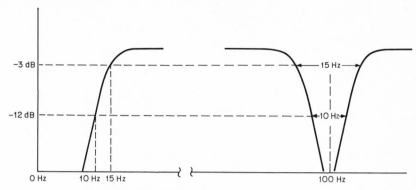

**Fig. 19-7**   High-pass band-reject relationship.

## RESPONSE FUNCTIONS

The previous section demonstrated the use of normalized low-pass curves. These curves represent transfer functions. By a change in the values and complexity of these transfer functions, the low-pass response can take on different shapes. This section discusses these various shapes and presents normalized low-pass frequency response curves.

### Butterworth Response

Butterworth low-pass filters are probably the most commonly used variety. The frequency response is very flat in the middle of the passband and somewhat rounded in the vicinity of cutoff. Beyond the 3-dB point the rate of attenuation increases and eventually reaches $n$ times 6 db per octave. For example, an $n = 3$ low-pass filter would increase its attenuation by 18 dB in the stop band every time the frequency was doubled. Butterworth filters are easy to manufacture because the resulting component values are more practical than most other types and are less critical to component tolerances.

Figure 19-8 provides a family of normalized attenuation curves for up to $n = 10$. Dual curves present expanded characteristics of the passband as well as the stop-band attenuation.

### Chebyshev Filters

An ideal normalized low-pass filter is illustrated in Fig. 19-9*a*. The filter has no attenuation over the range of direct current to 1 rad and infinite attenuation above 1 rad.

The Butterworth approximation to an ideal low-pass filter is shown in Fig. 19-9*b*. The approximation is good at frequencies far removed from 1 rad but poor near cutoff.

The Chebyshev curve of Fig. 19-9*c* is a better approximation to an ideal filter. The response in the region of cutoff is more rectangular, and the rate of descent to the stop band is steeper. These characteristics are obtained at the expense of allowing variations in the passband response referred to as *passband ripple.*

Chebyshev filters are more critical to manufacture than the Butterworth family and are more sensitive to component tolerances. The larger the ripple, the steeper the filter for a given order $n$, but the more critical the circuit becomes. During the first octave the attenuation exceeds $n$ times 6 dB per octave.

Figures 19-10 and 19-11 contain normalized attenuation curves for Chebyshev filters having ripples of 0.1 and 0.5 dB.

### Maximally Flat Delay

Butterworth and Chebyshev filters introduce varying amounts of delay to signals of different frequencies. The delay variation over the passband is referred to as *delay distortion.* This distortion increases with higher-order filters and increased ripple. If the

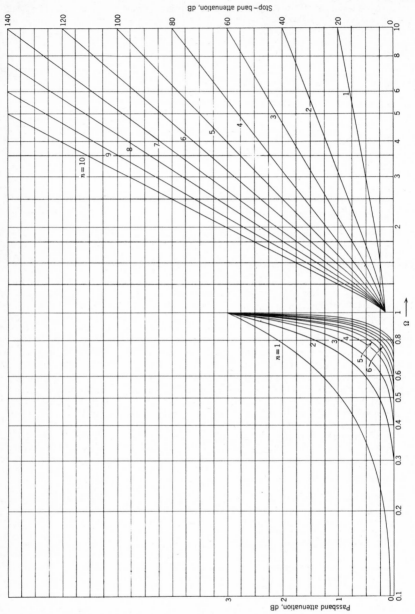

**Fig. 19-8** Attenuation characteristics for Butterworth filters. *(From A. I. Zverev, Handbook of Filter Synthesis, Wiley, New York, 1967.)*

input signal consists of multiple-frequency waveforms such as pulses or modulation, the output signal becomes distorted since the different frequency components are displaced in time.

The maximally flat delay low-pass filter family has a constant delay over the passband. However, the attenuation slope in the region near cutoff is very poor compared with a Butterworth or Chebyshev response. Even one or two octaves beyond cutoff attenuation is somewhat less than with the other types. These filters are therefore most useful where faithful signal reproduction is more important than attenuation characteristics.

The constant delay properties are not retained when the low-pass filter is transformed to a high-pass, bandpass, or band-reject configuration. Figure 19-12 depicts the normalized attenuation characteristics for this filter type.

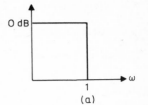
(a)

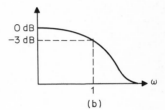
(b)

### Elliptic Function Family

Up to this point we have covered only all-pole low-pass filters. These filters provide infinite rejection only at infinite frequencies. Elliptic function filters contain zeros as well as poles in the transfer function. This results in infinite rejection at stop-band frequencies near cutoff. The passband has ripples similar to Chebyshev filters. The stop band has return lobes which are all equal in amplitude.

For a given filter order $n$, elliptic function filters have the steepest rate of descent to the stop-band theoretically possible. Figure 19-13 compares Butterworth, Chebyshev, and elliptic function filters of equivalent complexity.

The steeper rate of roll-off is obtained at the expense of return lobes. However, these return lobes are acceptable in most instances, providing they do not exceed the minimum attenuation required.

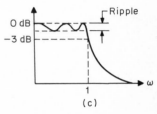
(c)

**Fig. 19-9** (*a*) Ideal low-pass filter; (*b*) Butterworth low-pass filter; (*c*) Chebyshev low-pass filter.

Elliptic function filter sections are more complex and critical than the other types. However, fewer sections are required for a given attenuation requirement than with other filter families. For moderate filter requirements the all-pole types are satisfactory. Where steep filter requirements are specified, elliptic function filters are a necessity.

Elliptic function low-pass filters are normalized so that the attenuation at 1 rad is equivalent to the ripple rather than 3 dB. Figure 19-14 illustrates a normalized elliptic function low-pass filter. A number of terms unique to these filters are used in Fig. 19-14 and are defined as follows:

$R_{dB}$ = passband ripple in decibels

$A_{dB}$ = minimum stop-band attenuation in decibels

$\omega_s$ = lowest stop-band frequency at which $A_{dB}$ occurs (in radians)

Elliptic function filters are categorized by these parameters in addition to order $n$. Since $R_{dB}$, $A_{dB}$, and $\omega_s$ define the passband and stop-band limits, normalized frequency-response curves are not required.

## 19.7 DESIGNING *LC* FILTERS USING TABLES

The design of *LC* filters is achieved by selecting the filter family and filter order $n$ in accordance with Sec. 19.5, looking up the corresponding normalized low-pass filter from the tables provided, and then modifying the tabulated values. If a high-pass, bandpass, or band-reject filter is required, the circuit configuration must be modified as well.

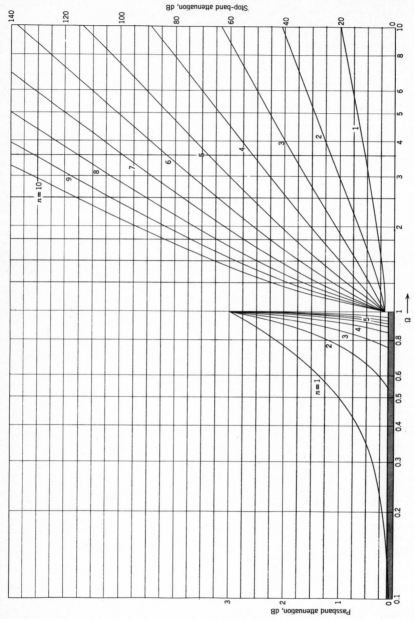

**Fig. 19-10** Attenuation characteristics for Chebyshev filter with 0.1-dB ripple. *(From A. I. Zverev, Handbook of Filter Synthesis, Wiley, New York, 1967.)*

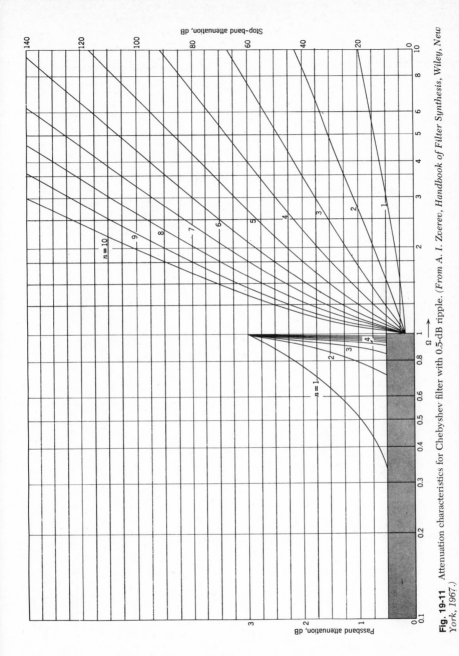

**Fig. 19-11** Attenuation characteristics for Chebyshev filter with 0.5-dB ripple. (*From A. I. Zverev, Handbook of Filter Synthesis, Wiley, New York, 1967.*)

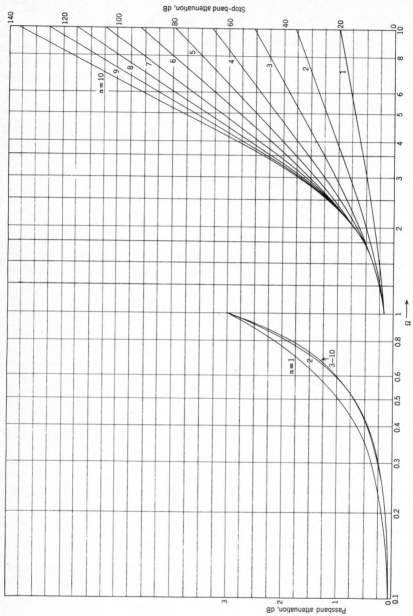

**Fig. 19-12** Attenuation characteristics for maximally flat delay (Bessel) filters. (*From A. I. Zverev, Handbook of Filter Synthesis, Wiley, New York, 1967.*)

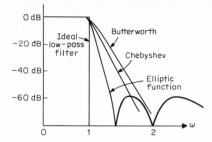

**Fig. 19-13**   Comparison of filter types.

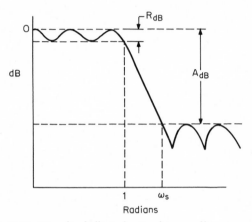

**Fig. 19-14**   Normalized elliptic function low-pass filter response.

## FREQUENCY AND IMPEDANCE SCALING

A filter can have its frequency response scaled (shifted) to a new frequency range if all the reactive element values (inductances and capacitances) are divided by a frequency-scaling factor (FSF). The FSF is the ratio of a particular reference frequency of the required scaled response to the frequency of the existing filter having the equivalent attenuation.

$$\text{FSF} = \frac{\text{reference frequency of scaled response}}{\text{reference frequency of existing response}} \qquad (19\text{-}12)$$

The reference frequency is usually the 3-dB points of low-pass or high-pass filters or the center frequency of bandpass or band-reject filters. Both numerator and denominator of the FSF must be expressed in identical units, whether hertz or radians, where the number of radians equals $2\pi F$.

When a filter is scaled to a different range, the new response can be obtained by multiplying all the numbers on the original frequency axis by the FSF. Figure 19-15 illustrates the effect upon the frequency response of frequency scaling a normalized low-pass filter by a factor of 10. By dividing all reactances by an FSF of 10, the frequency response is shifted higher in frequency by the same factor.

Let us assume that a low-pass filter is required having a 3-dB point at 1000 Hz and approximately 12 dB of attenuation at 2000 Hz. The normalized low-pass filter of Fig. 19-15 would satisfy this requirement when the 3-dB point is frequency-scaled from 1 rad to 1000 Hz. Since the FSF must be a ratio of identical units, 1000 Hz should be converted to radians in the following computation:

$$\text{FSF} = \frac{2\pi 1000 \text{ rad}}{1 \text{ rad}} = 6280 \qquad (19\text{-}12)$$

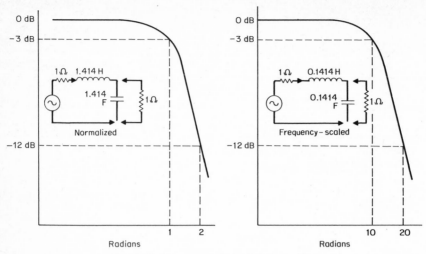

**Fig. 19-15** Frequency scaling of a low-pass filter.

It is apparent that a normalized low-pass filter can be denormalized (frequency scaled) by dividing all reactances by a simplified FSF defined as:

$$\text{FSF} = 2\pi F_c \qquad (19\text{-}13)$$

The low-pass filter of Fig. 19-16a is obtained by dividing the reactive element values of the normalized low-pass filter of Fig. 19-15 by the FSF. Although this filter satisfies the response requirements, the required inductor and capacitor values are highly impractical. This leads us to impedance scaling.

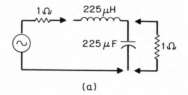

(a)

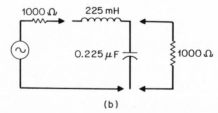

(b)

**Fig. 19-16** (a) Frequency-scaled low-pass filter; (b) frequency- and impedance-scaled low-pass filter.

If all impedances of a network are multiplied by a factor Z, the frequency response remains the same. Resistors and inductors are multiplied by Z. Capacitors, however, must be divided by Z for an impedance increase of the same factor. If the circuit of Fig. 19-16a is impedance-scaled by a Z of 1000, the circuit of Fig. 19-16b is obtained having practical values.

When filters are designed, frequency and impedance scaling is combined into one operation by using the following equations:

$$R' = Z \times R \tag{19-14}$$

$$L' = \frac{Z \times L}{FSF} \tag{19-15}$$

$$C' = \frac{C}{FSF \times Z} \tag{19-16}$$

where the primed components are the resulting values after the frequency and impedance scaling.

Both active and passive filters can be designed by scaling tabulated normalized values to the appropriate frequency and impedance ranges.

### LOW-PASS FILTER DESIGN

Low-pass filters are designed by first normalizing the requirements and selecting the appropriate filter from the normalized curves on pages 19-5 to 19-9. The corresponding values from Tables 19-1 to 19-6 are then frequency- and impedance-scaled.

**TABLE 19-1    Butterworth Low-Pass *LC* Element Values**

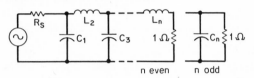

| $n$ | $R_s$ | $C_1$ | $L_2$ | $C_3$ | $L_4$ | $C_5$ | $L_6$ | $C_7$ |
|---|---|---|---|---|---|---|---|---|
| 2 | 1.000 | 1.4142 | 1.4142 | | | | | |
| 3 | 1.000 | 1.0000 | 2.0000 | 1.0000 | | | | |
| 4 | 1.000 | 0.7654 | 1.8478 | 1.8478 | 0.7654 | | | |
| 5 | 1.000 | 0.6180 | 1.6180 | 2.0000 | 1.6180 | 0.6180 | | |
| 6 | 1.000 | 0.5176 | 1.4142 | 1.9319 | 1.9319 | 1.4142 | 0.5176 | |
| 7 | 1.000 | 0.4450 | 1.2470 | 1.8019 | 2.0000 | 1.8019 | 1.2470 | 0.4450 |
| | $1/R_s$ | $L_1$ | $C_2$ | $L_3$ | $C_4$ | $L_5$ | $C_6$ | $L_7$ |

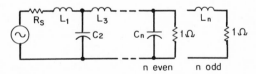

SOURCE: A. I. Zverev, *Handbook of Filter Synthesis*, Wiley, New York, 1967.

Low-pass filters are divided into two categories, all-pole and elliptic function. All-pole filters include Butterworth, Chebyshev, and maximally flat delay. Elliptic function filters are treated separately.

### All-Pole Type

Tables 19-1 to 19-4 contain the element values of normalized low-pass filters. The upper schematic corresponds to the column headings at the top of the table, whereas the lower schematic corresponds to the column headings at the bottom of the table. The upper configuration is normally used for odd $n$ low-pass filters since this circuit uses less inductors than the lower schematic.

**TABLE 19-2   0.1-dB Chebyshev Low-Pass *LC* Element Values**

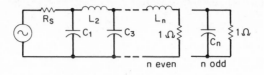

| $n$ | $R_s$ | $C_1$ | $L_2$ | $C_3$ | $L_4$ | $C_5$ | $L_6$ | $C_7$ |
|---|---|---|---|---|---|---|---|---|
| 2 | 1.3554 | 1.2087 | 1.6382 | | | | | |
| 3 | 1.000 | 1.4328 | 1.5937 | 1.4328 | | | | |
| 4 | 1.3554 | 0.9924 | 2.1476 | 1.5845 | 1.3451 | | | |
| 5 | 1.0000 | 1.3013 | 1.5559 | 2.2411 | 1.5559 | 1.3013 | | |
| 6 | 1.3554 | 0.9419 | 2.0797 | 1.6581 | 2.2473 | 1.5344 | 1.2767 | |
| 7 | 1.0000 | 1.2615 | 1.5196 | 2.2392 | 1.6804 | 2.2392 | 1.5196 | 1.2615 |
| | $1/R_s$ | $L_1$ | $C_2$ | $L_3$ | $C_4$ | $L_5$ | $C_6$ | $L_7$ |

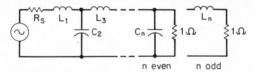

SOURCE: A. I. Zverev, *Handbook of Filter Synthesis*, Wiley, New York, 1967.

**TABLE 19-3   0.5-dB Chebyshev Low-Pass *LC* Element Values**

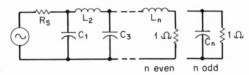

| $n$ | $R_s$ | $C_1$ | $L_2$ | $C_3$ | $L_4$ | $C_5$ | $L_6$ | $C_7$ |
|---|---|---|---|---|---|---|---|---|
| 2 | 1.9841 | 0.9827 | 1.9497 | | | | | |
| 3 | 1.0000 | 1.8636 | 1.2804 | 1.8636 | | | | |
| 4 | 1.9841 | 0.9202 | 2.5864 | 1.3036 | 1.8258 | | | |
| 5 | 1.0000 | 1.8068 | 1.3025 | 2.6914 | 1.3025 | 1.8068 | | |
| 6 | 1.9841 | 0.9053 | 2.5774 | 1.3675 | 2.7133 | 1.2991 | 1.7961 | |
| 7 | 1.0000 | 1.7896 | 1.2961 | 2.7177 | 1.3848 | 2.7177 | 1.2961 | 1.7896 |
| | $1/R_s$ | $L_1$ | $C_2$ | $L_3$ | $C_4$ | $L_5$ | $C_6$ | $L_7$ |

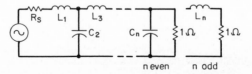

SOURCE: A. I. Zverev, *Handbook of Filter Synthesis*, Wiley, New York, 1967.

**TABLE 19-4 Maximally Flat Delay Low-Pass *LC* Element Values**

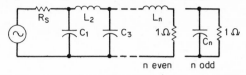

| $n$ | $R_s$ | $C_1$ | $L_2$ | $C_3$ | $L_4$ | $C_5$ | $L_6$ | $C_7$ |
|---|---|---|---|---|---|---|---|---|
| 2 | 1.0000 | 0.5755 | 2.1478 | | | | | |
| 3 | 1.0000 | 0.3374 | 0.9705 | 2.2034 | | | | |
| 4 | 1.0000 | 0.2334 | 0.6725 | 1.0815 | 2.2404 | | | |
| 5 | 1.0000 | 0.1743 | 0.5072 | 0.8040 | 1.1110 | 2.2582 | | |
| 6 | 1.0000 | 0.1365 | 0.4002 | 0.6392 | 0.8538 | 1.1126 | 2.2645 | |
| 7 | 1.0000 | 0.1106 | 0.3259 | 0.5249 | 0.7020 | 0.8690 | 1.1052 | 2.2659 |
| | $1/R_s$ | $L_1$ | $C_2$ | $L_3$ | $C_4$ | $L_5$ | $C_6$ | $L_7$ |

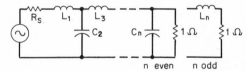

SOURCE: A. I. Zverev, *Handbook of Filter Synthesis*, Wiley, New York, 1967.

**problem 19.6**   A low-pass filter is specified having a 3-dB point at 3000 Hz and at least 25 dB of attenuation at 9000 Hz. The filter is required to operate from a source of 600 Ω and into a load of 600 Ω.

**theory**   The following procedure for the design of low-pass filters will serve as a guideline for proper use of the material presented in this chapter:
1. Normalize filter requirement by computing $A_s$ (pages 19-5 to 19-9).
2. Select the response function and corresponding filter (pages 19-9 to 19-11).
3. Frequency- and impedance-scale the normalized design to the required cutoff and impedance level after computing FSF and selecting Z (pages 19-11 to 19-17).
For even $n$ Chebyshev filters the normalized source and load resistances are unequal, and so the resulting circuit cannot operate between identical source and load resistances.

**solution**
   *a.* Compute $A_s$:

$$A_s = \frac{F_s}{F_c} = \frac{9000 \text{ Hz}}{3000 \text{ Hz}} = 3 \qquad (19\text{-}7)$$

   *b.* Select the response function and filter. When the normalized curves on pages 19-9 to 19-11 are used, an $n = 3$ Butterworth filter provides over 25 dB of attenuation at 3 rad. The corresponding normalized low pass from Table 19-1 is shown in Fig. 19-17*a*.
   Compute FSF and select Z:

$$\text{FSF} = 2\pi 3000 = 18{,}850 \qquad (19\text{-}13)$$

Since a source and load impedance of 600 Ω is required, Z = 600 is selected. Otherwise Z would be arbitrary but should be chosen to be such that the resulting $L$ and $C$ values are practical.
   Frequency- and impedance-scale the normalized circuit values:

$$C_1' = C_3' = \frac{1}{18{,}850 \times 600} = 0.0884 \ \mu\text{F} \qquad (19\text{-}16)$$

$$L_2' = \frac{600 \times 2}{18{,}850} = 63.7 \text{ mH} \qquad (19\text{-}15)$$
$$R' = 1 \times 600 = 600 \ \Omega \qquad (19\text{-}14)$$

The resulting circuit is shown in Fig. 19-17*b*.

**TABLE 19-5 Elliptic Function Low-Pass *LC* Element Values *N* = 3**

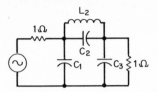

| | | | $R_{dB} = 1$ dB | | | |
|---|---|---|---|---|---|---|
| $\omega_s$ | $A_{dB}$ | $C_1$ | $C_2$ | $L_2$ | $\omega_2$ | $C_3$ |
| 1.295 | 20 | 1.570 | 0.805 | 0.613 | 1.424 | 1.570 |
| 1.484 | 25 | 1.688 | 0.497 | 0.729 | 1.660 | 1.688 |
| 1.732 | 30 | 1.783 | 0.322 | 0.812 | 1.954 | 1.783 |
| 2.048 | 35 | 1.852 | 0.214 | 0.865 | 2.324 | 1.852 |
| 2.418 | 40 | 1.910 | 0.145 | 0.905 | 2.762 | 1.910 |
| 2.856 | 45 | 1.965 | 0.101 | 0.929 | 3.279 | 1.965 |
| $\omega_s$ | $A_{dB}$ | $L_1$ | $L_2$ | $C_2$ | $\omega_2$ | $L_3$ |

| | | | $R_{dB} = 0.5$ dB | | | |
|---|---|---|---|---|---|---|
| $\omega_s$ | $A_{dB}$ | $C_1$ | $C_2$ | $L_2$ | $\omega_2$ | $C_3$ |
| 1.416 | 20 | 1.267 | 0.536 | 0.748 | 1.578 | 1.267 |
| 1.636 | 25 | 1.361 | 0.344 | 0.853 | 1.846 | 1.361 |
| 1.935 | 30 | 1.425 | 0.226 | 0.924 | 2.189 | 1.425 |
| 2.283 | 35 | 1.479 | 0.152 | 0.976 | 2.600 | 1.479 |
| 2.713 | 40 | 1.514 | 0.102 | 1.015 | 3.108 | 1.514 |
| $\omega_s$ | $A_{dB}$ | $L_1$ | $L_2$ | $C_2$ | $\omega_2$ | $L_3$ |

| | | | $R_{dB} = 0.1$ dB | | | |
|---|---|---|---|---|---|---|
| $\omega_s$ | $A_{dB}$ | $C_1$ | $C_2$ | $L_2$ | $\omega_2$ | $C_3$ |
| 1.756 | 20 | 0.850 | 0.290 | 0.871 | 1.986 | 0.850 |
| 2.082 | 25 | 0.902 | 0.188 | 0.951 | 2.362 | 0.902 |
| 2.465 | 30 | 0.941 | 0.125 | 1.012 | 2.813 | 0.941 |
| 2.921 | 35 | 0.958 | .0837 | 1.057 | 3.362 | 0.958 |
| 3.542 | 40 | 0.988 | .0570 | 1.081 | 4.027 | 0.988 |
| $\omega_s$ | $A_{dB}$ | $L_1$ | $L_2$ | $C_2$ | $\omega_2$ | $L_3$ |

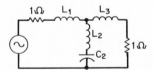

SOURCE: P. R. Geffe, *Simplified Modern Filter Design*, Rider, New York, 1963; by permission of Hayden Book Company, Inc.

**Elliptic Function Filters**  Tables 19-5 and 19-6 contain values for $n = 3$ and $n = 5$ elliptic function low-pass filters. The filters are tabulated for ripples of 0.1, 0.5, and 1 dB and for various $\omega_s$ and $A_{dB}$ as defined on pages 19-9 to 19-11.

As in the all-pole case the upper schematic corresponds to the column headings at the top of the table, whereas the lower schematic corresponds to the column headings at the bottom of the table. For low-pass filters the upper schematic should be used.

**TABLE 19-6   Elliptic Function Low-Pass *LC* Element Values *N*-5**

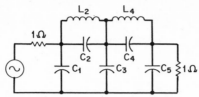

| $\omega_s$ | $A_{dB}$ | $C_1$ | $C_2$ | $L_2$ | $\omega_2$ | $C_3$ | $C_4$ | $L_4$ | $\omega_4$ | $C_5$ |
|---|---|---|---|---|---|---|---|---|---|---|
| | | | | | $R_{dB} = 1$ dB | | | | | |
| 1.145 | 35 | 1.783 | 0.474 | 0.827 | 1.597 | 1.978 | 1.487 | 0.488 | 1.174 | 1.276 |
| 1.217 | 40 | 1.861 | 0.372 | 0.873 | 1.755 | 2.142 | 1.107 | 0.578 | 1.250 | 1.427 |
| 1.245 | 45 | 1.923 | 0.293 | 0.947 | 1.898 | 2.296 | 0.848 | 0.684 | 1.313 | 1.553 |
| 1.407 | 50 | 1.933 | 0.223 | 0.963 | 2.158 | 2.392 | 0.626 | 0.750 | 1.459 | 1.635 |
| 1.528 | 55 | 1.976 | 0.178 | 0.986 | 2.387 | 2.519 | 0.487 | 0.811 | 1.591 | 1.732 |
| 1.674 | 60 | 2.007 | 0.141 | 1.003 | 2.660 | 2.620 | 0.380 | 0.862 | 1.747 | 1.807 |
| 1.841 | 65 | 2.036 | 0.113 | 1.016 | 2.952 | 2.703 | 0.301 | 0.901 | 1.920 | 1.873 |
| 2.036 | 70 | 2.056 | .0890 | 1.028 | 3.306 | 2.732 | 0.239 | 0.934 | 2.117 | 1.928 |
| $\omega_s$ | $A_{dB}$ | $L_1$ | $L_2$ | $C_2$ | $\omega_2$ | $L_3$ | $L_4$ | $C_4$ | $\omega_4$ | $L_5$ |

| $\omega_s$ | $A_{dB}$ | $C_1$ | $C_2$ | $L_2$ | $\omega_2$ | $C_3$ | $C_4$ | $L_4$ | $\omega_4$ | $C_5$ |
|---|---|---|---|---|---|---|---|---|---|---|
| | | | | | $R_{dB} = 0.5$ dB | | | | | |
| 1.186 | 35 | 1.439 | 0.358 | 0.967 | 1.700 | 1.762 | 1.116 | 0.600 | 1.222 | 1.026 |
| 1.270 | 40 | 1.495 | 0.279 | 1.016 | 1.878 | 1.880 | 0.840 | 0.696 | 1.308 | 1.114 |
| 1.369 | 45 | 1.530 | 0.218 | 1.063 | 2.077 | 1.997 | 0.627 | 0.795 | 1.416 | 1.241 |
| 1.481 | 50 | 1.563 | 0.172 | 1.099 | 2.300 | 2.113 | 0.482 | 0.875 | 1.540 | 1.320 |
| 1.618 | 55 | 1.559 | 0.134 | 1.140 | 2.558 | 2.188 | 0.369 | 0.949 | 1.690 | 1.342 |
| 1.782 | 60 | 1.603 | 0.108 | 1.143 | 2.847 | 2.248 | 0.291 | 0.995 | 1.858 | 1.449 |
| 1.963 | 65 | 1.626 | .0860 | 1.158 | 3.169 | 2.306 | 0.230 | 1.037 | 2.048 | 1.501 |
| 2.164 | 70 | 1.624 | .0679 | 1.178 | 3.536 | 2.319 | 0.182 | 1.078 | 2.258 | 1.521 |
| $\omega_s$ | $A_{dB}$ | $L_1$ | $L_2$ | $C_2$ | $\omega_2$ | $L_3$ | $L_4$ | $C_4$ | $\omega_4$ | $L_5$ |

| $\omega_s$ | $A_{dB}$ | $C_1$ | $C_2$ | $L_2$ | $\omega_2$ | $C_3$ | $C_4$ | $L_4$ | $\omega_4$ | $C_5$ |
|---|---|---|---|---|---|---|---|---|---|---|
| | | | | | $R_{dB} = 0.1$ dB | | | | | |
| 1.309 | 35 | 0.977 | 0.230 | 1.139 | 1.954 | 1.488 | 0.742 | 0.740 | 1.350 | 0.701 |
| 1.414 | 40 | 1.010 | 0.177 | 1.193 | 2.176 | 1.586 | 0.530 | 0.875 | 1.468 | 0.766 |
| 1.540 | 45 | 1.032 | 0.140 | 1.228 | 2.412 | 1.657 | 0.401 | 0.968 | 1.605 | 0.836 |
| 1.690 | 50 | 1.044 | 0.1178 | 1.180 | 2.682 | 1.726 | 0.283 | 1.134 | 1.765 | 0.885 |
| 1.860 | 55 | 1.072 | 0.0880 | 1.275 | 2.985 | 1.761 | 0.241 | 1.100 | 1.942 | 0.943 |
| 2.048 | 60 | 1.095 | 0.0699 | 1.292 | 3.328 | 1.801 | 0.192 | 1.148 | 2.130 | 0.988 |
| 2.262 | 65 | 1.108 | 0.0555 | 1.308 | 3.712 | 1.834 | 0.151 | 1.191 | 2.358 | 1.022 |
| 2.512 | 70 | 1.112 | 0.0440 | 1.319 | 4.151 | 1.858 | 0.119 | 1.225 | 2.619 | 1.044 |
| $\omega_s$ | $A_{dB}$ | $L_1$ | $L_2$ | $C_2$ | $\omega_2$ | $L_3$ | $L_4$ | $C_4$ | $\omega_4$ | $L_5$ |

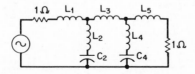

SOURCE: P. R. Geffe, *Simplified Modern Filter Design*, Rider, New York, 1963; by permission of Hayden Book Company, Inc.

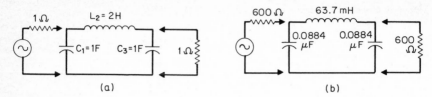

**Fig. 19-17**   (*a*) Normalized *n* = 3 Butterworth filter; (*b*) denormalized circuit.

**problem 19.7**   A low-pass filter is required having less than 1 dB of attentuation at 1000 Hz and more than 38 dB at 2800 Hz.

**theory**   Elliptic function low-pass filters are designed using the identical procedure as for all-pole low-pass filters and keeping in mind that the tables are normalized for an attenuation equal to the ripple at 1 rad (see Fig. 19-14).

**solution**

   *a.* Compute $A_s$:

$$A_s = \frac{F_s}{F_c} = \frac{2800 \text{ Hz}}{1000 \text{ Hz}} = 2.8 \qquad (19\text{-}7)$$

   *b.* Select the filter from Table 19-5 that achieves the required passband to stop-band transition with an $\omega_s$ not more than $A_s$. The ripple $R_{dB}$ and minimum attenuation $A_{dB}$ should comply to the specification.

   The filter for $n = 3$, $R_{dB} = 0.5$ dB, and $\omega_s = 2.713$ satisfies the requirements. The normalized low-pass filter is illustrated in Fig. 19-18*a*.

   *c.* Compute FSF and select Z:

$$\begin{aligned} \text{FSF} &= 2\pi 1000 = 6280 \\ \text{Let } Z &= 10{,}000 \ \Omega. \end{aligned} \qquad (19\text{-}13)$$

   *d.* Frequency- and impedance-scale the normalized circuit values:

$$C_1' = C_3' = \frac{1.514}{6280 \times 10{,}000} = 0.0241 \ \mu\text{F} \qquad (19\text{-}16)$$

$$C_2' = \frac{0.102}{6280 \times 10{,}000} = 0.00162 \ \mu\text{F} \qquad (19\text{-}16)$$

$$L_2' = \frac{10{,}000 \times 1.015}{6280} = 1.62 \text{ H} \qquad (19\text{-}15)$$

The circuit of Fig. 19.18*b* results.

   The resonant frequency $F_\infty$ of each parallel resonant circuit can be calculated by determining the product of $F_c \, \omega_\infty$ from the table:

$$F_\infty = F_c \omega_\infty \qquad (19\text{-}17)$$

So $F_\infty = 1000 \times 3.108 = 3108$ Hz.

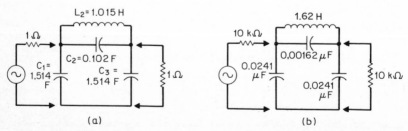

**Fig. 19-18**   (*a*) Normalized low-pass filter; (*b*) denormalized circuit.

## HIGH-PASS FILTER DESIGN

High-pass filters are designed directly from the normalized low-pass-filter values presented in Tables 19.1 to 19.6. The low-pass configuration must first be transformed into a high-pass filter which is also normalized to 1 rad. The normalized high-pass filter is then frequency- and impedance-scaled in accordance with the requirements.

**problem 19.8**    A high-pass filter is required with less than 3 dB of attenuation at 1000 Hz and more than 45 dB attenuation at 350 Hz.

**theory**    To design high-pass filters, the following sequence should be followed:

1. Normalize the high-pass requirement computing a high-pass steepness factor $A_s$ (pages 19-5 to 19-9).

2. Select the response function and corresponding filter (pages 19-9 to 19-11).

3. Transform the normalized low-pass filter circuit into a high-pass filter. This is achieved by replacing every capacitor with an inductor having an inductance equal to $1/C$ and by replacing every inductor with a capacitor having a capacitance equal to $1/L$.

$$\text{High-pass } C = \frac{1}{\text{low-pass } L} \tag{19-18}$$

$$\text{High-pass } L = \frac{1}{\text{low-pass } C} \tag{19-19}$$

This low-pass to high-pass transformation applies to any passive *LC* filter, whether an all-pole or elliptic function circuit. To minimize the number of inductors, use the lower schematic and corresponding values from Tables 19.1 to 19.6.

4. The resulting normalized high-pass filter can be frequency- and impedance-scaled after computing FSF and selecting Z (pages 19-11 to 19-17). If the filter is an elliptic function type, each series resonant frequency occurs at:

$$F_\infty = \frac{F_c}{\omega_\infty} \tag{19-20}$$

**solution**

*a.* Compute $A_s$:

$$A_s = \frac{F_c}{F_s} = \frac{1000 \text{ Hz}}{350 \text{ Hz}} = 2.86 \tag{19-8}$$

*b.* If an all-pole filter is desired, select a normalized low-pass filter from the curves on pages 19-9 to 19-11 having a minimum attenuation of 45 dB at 2.86 rad. For an elliptic function design, select an elliptic function low-pass filter from Table 19-5 or 19-6 having an $\omega_s$ of not more than 2.86 rad.

Let us select an all-pole design. Figure 19-8 indicates that an $n = 5$ Butterworth filter satisfies this requirement. Figure 19-19*a* shows the normalized low-pass filter using the lower schematic and column headings from Table 19-1.

To make a low-pass to high-pass transformation, replace each inductor with a capacitor having a value equal to $1/L$ and replace each capacitor with an inductor having a value $1/C$.

$$C_1 = C_5 = \frac{1}{0.618} = 1.618 \text{ F} \tag{19-18}$$

$$L_2 = L_4 = \frac{1}{1.618} = 0.618 \text{ H} \tag{19-19}$$

$$C_3 = \frac{1}{2} = 0.5 \text{ F} \tag{19-18}$$

Compute FSF and select Z:

$$\text{FSF} = 2\pi 1000 = 6280 \tag{19-13}$$

Let Z = 600 Ω.

$$C_1' = C_5' = \frac{1.618}{6280 \times 600} = 0.429 \text{ }\mu\text{F} \tag{19-16}$$

$$L_2' = L_4' = \frac{600 \times 0.618}{6280} = 0.059 \text{ H} \tag{19-15}$$

$$C_3' = \frac{0.5}{6280 \times 600} = 0.133 \, \mu\text{F} \qquad (19\text{-}16)$$

Figure 19-19c shows the resulting filter.

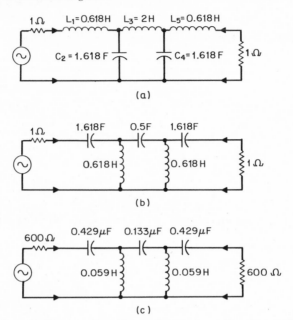

(a)

(b)

(c)

**Fig. 19-19** (a) Normalized low-pass filter; (b) normalized high-pass filter; (c) scaled high-pass filter.

## BANDPASS FILTER DESIGN

Bandpass filters are usually classified into two categories, narrow band and wide band. Generally if the ratio of the upper 3-dB frequency to the lower 3-dB frequency is more than 1.5, the filter is considered wide band.

### Wide Band

Wide-band bandpass filters are designed by separating the requirements into low-pass and high-pass specifications, designing a low-pass filter and a high-pass filter, and cascading the two filters. Although there may be some interaction between filters, these effects are usually of little consequence, especially for large ratios of upper cutoff to lower cutoff.

   **problem 19.9**   A bandpass filter is required having 3-dB points at 1000 and 3000 Hz, more than 45-dB attenuation at 350 Hz, and more than 25 dB at 9000 Hz. A source and load impedance of 600 Ω is specified.

   **theory**   Since the ratio of the upper 3-dB point to the lower 3-dB point is in excess of 1.5, the filter is a wide-band type. The filter can then be designed as a combined low-pass and high-pass filter.

   **solution**   The bandpass requirement is separated into the following low-pass and high-pass specifications:

Low-pass filter:   3 dB at 3000 Hz
25 dB minimum at 9000 Hz
High-pass filter:   3 dB at 1000 Hz
45 dB minimum at 350 Hz

   The low-pass filter can be designed as an all-pole type. The example of pages 19-17 to 19-22 satisfies this low-pass specification. Pages 19-23 to 19-24 illustrate the design of a high-pass filter that meets the high-pass requirement. The filters are cascaded, resulting in the circuit of Fig. 19-20.

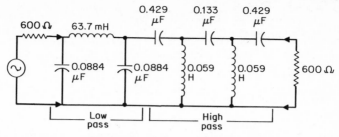

**Fig. 19-20**   Wide-band bandpass filter.

## Narrow Band

When the ratio of the upper 3-dB frequency to the lower 3-dB frequency is less than 1.5, the filter is designed as a narrow-band type. Pages 19-5 to 19-9 illustrate how a bandpass filter response is related to a low-pass filter characteristic. This section demonstrates how the bandpass filter circuit itself is directly derived from the normalized low-pass values.

**problem 19.10**   Design a bandpass filter centered about 1000 Hz with 3-dB points at 900 and 1100 Hz and having a minimum rejection of 15 dB at 800 and 1200 Hz. The source and load impedance should be 1000 Ω.

**theory**   Since the ratio of the upper 3-dB point to the lower 3-dB point is less than 1.5, a narrow-band design is required. The initial step is to normalize the bandpass requirements in accordance with pages 19-5 to 19-9. A satisfactory low-pass filter is selected using the normalized curves of pages 19-9 to 19-11 and values from Tables 19-1 to 19-6.

The normalized low-pass filter is then frequency- and impedance-scaled so that the cutoff frequency corresponds to the bandwidth of the bandpass filter. The impedance level is identical to the required bandpass source and load impedance.

This low-pass filter can then be transformed into a bandpass filter having a bandwidth equal to the low-pass cutoff. This is accomplished by resonating each low-pass inductor with a series capacitor and resonating each low-pass capacitor with a parallel inductor. Resonance in each circuit branch should occur at the filter geometric center frequency $F_0$, and so values are computed using the following relationship:

$$F_0 = \frac{1}{2\pi\sqrt{LC}} \tag{19-21}$$

$$L = \frac{1}{(2\pi F_0)^2 C} \tag{19-22}$$

Alternately,

$$C = \frac{1}{(2\pi F_0)^2 L} \tag{19-23}$$

### solution

*a.* Normalize the bandpass requirement and select an appropriate normalized low-pass filter by the following steps. Compute the geometric center frequency $F_0$:

$$F_0 = \sqrt{900 \times 1100} = 995 \text{ Hz} \tag{19-1}$$

Compute two pairs of geometrically related stop-band frequencies:

$$F_a = 800 \text{ Hz} \qquad F_b = \frac{995^2}{800} = 1237 \text{ Hz} \tag{19-9}$$

$$F_b - F_a = 437 \text{ Hz}$$

$$F_b = 1200 \text{ Hz} \qquad F_a = \frac{995^2}{1200} = 825 \text{ Hz} \tag{19-9}$$

$$F_b - F_a = 375 \text{ Hz}$$

Compute the band-pass steepness factor:

$$A_s = \frac{375 \text{ Hz}}{200 \text{ Hz}} = 1.88 \tag{19-10}$$

The normalized curves of Fig. 19-8 show that an $n = 3$ Butterworth filter provides at least 15 dB attenuation at 1.88 rad. The normalized low-pass filter from Table 19-1 is shown in Fig. 19-21a.

b. Frequency- and impedance-scale the normalized low-pass filter for a cutoff equal to the desired bandpass filter bandwidth of 200 Hz and required impedance level of 600 $\Omega$:

$$\text{FSF} = 1257 \qquad (19\text{-}13)$$

$$C_1' = C_3' = \frac{1}{1257 \times 600} = 1.33\,\mu\text{F} \qquad (19\text{-}16)$$

$$L_2' = \frac{600 \times 2}{1257} = 0.95\,\text{H} \qquad (19\text{-}15)$$

The scaled low-pass filter is illustrated in Fig. 19-21b.

c. Resonate each circuit branch to $F_0$ by adding a series capacitor to each inductor and a parallel inductor to each capacitor.

$$L = \frac{1}{(2\pi995)^2\,1.33 \times 10^{-6}} = 19.2\,\text{mH} \qquad (19\text{-}22)$$

$$C = \frac{1}{(2\pi995)^2\,0.95} = 0.0269\,\mu\text{F} \qquad (19\text{-}23)$$

The final bandpass filter of Fig. 19-21c results.

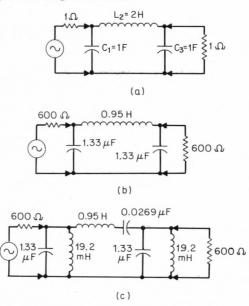

**Fig. 19-21**    (a) Normalized low-pass filter; (b) scaled low-pass filter; (c) bandpass filter.

## BAND-REJECT FILTER DESIGN

Band-reject filters are derived from a high-pass filter having a cutoff frequency equal to the desired band-reject filter bandwidth. The material on Normalizing Filter Requirements in Sec. 19.5 shows the interrelationship between low-pass, high-pass, and band-reject filters and defines procedures for normalization. Determination of the filter configuration and element values is illustrated in this section.

**problem 19.11**    A band-reject filter is required having 3-dB points at 900 and 1100 Hz and a minimum attenuation of 15 dB at 950 and 1050 Hz. A source and load impedance of 1000 $\Omega$ is required.

**theory**    To design a band-reject filter, a normalized low-pass filter must first be selected as described above, and then a normalized low-pass filter can be chosen from the normalized curves of Sec. 19.5. The corresponding low-pass circuit from Tables 19-1 to 19-6 is transformed to a band-reject filter in the following manner.

Transform the normalized low-pass filter into a normalized high-pass filter and then frequency- and impedance-scale the high-pass filter to a cutoff frequency equal to the bandwidth of the band-reject filter and to the required impedance level. Next transform the scaled high-pass filter into a band-reject filter by resonating each high-pass inductor with a series capacitor and resonating each high-pass capacitor with a parallel inductor. Resonance should occur at the geometric center frequency $F_0$.

### solution

*a.* Normalize the band-reject requirement and select an appropriate normalized high-pass filter by the following steps. Compute the geometric center frequency $F_0$:

$$F_0 = \sqrt{900 \times 1100} = 995 \text{ Hz} \tag{19-1}$$

Compute two pairs of geometrically related stop-band frequencies:

$$F_a = 950 \text{ Hz} \qquad F_b = \frac{995^2}{950} = 1042 \text{ Hz} \tag{19-9}$$

$$F_b - F_a = 92 \text{ Hz}$$

$$F_b = 1050 \text{ Hz} \qquad F_a = \frac{995^2}{1050} = 943 \text{ Hz} \tag{19-9}$$

$$F_b - F_a = 107 \text{ Hz}$$

Compute the band-reject steepness factor:

$$A_s = \frac{200 \text{ Hz}}{107 \text{ Hz}} = 1.87 \tag{19-11}$$

Examination of the normalized curves of Fig. 19-8 shows that an $n = 3$ Butterworth filter has over 15 dB of attenuation at 1.87 rad. The corresponding circuit from Table 19-1 is illustrated in Fig. 19-22*a*.

*b.* Convert the normalized low-pass filter into a normalized high-pass filter:

$$L_1 = L_3 = \frac{1}{1} = 1 \text{ H} \tag{19-19}$$

$$C_2 = \frac{1}{2} = 0.5 \text{ F} \tag{19-18}$$

The normalized high-pass filter is shown in Fig. 19-22*b*.

*(c)* Frequency- and impedance-scale the normalized high-pass filter for a cutoff frequency equivalent to the band-reject bandwith requirement and to the required impedance level.

$$\text{FSF} = 1257 \tag{19-13}$$

$$L_1' = L_3' = \frac{1000}{1257} = 0.796 \text{ H} \tag{19-15}$$

$$C_2' = \frac{0.5}{1257 \times 1000} = 0.398 \text{ }\mu\text{F} \tag{19-16}$$

Figure 19-22*c* shows the scaled high-pass filter.

*(d)* The final step in the design procedure is to resonate each branch to $F_0$ by adding a series capacitor to each inductor and a parallel inductor to each capacitor.

$$C = \frac{1}{(2\pi995)^2 \times 0.796} = 0.0321 \text{ }\mu\text{F} \tag{19-23}$$

$$L = \frac{1}{(2\pi995)^2 \times 0.398 \times 10^{-6}} = 64.3 \text{ mH} \tag{19-22}$$

The final circuit is shown in Fig. 19-22*d*.

## TUNING METHODS

Each resonant circuit can be adjusted to the exact resonant frequency required by varying the inductance. Most inductors provide a means for this adjustment such as ferrite slugs.

### Frequency-Response Method

Figure 19-23 contains a test circuit which can be used for adjusting parallel-tuned circuits. The load resistor is computed by:

$$R \approx \frac{2\pi FLQ}{10} \tag{19-24}$$

where $F$ is the frequency of resonance and $Q$ is the inductor quality factor.

Adjust inductor $L$ for a null on the VTVM with the sine-wave oscillator set for the desired resonant frequency. Each parallel-tuned circuit can be adjusted in this manner.

Adjust series resonance using the test circuit of Fig. 19-24 where:

$$R \approx \frac{20\pi FL}{Q} \tag{19-25}$$

A null will ocur on the VTVM at the resonant frequency.

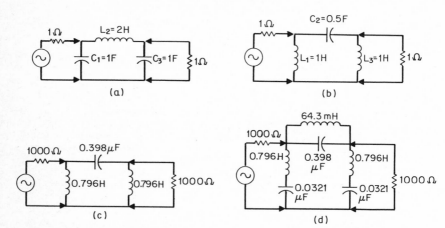

Fig. **19-22** (*a*) Normalized low-pass filter; (*b*) normalized high-pass filter; (*c*) scaled high-pass filter; (*d*) band-reject filter.

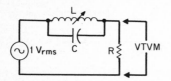

Fig. **19-23** Test circuit for parallel resonance.

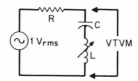

Fig. **19-24** Test circuit for series resonance.

### Lissajous Pattern Method

When inductor $Q$'s are very low (below 5), a very broad null is obtained using the method described above. An alternate scheme involves using an oscilloscope in the test circuits of Figs. 19-25 and 19-26. The vertical channel is connected to the oscillator output and the horizontal channel to the circuit output instead of using a VTVM. An ellipse will be obtained called a *Lissajous pattern*. At resonance this ellipse will collapse to a straight line (at a 45° angle). This straight line represents 0° phase shift which occurs only at resonance. Figure 19-27 shows typical Lissajous patterns for various phase shifts.

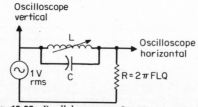

Fig. **19-25** Parallel-resonance Lissajous test circuit.

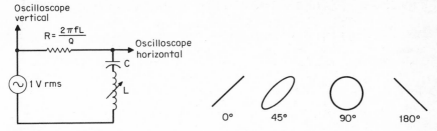

**Fig. 19-26**   Series-resonance Lissajous test circuit.

**Fig. 19-27**   Typical Lissajous patterns.

## EFFECTS OF COMPONENT SELECTION

The sucess or failure of a filter design depends a great deal upon component selection. Poor choice of component types can result in filter drift and large discrepancies between the measured and theoretical response.

### Capacitor Selection

The most commonly used capacitor types are polystyrene, Mylar polyester, and mica. For most applications below 100 kHz, polystyrene capacitors are recommended since they provide excellent performance at reasonable cost. For very low frequency filters where high-capacity values are required, Mylar polyester capacitors offer optimum space saving. Above 100 kHz mica capacitors should be used.

The following table lists the properties of the three basic capacitor types:

| Type | Temp. coefficient | Range | Temp. range |
|------|------|------|------|
| Polystyrene | −120 ppm/°C | 0.001 to 0.1 $\mu$F | −55 to + 85°C |
| Mylar | 400 ppm/°C | 0.001 to 10 $\mu$F | −55 to + 125°C |
| Mica | ±50 ppm/°C | 10 to 5000 pF | −55 to + 125°C |

### Inductor selection

The two critical parameters of inductors are $Q$ and stability. Low $Q$ results in rounding of the frequency response near cutoff and increased insertion loss. Figure 19-28 illustrates the effect of low inductor $Q$ on both low-pass and bandpass filters. As a rule, $Q$ requirements increase with filter complexity. Higher $Q$'s are required for narrow bandpass designs.

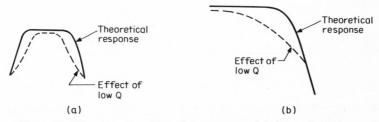

**Fig. 19-28**   Effect of low $Q$ on (a) bandpass response and (b) low-pass response.

Inductor stability and $Q$ depend on many factors such as the inductor core material, shape, and wire size. The most popular form of inductor is the ferrite potcore. Ferrite materials are available covering the range of subaudio to a few megahertz. Overlapping this frquency range are toroidal cores consisting of molybdenum Permalloy. Both materials can be supplied with positive temperature coefficients to compensate for the negative coefficient of polystyrene capacitors where high stability is required.

## 19.8   ACTIVE FILTER DESIGN

Active filters are composed of circuits containing operational amplifiers, resistors, and capacitors. Since they completely eliminate inductors, active filters offer cost savings and size reduction, especially at low frequencies where inductors can become prohibitively large and expensive.

### FREQUENCY AND IMPEDANCE SCALING

An active filter's frequency response can be shifted to a different frequency range by dividing the resistor or capacitor values by the FSF. As defined in Sec. 19.3, the FSF is the ratio of a particular reference frequency of the required scaled response to the corresponding frequency of the existing filter:

$$FSF = \frac{\text{reference frequency of scaled response}}{\text{reference frequency of existing response}} \qquad (19\text{-}12)$$

Figure 19-29$a$ shows a normalized low-pass active filter and its associated response. When the 3-dB point is scaled from 1 rad to 10 KHz, an FSF of 62,800 results in the circuit of Fig. 19-29$b$.

Although the circuit of Fig. 19-29$b$ has the desired response, the values are impractical. If the resistors are multiplied by a factor $Z$ and the capacitors are divided by this same factor, the circuit can be impedance-scaled without changing the frequency response. By use of a $Z$ of 10,000, the final circuit of Fig. 19-29$c$ results.

Frequency and impedance scaling can be combined, with the following formulas resulting:

$$C = \frac{C \text{ normalized}}{Z \times FSF} \qquad (19\text{-}26)$$

$$R = R \text{ normalized} \times Z \qquad (19\text{-}27)$$

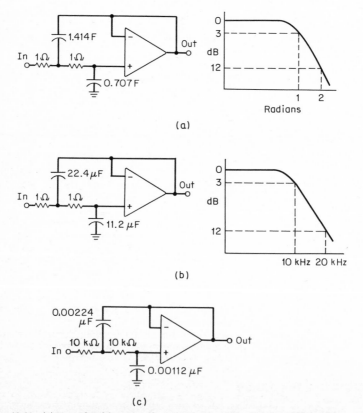

**Fig. 19-29**  ($a$) Normalized low-pass filter; ($b$) filter scaled to 10 kHz; ($c$) final filter.

## LOW-PASS FILTER DESIGN

To design an active low-pass filter, the requirements must first be normalized. A suitable response function is then selected from Sec. 19.5. Tables 19-7 to 19-11 contain normalized active low-pass filter values corresponding to each response function.

In a manner similar to that used for *LC* filters, the normalized low-pass filter is frequency- and impedance-scaled to the required cutoff frequency and to a convenient impedance level.

Active low-pass filters, like their *LC* counterpart, are divided into two categories, all pole and elliptic function. The all-pole family includes Butterworth, Chebyshev, and maximally flat delay.

**TABLE 19-7    Butterworth Normalized Active Low-Pass Values**

| Order $N$ | $C_1$ | $C_2$ | $C_3$ |
|:---:|:---:|:---:|:---:|
| 2 | 1.414 | 0.7071 | |
| 3 | 3.546 | 1.392 | 0.2024 |
| 4 | 1.082 | 0.9241 | |
|   | 2.613 | 0.3825 | |
| 5 | 1.753 | 1.354 | 0.4214 |
|   | 3.235 | 0.3090 | |
| 6 | 1.035 | 0.9660 | |
|   | 1.414 | 0.7071 | |
|   | 3.863 | 0.2588 | |
| 7 | 1.531 | 1.336 | 0.4885 |
|   | 1.604 | 0.6235 | |
|   | 4.493 | 0.2225 | |
| 8 | 1.020 | 0.9809 | |
|   | 1.202 | 0.8313 | |
|   | 1.800 | 0.5557 | |
|   | 5.125 | 0.1950 | |
| 9 | 1.455 | 1.327 | 0.5170 |
|   | 1.305 | 0.7661 | |
|   | 2.000 | 0.5000 | |
|   | 5.758 | 0.1736 | |
| 10 | 1.012 | 0.9874 | |
|   | 1.122 | 0.8908 | |
|   | 1.414 | 0.7071 | |
|   | 2.202 | 0.4540 | |
|   | 6.390 | 0.1563 | |

## All-Pole Filters

All-pole normalized low-pass filters consist of combinations of the two-pole and three-pole sections shown in Fig. 19-30. If the filter order $n$ is an even number, $n/2$ two-pole sections are used. If $n$ is odd, $(n-3)/2$ two-pole sections and a single three-pole section are needed.

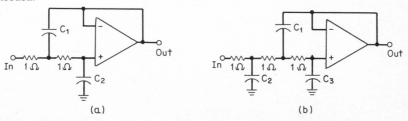

(a)                    (b)

**Fig. 19-30**  (*a*) Basic two-pole section; (*b*) basic three-pole section.

**TABLE 19-8    0.1-dB Ripple Chebyshev Normalized Active Low-Pass Values**

| Order $N$ | $C_1$ | $C_2$ | $C_3$ |
|---|---|---|---|
| 2 | 1.638 | 0.6955 | |
| 3 | 6.653 | 1.825 | 0.1345 |
| 4 | 1.900 | 1.241 | |
|   | 4.592 | 0.2410 | |
| 5 | 4.446 | 2.520 | 0.3804 |
|   | 6.810 | 0.1580 | |
| 6 | 2.553 | 1.776 | |
|   | 3.487 | 0.4917 | |
|   | 9.531 | 0.1110 | |
| 7 | 5.175 | 3.322 | 0.5693 |
|   | 4.546 | 0.3331 | |
|   | 12.73 | 0.08194 | |
| 8 | 3.270 | 2.323 | |
|   | 3.857 | 0.6890 | |
|   | 5.773 | 0.2398 | |
|   | 16.44 | 0.06292 | |
| 9 | 6.194 | 4.161 | 0.7483 |
|   | 4.678 | 0.4655 | |
|   | 7.170 | 0.1812 | |
|   | 20.64 | 0.04980 | |
| 10 | 4.011 | 2.877 | |
|    | 4.447 | 0.8756 | |
|    | 5.603 | 0.3353 | |
|    | 8.727 | 0.1419 | |
|    | 25.32 | 0.04037 | |

Each individual section has unity gain at direct current and may have a sharply peaked response in the passband. The composite response of all sections yields the required response funtion. The operational amplifier has nearly zero output impedance, and so sections can be directly cascaded.

**problem 19.12**    An active low-pass filter is required having a 3-dB cutoff of 100 Hz and 55-dB minimum rejection at 300 Hz.

**theory**    The following procedure should be followed for the design of active low-pass filters:
   1. Normalize the low-pass requirement by computing $A_s$ (pages 19-5 to 19-9).
   2. Select a satisfactory response function and a corresponding filter (pages 19-9 to 19-11).
   3. Frequency- and impedance-scale the normalized design to the desired cutoff and to a convenient impedance level.

**solution**
   *a.* Calculate the steepness factor $A_s$:

$$A_s = \frac{300 \text{ Hz}}{100 \text{ Hz}} = 3 \qquad (19\text{-}3)$$

   *b.* Select a response function and normalize the design. According to the normalized curves of Sec. 19.2.2, an $n = 5$, 0.5-dB Chebyshev filter has over 55-dB rejection at 3 rad. The normalized design from Table 19-9 is shown in Fig. 19-31*a*.
   *c.* Frequency- and impedance-scale the normalized filter:

$$\text{Compute FSF} = 2\pi 100 = 628 \qquad (19\text{-}13)$$

Choose $Z = 10,000$.
Then

$$C = \frac{C \text{ normalized}}{Z \times \text{FSF}} \qquad (19\text{-}26)$$

**TABLE 19-9  0.5-dB Ripple Chebyshev Normalized Active Low-Pass Values**

| Order $N$ | $C_1$ | $C_2$ | $C_3$ |
|---|---|---|---|
| 2 | 1.950 | 0.6533 | |
| 3 | 11.23 | 2.250 | 0.0895 |
| 4 | 2.582<br>6.233 | 1.300<br>0.1802 | |
| 5 | 6.842<br>9.462 | 3.317<br>0.1144 | 0.3033 |
| 6 | 3.592<br>4.907<br>13.40 | 1.921<br>0.3743<br>0.07902 | |
| 7 | 7.973<br>6.446<br>18.07 | 4.483<br>0.2429<br>0.05778 | 0.4700 |
| 8 | 4.665<br>5.502<br>8.237<br>23.45 | 2.547<br>0.5303<br>0.1714<br>0.04409 | |
| 9 | 9.563<br>6.697<br>10.26<br>29.54 | 5.680<br>0.3419<br>0.1279<br>0.03475 | 0.6260 |
| 10 | 5.760<br>6.383<br>8.048<br>12.53<br>36.36 | 3.175<br>0.6773<br>0.2406<br>0.09952<br>0.02810 | |

$$R = R \text{ normalized} \times Z \tag{19-27}$$

The resulting filter is shown in Fig. 19-31*b*.

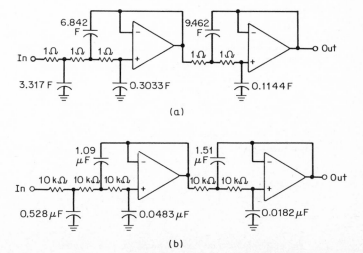

(a)

(b)

**Fig. 19-31**  (*a*) Normalized $n = 5$, 0.5-dB Chebyshev low-pass filter; (*b*) final low-pass filter.

**TABLE 19-10    Maximally Flat Delay Normalized Active Low-Pass Values**

| Order $N$ | $C_1$ | $C_2$ | $C_3$ |
|---|---|---|---|
| 2 | 0.9066 | 0.6800 | |
| 3 | 1.423 | 0.9880 | 0.2538 |
| 4 | 0.7351 | 0.6746 | |
|   | 1.012 | 0.3900 | |
| 5 | 1.010 | 0.8712 | 0.3095 |
|   | 1.041 | 0.3100 | |
| 6 | 0.6352 | 0.6100 | |
|   | 0.7225 | 0.4835 | |
|   | 1.073 | 0.2561 | |
| 7 | 0.8532 | 0.7792 | 0.3027 |
|   | 0.7250 | 0.4151 | |
|   | 1.100 | 0.2164 | |
| 8 | 0.5673 | 0.5540 | |
|   | 0.6090 | 0.4861 | |
|   | 0.7257 | 0.3590 | |
|   | 1.116 | 0.1857 | |
| 9 | 0.7564 | 0.7070 | 0.2851 |
|   | 0.6048 | 0.4352 | |
|   | 0.7307 | 0.3157 | |
|   | 1.137 | 0.1628 | |
| 10 | 0.5172 | 0.5092 | |
|   | 0.5412 | 0.4682 | |
|   | 0.6000 | 0.3896 | |
|   | 0.7326 | 0.2792 | |
|   | 1.151 | 0.1437 | |

## Elliptic-Function Filters

Elliptic-function low-pass active filters consist of the basic section shown in Fig. 19-32. The total number of sections required for a complete filter is $(n - 1)/2$ where $n$ is the order of the filter. Only odd-order active filters are provided in Table 19-11 since they utilize each operational amplifier most efficiently. Note that $R_5$ and $C_5$ appear on the output section only.

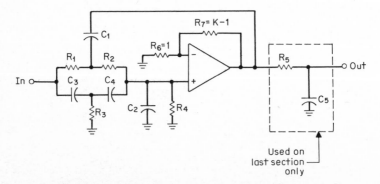

**Fig. 19-32**   Elliptic-function normalized low-pass filter section.

**TABLE 19-11  Elliptic Function Active Low-Pass Filter Values**

| $N$ | $\Omega_x$ | $\omega_s$ | $R_1$ | $R_2$ | $R_3$ | $R_4$ | $R_5$ | $C_1$ | $C_2$ | $C_3$ | $C_4$ | $C_5$ | $K$ |
|---|---|---|---|---|---|---|---|---|---|---|---|---|---|
| | | | | | | $R_{dB} = 0.01$ dB | | | | | | | |
| 3 | 5.241 | 41.00 | 0.3620 | 0.7240 | 2.805 | 12.62 | 1.000 | 2.340 | 0.5199 | 0.1342 | 0.0671 | 0.6193 | 1.206 |
| 3 | 3.628 | 31.14 | 0.3922 | 0.7844 | 1.481 | 6.662 | 1.000 | 2.183 | 0.4851 | 0.2570 | 0.1285 | 0.5968 | 1.343 |
| 3 | 2.459 | 20.4 | 0.4561 | 0.9121 | 0.8258 | 3.716 | 1.000 | 1.930 | 0.4290 | 0.4738 | 0.2369 | 0.5438 | 1.658 |
| 5 | 1.701 | 40.81 | 0.3866 | 0.7732 | 1.613 | 7.259 | 1.000 | 3.583 | 0.7963 | 0.3817 | 0.1908 | 1.039 | 1.050 |
| | | | 0.4848 | 0.9695 | 0.7114 | 3.201 | | 2.590 | 0.5756 | 0.7845 | 0.3923 | | 2.145 |
| 5 | 1.414 | 30.17 | 0.4239 | 0.8479 | 1.039 | 4.679 | 1.000 | 3.133 | 0.6962 | 0.5678 | 0.2839 | 0.9260 | 1.247 |
| | | | 0.5443 | 1.088 | 0.5733 | 2.580 | | 2.364 | 0.5253 | 0.9974 | 0.4987 | | 2.471 |
| 7 | 1.192 | 40.54 | 0.3909 | 0.7819 | 1.508 | 6.786 | 1.000 | 4.187 | 0.9305 | 0.4825 | 0.2412 | 1.229 | 1.039 |
| | | | 0.5157 | 1.031 | 0.6285 | 2.828 | | 2.846 | 0.6325 | 1.038 | 0.5190 | | 2.153 |
| | | | 0.5902 | 1.180 | 0.5107 | 2.298 | | 2.396 | 0.5325 | 1.230 | 0.6153 | | 2.837 |
| | | | | | | $R_{dB} = 0.1$ dB | | | | | | | |
| 3 | 3.628 | 40.77 | 0.3655 | 0.7311 | 2.522 | 11.34 | 1.000 | 3.166 | 0.7036 | 0.2040 | 0.1020 | 0.9897 | 1.313 |
| 3 | 2.559 | 31.13 | 0.3980 | 0.7961 | 1.367 | 6.152 | 1.000 | 2.924 | 0.6497 | 0.3783 | 0.1892 | 0.9484 | 1.465 |
| 3 | 1.788 | 20.53 | 0.4651 | 0.9301 | 0.7845 | 3.530 | 1.000 | 2.540 | 0.5645 | 0.6692 | 0.3346 | 0.8544 | 1.807 |
| 5 | 1.440 | 40.90 | 0.3873 | 0.7745 | 1.596 | 7.181 | 1.000 | 4.346 | 0.9659 | 0.4688 | 0.2344 | 1.501 | 1.215 |
| | | | 0.5048 | 1.009 | 0.6543 | 2.945 | | 2.776 | 0.6170 | 0.9518 | 0.4759 | | 2.318 |
| 5 | 1.236 | 30.59 | 0.4243 | 0.8485 | 1.036 | 4.666 | 1.000 | 3.756 | 0.8347 | 0.6831 | 0.3415 | 1.323 | 1.420 |
| | | | 0.5600 | 1.120 | 0.5490 | 2.470 | | 2.531 | 0.5625 | 1.147 | 0.5739 | | 2.626 |
| 7 | 1.155 | 46.24 | 0.3774 | 0.7547 | 1.905 | 8.571 | 1.000 | 5.264 | 1.169 | 0.4635 | 0.2318 | 1.823 | 1.139 |
| | | | 0.5058 | 1.011 | 0.6517 | 2.933 | | 3.115 | 0.6922 | 1.074 | 0.5372 | | 2.211 |
| | | | 0.5886 | 1.177 | 0.5124 | 2.306 | | 2.491 | 0.5536 | 1.272 | 0.6359 | | 2.860 |
| | | | | | | $R_{dB} = 0.28$ dB | | | | | | | |
| 3 | 2.924 | 39.48 | 0.3719 | 0.7438 | 2.142 | 9.638 | 1.000 | 3.538 | 0.7861 | 0.2730 | 0.1365 | 1.282 | 1.410 |
| 3 | 2.130 | 30.44 | 0.4069 | 0.8137 | 1.229 | 5.533 | 1.000 | 3.239 | 0.7198 | 0.4764 | 0.2382 | 1.221 | 1.578 |
| 3 | 1.556 | 20.58 | 0.4739 | 0.9479 | 0.7490 | 3.371 | 1.000 | 2.797 | 0.6215 | 0.7865 | 0.3933 | 1.093 | 1.927 |
| 5 | 1.305 | 39.17 | 0.3942 | 0.7884 | 1.439 | 6.477 | 1.000 | 4.572 | 1.015 | 0.5564 | 0.2782 | 1.831 | 1.360 |
| | | | 0.5280 | 1.056 | 0.6027 | 2.712 | | 2.757 | 0.6127 | 1.073 | 0.5368 | | 2.479 |
| 5 | 1.166 | 30.46 | 0.4284 | 0.8568 | 1.001 | 4.507 | 1.000 | 3.982 | 0.8848 | 0.7570 | 0.3785 | 1.623 | 1.551 |
| | | | 0.5781 | 1.156 | 0.5249 | 2.362 | | 2.531 | 0.5624 | 1.238 | 0.6194 | | 2.756 |
| 7 | 1.155 | 50.86 | 0.3696 | 0.7393 | 2.263 | 10.18 | 1.000 | 5.919 | 1.315 | 0.4297 | 0.2148 | 2.352 | 1.206 |
| | | | 0.4958 | 0.9916 | 0.6782 | 3.052 | | 3.274 | 0.7276 | 1.063 | 0.5319 | | 2.214 |
| | | | 0.5825 | 1.165 | 0.5195 | 2.338 | | 2.548 | 0.5661 | 1.270 | 0.6349 | | 2.843 |

**problem 19.13** An active low-pass filter is required having 0.5-dB maximum ripple up to 1000 Hz and 35-dB minimum rejection above 3000 Hz.

**theory** Active elliptic function low-pass filters are designed following a procedure similar to that used for the all-pole types.

1. Normalize the low-pass requirement by computing $A_s$ (pages 19-5 to 19-9).

2. Select a satisfactory filter from Table 19-11 having an $R_{dB}$ less than the required passband ripple, an $\omega_s$ less than the calculated steepness factor $A_s$, and an $A_{dB}$ in excess of the minimum required stop-band attenuation.

3. Frequency- and impedance-scale the normalized design to the desired cutoff and to a convenient impedance level.

**solution**

   $a$. Calculate the steepness factor $A_s$:

$$A_s = \frac{3000 \text{ Hz}}{1000 \text{ Hz}} = 3 \tag{19-3}$$

   $b$. Select a filter from Table 19-11:

$$R_{dB} = 0.28 \text{ dB}$$
$$N = 3$$
$$\omega_s = 2.924$$
$$A_{dB} = 39.48 \text{ dB}$$

The normalized design is shown in Fig. 19-33$a$.

   $c$. Frequency- and impedance-scale the normalized filter.

$$\text{Compute FSF} = 2\pi 1000 = 6280 \tag{19-13}$$

Choose $Z = 10,000$.
Then

$$C = \frac{C \text{ normalized}}{Z \times \text{FSF}} \tag{19-26}$$
$$R = R \text{ normalized} \times Z \tag{19-27}$$

The resulting filter is shown in Fig. 19-33$b$.

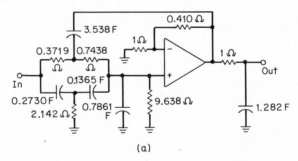

(a)

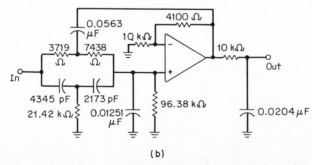

(b)

**Fig. 19-33**   ($a$) Normalized low-pass filter; ($b$) denormalized low-pass filter.

## HIGH-PASS FILTER DESIGN

Active high-pass filters are designed directly from the normalized low-pass circuit in a way similar to that for $LC$ high-pass filters. After transforming the normalized low-pass filter into a normalized high-pass filter, the circuit is then frequency- and impedance-scaled.

**problem 19.14** A high-pass filter is required having less than 3 dB of attenuation at 1000 Hz and more than 45 dB at 350 Hz.

**theory**

1. First normalize the high-pass requirement by computing the high-pass steepness factor $A_s$ (pages 19-5 to 19-9).

2. Then select a satisfactory response function from pages 19-9 to 19-11 and the appropriate active low-pass filter from Tables 19-7 to 19-11.

3. Next transform the normalized low-pass circuit into a normalized high-pass filter. This is accomplished by replacing each resistor $R$ by a capacitor of $1/R$ F. Each capacitor $C$ is replaced by a resistor of $1/C$ $\Omega$.

4. Finally frequency- and impedance-scale the normalized high-pass filter to the desired frequency cutoff and impedance level.

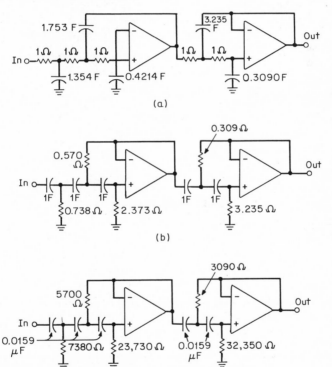

**Fig. 19-34** (a) Normalized low-pass filter; (b) transformed normalized high-pass filter; (c) final high-pass filter.

**solution**

a. Calculate the high-pass steepness factor $A_s$:

$$A_s = \frac{1000}{350} = 2.86 \tag{19-8}$$

b. Select a response function and a normalized low-pass filter. As the normalized curves on pages 19-9 to 19-11 indicate, an $n = 5$ Butterworth filter provides over 40 dB at 2.86 rad. Figure 19-34a shows the normalized low-pass filter from Table 19.7.

*c.* Transform the normalized low-pass filter into a normalized high-pass filter by replacing each resistor $R$ by a capacitor of $1/R$ farads and each capacitor $C$ by a resistor of $1/C$ $\Omega$. The resulting normalized high-pass filter is illustrated in Fig. 19-34*b*.

*d.* Frequency- and impedance-scale the normalized filter:

$$\text{Compute FSF} = 2\pi 1000 = 6280 \qquad (19\text{-}13)$$

Choose $Z = 10,000$.
Then

$$C = \frac{C \text{ normalized}}{Z \times \text{FSF}} \qquad (19\text{-}26)$$

$$R = R \text{ normalized} \times Z \qquad (19\text{-}27)$$

The resulting filter is shown in Fig. 19-34*c*.

## BANDPASS FILTER DESIGN

Active bandpass filters are classified as either wide band or narrow band. If the ratio of upper 3-dB frequency to lower 3-dB frequency is in excess of 1.5, the filter requirement falls into the wide-band category.

## Wide-Band Bandpass Filters

To design the filter as a wide-band type, separate the requirement into low-pass and high-pass specifications. A separate low-pass and high-pass filter can then be designed and cascaded with no interaction.

**problem 19.15**   Design a bandpass filter having 3-dB points at 1000 and 3000 Hz and having more than 25 dB attenuation at 300 and 9000 Hz.

**theory**   Because the ratio of the upper 3-dB point to the lower 3-dB point exceeds 1.5, the filter should be designed as a wide-band type.

**solution**

*a.* Separate the bandpass requirements into individual low-pass and high-pass specifications.

> Low-pass filter:    3 dB at 3000 Hz
> 25 dB minimum at 9000 Hz

> High-pass filter:    3 dB at 1000 Hz
> 25 dB minimum at 300 Hz

*b.* Calculate the steepness factor $A_s$ for both the low-pass and high-pass specifications:

$$\text{Low pass: } A_s = \frac{9000}{3000} = 3 \qquad (19\text{-}3)$$

$$\text{High pass: } A_s = \frac{1000}{300} = 3.33 \qquad (19\text{-}8)$$

*c.* Select normalized designs to satisfy the low-pass and high-pass requirements: According to the curves of Fig. 19-8, an $N = 3$ Butterworth design provides over 25 dB at $A_s = 3$ and $A_s = 3.33$.

*d.* Figure 19-35*a* shows the normalized low-pass and high-pass filters. The normalized high-pass filter was derived from a normalized $N = 3$ Butterworth low-pass filter by replacing every resistor with a capacitor having the value $1/R$ and by replacing each capacitor with a resistor $1/C$.

*e.* Scale the low-pass and high-pass to the required cutoffs and desired impedance levels. Compute FSF:

> Low pass:  FSF $= 2\pi 3000 = 18,850$     (19-13)
> High pass: FSF $= 2\pi 1000 = 6280$     (19-13)

With a $Z$ of 10,000, multiply all resistors by $Z$ and divide all capacitors by $Z \times$ FSF using the appropriate FSF for the high-pass and low-pass filters. Figure 19-35*b* shows the resulting bandpass filter.

## Narrow Bandpass filters

When the ratio of upper 3-dB cutoff to lower 3-dB cutoff is less than 1.5, a narrow-band design is required.

**problem 19.16**   Design a bandpass filter having a center frequency of 1000 Hz and a 3-dB bandwidth of 50 Hz. Determine the attenuation at a bandwidth of 200 Hz.

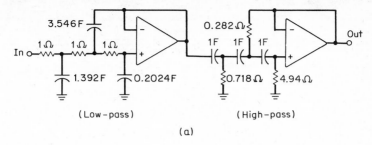

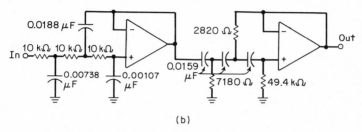

**Fig. 19-35**  (a) Normalized filter; (b) final bandpass filter.

**theory**    The circuit of Fig. 19-36 provides a narrow bandpass response up to a circuit $Q$ of 20. Filter gain is unity at the center frequency.

The response at any bandwidth $BW_x$ can be calculated by:

$$dB = 10 \log \left[ 1 + \left( \frac{BW_x}{BW_{3dB}} \right)^2 \right] \tag{19-28}$$

where $BW_{3dB}$ is $F_0/Q$. Equation (19-28) corresponds to $n = 1$ of Fig. 19-8 where $\Omega$ corresponds to $BW_x/BW_{3dB}$.

The design equations for the circuit of Fig. 19-36 are:

$$R_1 = \frac{Q}{2\pi F_0 C} \tag{19-29}$$

$$R_2 = \frac{R_1}{2Q^2 - 1} \tag{19-30}$$

$$R_3 = 2R_1 \tag{19-31}$$

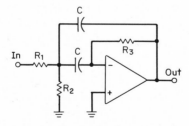

**Fig. 19-36**    Narrow bandpass filter circuit.

**solution**    Select $C = 0.01\ \mu F$. Then

$$R_1 = 318\ k\Omega \tag{19-29}$$
$$R_2 = 398\ \Omega \tag{19-30}$$
$$R_3 = 636\ k\Omega \tag{19-31}$$

Figure 19-37 illustrates the resulting filter. Resistor $R_2$ has been made adjustable for tuning purposes. Tuning can be accomplished by monitoring the phase shift between the input and output using an oscilloscope Lissajous pattern as described on pages 19-27 to 19-29. Adjust $R_2$ for a closed ellipse.

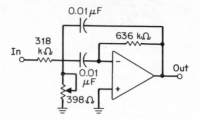

**Fig. 19-37**   Resulting filter of Problem 19.16.

## BAND-REJECT FILTERS

Wide-band band-reject filters have a ratio of upper 3-dB frequency to lower 3-dB frequency of 1.5 or more and are designed by combining a high-pass and low-pass filter using an additional operational amplifier. Notch networks are used for generating a narrow band-reject filter response characteristic.

### Wide-Band Band-Reject Filters

To design a wide-band band-reject filter, first separate the requirements into individual low-pass and high-pass specifications. Low-pass and high-pass filters are then designed and combined using an operational amplifier as shown in Fig. 19-38.

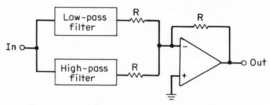

**Fig. 19-38**   Wide-band band-reject configuration.

**problem 19.17**   A band-reject filter is required having 3-dB points at 300 and 3000 Hz and more than 12-dB rejection between 600 and 1500 Hz.

**theory**   Since the ratio of the upper 3-dB frequency to the lower 3-dB frequency exceeds 1.5, a wide-band design is required.

**solution**   Separate the specification into individual low-pass and high-pass requirements.

Low-pass filter:   3 dB at 300 Hz
                   12 dB minimum at 600 Hz
High-pass filter:   3 dB at 3000 Hz
                   12 dB minimum at 1500 Hz

Calculate the steepness factor $A_s$ for both filter types:

$$\text{Low pass: } A_s = \frac{600}{300} = 2 \tag{19-7}$$

$$\text{High pass: } A_s = \frac{3000}{1500} = 2 \tag{19.8}$$

Figure 19.8 indicates that an $N = 2$ Butterworth design provides 12 dB at $\omega = 2$. This normalized filter satisfies both the low-pass and high-pass requirements.

The low-pass filter is designed in accordance with pages 19-31 to 19-36 and the procedure of pages 19-37 to 19-38 is followed for the high-pass filter. The resulting low-pass and high-pass filters are combined using the circuit of Fig. 19-38. The final filter is shown in Fig. 19-39.

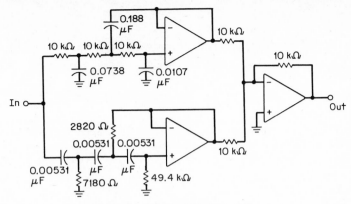

**Fig. 19-39** Wide-band band-reject filter.

## Notch Networks

A frequently used $RC$ null network is the "twin T." A deep null can be obtained at a particular frequency. However the circuit $Q$, which is defined as:

$$Q = \frac{\text{null frequency}}{\text{3-dB bandwidth}} \tag{19-32}$$

is only ¼. The circuit of Fig. 19-40 provides a means of increasing $Q$.

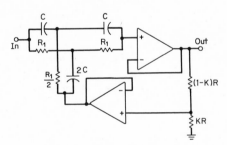

**Fig. 19-40**  Notch network.

**problem 19.18**  Design a filter having a notch at 1000 Hz and a 3-dB bandwidth of 50 Hz.

**theory**  The component values of Fig. 19-40 are computed by:

$$R_1 = \frac{1}{2\pi FC} \tag{19-33}$$

$$K = \frac{4Q - 1}{4Q} \tag{19-34}$$

$R$ and $C$ may be arbitrarily chosen.

**solution**

$$
\begin{aligned}
Q &= 20 &&(19\text{-}32)\\
K &= 0.9875 &&(19\text{-}34)\\
\text{Let } R &= 10\,\text{k}\Omega\\
C &= 0.01\,\mu\text{F} &&(19\text{-}33)\\
\text{Then } R_1 &= 15.9\,\text{k}\Omega
\end{aligned}
$$

The final circuit is shown in Fig. 19-41.

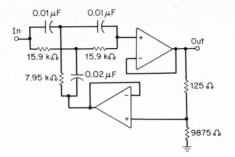

**Fig. 19-41**    Result of Problem 19.18.

## 19.8  BIBLIOGRAPHY

Christian, E., and Eisenmann, E.: *Filter Design Tables and Graphs,* Wiley, New York, 1966. Extensive tables for numerous filter types are provided.

Geffe, P.: *Simplified Modern Filter Design,* Rider, New York, 1963. Simplified design methods for *LC* filters are presented.

Giles, J. N. (ed.): *Linear Integrated Circuits Applications Handbook,* Fairchild Semiconductor Co., Mountainview, Calif., 1967. This book will prove useful to readers unfamiliar with operational amplifiers.

*Handbook of Operational Amplifier Active RC Networks,* Burr-Brown Research Corporation, Tucson, Ariz., 1966. This book consists of a collection of simple filter circuits and design formulas.

Huelsman, L. P.: *Theory and Design of Active RC Circuits,* McGraw-Hill, New York, 1968. A theoretical treatment and analysis of active *RC* circuits is presented.

Saal, R.: *Der Entwurf von Filtern mit Hilfe des Kataloges Normierter Tiefpasse,* TeleFunken Gmbh, Backhang, West Germany. This comprehensive collection of elliptic function *LC* filter tables is extremely useful for serious filter designers.

Weinberg, L.: *Network Analysis and Synthesis,* McGraw-Hill, New York, 1962. An excellent collection of *LC* filter design tables is provided along with useful theoretical material.

Williams, A. B.: *Active Filter Design,* Artech House Inc., Dedham, Mass., 1975. This book is directed toward the average engineer and provides many direct methods for the design of sophisticated active filters using simple formulas and tables.

Zverev, A. I.: *Handbook of Filter Synthesis,* Wiley, New York, 1967. A comprehensive overview of filter design is presented with emphasis on *LC* filters.

# Chapter 20

# Antennas

## LLOYD TEMES, P.E.

Associate Professor, Department of Electrical Technology, College
of Staten Island (CUNY)

## 20.1 INTRODUCTION

This chapter presents a number of practical problems relating to directional antennas. These include Hertz, multi-element, and long-wire antennas. Discussions are also presented of log-periodic and parabolic antennas.

The characteristics of single-element and multi-element antennas are covered. These include antenna gain, directivity, and impedance. The choice of an appropriate transmission line for an antenna is also discussed herein. (See Chapters 18 and 21 for additional information regarding transmission lines.) In addition, the important topics of line-of-sight transmission, effective radiated power (ERP), and radiation resistance are covered.

## 20.2 DESIGNING AN ANTENNA WHICH WILL RADIATE EQUALLY WELL IN ALL DIRECTIONS

**problem 20.1** An antenna is desired which will be effective in radiating and receiving a 100-MHz signal equally well in all directions. Consider the velocity factor to be 0.8.

**theory** A radiation pattern describes the effectiveness of an antenna in radiating energy in the various directions. A Marconi antenna has a circular radiation pattern as shown in Fig. 20-1. A Marconi antenna consists of a ¼-wavelength vertical conductor. Wavelength can be found from the formula:

$$f\lambda = kC \tag{20-1}$$

where $f$ = frequency
$\lambda$ = wavelength
$k$ = velocity factor*
$C$ = speed of electromagnetic waves in space

---

*Velocity factor $k$ is a decimal number which, when multiplied by the speed of electromagnetic waves in free space, will give the speed of such waves in a physical medium.

**Fig. 20-1**    Horizontal radiation pattern of a Marconi antenna.

**solution**    By Eq. (20-1):

$$f\lambda = kC$$
$$(100 \times 10^6)\lambda = (0.8)(3 \times 10^8)$$
$$\lambda = \frac{2.4 \times 10^8}{100 \times 10^6}$$
$$= 2.4 \text{ m}$$
$$\frac{\lambda}{4} = \frac{2.4}{4}$$
$$= 0.6 \text{ m}$$

## 20.3    DESIGNING AN ANTENNA WHICH RADIATES WELL IN TWO DIRECTIONS BUT IS INEFFECTIVE IN DIRECTIONS PERPENDICULAR TO THE OPTIMUM DIRECTIONS

**problem 20.2**    Design an antenna to be used at 60 MHz which will be equally effective in the north and south directions yet which will radiate very little energy in the east and west directions. Assume a velocity factor of 0.9.

**theory**    The radiation pattern of a Hertz antenna is that shown in Fig. 20-2a. Note that the antenna is most effective in two directions but is ineffective in the directions perpendicular to the most effective directions. The current and voltage distribution are shown in Fig. 20-2b. A Hertz antenna is cut to a length of ½ wavelength and is usually placed in a horizontal plane. See Fig. 20-3.

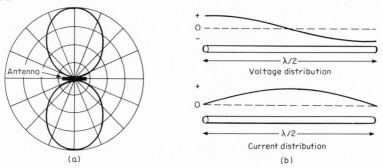

**Fig. 20-2**    (a) Radiation pattern of a Hertz dipole antenna. (b) Current and voltage distribution in dipole antenna.

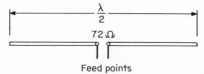

**Fig. 20-3**    A Hertz dipole antenna.

**solution**    By Eq. (20-1):

$$f\lambda = kC$$
$$60 \times 10^6\lambda = 0.9(3 \times 10^8)$$
$$\lambda = \frac{0.9 \times 3 \times 10^8}{60 \times 10^6}$$
$$= 4.5 \text{ m}$$

The length of a Hertz antenna is ½ wavelength; thus:

$$\frac{\lambda}{2} = 2.25 \text{ m}$$

See Fig. 20-4a for illustration of this solution.

Figure 20-4b is a graph giving the physical length of a ½-wave antenna having various length-to-diameter ratios. Radiation resistance is also given. A chart for converting electrical degrees to fractions of a wavelength is shown in Fig. 20-4c.

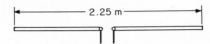

**Fig. 20-4**   (a) The Hertz dipole antenna for a 60-MHz signal.

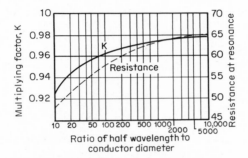

**Fig. 20-4**   (b) The solid curve shows the factor K by which the length of a half wave in free space should be multiplied to obtain the physical length of a resonant half-wave antenna having the length/diameter ratio shown along the horizontal axis. The broken curve shows how the radiation resistance of a half-wave antenna varies with the length/diameter ratio. (Courtesy A.R.R.L.)

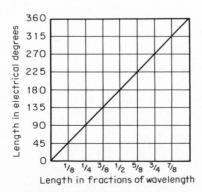

**Fig. 20-4**   (c) Chart for converting electrical degrees to fractions of a wavelength. (Courtesy A.R.R.L.)

## 20.4    DESIGNING AN ANTENNA HAVING HIGHLY DIRECTIVE PROPAGATION CHARACTERISTICS

**problem 20.3**    An antenna is necessary which will radiate in only the northerly direction and be relatively ineffective in the south, east, and west directions. The antenna is to be used for a 200-MHz signal. Design the antenna. Assume a velocity factor of 0.85 for the antenna elements and a velocity factor of 1.0 for the space between the elements.

**theory**    Directors and reflectors can be used in conjunction with a dipole antenna in order to produce an antenna with more directional characteristics than just the dipole alone. See Fig. 20-5 for a comparison of radiation patterns of a simple dipole antenna acting alone (Fig. 20-5a) and an antenna consisting of a dipole with reflectors and directors (Fig. 20-5b). This combination is called a *beam antenna* because it serves to focus the radiated energy into a narrower beam than would exist without reflectors and directors.

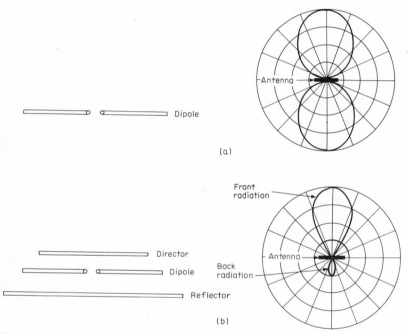

(a)

(b)

**Fig. 20-5**    (a) A dipole antenna and its radiation pattern. (b) A dipole antenna in conjunction with a reflector and a director, and its radiation pattern.

The reflectors and directors are called *parasitic elements* and the dipole is called the *driven element*. The reason for this nomenclature is that the dipole is connected to the transmission line while the directors and reflectors are not connected directly to the transmission line. Reflectors and directors get their excitation from the power radiated by the dipole. The combination is called a parasitic array.

A reflector is greater in length than the dipole while the director is shorter. Optimum spacing of the reflector from the driven element (dipole) is 0.15 wavelength if the reflector is 5% larger than the dipole. The optimum spacing between the driven element and the director is 0.1 wavelength when the length of the director is 5% smaller than the dipole. See Fig. 20-6.

### solution

$$f\lambda = kC \tag{20-1}$$

Using a velocity factor of 0.85 for the antenna elements yields

$$200 \times 10^6 \, \lambda = 0.85(3 \times 10^8)$$
$$= \frac{0.85(3 \times 10^8)}{200 \times 10^6} = \frac{0.85 \times 3}{2}$$
$$= 1.275 \text{ m}$$

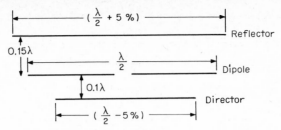

**Fig. 20-6**   The geometry of a three-element beam antenna.

The length of the dipole is thus:

$$\frac{\lambda}{2} = \frac{1.275}{2} = 0.6375 \text{ m}$$

Determining length of parasitic elements:

$$0.05\frac{\lambda}{2} = 0.05(0.6375)$$
$$= 0.319$$

Length of reflector $L_R$:

$$L_R = \frac{\lambda}{2} + 0.05\left(\frac{\lambda}{2}\right) = 0.6375 + 0.0319$$
$$= 0.6694 \text{ m}$$

Length of director:

$$L_D = \frac{\lambda}{2} - 0.05\left(\frac{\lambda}{2}\right) = 0.6375 - 0.0319$$
$$= 0.6056 \text{ m}$$

Now to calculate spacing. Calculate $\lambda$ in space with velocity factor of 1.0.

$$f\lambda = kC$$
$$200 \times 10^6 \lambda = 1.0(3 \times 10^8)$$
$$\lambda = \frac{1.0(3 \times 10^8)}{200 \times 10^6}$$
$$= 0.015 \times 10^2$$
$$= 1.5 \text{ m}$$

Driven element to reflector spacing:

$$0.15\lambda = 0.15(1.5)$$
$$\lambda = 0.225 \text{ m}$$

Driven element to director spacing:

$$0.1\lambda = 0.1(1.5)$$
$$= 0.15 \text{ m}$$

See Fig. 20-7.

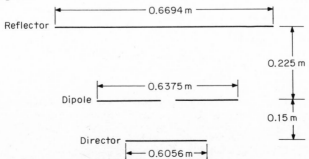

**Fig. 20-7**   Solution to Problem 20.3.

The previous problems involved the determination of element spacing in the design of a parasitic array for optimum conditions. There are times, however, when spacing other than 0.12 and 0.15λ is desirable.

Figure 20-8a shows how the gain of the antenna rolls off as the spacing from parasitic element to driven element varies from the optimum. These curves are each for an antenna having a driven element and one parasitic element, either a reflector or a director. Several examples of the spacing of multielement parasitic antennas are shown in Fig. 20-8b.

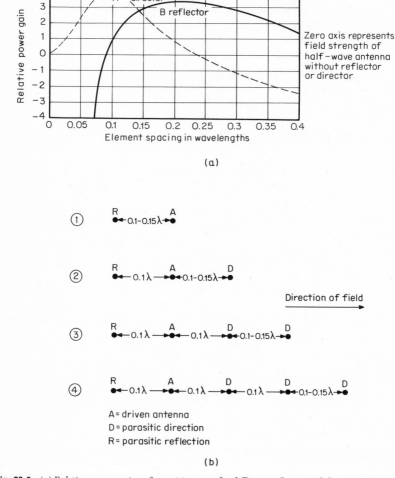

(a)

(1)   R          A
      0.1–0.15λ

(2)   R     A         D
      0.1λ  0.1–0.15λ

Direction of field

(3)   R     A      D         D
      0.1λ  0.1λ   0.1–0.15λ

(4)   R     A      D      D         D
      0.1λ  0.1λ   0.1λ   0.1–0.15λ

A = driven antenna
D = parasitic direction
R = parasitic reflection

(b)

**Fig. 20-8**   (a) Relative power gains of parasitic arrays for different reflector and director spacings. (b) Examples of multielement parasitic antennas.

Another effect of parasitic antenna design making use of parameters other than the optimum values used in previous illustrations is the change in radiation resistance of the antenna with variations in element spacing. This effect of element spacing on the radiation resistance of the antenna is shown in Fig. 20-9.

Front-to-back ratio is also affected by element spacing in a two-element parasitic array and is shown in Fig. 20-10.

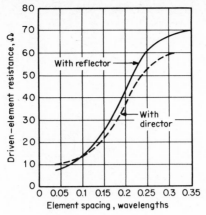

**Fig. 20-9**  Beam antenna radiation resistance as a function of parasitic element spacing. (*Courtesy A.R.R.L.*)

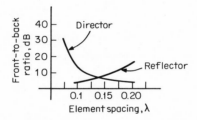

**Fig. 20-10**  Beam antenna front-to-back ratio as a function of parasitic element spacing.

## 20.5 DETERMINING MINIMUM HEIGHT OF A RECEIVING ANTENNA TO BE USED FOR RECEPTION OF LINE-OF-SIGHT WAVES

**problem 20.4**  How high must an antenna be in order to receive VHF commercial television broadcasts from a station whose transmitting antenna is 50 mi from the receiving antenna and is 1000 ft high. There are no interfering objects.

**theory**  VHF commercial television broadcasts are in the band of frequencies 54 to 216 MHz. Examining Fig. 20-11 shows that in this band of frequencies the only propagation which is effective is that due to line-of-sight waves.

Although ineffective at VHF commercial TV broadcast frequencies, two other modes of propagation exist, the ground wave and the sky wave. The ground wave is effective at lower frequencies, while the sky wave, also known as skip, is effective on a consistent basis only in the midfrequency range. See Fig. 20-12.

Within those frequencies where the line-of-sight wave is the only effective means of propagation, merely increasing the output power of a transmission will not necessarily increase the distance of effectiveness of the broadcast, since the line-of-sight wave travels essentially in a straight line. However, some bending of the wave travel does occur along the curvature of the earth's surface. This increases the optical line-of-sight distance by about 15%.

For the most part, in order to increase the effective reception distance, increasing the height of both the transmitting and receiving antenna is most effective. See Fig. 20-13a. (Note that doubling the height of the transmitting antenna is approximately equal to increasing the power by a factor of 5.)

The relationship between maximum distance of transmission and antenna height(s) is:

$$d = 1.41\sqrt{H(\text{ft})} \qquad (20\text{-}2)$$

where $H$ is the combined height of the transmitting and receiving antennas in feet, and $d$ is the line-of-sight transmission distance, including the effect of bending (refraction) along the earth's surface.

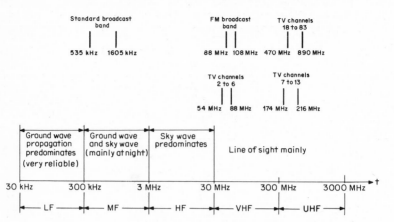

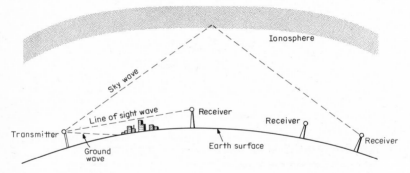

**Fig. 20-11**    Frequency bands and their predominant modes of propagation.

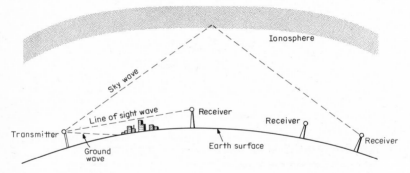

**Fig. 20-12**    The three modes of propagation of a radiated signal.

**solution**    From Eq. (20-2):

$$d = 1.41\sqrt{H\text{(ft)}}$$
$$50 = 1.41\sqrt{H}$$
$$\left(\frac{50}{1.41}\right)^2 = H$$
$$1259 \text{ ft} = H$$
$$1259 - 1000 = 259 \text{ ft} \quad \text{(receiving antenna height)}$$

Thus the minimum height of the receiving antenna must be 259 ft, in order to receive VHF transmissions broadcast by an antenna 50 mi distant and 1000 ft in height. The graph of Fig. 20-13$b$ shows line-of-sight reception distances. It gives both the optical distance and the distance as increased by refraction.

## 20.6    USING ANTENNA TRAPS TO CHANGE THE EFFECTIVE LENGTH OF AN ANTENNA

**problem 20.5**    Design a dipole antenna which will be equally effective at 50 and 30 MHz. Assume a velocity factor of 0.9.

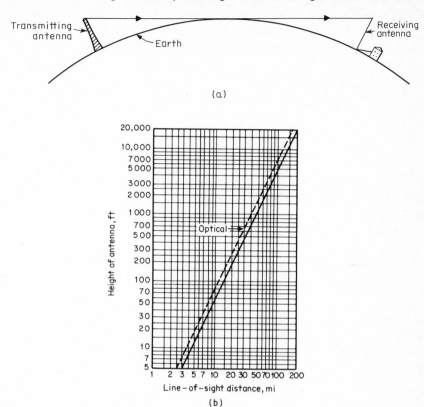

(a)

(b)

**Fig. 20-13** (*a*) Heights of both the transmitting and receiving antennas affect reception distance for line-of-sight waves. (*b*) Graph showing line-of-sight reception distance. Antenna height (*H*) is the combined height of the transmitting and receiving antennas. The solid line includes the effect of refraction. (*Courtesy A.R.R.L.*)

**theory**    The length of a dipole for transmission of a 50-MHz signal and the length required for optimum transmission of a 30-MHz signal are different. The lower-frequency signal will require a larger dipole than the higher-frequency signal.

Including a frequency-dependent circuit within each leg of the dipole which will act either as an open circuit or as a closed circuit will effectively lengthen or shorten the antenna as required. The frequency-dependent circuits used for this purpose are called *antenna traps*. See Fig. 20-14.

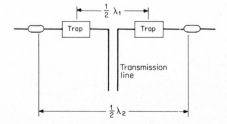

**Fig. 20-14**    Antenna traps being used to provide two separate tuned frequencies.

**solution**   First determine the wavelength of the 50-MHz signal.

$$f\lambda = kC$$
$$50 \times 10^6\lambda = 0.9(3 \times 10^8)$$
$$\lambda = \frac{0.9(3 \times 10^8)}{50 \times 10^6}$$
$$\lambda = 5.4 \text{ m}$$

(20-1)

The proper length for a 50-MHz dipole can then be found.

$$\frac{\lambda}{2} = \frac{5.4}{2} = 2.7 \text{ m}$$

Now determine the wavelength of the 30-MHz signal:

$$f\lambda =. kC$$
$$30 \times 10^6\lambda = 0.9(3 \times 10^8)$$
$$\lambda = \frac{0.9(3 \times 10^8)}{30 \times 10^6}$$
$$= 9 \text{ m}$$

The length of the dipole needed for the 30-MHz transmission can now be found from:

$$\frac{\lambda}{2} = \frac{9}{2} = 4.5 \text{ m}$$

The antenna meeting the needs specified above is shown in Fig. 20-15.

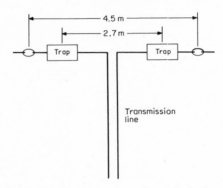

**Fig. 20-15**   Dipole antenna with antenna traps to be used at both 30 and 50 MHz.

## 20.7   DETERMINING THE Q OF AN ANTENNA

**problem 20.6**   An antenna is to be used to transmit an rf signal having a center frequency of 75 MHz and a bandwidth of 6 MHz. Determine the requirements for the Q of the antenna.

**theory**   Tuned circuits have an optimum frequency of operation, and they roll off away from this optimum as frequencies differ, either higher or lower. (See Fig. 20-16$a$). The responses for several values of $Q$ are shown.

As with tuned circuits, an antenna is said to have a resonant frequency (frequency for which its length was cut), a bandwidth (BW), and a $Q$.

$$\text{BW} = \frac{f_o}{Q}$$

(20-3)

In addition to appearing as a resistance, the antenna impedance also has a reactive component. As with lumped elements, the reactive portions of the impedance are the result of currents and voltages being out of phase in the power-consuming device, in this case the antenna. The reason that currents and voltages could be out of phase is that the antenna is not cut to the exact length called for by the type of antenna being considered. Actually an antenna is very much like a tuned circuit in that at its center frequency, that frequency at which its geometry is exactly correct, a maximum impedance which is purely resistive (radiation resistance plus ohmic resistance) is presented to the transmission line. As the frequency of the signal being presented to the antenna is changed, it becomes either higher or lower than the design frequency, and the impedance presented to the transmission line becomes reactive, just as with a tuned circuit. As a matter of fact, we speak of the bandwidth and $Q$ of an antenna as we do with tuned circuits, and these terms still maintain their original meaning. The relationship between the bandwidth and the $Q$ of an antenna is the same as that for tuned circuits.

$$\text{Bandwidth} = \frac{f_o}{Q} \qquad (20\text{-}3)$$

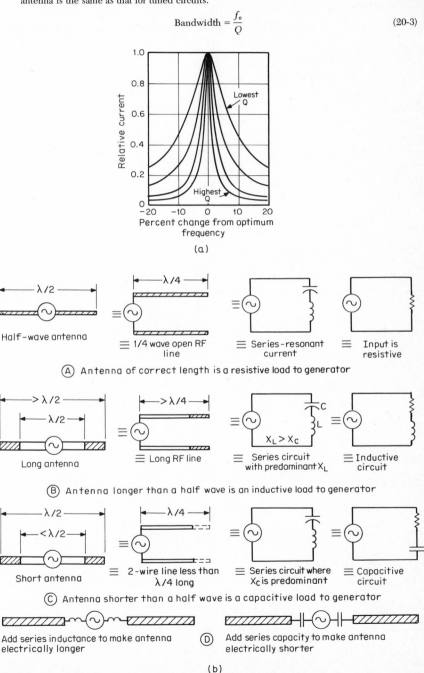

**Fig. 20-16**  (*a*) Optimum frequency response and roll off away from optimum design frequency, for several values of $Q$. (*b*) Effect of length on antenna impedance.

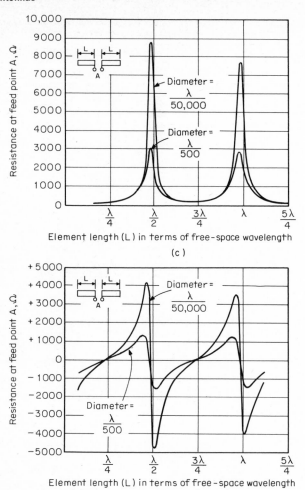

**Fig. 20-16 (cont.)**  $(c, d)$ Impedance curves for a center-fed antenna.

If for any reason the resonant frequency of an antenna is not exactly at the frequency to be transmitted, the antenna can be tuned. This can be done by using antenna tuners; these are for the most part tunable reactive circuits which are tuned so that when they are combined with the antenna, the reactive components of the antenna and the tuning circuit cancel out. Capacitive reactance can counterbalance inductive reactance, and inductive reactance can counterbalance capacitive reactance.

**solution**    Using Eq. (20-3) gives:

$$BW = \frac{f_o}{Q}$$

$$Q = \frac{f_o}{BW}$$

$$= \frac{75 \times 10^6}{6 \times 10^6}$$

$$= 12.5$$

This represents the maximum permissible $Q$ of the antenna since, according to Eq. (20-3), higher values of $Q$ appearing in the denominator of the equation would cause a reduction in bandwidth. A reduction in bandwidth would cause a loss of information since the total signal being transmitted would not pass from the antenna:

$$Q < 12.5$$

The effect of length on antenna impedance is illustrated graphically in Fig. 20-16$b$. Figure 20-16$c$ and $d$ shows the impedance curves for a center-fed antenna in terms of resistance and reactance.

## 20.8 CHOOSING THE BEAM WIDTH OF AN ANTENNA

**problem 20.7** A rotatable antenna is desired for point-to-point communication. Two antennas are available. The radiation patterns for both are shown in Fig. 20-17. Which of these antennas will be most effective for this service with regard to beam width?

**theory** The beam width of a directive antenna is the width, in degrees, of the major lobe between the two directions at which the relative radiated power is equal to one-half its value at the peak of the lobe. At these "half-power points" the field intensity is equal to 0.707 times its maximum value, or down 3 db from maximum.

**solution** On the basis of the above discussion, antenna A has a beam width of 40°, while antenna B has a beam width of 17°.

Antenna B, having the smaller beam width, would prove to be more desirable since we are interested in point-to-point communications using rotatable antennas. Using the antenna with the narrower beam width allows the same amount of energy to be concentrated into a narrower, more intense beam, allowing for a more penetrating and thus effective signal.

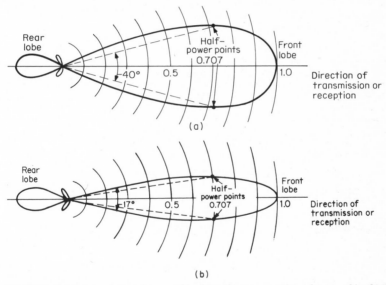

**Fig. 20-17** (*a*) Antenna A has a beam width of 40° (*b*) Antenna B has a beam width of 17°

## 20.9 CHOOSING AN ANTENNA ON THE BASIS OF FRONT-TO-BACK RATIO

**problem 20.8** A choice is to be made as to which of two antennas to purchase for use in a base station for a ship-to-shore communication system. One antenna has a front-to-back ratio of 3 dB while the second has a front-to-back ratio of 10 dB.

**theory** Front-to-back ratio is the ratio of power radiated in the maximum direction to power radiated in the opposite direction. Front-to-back ratio is usually expressed in decibels.

An antenna which is to serve as the base station in ship-to-shore service must usually radiate minimally in the direction of land and radiate most efficiently in the direction of water. See Fig. 20-18.

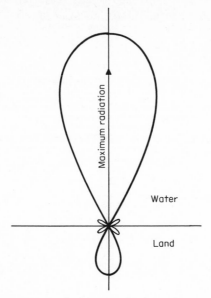

**Fig. 20-18**   Desired radiation pattern for ship-to-shore service.

**solution**   The antenna with a front-to-back ratio of 10 db is preferred over the antenna which has a front-to-back ratio of 3 dB because of the greater directional properties and, therefore, less waste of energy being radiated from the coast, inland. The antenna with the greater front-to-back ratio serves to focus the energy in the direction it is needed, out to sea.

## 20.10   CHOOSING AN APPROPRIATE TRANSMISSION LINE FOR AN ANTENNA

**problem 20.9**   A 50-Ω Marconi antenna is to feed a transmission line. Choose an appropriate transmission line* for this situation. The environment through which the transmission line passes is very noisy electrically.

**theory**   The primary concern when choosing an antenna is that the characteristic impedance of the transmission line be equal to the radiation resistance of the antenna. This is to minimize the development of standing waves on the line and the concomitant loss of power on the transmission line.

Other considerations include whether the transmission line should be a balanced or unbalanced line and whether shielding is necessary. Shielding of the transmission line is significant primarily when dealing with a receiving system in which reception is marginal and which is located in an electrically noisy environment.

The use of shielded cable also becomes highly important when the transmission line for a transmitter installation passes in close proximity to delicate instrumentation which can be affected by the radiation coming off the cable.

**solution**   A transmission line having a characteristic impedance of approximately 50 Ω will satisfy the characteristic impedance requirements for the desired transmission line. Owing to the noise conditions described in the problem, a shielded cable is desired.

Since connection to a Marconi antenna requires one leg to feed the antenna element and the other to be grounded, an unbalanced line is desired. A 50-Ω coaxial cable will thus fill the bill. RG 8A/U is a 53-Ω coaxial cable meeting all the above requirements. (See Table 18-2.)

*See the Appendix at the end of Chap. 18 for tables of transmission line characteristics.

## 20.11   DETERMINING RADIATION RESISTANCE AND DUMMY LOAD REQUIREMENTS

**problem 20.10**   An antenna which is matched to its transmission line radiates 750 W when the antenna feed carries 3.5 A. Determine the radiation resistance of the antenna and specify the necessary requirements for a dummy load to be used in lieu of the antenna when adjusting the transmitter.

**theory**   Radiation resistance is an intangible resistance that is the measure of the actual radiation of energy into space. It is defined as the ratio of the total power radiated to the square of the effective value of the maximum current in the radiating system. For a half-wave antenna, this resistance is approximately 73 Ω and equals the input resistance of the center of the antenna. In general, however, the input resistance and the radiation resistance are not equal.

Radiation resistance can be found from:

$$R_{rad} = \frac{P}{I^2} \tag{20-4}$$

When a transmitter is adjusted, the antenna is disconnected so that radiations do not occur. If nothing is used to replace the antenna, a mismatch would result, and the output stage of the transmitter would be damaged. In place of the antenna, a resistor is used having a resistance equal to the radiation resistance of the antenna and having a power-handling capability to allow it safely to dissipate the power that would be taken by the antenna.

**solution**   By Eq. (20-4):

$$R_{rad} = \frac{P}{I^2}$$
$$= \frac{750}{(3.5)^2}$$
$$= 61.22 \ \Omega$$

Specifications of dummy load:

$$R = 61.22 \ \Omega$$
$$P_{rated} > 750 \ W$$

*Note:* In practice, the power rating is usually doubled, for safe and reliable operation of the dummy load. In this case a power rating of about 1500 W should be used.

Figure 20-19a shows how the radiation resistance varies with antenna length for an antenna in free space. For a half-wave antenna the radiation resistance is approximately 73.2 Ω, measured at the current maximum which is at the center of the antenna. For a quarter-wave antenna measured at its current maximum, the radiation resistance is approximately 36.6 Ω. The radiation resistance is also affected somewhat by the height of the antenna above ground and by its proximity.

Figure 20-19b shows the variation of radiation resistance vs. antenna height, for a vertical antenna over a perfectly conducting ground.

## 20.12   CALCULATING THE GAIN OF AN ANTENNA

**problem 20.11**   A receiving antenna intercepts a 25-μV signal. The directional properties of the transmitting antenna are changed by the broadcaster, causing the receiving antenna under the same conditions to intercept a 50-μV signal. Determine the gain of the second transmitting antenna over the first.

**theory**   Antenna gain is a comparison of the output in a particular direction of the antenna in question and a reference antenna. Unless otherwise specified, the reference antenna is either an omnidirectional antenna (radiates equally in all directions) or a dipole. The increased power being radiated in a particular direction is obtained at the expense of the other directions. The antenna *does not* produce additional power.

**solution**   In this problem two transmission antennas are being compared with each other. This eliminates the necessity of specifying a reference antenna.

$$A_{dB} = 20 \log \frac{V2}{V1}$$
$$= 20 \log \frac{50}{25}$$
$$= 20 \log 2 \tag{20-5}$$
$$= 20(0.3)$$
$$= 6 \ dB$$

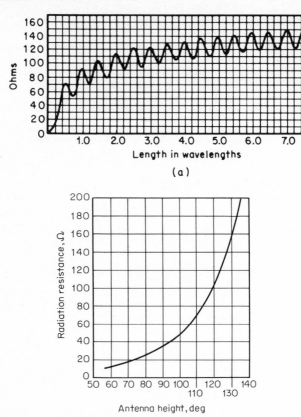

**Fig. 20-19** (*a*) Radiation resistance of antennas measured at current maxima for various lengths. (*b*) Radiation resistance vs. antenna height in degrees, for a vertical antenna over perfectly conducting ground or over a highly conducting ground plane. (*Courtesy A.R.R.L.*)

Thus the antenna gain of the second antenna over the first is 6 dB. Figure 20-20 is a chart of antenna voltage gain or loss in terms of decibels vs. power or voltage.

## 20.13   FINDING THE EFFECTIVE RADIATED POWER OF AN ANTENNA

**problem 20.12**   Determine the effective radiated power of an antenna which receives 1 kW and has an antenna power gain of 3 dB in the direction in question.

**theory**   Effective radiated power (ERP) is the product of antenna input power and antenna power gain.

**solution**

$$A_{dB} = 10 \log \frac{P_2}{P_1}$$

$$3 = 10 \log \frac{\text{ERP}}{1000}$$

$$2 = \frac{\text{ERP}}{1000} \tag{20-6}$$

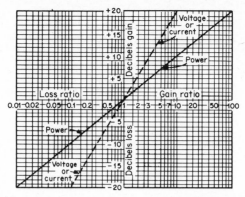

**Fig. 20-20** Chart of decibels vs. power or voltage gain or loss. When the voltage curve is used, the voltages must be measured across identical impedances. The range of the chart can be extended by adding (or subtracting if a loss) 10 dB each time the power ratio is multiplied or divided by 10, or 20 dB each time the voltage is multiplied or divided by 10. *(Courtesy A.R.R.L.)*

$$\text{ERP} = 2000 \text{ W}$$
$$= 2 \text{ kW in the direction in question}$$

*Note:* Three decibels equals a power gain of 2 (approximately). Thus $2 \times 1000 = 2000$ W (approximately).

## 20.14 DETERMINE THE RADIATION RESISTANCE OF A LONG-WIRE ANTENNA

**problem 20.13**   Calculate the radiation resistance of a resonant long-wire antenna whose length is six times the wavelength of the signal handled.

**theory**   A long-wire antenna is one made up of a length of wire or wires whose length is long as compared with a half-wavelength of the signal being received or transmitted. Long-wire antennas are classified as either resonant or nonresonant. A resonant antenna is one whose length is an even multiple of a half-wavelength of the signal being handled.

The radiation resistance of a resonant long-wire antenna can be determined from:

$$R_{\text{rad}} = 73 + 69 \log n \tag{20-7}$$

where $n$ is the number of half wavelengths of the antenna.

**solution**   Using the equation for the radiation resistance of a resonant long-wire antenna yields:

$$R_{\text{rad}} = 73 + 69 \log n \tag{20-7}$$

$n = 6 \times 2 = 12$ half-wavelengths
Substituting yields:

$$R_{\text{rad}} = 73 + 69 \log 12$$
$$= 73 + 69(1.079)$$
$$= 147.45 \ \Omega$$

## 20.15 LOG-PERIODIC ANTENNAS

A class of antenna has been developed known as *log-periodic antennas.* These antennas have very wide band characteristics and are therefore also referred to as *frequency independent antennas.* This class of antenna consists of a number of different types and configurations. A few of the antenna types in the log-periodic or frequency-independent class are shown in Fig. 20-21. The simplest antenna of this class to describe is the log-periodic dipole array shown in Fig. 20-22*a.*

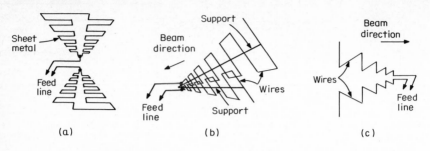

(a)          (b)          (c)

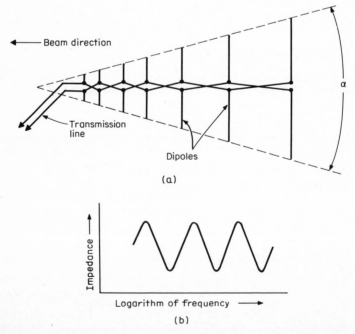

(d)          (e)

**Fig. 20-21**    Five examples of log-periodic antennas.

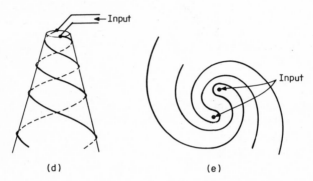

**Fig. 20-22**    (a) The log-periodic dipole array. (b) The periodic nature of antenna impedance when plotted on a logarithmic scale.

The term *log periodic* does not, as is sometimes mistakenly supposed, refer to the relative lengths of the elements or the relationship between their distances. Instead, it refers to their electrical characteristics. If a graph were made showing the antenna impedance against the logarithm of the frequency, it would be a curve similar to the one shown in Fig. 20-22*b*. Note that the maximum value of impedance on this curve repeats itself *periodically* on the *logarithmic scale,* and hence the name *log periodic.* This repetition is true not only for the impedance of these antennas but also for other electrical characteristics.

An important feature of log-periodic antennas is the *geometric* relationship between the spacing of the elements and the length of the elements. Characteristically, these spacings and lengths vary in a geometric progression which results in the antenna getting larger and larger as the distance from a theoretical point—called the *apex*—increases. In a similar manner, the length of the elements also increases, as can be demonstrated by the log-periodic antenna shown in Fig. 20-22*a*. This particular type of log periodic is sometimes referred to as a *dipole array* log-periodic antenna because it consists of a number of dipoles which vary in distances and lengths. The antenna is designed in such a way that the distances between the dipoles and the length of the dipoles are in a constant ratio to each other. The antenna is said to have a scale factor $\tau$ such that (because of the fixed ratio of growth) the tips of the antenna elements determine straignt lines. The two straight lines thus determined define an angle $\propto$, as shown in Fig. 20-22*a*.

When a station broadcasts a signal, and the log-periodic antenna is pointed in the direction necessary to receive this signal, it can be shown that only one or two of the dipole elements in the antenna will react to that frequency. All the other elements are inactive at that particular frequency; however, at some other frequency, they will become active. In other words, for any particular frequency being received, only one or two of the log-periodic elements are considered active.

## 20.16  PARABOLIC ANTENNAS

When a multielement broadside array is excited, the E field which exists in front of the antenna will be in a single plane as shown in Fig. 20-23*a* rather than in an arc as in the case of a single half-wave element. The larger the dimensions of this plane in terms of wavelengths, the greater is the directivity of the antenna system and the narrower the radiated beam.

Although a multielement broadside array gives good results, it is quite complicated in structure. In it every element must be driven, and all spacings and dimensions must be quite exact. A much simpler device for producing an electric field in a single plane is the parabolic reflector. As you can see in Fig. 20-23*b*, the parabola has its focal point at *F*. If a single antenna is placed at *F* and caused to radiate a field, the electric field will leave the antenna in all directions at the same rate in the form of an arc as indicated at point *A*. As each part of the wavefront reaches the reflecting surface, it is shifted 180° in phase and sent outward at an angle of reflection that is equal to the angle of incidence. All parts of the field will arrive at line *BB'* at the same time after reflection because all paths from *F*, to the reflector, to line *BB'* are equal in length. Thus you see that with only one antenna and a specially shaped reflector, it is possible to produce a large electric field in a single plane. Or, looking at it another way, all parts of the field travel in parallel paths after reflection from the parabola in such a way that the rays are focused as are the headlight beams from an automobile.

Like the broadside array, high directivity is not obtained until the diameter of the parabolic reflector is made many wavelengths long. This prohibits the use of the parabolic reflector at low frequencies, but for 3- and 10-cm radar equipment they are very practical.

In the illustration of Fig. 20-23*c* showing the exciting antenna for a parabolic reflector, the half-wave dipole is mounted a quarter-wave back from the short on the coaxial line. To sharpen the focal point, the antenna is physically less than a half-wave long. However, the balls at the end make it electrically a half-wave long. This broadens the band of frequencies it will handle. The airtight cylinder in which the antenna is enclosed permits the coaxial line to be pressurized. The inner surface of half the cylinder, that is, the side away from the parabolic reflector, is coated with a reflecting foil. This reflecting surface directs energy from that side of the antenna into the large reflector. Without this reflector half of the antenna radiation would be nondirectional.

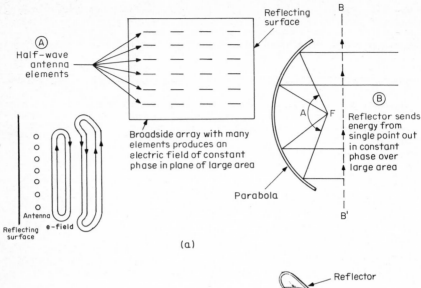

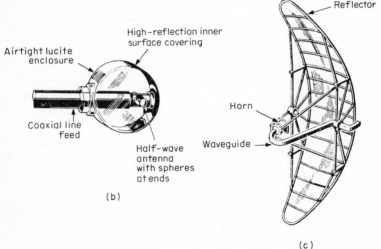

**Fig. 20-23** (*a, b*) Development of parabolic reflector. (*c*) Dipole for exciting parabolic reflector in 10-cm system. (*d*) Orange-peel parabola with waveguide feed.

The natural directivity of a dipole causes the pattern from a parabolic reflector to be somewhat sharper in the plane containing the dipole than in the other plane. For this reason, the dipole is erected horizontally for maximum azimuth accuracy in radar systems. If vertical accuracy is of primary importance, the dipole is mounted vertically.

Other methods of preventing the direct forward radiation use a parasitic reflector with the half-wave dipole and a disk which is placed in front of the dipole.

When a waveguide is used in the rf system, it is possible to send the energy into the parabolic reflector with a horn radiator. In this case an orange-peel reflector, or section of a complete circular paraboloid as shown in Fig. 20-23*d*, may be used as the reflector. The feed is by a waveguide and a horn-type radiator. The horn just about covers the shape of the reflector and prevents very little rf energy from escaping at the sides. This arrangement is highly directive in the vertical plane. It is used principally to determine the altitude of airplanes.

One form of parabolic reflector is known as the *paraboloid of revolution* or *rotational parabola*. This is the surface generated by the revolution of a parabola about its axis and somewhat resembles an eggshell cut in half. Figure 20-24a shows a cross-section view of a rotational parabola which is excited by a vertical antenna located at the focal point inside the parabola. A hemispherical shield is used to direct all the radiation back toward the parabolic surface. By this means direct radiation is eliminated, the beam is made sharper, and power is saved. Without the shield some of the power would leave the radiator directly. Since it would not be reflected, it would not become a part of the main beam and thus could serve no useful purpose.

Another method of accomplishing the same result is through the use of a parasitic array to direct the power back to the reflector. The radiation pattern of a rotational parabola contains a major lobe, which is directed along the axis of revolution, and several minor lobes (Fig. 20-24b). Very narrow beams are possible with this type of reflector.

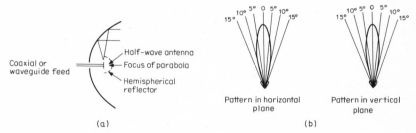

**Fig. 20-24**   (a) Paraboloid or rotational parabola-cross-section view. (b) Typical directional pattern obtainable with rotational parabola antenna of Fig. 20-24a.

Another form of parabolic reflector is the cylindrical parabola with open or closed ends (Fig. 20-25). Cylindrical parabolic reflectors have a parabolic curvature in one plane, usually the horizontal plane, and no curvature in any plane perpendicular to this horizontal plane. This type of parabola normally is excited by an antenna placed parallel to the cylindrical surface and located at the axis of the parabola. The parabola should be so designed that the focus lies well within its mouth, in order that most of the radiated energy will be intercepted by the reflecting surfaces.

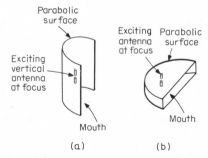

**Fig. 20-25**   Cylindrical parabolas with open and closed ends.

## 20.17  BIBLIOGRAPHY

Adam, S. F.: *Microwave Theory and Applications,* Prentice Hall, Englewood Cliffs, N.J., 1969.
Belden Corp. staff: *Electronic Cable Handbook,* Sams, Indianapolis, Ind., 1966.
Blake, L. V.: *Antennas,* Wiley, New York, 1966.
DeFrance, J. J.: *Communications Electronics Circuits,* Holt, New York, 1966.
DeMaw, Doug (ed.): *The Radio Amateur's Handbook,* American Radio Relay League, Hartford, Conn. 1976.

Department of the Air Force, *Antenna Systems*, AF Manual 52-19, 1953.

Kaufman, Milton: *Radio Operator's License Q & A Manual*, Hayden, New York, 1975.

Langford and Smith: *Radiotron Designer's Handbook*, Radio Corporation of America, Harrison, N.J., 1953.

McKenzie, Alexander: *Radiotelephone Examination Key and Answers*, McGraw-Hill, New York, 1972.

Schure, Alexander: *R-F Transmission Lines*, Rider, New York, 1956.

Shrader, R. L.: *Electronic Communication*, McGraw-Hill, New York, 1967.

Stewart, J. L.: *Circuit Analysis of Transmission Lines*, Wiley, New York, 1958.

Temes, Lloyd: *Communication Electronics for Technicians*, McGraw-Hill, New York, 1974.

Westman, H. P. (ed.): *Reference Data for Radio Engineers*, Sams, Indianapolis, 1970.

# Chapter 21

# Microwaves

## NICK KUHN

**Senior Applications Engineer, Hewlett-Packard Co., Palo Alto, Calif.**

## INTRODUCTION

The microwave frequency range includes frequencies so high that circuits and interconnections become comparable in size with a wavelength. The low-frequency boundary is sometimes given as 1 GHz, although microwave techniques are commonly used down to 100 MHz and sometimes even lower. On the upper-frequency end, the microwave frequency range goes up to the infrared frequencies at about 400 GHz. It is estimated that 99% of the microwave devices operate below 100 GHz, 90% below 40 GHz, and 75% below 12.4 GHz.

At microwave frequencies, the velocity of propagation and time delay greatly affect most designs and observations. The concept of reflection largely replaces the concept of impedance. Consequently many of the calculations discussed in this chapter are concerned with the velocity of propagation and the interpretation of reflection coefficient in terms of impedance and vice versa.

The calculations examined are those most likely to be required of a technician who is working at microwave frequencies. The subjects discussed are microwave transmission lines, simple matching techniques for reducing reflections, the concept of power flow, and those basic measurement principles and techniques that require the processing of data at the technician level. Measurement techniques in general are not discussed.

Problems and review theory are presented with the goal of understanding the physical principles that underlie the calculation. Thus the logic behind the calculation is stressed as well as the actual calculation.

In a section of this size, only a small fraction of the possible calculations are discussed. Those that are included in the design of equipment, but not in its use, are not covered here. For this reason cavity wavemeters and filters are not considered. With the exception of simple impedance matching, design calculations are not discussed because they are normally performed by a microwave specialist at a much more advanced level than a technician or bachelor's degree in engineering. Many microwave techniques involve calculations which are so difficult or so approximate that, in practice, the problem is solved by trial and error. Calculations for such problems are also not discussed here.

## 21.1 CALCULATING THE WAVELENGTH AND VELOCITY OF A TRAVELING WAVE

**problem 21.1** At a frequency 10 GHz find the wavelength and velocity of a wave traveling through snow.

**theory** Frequency $f$, velocity $v$, and wavelength $\lambda$ are related by

$$f = \frac{v}{\lambda} \tag{21-1}$$

The velocity of a wave in free space $c$ is

$$c = 2.9979 \times 10^8 \text{ m/s} \tag{21-2}$$

The velocity in other media is related to that in free space by

$$v = \frac{c}{\sqrt{\epsilon_r}} \tag{21-3}$$

where $\epsilon_r$ is the relative dielectric constant, also called the *relative permittivity*, of the medium. The relative dielectric constants for some common materials are given in Table 21-1.

**TABLE 21-1 Relative Dielectric Constant $\epsilon_r$, or Relative Permittivity, of Some Common Dielectric Materials**

| Material | $\epsilon_R$ | Material | $\epsilon_R$ |
|---|---|---|---|
| Air | 1.0006 | Polystyrene | 2.53 |
| Alcohol, ethyl | 25 | Porcelain (dry process) | 6 |
| Aluminum oxide | 8.8 | Pyrex glass | 5 |
| Amber | 2.8 | Pyranol | 4.4 |
| Asbestos fiber | 4.8 | Quartz (fused) | 3.8 |
| Bakelite | 4.75 | Rubber | 2.5–3 |
| Barium titanate | 1200 | Silica (fused) | 3.8 |
| Carbon dioxide | 1.001 | Sodium chloride | 5.9 |
| Carbon tetrachloride | 2.2 | Snow | 3.3 |
| Glass | 4–7 | Soil (dry) | 2.8 |
| Glycerine | 40 | Steatite | 5.8 |
| Ice | 4.2 | Styrofoam | 1.03 |
| Mica | 5.4 | Teflon | 2.03 |
| Neoprene | 6.7 | Titanium dioxide | 100 |
| Nylon | 4 | Turpentine | 2.2 |
| Paper | 2–4 | Water (distilled) | 80 |
| Plexiglass | 3.45 | Wood (dry) | 1.5–4 |
| Polyethylene | 2.26 | | |
| Polypropylene | 2.25 | | |

SOURCE: W. H. Hayt, *Engineering Electromagnetics*, McGraw-Hill, New York, 1974.

**solution** From Table 21-1, $\epsilon_r$ is seen to be 3.3. Substituting this value into Eq. (21-3) and using $c = 2.9979 \times 10^8$ m/s yields:

$$v = 2.9979 \times \frac{10^8}{\sqrt{3.3}} = 1.65 \times 10^8 \text{ m/s}$$

## 21.2 CALCULATING THE PROPERTIES AND DIMENSIONS OF COAXIAL TRANSMISSION LINE

**problem 21.2** A coaxial line of 50-$\Omega$ characteristic impedance is to be built with a center conductor diameter of 3 mm and with Teflon dielectric. Find the inside diameter of the outer conductor, the maximum frequency of reliable transmission, the velocity of propagation, and the time delay in nanoseconds per meter (ns/m).

**theory** The characteristic impedance of a coaxial line is the ratio of voltage to current for a single wave traveling in one direction on the transmission line. For coaxial line the characteristic impedance is given by

$$Z_o = \frac{138}{\sqrt{\epsilon_r}} \log \frac{d_o}{d_i} \tag{21-4}$$

where $d_o$ and $d_i$ are the diameters of the outer and inner conductors, respectively.

The velocity of propagation is calculated in the same way as in Problem 21.1.

$$v = \frac{c}{\sqrt{\epsilon_r}}$$ (21-5)

The frequency of reliable transmission is limited by the lowest frequency at which non-TEM or waveguide type propagation can take place. This occurs when the average circumference is roughly one wavelength or

$$\lambda = \frac{v}{f} = \left(\frac{d_o + d_i}{2}\right)\pi$$ (21-6)

**solution**   The dielectric constant of Teflon is 2.03 (from Table 21-1). Equation (21-4) must be solved for $d_o$. This is done by first solving for log $(d_o /d_i)$ and then raising each side to a power of 10 giving

$$10^{\log (d_o/d_i)} = \frac{d_o}{d_i} = 10^{Z_o(\epsilon_r)^{1/2}/138} = 3.28$$ (21-7)

Since $d_i$ is given as 3 mm, $d_o$ is 3(3.28) or 9.84 mm. The velocity of propagation is found from Eq. (21-5):

$$v = \frac{c}{\sqrt{\epsilon_r}} = \frac{2.9979 \times 10^8}{\sqrt{2.03}} = 2.104 \times 10^8 \text{ m/s}$$

The time delay for each meter of transmission line is equal to $1/v$ or 4.75 ns/m.

The lowest frequency for non-TEM propagation is found from Eq. (21-6), after converting the measurements of diameters to meters from millimeters:

$$f = \frac{2v}{\pi} \frac{1}{(d_o + d_i)}$$

$$= \frac{2(2.104 \times 10^8)}{\pi(0.003 + 0.00984)} = 10.43 \text{ GHz}$$

## 21.3   CALCULATING THE PROPERTIES OF MICROSTRIP TRANSMISSION LINE

**problem 21.3**   A 50-$\Omega$ transmission line is to be constructed on glass-epoxy (type G-10) printed circuit board that is 0.060 in. thick. The line is to be one-quarter wavelength long at 3 GHz. What are the dimensions of the transmission line?

**theory**   The configuration of transmission line on one side of a printed circuit board, with a ground plane on the other side, is known as *microstrip* and is shown in Fig. 21-1. Propagation takes place partially in the dielectric region and partially in the free space above the dielectric. Both the characteristic impedance and the wave velocity fall between the values that would be proper for all free space or all dielectric. The relationship between the line dimensions, actually width to height ($w/h$), and the transmission-line characteristic impedance $Z_o$ are given in the graph of Fig. 21-1$b$. Also shown is the relationship between $w/h$ and the wave velocity relative to that of free space. What is not indicated in Fig. 21-1$b$ is that $Z_o$ and $v_r$ also depend somewhat on frequency and the presence of other conductors. Figure 21-1$b$ should be accurate enough for most situations where microstrip is used.

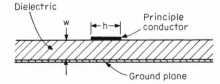

**Fig. 21-1**   ($a$) Cross section of microstrip transmission line.

**solution**   From Fig. 21-1 a 50-$\Omega$ line on G-10 dielectric ($\epsilon_r = 4.55$) requires a width-to-height ratio $w/h$ of 1.88. For a dielectric thickness of 0.060 in, the width should be

$$w = \left(\frac{w}{h}\right) h = 1.88(0.060) = 0.113 \text{ in}$$ (21-8)

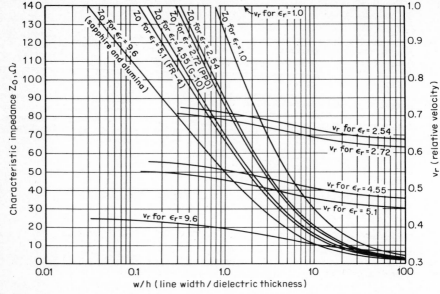

**Fig. 21-1 (cont.)** (*b*) Characteristic impedance $z_0$ and relative phase velocity $V_r$ versus transmission line width to thickness ratio (*w/h*) for microstrip. (*From equations given in I. J. Bahl and D. K. Trivedi, "A Designer's Guide to Microstrip Line," Microwaves, pp. 174–182, May 1977.*)

Once *w/h* is set to 1.88, the relative velocity is found from Fig. 21-1*b* to be 0.54. The actual velocity is, therefore,

$$v = v_r c = 0.54(3 \times 10^8) = 1.62 \times 10^8 \text{ m/s} \qquad (21\text{-}9)$$

The relationship between wavelength, frequency, and velocity, already covered in Eq. (21-1), is

$$\lambda = \frac{v}{f} = \frac{1.62 \times 10^8}{3 \times 10^9} = 0.054 \text{ m}$$

Since this transmission line is to be one-quarter wavelength long,

$$l = \frac{\lambda}{4} = \frac{0.054}{4} = 0.0135 \text{ m}$$
$$= 1.35 \text{ cm} = 0.532 \text{ in}$$

## 21.4    CALCULATING BANDWIDTH LIMITATIONS, WAVELENGTH, AND VELOCITY OF PROPAGATION IN RECTANGULAR WAVEGUIDE

**problem 21.4**    For a rectangular waveguide of inside dimensions 0.9 and 0.4 in and of free-space dielectric, calculate the cutoff frequency of the dominant mode and the next highest mode. For a frequency of 10 GHz, calculate the guide wavelength, phase velocity, and group velocity.

**theory**    Propagation in a rectangular waveguide may be compared with propagation of light down a hallway of mirrors. The wave is guided down the dielectric (usually air) by reflections off the walls. Propagation is possible in various modes. Each mode corresponds to a different zigzag pattern of propagation. Each mode acts like a high-pass filter; it can propagate only at frequencies above the cutoff frequency $f_c$ given by

$$f_c = \frac{v}{2} \sqrt{\left(\frac{m}{a}\right)^2 + \left(\frac{n}{b}\right)^2} \qquad (21\text{-}10)$$

where $v$ is the velocity of light in the dielectric ($v = c/\sqrt{\epsilon_r}$), $m$ and $n$ are integers that give the mode number, and $a$ and $b$ are the cross-sectional dimensions of the waveguide. Either $m$ or $n$ can be zero but not both.

If $a$ is the larger cross-sectional dimension, the lowest possible cutoff frequency occurs when $m$ is 1 and $n$ is zero. This mode, called the $TE_{10}$ mode, therefore has a cutoff frequency given by

$$f_c\bigg|_{TE_{10}} = \frac{v}{2a} \tag{21-11}$$

Because this mode has a frequency range where no other modes can propagate, it is called the *principal mode* or *dominant mode*. Depending on the relative waveguide dimension $b/a$, the next highest cutoff frequency occurs either for $m = 0$ and $n = 1$ or for $m = 2$ and $n = 0$.

The wavelength in waveguide is said to be longer than in free space. This is because of the zigzag nature of propagation. In Fig. 21-2, the distance between wave crests along the center of

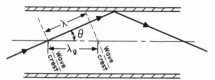

**Fig. 21-2**   Propagation down rectangular waveguide shown as reflection off the waveguide walls.

waveguide is given by $\lambda_g$. The distance between crests along a ray of the wave is still $\lambda$. From Fig. 21-2 the angle of zigzag $\theta$ is

$$\sin\theta = \frac{\lambda}{\lambda_g} \tag{21-12}$$

It is easily seen that $\lambda_g$ is larger than $\lambda$. The guide wavelength $\lambda_g$ is related to the cutoff frequency and the free-space wavelength $\lambda$ by

$$\lambda_g = \frac{\lambda}{\sqrt{1 - (f_c/f)^2}} \tag{21-13}$$

If the guide wavelength and frequency are used to calculate velocity from Eq. (21-1), the result is called the phase velocity $v_p$:

$$v_p = \lambda_g f \tag{21-14}$$

Although the phase velocity is larger than the velocity of light, it must be remembered that $v_p$ is only calculated; no matter or energy is actually moving that fast. The velocity of a single wave crest down the waveguide is called the *group velocity* and can be found from Fig. 21-2 by considering the component of the velocity of the ray that aims down the waveguide:

$$v_g = v\sin\theta \tag{21-15}$$

This velocity is less than the velocity of light in free space.

**solution**    The cutoff frequency is given by Eq. (21-11), but first $a$ and $v$ must be in compatible units.

$$a = 0.9(0.0254) = 0.02286 \text{ m}$$

The cutoff frequency is, therefore,

$$f_c\bigg|_{TE_{10}} = \frac{3 \times 10^8}{2(0.02286)} = 6.562 \text{ GHz}$$

The mode corresponding to $m = 2$ and $n = 0$ has a cutoff frequency, calculated from Eq. (21-10), of

$$f_c\bigg|_{TE_{20}} = \frac{(3 \times 10^8)2}{2(0.02286)} = 13.12 \text{ GHz}$$

The mode corresponding to $m = 0$ and $n = 1$ has a cutoff frequency, calculated from Eq. (21-10), of

$$f_c = \left(\frac{3 \times 10^8}{2}\right)\frac{1}{0.4(0.0254)} = 14.76 \text{ GHz}$$

Thus the next highest cutoff frequency is 13.12 GHz.

Before the guide wavelength $\lambda_g$ is calculated from Eq. (21-13), the free-space wavelength is needed from Eq. (21-1).

$$\lambda = \frac{3 \times 10^8}{10 \times 10^9} = 0.03 \text{ m} = 3 \text{ cm}$$

Then from Eq. (21-13) the guide wavelength is

$$\lambda_g = \frac{0.03}{\sqrt{1 - (6.562/10)^2}} = \frac{0.03}{\sqrt{1 - 0.4306}} = \frac{0.03}{0.7546}$$
$$= 0.0398 \text{ m} = 3.98 \text{ cm}$$

From Eq. (21-14), the phase velocity is

$$v_p = 0.0398 \times 10^{10} = 3.98 \times 10^8 \text{ m/s}$$

Before calculating the group velocity, $\sin \theta$ is calculated from Eq. (21-12):

$$\sin \theta = \frac{\lambda}{\lambda_g} = \frac{0.03}{0.0398} = 0.7538$$

The group velocity can now be calculated from Eq. (21-15):

$$v_g = 3 \times 10^8 (0.7538) = 2.26 \times 10^8$$

## 21.5  CALCULATING BANDWIDTH, WAVELENGTH, AND VELOCITY OF PROPAGATION IN CYLINDRICAL WAVEGUIDE

**problem 21.5**  Find the radius of a cylindrical waveguide which has only dominant-mode propagation below 12.5 GHz. What is the phase velocity $v_p$ and guide wavelength at a frequency of 10 GHz?

**theory**  For a cylindrical waveguide the concept of zigzag propagation is too confusing to be of much value. Still there are various modes of propagation, each corresponding to a different interference pattern and having a certain cutoff frequency. Once the cutoff frequency is determined, expressions for guide wavelength and velocity from rectangular waveguide still hold; that is, Eqs. (21-13) and (21-14) are valid. The cylindrical waveguide modes that are of greatest interest are named the $TE_{11}$, $TM_{01}$, and $TE_{01}$.

The $TE_{11}$ mode is the dominant mode and, therefore, has the lowest cutoff frequency given by

$$f_c = 0.293 \frac{v}{a} \tag{21-16}$$

where $a$ is the inside radius and $v$ is the velocity of light in the dielectric. Usually the dielectric is free space or air, and $v$ is $3 \times 10^8$ m/s.

The $TM_{01}$ mode is of interest because it has the next lowest cutoff frequency given by

$$f_c = 0.383 \frac{v}{a} \tag{21-17}$$

This mode is of interest because above this frequency, propagation can normally take place in more than one mode simultaneously. This means that there are two velocities of propagation for the same signal and, therefore, the signal becomes garbled.

The $TE_{01}$ mode is sometimes of special interest because it has very low loss. This mode, however, is not dominant, and special precautions and structures are used to prevent other modes from propagating. The cutoff frequency is given by

$$f_c = 0.609 \frac{v}{a} \tag{21-18}$$

**solution**  For only dominant-mode propagation below 12.5 GHz, the $TM_{01}$ mode should have its cutoff frequency at that frequency. Using Eq. (21-17) gives:

$$a = 0.383 \frac{v}{f_c}$$
$$= \frac{0.383(3 \times 10^8)}{12.5 \times 10^9}$$
$$= 9.19 \times 10^{-3} \text{ m} = 0.919 \text{ cm}$$

The cutoff frequency for the dominant mode ($TE_{11}$) is calculated from Eq. (21-16):

$$f_c = \frac{0.293(3 \times 10^8)}{9.19 \times 10^{-3}}$$
$$= 9.56 \text{ GHz}$$

At 10 GHz, the free-space wavelength is

$$\lambda = 3 \times 10^8 / 10^{10} = 3 \times 10^{-2} \text{ m}$$
$$= 3 \text{ cm}$$

By use of this and Eq. (21-13), the guide wavelength can be found:

$$\lambda_g = \frac{3 \times 10^{-2}}{\sqrt{1 - (9.56/10)^2}} = 0.102 \text{ m}$$
$$= 10.2 \text{ cm}$$

If this guide wavelength is used to calculate the velocity according to Eq. (21-1), the result is the phase velocity:

$$v_p = f\lambda_g = 10^{10}(0.102) = 1.02 \times 10^9 \text{ m/s}$$

As a matter of interest, the group velocity $v_g$ is related to the phase velocity in many waveguide structures, including rectangular and cylindrical waveguides by

$$v_g \cdot v_p = v^2 \qquad (21\text{-}19)$$

So in this case the group velocity at 10 GHz is

$$v_g = \frac{(3 \times 10^8)^2}{1.02 \times 10^9} = 8.82 \times 10^7 \text{ m/s}$$

## 21.6  CALCULATING THE ATTENUATION OF WAVEGUIDE BEYOND CUTOFF

**problem 21.6**  A cylindrical, waveguide-beyond-cutoff attenuator is used with coax transmission lines as shown in Fig. 21-3. If the distance between the coax transmission lines is increased from $z$ to $(z + 1)$ inches, calculate the increase in attenuation at 10, 100, 1000, and 10,000 MHz.

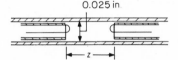

0.025 in

**Fig. 21-3**  A waveguide-beyond-cutoff attenuator.

**theory**  In waveguide beyond cutoff, the attenuation (in decibels) for a length $z$ is

$$\text{Loss} = 8.686\,(\alpha z) \qquad (21\text{-}20)$$

where $\alpha$ is the attenuation constant given by

$$\alpha = \frac{2\pi}{\lambda_c}\sqrt{1 - \left(\frac{f}{f_c}\right)^2} \qquad (21\text{-}21)$$

where $\lambda_c$ and $f_c$ are the cutoff wavelength in free space and cutoff frequency of the waveguide.

**solution**  The smaller the value of $\alpha$, the smaller the attenuation and the stronger the signal; the mode with the smallest $\alpha$ determines the signal level. From Eq. (21-21), the smallest $\alpha$ occurs for the dominant mode. For cylindrical waveguide, the dominant $TE_{11}$ mode has a cutoff frequency calculated from Eq. (21-16) after converting the 0.25-in diameter to a radius of 3.175 mm:

$$f_c = 0.293\,\frac{3 \times 10^8}{3.175 \times 10^{-3}} = 27.68 \text{ GHz}$$

The cutoff wavelength is, from Eq. (21-1),

$$\lambda_c = \frac{v}{f_c} = \frac{3 \times 10^8}{27.68 \times 10^9} = 0.0108 \text{ m}$$

The $\alpha$ for each frequency can now be calculated from Eq. (21-21). At 10 GHz:

$$\alpha = \frac{2\pi}{0.0108}\sqrt{1 - \left(\frac{10}{27.68}\right)^2} = \frac{2\pi}{0.0108}(0.934)$$
$$= 582(0.932) = 542/\text{m}$$

At 1 GHz:

$$\alpha = \frac{2\pi}{0.0108}\sqrt{1 - \left(\frac{1}{27.68}\right)^2} = 582(0.9994)$$
$$= 581.4/\text{m}$$

At frequencies of 100 MHz and lower the quantity under the square root is, for practical purposes, equal to 1. So at 100 and 10 MHz:

$$\alpha = \frac{2\pi}{0.0108} = 582/\text{m}$$

Now that $\alpha$ has been calculated, the increased attenuation for a length increase of 1 in (0.0254 m) can be calculated. At 10 GHz:

$$\text{Loss} = 8.686(0.0254)543 = 119.6 \text{ dB}$$

At 1 GHz:

$$\text{Loss} = 8.686(0.0254)581.4 = 128.3 \text{ dB}$$

At 10 MHz and 100 MHz:

$$\text{Loss} = 8.686(0.0254)582 = 128.4 \text{ dB}$$

## 21.7  CALCULATING THE RELATIVE SIZES OF WAVES ON A TRANSMISSION LINE FOR A PARTICULAR LOAD

**problem 21.7**  A wave, called the incident wave, of amplitude 10 V is propagating down a 50-$\Omega$ transmission line toward a load. For loads of resistance 0 (a short circuit), 25, 50, and $\infty$ (an open circuit) $\Omega$, calculate the voltage amplitude of the reflected wave, the current associated with each wave, and the total load current and voltage.

**theory**  The voltage $V_i$ and current $I_i$ of a wave traveling from a generator toward the load, called the incident wave, are related by

$$\frac{V_i}{I_i} = Z_o \tag{21-22}$$

Yet the voltage across the load and current through the load are related by Ohm's law:

$$\frac{V_l}{I_l} = Z_l \tag{21-23}$$

If $Z_l$ is not equal to $Z_o$, Eqs. (21-22) and (21-23) are different. Yet both equations must be satisfied. This happens by means of a reflected wave that travels from the load toward the generator. Thus there are generally two waves on a transmission line, an incident wave and a reflected wave. The voltage and current of the reflected wave are related and are the same as the incident wave:

$$\frac{V_r}{I_r} = Z_o \tag{21-24}$$

The reference directions of current flow and of voltage polarity are shown in Fig. 21-4. At any point along the transmission line, the net voltage between the two conductors, that is, the voltage that would be measured by a voltmeter, is the sum of the incident and reflected voltages at that point. This is also true at the load, where the result is

$$V_i + V_r = V_l \tag{21-25}$$

The total current is also the superposition of the currents from the incident and reflected waves.

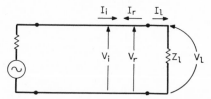

**Fig. 21-4**  The various voltages and currents on a transmission line.

But the arrowhead of $I_r$ is opposite to that of $I_i$ and $I_l$. Using Kirchhoff's current law at the load gives

$$I_i - I_r = I_l \tag{21-26}$$

Using these equations, the ratio of $V_r$ to $V_i$, called the *reflection coefficient*, is found to be

$$\frac{V_r}{V_i} = \frac{Z_l - Z_o}{Z_l + Z_o} \tag{21-27}$$

**solution**   The results of using Eqs. (21-27), (21-22), and (21-24) to (21-26) to solve are shown below. For $Z_l = 0$:

$$V_r = V_i \frac{Z_l - Z_o}{Z_l + Z_o} = 10 \frac{0 - 50}{0 + 50} = -10 \text{ V}$$

$$I_i = \frac{V_i}{Z_o} = \frac{10}{50} = 0.2 \text{ A}$$

$$I_r = \frac{V_r}{Z_o} = \frac{-10}{50} = -0.2 \text{ A}$$
$$V_l = V_i + V_r = 10 - 10 = 0 \text{ V}$$
$$I_l = I_i - I_r = 0.2 - (-0.2) = 0.4 \text{ A}$$

Note: $V_l/I_l = Z_l = 0$ as expected.
For $Z_l = 25$:

$$V_r = V_i \frac{Z_l - Z_o}{Z_l + Z_o} = 10 \frac{25 - 50}{25 + 50} = -\frac{10}{3} \text{ V}$$

$$I_i = \frac{V_i}{Z_o} = \frac{10}{50} = 0.2 \text{ A}$$

$$I_r = \frac{V_r}{Z_o} = \frac{-\dfrac{10}{3}}{50} = -\frac{1}{15} \text{ A}$$

$$V_l = V_i + V_r = 10 - \frac{10}{3} = \frac{20}{3} \text{ V}$$

$$I_l = I_i = I_r = \frac{2}{10} - \left(-\frac{1}{15}\right) = \frac{4}{15} \text{ A}$$

Note: $V_l/I_l = (20/3) \times (15/4) = 25 \ \Omega$ as expected.
For $Z_l = 50$:

$$V_r = V_i \frac{Z_l - Z_o}{Z_l + Z_o} = 10 \times \frac{50 - 50}{50 + 50} = 0 \text{ V}$$

$$I_i = \frac{V_r}{Z_o} = \frac{10}{50} = 0.2 \text{ A}$$

$$I_r = \frac{V_r}{Z_o} = \frac{0}{50} = 0 \text{ A}$$
$$V_l = V_i + V_r = 10 + 0 = 10 \text{ V}$$
$$I_l = I_i - I_r = 0.2 - 0 = 0.2 \text{ A}$$

Note: $V_l/I_l = 10/0.2 = 50 \ \Omega$ as expected.
For $Z_l = \infty$:

$$V_r = V_i \frac{Z_l - Z_o}{Z_l + Z_o} = 10 \frac{\infty - 50}{\infty + 50} = 10 \text{ V}$$

$$I_i = \frac{V_i}{Z_o} = \frac{10}{50} = 0.2 \text{ A}$$

$$I_r = \frac{V_r}{Z_o} = \frac{10}{50} = 0.2 \text{ A}$$
$$V_l = V_i + V_r = 10 + 10 = 20 \text{ V}$$
$$I_l = I_i - I_r = 0.2 - 0.2 = 0 \text{ A}$$

Note: $V_l/I_l = 20/0 = \infty$ as expected.

## 21.8   CALCULATING THE IMPEDANCE ALONG THE LENGTH OF A TRANSMISSION LINE

**problem 21.8**   A 50-$\Omega$ transmission line is terminated in a load impedance of $75-j100 \ \Omega$. A wavelength on the transmission line is 1.5 m. Find the apparent load impedance looking toward the load through 10 m of the line.

**theory**    The total voltage and total current vary from point to point along a transmission. This is because they are made of two waves, one traveling to the right and one traveling toward the left. The ratio of total voltage to total current, called impedance, also varies from point to point along the transmission line according to

$$Z = Z_o \frac{Z_l \cos{(2\pi l/\lambda)} + jZ_o \sin{(2\pi l/\lambda)}}{Z_o \cos{(2\pi l/\lambda)} + jZ_l \sin{(2\pi l/\lambda)}} \tag{21-28}$$

where $2\pi l/\lambda$ is the distance in radians from the load to the point where $Z$ is being calculated. The term $2\pi/\lambda$ occurs so frequently that it is often given the separate symbol $\beta$.

**solution**    The angle to be used in the formula is first found in radians and then converted to degrees:

$$\theta = \frac{2\pi 10}{1.5} \, \text{rad} \cdot \frac{360°}{2\pi \, \text{rad}} = 2400° \tag{21-29}$$

To find the cosine and sine of this angle, six complete revolutions, or 2160°, may be subtracted and still yield the same trigonometric functions. Therefore

$$\cos\theta = \cos 2400° = \cos{(2400° - 2160°)} = \cos 240° = -0.5$$
$$\sin\theta = \sin 240° = -0.866$$

These values are substituted into Eq. (21-28) to calculate the desired impedance:

$$Z = 50 \frac{(75 - j100)(-0.5) + j50(-0.866)}{50(-0.5) + j(75 - j100)(-0.866)}$$
$$= 50 \frac{-37.5 + j50 - j43.3}{-25 - j65 - 86.6} = 50 \frac{-37.5 + j6.7}{-111.6 - j65}$$
$$= 50 \frac{38/\underline{169.87}}{129/\underline{-149.8}} = 50(0.295)/\underline{319.7}$$
$$= 14.75/\underline{319.7} = 11.2 - j9.5 \, \Omega$$

## 21.9   CALCULATING THE LENGTH AND CHARACTERISTIC IMPEDANCE OF A TRANSMISSION LINE WHICH WILL CHANGE ONE REAL IMPEDANCE INTO ANOTHER REAL IMPEDANCE

**problem 21.9**    Find the length and characteristic impedance of an air transmission line which will make a load resistance of 85 $\Omega$ appear as 50 $\Omega$ at a frequency of 1 GHz.

**theory**    Impedances are changed according to Eq. (21-28). For the special case of a quarter-wavelength line ($\cos 2\pi l/\lambda = 0$; $\sin 2\pi l/\lambda = 1$), that equation shows how a transmission line transforms $Z_l$ into another impedance $Z$. The result is

$$Z = \frac{Z_o^2}{Z_l} \tag{21-30}$$

**solution**    At 1 GHz, wavelength is given by Eq. (21-1):

$$\lambda = \frac{v}{f} = \frac{3 \times 10^8}{10^9} = 0.3 \, \text{m} = 30 \, \text{cm}$$

The transmission line should be ¼ wavelength or

$$l = \frac{1}{4} \, 30 \, \text{cm} = 7.5 \, \text{cm}$$

To find the characteristic impedance $Z_o$, Eq. (21-30) is solved for $Z_o$:

$$Z_o = \sqrt{ZZ_l} = \sqrt{50 \cdot 85} = 65.2 \, \Omega$$

## 21.10   CALCULATING THE LENGTH AND CHARACTERISTIC IMPEDANCE OF A TRANSMISSION LINE WHICH MAKES A COMPLEX IMPEDANCE APPEAR REAL

**problem 21.10**    At a frequency of 500 MHz, a complex load impedance of $(35 - j60)$ $\Omega$ is to be transformed into a real impedance with an air transmission line. Find the length and characteristic impedance of a transmission line that accomplishes this as well as the real impedance.

**theory**    For the particular case of a one-eighth wavelength line the sines and cosines in Eq. (21-28) take on the value of 0.707 so that the equation becomes

$$Z = Z_o \frac{Z_l + jZ_o}{Z_o + jZ_l} \tag{21-31}$$

In this equation $Z_l$ can be complex ($Z_l = R_l + jX_l$). With algebra, it can be shown that if

$$Z_o = \sqrt{R_l^2 + X_l^2} \tag{21-32}$$

then the imaginary part of Z is zero. Z is, therefore, purely real and has the value

$$Z = \frac{R_l Z_o}{Z_o - X_l} \tag{21-33}$$

**solution**    The transmission line is to be $\lambda/8$ long at 500 MHz where $\lambda$ is given by Eq. (21-1):

$$\lambda = \frac{v}{f} = \frac{3 \times 10^8}{5 \times 10^8} = 0.6 \text{ m} = 60 \text{ cm}$$

Thus the line is to be 60/8 = 7.5 cm long. The $Z_o$ of the line, from Eq. (21-32), is

$$Z_o = (R_l^2 + X_l^2)^{1/2} = \sqrt{(35)^2 + (-60)^2}$$
$$= \sqrt{1225 + 3600} = \sqrt{4825} = 69.46 \ \Omega$$

According to Eq. (21-33), with this 7.5 cm length of 69.46-$\Omega$ line, the impedance of $(35 - j60) \ \Omega$ is transformed to

$$Z = \frac{35(69.46)}{69.46 + 60} = 18.8$$

## 21.11   CALCULATING REFLECTION COEFFICIENT FROM IMPEDANCE AND IMPEDANCE FROM REFLECTION COEFFICIENT

**problem 21.11**    For a transmission line of $Z_o = 50 \ \Omega$, find the reflection coefficients corresponding to impedances of 0, 50, 1000, 10, $j50$ and $(15 + j30) \ \Omega$. Also find the impedance that corresponds to a reflection coefficient of $0.5\underline{/45°}$.

**theory**    The relationship between impedance Z and the ratio of the reflected wave to the incident wave $V_r/V_i$ was found in Eq. (21-27). That ratio, called the *reflection coefficient* and symbolized by $\Gamma$, is:

$$\Gamma = \frac{V_r}{V_i} = \frac{Z - Z_o}{Z + Z_o} \tag{21-34}$$

In order to convert from reflection coefficient to impedance, Eq. (21-34) should be solved for Z:

$$Z = Z_o \frac{1 + \Gamma}{1 - \Gamma} \tag{21-35}$$

**solution**    From Eq. (21-34) we can calculate the reflection coefficient.

For $Z = 0$:

$$\Gamma = \frac{-50}{50} = -1$$

For $Z = 50$:

$$\Gamma = \frac{50 - 50}{50 + 50} = 0$$

For $Z = 1000$:

$$\Gamma = \frac{1000 - 50}{1000 + 50} = \frac{950}{1050} = 0.905$$

For $Z = 10^6$:

$$\Gamma = \frac{10^6 - 50}{10^6 + 50} = 0.99990$$

For $Z = j50$:

$$\Gamma = \frac{j50 - 50}{j50 + 50} = \frac{70.7\underline{/135°}}{70.7\underline{/45°}} = 1\underline{/90°} = 0 + j1$$

For $Z = 15 + j30$:

$$\Gamma = \frac{15 + j30 - 50}{15 + j30 + 50} = \frac{-35 + j30}{65 + j30} = \frac{46.1\underline{/139.4}}{71.6\underline{/24.8}}$$
$$= 0.64\underline{/114.6}$$

From Eq. (21-35), for $\Gamma = 0.5\underline{/45°}$:

$$Z = 50\frac{1 + 0.5\underline{/45°}}{1 - 0.5\underline{/45°}} = 50\frac{1 + 0.3536 + j0.3536}{1 - 0.3536 - j0.3536}$$
$$= 50\frac{1.3536 + j0.3536}{0.6464 - j0.3536} = 40\frac{1.4\underline{/14.6}}{0.7368\underline{/-28.7}}$$
$$= 94.9\underline{/43.3} = 69.07 + j65.12$$

## 21.12    USING A SMITH CHART TO TRANSFORM BETWEEN IMPEDANCE AND REFLECTION COEFFICIENT

**problem 21.12**    For reflection coefficients of $-1, 0, 1, 0.5\underline{/90°}$ find the corresponding impedances that terminate a 50-$\Omega$ transmission line. What are the impedances for a 75-$\Omega$ transmission line?

**theory**    The Smith chart, Fig. 21-5, is a graphical and very convenient solution to Eqs. (21-34) and (21-35) for transforming back and forth between reflection coefficient $\Gamma$ and impedance $Z$.

The center of the Smith chart (label no. 1 of Fig. 21-5) is also the origin for $\Gamma$ and, therefore, represents $\Gamma = 0$. The magnitude of $\Gamma$, expressed by the Greek letter $\rho$, is given by distance from the origin. A magnitude of 1 is the outer circle (label no. 4). A convenient scale for measuring $\rho$, or $|\Gamma|$, is at the bottom of the chart (label no. 10). The phase angle of $\Gamma$, expressed by the Greek letter $\phi$, is measured from the positive $x$ axis with a positive angle referring to progress in the counterclockwise direction. There is a convenient degree scale along the outside edge (label no. 6) of the Smith chart. Thus point no. 8 represents a reflection coefficient of $0.45\underline{/-63.4°}$.

All the circles and arcs inside the Smith chart represent coordinates for reading off the corresponding value of $Z$. All the impedances on the chart are actually normalized impedances $Z_n$ that are related to the true impedances $Z$ by

$$Z_n = \frac{Z}{Z_o} = R_n + jX_n \tag{21-36}$$

where $Z_o$ is the characteristic impedance of the transmission line used to define $\Gamma$. The horizontal diameter of the Smith chart also lies on the horizontal diameter of a whole family of circles. Those circles represent constant values of normalized resistance $R_n$. The values of $R_n$ are labeled along the principal axis (label no. 12) and along the two arcs that intersect the top and bottom of the chart (labels nos. 5 and 7).

The family of arcs that intersects the outer edge of the Smith chart and intersects the extreme right-hand point consists of lines of constant reactance. The arcs in the upper half of the Smith chart represent inductive (positive) reactance. The arcs in the lower half of the Smith chart represent capacitive (negative) reactance. The values of normalized reactance are labeled along the intersection with the outer edge and along the $R_n = 1$ circle (labels nos. 2 and 9). Thus point no. 11 represents a normalized impedance of $0.5 - j0.3\ \Omega$ or a true impedance, from Eq. (21-36), of

$$Z = Z_o Z_n = 50(0.5 - j0.3) = 25 - j15\ \Omega$$

**solution**    $\Gamma = -1$ corresponds to $\Gamma = 1\underline{/180°}$. This is the extreme left-hand point of the chart (point no. 13). Reading the normalized impedance of this point shows that $R_n = 0$ and $X_n = 0$. Thus $Z_n = 0$ and $Z = 0$. This corresponds to a short circuit.

$\Gamma = 0$ corresponds to zero radius or the origin of the Smith chart (point no. 1). The normalized impedance of this point is $1 + j0$. Therefore, using Eq. (21-36) yields:

$$Z = Z_o Z_n = 50(1 + j0) = 50\ \Omega$$

$\Gamma = 1$ corresponds to $\Gamma = \underline{/10°}$ and is represented by the extreme right-hand edge of the chart. The normalized impedance of this point is found by recognizing that the values of the $R_n$ and $X_n$ scales are approaching infinity. Therefore

$$Z = Z_o Z_n = 50(\infty + j\infty) = \infty$$

This corresponds to an open circuit.

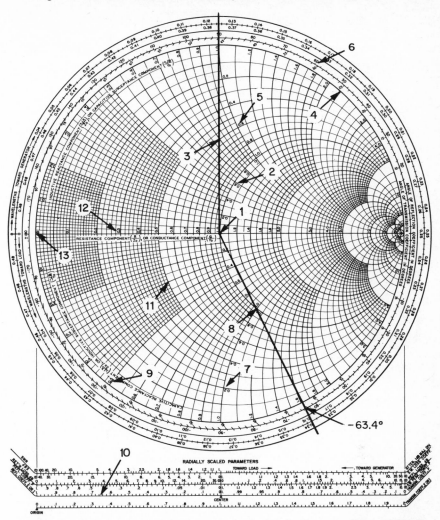

**Fig. 21-5** The Smith chart for converting between impedance and reflection coefficient. [*Copyright 1949 by Kay Electric Co.; renewed 1977 by P. H. Smith. Reproduced by permission of copyright owner. (P. H. Smith*, Electronic Applications of the Smith Chart, *McGraw-Hill Book Co., 1969).]*

$\Gamma = 0.5\underline{/90°}$ is shown as point no. 3. The normalized impedance is read off directly as $Z_n = 0.6 + j0.8$ or

$$Z = Z_o Z_n = 50(0.6 + j0.8) = 30 + j40 \ \Omega$$

If the transmission line were 75 $\Omega$, the same values of $\Gamma$ would yield the same values of $Z_n$. The true impedances would be:
For $\Gamma = -1$:

$$Z = 75(0) = 0$$

For $\Gamma = 0$:

$$Z = 75(1 + j0) = 75 \ \Omega$$

For $\Gamma = 1$:

$$Z = 75(\infty + j\infty) = \infty \ \Omega$$

For $\Gamma = 0.5/\underline{90°}$:

$$Z = 75(0.6 + j0.8) = 45 + j60 \ \Omega$$

## 21.13 USING THE SMITH CHART TO CALCULATE THE REFLECTION COEFFICIENT AND IMPEDANCE ALONG THE LENGTH OF A TRANSMISSION LINE

**problem 21.13** A 50-$\Omega$ transmission line that is 0.87 wavelength long is terminated in a load of $35 + j70 \ \Omega$. Find the impedance and reflection coefficient as viewed from the input end of the transmission line.

**theory** The difficult algebraic Eq. (21-28) is replaced by a rather simple graphical process on the Smith chart. The reflection coefficient of the load $\Gamma_l$, normalized to the $Z_0$ of the line being used, is plotted on the Smith chart. As the observer moves back toward the generator, the angle of

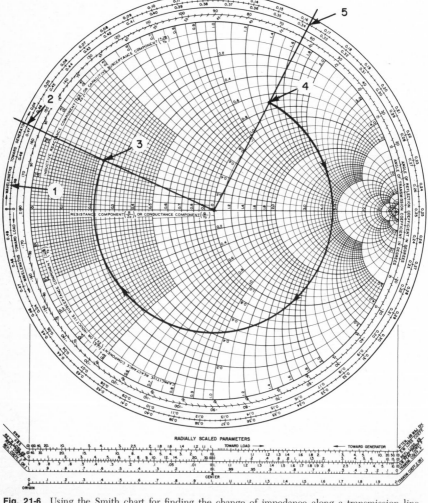

**Fig. 21-6** Using the Smith chart for finding the change of impedance along a transmission line. (*Copyright 1949 by Kay Electric Co.; renewed 1977 by P. H. Smith. Reproduced by permission of copyright owner.*)

the reflection coefficient moves in the negative or clockwise direction. For a low-loss transmission line, the magnitude of reflection coefficient remains the same. A reminder of the direction to move is usually printed at the outer left edge of the Smith chart (label no. 1 of Fig. 21-6). The reminder, *Wavelengths Toward Generator*, also shows that the outer scale of the chart is graduated directly in wavelengths. A movement all the way around the Smith chart corresponds to a movement of position of a half wavelength.

**solution**    $Z_l = 35 + j70$ means that the normalized value is $Z_n = Z_l/Z_o = 0.7 + j1.4$ (label no. 4 in Fig. 21-6) which corresponds to $\Gamma = 0.65\underline{/63°}$. Clockwise movement of 0.87 $\lambda$, corresponding to movement toward the generator, is a full revolution plus another 0.37 wavelength. The reflection coefficient begins at the 0.163 wavelength position (label no. 5 in Fig. 21-6) so that another 0.37 $\lambda$ means that it should finally be 0.533 $\lambda$, but the scale goes only to 0.5 $\lambda$ and begins over. The angle desired, therefore, corresponds to $(0.533 - 0.5) \lambda = 0.033 \lambda$ (label no. 2). The reflection coefficient at the input is $\Gamma_{in} = 0.65\underline{/156°}$, and the normalized input impedance is $0.22 + j0.2$. The actual input impedance is, therefore,

$$Z_{in} = Z_o(0.22 + j0.2) = 11.0 + j10$$

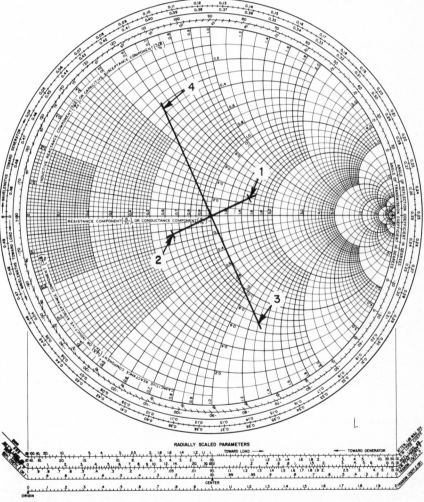

**Fig. 21-7**  Using the Smith chart to convert between impedance and admittance. (*Copyright 1949 by Kay Electric Co.; renewed 1977 by P. H. Smith. Reproduced by permission of copyright owner.*)

## 21.14    USING THE SMITH CHART TO CONVERT BETWEEN IMPEDANCE AND ADMITTANCE

**problem 21.14**    Using the Smith chart, find the admittances that correspond to impedances of $75 + j15$ and $35 - j65 \ \Omega$.

**theory**    Because admittance is the reciprocal of impedance, $Y = 1/Z$, Eq. (21-35) can be written in terms of $Y$ instead of $Z$:

$$Z = Z_o \frac{1 + \Gamma}{1 - \Gamma} = \frac{1}{Y} \tag{21-37}$$

or

$$Y = \frac{1}{Z_o} \frac{1 - \Gamma}{1 + \Gamma} = Y_o \frac{1 + (-\Gamma)}{1 - (-\Gamma)} \tag{21-38}$$

Equation (21-38) means that if $-\Gamma$ is plotted on the Smith chart, then the normalized admittance, $Y/Y_o = YZ_o$, may be read off the arcs and circles. $-\Gamma$ is the point diametrically opposite $\Gamma$.

**solution**    If $Z = 75 + j15$, the normalized impedance is

$$Z_n = \frac{1}{50}(75 + j15) = 1.5 + j0.3$$

(label no. 1 in Fig. 21-7). The diametrically opposite point (label no. 2) gives $Y_n = 0.65 - j0.13$. The total admittance is:

$$Y = Y_o Y_n = \frac{Y_n}{Z_o} = \frac{1}{50}(0.65 - j0.13)$$
$$= 13 - j2.6 \text{ millisiemens (mS)}$$

For the second part of the problem, the normalized impedance is

$$Z_n = \frac{1}{50}(35 - j65) = 0.7 - j1.3$$

(label no. 3 in Fig. 21-7). The diametrically opposite point (label no. 4) gives $Y_n = 0.32 + j0.59$. The total admittance is

$$Y = Y_o Y_n = \frac{1}{50}(0.32 + j0.6) = 6.4 + j12 \text{ mS}$$

## 21.15    USE OF A SLOTTED LINE FOR FINDING A STANDING-WAVE RATIO AND REFLECTION-COEFFICIENT MAGNITUDE

**problem 21.15**    Explain how to set up and use a slotted line to measure a standing-wave ratio of 4:1 and then calculate the load reflection coefficient.

**theory**    The incident and reflected waves on a transmission line form an interference pattern. If, for example, the load were a short circuit (see Fig. 21-8a), the incident and reflected voltages at the load are equal in magnitude but opposite in sign during every part of the high-frequency cycle. The result is that there is no voltage at the load. The two waves add together destructively. Now consider a position a quarter wavelength toward the generator. The incident wave is 90° ahead of the incident wave at the load. The reflected wave is 90° behind the reflected wave at the load. Thus at this position, the two waves are equal in magnitude and in phase during the entire high-frequency cycle. The two waves add together constructively. Another quarter wavelength toward the generator the waves will be found to add destructively again. Where two waves in opposite directions add constructively, the total voltage on the line is $V_{max} = V_i + V_r$. Where they add destructively, the total voltage is $V_{min} = V_i - V_r$. The ratio of these is the standing-wave ratio (SWR), sometimes called the voltage standing-wave ratio (VSWR).

$$\text{VSWR} = \frac{V_{max}}{V_{min}} = \frac{V_i + V_r}{V_i - V_r} = \frac{1 + (V_r/V_i)}{1 - (V_r/V_i)}$$
$$= \frac{1 + |\Gamma|}{1 - |\Gamma|} = \frac{1 + \rho}{1 - \rho} \tag{21-39}$$

**solution**    A slotted line (Fig. 21-8b) is an instrument in which a probe, that is a small antenna, can be moved along the length of a transmission line to sense the total signal amplitude at various positions. The probe output is rectified by a crystal detector, and the rectified output is displayed on a meter. First the probe is moved along the line to a minimum meter indication. The probe

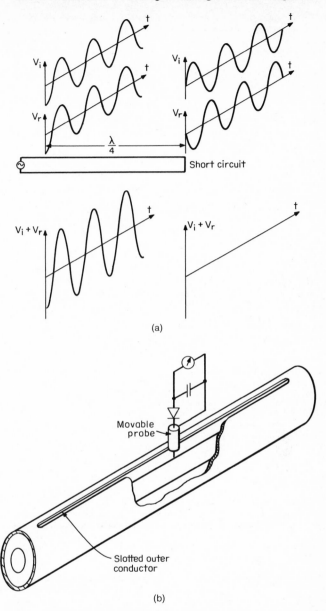

**Fig. 21-8**  (*a*) Incident, reflected, and total voltage along a short-circuited transmission line. (*b*) Principal parts of a coaxial slotted line.

penetration should be adjusted for a small-signal level so that the disturbance to waves on the line is a minimum. The meter reading is noted, and then the probe is moved to a position where the total signal is a maximum. The meter reading is again noted. Crystal rectifiers, at these small-signal levels, usually give outputs that are proportional to the square of the high-frequency voltage. To find the VSWR, the square root of the ratio of the maximum to minimum output indication is taken. Commercial instruments, called *standing-wave amplifiers,* have meter scales that are already graduated in terms of square root. Furthermore the commercial instruments have a gain adjustment. At the maximum position, the gain is adjusted to full scale, and the meter reads 1.0. At the minimum, the needle moves down the scale to increasing scale numbers that give VSWR directly. In this problem the answer is 4.

Once VSWR is measured, Eq. (21-39) can be solved for $\rho$:

$$\rho = \frac{SWR - 1}{SWR + 1} = \frac{4 - 1}{4 + 1} = \frac{3}{5} = 0.6$$

## 21.16    DETERMINING THE COMPLEX REFLECTION COEFFICIENT AND IMPEDANCE FROM SLOTTED-LINE MEASUREMENTS

**problem 21.16**    Explain the method for determining the magnitude and phase-of-reflection coefficient. What are the expected measurement quantities for a load of $15 - j10$ $\Omega$ on a 50-$\Omega$ transmission line at a frequency of 4 GHz?

**theory**    The phase-of-reflection coefficient $\phi$ is found by locating the position of the minimum of the standing-wave pattern and comparing it with the position of the minimum for a short circuit.

The total voltage at a load is the vector sum of (1) the voltage associated with the incident wave, and (2) the voltage associated with the reflected wave. To solve the problem, the incident voltage $V_i$ is considered the reference and is normalized to $\underline{/10°}$, and the reflected voltage is considered in relation to $V_i$ by the reflection coefficient:

$$\Gamma = \frac{V_r}{V_i} \tag{21-40}$$

The two vectors are drawn in Fig. 21-9a for an example load reflection coefficient of $0.5\underline{/135°}$.

As the observing position is moved toward the generator by distance $l$ (or $l/\lambda$ wavelengths), the phase of $V_r$ with respect to $V_i$ is delayed by $2l/\lambda$ complete rotations. The result is $2l/\lambda$ because one $l/\lambda$ is due to change in the incident wave and the other $l/\lambda$ is due to change in the reflected wave. This means that the vector $V_r$ should be rotated clockwise by

$$\phi = 360 \frac{2l}{\lambda} \text{ degrees} \tag{21-41}$$

At any position of the slotted-line probe, the detector output is proportional to magnitude of the vector sum $(V_i + V_r)$. This is the third side of the triangle formed by the $V_i$ and $V_r$ vectors.

When $V_i$ and $V_r$ are aligned in the same direction, the detector output is proportional to $|V_i| + |V_r|$, and the probe is at a maximum of the standing-wave pattern. When $V_i$ and $V_r$ are aligned in opposite directions, the detector output is proportional to $|V_i| - |V_r|$, and the probe is at a minimum of the standing-wave pattern.

For a short-circuit load, $V_i$ and $V_r$ are aligned in opposite directions at the load and at every multiple of one-half wavelength toward the generator from the load.

The triangle formed by $V_i$, $V_r$, and $(V_i + V_r)$ can be placed on the Smith chart. The most convenient position is that at which $V_i$ lies along the principal axis and the arrow ends at the center of the chart. Then the vector $V_r$ exactly coincides with the reflection-coefficient vector.

**solution**    The first part of the technique is to measure the position of the minimum for a short-circuit load. Slotted lines generally have a fixed scale attached for locating the minimums. Minimums are used instead of maximums because the minimums are sharper and therefore easier to define than the maximums. The probe should be adjusted for the smallest penetration that will still yield the information needed. At 4 GHz a wavelength is, from Eq. (21-1), $(3 \times 10^8)/(4 \times 10^9) = 0.075$ m, or 7.5 cm. One-half wavelength is 3.75 cm. For a short circuit, minimums will occur at the load and at every 3.75-cm point toward the generator. The "zero" position of the scale on the slotted line does not usually correspond with the exact position of the short. Suppose in this case that a minimum is observed at the 9.80-cm position. Then other minimums may be observed at the 13.55-, 17.30-cm, etc., positions. The $V_i$ and $V_r$ (labeled as $V_{rs}$) vectors for the short circuit are plotted on the Smith chart of Fig. 21-9b.

It is good practice to slide the slotted-line probe along the carriage to observe whether the standing-wave pattern is well behaved. Bad behavior includes maximums that are not midway between the minimums and secondary maximums and minimums. Such anomalies are due to other frequencies on the transmission line. These frequencies, like the second or third harmonic of the desired generator output, give erroneous data.

The next step in the process is to attach the unknown load to the slotted line and to measure the standing-wave ratio. The load for this problem corresponds to a normalized impedance of, from Eq. (21-36),

$$Z_i = \frac{Z}{Z_o} = \frac{15 - j10}{50} = 0.3 - j0.2$$

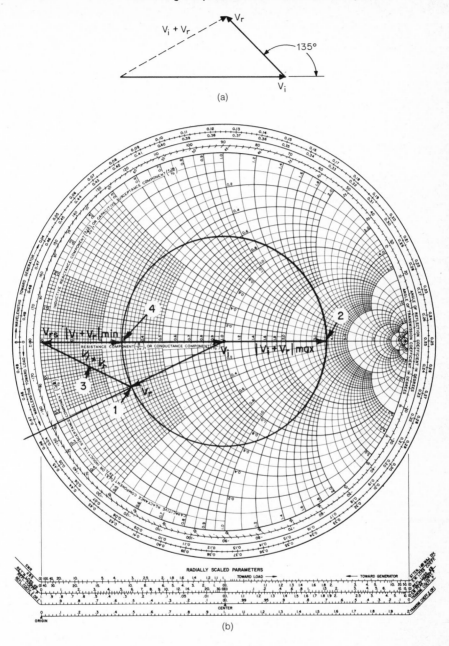

**Fig. 21-9**  (*a*) The vector addition of incident and reflected wave voltages to form the total voltage. (*b*) Reflected and incident wave voltages plotted on a Smith chart. *(Copyright 1949 by Kay Electric Co.; renewed 1977 by P. H. Smith. Reproduced by permission of copyright owner.)*

This is plotted (label no. 1) on the Smith chart of Fig. 21-9b. It corresponds to a reflection coefficient of $0.55\underline{/-155°}$.

The standing-wave ratio is measured by the same technique as in Problem 21.15. The measured value should be, according to Eq. (21-39),

$$\text{SWR} = \frac{1 + \rho}{1 - \rho} = \frac{1 + 0.55}{1 - 0.55} = 3.45$$

One of the convenient characteristics of the Smith chart is that the VSWR can be plotted directly on the principal axis (label no 2). A circle, with center at the origin, drawn through the measured VSWR point of the principal axis, represents all the possible $V_r$'s as the observing position is moved along the line.

At the position of the load, the $V_r$ vector is drawn from the center of the Smith chart to the reflection-coefficient point. The load voltage vector, $V_i + V_r$ (label no. 3 of Fig. 21-9b), is the third side of the $V_i$, $V_r$ triangle. The length of that vector represents the amplitude of the slotted-line-detector output if the probe could be placed at the load. That same output occurs each half wavelength away from the load, that is, wherever there is a minimum when the short circuit is connected.

As the probe position is moved a little closer to the generator, $V_r$ progresses clockwise from label no. 1. The $(V_i + V_r)$ vector gets somewhat shorter until it becomes a minimum (label no. 4). Probe movement from point no. 1 to point no. 4 is 0.035 wavelength (0.263 cm) toward the generator as read off the outside of the Smith chart. With the unknown load connected, minimums in the standing-wave pattern would occur 0.263 cm toward the generator from every point where a minimum existed for the short circuit.

## 21.17 CALCULATING POWER FLOW

**problem 21.17** A 10-V wave is flowing from a generator down a 50-$\Omega$ transmission line toward an antenna of impedance $40 - j10$ $\Omega$. Calculate the power flowing toward the antenna, the power reflected from the antenna, and the power absorbed by the antenna.

**theory** The power associated with a wave on a transmission line is the product of voltage and current for that wave. Thus the incident power can be written

$$P_i = V_i I_i = V_i \frac{V_i}{Z_o} = \frac{V_i^2}{Z_o} \tag{21-42}$$

where $I_i = V_i/Z_o$ from Eq. (21-22). In the same way, reflected power can be written

$$P_r = V_r I_r = \frac{V_r^2}{Z_o} \tag{21-43}$$

Note that unless otherwise stated, voltage and current amplitude refer to rms values, not peak values.

Only two things can happen to power flowing toward a load; it is either reflected from the load or dissipated in the load. Dissipated power is usually converted into another form of energy. Resistors convert the power into heat, antennas convert the power into waves propagating in space (receiving antennas convert power in space into a transmission-line wave), and mixers convert the power to a different frequency. The relationship between incident, reflected, and dissipated power is

$$P_d = P_i - P_r \tag{21-44}$$

**solution** The incident power is given directly by Eq. (21-42):

$$P_i = \frac{V_i^2}{Z_o} = \frac{10^2}{50} = 2\text{W}$$

The reflection coefficient of the antenna can be found from a Smith chart or calculated from Eq. (21-34):

$$\Gamma = \frac{Z_l - Z_o}{Z_l + Z_o} = \frac{40 - j10 - 50}{40 - j10 + 50} = \frac{-10 - j10}{90 - j10} = \frac{14.14}{90.55} \frac{\underline{/-135}}{\underline{/-6.34}}$$
$$= 0.156 \underline{/-128.7}$$

The voltage of the reflected wave is, therefore,

$$V_r = \Gamma V_i = (0.156 \quad \underline{/-128.7})10 = 1.56 \quad \underline{/-128.7}$$

The reflected power is calculated using Eq. (21-43):

$$P_r = \frac{V_r^2}{Z_o} = \frac{1.56^2}{50} = 0.049 \text{ W}$$

One way of finding the dissipated power is by Eq. (21-44):

$$P_d = P_i - P_r = 2 - 0.049 = 1.95 \text{ W}$$

The other way is to first calculate the load voltage and current from Eqs. (21-25) and (21-23):

$$V_l = V_i + V_r = 10 + 1.56 \; \underline{/-128.7} = 10 - 0.975 - j1.217$$
$$= 9.025 - j\,1.217$$
$$I_l = \frac{V_l}{Z_l} = \frac{9.025 - j1.217}{40 - j10} = \frac{9.106 \; \underline{/-7.683}}{41.23 \; \underline{/-14.036}} = 0.221 \; \underline{/6.35}$$

The dissipated power is the product of voltage, current, and the cosine of the angle between them:

$$P_l = V_l I_l \cos \phi = 9.106(0.221) \cos (-7.683 - 6.353)$$
$$= 9.106(0.221) \cos (-14.036) = 1.95 \text{ W}$$

## 21.18  CALCULATING dB, RETURN LOSS, MISMATCH LOSS, ATTENUATION, AND dBm

**problem 21.18**  A 20-mW signal is incident upon a 20-dB attenuator that has a reflection coefficient at its input port of $0.1 \; \underline{/\,32°}$. Calculate the return loss and mismatch loss for the attenuator. Express the incident power and the power emerging from the output port of the attenuator in terms of decibels above 1 mW (dBm) (see below).

**theory**  Decibels (dB) are defined as

$$dB = 10 \log \frac{P}{P_{\text{ref}}} \tag{21-45}$$

where $P$ and $P_{\text{ref}}$ are powers. If the ratio is greater than 1, the number of dB is positive. If the ratio is less than 1, the number of dB is negative. Sometimes, when the number of dB turns out negative, a second negative term such as loss or attenuation is used to make the result positive. Thus a loss of 10 dB is a gain of $-10$ dB.

Frequently each of the powers in Eq. (21-45) is written in terms of voltage and impedance. If, for example,

$$P = |V^2/Z_o| \text{ and } P_{\text{ref}} = |V_{\text{ref}}^2/Z_o|, \text{ then Eq. (21-45)}$$

becomes

$$dB = 10 \log \left| \frac{V^2}{Z_o} \cdot \frac{Z_o}{V_{\text{ref}}^2} \right| = 10 \log \left| \frac{V}{V_{\text{ref}}} \right|^2$$
$$= 20 \log \left| \frac{V}{V_{\text{ref}}} \right| \tag{21-46}$$

Return loss (RL) is defined as the ratio, expressed in dB loss, of the reflected power to the incident power.

$$RL = -10 \log \frac{P_r}{P_i} = -10 \log \left| \frac{V_r}{V_i} \right|^2 = -10 \log |\Gamma|^2$$
$$= -20 \log |\Gamma| = -20 \log \rho \tag{21-47}$$

Mismatch loss (ML) is defined as the ratio, expressed in dB loss, of the power transmitted through the mismatch to the incident power.

$$ML = -10 \log \frac{P_d}{P_i} = -10 \log \frac{P_i - P_r}{P_i}$$
$$= -10 \log \left( 1 - \frac{P_r}{P_i} \right) = -10 \log (1 - \rho^2) \tag{21-48}$$

Mismatch loss shows the amount of power that was lost due to reflection.

Attenuation is defined as the ratio, expressed as dB loss, of the emerging power from the output port to the incident power at the input port.

$$\text{Attenuation} = -10 \log \frac{P_{out}}{P_i} \tag{21-49}$$

The term *insertion loss* is often used interchangeably with attenuation.

Popular usage has added another convenient unit, the dBm, meaning decibels above one milliwatt; a negative number of dBm means decibels below one milliwatt. The formula for dBm is like Eq. (21-45) except that the denominator $P_{ref}$ is always one milliwatt:

$$\text{dBm} = 10 \log \frac{P}{1 \text{ mW}} \tag{21-50}$$

In this equation, $P$ is the only variable, and so dBm is used as a measure of absolute power.

If the number of dB or dBm is known, then the absolute ratio may be calculated by solving any of the above equations backwards. The key to such a solution is that

$$10^{\log x} = x \tag{21-51}$$

By use of this relationship, Eq. (21-45) can be solved for the ratio

$$\frac{P}{P_{ref}} = 10^{dB/10} \tag{21-52}$$

**solution**    Return loss is calculated directly from the definition in Eq. (21-47).

$$\text{RL} = -20 \log \rho = -20 \log 0.1$$
$$= -20(-1) = 20 \text{ dB}$$

Mismatch loss is calculated directly from the definition in Eq. (21-48).

$$\text{ML} = -10 \log (1 - \rho^2) = -10 \log (1 - 0.01)$$
$$= -10 \log 0.99 = 0.044 \text{ dB}$$

The incident power is expressed in dBm by using Eq. (21-50):

$$P_i(\text{dBm}) = 10 \log \frac{20 \text{ mW}}{1 \text{ mW}} = 10 \log 20 = 10(1.30)$$
$$= 13 \text{ dBm}$$

Calculation of the output level can be done the long way by calculating $P_{out}$ from Eq. (21-49) and then calculating dBm from Eq. (21-50). The advantage of dB, however, is that multiplication and division of absolute ratios is replaced by addition and subtraction of dB. Thus the output emerging wave $P_{out}$ is 20 dB lower than the incident wave $P_i$. But $P_i$ was just calculated to be 13 dBm, and so

$$P_{out}(\text{dBm}) = P_i(\text{dBm}) - \text{attenuation}$$
$$= 13 - 20 = -7 \text{ dBm} \tag{21-53}$$

## 21.19    MEASURING REFLECTION COEFFICIENT BY FIRST MEASURING RETURN LOSS ON A SWEPT REFLECTOMETER

**problem 21.19**    Describe a reflectometer for measuring return loss. What reflection coefficient magnitude corresponds to a return loss of 40 dB? of 30, 26, 20, and 10 dB?

**theory**    A reflectometer operates by separating the incident wave from the reflected wave, measuring each one, then finding and displaying the ratio.

The components used to separate waves traveling in different directions on the same transmission line are called *directional couplers*. A directional coupler usually consists of three ports. Two ports, ports 1 and 2 of Fig. 21-10a, form the input and output of the main transmission line. Another transmission line, called the *auxiliary line*, samples or taps off some of the energy from the main line. The sample of the wave propagating from port 1 to port 2 emerges from port 3. The sample of the wave propagating from port 2 to port 1 does not emerge from port 3 but is dissipated in an internal load. The arrows in the schematic of the coupler (Fig. 21-10a) show the direction of propagation of the sampled waves.

The magnitude of the coupled wave is given by the coupling factor $C$.

$$C = \frac{V_{o3}}{V_{i1}} \tag{21-54}$$

$V_{o3}$ is the voltage associated with the wave traveling out of port 3, and $V_{i1}$ is the voltage associated with the wave traveling into port 1. Usually the coupling factor is given in decibels:

$$C(\text{dB}) = -20 \log \frac{V_{o3}}{V_{i1}} \tag{21-55}$$

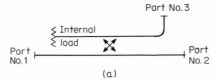

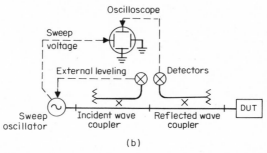

**Fig. 21-10**  (*a*) Schematic diagram of a directional coupler. (*b*) Schematic diagram of a swept reflectometer (low-frequency lines are shown dashed).

The attenuation between ports 1 and 3 is identical to the coupling.

Directional couplers, unfortunately, couple some power out of the auxiliary arm from reverse traveling waves on the main line; the directional coupler is not perfectly directional. Some of the power entering the main line at port 2 is sampled and emerges from port 3 of the coupler. This undesired coupling is lower than the desired coupling $C$ by the coupler directivity $D$. In equation form, for a wave traveling backwards on the main line,

$$CD = \frac{V_{o3}}{V_{i2}} \qquad \text{or} \qquad D = \frac{1}{C}\frac{V_{o3}}{V_{i2}} \tag{21-56}$$

where $V_{i2}$ is the voltage associated with the wave traveling into port 2, and $V_{o3}$ is the voltage associated with the undesired coupled wave out of port 3. The attenuation between ports 2 and 3 is equal to $CD$ (and not $D$ alone as is often thought). Directivity is usually expressed in decibels, and so Eq. (21-56) becomes

$$D = -10 \log \left(\frac{V_{o3}}{V_{i2}}\right)^2 - C \qquad \text{(dB)} \tag{21-57}$$

If a directional coupler has a coupling factor of 20 dB and a directivity of 30 dB, then the auxiliary line output (at port 3 of Fig. 21-10*a*) for a reverse main line input (at port 2) is 50 dB down.

Return loss was already defined in Eq. (21-47) as

$$RL = -10 \log \frac{P_r}{P_i} = -20 \log \frac{V_r}{V_i} = -20 \log \rho \tag{21-58}$$

A high return loss means that the reflection coefficient is small.

**solution**  The fundamental parts of a swept reflectometer are shown in Fig. 21-10*b*. The sweep oscillator repetitively changes its microwave frequency output from a start frequency to a stop frequency. The sweep output voltage is used to drive the horizontal plates of an oscilloscope.

The microwave power output from the sweep oscillator is sampled by the incident wave coupler and detected at the auxiliary arm output port. The detected output is fed back to the sweep oscillator. Leveling circuits in the sweep oscillator then adjust the output power to keep the detected feedback signal a constant. Thus the emerging wave from the sweep oscillator is ideally the same amplitude at every frequency.

The reflected wave from the device under test (DUT) is sampled by the reflected wave coupler and detected at the auxiliary arm output port. The detected reflected wave is fed to the vertical plates of the oscilloscope.

The entire system is calibrated by first measuring a short circuit and noting the vertical deflection on the oscilloscope. When the unknown is connected, the ratio of the vertical deflection for the DUT to the vertical deflection for the short circuit indicates the size of the reflection coefficient.

Several display methods are available that provide convenience and accuracy in interpreting the vertical display on the oscilloscope. Some test sets have logarithmic amplifiers so that the vertical display is read directly in return loss. Some of these also have specially shaped amplifiers to account for detector nonlinearities. Other systems, called *microwave network analyzers*, replace the detectors and oscilloscope with instruments that measure the complex ratio of the two sampled waves. Thus the microwave network analyzer can measure the phase as well as the magnitude of reflection coefficient.

If an ordinary oscilloscope is used, the vertical scale must be interpreted properly. For waveguide reflectometers, the reflected detector output can be scaled by means of a precision variable attenuator installed between the reflected-wave-coupler auxiliary port and the detector. With the short circuit as the DUT and 0 dB on the attenuator, the oscilloscope display corresponds to 0-dB return loss. If the attenuator is adjusted to 10 dB, then the oscilloscope display corresponds to a 10-dB return loss. The oscilloscope level can be noted, or a 10-dB return-loss line can be sketched by grease pencil on a protective cover placed over the cathode-ray screen. Similar lines can be drawn for 20-dB, 40-dB, or any desired level of return loss.

In coaxial systems, where such precision attenuators are not available, the sweep oscillator output must be adjusted so that the sample of the reflected wave reaching the reflected-wave detector is below $-20$ dBm (10 $\mu$W) for the short circuit as the DUT. This is sufficient to assure that the detector is in the square-law range; this means that the detector voltage output is proportional to the microwave power input. Note that power is proportional to the square of the voltage—hence the term *square law*. If the vertical voltage for the short circuit is called $V_s$ and for the DUT it is called $V_{DUT}$, the reflection coefficient magnitude is given by

$$\rho = \sqrt{\frac{V_{OUT}}{V_s}} \qquad (21\text{-}59)$$

The reflection coefficient magnitude corresponding to a particular return loss is found by solving for $\rho$ in Eq. (21-58).

$$\rho = 10^{-RL/20} \qquad (21\text{-}60)$$

For RL = 40 dB:

$$\rho = 10^{-2} = 0.01$$

For RL = 30 dB:

$$\rho = 10^{-1.5} = 0.0316$$

For RL = 26 dB:

$$\rho = 10^{-1.3} = 0.050$$

For RL = 20 dB:

$$\rho = 10^{-1} = 0.1$$

For RL = 10 dB:

$$\rho = 10^{-0.5} = 0.316$$

This explanation of good reflectometer technique is greatly simplified. Consult the measurement references for a more complete treatment.

## 21.20 CALCULATING THE LIMITS OF POWER INCIDENT FROM A GENERATOR TO A LOAD—MISMATCH UNCERTAINTY

**problem 21.20** A microwave generator is specified to have an output power of 10 mW into a 50-$\Omega$ load. The generator is found to have a reflection coefficient magnitude of 0.5. A load with reflection coefficient of 0.3 is connected by means of a 50-$\Omega$ transmission line of variable length. What is the maximum and minimum power that is incident upon the load as the transmission line length is varied? What is the power dissipated in the load for each case? Find the mismatch uncertainty limits.

**theory** Microwave generators are usually rated by the power they dissipate in a reference load, usually 50 $\Omega$. Let this power be called $P_{gZ_0}$. With other loads, however, the generator may put out more or less net power. The incident wave upon an arbitrary load generally reflects power back toward the generator. Upon reaching the generator, the reflected wave re-reflects according to the generator reflection coefficient. The re-reflected wave adds to what is then being generated by the generator. The addition of waves could be in phase, out of phase, or anywhere in between. The phase depends on (1) the phase angle of $\Gamma_l$, (2) the phase angle of $\Gamma_g$, and (3) the round-trip phase shift along the transmission line. Seldom are the three phase angles known, and so the best that can be done is to calculate the limits of the incident power. These are given by

$$\frac{P_{gZ_0}}{(1 + |\Gamma_l\Gamma_g|)^2} \leq P_i \leq \frac{P_{gZ_0}}{(1 - |\Gamma_l\Gamma_g|)^2} \qquad (21\text{-}61)$$

In actual practice, $P_i$ is measured with a power meter, and then the limits $P_{g_{Z_0}}$ are calculated from that measurement.

$$P_i(1 - |\Gamma_l\Gamma_g|)^2 \leqslant P_{g_{Z_0}} \leqslant P_i(1 + |\Gamma_l\Gamma_g|)^2 \qquad (21\text{-}62)$$

The limits of $P_{g_{Z_0}}/P_i$, expressed in decibels, are called the *mismatch uncertainty limits* and are given by

$$M_{\mu,\text{max}} = 10 \log (1 + |\Gamma_l\Gamma_g|)^2 \qquad (21\text{-}63)$$
$$M_{\mu,\text{min}} = 10 \log (1 - |\Gamma_l\Gamma_g|)^2 \qquad (21\text{-}64)$$

**solution**   From Eq. (21-61) the maximum possible value of incident power for this generator and load can be calculated:

$$P_{i,\text{max}} = \frac{P_{g_{Z_0}}}{(1 - |\Gamma_l\Gamma_g|)^2} = \frac{10 \text{ mW}}{(1 - 0.5 \cdot 0.3)^2} = \frac{10}{0.85^2}$$
$$= 13.84 \text{ mW}$$

The minimum power, from the same equation, is

$$P_{i,\text{min}} = \frac{P_{g_{Z_0}}}{(1 + |\Gamma_l\Gamma_g|)^2} = \frac{10 \text{ mW}}{(1 + 0.5 \cdot 0.3)^2} = \frac{10}{1.15^2}$$
$$= 7.56 \text{ mW}$$

The power reflected from the load can be found from the definition of reflection coefficient:

$$P_r = \frac{|V_r|^2}{Z_o} = \frac{|\Gamma_l V_i|^2}{Z_o} = \rho_l^2 P_i$$

so that

$$P_{r,\text{max}} = 0.09(13.84) = 1.25 \text{ mW}$$
$$P_{r,\text{min}} = 0.09(7.56) = 0.68 \text{ mW}$$

The power dissipated in the load is the difference between the incident and reflected power.

$$P_{d,\text{max}} = 13.84 - 1.25 = 12.59 \text{ mW}$$
$$P_{d,\text{min}} = 7.56 - 0.68 = 6.88 \text{ mW}$$

The mismatch uncertainty limits are, according to Eqs. (21-63) and (21-64),

$$M_{\mu,\text{max}} = 10 \log (1 + 0.5 \cdot 0.3)^2 = 10 \log 1.15^2 = 1.21 \text{ dB}$$
$$M_{\mu,\text{min}} = 10 \log (1 - 0.5 \cdot 0.3)^2 = 10 \log 0.85^2 = -1.41 \text{dB}$$

## 21.21   POWER TRANSFER THROUGH AN ATTENUATOR

**problem 21.21**   A 0 to 11-dB-step attenuator is inserted between a generator and a power sensor (Fig. 21-11). The generator dissipates 10 mW in a $Z_o$ load and has $\rho_g \leqslant 0.33$. The power sensor has a $\rho_l \leqslant 0.2$. The attenuator, known to have an insertion loss of 1.5 dB, is set to 6 dB. The attenuator reflection coefficients are less than 0.2. Find the range of the power that is incident upon the power sensor.

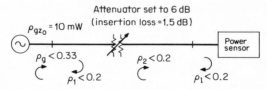

**Fig. 21-11**   Schematic diagram for finding the power through an attenuator and incident upon the load.

**theory**   The principal cause of ambiguity to the incident power is the mismatch uncertainty. In the case of a two-port device, like an attenuator, there are more sources of reflection and re-reflection than discussed for the previous problem. There are actually three re-reflection phenomena: (1) at the generator/attenuator connection, (2) at the attenuator/power sensor (or load) connection, and (3) between the generator and power sensor by passing through the attenuator. This last source of re-reflected waves is reduced by the attenuation and can often be ignored. The formula for mismatch uncertainty for a two-port inserted between a generator and load is

$$M_\mu = 10 \log [(1 \pm \rho_g\rho_1)(1 \pm \rho_l\rho_2) \pm \rho_l\rho_g|T|^2]^2 \qquad (21\text{-}65)$$

$\rho_1$ and $\rho_2$ refer to the reflection coefficients at the input and output of the two-port, and $T$ refers to the transmission coefficient of the two-port. The plus signs refer to the maximum value of $M_\mu$ (that is, $M_{\mu,\text{max}}$ ), and the minus signs refer to the minimum value of $M_\mu$ (that is, $M_{\mu,\text{min}}$ ). The magnitude of the transmission is often symbolized by $\tau$.

**solution**    The insertion loss of a two-port device is defined as the transmission coefficient in decibels of the two-port when it is inserted between a generator with zero reflection coefficient and a load of zero reflection coefficient. Insertion loss, therefore, refers to the case of no mismatch uncertainty. For a variable attenuator, insertion loss is the attenuation when the attenuator is set to 0 dB. In this case the loss at the 0 dB setting is 1.5 dB.

The attenuation of a variable attenuator usually refers to the incremental attenuation, that is, the attenuation over and above the 0-dB value. The attenuator in this problem, therefore, has a total loss of $6 + 1.5 = 7.5$ dB. This refers to the loss considered for a reflectionless generator and load. If $\rho_l$ and $\rho_g$ were both zero in Eq. (21-65), $M_\mu$ would have the value 1 for both maximum and minimum—there is no mismatch uncertainty. For that special case the power to the load is 7.5 dB less than the generated power. But the generated power is 10 mW or 10 dBm [from Eq. (21-50)], and so the power to the load is $10 - 7.5 = 2.5$ dBm.

For the usual case where mismatch uncertainty is present, this value of 2.5 dBm can increase or decrease by the amount given by Eq. (21-65). The magnitude of the transmission coefficient is given by

$$\text{dB} = 10 \log \tau^2 = 20 \log \tau$$

or

$$\log \tau = \frac{\text{dB}}{20}$$
$$10^{\log \tau} = \tau = 10^{\text{dB}/20}$$

For a loss of 7.5 dB

$$\tau = 10^{-7.5/20} = 0.422$$
$$\tau^2 = 0.178$$

Therefore

$$\begin{aligned}
M_{\mu,\text{max}} &= 20 \log \left[ (1 + 0.2\times0.33)(1+0.2\times0.2) + 0.33\times0.2\times0.178 \right] \\
&= 20 \log \left[ 1.066\times1.04 + 0.0117 \right] \\
&= 20 \log (1.120) \\
&= 0.99 \text{ dB}
\end{aligned}$$

Thus the maximum power to the load is 0.99 dB more than 2.5 dBm or a total of 3.49 dBm. Solving Eq. (21-50) for absolute power gives

$$\begin{aligned}
P_{i,\text{max}} &= 3.49 \text{ dBm} \\
&= 10^{3.49/10} = 2.23 \text{ mW}
\end{aligned}$$

Also

$$\begin{aligned}
M_{\mu,\text{min}} &= 20 \log \left[ (1-0.33\times0.2)(1-0.2\times0.2) - 0.33\times0.2\times0.178 \right] \\
&= 20 \log \left[ 0.934\times0.96 - 0.0117 \right] \\
&= 20 \log 0.885 \\
&= -1.06 \text{ dB}
\end{aligned}$$

The minimum power to the load is 1.06 dB less than 2.5 dBm or a total of 1.44 dBm. Solving Eq. (21-50) for absolute power gives

$$\begin{aligned}
P_{i,\text{min}} &= 1.44 \text{ dBm} \\
&= 10^{1.44/10} = 1.39 \text{ mW}
\end{aligned}$$

## 21.22    DETERMINING THE SOURCE REFLECTION COEFFICIENT OF A LEVELED SIGNAL GENERATOR

**problem 21.22**    A microwave generator is rated as having an output of 20 mW into a 50-$\Omega$ load and a reflection coefficient magnitude of 0.5. The generator is to be externally leveled with a directional coupler, crystal detector, and high-gain amplifier. The coupler has 30-dB directivity, and the reflection coefficient on the main line is zero. What is the source-reflection coefficient of the leveled generator?

**theory**    The coupler is connected to the generator as in Fig. 21-12. The power emerging from the generator (incident upon the load) is sampled in the auxiliary arm of the coupler and rectified by the crystal detector. A coupler detector combination is called a *directional detector*. The detector output is compared with a reference voltage in a differential amplifier. The amplified difference or error voltage is fed to the generator circuits that control the size of the wave emerging from the generator.

Now if a wave reflected from a load re-reflects off the generator and increases the incident power to the load, the increase is detected and creates an error voltage, and the generator decreases its internally generated power by the appropriate amount. The net result is that re-reflections from the generator are completely corrected. There are two remaining possible sources of effective re-reflection. One is the main line of the directional coupler near the output end, from which re-reflection may possibly not be sampled by the auxiliary arm. The second is the directivity of the coupler; the reflection from the load is inadvertently sampled by the coupler because of its directivity, and this signal masquerades as a change in the incident power and changes the generator output.

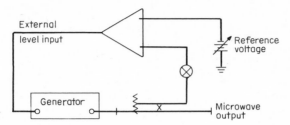

**Fig. 21-12**   Schematic diagram of an externally leveled generator.

**solution**   Any reflections from the generator are monitored by the directional detector and completely corrected. A wave incident upon the generator $V_{ig}$, is sampled and changes the generator output by $DV_{ig}$ where $D$ is the coupler directivity (expressed as a ratio instead of in decibels). The apparent generator-reflection coefficient is the change in the total generator output wave for a change in the input wave (incident upon the generator).

$$\Gamma_g = \frac{\Delta V_o}{V_{ig}} = \frac{DV_{ig}}{V_{ig}} = D \qquad (21\text{-}66)$$

But from the definition of dB

$$D_{dB} = -10 \log (D)^2 = -20 \log D \qquad (21\text{-}67)$$

Solving for $D$ gives

$$\Gamma_g = D = 10^{-D_{dB}/20} = 0.031$$

Thus leveling reduces the generator-reflection coefficient from 0.5 to 0.031.

## 21.23   THE RELATIONSHIP BETWEEN PULSE POWER AND AVERAGE POWER

**problem 21.23**   The power from a radar transmitter is measured by connecting a power sensor and average-reading power meter to the auxiliary arm of a 40-dB directional coupler. The main line of the coupler is inserted between the transmitter and the antenna. The radar pulses are also known to be 2-$\mu$s wide, and they occur at a 1-kHz rate. If the power meter reads 10 mW, what is the power of each radar pulse?

**theory**   Most average-reading power meters convert the microwave energy to heat in a resistor and then measure the temperature of the resistor or the heat flow. Such heat-sensing methods are too slow to respond to each radar pulse. Instead they average the power over periods of time from several hundredths of a second to several seconds. Radar pulses are usually rectangular, and so the relation of average power to the pulse power is given by the duty cycle:

$$P_p = \frac{P_{avg}}{\text{duty cycle}} \qquad (21\text{-}68)$$

where the duty cycle is the fraction of total time that the pulse is present.

$$\text{Duty cycle} = \frac{\tau}{T_o} \qquad (21\text{-}69)$$

where $\tau$ is the pulse width and $T_o$ is the period of radar cycle (equal to the reciprocal of the repetition rate).

**solution**   From Eq. (21-69) the duty cycle is

$$\text{Duty cycle} = \frac{2 \times 10^{-6}}{1/1000} = 0.002$$

The pulse power delivered to the sensor is calculated from Eq. (21-68):

$$P_p = \frac{P_{\text{avg}}}{\text{duty cycle}} = \frac{10 \text{ mW}}{0.002} = 5 \text{ W}$$

But this is the power in the secondary arm of the directional coupler. The power in the main line is 40 dB ($10^4$ times) stronger. Thus the transmitter power that is incident on the antenna is

$$P_{\text{ant}} = 5 \times 10^4 \text{ W} = 50 \text{ kW}$$

## 21.24    CALCULATING THE PULSE RESPONSE OF A MICROWAVE DETECTOR

**problem 21.24**    The output of a diode detector, whose equivalent circuit is shown in Fig. 21-13, is connected to the vertical input of an oscilloscope with 3 ft of cable. The cable capacity is 30 pF/ft. Find the 10 to 90% rise time of the detector cable combination. What would the rise time be if a 50-$\Omega$ resistor were placed across the oscilloscope terminals (in parallel with the cable)?

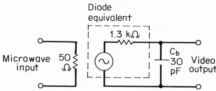

**Fig. 21-13**   Equivalent circuit of a diode detector.

**theory**    Detector circuits can usually be represented as a voltage generator connected to an $RC$ circuit. The response of such a circuit to a step function from the generator is

$$V_o = V_g(1 - e^{-t/RC}) \tag{21-70}$$

where $R$ is the resistance through which the total capacity $C$ must charge.

**solution**    The resistance for Eq. (21-70) is 1.3 k$\Omega$. The capacity is the total capacity, 30 pF for the bypass capacitor, and 90 pF for the cable to the oscilloscope. Thus the $RC$ time constant is given by

$$\tau = RC = (1.3 \times 10^3)(120 \times 10^{-12}) = 156 \times 10^{-9} \text{ s} \tag{21-71}$$

The final value of $V_o$ from Eq. (21-70), evaluated by considering $t = \infty$, is $V_g$. To find the elapsed time until the output becomes 10% of its final value, Eq. (21-70) must be solved as follows:

$$V_o|_{t=t_{10}} = 0.1V_g = V_g(1 - e^{-t_{10}/\tau})$$

$$e^{-t_{10}/\tau} = 1 - 0.1 = 0.9$$

or
$$\ln(e^{-t_{10}/\tau}) = \frac{t_{10}}{\tau} = \ln 0.9$$

$$t_{10} = -\tau \ln 0.9 = 0.11 \tau$$

In a similar manner, to find the elapsed time until the output becomes 90% of its final value, Eq. (21-70) is solved as follows:

$$V_o|_{t=t_{90}} = 0.9 V_g = V_g(1 - e^{-t_{90}/\tau})$$
$$e^{-t_{90}/\tau} = 1 - 0.9 = 0.1$$
$$\ln(e^{-t_{90}/\tau}) = \frac{t_{90}}{\tau} = \ln 0.1$$
$$t_{90} = -\tau \ln 0.1 = 2.3 \tau$$

Thus the 10 to 90% rise time is

$$t_r = t_{90} - t_{10} = 2.3 \tau - 0.11 \tau = 2.2 \tau$$

The rise time for the diode/cable combination is, therefore,

$$t_r = 2.2(156 \times 10^{-9}) = 343 \text{ ns}$$

If a 50-$\Omega$ resistor were placed at the output of the cable, the equivalent resistance that contributes to the time constant is 50 $\Omega$ in parallel with the 1.3-k$\Omega$ source resistance. Therefore

$$R_{eq} = \frac{50(1300)}{50 + 1300} = 48 \ \Omega$$

Now the time constant is

$$\tau = R_{eq}C = 48(120 \times 10^{-12}) = 5.76 \times 10^{-9} \text{ s}$$

The rise time for the diode/cable/resistor combination is

$$t_r = 2.2 \ \tau = 2.2(5.76 \times 10^{-9}) = 12.7 \text{ ns}$$

The response is speeded up by a factor of 343/12.7 or 27 times. There is a cost for the increased speed—the 50-$\Omega$ resistor lowers the voltage to the oscilloscope. The voltage to the vertical amplifier now comes from a voltage divider of 50 $\Omega$ in series with 1.3 k$\Omega$. The input voltage is, therefore, reduced by a factor of 1350/50 or 27.

## 21.25  CALCULATING THERMAL NOISE POWER

**problem 21.25**  Calculate the thermal noise power available in dBm from a resistor at a temperature of 17°C (290 K). The power is to be measured first over a 1-MHz bandwidth, then over a 1-Hz bandwidth.

**theory**  Boltzmann's constant is symbolized by $k$ and has a value $1.38 \times 10^{-23}$ J/°C. Boltzmann's constant gives the thermal energy of a particle due to its temperature. In the case of resistors, this thermal energy in the form of electron movement can be transmitted electrically. It is natural that the greater the bandwidth for transmitting the thermal energy from the electrons, the greater the amount of energy that can be transferred per second. Furthermore, the hotter the resistor, the greater the power transmitted. The noise power available from a resistor is symbolized by $kTB$, where $T$ is the absolute temperature and $B$ is the bandwidth.

**solution**  The noise power available from the resistor for $T = 290$ K and $B = 1$ MHz is

$$kTB = 1.38 \times 10^{-23} \frac{\text{J}}{\text{K}} \times 290 \text{ K} \times \frac{10^6}{\text{s}}$$
$$= 4.002 \times 10^{-15} \text{ J/s} = 4 \times 10^{-15} \text{ W} \qquad (21\text{-}72)$$

The power in dBm [from Eq. (21-50)] is

$$\text{dBm} = 10 \log \frac{4 \times 10^{-15} \text{ W}}{10^{-3} \text{ W}} = -114 \text{ dBm} \qquad (21\text{-}73)$$

for a 1-MHz bandwidth.

For a 1-Hz bandwidth, the result would be $10^6$ smaller, or 60 dB smaller. Therefore the noise power available is

$$\text{dBm} = (114 - 60) \text{ dBm} = -174 \text{ dBm} \qquad (21\text{-}74)$$

for a 1-Hz bandwidth.

## 21.26  CALCULATING THE NOISE FIGURE AND EFFECTIVE INPUT NOISE TEMPERATURE FROM MEASURED Y FACTOR DATA

**problem 21.26**  The noise power is measured at the output of an amplifier under two conditions. In the first, the input to the amplifier is a noise source in the "turned-on" condition. The measured output power is $-52.55$ dBm. In the second, the noise source is "turned off," and the measured output power is $-58.25$ dBm. The excess noise ratio of the noise source is specified as 15.3 dB. Find the $Y$ factor, the amplifier noise figure, and the effective input noise temperature.

**theory**  The noise power at the output of a two-port device is the sum of two major components: (1) the noise power at the input of the amplifier (noise of the source) increased by the gain of the device under test, and (2) the noise added by the two-port under test. If the total power output is measured for two different levels of input noise, it is then possible to calculate the noise power added by the two-port under test.

In most discussions about noise, the "standard" temperature is 290° above absolute zero (290 K or 17°C) and is symbolized by $T_o$. Even though room temperature may be somewhat different than 17°C, the errors, in assuming that the "off" or "cold" source impedance is at 17°C, are small enough compared with other sources of error that they can be ignored. The excess noise ratio

(ENR) of a noise source is a fraction that is usually expressed in dB. The numerator of the fraction is the noise power output over and above $kT_oB$; it is noise power output in excess of $kT_oB$. The denominator is $kT_oB$. Therefore

$$\text{ENR} = 10 \log \left( \frac{kT_{\text{HOT}}B - kT_oB}{kT_oB} \right)$$

$$= 10 \log \left( \frac{T_{\text{HOT}}}{T_o} - 1 \right) \tag{21-75}$$

$Y$ factor can be defined according to

$$Y = \frac{N_{TOH}}{N_{TOC}} \tag{21-76}$$

where $N_{TOH}$ is the noise power transferred from the output of the unit under test for a hot noise source connected to the input, and $N_{TOC}$ is the noise power transferred from the output of the unit under test for a cold noise source connected to the input. The cold temperature is usually room temperature. $Y$ factor is frequently expressed in decibels.

The definition of noise figure, expressed as a ratio, is

$$F = \frac{N_{TO} \big|_{T_s=290 \text{ K}}}{G_a kT_o B} \tag{21-77}$$

where $B$ is the bandwidth of the unit under test, $G_a$ is the available gain of the unit under test, and the numerator is the noise power available at the output of the two-port for the source at the standard temperature $T_o$. The numerator includes both the amplified thermal noise from the source and the noise added by the unit under test. The denominator is only the amplified thermal noise of the source. The relationship of $F$ to the $Y$ factor can be shown to be

$$F = \frac{\text{ENR}}{Y - 1} \tag{21-78}$$

where ENR, $F$, and $Y$ are all simple ratios rather than quantities of decibels. On specification sheets, $F$ is usually written in terms of decibels.

The effective input noise temperature is defined as the temperature the source would have to be in order to produce, at the output of the unit under test, a power equal to the noise power added by the unit under test. Effective input noise temperature $T_e$ is related to noise figure $F$ by

$$T_e = (F - 1)T_o \tag{21-79}$$

**solution**    The $Y$ factor, from Eq. (21-76), is simply the ratio of the two powers given in the problem.

$$Y_{\text{dB}} = 10 \log (N_{TOH}) - 10 \log (N_{TOC})$$
$$= -52.55 \text{ dBm} - (-58.25 \text{ dBm})$$
$$= 5.7 \text{ dB}$$

As a simple ratio

$$Y = 10^{(5.7/10)} = 3.72$$

The excess noise ratio, converting from decibels, is

$$\text{ENR} = 10^{(15.3/10)}$$
$$= 33.88$$

The noise figure is calculated from Eq. (21-78):

$$F = \frac{\text{ENR}}{Y - 1} = \frac{33.88}{3.72 - 1}$$
$$= 12.46$$

or in decibels:    $$F(\text{dB}) = 10 \log 12.46 = 10.95 \text{ dB}$$

The noise figure also shows the deterioration of the signal-to-noise ratio in going through the unit under test. For this problem, therefore, the signal-to-noise ratio at the output is 10.95 dB worse than at the input of the unit under test. Thus if the signal were 10.95 dB larger than the noise at the input, it would be equal to noise at the output.

The effective input noise temperature is calculated from Eq. (21-79):

$$T_e = (F - 1)T_o = (12.46 - 1)290$$
$$= 3323 \text{ K}$$

## 21.27    CALCULATING THE OVERALL NOISE FIGURE OF CASCADED DEVICES

**problem 21.27**    The noise figure of an input amplifier is 3.2 dB, and the available gain is 6.7 dB. The noise figure of a second-stage amplifier is 7 dB, and the gain is 12 dB. The output stage has a noise figure of 15 dB. Find the noise figure of the three amplifiers connected in cascade.

**theory**    The noise figure of cascaded devices is given by

$$F = F_1 + \frac{F_2 - 1}{G_1} + \frac{F_3 - 1}{G_1 G_2} + \cdots + \frac{F_n - 1}{G_1 G_2 \cdots G_{n-1}} \tag{21-80}$$

where the subscripts 1, 2, 3, etc., refer to individual devices. The noise figures and gains of Eq. (21-80) are all in simple ratios rather than decibels.

**solution**    The individual gains and noise figures must first be converted from decibels to simple ratios.

$$F_1 = 10^{3.2/10} = 2.09$$
$$G_1 = 10^{6.7/10} = 4.68$$
$$F_2 = 10^{7/10} = 5.01$$
$$G_2 = 10^{12/10} = 15.85$$
$$F_3 = 10^{15/10} = 31.62$$

The overall noise figure of the three cascaded units is found from Eq. (21-80):

$$F = 2.09 + \frac{5.01 - 1}{4.68} + \frac{31.62 - 1}{(15.85)(4.68)}$$
$$= 2.09 + 0.86 + 0.41$$
$$= 3.36$$

In decibels: $\qquad\qquad\qquad\qquad F = 5.26 \text{ dB}$

This problem demonstrates that the overall noise figure is mainly determined by the noise figure of the first stage. This is true so long as the gain of the first stage is large enough to reduce the contribution of the second stage to the overall noise figure.

## 21.28    CALCULATING RECEIVER NOISE FIGURE CONSIDERING THE IMAGE RESPONSE

**problem 21.28**    A 10-GHz receiver is composed of an ordinary mixer followed by an i-f amplifier at 30 MHz. The noise figure of the receiver is measured, by measuring the $Y$ factor as in the preceding problem, and found to be 5.6 dB. What is the actual noise figure of the receiver?

**theory**    Ordinary mixers convert all signals that satisfy the relation

$$f_{\text{sig}} = f_{\text{lo}} \pm f_{if} \tag{21-81}$$

where $f_{\text{lo}}$ is local oscillator frequency and $f_{if}$ is the i-f frequency. This shows there are two possible frequency bands that can create a response at the output of the i-f amplifier, one above $f_{\text{lo}}$ and the other below $f_{\text{lo}}$. When a noise figure is measured by means of a broad-band noise source, the noise from the source over both those bands is converted by the mixer and enters the i-f amplifier. If, in its normal operation, the receiver does not utilize the full passband for signal information (most receivers use only one sideband), its operating noise figure will be higher than the measured noise figure. The operating noise figure can be calculated by the equation

$$F \text{ (operating)} = F \text{ (measured)} \cdot \frac{B \text{ (total)}}{B \text{ (utilized)}} \tag{21-82}$$

where $F$ is expressed as a simple ratio instead of in decibels.

**solution**    When the receiver is tuned to 10 GHz, the local oscillator frequency is at either 10.03 or 9.97 GHz. Assume that it is at 10.03 GHz. Then, using Eq. (21-81), the receiver will also respond to signals at 10.06 GHz unless the mixer and its circuitry are considerably more complicated than the "ordinary" mixer specified for this problem. Wherever this receiver is used, the 10.06-GHz band probably has no signals to give confusing outputs. But when the noise source was used, 10.06-GHz noise signals were present. If the mixer is "ordinary," the response at 10.06 GHz can be considered identical to that at 10 GHz. This means that $B$(total) is actually twice the if bandwidth. Yet $B$(utilized) is the actual i-f bandwidth giving a bandwidth ratio of 2 in Eq. (21-82). Equation (21-82) can be written in decibels:

$$10 \log F_{op} = 10 \log F_{mea} + 10 \log \frac{B_t}{B_u}$$
$$F_{op}(dB) = 5.6 + 10 \log (2)$$
$$= 5.6 + 3$$
$$= 8.6 \text{ dB}$$

## 21.29  BIBLIOGRAPHY

Adam, S. F.: *Microwave Theory and Applications,* Prentice-Hall, Englewood Cliffs, N.J., 1969.
Bahl and Trivedi: "A Designer's Guide to Microstrip Line," *Microwaves,* pp. 174–182, May 1977.
Beatty, R. W.: "Intrinsic Attenuation," *IEEE Transactions on Microwave Theory and Techniques,* vol. 11, no. 3, pp. 179–182, May 1963.
Ely, P. C.: "Swept-Frequency Techniques," *Proceedings of the IEEE,* vol. 55, no. 6, pp. 991–1002, June 1967.
Hayt, W. H.: *Engineering Electromagnetics,* 3d ed., McGraw-Hill, New York, 1974.
Hewlett-Packard Company: "Noise Figure Primer," *Application Note 57.*
———: "Fundamentals of RF and Microwave Power Measurements," *Application Note 64-1.*
———: "High-Frequency Swept Measurements," *Application Note 183.*
Lance, A. L.: *Introduction to Microwave Theory and Measurements,* McGraw-Hill, New York, 1964.
Moreno, T.: *Microwave Transmission Design Data,* Dover, New York, 1958.
Mumford and Scheibe: *Noise Performance Factors in Communications Systems,* Horizon House—Microwave, Inc., 1968.
*Reference Data for Radio Engineers,* Sams, Indianapolis, Ind., 1975.
Smith, P. H.: *Electronic Applications of the Smith Chart,* McGraw-Hill, New York, 1969.

# Chapter 22

# Communications Systems

## HAROLD E. ENNES

Consulting electronics writer; formerly Assistant Chief Engineer for
Maintenance, WTAE-TV; Charter Senior Member and Fellow,
Society of Broadcast Engineers

## 22.1 INTRODUCTION

The imparting or interchange of thought, opinion, or information by speech, music, writing, signs, or symbols is the act of communication. The modern form of communication is via applied technology. This technology has become quite complex since telegraphy signals were first transmitted. It now includes systems for the transmission of not only telegraphy, but also AM and FM broadcasting, as well as TV and digital communications systems.

This chapter presents practical problems involving important calculations and measurements in AM, FM, digital, and radio-relay systems. In addition, a number of useful tables are included.

## 22.2 CALCULATING SENSITIVITY RATINGS OF MICROPHONES

**problem 22.1** Compare the sensitivity rating in dBm of the following microphones:
1. Low-impedance type (150 Ω) with output rating of −60 dBm at 10 dynes/cm².
2. Low-impedance type (150 Ω) with EIA rating of −150 dBm.
3. High-impedance type (40,000 Ω) with −60 dBv output rating. (Reference: 1 dyne/cm².)

**theory** A microphone specification sheet may express the effective output rating in terms of voltage or power. The output is obviously dependent on the sound level. Therefore, the specified output is related to the acoustical sound pressure (dynes) at which the output level is given.

Table 22-1 shows typical sound levels in dynes per square centimeter correlated with the decibel scale (where 0 dB = 0.0002 dyne/cm² = threshold of hearing in young persons at 1000 Hz). The effective output levels for microphones other than those with an EIA rating are given for a specified pressure of either 1 or 10 dynes/cm². Note that a 10-times increase in acoustical pressure is equal to a 20-dB gain. Thus the following conversion rules apply:

To convert from 1 dyne/cm² to 10 dynes/cm², add +20 dB.

To convert from 10 dynes/cm² to 1 dyne/cm², add −20 dB.

Most broadcast-type microphones are rated directly in terms of power output (dBm) at a stated sound pressure, usually 10 dynes/cm². Note from Table 22-1 that this rating is near the upper region of the program sound pressure encountered in practice.

The EIA microphone system rating uses a ratio in decibels relative to 1 mW and 0.0002 dyne/cm² (threshold of hearing). Such a rating is given by:

$$G_M = \left(20 \log \frac{V}{P} - 10 \log R_M\right) - 50 \text{ dB} \qquad (22\text{-}1)$$

where $G_M$ = microphone system rating (sensitivity), dBm
$V$ = open-circuit voltage generated by the microphone
$P$ = sound pressure, dynes/cm²
$R_M$ = microphone rating impedance (see Table 22-2)

When a microphone output is specified in terms of voltage, the reference is 1 V (open circuit, or $V_{oc}$), usually at 1 dyne/cm². Thus with 0 dBv = 1 V, −60 dBv, for example, indicates 60 dB below 1 V, open circuit.

The effective output level in milliwatts of a microphone connected to a matching impedance is given by:

$$P_o = 1000 \frac{V_g^2}{4R_M} \qquad (22\text{-}2)$$

where $P_o$ = output level, mW
$V_g$ = open circuit output, V
$R_M$ = nominal microphone impedance, Ω
$P_o$ can then be converted to dBm.

**TABLE 22-1   Typical Sound Levels in decibels and dynes/cm²**

|  | dB | dynes/cm² |  |
|---|---|---|---|
|  | 140 | 2000 |  |
|  | 130 |  | Threshold of pain (130 dB) |
|  | 120 | 200 |  |
|  |  |  | Jet engine control room (115 dB) |
| ——100 | 100 | 20 | Riveter at 35 ft (97 dB) |
| ↑ | 94 | 10 | Inside motor bus (90 dB) |
| Music | 80 | 2 | Factory (78 dB) |
| range | 74 | 1 |  |
| ——68 | 60 | 0.2 | Restaurant (60 dB) |
|  |  |  | Small store (52 dB) |
| ↓  Speech | 40 | 0.02 | House, large city (40 dB) |
| —30  range ——30 |  |  | House, country (30 dB) |
| ↓ | 20 | 0.002 | Average whisper at 4 ft (20 dB) |
| ——16 |  |  | Quiet whisper at 5 ft (10 dB) |
|  | 0 | 0.0002 | Threshold of hearing* |

*0 dB = 0.0002 dyne/cm² (threshold of hearing at 1000 Hz).
*Source:* H. F. Olson, *Elements of Acoustical Engineering,* 2d ed., Van Nostrand, New York, 1947: H. M. Tremaine, *Audio Cyclopedia,* 2d ed., Sams, Indianapolis, Ind., 1969, p. 28.

**TABLE 22-2   Microphone and $R_M$ Impedance**

| Microphone Z, Ω | $R_M$, Ω |
|---|---|
| 19–75 | 38 |
| 75–300 | 150 |
| 300–1200 | 600 |
| 1200–4800 | 2400 |
| 4800–20,000 | 9600 |
| 20,000–80,000 | 40,000 |
| 80,000 or greater | 100,000 |

**solution**   (*a*) The output is stated directly as −60 dBm at a sound pressure of 10 dynes/cm². The other two microphones to be compared must consider what the output would be at this reference acoustical pressure.

(*b*) The EIA microphone rating is based upon 0.0002 dyne/cm² as reference level. The comparison microphone (*a*) is rated at 10 dynes/cm². The ratio of 10 to 0.0002 is 50,000/1, or 94 dB. Thus, to convert from $G_M$ rating to effective output at 10 dynes/cm², add +94 dB to −150 dB, giving −56 dBm.

(*c*) Since the output is −60 dBv at 1 dyne/cm², the output at 10 dynes/cm² is −60 +20 = −40 dBv. Then, either Eq. (22-1) or (22-2) may be used. First, note that −40 dBv = 0.01 V. Then using Eq. (22-1):

$$G_M = \left(20 \log \frac{0.01}{0.0002} - 10 \log 40{,}000\right) - 50 \text{ dB}$$
$$= (20 \log 50 - 10 \log 40{,}000) - 50 \text{ dB}$$
$$= (20)(1.7) - (10)(4.6) - 50 \text{ dB}$$
$$= 34 - 46 - 50$$
$$= -62 \text{ dBm}$$

Equation (22-2) does not involve logarithms. Thus:

$$P_o = 1000 \frac{0.01^2}{4(40{,}000)}$$

$$= 1000 \frac{0.0001}{160{,}000}$$
$$= (1000) \, 6.25(10^{-10})$$
$$= 6.25(10^{-7}) \text{ mW}$$

and $6.25(10^{-7})$ mW = −62 dBm.

To summarize the answers:

(*a*) −60 dBm
(*b*) −56 dBm
(*c*) −62 dBm

## 22.3   DESIGNING A MATCHING *H* PAD

**problem 22.2**   A line amplifier with 150-Ω balance input and maximum input rating of −35 dBm is to be used in a line with typical input level of −30 dBm. Design the required matching pad.

**theory**   Figure 22-1 illustrates an audio arrangement with a multiple sound source mixer that has a typical output of −30 dBm. Since the maximum input level to most line amplifiers is from −32 to −35 dBm, the pad is designed for a safety margin in this example so that typical input level will be −40 dBm. The line amplifier gain is normally adjusted to provide an output of +10 dBm in a 600-Ω load. In this example, the input circuit is 150 Ω on each side of the pad. Owing to shunt capacitance effects across large values of resistors, it is not practical to obtain attenuation of more than 40 dB in a single pad. If greater attenuation is necessary, pads are connected in tandem.

For other than laboratory precision, the nearest EIA value of resistance to the theoretical absolute value can be used.

The drawings at the top of Table 22-3 show the *T* and *H* pad configurations. The *T* pad is used where one side of the circuit is grounded or provides a common return. The *H* pad is used for balanced-to-ground circuitry.

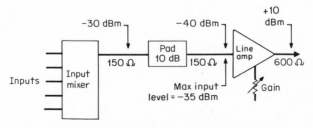

**Fig. 22-1**   Typical audio arrangement.

**solution**   Table 22-3 lists the closest EIA values of resistors to be used when the input and output impedances equal 600 Ω. For other than 600 Ω (but equal impedances), multiply all resistance values by the factor $Z_x/600$. Thus for 150 Ω, the factor is 0.25.

Note that the values of the series resistors in the $H$ pad are essentially one-half the values of those in the $T$ pad, since the two sides of the circuit must total the desired value; $R_2$ is the same in either case, for a given impedance and attenuation.

Since the problem here calls for a balanced pad, the $H$ configuration is used. Also, since the impedance is 150 $\Omega$, the values of all resistors in the table are multiplied by the factor 0.25. Thus for a 10-dB pad:

$$R_2 = (430)(0.25) = 107.5 \ \Omega \quad \text{(nearest EIA value} = 110 \ \Omega)$$
$$R_3 = (160)(0.25) = 40 \ \Omega \quad \text{(nearest EIA value} = 39 \ \Omega)$$

The "stack up" of tolerance in departure from absolute values of resistors in this example will give an attenuation slightly different than 10 dB. This is insignificant in communication circuitry concerned with speech and program waves.

**TABLE 22-3   Design Data for Fixed Pads**

For the case where $Z_i$ and $Z_o$ = 600 $\Omega$. For other than 600 $\Omega$ (but equal impedances), multiply all resistance values by factor $Z_x/600$ (0.41 for 250 $\Omega$, 0.25 for 150 $\Omega$, 0.083 for 50 $\Omega$).

EIA Values Nearest to Exact Values

| Loss, dB | $R_1, \Omega$ | $R_2, \Omega$ | $R_3, \Omega$ |
|---|---|---|---|
| 0.5 | 18 | 10,000 | 8.2 |
| 1.0 | 36 | 5100 | 18 |
| 2 | 68 | 2700 | 36 |
| 3 | 100 | 1800 | 51 |
| 4 | 130 | 1200 | 68 |
| 5 | 160 | 1000 | 82 |
| 6 | 200 | 820 | 100 |
| 7 | 220 | 680 | 110 |
| 8 | 270 | 560 | 130 |
| 9 | 300 | 470 | 150 |
| 10 | 300 | 430 | 160 |
| 11 | 330 | 360 | 160 |
| 12 | 360 | 330 | 180 |
| 13 | 390 | 270 | 200 |
| 14 | 390 | 240 | 200 |
| 15 | 430 | 220 | 200 |
| 16 | 430 | 200 | 220 |
| 17 | 470 | 180 | 220 |
| 18 | 470 | 150 | 240 |
| 19 | 470 | 130 | 240 |
| 20 | 510 | 120 | 240 |
| 22 | 510 | 100 | 270 |
| 24 | 510 | 75 | 270 |
| 26 | 560 | 62 | 270 |
| 28 | 560 | 47 | 270 |
| 30 | 560 | 39 | 270 |
| 32 | 560 | 30 | 300 |
| 34 | 560 | 24 | 300 |
| 36 | 560 | 18 | 300 |
| 38 | 560 | 15 | 300 |
| 40 | 560 | 12 | 300 |

## 22.4   DESIGNING A BRIDGING PAD FOR A 600-$\Omega$ LINE

**problem 22.3**   It is necessary to bridge a terminated 600-$\Omega$ line (balanced) for additional use. Level is +10 dBm. The balanced input amplifier to be bridged across the line must have an input level of −24 dBm to obtain required output. Design the required bridging pad, and calculate the number of such bridges possible.

**theory**    A bridging pad must provide a high impedance to the bus to be bridged and a matching impedance to the input of the amplifier to be used. The bridging impedance should be at least 10 times that of the line to be bridged to avoid attenuation of the bridged circuit due to loading, and so that any effect of shorting or oscillations which could occur on the bridged branch will not disturb the original circuit. A 600-Ω line is the most common impedance to be bridged. 22.5

Figure 22-2 shows the three most common bridging arrangements in practice. When bridging pads are used, a loss of about 28 dB is the lowest practical value, although less loss can be tolerated when only one bridge is to be made and less isolation can be tolerated. The most efficient bridging arrangement is by use of a bridging transformer with typical loss of 18 to 20 dB. These transformers provide a very high impedance to the bridged circuit, and the quantity used is limited only by the effect of shunt capacitances on frequency response of a given system. Bridging amplifiers are also available. Resistive pads are the most economical.

Table 22-4 shows the value of bridging resistor ($R_1$) vs. bridging attenuation in decibels, when bridging a 600-Ω source and feeding a 600-Ω amplifier. The loss for other values of $R_1$ and $R_2$ may be related as:

$$dB = 10 \log \left( \frac{R_1}{R_2} \right) + 20 \log \left( \sqrt{\frac{R_1}{R_2}} + \sqrt{\frac{R_1}{R_2} - 1} \right) \qquad (22\text{-}3)$$

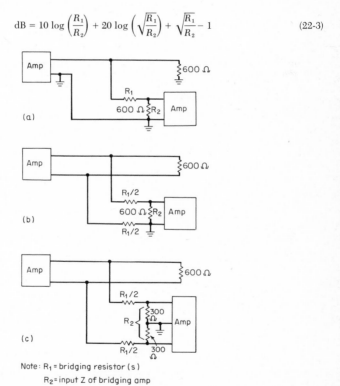

Note: $R_1$ = bridging resistor (s)

$R_2$ = input Z of bridging amp

**Fig. 22-2**    (*a*) Unbalanced-to-unbalanced bridge. (*b*) Balanced-to-unbalanced bridge. (*c*) Balanced-to-balanced bridge.

**solution**    The necessary bridging loss is 34 dB. From Table 22-4, this requires a bridging resistor which totals 15,000 Ω. Since this is a balanced line, $R_1/R_2$ is 7500 Ω in each leg. $R_2$ is two 300-Ω resistors center tapped to the center connection of the balanced amplifier input. It may or may not be connected to physical ground.

The total number of such bridges should be such that the total bridging resistance is at least 10 times 600 Ω, or 6000 Ω. Two 15,000-Ω bridges (paralleled) would be an effective bridging resistance of 15,000/2 = 7500 Ω. Three such bridges would total 15,000/3 = 5000 Ω, which is less than ideal isolation. Therefore, two bridges should be considered as maximum for this problem, when resistive bridges are used.

TABLE 22-4    Value of
Bridging Resistor ($R_1$) for
600-$\Omega$ Circuit

| Loss, dB | $R_1$, $\Omega$ |
|---|---|
| 40.1 | 30,000 |
| 38.5 | 25,000 |
| 36.8 | 20,000 |
| 34.2 | 15,000 |
| 30.7 | 10,000 |
| 28.3 | 7500 |
| 24.9 | 5000 |
| 20.8 | 3000 |

*Note:* The given dB loss is accurate only when bridging a
600-$\Omega$ circuit and feeding a 600-$\Omega$ input bridged amplifier.

## 22.5    FIND THE PROPER READING OF A VU METER ACROSS 150-$\Omega$ INPUT

**problem 22-4**    A standard volume unit (VU) meter is to be used across a 150-$\Omega$ input to a
crossbar switcher with a desired input level of +10 dBm. Find the proper reading.

**theory**    The external circuit for the standard VU meter is illustrated by Fig. 22-3. The network is
as follows:

A. Zero adjuster, 800 to 1000 $\Omega$ typical.

B. Fixed resistor, approximately 3200 $\Omega$, selected so that with A at midposition, $A + B = 3600$
$\Omega$.

C. Meter multiplier $T$ attenuator, 3900-$\Omega$ input and output.

The meter input impedance as seen by a program line is 7500 $\Omega$ except when a "test" position is
provided (Problem 22.7). This is not used in monitoring communications lines.

When the meter is bridged across the program line as in Fig. 22-3 (total bridging impedance of
7500 $\Omega$), the maximum sensitivity is +4 dBm for 0 VU deflection. Most program lines are fed with
+8 or +10 dBm. In this case, the meter multiplier is set to +8 or +10, and the meter indicates +8
or +10 dBm at 0 VU reference.

The VU meter is properly calibrated *only* when connected across 600 $\Omega$. For any other
impedance, the reading must be corrected by adding 10 log (600/Z) where Z is the actual
impedance in ohms. Fig. 22-4 shows this correction factor for 10 to 10,000 $\Omega$.

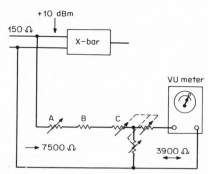

**Fig. 22-3**    When the standard VU meter is used across other than 600 $\Omega$, a correction factor must
be applied as in Fig. 22-4.

**solution**    In Fig. 22-3, the VU meter is connected across 150 $\Omega$. The correction is +6 dB by
using Fig. 22-4 or the relationship 10 log (600/150) = 10 log 4 = (10)(0.6021) = +6 dB. Since 0 VU
= +4 dBm, (+6)+(+4) = +10 dBm. Thus the VU meter will read 0 VU deflection (100%) for a
+10-dBm level in 150 $\Omega$.

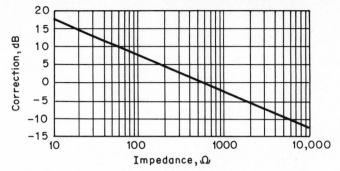

**Fig. 22-4** Correction factor for VU meter connected across various impedances. Correction factor is 10 log (600/Z) where Z = actual impedance in ohms.

## 22.6 CALCULATING INPUT LEVEL AND S/N RATIO FOR A PREAMP

**problem 22.5** A microphone preamplifier with fixed gain of 40 dB is rated as having a noise level of −80 dBm referred to output. Find the minimum input level required to maintain a 60-dB signal-to-noise (S/N) ratio.

**theory** The S/N ratio of a well-designed preamp is largely determined by the input level. A "well-designed" amplifier for a microphone input normally requires that the input stage transistor be operated at very low collector current with low amplification, for best S/N ratio performance.

**solution** Since the noise level is −80 dBm referred to output, the output level required for a 60-dB S/N ratio is −80 + 60 = −20 dBm. Then the required input level for a gain of 40 dB and an output of −20 dBm is −20 + (−40) = −60 dBm.

## 22.7 FIND MAXIMUM INPUT LEVEL AND NEW S/N RATIO

**problem 22.6** The preamp of Problem 22.5 is rated at maximum output level of +18 dBm. Calculate: (a) the maximum input level and (b) the S/N ratio at maximum input level.

**theory** Many preamplifiers are quite flexible in design so that the input can be very low as from a microphone, or at a somewhat higher level (maximum input and/or output is normally specified) for other applications. This example illustrates that the S/N ratio for a given amplifier is largely dependent upon the input level.

**solution**
(a) Maximum input level = +18 − 40 = −22 dBm.
(b) S/N ratio = +18 − (−80) = 98 dB.

## 22.8 MAKING AM FREQUENCY RESPONSE, NOISE, AND DISTORTION MEASUREMENTS

**problem 22.7** You are charged with the responsibility of running frequency response, noise, and distortion measurements on an AM system. Give procedure and calculations.

**theory** The overall performance of any given communications system can be determined only by complete frequency response, noise, and distortion measurements employing sine waves. For broadcast and certain other communications systems, FCC minimum requirements must be met. Where this is not required, the performance should be compared with EIA or manufacturers' specifications.

An overall check must include everything from the microphone input to the transmitter output. In general, performance can be affected by degree of transmitter modulation, and checks are normally made at selected percentages of modulation.

Figure 22-5a shows the output section of a typical signal generator which normally has a low-impedance output (50 to 600 Ω) that can be adjusted to match the load over this range. In this generator, the dBm meter is always loaded by the constant impedance of the variable attenuators—tens, units, and tenths—so that precise measurements can be made.

The actual signal generator output is the reading of the dBm meter minus the setting of the calibrated attenuator. For example, to feed a microphone preamplifier, the generator gain might be adjusted to give a meter reading of +10 dBm and the attenuator set to 60 dB. The actual output is then 10 − 60 = −50 dBm, a typical level for the input circuit of a microphone preamplifier.

When the generator output is matched to the load impedance, the actual dBm input to the system is independent of the value of the load. No conversion in dBm is necessary regardless of the input impedance, providing the generator is matched to the load.

Since a signal generator often provides only a range of lower values of impedance at the output, a matching transformer (Fig. 22-5b) must be used if high-impedance microphones are employed (microphone preamp input of high impedance). This transformer must have a frequency response better than system response. Power loss is considered negligible. The primary matches the generator output, and the secondary matches the input impedance of the preamp.

Figure 22-6 illustrates a simplified block diagram of a typical noise-distortion (N/D) meter which is also used for the frequency response measurements. In measuring total harmonic distortion, the following action occurs:

1. The amplitude of a single-frequency sine wave is measured.

2. A tuning circuit is adjusted to suppress the fundamental frequency. This is done by a sharply tuned circuit and bridge-balance control to get at least 80 dB of suppression at the fundamental frequency.

3. The remaining measured amplitude is the total harmonic distortion.

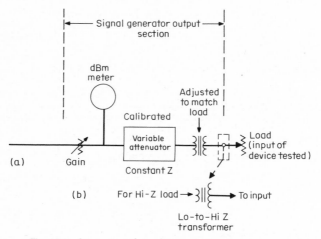

**Fig. 22-5**   Connection of signal generator to system being tested.

**solution**   The first requirement is to draw a block diagram as typified by Fig. 22-7.

The second requirement is to run back-to-back measurements on the audio generator and N/D meter. Be sure this performance meets the manufacturer's specifications. Normally, frequency response is flat within 2 dB from 50 to 15,000 Hz. Distortion should measure less than 0.2% from 50 Hz to 20 kHz, and noise should measure at least 80 dB below the output level of the audio oscillator.

For frequency response runs, remember to apply any necessary correction factor from the back-to-back runs. For example, if the back-to-back response is down 2 dB (relative to a 1000-Hz reference frequency) at 5000 Hz, then feed an input 2 dB higher at this frequency for the test run, or compensate on the tabulated data explained below.

Be sure that all the notations of Fig. 22-7 are strictly adhered to.

## Steps for Frequency Response Measurements

Figure 22-8 shows a typical form for recording frequency response data.

1. Set the audio generator at +10 dBm output and adjust the calibrated attenuator to 60 dB. (This is −50 dBm to the microphone input.) Record the 60 all the way across in row 1 on the form. Copy the 60 also in row 2 for 1000 Hz.

2. Adjust the microphone fader and master gain (where used) for reference modulation of the transmitter. Start with 100% modulation.

3. Place the signal generator on 50 Hz and readjust the generator attenuator (if necessary) for 100% modulation. Record the new attenuation figure in row 2 under 50 Hz. Repeat this procedure for the other frequencies listed.

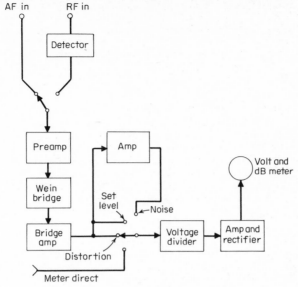

**Fig. 22-6** Simplified block diagram of typical N/D meter.

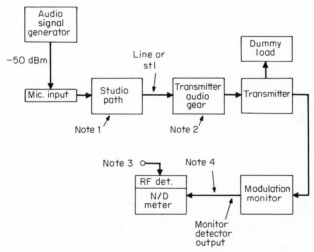

**Fig. 22-7** Typical block diagram for system proof of performance.

% Modulation

| Hz | 50 | 100 | 400 | 1000 | 5000 | 7500 |
|---|---|---|---|---|---|---|
| 1.(Ref) | 60 | 60 | 60 | 60 | 60 | 60 |
| 2. | 58.5 | 59.0 | 59.6 | 60 | 60.5 | 61.2 |
| 3. | +1.5 | +1.0 | +0.4 | 0 | −0.5 | −1.2 |

**Fig. 22-8** Typical form for recording frequency response data.

4. Fill in row 3 by subtracting the readings in row 2 from those in row 1. This is a record of the response variation and may be used to plot a graph if required.

5. Repeat the entire process for lower percentages of modulation. Use a separate form for each modulation percentage.

When a system fails to meet response specifications, determine whether the trouble is at the originating point, in the line (or stl), or at the transmitter. Where the transmitter is remote from the studio or originating point, feed the oscillator directly into the line or stl. This provides the appropriate clue for further trace of the problem.

### Measuring Noise

The reference frequency is 1000 Hz, and this measurement may actually be performed after step 2 of the preceding procedure. Remove the oscillator input and terminate the microphone input with a resistor equal to the microphone impedance. Leave all gains as originally set, but be sure that no other sources are being fed (all other faders should be off). The N/D meter is then placed in the noise mode, and sensitivity is increased to obtain the noise level in the unmodulated carrier. Input stages of preamplifiers are the most common cause of noise and most susceptible to hum pickup.

### Measuring Distortion

1. Obtain the reference modulation at 1000 Hz. Place the N/D meter in the set level position and adjust the meter to 100% calibration.

2. Place the N/D meter in the distortion position and adjust both tuning and bridge balance controls for minimum reading. Increase the sensitivity in steps of 10 dB and adjust these controls for minimum each time.

3. Read the percentage distortion directly on the most sensitive scale possible to obtain a minimum reading by use of the tuning and balance controls.

4. Repeat the procedure at all frequencies used in the frequency response run.

### Carrier Shift

This is a change in the average value of the modulated rf carrier compared with the average value of the unmodulated carrier. Excessive carrier shift results in unwanted harmonics and spurious sideband frequencies with consequent interference on adjacent channels.

Assume, for example, that a rectified unmodulated carrier measures 1 V. Under 100% modulation (usually with a 400-Hz sine wage), the voltage drops to 0.95V. The difference voltage is 0.05 V. This is a 5% carrier shift. In general, this is the maximum carrier shift that should be allowed.

Probable causes of excessive carrier shift are:
1. Overmodulation (check accuracy of modulation monitor)
2. Improper bias or poor bias supply regulation
3. Defective power supply filters
4. Faulty neutralization
5. Improper rf excitation

## 22.9   MAKING FM FREQUENCY RESPONSE, NOISE, AND DISTORTION MEASUREMENTS

**problem 22.8**   You are assigned the responsibility of running frequency response, noise, and distortion measurements on an fm system. Give procedure and calculations.

**theory**   The only difference between this and Problem 22.7 is (usually) the broadened frequency range and the modulation-demodulation (modem) characteristics. Preemphasis is normally employed at the transmitter for a good S/N ratio, with matching deemphasis at the receiver.

Figure 22-9 shows the standard preemphasis curve, the use of which is described under Solution below. Some limited number of general communications fm systems employ a different preemphasis, and this must be obtained from the manufacturer.

**solution**   Figure 22-10 illustrates a form with typical readings for recording fm system data. Note that as the frequency is increased, the attenuator settings must be increased to decrease the modulation owing to preemphasis. Otherwise, the procedure is the same as for Fig. 22-8.

The data of the initial run are plotted as open circles on Fig. 22-9. Note that certain frequencies are outside the curve. In this example, the axis can be shifted −1 dB (closed circles) to fit under the curve. If no correction factor can be found to allow all dots to fall within the curve, remedial measures are required.

Noise and distortion measurements are identical to those of Problem 22.7 except that such measurements must be made through a standard deemphasis demodulator.

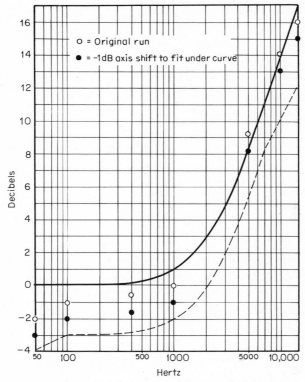

**Fig. 22-9**   Standard 75-μs preemphasis curve with initial and compensated measurements. Initial measurements are those of Fig. 22-10.

% Modulation

| Hz | 50 | 100 | 400 | 1000 | 5000 | 10,000 | 15,000 |
|---|---|---|---|---|---|---|---|
| 1. | 60 | 60 | 60 | 60 | 60 | 60 | 60 |
| 2. | 58 | 59 | 59.4 | 60 | 69.2 | 74 | 76 |
| 3. | +2 | +1 | +0.6 | 0 | −9.2 | −14 | −16 |

**Fig. 22-10**   Typical response data for fm showing preemphasis prior to modulation.

## 22.10   CHECKING THE ACCURACY OF AN AM MODULATION MONITOR

**problem 22.9**   You are given the responsibility of checking the accuracy of an AM modulation monitor. Give procedure and calculations.

**theory**   The typical AM modulation monitor is a semipeak indicating meter, with switch-reversal provision to monitor either the positive or negative side of the modulated envelope individually.

Figure 22-11a shows the pattern of an am envelope modulated 100% by a sine wave. The negative peaks just reach zero carrier without clipping, and the positive peaks reach a certain reference level representing 100% modulation. Figure 22-11b shows the result of overmodulation by a sine wave. The negative peak is clipped, representing carrier cutoff for the duration of the negative peak. This represents overmodulation in the worst form; carrier clipping results in the generation of spurious frequencies exceeding the bandwidth of the am channel and causing interference to adjacent channels (an FCC violation). Overmodulation in the positive direction does not cause adjacent channel interference.

The degree of modulation is given by the following equation:

$$m = \frac{V_{av} - V_{min}}{V_{av}} \tag{22-4}$$

where $m$ = degree of modulation
$V_{av}$ = average envelope amplitude
$V_{min}$ = minimum envelope amplitude

For sine-wave modulation, the peaks and troughs of the envelope will be equal. When the minimum envelope amplitude (negative peak) is zero in Eq. 22-4, $m$ is 1, and the degree of modulation is complete, or 100% expressed in percentage modulation.

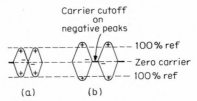

**Fig. 22-11**   AM envelope pattern. (*a*) 100% modulation. (*b*) Overmodulation.

**solution**   Couple an oscilloscope to the final rf stage as shown by Fig. 22-12. If the oscilloscope vertical amplifier has a frequency response suitable to the carrier frequency, the vertical input of the oscilloscope can be used. If the carrier frequency is higher than the oscilloscope amplifier response, a direct connection to the vertical plates of the CRT must be used. In this case, sufficient turns in the link and tightness of coupling must be employed to obtain proper voltage deflection.

Modulate the transmitter and arbitrarily adjust the gain for suitable deflection, such as six graticule divisions for the positive peaks (Fig. 22-13). The percent modulation (%M) is:

$$\%M = \frac{max - min}{max + min}(100) \tag{22-5}$$

For the example of Fig. 22-13:

$$\%M = \frac{6 - 2}{6 + 2}(100) = \frac{4}{8}(100) = 0.5(100) = 50\%$$

Compare the modulation monitor reading with this value on both positive and negative peak position of the selector switch. Run similar checks at every 10% point from 10 to 100% modulation levels. The modulation meter should track these readings on both positive and negative peak values for sine-wave modulation. The oscilloscope graticule vertical calibration line normally has five divisions between major horizontal lines, thus providing sufficient accuracy.

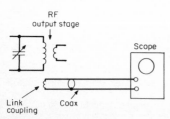

**Fig. 22-12**   Coupling oscilloscope to final rf stage for modulation pattern display.

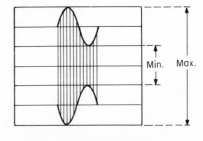

Max. = 6 divisions
Min. = 2 divisions

**Fig. 22-13**   Measuring percentage of modulation on an oscilloscope.

## 22.11    CHECKING THE ACCURACY OF AN FM MODULATION MONITOR

**problem 22.10**    You are given the responsibility of checking the accuracy of an fm modulation monitor. Describe the procedure and provide calculations.

**theory**    The modulated fm wave can be resolved into its carrier and sideband-frequency components, the amplitudes of which vary as Bessel functions of the modulation index according to this equation:

$$m = \frac{\Delta F}{f} \tag{22-6}$$

where $m$ = modulation index
$\Delta F$ = peak deviation, Hz
$f$ = modulating frequency, Hz

Therefore the required modulating frequency $f$ to achieve a given modulation index is:

$$f = \frac{\Delta F}{m} \tag{22-7}$$

At certain values of $m$, the carrier amplitude becomes zero (is crossing the zero axis), and all the energy is transmitted in the sidebands. Figure 22-14 gives the modulation indexes necessary to produce successive carrier nulls. If the modulating frequency $f$ is properly chosen, one of the carrier null points can be made to correspond to the desired deviation.

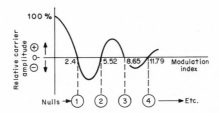

**Fig. 22-14**    Bessel functions of the modulation index.

**solution**    See Fig. 22-15. Tune the communications receiver (with narrow i-f selectivity) to the unmodulated carrier. Peak the beat-frequency oscillator (BFO) in the receiver to the unmodulated carrier.

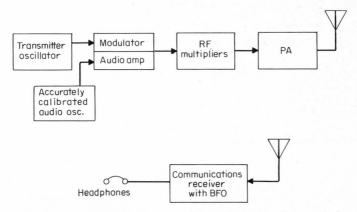

**Fig. 22-15**    Measurement of frequency deviation by the Bessel-zero method.

**example 1**    Assume that the frequency deviation is ±15 kHz for 100% modulation. To use the first null ($m$ = 2.4, Fig. 22-14), the modulating frequency $f$ must be [according to Eq. (22-7)]:

$$f = \frac{15 \text{ kHz}}{2.4} = 6.25 \text{ kHz for 100\% modulation}$$

Set the audio oscillator to 6.25 kHz and, starting with zero amplitude, increase the output until the beat note in the headphones disappears. The first null exists when the beat note first disappears. The fM modulation monitor should now indicate 100% modulation.

Check at lower modulation levels. In this example, a 50% modulation (±7.5 kHz) occurs at the first null when the frequency from the audio oscillator is 3.125 kHz, etc.

**example 2**    Assume that the frequency deviation is ±75 kHz for 100% modulation. To use the first null, the modulating frequency would need to be 31.25 kHz, which is outside the passband of the audio system. The second null can be used, where $f = 75$ kHz/5.52 = 13.586 kHz, which is practical. Or the third null could be used: $f = 75/8.65 = 8.67$ kHz.

Accuracy of measurement with this technique is generally within 5% and depends largely upon the accuracy of the audio oscillator calibration.

## 22.12    RELATIONSHIP BETWEEN SINE WAVES AND SPEECH FOR VU READINGS

**problem 22.11**    A sine wave is often used to "check levels" in communications systems. Give the relationship between sine wave and actual speech or music for VU-meter readings.

**theory**    The rms value of a sine wave is 0.3535 of the peak-to-peak value. The standard VU meter is a full-wave rms device. It is characteristically damped such that relatively sharp peaks in the short bursts of energy in program audio cannot be indicated. Basically, the meter shows a kind of average of program audio, and actual program voltage peaks exceed the meter indication. It is a close approximation of the sound level emanating from the loudspeaker but does not approximate modulation levels of a transmitter.

Even a so-called "peak" meter as used in modulation monitors does not truly indicate peak modulation. Because of this, peak "flashers" are used which respond to sharp peaks of program waves to indicate when 100% modulation is reached or exceeded.

Figure 22-16 illustrates an oscilloscope display of a sine wave arbitrarily adjusted for two divisions on the oscilloscope graticule. This sine wave was fixed at 0 VU on the studio console. Superimposed on this sine wave is the program audio set to trigger the oscilloscope when the VU meter hits 0 VU. The peak-to-peak excursion of the program signal in Fig. 22-16 covers six divisions, a voltage ratio of 3:1, or about 9.6 dB. In practice, this ratio of program peaks to sine-wave voltage at 0 VU deflection can be anywhere from 9 to 12 dB, with 10 dB an average value.

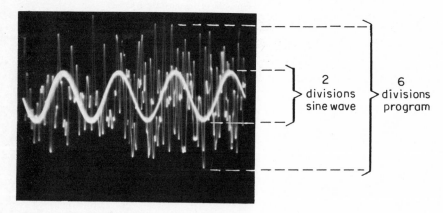

**Fig. 22-16**    Sine wave vs. program wave at 0 VU deflection.

**solution**    If a sine wave is being sent from the control point at 0 VU meter indication, and the transmitter gain control is adjusted for 100% modulation, the application of speech or music monitored by the same VU meter will severely overmodulate the transmitter. If a calibrated gain control is available at the transmitter, first adjust this to get 100% modulation, then reduce the gain by at least 10 dB prior to program modulation. Normally, this gain control is left on the position which previous experience has dictated for normal program operation.

Thus if the gain control is left in normal program position, a sine wave at 0 VU from the control point will modulate the transmitter about 10 dB under 100% modulation, or around 32%.

## 22.13  CALCULATING REQUIREMENTS OF A RADIO RELAY SYSTEM

**problem 22.12**   A radio relay at a 950-MHz operating frequency covers a 20-mi distance. (See Fig. 22-17.) The transmitter power output is 6 W, and the receiver sensitivity requires −115 dBW for a 60 dB S/N ratio. Total transmission line loss is 10 dB, and antenna gain at each end of the path is 15 dB. Specify requirements and show calculations. Evaluate the equipment specifications for this application.

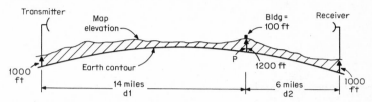

**Fig. 22-17**   Find the necessary height of antennas (assuming equal heights) and evaluate the system for the equipment specifications given (see Problem 22.12).

**theory**   The effectiveness of a radio relay is dependent on the path distance (for a given terrain), the transmitter power output, receiver sensitivity, and signal fading margin allowed.

For analysis of radio relays in general, we shall use the data of Fig. 22-17 as an example. In this figure, a distance of 14 mi from the transmitter (6 mi from the receiver) is found to provide the highest path obstruction. The bulge ($h$) of the earth in feet at a distance $d1$ miles from the near end and $d2$ miles from the far end of the path is:

$$h = 0.5 \; d1 \times d2 \tag{22-8}$$

Energy is characterized into *Fresnel zones*. The primary energy is contained in the first Fresnel zone, and energy contained in even-numbered zones is phase canceling; thus it is desirable to obstruct the energy contained in all but the first Fresnel zone. The first zone must provide ample clearance. A value of 0.6 times the radius of the first Fresnel zone normally is taken as the minimum clearance.

The radius of the first Fresnel zone at the point of major obstruction in the path is:

$$R = 72 \; \sqrt{\frac{AB}{Pf}} \tag{22-9}$$

where $R$ = radius, ft
$A$ = distance from one end of the path to the point of obstruction, mi
$B$ = distance from the other end of the path to the point of obstruction, mi
$P$ = total path length, mi
$f$ = frequency, GHz

The field strength at the receiver is dependent on the free-space path loss (Table 22-5), the total line loss, the total antenna gain (sum of transmitter and receiver gains), and the transmitter output in dBW (decibels above 1 watt where 1 watt = 0 dBW).

The values of free-space loss in Table 22-5 are based on the following relationship:

$$A' = 37 + 20 \log f + 20 \log D \tag{22-10}$$

where $A'$ = free-space loss, dB
$f$ = operating frequency, MHz
$D$ = distance, mi

The *net* path loss is the combined transmission line loss and antenna gain, minus the free-space loss:

$$A = A' - G_t \tag{22-11}$$

where $A$ = net path loss
$A'$ = free-space loss
$G_t$ = combined line loss and antenna gain

The receiver power input is:

$$P_r = P_t - A \tag{22-12}$$

where $P_r$ = receiver power input, dBW
$P_t$ = transmitter power output, dBW
$A$ = net path attenuation

Adequate *fading margin* must be provided to obtain a given reliability factor as shown by Table 22-6.

**TABLE 22-5  Approximate Free-Space Field Loss vs. Frequency***

| Path length, mi | Loss, dB | | | |
|---|---|---|---|---|
| | 950 MHz | 2000 MHz | 7000 MHz | 13,000 MHz |
| 5 | 110 | 117 | 127 | 133 |
| 10 | 115 | 123 | 133 | 140 |
| 15 | 118 | 126 | 136 | 143 |
| 20 | 121 | 129 | 139 | 146 |
| 25 | 124 | 131 | 141 | 147 |
| 30 | 126 | 132 | 142 | 148 |

*Actual free-space loss in dB = $A' = 37 + 20 \log f + 20 \log D$

where $A'$ = free-space attenuation, dB
$\quad f$ = operating frequency, MHz
$\quad D$ = distance, mi

**TABLE 22-6  Approximate Fading Allowance vs. Reliability**

| Path length, mi | dB fade allowance for reliability of: | | |
|---|---|---|---|
| | 99% | 99.9% | 99.99% |
| 10 | 5 | 7.5 | 10 |
| 20 | 14 | 24 | 32 |
| 30 | 20 | 30 | 40 |

*Source:* Average of composite data from various radio relay manufacturers.

**solution**  To find the necessary equal height requirements for the antennas, begin by finding the first Fresnel zone clearance at the major point of obstruction by using Eq. (22-9):

$$R = 72 \sqrt{\frac{(14)(6)}{(20)(0.95)}} \quad \text{Note: 950 MHz = 0.95 GHz}$$

$$= 72 \sqrt{\frac{84}{19}}$$

$$= 72 \sqrt{4.42}$$
$$= (72)(2.1) = 151.2 \text{ ft radius}$$

Then the minimum clearance = $(0.6)(151.2) = 90.7$ ft or approximately 91 ft.
From Eq. (22-8):

$$h = 0.5(14)(8) = 0.5(112) = 56 \text{ ft}$$

Note from Fig. 22-17 that the transmitting and receiving points are the same height above sea level. But point $P$ is 200 ft + 100 ft = 300 ft above this value.
The total is computed as follows:

| | |
|---|---|
| Earth curvature | 56 |
| Minimum Fresnel zone clearance | 91 |
| Terrain and building | 300 |
| Total | 447 ft |

Thus 447 ft is the required height of the sending and receiving antennas to meet minimum requirements.

To evaluate the system per equipment specifications given in the problem, note that a transmitter power output of 6 W is 8 dBW (approx.). The total antenna gain (15 dB at each end) is 30 dB. The effective antenna gain is 30 dB minus total transmission line loss of 10 dB, or 20 dB. Then the net path loss from Table 22-5 and Eq. (22-11) is:

$$A = 121 - 20 = 101 \text{ dB}$$

The receiver power input from Eq. (22-12) is:

$$P_r = 8 - 101 = -93 \text{ dBW}$$

This allows a fade margin of only $115 - 93 = 22$ dB. Note from Table 22-6 that this does not quite provide a 99.9% reliability. To obtain a 99.99% reliability, a fade margin of 32 dB is required for a 20-mi path. In practice, the engineer would recommend a higher-gain antenna system to obtain greater reliability.

## 22.14  CALCULATING THE PLATE CURRENT OF AN RF FINAL AMPLIFIER

**problem 22.13**  Plate voltage of a transmitter final stage is 3000 V. Efficiency factor $F$ as supplied by the manufacturer is 0.7. Find the plate current that must be "loaded into" the final stage, if the power supplied to the transmission line is to be 1000 W.

**theory**  The power output of a transmitter is:

$$P_o = (V_p)(I_p)(F) \tag{22-12}$$

where $P_o$ = power output, W
$V_p$ = plate voltage on final stage
$I_p$ = plate current of final stage
$F$ = efficiency factor furnished by the manufacturer
Therefore, the current that must be drawn by the final stage by loading adjustment is:

$$I_p = \frac{P_o}{V_p(F)} \tag{22-13}$$

**solution**

$$I_p = \frac{1000}{3000(0.7)} = \frac{1000}{2100} = 0.476 \text{ A}$$

## 22.15  CALCULATING POWER INPUT TO FINAL STAGE OF SSSC TRANSMITTER

**problem 22.14**  Assume that Problem 22.13 concerned a double-sideband AM transmitter. Calculate the power input to the final stage of a single-sideband suppressed carrier (SSSC) transmitter to produce the same sideband power as that obtained in Problem 22.13. Assume the same value for $F$.

**theory**  Figure 22-18 illustrates the block diagram of a typical SSSC modem, with frequency designations.

In the transmitter, the low-level oscillator is typically 20 kHz for voice band frequency transmission of 300 to 3000 Hz. The audio and 20-kHz carrier are both canceled in the balanced modulator, leaving only the sideband frequencies plus and minus the carrier frequency. The sideband filter passes only the upper sideband which is multiplied in frequency by the first mixer-oscillator combination. In this example, the frequency is again increased to the required carrier frequency which is only one sideband. As many multiplier stages as necessary are used to reach the final transmission frequency. The receiver section of the modem reverses the process to extract the original signal.

In a double-sideband transmission, one-half the carrier power is distributed in the sidebands for 100% modulation. Since there are two sidebands, each sideband contains one-fourth the carrier power. A single sideband carries all the necessary information of the initial signal.

Bandwidth requirements for an SSSC transmitter are slightly less than one-half that required for a full-carrier, double-sideband transmission. The power input to the final stage is only one-fourth of that required for double-sideband. The narrower bandwidth and increased sideband power for a given power input improve the S/N ratio. The disadvantage is that the transmitters and receivers are more complex. The reinserted carrier in the receiver must be within ±10 Hz of that employed at the transmitter to avoid distortion.

**solution**  The double-sideband transmitter of 1000-W output produces two sidebands of 500 W total, or 250 W in a single sideband. The equivalent SSSC transmitter need produce only 250 W output, or one-fourth the power. This could be obtained by halving both the plate voltage $V_p$ and plate current $I_p$ of Problem 22.13 for values of:

$$V_p = 1500 \text{ V}$$
$$I_p = 0.238 \text{ A}$$

Then [using Eq. (22-12)]:

$$P_o = (1500)(0.238)(0.7) = 250 \text{ W}$$

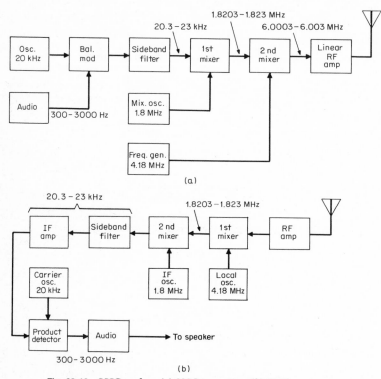

(a)

(b)

**Fig. 22-18**    SSSC modem. (*a*) SSSC transmitter. (*b*) SSSC receiver.

## 22.16   CALCULATING THE EFFECTIVE RADIATED RF POWER

**problem 22.15**   The power output of an fm transmitter is 10 kW. Transmission line efficiency (for type, operating frequency, and length of line required) is 80%. Antenna power gain is 3. Calculate the effective radiated power (erp). Solve by two methods, percent and decibel.

**theory**   The effective radiated power of a transmitter is the product of power output, transmission line efficiency, and antenna gain. If a harmonic filter is used, the loss (if any) must be added to transmission line loss. In television, losses in diplexer and sideband filter (if used) must be added to the transmission line loss. The relationship is:

$$\text{erp} = P_o \times T_e \times G_A \tag{22-14}$$

where $P_o$ = transmitter power output
$\quad T_e$ = all transmission losses
$\quad G_A$ = antenna gain

Power may be referred to in terms of dBk, or dB above or below 1 kW of power. See the graph of Fig. 22-19a. Figure 22-19b provides a convenient scale to convert transmission line efficiency to dB attenuation, or vice versa. Advantages of using the dB scale are:

1. Transmission losses may be subtracted directly from transmitter power level in dB above or below 1 kW.

2. Antenna gain in dB may be added directly to transmitter power level in dBk.

3. An increase of 1 dB at the transmitter results in an increase of 1 dB in received field strength.

**solution**   By the percent method:

$$\text{erp} = (10,000)(0.8) = 8000$$
$$(8000)(3) = 24,000 \text{ W erp}$$

By the decibel method:

$$10,000 \text{ W} = +10 \text{ dBk (Fig. 22-19}a)$$
$$80\% \text{ efficiency} = 1 \text{ dB (Fig. 22-19}b)$$
$$\text{Power gain of 3} = 4.8 \text{ dB}$$
$$\text{erp} = +10 - 1 + 4.8 = +13.8 \text{ dBk}$$

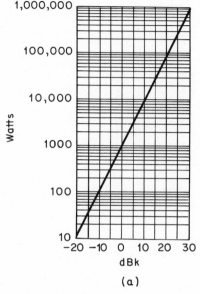

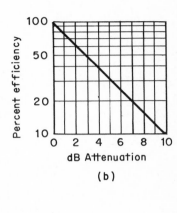

**Fig. 22-19** Useful data for transmission measurements. (*a*) dBk and power in watts. (*b*) Percent efficiency and dB attenuation.

## 22.17 CALCULATING EFFECTIVE RADIATED POWER WITH FCC CHART

**problem 22.16** Calculate the erp required from an fm station to produce 100 $\mu$V/m at 60 mi from an antenna 500 ft high, using an FCC chart.

**theory** The field strength is conveniently stated in terms of dBu, where 0 dB = 1 $\mu$V/m. Figure 22-20 shows this relationship, which is also shown in scale form on the FCC chart, Fig. 22-21. This chart is used to determine either a given field strength contour or the required erp to produce a given field strength at a certain distance, for a given antenna height. The chart is based on a receiving antenna height of 30 ft. As an example of its use to estimate the approximate radius of the area within a given field-strength contour, assume that the erp is 1 kW and effective antenna height above average terrain is 200 ft. Find the distance to the 1000-$\mu$V/m contour.

Note that 1000 $\mu$V/m = +60 dBu. Place a straightedge from 60 on the right edge to 60 on the left edge of the chart. Mark the point at which the vertical line representing 200 ft (noted along the top and bottom of the graph) intersects the straightedge. In this example, the distance to the 1000-$\mu$V/m contour is 9 mi.

Already drawn on the chart is a reference line connecting the 40-dBu (100-$\mu$V/m) points at the left and right edges. This line intersects the theoretical contour limits for 100 $\mu$V/m (40 dBu) at various antenna heights.

The FCC graph is drawn on the assumption of 1 kW (0 dBk) power. To use the chart for any other power, subtract the number of dBk for the power used from the number of dBu corresponding to the desired signal strength; then use the resulting number in the manner just described.

**solution** Locate the 60-mi point on the 500-ft antenna height line and place a straightedge (preferably transparent) across the chart through this point parallel with the bottom line. The field strength resulting from 1-kW erp is read on the left edge of the chart. In this case, it is 26 dBu. The required field strength is 100 $\mu$V/m = 40 dBu. The difference, 40 − 26 = 14 dB, is added to 0 dBk to obtain + dBk erp required. This is 25,120 W.

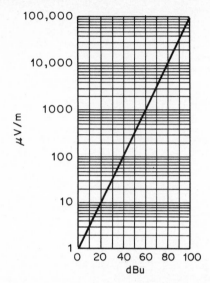

**Fig. 22-20**  Graph relating dBu to microvolts per meter.

## 22.18  CALCULATION OF FIELD STRENGTH WITH A REPLACEMENT ANTENNA

**problem 22-17**  An antenna with power gain of 3 dB results in a monitor point reading of 100 $\mu$V/m. This antenna is replaced by an antenna with a power gain of 8 dB. Calculate the expected new field strength at the same monitor point.

**theory**  Antenna gains are generally given relative to the "standard half-wave dipole" which gives a field intensity of 137 mV/m at 1 mi for a power of 1000 W. In free space, field intensity is independent of the frequency (wavelength) of operation. If an antenna gain is given relative to an isotropic radiator, the gain of a standard half-wave dipole is 1.64 that of the isotrope.

If a given antenna results in a free-space field intensity of 137 mV/m at 1 mi with only 100 W of power, the power gain (relative to a half-wave dipole) is 1000/100 = 10. When the power gain is known or specified, the field gain is the square root of the power gain: $E_G = \sqrt{P_G}$.

**solution**  The new antenna has an effective gain of 8 − 3 = 5 dB over the old antenna. For the old antenna, 100 $\mu$V/m = 40 dBu. (See Fig. 22-20.) Then 40 dBu + 5 dB = 45 dBu, and 45 dBu = 200 $\mu$V/m (approximate).

To be more exact, a power gain of 5 dB = 3.162. Then the field gain = $\sqrt{3.162}$ = 1.778. Therefore (100)(1.778) = 177.8 $\mu$V/m.

## 22.19  CALCULATING CHANNEL CAPACITY IN BITS PER SECOND

**problem 20.18**  A digital communication signal is to be transmitted on a line with bandwidth of 3000 Hz and 20-dB S/N ratio. Calculate maximum theoretical channel capacity in binary digits (bits) per second.

**theory**  The Hartley-Shannon law is the basic principal of information theory. This law relates the *rate* of transmission of information, or *capacity,* of the communication channel as:

$$C = B \log_2 (1 + \text{S/N}) \qquad \text{bits/s} \qquad (22\text{-}15)$$

where  $C$ = channel capacity, bits/s
  $B$ = bandwidth, Hz
  S/N = signal-to-noise average power ratio

Note that for bits the logarithm to base 2 is used rather than the more conventional base 10. $\log_2$ tables are not readily available. Figure 22-22 may be used for a very close estimate, or base 10 (common) logarithms can be used to obtain the answer in *hartleys.* This answer, multiplied by 3.32, gives the solution in bits/s.

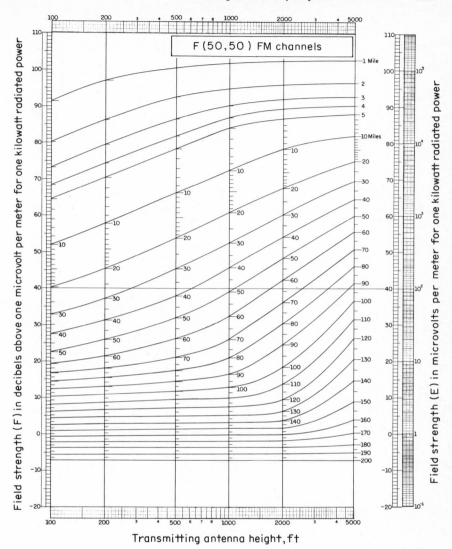

FM channels    Estimated field strength exceeded at 50 percent of the
potential receiver locations for at least 50 percent of
the time at a receiving antenna height of 30 ft

**Fig. 22-21**    FCC chart for fm broadcast channels.

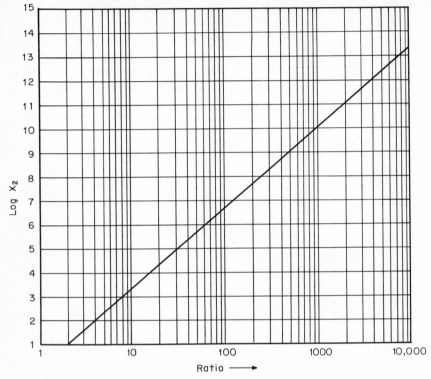

**Fig. 22-22** Log $_2$ graph.

**solution** To get the answer directly in bits/s, $\log_2$ is used:

$$C = 3000 \log_2 (1 + 100) \qquad \text{Note: 20-dB power} = 100/1 \text{ ratio.}$$
$$= 3000 \log_2 (100) \qquad \text{The 1 can be dropped.}$$
$$= 3000 \,(6.64) \qquad \text{From the graph of Fig. 22-22.}$$
$$= 19{,}920 \text{ bits/s}$$

Note carefully that the S/N ratio is converted to actual power ratio. Also with such high values of power ratio, the 1 is insignificant. This becomes significant at low power ratios.

To solve first by using common logarithms:

$$C = 3000 \log 100$$
$$= 3000 \,(2)$$
$$= 6000 \text{ hartleys}$$

Then to convert hartleys to bits/s:

$$(6000)(3.32) = 19{,}920$$

## 22.20 CALCULATING SIGNAL-TO-NOISE RATIO FOR A DIGITAL SYSTEM

**problem 22.19** For a digital communication system with capacity of 1 megabit/s, find the minimum S/N ratio required for a transmission path of 200-kHz bandwidth.

**theory** Note from Eq. (22-15) that the minimum S/N power ratio necessary for a binary channel of known bit rate and bandwidth can be found. The simplified relationship in terms of the S/N ratio is:

$$\text{S/N} = 2^{C/B} - 1 \tag{22-16}$$

**solution**

$$S/N = 2^{106/2(10^5)} - 1$$
$$= 2^5 - 1$$
$$= 32 - 1$$
$$= 31$$

To convert a power ratio of 31 to decibels:

$$dB = 10 \log 31$$
$$= 10 (1.4914)$$
$$= 14.9 \text{ dB, or approximately 15 dB}$$

## 22.21  BIBLIOGRAPHY

Ennes, H. E.: *AM-FM Broadcasting; Equipment, Operations and Maintenance,* Sams, Indianapolis, Ind., 1974.

——: *Television Broadcasting; Equipment, Systems, and Operating Fundamentals,* Sams, Indianapolis, Ind., 1971.

Everitt, *Communication Engineering,* McGraw-Hill, New York.

Jakes, W. C., Jr.: *Microwave Mobile Communications,* Wiley, New York, 1974.

Jordan, E. C.: *Electromagnetic Waves and Radiating Systems,* Prentice-Hall, Englewood Cliffs, N.J. 1950.

Miller, J. R. (ed.): *Communications Handbook,* Pt. I and Pt. II, Texas Instruments, Inc., Dallas, 1965.

Panter, P. F.: *Communication Systems Design,* McGraw-Hill, New York, 1972.

Pender, Harold, and McIlwain, Know: *Electrical Engineers Handbook (Communication and Electronics),* 4th ed., Wiley, New York, 1950.

*Reference Data for Radio Engineers,* 4th ed., International Telephone and Telegraph Corp., New York, 1956.

Terman, F. E.: *Electronic and Radio Engineering,* 4th ed., McGraw-Hill, New York, 1955.

Chapter **23**

# Measurements

## SOL PRENSKY

President, Radiolab Publishing Co. Formerly, Professor of Electrical
Technology, Fairleigh Dickinson College

## TED M. HEINRICH

Manager, Industrial Controls, Research & Development Center,
Westinghouse Electric Corp., Pittsburgh.

### 23.1 INTRODUCTION

Numerous measurements and the principles involved are to be found in many of the
individual chapters of this handbook. For example, measurements relating to components
of a communications system will be found in Chapter 22. Measurements relating to
transmission lines and antennas will be found in Chapters 18, 20, and 21. And so on.

In this chapter, measurement techniques involving meters, an oscilloscope, and a
Wheatstone bridge are presented. These include common and special measurements and
the determination of the accuracy of some of the measurements. Also included is the
measurement of a square wave using two different meters. The method of determining
the accuracy of a digital frequency meter and a digital voltmeter is illustrated. Another
important measurement explained here is the error introduced by any oscilloscope when
measuring the rise time of a pulse.

### 23.2 DETERMINING METER SENSITIVITY

**problem 23.1** Find the basic current sensitivity of a moving-coil (analog type) meter.

**theory** In determining the sensitivity of the common analog or d'Arsonval type of meter (where
a moving coil causes a pointer deflection, as opposed to the digital type), the fundamental factor
that determines the meter reading is the amount of direct current required for full-scale deflection
(FSD) of the pointer. This is true no matter how the meter is designated, whether as a multimeter
(MM) or volt-ohm-milliammeter (VOM). The position of the pointer on any particular scale
depends on how much direct current is actually flowing through the moving coil of the meter.

Consequently, the first value we wish to know about any such meter is the full-scale deflection.
The most common general-purpose analog meter now used is a 0- to 50-$\mu$A unit.

On many meters, the value for full-scale deflection is shown in small type at the bottom of the meter face. This is generally the case when the meter circuitry contains series resistors (for voltage ranges) or shunts (for additional current ranges). In order to measure the full-scale deflection directly, it is necessary to connect the terminals of the moving-coil in series with a variable power source and an accurate standard meter. By varying the voltage applied to the two meters in series until the unknown meter is deflected to full scale, we determine its full-scale current value from the reading of the standard meter. This basic value can then be the basis for other scales, for whatever other functions we may wish the basic meter to provide.

## 23.3   CONVERTING THE BASIC METER INTO A VOLTMETER

**problem 23.2**   Starting with a basic 0- to 50-$\mu$A moving-coil meter (full scale = 50 $\mu$A), devise a circuit that will convert the basic meter into a voltmeter having a desired voltage range of 0 to 10 V direct current.

**theory**   Although the basic meter is a current meter, it is arranged to act as a voltmeter if the proper resistor (also called a multiplier) is used in series with it, as shown by $R_s$ in Fig. 23-1. The proper value of this resistor, which will limit the full-scale current to 50 $\mu$A when 10 V direct current is applied, is obtained from Ohm's law as follows.

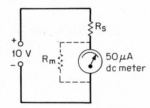

**Fig. 23-1**   A 50-$\mu$A dc meter employed as a 0- to 10-V dc voltmeter.

Calling the total resistance $R_t$ ($R_s + R_m$) and the resistance of the meter $R_m$, we have

$$50 \ \mu\text{A} = \frac{10 \text{ V}}{R_t} \quad \text{and} \quad R_t = R_s + R_m$$

$$R_t = \frac{10 \text{ V}}{0.000050 \text{ A}} \tag{23-1}$$

or

$$R_t = \frac{10 \text{ V}}{50 \ (10^{-6}) \text{ A}}$$

Then, $R_t = 0.2(10^6) = 200{,}000 \ \Omega$, or 200 k$\Omega$.

And

$$\begin{aligned} R_s &= R_t - R_m \\ &= 200{,}000 - 1000 \\ &= 200 \text{ k}\Omega \quad \text{(neglecting } R_m)^* \end{aligned} \tag{23-2}$$

## 23.4   CALCULATING THE OHMS/VOLT RATING FOR A DC VOLTMETER

**problem 23.3**   Calculate the ohms/volt rating for the dc voltmeter of Sec. 23.2. (This method applies generally for all dc voltmeters.)

**theory**   The expression *ohms/volt* refers to the sensitivity of the current meter when used in measuring volts, and its value indicates the amount of resistance required to produce full-scale current when a unit input is applied as shown in Fig. 23-2.

$$\Omega/\text{V} = \frac{1 \text{ V}}{\text{full-scale current}} \tag{23-3}$$

*Since $R_m$ (usually around 1 k$\Omega$) is so much smaller than the 200 k$\Omega$ of $R_s$, we can neglect this small percentage contributed by $R_m$ (1/200 or ½%) and still be well within the usual ±2% tolerance of the meter. This simplification of neglecting $R_m$ (while keeping within the ±2% tolerance of the meter) will hold as long as $R_s$ is about 50 times greater than $R_m$.

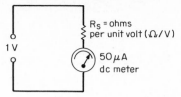

**Fig. 23-2**   Voltmeter sensitivity expressed as ohms/volt = 20 kΩ/V.

**solution**   The meter used in Sec. 23.2 has a full-scale deflection current of 50 μA. Therefore,

$$\Omega/V = \frac{1\ V}{50 \times (10^{-6})}$$

$$= \frac{1,000,000}{50}$$

$$= 20,000 \text{ or } 20 \text{ k}\Omega/V^*$$

## 23.5   USING AN EXISTING LOW-VOLTAGE METER SCALE TO ACCURATELY MEASURE A HIGHER VOLTAGE

**problem 23.4**   A multimeter or volt-ohm meter (VOM) is the only meter available and is to be used to measure the state of charge of a 12-V lead-acid battery. When fully charged and with a light load (say, 0.5 A), the voltage across the battery terminals is 12 V.

When the battery is discharged, under the same load condition, the battery voltage will be 10.8 V. The problem is that the VOM has ranges (scales) of 2.5, 10, 50, 250, and 1000 V. Thus, to read the state of charge of the battery, the 50-V scale must be used. However, the meter's accuracy is guaranteed only at full scale. The meter's accuracy is 1.5% of full scale. On the 50-V scale, this means that the reading may be in error by 1.5% of 50 V, or 0.75 V. In measuring 12 V on the 50-V range, an error of 0.75 V is equal to 6.25% of 12 V. Since the terminal voltage difference between a fully charged and a discharged battery is only 1.2 V, this accuracy of 6.25% is unacceptable. However, we can employ the voltmeter's 2.5-V scale to advantage and approach the inherent 1.5% accuracy of the meter. A resistive divider may be assembled to scale down the battery voltage to about 2.5 V so that the voltage measurement may be made on the 2.5-V scale. Figure 23-3 shows this arrangement. If the meter resistance is 20,000 Ω/V ± 0.5%, what value should resistor $R_1$ be set to for maximum accuracy? If $R_1$ can be set to within 0.1% of the desired value, what will be the accuracy of the measurement?

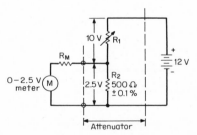

**Fig. 23-3**   Using a 2.5-V meter to check a 12-V battery. *Note:* The voltages across $R_1$ and $R_2$ would seem to add to more than the 12-V battery voltage. However, this is an assumption to simplify calculations (see text).

---

*When the ohms/volt figure is known, to mentally calculate the series resistance required for any desired range of voltage becomes very simple. Thus, in the previous problem for a voltage range of 0 to 10 V direct current, the figure of 20 kΩ/V is simply multiplied by the desired range of 10 V to obtain 20 kΩ/V × 10 = 200 kΩ for the series resistor. It should be kept in mind, however, that this figure is the total resistance $R_b$ which is the same as series resistance $R_s$ only when the meter resistance $R_m$ can be neglected, as was explained in Sec. 23.4. Note also that all the above ohms/volt calculations were concerned only with a dc voltmeter.

**solution**    The meter resistance is calculated by the full-scale voltage rating multiplied by the ohms/volt constant; therefore, the meter resistance $R_M$ is:

$$R_M = 2.5 \times 20{,}000$$
$$= 50{,}000 \ \Omega \qquad (23\text{-}4)$$

The parallel combination of the meter resistance and the 500-$\Omega$ resistor $R_2$ ($R_p$), can now be calculated.

$$R_p = \frac{R_M \times R_2}{R_M + R_2}$$

$$= \frac{R_M \times 500}{R_M + 500}$$

$$= \frac{25 \times 10^6}{50.5 \times 10^3} \qquad (23\text{-}5)$$

$$= 495.05 \ \Omega$$

At this point we shall calculate the minimum and maximum values for $R_p$, since they are required to calculate the overall tolerance. The value of $R_p$ will be maximum when both $R_M$ and $R_2$ have their maximum values and minimum when both $R_M$ and $R_2$ have their minimum values. Repeating the above calculations for the maximum and minimum values of $R_p$, we obtain

$$R_{p,\max} = 495.56$$

and
$$R_{p,\min} = 494.53$$

Note that shunting the meter resistance which has a tolerance of $\pm 0.5\%$ with a tighter tolerance resistor ($R_2$) of much lower resistance has the effect of improving the tolerance to essentially that of the lower resistance, namely $\pm 0.1\%$.*

Now the resistance $R_1$ is calculated to provide a full-scale deflection of 2.5 V of the meter, from the 12-V battery.

At this point it is wise to consider the ease of reading the meter. It is more convenient for 12.5 V to produce a 2.5-V reading on the meter, since then it is necessary merely to multiply the meter reading by 5 rather than by some irrational number. We may now calculate the nominal value of $R_1$. Since the current through $R_p$ must equal that through $R_1$ (see Fig. 23-3), we may write:

$$I = \frac{12.5 \ \text{V} - 2.5 \ \text{V}}{R_1} \qquad (23\text{-}6)$$

and also
$$I = \frac{2.5 \ \text{V}}{R_p}$$

Combining the above two terms yields:

$$R_1 = \frac{10 \ \text{V} \times R_p}{2.5 \ \text{V}}$$

$$= 4 \times R_p$$
$$= 4 \times 495.05 \qquad (23\text{-}7)$$
$$= 1980.2 \ \Omega$$

Now the worst-case accuracy of the resistive divider ratio may be calculated. The ratio has its lowest value of accuracy when the value of $R_1$ is greatest and that of $R_p$ is least. Conversely, the ratio has its highest value of accuracy when the value of $R_1$ is least and that of $R_p$ is greatest.

Therefore
$$k_{\min} = \frac{R_{p,\min}}{R_{1,\max} + R_{p,\min}}$$

$$= \frac{494.53}{1980.2 \times 1.001 + 494.53} \qquad (23\text{-}8)$$

$$= 0.1997$$

Conversely,
$$k_{\max} = \frac{R_{p,\max}}{R_{1,\min} + R_{p,\max}}$$

$$= \frac{495.56}{1980.2 \times 0.999 + 495.56} \qquad (23\text{-}9)$$

$$= 0.2003$$

*This technique lowers the sensitivity of the meter and should be used only when the internal resistance of the voltage source is low (less than 1.0 $\Omega$). For a lead-acid battery, the internal resistance is approximately 0.01 $\Omega$.

The *nominal* value of $k = 0.200$, and the tolerance of $k$ is $\pm 0.0003$. Therefore

$$\% \text{ tolerance} = \pm \frac{0.0003}{0.200} \times 100$$

$$= \pm 0.15\% \tag{23-10}$$

The overall tolerance is the sum of the tolerance of the meter and the voltage divider ratio, which is $\pm 1.65\%$. The approximate voltage error this represents at the 12-V level is $\pm 0.198$ V, rather than the $\pm 0.75$ V which would have resulted had the 50-V scale been used.

## 23.6  MODIFYING THE BASIC DC METER TO PROVIDE HIGHER CURRENT RANGES

**problem 23.5a**   It is required to produce a new current range 25 times larger than the original 0-to1-mA direct current of a basic dc meter. The resistance of the meter $R_m$ is 120 Ω. Find the value of the shunt resistance $R_{sh}$.

**theory**   The amount of current required for full-scale deflection of this basic meter, 0 to 1 mA, is the starting point, together with the meter resistance of 120 Ω. In order to change the range to 0 to 25 mA, it is necessary to connect a resistor, called a *shunt*, across the meter terminals, so that when 25 mA is applied to the parallel combination, only 1 mA flows through $R_m$ of the meter, while the additional 24 mA flows through the shunt resistor $R_{sh}$ in Fig. 23-4a. By inspection, in this simple case, the required value of the shunt resistor $R_{sh}$ must be $\frac{1}{24}$ of $R_m$, or

$$R_{sh} = \frac{R_m}{24}$$

$$= \frac{120}{24} \tag{23-11}$$

$$= 5 \text{ Ω}$$

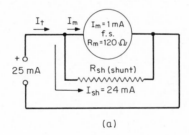

(a)

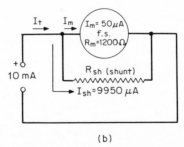

(b)

**Fig. 23-4**   (*a*) Obtaining a 25-mA full-scale range, with a 0- to 1-mA dc meter. (*b*) Obtaining a 10-mA full-scale range with a 0- to 50-μA dc meter.

For the more general situation, the *current-division* formula is used to find the value required for the shunt resistor $R_{sh}$. This formula is given in terms of the full-scale current $I_m$ of the basic meter and the total current $I_t$ for the desired range:

$$I_m = I_t \frac{R_{sh}}{R_{sh} + R_m} \tag{23-12}$$

Rearranged to solve for shunt resistance $R_{sh}$, this formula is

$$R_{sh} = \frac{I_m R_m}{I_t - I_m}$$

$$= \frac{1 \text{ mA} (120)}{25 \text{ mA} - 1 \text{ mA}}$$

$$= \frac{120}{24} \quad \text{as before}$$

$$= 5 \,\Omega$$

(23-13)

**problem 23.5b**   Using the more common 0- to 50-$\mu$A basic meter as in Fig. 23-4b, find the shunt resistance $R_{sh}$ required to produce a range of 10 mA. The meter resistance $R_m$ is 1200 $\Omega$.

**solution**   Using the above formula to find $R_{sh}$, and expressing both currents ($I_t$ and $I_m$) in microamperes yields:

$$R_{sh} = \frac{50 \,\mu\text{A} (1200)}{10,000 \,\mu\text{A} - 50 \,\mu\text{A}}$$

$$= \frac{50(1200)^*}{10,000}$$

$$= \frac{60,000}{10,000}$$

$$= 6 \,\Omega$$

## 23.7   USING A ZENER DIODE IN CONJUNCTION WITH A METER SHUNT TO MEASURE A SPECIFIC VOLTAGE

**problem 23.6**   One way in which a battery voltage may be measured accurately is with the arrangement shown in Fig. 23-5. The zener diode will conduct current as long as the battery voltage is above 10 V and the meter will indicate the value of current which is conducted by the zener diode. This current is proportional to the difference of the battery voltage and the zener diode voltage so that the indication given on the ammeter is extremely sensitive to small variations in the battery terminal voltage. The purpose of the shunt resistor may not seem obvious. Why not use a series resistor so that 12 V on the battery produces 50 $\mu$A into the 10-V zener diode? This is not practical because the zener voltage is not well defined when only 50 $\mu$A is conducted. In order to operate the zener "over its knee" or in its well-defined region, about 1 mA must be supplied to the zener, and hence the meter must operate with a shunting resistor. What value should this resistor have in order to provide full-scale indication when the battery is fully charged?

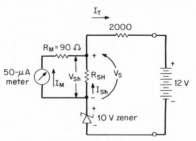

**Fig. 23-5**   A method of measuring the battery voltage using a microammeter, a zener diode, and a shunt resistor.

**theory**   The current that flows into the parallel combination of the meter shunt and meter coil produces a voltage drop of $V_{sh}$, and the current $I_T$ must equal the sum of $I_M$ and $I_{sh}$:

$$I_T = I_M + I_{sh}$$

(23-14)

*We are ignoring the 50 $\mu$A in comparison with 10,000 $\mu$A, since the error is very small. If we subtract the 50 $\mu$A, the value of $R_{sh}$ is 6.03 $\Omega$.

$$I_M = \frac{V_M}{R_M} \tag{23-15}$$

$$I_{sh} = \frac{V_M}{R_S} \tag{23-16}$$

$$= \frac{I_M R_M}{R_{sh}}$$

Substituting Eq. (23-16) into Eq. (23-14) gives

$$I_T = I_M + \frac{I_M R_M}{R_{sh}} \tag{23-17}$$

$$= I_M \frac{1 + R_M}{R_{sh}}$$

By rearranging terms we obtain

$$I_M = I_T \frac{R_{sh}}{R_M + R_{sh}} \tag{23-18}$$

This equation indicates that the current $I_T$ can be scaled down by choosing a suitably small value of $R_{sh}$.

**solution**    The total current which will flow in the zener diode is given by the difference of voltage between a fully charged 12-V battery and the 10-V zener voltage, divided by the series combination of 2000 $\Omega$ and the parallel combination of the meter resistance and shunt resistance. For the moment, let us assume the parallel resistance of the meter and shunt is so small that the series combination with the 2000-$\Omega$ resistor is practically 2000 $\Omega$. The current flowing in the zener diode when the battery is fully charged is

$$I_T = \frac{12 - 10}{2000} \tag{23-19}$$
$$= 1 \text{ mA}$$

Using Eq. (23-18), we may now calculate the value of $R_{sh}$ to provide full-scale 50-$\mu$A current into the meter:

$$I_M = \frac{I_T \times R_{sh}}{R_M + R_{sh}} \tag{23-20}$$

or

$$50 \times 10^{-6} = \frac{1 \cdot 10^{-3} \times R_{sh}}{R_M + R_{sh}}$$

and

$$R_M + R_{sh} = 20 \times R_{sh}$$

Therefore

$$R_{sh} = \frac{R_M}{19}$$

$$= \frac{90}{19}$$

$$= 4.74 \ \Omega$$

The parallel combination of $R_{sh}$ and $R_M$ is

$$R_p = \frac{R_M \times R_{sh}}{R_M + R_{sh}}$$

$$= \frac{90 \times 4.74}{94.74} \tag{23-21}$$

$$= 4.50 \ \Omega$$

This is almost negligible compared with 2000 $\Omega$ as was originally assumed, but if a more exact solution is required, the current $I_T$ can be recalculated with the 4.50 $\Omega$ added in series with the 2000 $\Omega$:

$$I_T = \frac{2}{2004.5}$$

$$= 0.9978 \text{ mA} \tag{23-22}$$

Using Eq. (23-20) yields

$$50 \times 10^{-6} = \frac{0.9978 \times 10^{-3} \, R_{sh}}{R_M + R_{sh}} \qquad\qquad (23\text{-}20)$$

and

$$R_{sh} = \frac{R_M}{18.96}$$

$$= \frac{90}{18.96}$$

$$= 4.75 \, \Omega$$

## 23.8   CALCULATING THE READINGS ON TWO DIFFERENT METERS WHEN MEASURING A SQUARE WAVE

**problem 23.7**   A square-wave inverter is used as an auxiliary battery-powered ac supply in a recreational vehicle. The output of this inverter is a square wave as shown in Fig. 23-6 with a peak value of 108 V. Suppose that a permanent-magnet moving-coil (PMMC) meter is used to measure its output. What would the reading be? What would the reading be with a true rms meter?

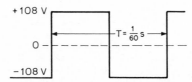

**Fig. 23-6**   Output waveform of the inverter.

**theory**   The unfiltered square-wave inverter is commonly encountered as an auxiliary power supply. The square wave is scaled to the value shown because that is the voltage required to make the square-wave *average* voltage equal to the average voltage of a 120-V rms sine wave. Magnetic devices such as motors or transformers are sensitive to average voltage and saturate or draw excessive current when their average voltage rating is exceeded. The PMMC meter is basically a dc-meter movement with a rectifier converting an ac input signal to a dc level for measurement. As such the PMMC meter responds to average value. However, manufacturers recognize that most of the time a PMMC meter will be used with a sine-wave input and, consequently, have calibrated the ac scale in rms volts corresponding to the pointer deflection produced by a sine wave. Table 23-1 lists the conversion factors for various types of waveforms and the meters of differing basic sensitivities to allow correction of the meter reading when the proper sensitivity meter is not available. The only caution one should observe when using this table is that errors may be introduced when the waveform measured is not ideally the same as that shown in the table. For example, the square wave may be asymmetrical, contain a small dc component, or be slightly "rounded off" at the corners. The best practice is to obtain the proper type of meter sensitivity and use correction factors only when necessary.

**solution**   The PMMC meter is referred to as a type-I meter in the table; i.e., it is an average responding meter and is calibrated rms for sinusoidal waves. A square wave after full-wave rectification appears as a dc level of amplitude equal to the peak value of square wave $V_{pl}$ and by definition this dc level is the average value $V_{av}$ of the square wave:

$$V_{av} = V_{pk}$$
$$= 108 \, V$$

From the table, a square wave should have the meter reading multiplied by $2\sqrt{2}/\pi$ to obtain the average value:

$$V_{av} = 108 \qquad\qquad (23\text{-}23)$$

$$= (\text{reading}) \times \frac{2\sqrt{2}}{\pi} \, V$$

We can now solve for the reading which is

$$\text{Reading} = \frac{108 \times \pi}{2\sqrt{2}}$$

$$= 120 \, V$$

Also from Table 23-1, we see that the factor for converting the true rms meter reading, type III, to an average reading, is 1. Therefore, the true rms meter will read 108 V.

## 23.9  CALCULATING THE PERCENTAGE ERRORS ON TWO DIFFERENT SCALES OF AN ANALOG METER

**problem 23.8**   A 20-V dc source is to be measured by a multimeter having a stated tolerance of ±2% full scale. The percent error will be different, depending upon the full-scale range employed for the reading. Find the percent error when each of the following ranges is used to measure the 20-V direct current: (*a*) the 0- to 100-V range; (*b*) the 0- to 25-V range.

**theory**   In either case, the percent error for each range is given by

$$\% \text{ error} = \frac{\% \text{ tolerance} \times \text{full-scale reading (V)}}{\text{volts being measured}} \times 100 \qquad (23\text{-}24)$$

**solution**
(*a*) ±2% tolerance of full scale on the 0- to 100-V range allows ±2 V tolerance in 20 V:

$$\% \text{ error} = \frac{2 \text{ V}}{20 \text{ V}} \times 100$$
$$= 10\% \text{ error*}$$

(*b*) The same ±2% tolerance of full scale on the 0- to 25 V range allows ±0.5 V tolerance in 20 V:

$$\% \text{ error} = \frac{0.5 \text{ V}}{20 \text{ V}} \times 100$$
$$= 2.5\% \text{ error*}$$

## 23.10  CALCULATING THE PERCENT ERROR OF A DIGITAL VOLTMETER

**problem 23.9**   A 3½ digit voltmeter is used on its 0- to 20-V dc range to measure a source of 5.00 V. (The 3½ digit model provides a reading up to 19.99 V rather than only to 9.99 V, as for a 3-digit model.) Its accuracy specification is given as ± (0.1% of reading +1 digit).† Find the allowable limit of percent error.

**solution**   Allowable error from ±0.1% of nominal 5-V reading equals ±0.005 V, and allowable error from one digit equals 0.01/5.00 or ±0.002 V; therefore, the total equals ±0.007 V.

$$\% \text{ error} = \frac{0.007 \text{ V}}{5.00 \text{ V}} \times 100 = 0.14\%$$

## 23.11  CALCULATING THE ACCURACY OF A DIGITAL FREQUENCY METER

**problem 23.10**   A digital frequency meter specifies its internal reference (quartz crystal) frequency to be correct within 20 ppm (parts per million). (*a*) Express 20 ppm as percent accuracy. (*b*) What ppm figure denotes an accuracy percentage of 0.005% (as required for transmission frequencies in the citizens' band at 27 MHz)?

**theory**   The relation between ppm and % accuracy can be seen by expressing ppm in the notation of powers of 10, where 1 ppm = 0.000001 or $1 \times 10^{-6}$. The corresponding percentage figure can then be obtained from the decimal notation.

**solution**
(*a*) 20 ppm = 20 parts in $10^{-6}$
$= 20 \times 0.000001$
$= 0.000020$
$= 0.002\%$
(*b*) 0.005% = 0.000050
$= 50 \times 10^{-6}$
$= 50$ ppm

---

*In the usual case, where the measurement tolerance of an analog meter is given as a percentage of full scale, it is good practice to select a range so that the pointer falls in the highest third of the scale. This provides the lowest percent error for that reading, as shown above.
†With digital meters, the tolerance percentage is given in *percent of reading*, rather than in *percent of full scale* as with analog meters.

**TABLE 23-1 Multiplier (M) for Voltage and Current Calculations**

| Waveform | | I | II | III | IV | V | VI |
|---|---|---|---|---|---|---|---|
| | | | | | Voltmeter or ammeter response (see Notes) | | |
| Sine: | RMS | 1 | 1 | 1 | $\pi/2\sqrt{2}$ | $1/\sqrt{2}$ | $1/2\sqrt{2}$ |
| | AUG | $2\sqrt{2}/\pi$ | $2\sqrt{2}/\pi$ | $2\sqrt{2}/\pi$ | 1 | $2/\pi$ | $1/\pi$ |
| | O-P | $\sqrt{2}$ | $\sqrt{2}$ | $\sqrt{2}$ | $\pi/2$ | 1 | $1/2$ |
| | P-P | $2\sqrt{2}$ | $2\sqrt{2}$ | $2\sqrt{2}$ | $\pi$ | 2 | 1 |
| Full-wave-rectified sine: | RMS | 1 | 1 | 1 | $\pi/2\sqrt{2}$ | $1/\sqrt{2}$ | $1/\sqrt{2}$ |
| | AUG | $2\sqrt{2}/\pi$ | $2\sqrt{2}/\pi$ | $2\sqrt{2}/\pi$ | 1 | $2/\pi$ | $2/\pi$ |
| | O-P | $\sqrt{2}$ | $\sqrt{2}$ | $\sqrt{2}$ | $\pi/2$ | 1 | 1 |
| | P-P | $2\sqrt{2}$ | $2\sqrt{2}$ | $2\sqrt{2}$ | $\pi/2$ | 1 | 1 |
| Half-wave-rectified sine: | RMS | $\sqrt{2}$ | $1/\sqrt{2}$ | 1 | $\pi/2$ | $1/2$ | $1/2$ |
| | AUG | $2\sqrt{2}/\pi$ | $\sqrt{2}/\pi$ | $2/\pi$ | 1 | $1/\pi$ | $1/\pi$ |
| | O-P | $2\sqrt{2}$ | $\sqrt{2}$ | 2 | $\pi$ | 1 | 1 |
| | P-P | $2\sqrt{2}$ | $\sqrt{2}$ | 2 | $\pi$ | 1 | 1 |
| Sine pulse: (Notes 1 and 2) | RMS | $\sqrt{T/t}$ | $\sqrt{t/T}$ | 1 | $\pi\sqrt{T}/2\sqrt{2}t$ | $\sqrt{t/2T}$ | $\sqrt{t/2T}$ |
| | AUG | $2\sqrt{2}/\pi$ | $2\sqrt{2}t/\pi T$ | $2\sqrt{2}t/\pi\sqrt{T}$ | 1 | $2t/\pi T$ | $2t/\pi T$ |
| | O-P | $\sqrt{2}T/t$ | $\sqrt{2}$ | $\sqrt{2T}/t$ | $\pi T/2t$ | $t$ | 1 |
| | P-P | $\sqrt{2}T/t$ | $\sqrt{2}$ | $\sqrt{2T}/t$ | $\pi T/2t$ | $t$ | 1 |
| Segmental sine: (Note 3) | RMS | $2E/C$ | $E$ | 1 | $\sqrt{\pi}E/\sqrt{2}C$ | $E/\sqrt{2}$ | $E/\sqrt{2}$ |
| | AUG | $2\sqrt{2}/\pi$ | $\sqrt{2}C/\pi$ | $\sqrt{2}C/\pi E$ | 1 | $C/\pi$ | $C/\pi$ |
| | O-P | $2\sqrt{2}F/C$ | $\sqrt{2}E$ | $\sqrt{2}F/E$ | $\pi F/C$ | $F$ | $F$ |
| | P-P | $4\sqrt{2}F/C$ | $2\sqrt{2}F$ | $2\sqrt{2}F/E$ | $2\pi F/C$ | $F$ | $F$ |

| Waveform | | | | | | | |
|---|---|---|---|---|---|---|---|
| Full-wave-rectified segmental sine: (Note 2) | RMS | $2E/C$ | $E$ | $1$ | $\pi E/\sqrt{2}C$ | $E/\sqrt{2}$ | $E/\sqrt{2}$ |
| | AUG | $2\sqrt{2}$ | $\sqrt{2}C/\pi$ | $\sqrt{2}C/\pi E$ | $1$ | $C/\pi$ | $C/\pi$ |
| | O-P | $2\sqrt{2}F/C$ | $\sqrt{2}F$ | $\sqrt{2}F/E$ | $\pi F/C$ | $F$ | $F$ |
| | P-P | $2\sqrt{2}F/C$ | $\sqrt{2}F$ | $\sqrt{2}F/E$ | $\pi F/C$ | $F$ | $F$ |
| Half-wave-rectified sine: (Note 4) | RMS | $2\sqrt{2}E/C$ | $E/\sqrt{2}$ | $1$ | $\pi E/C$ | $E/2$ | $E/2$ |
| | AUG | $2\sqrt{2}/\pi$ | $C/\pi\sqrt{2}$ | $C/\pi E$ | $1$ | $C/2\pi$ | $C/2\pi$ |
| | O-P | $4\sqrt{2}F/C$ | $\sqrt{2}F$ | $2F/E$ | $2\pi F/C$ | $F$ | $F$ |
| | P-P | $4\sqrt{2}F/C$ | $\sqrt{2}F$ | $2F/E$ | $2\pi F/C$ | $F$ | $F$ |
| Sine squared: (Notes 1 and 2) | RMS | $2\sqrt{3T}/\sqrt{t}$ | $\sqrt{3t}/2\sqrt{T}$ | $t$ | $\sqrt{3T}/2t$ | $\sqrt{3T}/2\sqrt{2T}$ | $\sqrt{3T}/2\sqrt{2T}$ |
| | AUG | $2\sqrt{2}/\pi$ | $t/\sqrt{2T}$ | $\sqrt{2t/3T}$ | $1$ | $t/2T$ | $t/2T$ |
| | O-P | $4\sqrt{2T}/\pi t$ | $\sqrt{2}$ | $2\sqrt{2T}/3t$ | $2T/t$ | $1$ | $1$ |
| | P-P | $4\sqrt{2T}/\pi t$ | $\sqrt{2}\,t$ | $2\sqrt{2T}/3t$ | $2T/t$ | $1$ | $1$ |
| Fractional sine pulse: (Notes 1 and 2) | RMS | $2B/\sqrt{\pi}A$ | $B/\sqrt{\pi}$ | $1$ | $\sqrt{\pi}B/\sqrt{2}A$ | $B/\sqrt{2\pi}$ | $B/\sqrt{2\pi}$ |
| | AUG | $2\sqrt{2}/\pi$ | $\sqrt{2}A/\pi$ | $\sqrt{2}A/\sqrt{\pi}B$ | $1$ | $A/\pi$ | $A/\pi$ |
| | O-P | $2\sqrt{2}A$ | $\sqrt{2}$ | $\sqrt{2\pi}/B$ | $1$ | $1$ | $1$ |
| | P-P | $2\sqrt{2}A$ | $\sqrt{2}$ | $\sqrt{2\pi}/B$ | $\pi/A$ | $1$ | $1$ |
| Triangle or sawtooth: | RMS | $4\sqrt{2}/\sqrt{3}\pi$ | $\sqrt{2/3}$ | $1$ | $2\sqrt{3}$ | $1/\sqrt{3}$ | $1/2\sqrt{3}$ |
| | AUG | $2\sqrt{2}/\pi$ | $1/\sqrt{2}$ | $\sqrt{3}/2$ | $1$ | $1/2$ | $1/4$ |
| | O-P | $4\sqrt{2}/\pi$ | $\sqrt{2}$ | $\sqrt{3}$ | $2$ | $1$ | $1/2$ |
| | P-P | $8\sqrt{2}/\pi$ | $\sqrt{2}$ | $2\sqrt{3}$ | $4$ | $2$ | $1$ |
| Full-wave-rectified triangle or sawtooth: | RMS | $4\sqrt{2}/\sqrt{3}\pi$ | $\sqrt{2/3}$ | $1$ | $2\sqrt{3}$ | $1/\sqrt{3}$ | $1/\sqrt{3}$ |
| | AUG | $2\sqrt{2}/\pi$ | $1/\sqrt{2}$ | $\sqrt{3}/2$ | $1$ | $1/2$ | $1/2$ |
| | O-P | $4\sqrt{2}/\pi$ | $\sqrt{2}$ | $\sqrt{3}$ | $2$ | $1$ | $1$ |
| | P-P | $4\sqrt{2}/\pi$ | $\sqrt{2}$ | $\sqrt{3}$ | $2$ | $1$ | $1$ |

**TABLE 23-1 Multiplier (M) for Voltage and Current Calculations** (*Continued*)

| Waveform | | Voltmeter or ammeter response (see Notes) | | | | | |
|---|---|---|---|---|---|---|---|
| | | I | II | III | IV | V | VI |
| Half-wave-rectified triangle or sawtooth: | RMS | $8/\pi\sqrt{3}$ | $1/\sqrt{3}$ | $1$ | $2\sqrt{2/3}$ | $3\sqrt{6}$ | $1/\sqrt{6}$ |
| | AUG | $2\sqrt{2}/\pi$ | $\sqrt{2}/4$ | $\sqrt{3}/2\sqrt{2}$ | $1$ | $1/4$ | $1/4$ |
| | O-P | $8\sqrt{2}/\pi$ | $\sqrt{2}$ | $\sqrt{6}$ | $4$ | $1$ | $1$ |
| | P-P | $8\sqrt{2}/\pi$ | $\sqrt{2}$ | $\sqrt{6}$ | $4$ | $1$ | $1$ |
| Triangle or sawtooth pulse: Notes 1 and 2 | RMS | $4\sqrt{2T}/\pi\sqrt{3t}$ | $\sqrt{2t/3T}$ | $1$ | $2\sqrt{T}/\sqrt{3T}$ | $\sqrt{t}/\sqrt{3T}$ | $\sqrt{t}/\sqrt{6}$ |
| | AUG | $2\sqrt{2}/\pi$ | $t/\sqrt{2T}$ | $\sqrt{3t}/2\sqrt{T}$ | $1$ | $t/2T$ | $t/2T$ |
| | O-P | $4\sqrt{2T}/\pi t$ | $\sqrt{2}$ | $\sqrt{3T}/t$ | $2T/t$ | $1$ | $1$ |
| | P-P | $4\sqrt{2T}/\pi t$ | $\sqrt{2}$ | $\sqrt{3T}/t$ | $2T/t$ | $1$ | $1$ |
| Square: | RMS | $2\sqrt{2}/\pi$ | $\sqrt{2}$ | $1$ | $1$ | $1$ | $1/2$ |
| | AUG | $2\sqrt{2}/\pi$ | $\sqrt{2}$ | $1$ | $1$ | $1$ | $1/2$ |
| | O-P | $2\sqrt{2}/\pi$ | $\sqrt{2}$ | $1$ | $1$ | $1$ | $1/2$ |
| | P-P | $4\sqrt{2}/\pi$ | $2\sqrt{2}$ | $2$ | $2$ | $2$ | $1$ |
| DC and full-wave-rectified square: | RMS | $2\sqrt{2}/\pi$ | $\sqrt{2}$ | Trivial case (RMS = AUG = 0-P = P-P ≡ DC) | | | |
| | AUG | $2\sqrt{2}/\pi$ | $\sqrt{2}$ | | | | |
| | O-P | $2\sqrt{2}/\pi$ | $\sqrt{2}$ | | | | |
| | P-P | $2\sqrt{2}/\pi$ | $\sqrt{2}$ | | | | |
| Half-wave-rectified square: | RMS | $4/\pi$ | $1$ | $1$ | $\sqrt{2}$ | $\sqrt{2}$ | $1/\sqrt{2}$ |
| | AUG | $2\sqrt{2}/\pi$ | $1/\sqrt{2}$ | $1/\sqrt{2}$ | $1/2$ | $1/2$ | $1/2$ |
| | O-P | $4\sqrt{2}/\pi$ | $\sqrt{2}$ | $\sqrt{2}$ | $1$ | $2$ | $1$ |
| | P-P | $4\sqrt{2}/\pi$ | $\sqrt{2}$ | $\sqrt{2}$ | $1$ | $2$ | $1$ |

| Rectangular pulse: | | | | | | | |
|---|---|---|---|---|---|---|---|
| | RMS | $2\sqrt{2T}/\pi\sqrt{t}$ | $\sqrt{2}t/\sqrt{T}$ | 1 | $\sqrt{T/t}$ | $\sqrt{t/T}$ | $\sqrt{t/T}$ |
| Notes 1 | AUG | $2\sqrt{2}/\pi$ | $\sqrt{2}t/T$ | $\sqrt{t/T}$ | 1 | $t/T$ | $t/2T$ |
| and 2 | O-P | $2\sqrt{2}/\pi$ | $\sqrt{2}$ | $\sqrt{T/t}$ | $T/t$ | 1 | 1 |
| | P-P | $2\sqrt{2}/\pi$ | $\sqrt{2}$ | $\sqrt{T/t}$ | $T/t$ | 1 | 1 |
| Exponential pulse (critically damped): | RMS | $\sqrt{2T}/\pi e\sqrt{t}$ | $e\sqrt{t}/\sqrt{2T}$ | 1 | $\sqrt{T}/2e\sqrt{t}$ | $e\sqrt{t}/2\sqrt{T}$ | $e\sqrt{t}/2\sqrt{T}$ |
| | AUG | $\sqrt{2}/\pi$ | $\sqrt{2}et/T$ | $2e\sqrt{t}/\sqrt{T}$ | 1 | $et/T$ | $et/T$ |
| | O-P | $2\sqrt{2T}/\pi et$ | $\sqrt{2}$ | $2\sqrt{T}/e\sqrt{t}$ | $T/et$ | 1 | 1 |
| | P-P | $2\sqrt{2T}/\pi et$ | $\sqrt{2}$ | $2\sqrt{T}/e\sqrt{t}$ | $T/et$ | 1 | 1 |

NOTES:
1. Full-wave-rectified pulse train $t/T = 1$
2. Half-wave-rectified pulse train $t/T = 1/2$
3. Assumes theoretical duty cycle of 1. Actual duty cycle is function of $\theta_c$ and $t/T$
4. Assumes theoretical duty cycle of 1/2. Actual duty cycle is function of $\theta_c$ and $t/T$

$A = [(\sin \alpha \cos \alpha)/(1 - \cos \alpha)]^2$

$B = [2\alpha + \alpha \cos 2\alpha - (3/2) \sin 2\alpha]/(1 - \cos \alpha)^2$ } $\alpha = \pi t/T$ Radians

$C = 1 - \cos \alpha_c$

$E = [(\theta_c/180) - <(\sin 2\theta_c)/2\pi>]^{1/2}$ } $\theta_c$ = conduction angle (degrees)

$F = \sin \theta_c$ for $0° \le \theta_c \le 90°$

$F = 1$ for $90° \le \theta_c \le 180°$

$\pi = 3.14159 \ldots \epsilon = 2.71828 \ldots$

AMMETER OR VOLTMETER RESPONSE

I = average responding, calibrated rms for sines
II = peak responding, calibrated rms for sines
III = true rms
IV = true average
V = true peak
VI = peak-to-peak

SOURCE: EDN, pp. 117–118, Caliner Publishing, Sept. 20, 1977.

## 23.12    DETERMINING WHAT TYPES OF METERS SHOULD BE USED FOR MEASUREMENTS AT THE INPUT AND OUTPUT OF A DIFFERENTIAL AMPLIFIER

**problem 23.11**    Figure 23-7 is the schematic diagram of a differential transistor amplifier stage. A variable-frequency sine-wave signal generator with an output impedance of 50 $\Omega$ is connected to the input as shown. The small signal bandwidth of the amplifier is to be determined; also, the amplifier gain is to be determined at 1000 Hz to an accuracy of 1%. What types of meters should be used at the input and output of the amplifier for the measurements?

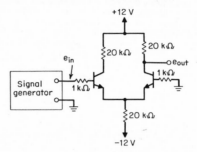

**Fig. 23-7**    Differential amplifier circuit employed to determine the selection and use of applicable instruments.

**theory**    Factors which must be considered when selecting an instrument are the following:

1. What is the intended use of the data which the measurement generates? Answering this question provides insight to the accuracy, number of data points, etc., when these specifications are not given.

2. What accuracy is required? In general, accuracies are specified as percentage deviation of indicated value from the true value, when measuring at full scale. As an example, a 10-V meter with an accuracy of 1% will have a maximum error of $\pm$ 0.1 V *at any point on the scale*. So if 1.0 V was measured on this scale, the error could be as much as $\pm 10\%$. Needless to say, a meter should be used which reads as close as possible to the full-scale value. Many measurements require that more than one reading be taken, e.g., the measurement of gain, attenuation, relative phase shifts, efficiency. In such cases, the individual measurement errors should be assumed to be the worst case, when the final calculation is made to determine the overall accuracy of the measurement. In the case where the ratio of two measurements is required to calculate the desired quantity, as in the measurement of gain, for instance, the overall percent error $E$ is given by

$$E = \frac{(\text{measured value}) - (\text{true value})}{(\text{true value})} \times 100\% \tag{23-25}$$

The measured value of the gain $G_M$ is

$$G_M = \frac{e_o(1 \pm x_o/100)}{e_i(1 \pm x_i/100)} \tag{23-26}$$

where $e_o$ = true value of the output voltage
$e_i$ = true value of the input voltage
$x_o$ = percent accuracy of the instrument used to measure the output voltage
$x_i$ = percent accuracy of the input measuring instrument

Substituting Eq. (23-26) into Eq. (23-25) gives

$$E = \left(\frac{100 \pm x_o}{100 \pm x_i} - 1\right) \times 100\% \tag{23-27}$$

The worst-case errors then are

$$E = \left(\frac{100 + x_o}{100 - x_i} - 1\right) \times 100\% \tag{23-28}$$

or

$$E = \left(\frac{100 - x_o}{100 + x_i} - 1\right) \times 100\% \tag{23-29}$$

3. The input impedance of the measuring instrument must be high enough that the measured signal is not affected by instrument loading. An approximate expression for the meter input impedance $Z_M$, required to achieve an accuracy of $A$ percent, is

$$Z_M \geq Z_o \times \frac{100}{A} \, \Omega \tag{23-30}$$

where $Z_o$ equals the output impedance of the circuit to be measured (in ohms).

4. The bandwidth (of frequency response) of the measuring instrument should be sufficiently wide that the frequency components of the signal to be measured are not attenuated. It is difficult to make a general statement concerning the effect of bandwidth on accuracy; however, most manufacturers of instruments will provide information as to the range of frequencies over which the accuracy is guaranteed.

5. Basic sensitivity should be considered when selecting an instrument. Some instruments are inherently sensitive to average values; others to the heating value or rms value, others to the peak value. Table 23-2 lists some common instrument types and the value to which each responds.

**TABLE 23-2    Sensitivity of Meters and Oscilloscope**

| Meter description | Type of sensitivity |
|---|---|
| Volt-ohm-ammeters (VOMs) with permanent-magnet, moving-coil movements (PMMC movements commonly called D'Arsonval movements) | Average |
| Electrodynammeter movements—voltmeter or ammeter | RMS |
| Moving-iron movements—ammeters or voltmeters | RMS |
| Thermocouple-type meters | RMS |
| Digital meters utilizing a dual slope analog-to-digital converter | Average |
| Oscilloscope | Peak or peak-to-peak |

Table 23-1, as noted earlier, presents the relationship between peak, average, and rms for some commonly encountered waveforms, such that corrections may be made when the correct meter type is not available for the required measurement.

6. Resolution of the instrument should be compatible with the requirements of the measurement. With moving-pointer meters, this merely means that the meter should not be used beyond its capability to be read accurately. In the case of meters with digital readouts, the resolution is determined by the weighting of the least significant digit on the display. This digit has a rounding or quantizing error of at least plus or minus one-half count and quite often an error of plus or minus one count in addition to the basic accuracy of the meter.

**solution**    To obtain the bandwidth of the differential amplifier requires that the gain be determined as a function of frequency. The frequencies at which the gain decreases by 3 dB, that is, becomes 0.707 at the low- and high-frequency ends, determine the amplifier's bandwidth. The most straightforward way to measure the bandwidth would be to hold the input voltage constant while measuring the output voltage and varying the frequency of the signal generator. When the output voltage of the amplifier decreases by 3 dB at the low- and high-frequency ends, the frequency difference of the signal generator settings is the bandwidth. Since we do not know what the bandwidth of this amplifier might be, it would be most convenient to utilize a widebandwidth instrument such as an oscilloscope. It is not difficult to obtain an oscilloscope with a bandwidth of greater than 10 MHz and with an input impedance of greater than a megohm. However, the ability to read an amplitude from the face of an oscilloscope cathode-ray tube (CRT) limits the accuracy of this type of device. Suppose the CRT has six major divisions each with five minor divisions for a total of 30 divisions. This means that the vertical deflection of the CRT trace, which represents the amplitude of the output voltage to be measured, can be resolved to only 1 part in 30. Perhaps by estimating the distance between minor divisions a resolution of 1 part in 60 or 1.7% can be obtained. Is this resolution of 1.7% good enough?

We must now consider the intended purpose of the data the measurement will provide. Bandwidth measurements are usually made to determine whether an element is capable of passing a certain frequency component without appreciable attenuation. Furthermore, the requirement for an accurate measurement at 1000 Hz indicates that this is the frequency of interest. Assuming that this is the case, an oscilloscope will do for the measurement of the bandwidth. The gain measurement at 1000 Hz is a different story, however.

Applying Eqs. (23-28) and (23-29) with meters that have a 1% accuracy gives worse-case errors of

$$E = \left(\frac{100 \pm x_o}{100 \pm x_i} - 1\right) \times 100\% \tag{23-31}$$

$$E = \left(\frac{100 + 1}{100 - 1} - 1\right) \times 100 = +2.02\% \tag{23-32}$$

or

$$E_2 = \left(\frac{100 - 1}{100 + 1} - 1\right) \times 100 = 1.98\% \tag{23-33}$$

This does not meet the accuracy requirement of 1%. Repeating the above calculations with meters of 0.5% accuracy yields the desired 1% accuracy for the gain measurement. The same meter may be used to measure both the input and output voltages if the meter is a multiscale type, provided that the requisite 0.5% accuracy applies to both scales.

The input impedance required for the meter when connected to the output voltage should be, in accordance with Eq. 23-30,

$$Z_M \geq Z_o \times \frac{100}{A} \; \Omega$$

$$\geq 20 \times 10^3 \times \frac{100}{0.5} \tag{23-30}$$

$$\geq 4 \times 10^6 \; \Omega$$

The output impedance $Z_o$ of the differential amplifier is actually less than 20 k$\Omega$, but even if it were less by a factor of 20, the calculation indicates that the instrument must be a high-input impedance type such as a vacuum-tube voltmeter (VTVM) or a field-effect transistor (FET) buffered meter.

In this particular case, the basic sensitivity of the meters is not important as long as both instruments used for the input and output readings have the same type of sensitivity, since the measurement requires the ratio of the two readings. However, care should be taken to be sure that no appreciable distortion is introduced into the output voltage by overdriving the amplifier and that the meters do not appreciably load the circuits (see Sec. 23.12).

## 23.13    CALCULATING THE "LOADING" ERROR OF THREE DIFFERENT VOLTMETERS

**problem 23.12**   In measuring the voltage drop across a load resistor $R_L$ in Fig. 23-8, the reading obtained by one voltmeter $M_1$ (at 1000 $\Omega$/V) is consistently lower than the theoretical voltage (true voltage before the meter is connected). This lowered voltage reading (on a correctly calibrated meter) is explained by the "loading" error, i.e., the effect of the nonnegligible current drawn by the voltmeter. Find the loading error for the following voltmeters, all used on their 0- to 6-V range:

(*a*) Meter 1, of 1000 $\Omega$/V
(*b*) Meter 2, of 20 k$\Omega$/V
(*c*) Digital voltmeter 3, with $Z_{in}$ = 10 M$\Omega$

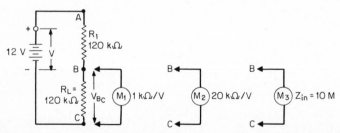

**Fig. 23-8**   Circuit to show loading error with voltmeters having widely different values of input impedance $Z_{1n}$.

**theory**   Ideally, the connection of a voltmeter across a source of voltage should not alter the voltage. Accordingly, in practice, one should use a voltmeter with a high-enough input impedance so that any current drawn by the meter can be considered negligible. Otherwise, a "loading" error results, because of the undesirable current flowing through the relatively low input impedance of the meter.

When a meter is connected across $R_L$ in Fig. 23-8, we have a new value of resistance from $B$ to $C$. This is the parallel combination of $R_L$ and the input impedance of the meter. By voltage divider action, the voltmeter reading across $R_{BC}$ ($V_{BC}$) is found by the voltage division relation:

$$V_{BC} = \frac{R_{BC}}{R_1 + R_{BC}} \times V \tag{23-34}$$

where $R_{BC}$ equals parallel resistance of $R_L$ and the meter resistance, and $V$ equals the battery voltage.

## solution

For voltmeter 1 (1000 Ω/V), the resistance $R_{M1}$ is

$$R_{M1} = 1000 \times 6 \text{ V (range)}$$
$$= 6 \text{ k}\Omega$$

Therefore, the parallel resistance $R_{BC}$ is

$$R_{BC} = \frac{R_L \times R_{M1}}{R_L + R_{M1}}$$
$$= \frac{120 \text{ k}\Omega \times 6 \text{ k}\Omega}{120 \text{ k}\Omega + 6 \text{ k}\Omega} \tag{23-35}$$
$$= 5.7 \text{ k}\Omega$$

Then, from Eq. (23-34)

$$V_{BC} = \frac{5.7 \text{ k}\Omega}{125.7 \text{ k}\Omega} \times 12 \text{ V}$$
$$= 0.544 \text{ V}$$

Thus, the nominal 6 V is reduced (on the meter) to 0.54 V. This means that the loading error of the 1000 Ω/V meter is excessive when measuring across this relatively high impedance circuit.

(b) For voltmeter 2 (20 kΩ/V), the resistance $R_{M2}$ on its 6-V range is

$$R_{M2} = 20 \text{ k}\Omega \times 6 \text{V (range)}$$
$$= 120 \text{ k}\Omega$$

and the parallel resistance $R_{BC}$ is

$$R_{BC} = \frac{120 \text{ k}\Omega \times 120 \text{ k}\Omega}{120 \text{ k}\Omega + 120 \text{ k}\Omega}$$
$$= 60 \text{ k}\Omega$$

Then from Eq. (23-34)

$$V_{BC} = \frac{60 \text{ k}\Omega}{120 \text{ k}\Omega + 60 \text{ k}\Omega} \times 12 \text{ V}$$
$$= 4 \text{ V}$$

Thus, the 20 kΩ/V meter is also not suitable for measuring voltage in this relatively high impedance circuit. It causes an error of

$$\frac{2 \text{ V}}{6 \text{ V}} \times 100 = 33.33\%$$

(c) For voltmeter 3 (10 MΩ input impedance), the resistance is the same on all ranges: The parallel resistance $R_{BC}$ is

$$R_{BC} = \frac{0.12 \text{ M}\Omega \times 10 \text{ M}\Omega}{0.12 \text{ M}\Omega + 10 \text{ M}\Omega}$$
$$= 0.119 \text{ M}\Omega$$
$$= 119 \text{ k}\Omega$$

Then from Eq. (23-34)

$$V_{BC} = \frac{119 \text{ k}\Omega}{120 \text{ k}\Omega + 119 \text{ k}\Omega} \times 12 \text{ V}$$
$$= 5.97 \text{ V}$$

This is within 0.5% of the actual 6 V and is generally acceptable for a loading error, except for very high precision measurements.

*Note:* As a rule-of-thumb method for keeping the loading error to less than 5%, be sure that the input impedance of the voltmeter is at least 20 times greater than the resistance across which it is connected.

## 23.14 CALCULATING THE TRUE VALUE OF A QUANTITY BY THE MEAN VALUE AND ALSO BY STANDARD DEVIATION

In determining the true value of a quantity by means of data made up of a number of slightly different readings, an average or mean value is used, as calculated in Problem 23.13a. Another method, the standard deviation ($\sigma$), is used in statistical analysis for obtaining the best approximation to the true value, as calculated in Problem 23.13b.

**problem 23.13a (Arithmetic Mean)**    The five readings ($x$) in Table 23-3 were obtained by measuring a nominal 15-$\Omega$ resistor. From the listing of the readings in Table 23-3, calculate: (1) the arithmetic mean ($\bar{x}$); (2) the standard deviation ($\sigma$).

**TABLE 23-3    Data for Problem 23.13**

| Reading, $x$ | Deviation from mean, $d$ | $d^2$ |
|---|---|---|
| 15.25 | +0.15 | 225($10^{-4}$) |
| 15.15 | +0.05 | 25($10^{-4}$) |
| 14.95 | −0.15 | 225($10^{-4}$) |
| 15.20 | +0.10 | 100($10^{-4}$) |
| 14.95 | −0.15 | 225($10^{-4}$) |
| $\Sigma x = 75.50$ (mean = 5.10) | $\Sigma d = 0^*$ | $\Sigma d^2 = 800(10^{-4})$ |

*The algebraic sum of the positive and negative deviations should equal zero.

**theory**    The arithmetic mean ($\bar{x}$) (or average) is defined as the sum of the readings ($\Sigma x$) divided by the number of readings ($n$). This mean is expressed by the equation:

$$\bar{x} = \frac{\Sigma x}{n} \tag{23-36}$$

**solution**    For the five readings of $x$, the arithmetic mean is, from Eq. (23-36):

$$\bar{x} = \frac{75.50}{5}$$
$$= 15.10$$

**problem 23.13b (Standard Deviation)**    Standard deviation ($\sigma$) is the root-mean-square value of the deviations, defined as the square root of the quantity; the sum of the squares of the deviations ($\Sigma d^2$) divided by 1 less than the number of readings ($n - 1$):

$$\sigma = \sqrt{\frac{\Sigma d^2}{(n-1)}} \tag{23-37}$$

**solution**    Using the values from Table 23-3 and Eq. 23-37, we have

$$\sigma = \frac{800 \times 10^{-4}}{4}$$
$$= 14.15 \times 10^{-2}$$
$$\approx 0.142$$

*Note:* Analyzing random errors is a practical use for the standard deviation $\sigma$. In the normal distribution, 95% of all such errors will be found to lie between $\pm 2\sigma$, or in this case, between $\pm 0.284$ from the mean of 15.10, the average reading.

## 23.15 CALIBRATING AN OSCILLOSCOPE AND ITS PROBE

**problem 23.14** In order to maintain an oscilloscope at its inherent accuracy, the oscilloscope and its probe must be calibrated prior to use in any critical measurement. What is the procedure for calibrating the vertical deflection amplifier and the oscilloscope probe?

**solution** Most oscilloscopes have available on the user panel a calibration voltage or calibrator with a range of selectable voltages. The oscilloscope probe may be contacted to one of these calibration voltage points. The calibration voltage should be preferably of the same order of magnitude as the voltage to be measured. If the oscilloscope probe is of the "×10" variety, that is, if the probe attenuates the measured signal by a factor of 10, it must be compensated prior to calibration.

Figure 23-9 illustrates the equivalent circuit of the ×10 oscilloscope probe and the input of vertical amplifier. Note that the oscilloscope input impedance consists of a 1-MΩ resistor in

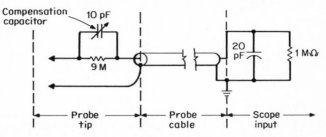

**Fig. 23-9** Equivalent circuit of an oscilloscope probe and input to the oscilloscope.

parallel with a 20-pF capacitor. To attenuate the input signal to the oscilloscope so that larger signals may be measured, a probe with a 9-MΩ resistor is used which has the added advantage of increasing the input resistance of the oscilloscope to 10 MΩ. However, this resistor in combination with the input capacitance of the oscilloscope causes the signal seen by the oscilloscope to become distorted because of the high-frequency filtering effect of the input capacitance. To compensate for this, oscilloscope manufacturers provide a variable capacitor in parallel with the probe resistance as is shown in Fig. 23-9. Figure 23-10 illustrates the traces that result when the probe is under- or overcompensated, when the calibrator square-wave voltage is viewed. The compensation capacitor is generally located on the probe tip, although sometimes it is integral with the connector to the oscilloscope. The procedure is simply to adjust the compensating capacitor until the corners of the square-wave trace are as square as possible.

Once this procedure has been completed, the sensitivity of the vertical amplifier, labeled *vertical gain adjust* (or some similar term), is adjusted until the proper deflection is obtained. For

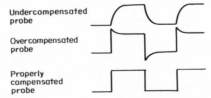

**Fig. 23-10** Waveforms showing conditions of compensation for an oscilloscope probe.

example, with a deflection sensitivity of 2 V/cm and a 2-V square wave from the calibrator, the vertical sensitivity is adjusted until a 1-cm (one major division) deflection is obtained. The scope is now ready for use.

## 23.16 CALCULATING PHASE-ANGLE FROM OSCILLOSCOPE PATTERN

**problem 25.15** An oscilloscope pattern as shown in Fig. 23-11 is obtained from two same-frequency, equal-amplitude signals, connected, respectively, to the $X$ and $Y$ inputs of an oscilloscope. Find the phase-angle between the two signals.

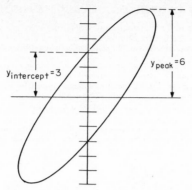

**Fig. 23-11**  Obtaining phase-angle $\phi$ from an oscilloscope pattern.

**theory**  With the oscilloscope properly adjusted to display equal amplitudes on the $X$ and $Y$ axes, the sine of the phase angle $\phi$ between the two signals is obtained from the equation:

$$\text{Sin } \phi = \frac{y \text{ intercept}}{y \text{ peak}} \qquad (23\text{-}38)$$

**solution**  By measurement, the following ratio is obtained, as shown in Fig. 23-12 and using Eq. (23-38):

$$\text{Sin } \phi = \frac{3 \text{ units}}{6 \text{ units}}$$
$$= 0.5$$

From a trigonometry table, the angle $\phi$ whose sine is 0.5 is found to be

$$\phi = 30°$$

## 23.17  CALCULATING THE ERROR INTRODUCED BY ANY OSCILLOSCOPE WHEN MEASURING THE RISE TIME OF A PULSE

**problem 23.16**  An oscilloscope has been used to measure the rise time of a voltage pulse, and the observed waveform is shown in Fig. 23-12. If the oscilloscope used to make the measurement has a bandwidth of 35 mHz and the oscilloscope probe has a rise time of 5 ns, what percent error exists in the measurement due to the oscilloscope's response time?

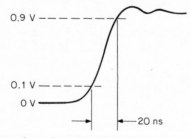

**Fig. 23-12**  Observed waveform of a voltage pulse, as seen on an oscilloscope.

**theory**  The response time, or the rise time $T_R$, of an oscilloscope is inversely proportional to its bandwidth and given by the following equation:

$$T_R = \frac{\ln 9}{2\pi \times \text{BW}}$$
$$= \frac{0.35}{\text{BW}} \text{ s} \qquad (23\text{-}39)$$

where BW equals the bandwidth in hertz.

Furthermore, the rise time observed on the oscilloscope is approximately equal to the square root of the sum of the squares of the individual rise times of the measured pulse, probe, and scope.

$$\text{Rise time observed} = \sqrt{T_{R_1}^2 + T_{R_2}^2 + T_{R_3}^2} \qquad (23\text{-}40)$$

where $T_{R_1}$ = rise time of measured pulse

$T_{R_2}$ = rise time of probe

$T_{R_3}$ = rise time of oscilloscope

**solution**    The rise time of the oscilloscope as well as the bandwidth is usually specified by the manufacturer. However, the rise time of the oscilloscope $T_{R_3}$ can be calculated from Eq. (23-39).

$$T_{R_3} = \frac{0.35}{\text{BW}}$$

$$= \frac{0.35}{35 \times 10^6}$$

$$= 10 \text{ ns}$$

We now know everything except the approximate rise time of the measured waveform, which can be found by using Eq. 23-40:

$$\text{Rise time observed} = \sqrt{T_{R_1}^2 + T_{R_2}^2 + T_{R_3}^2}$$

$$20 \times 10^{-9} = \sqrt{T_{R_1}^2 + (5 \times 10^{-9})^2 + (10 \times 10^{-9})^2}$$

$$400 \times 10^{-18} = T_{R_1}^2 + 25 \times 10^{-18} + 100 \times 10^{-18}$$

$$T_{R_1}^2 = 275 \times 10^{-18}$$

$$T_{R_1} = 16.6 \text{ ns}$$

The percent error is

$$\% \text{ error} = \frac{20 - 16.6}{16.6} \times 100 = 20.5\%$$

## 23.18    USING A WHEATSTONE BRIDGE TO ACCURATELY MEASURE RESISTANCE WITHIN 0.1%

**problem 23.17**    The classic method of accurately measuring resistance is by the use of a Wheatstone bridge. How would this bridge be used to make such a measurement of a variable resistor which is supposed to be set to 1980 $\Omega$?

**theory**    Figure 23-13 shows the *basic* Wheatstone bridge arrangement. Resistors $R_1$, $R_2$, and $R_3$ are precision adjustable resistors. $R_X$ is the unknown resistor. A sensitive galvanometer or microammeter is placed across two arms of the bridge as shown, and the opposite corners of the bridge are connected to a dc supply or battery. The resistors are adjusted to "balance" the bridge, that is to say, until the microammeter indicates no current. Since the meter current is zero, the voltage across the meter is zero, and the voltage across $R_1$ and $R_2$ must be equal. Likewise the voltage across $R_X$ and $R_3$ must be the same.

$$V_1 = V_2$$
$$V_X = V_3 \qquad (23\text{-}41)$$

Furthermore, at balance the current through $R_1$ equals the current in $R_X$ and the current through $R_2$ equals that of $R_3$.

$$\frac{V_1}{R_1} = \frac{V_X}{R_X} \qquad (23\text{-}42)$$

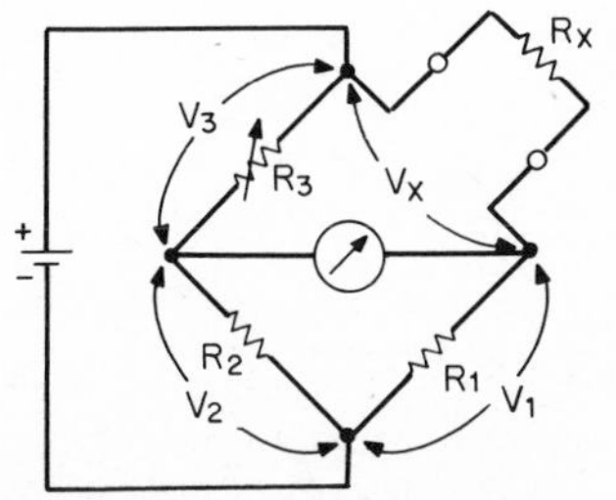

**Fig. 23-13**    Using the Wheatstone bridge to measure an unknown resistance $R_X$.

$$\frac{V_2}{R_2} = \frac{V_3}{R_3} \tag{23-43}$$

Substituting for $V_2$ and $V_3$ in Eq. (23-43) gives

$$\frac{V_1}{R_2} = \frac{V_x}{R_3}$$

$$V_1 = V_x \frac{R_2}{R_3} \tag{23-44}$$

and substituting in Eq. (23-44) gives

$$\frac{R_2}{R_1} = \frac{R_3}{R_x} \tag{23-45}$$

or

$$R_x = \frac{R_1}{R_2} \times R_3 \tag{23-46}$$

The resistor ratio $R_1/R_2$ is usually a decade multiplier, that is, $R_1/R_2 = 1/10$, 1, 10, etc., and $R_3$ is adjusted for balance. The resistor value is then equal to the decade multiplier setting times the value $R_3$.

**solution**   Using the bridge to set the variable resistor to 1980 Ω is quite simple. In this case, the value of $R_3$ is set to 1980 Ω, the decade multiplier $R_1/R_2$ set to 1, and the variable resistor is connected to the $R_x$ terminals. The microammeter usually has sensitivity adjustments, and the least sensitive setting should be used initially while the variable resistor is adjusted for null. The final "touch up" of the resistor should be done with the most sensitive setting of the galvanometer.

*Note:* The procedure to use when measuring the value of any unknown resistor is similar. The unknown resistance is connected to the $R_x$ terminals, and $R_3$ is set to a convenient value, say 1000. With the galvanometer set to its least sensitive position, the decade multiplier is switched from its highest setting down the scale. When the galvanometer changes polarity, the value of $R_3$ is adjusted to balance the bridge, using the highest galvanometer sensitivity. The unknown resistance is given by the product of $(R_1/R_2) \times R_3$.

## 23.19  BIBLIOGRAPHY

Dalton, O.: "Boosting Your HF Scope's Utility," pp. 83–89, *EDN*, Caliner Publishing, Nov. 5, 1977.

Fink, D. G. (ed.): *Electronic Engineers Handbook*, McGraw-Hill, New York, 1975, pp. 17-1 to 17-40.

Kaufman, M., and Seidman, A. H. (eds.), *Handbook for Electronics Engineering Technicians*, McGraw-Hill, New York, 1976, pp. 7-1–7-33.

Kraengel, W. D., Jr.: "Table for AC Waveform/Meter Conversion Factors," pp. 115–117, *EDN*, Caliner Publishing, Sept. 20, 1977.

LePage, W. R.: *Analysis of AC Circuits*, McGraw-Hill, New York, 1952, pp. 47–70.

Stout, M. B.: *Basic Electrical Measurements*, Prentice-Hall, Englewood Cliffs, N.J., 1950.

Westman, H. P.: *Reference Data for Radio Engineers*, Sams, Indianapolis, 1970, pp. 11-1 to 11-7.

Chapter **24**

# Thick-Film Technology

## WILLIAM G. DRYDEN

**Production Manager, Electro Materials Corp. of America, Mamaroneck, N.Y.**

### 24.1 INTRODUCTION

Figure 24-1 shows a typical thick-film hybrid microelectronic circuit. This is a relatively complex circuit which is made up of the following components: substrate, thick-film components, chip components, flywire interconnections, and package.

### Substrates

The substrate is a supporting structure for the circuit. It acts as a depository for the thick-film components and provides a base for mechanical support of the chip components. It must be an electrical insulator to provide isolation for the circuit components, and it must have good thermal conductivity for heat removal.

The most popular substrate material for thick-film work is 96% aluminum oxide ($Al_2O_3$), or *alumina*. It is a ceramic capable of withstanding the high temperatures required for thick-film processing. Alumina is available in sizes ranging from tiny chips to large circuit boards. It can be fabricated in a wide variety of shapes, and the fired sheets can be drilled or cut with diamond tools or a laser beam.

Two other materials occasionally used in thick-film circuits are 99% aluminum oxide and beryllium oxide (BeO), or beryllia. The 99% alumina has a smoother surface than the 96% material, and it is used primarily for thin-film deposition. It is used for thick-film work only when thin and thick films have to be deposited on the same substrate. Beryllia has a high thermal conductivity and is used as a replacement for alumina in critical thermal design situations. A summary of the electrical and mechanical properties of substrates is given in Table 24-1.

### Thick Films

Thick-film components are conductive, resistive, and dielectric film patterns on a substrate surface. These materials are in the form of an ink composed of various metal, oxide, ceramic, and glass powders suspended in an organic vehicle. The inks are deposited on the substrate surface by screen printing through stainless steel masks. The films are dried and then fired in a belt kiln to an exact time-and-temperature profile. During firing, the vehicle is driven off and the powders melt and sinter together to form a hard, durable film in intimate contact with the substrate.

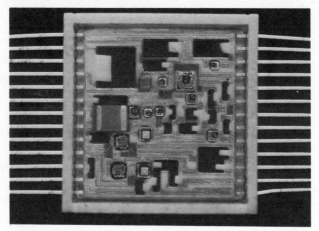

**Fig. 24-1**   Example of a thick-film hybrid circuit. *(Courtesy EMCA.)*

**TABLE 24-1   Properties of Ceramic Substrate Materials**

| Properties | $Al_2O_3$ 96% | $Al_2O_3$ 99.5% | BeO 99.5% |
|---|---|---|---|
| Specific gravity | 3.7 | 3.85 | 2.88 |
| Rockwell hardness | 78 | 80 | 62 |
| Thermal expansion, per °C | $7.9 \times 10^{-6}$ | $7.8 \times 10^{-6}$ | $8.5 \times 10^{-6}$ |
| Tensile strength, psi | 25,000 | 28,000 | 23,000 |
| Flexural strength, psi | 46,000 | 48,000 | 33,000 |
| Thermal conductivity, (cal)(cm)/ (s)(cm²)(°C) at 25°C | 0.084 | 0.088 | 0.59 |
| Dielectric constant, 1 MHz at 25°C | 9.3 | 9.8 | 6.5 |
| Dissipation factor, 1 MHz at 25°C | 0.0003 | 0.0001 | 0.0002 |
| Loss factor, 1 MHz at 25°C | 0.0028 | 0.0010 | 0.0013 |

## Conductives

The two most popular conductive systems used in thick-film work are silver-based and gold-based systems. The advantage of using noble metals is that they can be fired in air. Base-metal systems, like copper or nickel, must be fired in inert or reducing atmospheres.

Among the silver-based conductors, three major systems are in use: Silver, palladium-silver, and platinum-silver. Plain silver conductors are used primarily as ground planes. Since silver tends to migrate and to tarnish, it is not a good material for circuit conductors or resistor electrodes. Silver has fair solderability, but its leaching (dissolving during soldering) characteristics are poor. It will not accept gold thermocompression flywire bonding, but it provides a fair base for ultrasonic attachment of aluminum flywires.

Palladium-silver is a more stable conductor than plain silver. The addition of palladium ties up the silver and makes the conductor less active. It is an excellent material for circuit conductors and for resistor electrodes. It solders well and has good leaching characteristics. It does not accept thermocompression gold flywire bonding, but it does have good ultrasonic aluminum flywire bonding capabilities. Because of its low-cost and good electrical and mechanical properties, it is the most widely used of all thick-film conductors.

Platinum-silver is a compromise material. It was developed as a low-cost substitute for palladium-silver during a period when palladium prices were high. The addition of platinum tends to stabilize the silver but not to the same extent as palladium. Platinum-silver has excellent flywire bonding characteristics, but its other properties are not as good as those of palladium-silver.

In the gold family, the most popular conductors are: gold, platinum-gold, and palladium-gold. Gold is an excellent all-around material for thick-film use. Despite its cost, which is approximately six to seven times that of palladium-silver, it enjoys a wide popularity among hybrid designers and manufacturers. It does not tarnish or migrate, which makes it very suitable for circuit conductors and resistor electrodes. It has excellent flywire bonding properties and is the only thick-film conductor which accepts eutectic bonding of silicon devices. Its only drawback is that it cannot be used in soldering processes since gold dissolves readily in tin.

Platinum-gold conductors were developed as a complement to thick-film gold. The addition of platinum prevents the gold from leaching, thus giving the material excellent solderability. Unfortunately, platinum-gold has poor die and wire bond characteristics, and its use is generally limited to spot soldering applications in gold systems.

## Resistives

Most thick-film resistor systems are now based on a ruthenium-oxide metallurgy. This is a relatively inexpensive system which can be formulated in a range of sheet resistivity values from 10 $\Omega$/square to 10 M$\Omega$/square. In the midrange, from 100 $\Omega$/square to 100 k$\Omega$/square, temperature coefficients as low as $\pm 50$ ppm/°C can be achieved. In the extended ranges, temperature coefficients of $\pm 100$ ppm/°C are available in resistivities as low as 10 $\Omega$/square and as high as 1 M$\Omega$/square.

In the "as-fired" state, thick-film resistors normally exhibit a tolerance of $\pm 20\%$. To achieve precise values, the resistors are adjusted or "trimmed" to value. In the trimming process, a portion of the resistor body is systematically cut away with an air-abrasion tool or a laser beam. This has the effect of increasing the resistor length with a concomitant increase in resistance. Thick-film resistors can be easily trimmed in production to tolerances of $\pm 1\%$; with some care, tolerances of $\pm 0.5\%$ may be realized. A summary of thick-film resistor properties is given in Table 24-2.

**TABLE 24-2   Properties of Thick-Film Resistive Materials**

| Resistivity, per square | 10 $\Omega$ | 100 $\Omega$ | 1 k$\Omega$ | 10 k$\Omega$ | 100 k$\Omega$ | 1 M$\Omega$ | 10 M$\Omega$ |
|---|---|---|---|---|---|---|---|
| TCR, ppm/°C | $\pm 100$ | $\pm 50$ | $\pm 50$ | $\pm 50$ | $\pm 50$ | $\pm 100$ | $\pm 300$ |
| Noise, dB | N.A. | $-25$ | $-15$ | $-10$ | $+10$ | $+20$ | N.A. |
| Resistivity, % tolerance | $\pm 15$ | $\pm 10$ | $\pm 10$ | $\pm 10$ | $\pm 10$ | $\pm 15$ | $\pm 20$ |

## Dielectrics

The most widely used thick-film dielectrics are overglaze, multilayer dielectric, and capacitor dielectric. Overglaze is a low-melting vitreous material. It is generally used as a resistor overcoat to passivate thick-film resistors and to protect them from mechanical damage. It is also used as a conductor insulator and solder barrier.

A multilayer dielectric is a mixture of ceramic and devitrifying glasses. It is used as an insulating barrier between conductors for crossover and multilayer work. These materials have a dielectric constant of 8 to 14 and a breakdown voltage greater than 500 V/mil.

## Capacitor Dielectrics

Thick-film capacitors are an offshoot of thick-film multilayer technology. Materials are available with dielectric constants ranging from 10 to 2000. Table 24-3 provides a summary of the characteristics of capacitor dielectrics.

**TABLE 24-3   Properties of Thick-Film Capacitor Dielectrics***

| Dielectric constant | 10 | 30 | 400 | 1100 | 2000 |
|---|---|---|---|---|---|
| TCC | $\pm 50$ ppm | $\pm 30$ ppm | $-10\%$ | $-15\%$ | $-18\%$ |
| Dissipation factor, % | 0.001 | 0.09 | 1.5 | 1.5 | 1.6 |
| Insulation resistance, $\Omega$/square | $1 \times 10^{12}$ | $1 \times 10^{12}$ | $5 \times 10^{11}$ | $1 \times 10^{12}$ | $5 \times 10^{11}$ |
| Power factor/in$^2$ | 2000 | 6000 | 55,000 | 175,000 | 240,000 |
| Dielectric strength, V/mil | 800 | 500 | 500 | 500 | 500 |

*Tested on Pd-Ag with prefired bottom electrode and top electrode cofired at 1000°C.

## Chip Components

The term *chip components* describes uncased active and passive circuit elements designed for use in hybrid microelectronic circuitry. Included in this category are capacitors, resistors, transistors, diodes, and integrated circuits. Chip selection guides for these components are provided in Tables 24-4 to 24-8.

**TABLE 24-4    Chip Selection Guide for Signal Amplifier Applications**

| Device type | Frequency, $f_T$, MHz | Gain, @ 1 kHz | Breakdown voltage, V | Max. power dissipation, mW |
|---|---|---|---|---|
| 2N918 | 600 | 20 | 15 | 360 |
| 2N2369A | 500 | 40 | 15 | 360 |
| 2N3723 | 300 | 25 | 80 | 500 |
| 2N2219A | 250 | 75 | 40 | 500 |
| 2N2907A* | 200 | 50 | 60 | 400 |
| 2N3798* | 100 | 150 | 90 | 360 |
| 2N2484 | 60 | 175 | 60 | 360 |

*pnp; all others are npn.

**TABLE 24-5    Chip Selection Guide for Switching Applications**

| Device type | Switching speed | | Breakdown voltage | Maximum power dissipation, mW |
|---|---|---|---|---|
| | Max. $t_{on}$, ns | Max $t_{off}$, ns | | |
| 2N2369A | 23 | 28 | 15 | 360 |
| 2N3725 | 35 | 60 | 50 | 500 |
| 2N2219A | 35 | 300 | 40 | 500 |
| 2N3467* | 40 | 90 | 40 | 600 |
| 2N2907A* | 45 | 200 | 60 | 400 |

*pnp; all others are npn.

**TABLE 24-6    Commonly Used Diode Chips**

| Type | Description |
|---|---|
| 1N485B | Low leakage |
| 1N645 | High conductance |
| 1N914 | Fast switching |

## Flywire Attachments

Fine aluminum or gold wires called *flywires* are used to make interconnections between faceup bonded chips and substrate conductors or between substrate conductors and package leads. The wires range from 0.0018 to 0.038 cm in diameter; most work is done with 1-mil (0.0025-cm) diameter wire.

Gold flywires are bonded with a heat and pressure technique called *thermocompression bonding*. This is a very reliable method requiring minimal control. Its only disadvantage is that the substrate must be kept at a temperature of 200 to 250°C during the bonding cycle which may adversely affect temperature-sensitive components.

Aluminum flywires are bonded by ultrasonic action to form a molecular bond between the wire and its base metalization. Ultrasonic bonding has the advantage of being a cold process, but it requires very precise control over machine parameters such as power, pressure, and time.

## Packaging

The circuit package has three functions; (1) It protects the thick-film and semiconductor components from mechanical damage and facilitates handling. (2) It provides a means of electrical interconnection between the hybrid circuit and the next-higher assembly, such as a printed circuit board. (3) It protects the circuit components from exposure to moisture, gases, etc.

**TABLE 24-7   Commonly Used Digital Microcircuits**

110tRTL: Available from Motorola (−55 to 125°C)

Dual 2 input gate
Dual 3 input gate
Buffer
J-K flip flop

MRTL: Availablee from Motorola (−55 to 125°C and 0 to 85°C)

Quad 2 input gate
Quad 2 input expander
Dual 2 gate, dual 2 expander
Dual flipflop
Dual buffer, dual 3 input gate

DTL: Available from Fairchild and ITT (0 to 70°C)

Dual J-K
Dual buffer
Dual 4 input buffer
Dual 4 input gate
Triple 3 input gate
Quad 2 input gate
Dual extender
One shot

TTL: Available from Texas Instruments, Inc.

Low Power:
Quad 2 input gate
Triple 3 input gate
Dual 4 input gate
8 input gate
Dual AND-OR-invert gate
J-K flipflop
Dual J-K flipflop
Medium Power:
Quad 2 input gate
Single 8 input gate
Dual 4 input buffer
4 wide 2 input AND-OR-invert gate
Dual J-K flipflop
4-bit shift register
High Power—High Speed:
Quad 2 input gate
Single 8 input gate
Dual 4 input buffer
J-K flipflop

**TABLE 24-8   Linear Microcircuits**

| Type | Description |
| --- | --- |
| 709 | Operational amplifier |
| 710 | Comparator |
| 711 | Dual comparator |
| 723 | Voltage regulator |
| 733 | Differential video amplifier |
| 741 | Operational amplifier |
| 748 | Operational amplifier |

There are three general configurations of hybrid circuit packages:

1. The transistor outline (TO) package was designed to contain single-chip components but may be used for substrate assemblies.

2. The flatpack is a low-profile square or rectangular package with leads extending through the sides.

3. The dual-in-line (DIP) is a rectangular box-shaped package. It gets its name from two rows of leads extending from the base.

In addition to the above standard package types, many other packaging configurations are used. One of the most flexible and popular is shown in Fig. 24-2. A lead frame is soldered directly to the substrate conductor pads, and the substrate is either conformally coated or plastic molded. By use of this method, custom flatpack or DIP units can be designed.

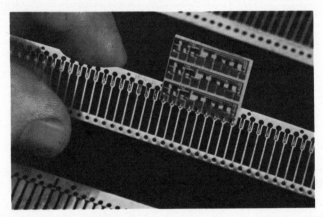

**Fig. 24-2**  Thick-film hybrid circuit with lead frame attachment. *(Courtesy EMCA.)*

## 24.2  ASPECT RATIO CALCULATIONS

**problem 24.1**  Consider the thick-film resistor shapes shown in Fig. 24-3. What is the aspect ratio of each of these resistors? Are these good designs or not? If not, how would you improve them?

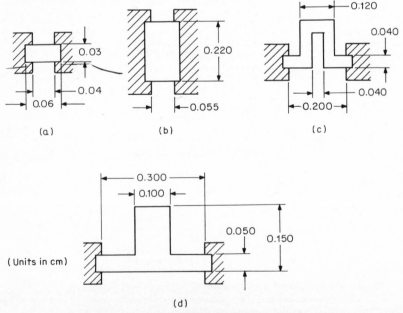

**Fig. 24-3**  Examples of thick-film resistor shapes (Problem 24.1).

**theory**  A geometric representation of a thick-film resistor is shown in Fig. 24-4. The basic equation for the resistance of this piece of material is:

$$R = P_s \frac{L}{Wt} = P_s \frac{L}{W}\frac{1}{t} \tag{24-1}$$

where $R$ = resistance measured from $A$ to $B$
  $P_s$ = sheet resistivity constant
  $W$ = resistor width
  $L$ = resistor length
  $t$ = film thickness

In the equation, $L/W$ is a dimensionless ratio called the *aspect ratio* (AR) of the resistor. Aspect ratio is really just a term used to define the number of square-shaped areas contained in the resistor body. For example, a resistor 0.15 cm long and 0.075 cm wide would have an $L/W$ or aspect ratio of 2 and would contain two square-shaped areas. It would be called a *two-square* resistor.

Manufacturing experience has shown that thick-film resistors with aspect ratios of less than 10:1 and greater than 1:3 have the best geometric characteristics for screen printing. Note that resistor length is measured between the electrodes and does not include the resistor/conductor overlap.

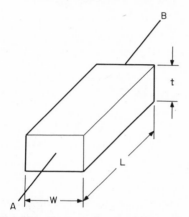

**Fig. 24-4**  Geometrical representation of a thick-film resistor.

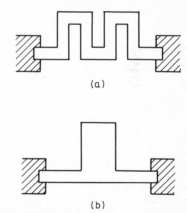

(a)

(b)

**Fig. 24-5**  Special-purpose resistor shapes. (*a*) Bent resistor. (*b*) Hat resistor.

There are two special-purpose resistor designs in which adjustments must be made in the calculations. Consider first the *bent (folded)* resistor configuration of Fig. 24-5a. This is a useful design when a large number of squares is required in a small area. The aspect ratio of the bent resistor is calculated in the same manner as that of a rectangular resistor, except that the corner areas are treated as one-half squares.

The *hat* resistor shown in Fig. 24-5b is used in cases where trimming adjustments have to be made while the circuit is operating to produce a required frequency response or to achieve an optimum voltage level. Hat resistors can be trimmed to aspect ratios of greater than 10:1. In practical application, the hat portion contributes little to the untrimmed value owing to the nonuniform voltage gradient in the hat region. When calculating its aspect ratio, the resistor is treated as if the hat were cut off except that the aspect ratio in the hat region is multiplied by 0.6.

**solution**  For Fig. 24-3a the aspect ratio is:

$$AR = \frac{L}{W} = \frac{0.04}{0.03} = 1.33 \text{ squares}$$

This is a good design according to our guidelines.
  For Fig. 24-3b the aspect ratio is:

$$AR = \frac{0.055}{0.220} = 0.25 \text{ squares}$$

This is less than our guideline minimum of 0.33 squares. The resistor should be redesigned by increasing the aspect ratio to 0.33 or 0.5 and printing with a lower-value paste. When a very low value is required, you may need to design several resistors in parallel. In this case, the technique shown in Fig. 24-6 can be very helpful in saving space.

In Fig. 24-3c the aspect ratio is determined by redrawing the resistor and dividing it into squares and half-squares (Fig. 24-7). The aspect ratio is:

$$AR = 7(1) + 4(1/2) = 9 \text{ squares}$$

This is within our guidelines and is an acceptable design.

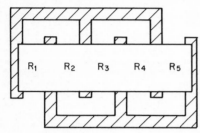

**Fig. 24-6**  Resistors in parallel.          **Fig. 24-7**  Bent resistor aspect ratio.

In Fig. 24-3d the expected aspect ratio would be:

$$AR = \frac{0.100}{0.050} + \frac{0.100}{0.150} + \frac{0.100}{0.050}$$
$$= 4.67 \text{ squares}$$

However, as noted in the theory portion of this problem, the practical aspect ratio is:

$$AR = \frac{0.100}{0.050} + 0.6 \left(\frac{0.100}{0.050}\right) + \frac{0.100}{0.050}$$
$$= 5.2 \text{ squares}$$

This is an untrimmed or unadjusted aspect ratio for the resistor. It establishes only a minimum value for the resistor. The hat resistor is designed to be trimmed to an aspect ratio of greater than 10:1.

## 24.3  SHEET RESISTIVITY CALCULATIONS

**problem 24.2**  (a) A thick-film resistive paste is rated by the manufacturer at 100 $\Omega$/square/25 $\mu$m dried print thickness. What would be the value of a resistor 0.13 cm long by 0.39 cm wide printed with this paste to a dried thickness of 23 $\mu$m? What would its value be if it was printed at 27 $\mu$m? (b) A thick-film resistor 0.3 cm long by 0.15 cm wide has a dried print thickness of 26 $\mu$m. The resistance after firing is 386 $\Omega$. What is the sheet resistivity of the paste expressed in ohms/square/25 micrometers?

**theory**  The basic equation, Eq. (24-1), for the resistance of a film resistor is $R = P_s (L/W)(1/t)$. This can be rewritten as:

$$P_s = \frac{R}{(L/W)/t} \tag{24-2}$$

The sheet resistivity constant $P_s$ is the standard used to describe the value of a thick-film resistive paste. The dimensions of $P_s$ are given in ohms/square/unit of film thickness.

The term *square* refers to a film of equal length and width. The "square" dimension in the sheet resistivity constant serves to normalize the resistance of any size resistor to the resistance of a one-square section of the resistive film. For example, a resistor 0.3 cm long by 0.1 cm wide having a resistance of 450 $\Omega$ has a sheet resistivity of $P_s = 450 \Omega/(0.3/0.1) = 150 \Omega$/square.

The thickness dimension is the most important component of the sheet resistivity constant. Control of print thickness is of major concern in thick-film processing. From Eq. (24-1), resistance varies inversely with the thickness of the film. Since the films themselves are so thin, generally 0.3 to 1 mil, minor variations in thickness can cause major variations in resistance. For example, a 5-$\mu$m variation in a resistor nominally printed at a 25-$\mu$m (1 mil) thickness causes a 20% shift in resistance.

Resistor thickness is generally measured after the print has been dried. There are two very good, practical reasons for standardizing on dried rather than fired print thickness.

First, print thickness is usually measured with optical equipment, such as a light-section microscope. The matte finish of a dried print allows a more accurate measurement than the glossy finish of a fired resistor.

Second, accurate set up of a thick-film printer is a trial-and-error process, generally requiring several adjustments. It takes only a few minutes to print, dry, and measure a thick-film resistor. If the firing cycle is added, however, the time required for printing adjustments becomes lengthy and impractical for production work.

The standard unit of dried-film thickness used for resistor measurement is 25 $\mu$m (or 1 mil). Over a limited range, approximately ±3 $\mu$m, resistance varies linearly with dried thickness. Beyond this range, the material characteristics of the films coupled with firing effects tend to exaggerate the resistance/thickness relationship. This is especially noticeable in higher-value inks, i.e., above 100 k$\Omega$. Good manufacturing practice dictates that resistor thickness be controlled within ±2 $\mu$m for best results.

**solution**    (a) From Eq. (24-1) it is seen that if the resistor is printed at 25 $\mu$ and fired, its value is:

$$R = 100 \ \Omega/\text{square}/25 \ \mu\text{m} \times \frac{0.13 \ \text{cm}}{0.39 \ \text{cm}} \times 1/25 \ \mu\text{m}$$

$$= 33.3 \ \Omega$$

If the dried-print thickness of the film is 23 $\mu$m, the resistance increases linearly as follows:

$$R = 33.3 \ \Omega \ (25 \ \mu\text{m}/23 \ \mu\text{m}) = 36.2 \ \Omega$$

Similarly, if the thickness is increased to 27 $\mu$m, the resistance would be:

$$R = 33.3 \ \Omega \times (25 \ \mu\text{m}/27 \ \mu\text{m}) = 30.8 \ \Omega$$

(b) To determine the sheet resistivity of any size and value, we use Eq. (24-2). Substituting the values given for resistance, length, and width into the equation, we can obtain a sheet resistivity for the film printed at 26 $\mu$m:

$$P_s = \frac{386 \ \Omega}{(0.3 \ \text{cm}/0.15 \ \text{cm})/26 \ \mu\text{m}}$$

$$= 193 \ \Omega/\text{square}/26 \ \mu\text{m}. \ \text{Correcting for thickness we get:}$$

$$P_s = 193 \ \Omega/\text{square}/26 \ \mu\text{m} \times \frac{26 \ \mu\text{m}}{25 \ \mu\text{m}}$$

$$= 201 \ \Omega/\text{square}/25 \ \mu\text{m}$$

## 24.4   RESISTOR DESIGN CALCULATIONS, GEOMETRY EFFECTS, AND TRIMMING ALLOWANCE

**problem 24.3**   A resistor test pattern is used to characterize a 10-k$\Omega$/square resistive ink. The ink is tested with gold electrodes and with palladium-silver electrodes. The mean resistor values obtained on each conductor type are as follows:

| Resistor | Width, mils | Length, mils | Resistance | |
|---|---|---|---|---|
| | | | On Au, k$\Omega$ | On Pd-Ag, k$\Omega$ |
| $R_1$ | 40 | 20 | 3.5 | 3.0 |
| $R_2$ | 40 | 40 | 10.0 | 10.0 |
| $R_3$ | 40 | 80 | 24.0 | 28.0 |
| $R_4$ | 40 | 120 | 37.5 | 45.3 |
| $R_5$ | 40 | 160 | 52.0 | 62.0 |
| $R_6$ | 40 | 200 | 66.0 | 77.5 |
| $R_7$ | 40 | 240 | 79.2 | 93.0 |

Owing to processing variables, the mean values have a tolerance of ±20%. (a) How is this information used to develop design parameters for the 10-k$\Omega$/square ink? (b) How would you design a 44-k$\Omega$ resistor terminated with gold electrodes? The resistor may be trimmed to final value.

**theory**   To this point, we have treated sheet resistivity as a constant for each thick-film resistive paste. We have assumed that a resistive paste rated at 100 $\Omega$/square yields 100 $\Omega$/square under all design conditions. Unfortunately, this is not the case. Actually, sheet resistivity is dependent upon resistor size and upon the type of conductor used for the resistor electrodes. This dependence is variously described as *contact resistance*, *termination resistance*, and *geometry effect*.

It is hard to categorize these effects because they vary greatly with the type and value of resistive ink used, the type of conductive ink used, and the size of the resistor. In general, however, the geometry effect has the following characteristics:

1. Small resistors have lower sheet resistivities than large resistors. For example, a one-square resistor 20 by 20 mils in size printed with a 100-$\Omega$/square ink may have a resistance of only 45 $\Omega$.

In contrast, a one-square resistor 80 by 80 mils in size, printed with the same ink, may have a resistance of 130 Ω.

2. Resistors printed on active conductors, such as silver and platinum-silver, have lower sheet resistivities than resistors printed on less active conductors like gold and palladium-silver. For example, a 40-by-40-mil resistor printed with 100-Ω/square ink on platinum-silver electrodes may have a resistance of 80 Ω. The same resistor printed on gold electrodes may have a value of 130 Ω.

3. Geometry effects become more pronounced with higher-value resistive inks. If the size of a resistor made with 1-kΩ/square ink is doubled, the resistivity may increase by 50%. If the same procedure is tried with a resistor made with 100-kΩ/square ink, the resistivity may increase by over 100%.

4. Geometry effects are very much dependent upon the chemical formulations of the resistive and conductive inks. For example, the same resistor printed on two types of gold electrodes may show a difference in value of 20 to 30%.

The explanation for these phenomena is quite simple. During the firing process there is a transfer of metal between a thick-film resistor and its adjacent electrodes. Metal from the electrodes diffuses into the resistor, creating metal-enriched zones in the body of the resistor adjacent to the electrodes. This, of course, causes a drop in resistance. Since the diffusion rate is constant for each resistor/conductor combination, small resistors pick up a greater percentage of transferred metal than large resistors.

A relationship between sheet resistivity and resistor size can be empirically developed for any resistor/conductor combination. This relationship can be presented in many ways. The most

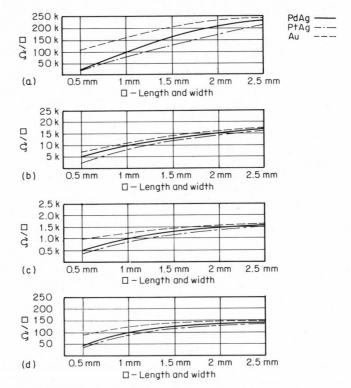

**Fig. 24-8** Resistor design curves for palladium-silver, platinum-gold, and gold terminations. *(Courtesy EMCA.)*

useful presentation is to put the information in the form of a design curve plotting sheet resistivity in ohms/square vs. resistor size.

A series of sample curves developed by manufacturers is given in Fig. 24-8. These design curves present a mean value of resistivity that can be expected under ideal manufacturing conditions. They do not reflect the day-to-day variations in resistivity that are caused by processing variables, such as print thickness and firing cycle. In order for the curves to be really useful, the manufacturing tolerances must be superimposed upon the mean values.

In a tightly controlled operation, these tolerances may average ±10 to ±12%. In a poorly controlled operation they may run in excess of ±30%. A good manufacturing facility with normal process controls will average about ±20%. Some thick-film manufacturing operations have taken extensive resistor data and processed them in a computer to create design curves with a 3-sigma probability.

A common mistake in thick-film resistor design is to minimize resistor size. In general, it is not a good practice to design resistors smaller than 40 mils (1 mm) square. Most resistor patterns are printed through a 200-mesh screen which has a wire size of 0.004 cm and an opening size of 0.0086 cm. A 20-mil-square resistor, for example, is squeezed through only four openings. Very small resistors have poorly defined shapes with a concomitant decrease in accuracy and consistency.

Another factor that must be taken into account in resistor design is the trimming allowance. As long as a resistor does not break down, it can be trimmed to as high a value as desired. Using laser trimming techniques, for example, it is possible to trim a 100-Ω resistor to over 1 MΩ. Such extreme trimming, however, requires that the resistor be cut to extremely thin sections. This creates hot spots and affects stability. Most resistor stability data are based on a 60% maximum cut depth which causes an increase in resistance of up to twice the as-fired value. With this limitation, excellent stability characteristics can be achieved.

**solution**    ($a$) First obtain the aspect ratio for each resistor by dividing length by width. Then obtain the sheet resistivity in ohms/square for each resistor by dividing resistance by the aspect ratio. This yields the following values:

|  |  | Sheet resistivity | |
|---|---|---|---|
| Resistor | No. of squares | Au, kΩ/square | Pd-Ag, kΩ/square |
| $R_1$ | 0.5 | 7.0 | 6.0 |
| $R_2$ | 1.0 | 10.0 | 10.0 |
| $R_3$ | 2.0 | 12.0 | 14.0 |
| $R_4$ | 3.0 | 12.5 | 15.1 |
| $R_5$ | 4.0 | 13.0 | 15.5 |
| $R_6$ | 5.0 | 13.2 | 15.5 |
| $R_7$ | 6.0 | 13.2 | 15.5 |

Next, obtain the maximum and minimum tolerances for each resistor. This is done by multiplying the mean resistivity of each resistor by 1.2 and by 0.8, respectively. Now reassemble the data as follows:

|  | | Sheet resistivity | | | |
|---|---|---|---|---|---|
|  |  | On Au | | On Pd-Ag | |
|  | No. of squares | Max., kΩ/square | Min., kΩ/square | Max., kΩ/square | Min., kΩ/square |
| $R_1$ | 0.5 | 8.4 | 5.6 | 7.2 | 4.8 |
| $R_2$ | 1.0 | 12.0 | 8.0 | 12.0 | 8.0 |
| $R_3$ | 2.0 | 14.4 | 9.6 | 16.8 | 11.2 |
| $R_4$ | 3.0 | 15.0 | 10.0 | 18.1 | 12.1 |
| $R_5$ | 4.0 | 15.6 | 10.4 | 18.6 | 12.4 |
| $R_6$ | 5.0 | 15.8 | 10.6 | 18.6 | 12.4 |
| $R_7$ | 6.0 | 15.8 | 10.6 | 18.6 | 12.4 |

Plot the design curves showing maximum and minimum sheet resistivity as a function of resistor length for each material, as shown in Figs. 24-9 and 24-10.

($b$) In order to design a resistor valued at 44 kΩ, we must keep in mind the maximum and minimum values of the design curve and the trimming allowance. The final trimmed value of the resistor must be less than the maximum untrimmed value as given by the design curve. This is because trimming is a one-way process. A resistor can be trimmed up in value but not down. Also, the minimum untrimmed value of the resistor as given by the design curve must be greater than 50% of the final trimmed value in order to fall within the allowable trimming requirements.

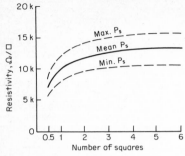

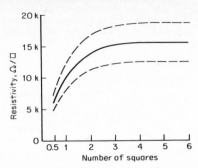

**Fig. 24-9**  Resistor design curve, 10-kΩ/square resistor on gold terminations.

**Fig. 24-10**  Resistor design curve, 10-kΩ/square resistor on palladium-silver terminations.

To determine the resistor dimensions, refer to Fig. 24-9 and set up the following table:

| No. of squares | Design value | | From curve | |
| --- | --- | --- | --- | --- |
| | $P_s$, kΩ/square | $0.5\,P_s$, kΩ/square | Max. $P_s$, kΩ/square | Min. $P_s$, kΩ/square |
| 1 | 44.0 | 22.0 | 12.0 | 8.0 |
| 2 | 22.0 | 11.0 | 14.4 | 9.6 |
| 3 | 14.7 | 7.4 | 15.0 | 10.0 |
| 4 | 11.0 | 5.5 | 15.6 | 10.4 |

By inspection, we see that a two-square design is unsuitable because the minimum value of 9.6 kΩ/square is below the lowest allowable trim value 11.0 kΩ/square. The three-square resistor is also unsuitable because its maximum value of 15.0 kΩ/square is above the design value of 14.7 kΩ/square. The solution, therefore, must lie between two and three squares. If we interpolate the data at 2.5 squares, we get:

| No. of squares | Design value | | From curve | |
| --- | --- | --- | --- | --- |
| | $P_s$ kΩ/square | $0.5\,P_s$ kΩ/square | Max. $P_s$ kΩ/square | Min. $P_s$ kΩ/square |
| 2.5 | 17.6 | 8.8 | 14.8 | 9.8 |

From our criteria, a 2.5-square resistor 40 mils wide by 100 mils long will work in this design. It will have a maximum untrimmed value of 37 kΩ which can be trimmed to 44 kΩ. Its minimum value is 24.5 kΩ, which is greater than the minimum allowable value of 24 kΩ.

For example purposes, the resistor width was kept constant at 40 mils. In an actual design, width would be used as an additional design variable.

## 24.5  TEMPERATURE COEFFICIENT OF RESISTANCE, TCR LINEARITY, AND TRACKING CALCULATIONS

**problem 24.4**  Two thick-film resistors are operated over the temperature range $-55$ to $+125°C$. The following data have been taken:

| Temperature, °C | $R_1$, Ω | $R_2$, Ω |
| --- | --- | --- |
| $-55$ | 100.240 | 300.480 |
| $-25$ | 100.125 | 300.225 |
| 0 | 100.050 | 300.075 |
| $+25$ | 100.000 | 300.000 |
| $+50$ | 100.138 | 300.375 |
| $+75$ | 100.250 | 300.825 |
| $+100$ | 100.338 | 301.350 |
| $+125$ | 100.500 | 301.950 |

(a) What is the temperature coefficient of resistance (TCR) of each resistor over the range −55 to +125°C? What is the TCR over the range + 25 to +125°C? (b) What is the TCR linearity over the hot and cold temperature ranges? How well do the two resistors track?

**theory**    The TCR is a measure of the way a resistor varies with increasing or decreasing temperature. It is defined as follows:

$$\text{TCR} = \left[ \frac{R_2 - R_1}{R_1(T_2 - T_1)} \right] \times 10^6 \qquad (24\text{-}3)$$

where TCR = temperature coefficient of resistance, parts per million per °C (ppm/°C)
$R_1$ = resistance at room or reference temperature
$R_2$ = resistance at operating ambient temperature
$T_1$ = room or reference temperature; almost always given as 25°C
$T_2$ = operating ambient temperature, °C
Generally, a "hot" and a "cold" value are given for TCR. The "hot" value is measured from +25°C, or room temperature, to the high end of the operating range, usually +125°C. The "cold" value is measured from +25°C to the low end of the operating range, usually −55°C. If resistance increases over the hot portion of the range, the TCR is positive; if it decreases, the TCR is negative. Conversely, if the resistance increases over the cold portion of the range, the TCR is negative; if it decreases, the TCR is positive. Two temperature ranges are necessary because TCR varies nonlinearly with temperature.

Pure metals tend to exhibit a very positive TCR. Heat increases the activity of their electrons, causing an increase in resistance. Metal oxides may have positive or negative TCRs. Ruthenium oxide, which forms the base for many resistor systems, has a positive TCR. The glasses, by themselves, have a slightly positive TCR. During the firing process, however, some of the active materials become dissolved in the glass. This causes the glass to behave like a semiconductor, resulting in a negative TCR. (For a more complete discussion, see the works by Sargent and by Holmes and Loasby in the Bibliography.)

The resistivity of a family of thick-film pastes is regulated by controlling the amount of active materials (metals and metal oxides) in the paste. The more active material, the lower the sheet resistivity. Inks with lower sheet resistivities tend to take on metallic characteristics and exhibit positive TCRs. Inks with higher sheet resistivity take on semiconductor characteristics and exhibit negative TCRs.

TCR, like sheet resistivity, is influenced by geometry and by processing variations. It is difficult to quantify these variations since they depend on specific usage. However, we can make the following general observations:

1. Short resistors tend to have more positive TCRs than long resistors.
2. Resistors terminated with active conductive materials, such as silver and platinum-silver, have more positive TCRs than resistors terminated with less active conductors, such as gold or palladium-silver.
3. Thin resistors have more negative TCRs than thick resistors.

Two important paremeters are TCR linearity and TCR tracking. *TCR linearity* is the maximum deviation of TCR from a straight-line function as measured over the temperature range. *TCR*

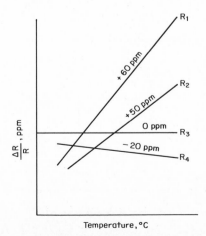

**Fig. 24-11**   TCR tracking.

*tracking* is the maximum difference in TCR between two resistors as measured over the temperature range. TCR tracking is illustrated in Fig. 24-11 where the tracking between $R_1$ and $R_2$ is:

$$|60 - 50| = 10$$

Tracking is always expressed as an absolute value, i.e., without sign. So the TCR tracking between $R_3$ and $R_4$ is:

$$|0 - 20| = 20$$

**solution**    (*a*) By Eq. (24-3) and the given values for $R_1$ and $R_2$ the hot and cold TCRs at the extreme temperature points of $-55$ and $+125°C$ are:

For $R_1$:

$$\text{at } -55°C: \text{TCR} = \frac{100.240 - 100.000}{100(-55 - 25)} \times 10^6$$
$$= -30 \text{ ppm/°C}$$
$$\text{at } +125°C: \text{TCR} = \frac{100.500 - 100.000}{100(125 - 25)} \times 10^6$$
$$= +50 \text{ ppm/°C}$$

For $R_2$:

$$\text{at } -55°C: \text{TCR} = \frac{300.480 - 300.000}{300(-55 - 25)} \times 10^6$$
$$= -20 \text{ ppm/°C}$$
$$\text{at } +125°C: \text{TCR} = \frac{301.950 - 300.00}{300(125 - 25)} \times 10^6$$
$$= +65 \text{ ppm/°C}$$

In a similar manner, the TCRs at other points are found and are summarized below.

|  | TCR | |
|---|---|---|
| Temp., °C | $R_1$, ppm/°C | $R_2$, ppm/°C |
| $-55$ | $-30$ | $-20$ |
| $-25$ | $-25$ | $-15$ |
| $0$ | $-20$ | $-10$ |
| $+25$ | | |
| $+50$ | $+55$ | $+50$ |
| $+75$ | $+50$ | $+55$ |
| $+100$ | $+45$ | $+60$ |
| $+125$ | $+50$ | $+65$ |

(*b*) The TCR linearity over the hot and cold ranges is found by subtracting the maximum or minimum TCR from the straight-line end-point TCR. It is the absolute value of the greatest deviation over the temperature range. The calculations are as follows:

For $R_1$:
  $+25$ to $-55°C$: Linearity $= |-30 - (-20)| = 10$ ppm/°C max.
  $+25$ to $+125°C$: Linearity $= |55 - 45| = 10$ ppm/°C max.
For $R_2$:
  $+25$ to $-55°C$: Linearity $= |-20 - (-10)| = 10$ ppm/°C max.
  $+25$ to $+125°C$: Linearity $= |65 - 50| = 15$ ppm/°C max.

TCR tracking is found by subtracting the TCR of $R_1$ from the TCR of $R_2$ at each data point over the temperature range as follows:

|  | TCR | | |
|---|---|---|---|
| Temp., °C | $R_1$ | $R_2$ | $\Delta$TCR |
| $-55$ | $-30$ | $-20$ | $10$ |
| $-25$ | $-25$ | $-15$ | $10$ |
| $0$ | $-20$ | $-10$ | $10$ |
| $+25$ | | | |
| $+50$ | $+55$ | $+50$ | $5$ |
| $+75$ | $+50$ | $+55$ | $5$ |
| $+100$ | $+45$ | $+60$ | $15$ |
| $+125$ | $+50$ | $+65$ | $15$ |

In summary, for the cold range of +25 to $-55°C$, $R_1$ and $R_2$ track within 10 ppm/°C. Over the hot range of +25 to +125°C, $R_1$ and $R_2$ track within 15 ppm/°C.

## 24.6  VOLTAGE COEFFICIENT OF RESISTANCE CALCULATIONS

**problem 24.5**  A precision thick-film resistor is required to have a value of 125,000 $\Omega$ with a tolerance of ±0.25%. The resistor can be laser trimmed to a tolerance of ±0.1%. The voltage of the trimming bridge is 28 V, and the operating voltage of the resistor is 5 V. The resistive material has a voltage coefficient of resistance of $-200$ ppm/V. Will the resistor be within tolerance at its operating voltage? If not, what adjustments must be made to the trimming process?

**theory**  The resistivity of high-value thick-film pastes, 10 k$\Omega$/square and up, varies with voltage in much the same way that it varies with temperature. This effect is called the voltage coefficient of resistance (VCR), and it is defined in the same manner as TCR:

$$\text{VCR} = \left[\frac{R_2 - R_1}{R_1(V_2 - V_1)}\right] \times 10^6 \tag{24-4}$$

where VCR = voltage coefficient of resistance, ppm/V
  $V_1$ = reference voltage
  $V_2$ = applied voltage
  $R_1$ = resistance at $V_1$
  $R_2$ = resistance at $V_2$

The magnitude of VCR varies widely among thick-film resistor materials. Like sheet resistivity and TCR, it is dependent upon the type of termination material used and the geometry of the resistor. Within these constraints, however, it is linear, negative, and reproducible.

VCR becomes important when we deal with high-value, close-tolerance resistors. The different voltage levels applied by trimming bridges and digital ohmmeters coupled with a high VCR can often result in measurement errors with concomitant high-yield losses.

**solution**  Since the VCR effect causes resistivity to shift downward with increasing voltage, the resistivity at the trimming voltage (28 V) will be lower than the resistivity at operating voltage. Because of this, the worst case occurs when the resistor is adjusted to the maximum extent of its trim tolerance. In this case, its trimmed resistance is:

$$R_1 = 125,000 + 0.001\ (125,000)$$
$$= 125,125\ \Omega$$

The resistance at operating voltage $R_2$, by Eq. (24-4), is:

$$\text{VCR} = \left[\frac{R_2 - R_1}{R_1(V_2 - V_1)}\right] \times 10^6$$

Then
$$R_2 = R_1\,[1 + (\text{VCR})(V_2 - V_1) \times 10^{-6}]$$
$$= 125,125\ [1 + (-200)(5 - 28) \times 10^{-6}]$$
$$= 125,700\ \Omega$$

Since the allowable operating tolerance is ± 0.25%, the maximum value we can have for $R_2$ is:

$$R_2 = 125,000 + 0.0025(125,000)$$
$$= 125,312\ \Omega$$

The resistor will be greater than its allowable value at operating voltage. To remedy this, we may take two courses of action: First, we can lower the VCR of the resistor. Requiring a redesign of the resistor, this may be done by using a lower-value resistive paste and increasing the aspect ratio, or by keeping the same aspect ratio and increasing the physical size of the resistor.

To find the required VCR, we substitute our maximum allowable operating resistance and minimum trimmed resistance into Eq. (24-4):

$$\text{VCR} = \frac{125,312 - 124,875}{124,875(5 - 28)} \times 10^6$$
$$= 152\ \text{ppm/V}$$

Second, if design conditions will not permit an increase in resistor size, we can adjust the trimmed resistance to bring the operating resistance into value. To do this, Eq. (24-4) is solved for $R_1$ with $R_2$ equal to the nominal allowable operating resistance:

$$R_1 = \frac{R_2}{1 + (\text{VCR})(V_2 - V_1) \times 10^{-6}}$$

$$= \frac{125,000}{1 + (-200)(-23) \times 10^{-6}}$$
$$= 124,428\ \Omega$$

Since the operating tolerance of ±0.25% is greater than the trimming tolerance of 0 ±0.10%, we can see that the resistor will operate within tolerance at its maximum and minimum trimmed values.

## 24.7   BLENDING RESISTOR INKS FOR RESISTIVITY AND TCR

**problem 24.6**   Two thick-film resistor inks are available for use. Their nominal values are 100 Ω/square and 1 kΩ/square. When printed to a 25-μm dried thickness, the actual values, after firing, are found to be: 95 Ω/square and 980 Ω/square. How would you blend these inks to make 80 g of a 500-Ω/square ink?

**theory**   Because chemical formulations of inks are complex, catastrophic results can occur when inks of different chemical bases are mixed together. Within a particular group or family of inks, however, blending can be done successfully. In general, the blending relationship for a compatible ink series conforms to the following empirical formula:

$$\log R_B = \log R_L + \frac{W_H \pm 0.4\,W_L W_H}{W_B}(\log R_H - \log R_L) \tag{24-5}$$

where $R_B$ = resistivity of blend
$\quad R_H$ = resistivity of high component
$\quad R_L$ = resistivity of low component
$\quad W_H$ = weight of high component
$\quad W_L$ = weight of low component
$\quad W_B$ = weight of blend

Fortunately, this relationship can be presented graphically as shown in Fig. 24-12. This is known as a *blending curve*, and it forms the basis for mixing any thick-film resistor inks. The blending curve is drawn on a semilogarithmic scale. The $y$, or log, axis denotes resistivity in ohms/square. The $x$ axis represents the percentage proportions by weight of each resistive ink. At the far left point there is 100% of the low component and 0% of the high component. At the far right there is 100% of the high component and 0% of the low component. Since blending is not an exact science, the blending curve usually exhibits some sag or bulge. Before a blend is attempted, a small sample mix should be made to determine the actual shape of the curve.

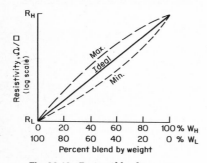

**Fig. 24-12**   Resistor blending curve.

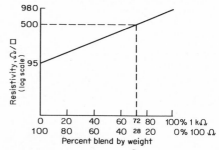

**Fig. 24-13**   Blending curve for 100-Ω/square and 1-kΩ/square inks.

Blending for specific TCR values is difficult and should be avoided if possible. Paste manufacturers have developed additives for their inks to bias TCR in a positive or negative direction. Since these materials are proprietary, the manufacturers are generally unwilling to sell them as a product. They do, nowever, offer TCR blending as a service to their customers.

Before closing this discussion, we should observe two practical cautions. First, for accurate blending, the high and low components should be no more than a decade apart. The wider apart the two components are in value, the more inaccurate the blending curve becomes. Second, proper mixing of the ink is necessary to obtain reproducible results. Many blends give poor results simply because they were not mixed thoroughly. Be sure to get mixing recommendations from the ink manufacturer before attempting a blend.

**solution**   First, construct the resistivity blending curve shown in Fig. 24-13. From this figure the percentages of ink required to made a 500-Ω/square blend are found to be 28% 100 Ω/square and 72% 1 kΩ/square. To make 80 g, the amount required is:

$$80 \times 0.28 = 22.4 \text{ g of } 100 \text{ Ω/square}$$
$$80 \times 0.72 = 57.6 \text{ g of } 1 \text{ kΩ/square}$$

## 24.8   THERMAL DESIGN CALCULATIONS

**problem 24.7**   A power transistor chip, 0.254 cm wide by 0.254 cm long by 0.178 cm thick, is eutectically bonded to a 0.064-cm-thick 96% alumina substrate. The substrate is backed with palladium-silver and soldered into a Kovar flatpack with an 0.05-cm-thick base. The package is bonded with a BeO-filled epoxy to a copper heat sink which is maintained at 95°C. The epoxy bond is 3 mils thick.

What is the junction temperature of the transistor when it is dissipating 7 W? What would the junction temperature be if the substrate is made from beryllia?

**theory**   The basic problem in thermal design of electronic assemblies is to cool the high-heat-dissipating components by providing a path for the removal of thermal energy from the high-temperature heat source to a low-temperature heat sink. This may be done in three ways: conduction, radiation, and convection.

A well-designed electronic assembly makes the maximum use of all three methods of heat removal. Copper strips are added for conduction; finned aluminum extrusions aid convection; and light-colored surfaces emit radiant energy. In the case of the hybrid microelectronic circuit, most of the thermal energy is removed by conduction. The amount of convective and radiation cooling is generally negligible.

In conduction, heat in the form of thermal energy flows from a high-temperature heat source through the thermal path to a low-temperature heat sink where it is absorbed. The heat source may be a resistor or transistor; the thermal path may be a substrate, a solder joint, or a package lead; and the heat sink may be a copper strip on a printed circuit board or a finned extrusion. The basic equation for the conduction of thermal energy is:

$$Q = \frac{KA}{L} \Delta T = \frac{KA}{L}(T_1 - T_2) \qquad (24\text{-}6)$$

where $Q$ = heat flow per unit of time, W
  $K$ = thermal conductivity constant, W/(cm)(°C)
  $A$ = area of thermal path, cm$^2$
  $L$ = length of thermal path, cm
  $T_1$ = temperature of heat source, °C
  $T_2$ = temperature of heat sink, °C

A representative listing of the thermal conductivities of commonly used hybrid materials is given in Table 24-9.

**TABLE 24-9   Thermal Conductivity of Hybrid Circuit Materials at 20°C**

| Material | Thermal conductivity, | |
|---|---|---|
| | W/in/°C | W/cm/°C |
| Silver | 10.6 | 4.17 |
| Copper | 9.6 | 3.78 |
| Au-Si eutectic | 7.5 | 2.95 |
| Gold | 7.5 | 2.95 |
| Beryllia | 6.58 | 2.6 |
| Aluminum | 5.52 | 2.17 |
| Nickel | 2.29 | 0.9 |
| Silicon | 2.13 | 0.84 |
| Alumina (99%) | 0.93 | 0.36 |
| Solder (60-40) | 0.91 | 0.358 |
| Alumina (96%) | 0.89 | 0.35 |
| Kovar | 0.49 | 0.192 |
| Multilayer dielectric | 0.55 | 0.216 |
| Steatite | 0.15 | 0.06 |
| Epoxy adhesive, BeO-filled | 0.088 | 0.035 |
| Silicon RTV, BeO-filled | 0.066 | 0.026 |
| Epoxy, silver-filled | 0.040 | 0.016 |
| Glass, lead borosilicate | 0.026 | 0.010 |
| Epoxy | 0.004 | 0.0016 |

Equation (24-6) can be rewritten as:

$$Q = \frac{T_1 - T_2}{L/KA} \qquad (24\text{-}7)$$

or
$$Q = \frac{T_1 - T_2}{\theta} \qquad (24\text{-}8)$$

where $\theta$ equals $L/KA$.

Element $\theta$ is the *thermal resistance* of the conductive path and is given in watts per degree Celsius.

Figure 24-14 illustrates heat flow through a simple series conductive path. Heat flows from a transistor heat source at surface temperature $T_1$ to a copper heat sink maintained at temperature

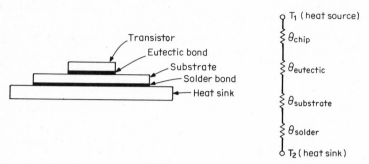

**Fig. 24-14**   Series thermal conductive path.

$T_2$. The thermal path consists of the silicon chip, the eutectic bond, the substrate, and the solder joint. The thermal resistance of the path is equal to the sum of the thermal resistances of the individual elements:

$$\theta_{\text{total}} = \theta_1 + \theta_2 + \theta_3 + \cdots \qquad (24\text{-}9)$$

and the heat flow can be calculated using Eq. (24-7):

$$Q = \frac{T_1 - T_2}{\theta_{\text{total}}}$$

Parallel thermal resistances are totaled in the same manner as parallel electrical resistances:

$$\frac{1}{\theta_{\text{total}}} = \frac{1}{\theta_1} + \frac{1}{\theta_2} + \frac{1}{\theta_3} + \cdots \qquad (24\text{-}10)$$

For example, consider a transistor heat source bonded to a substrate. The substrate is connected to a printed circuit board heat sink by two sets of leads. In this case, there are two equal parallel paths to the heat sink, as shown in Fig. 24-15. The total thermal resistance is:

$$\theta_{\text{total}} = \theta_{\text{chip}} + \theta_{\text{bond}} + (\theta_{\text{substrate}} + \theta_{\text{lead}}/3)/2$$

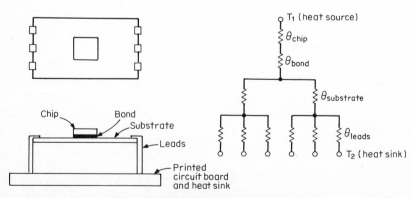

**Fig. 24-15**   Parallel thermal conductive paths.

In calculating the thermal resistance of a thermal network, there are certain elements which can be treated as isotherms. That is, there is no temperature drop across the element; its thermal resistance, therefore, is zero. Thick-film resistors and conductors may be treated as isotherms because of their thin cross sections and intimate bonding to the substrate. Other circuit elements such as eutectic die bonds and thin solder joints may also be considered isothermal.

The thermal path of a small surface area heat source through a large surface area conductor is not easily measured because the heat tends to spread out through the conductive material. For thin sections such as silicon chips or mechanical bonds, we can assume that the heat travels in a straight path. In larger sections, however, this approach is too conservative for optimum design, and thermal spreading must be considered.

For a square-shaped source:

$$\theta_{\text{square}} = \frac{L}{Ka(a + 2L)} \qquad (24\text{-}11)$$

where $a$ equals length of side.

For a rectangular source:

$$\theta_{\text{rectangle}} = \frac{1}{2K(a - b)}\left[\ln\left(\frac{a}{b}\right)\frac{2L + b}{2L + a}\right] \qquad (24\text{-}12)$$

where $a$ and $b$ are lengths of sides.

For a circular source:

$$\theta_{\text{circle}} = \frac{L}{K\pi(r^2 + rL)} \qquad (24\text{-}13)$$

where $r$ equals the radius of the circle.

In the planning of a thick-film hybrid assembly, the following guidelines should be observed for good thermal design:

1. Try to spread component heat evenly over the substrate. Do not group high power dissipating components together.

2. In mechanically bonding high-power components to substrates conductors, select high-conductivity materials. Use eutectic or solder joining rather than epoxy bonding.

3. Keep the thermal path of high-power components as short as possible.

4. When designing high-power thick-film resistors, be sure to consider area reduction due to trimming adjustments.

5. Do not locate passive components such as chip capacitors over heat-dissipating thick-film resistors.

6. When using pins or leads as a thermal path, select high-conductivity materials (e.g., copper rather than Kovar). Also try to maximize cross-sectional area and minimize length.

**solution**    First, we must determine the thermal resistance of the path from the heat source to the heat sink. This is a straight series path (Fig. 24-16). We can assume that the eutectic bond and the substrate solder joint are isothermal. We also assume that the thermal path through the transistor and through the epoxy joint is straight. We shall calculate only thermal spreading in the substrate and the Kovar package.

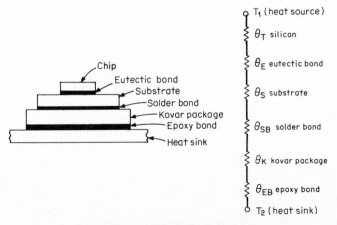

**Fig. 24-16**   Thermal path for Problem 24.7.

By Eqs. (24-8) and (24-11), we calculate the individual thermal resistance:

$$\theta_T = \frac{0.0178}{0.84 \times 0.254 \times 0.254} = 0.33°C/W$$

$$\theta_S = \frac{0.064}{0.35 \times 0.254(0254 + 2 \times 0.064)} = 1.87°C/W$$

$$\theta_K = \frac{0.05}{0.192 \times 0.38(0.38 + 2 \times 0.05)} = 1.43°C/W$$

$$\theta_{EB} = \frac{0.0076}{0.035 \times 0.48 \times 0.48} = 0.94 °C/W$$

The total thermal resistance is:

$$\begin{aligned}\theta_{total} &= 0.33 + 1.87 + 1.43 + 0.94 \\ &= 4.57°C/W\end{aligned}$$

Substituting this value into Eq. (24-7), we can solve for $T_1$:

$$7 W = \frac{T_1 - 95°C}{4.57°C/W}$$
$$T_1 = 127°C$$

This is a little too high for the best operating condition. We would like to reduce $T_1$ to less than 125°C by using a beryllia substrate. The thermal resistance of the substrate would be:

$$\theta_S = \frac{0.064}{2.6 \times 0.254(0.254 + 2 \times 0.064)} = 0.25°C/W$$

The total thermal resistance is now:

$$\begin{aligned}\theta_{total} &= 0.33 + 1.43 + 0.94 + 0.25 \\ &= 2.95°C/W\end{aligned}$$

By Eq. (24-7),

$$7W = \frac{T_1 - 95°C}{2.95°C/W}$$
$$T_1 = 115.65°C$$

## 24.9    DESIGN LAYOUT OF A THICK-FILM HYBRID CIRCUIT

**problem 24.8**    Figure 24-17 shows the schematic diagram of a switch driver circuit. This circuit has been in use for some time as a printed wiring-board assembly built with discrete semiconductor components and carbon resistors. How would you convert this design to a hybrid circuit configuration?

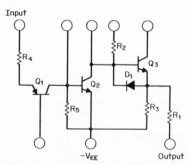

**Fig. 24-17**    Schematic diagram of a switch driver circuit (Problem 24.8).

**theory**    The design of thick-film circuitry is similar in many ways to the design of printed wiring-board circuitry. There are basic guidelines to be followed regarding line widths, spacing, edge clearance, component mounting, pad configurations, etc. Keep in mind that these are conventions for good conservative design; they are not ironclad rules. As in anything else, rules are made to be broken, and thick-film design is no exception.

**Substrates**    Substrates can be obtained in almost any shape in sizes up to 10 by 10 cm with thickness ranging from 0.038 to 0.1 cm. The standard design parameters are:
1. *Length and width tolerances:* ±½% to a minimum of ±0.0076 cm.
2. *Thickness tolerance:* ±10% to a minimum of 0.0025 cm.
3. *Flatness:* ±0.1 cm/cm as fired. Grinding or lapping may be used to reduce the tolerance to ±0.0025 cm/cm. This is expensive and should be avoided unless absolutely necessary.
4. *Hole diameters:* No smaller than two-thirds substrate thickness with an absolute minimum of 0.05 cm. Allow at least 0.0125 cm for riser pin clearance. The minimum hole-to-hole or hole-to-substrate edge distance must not be less than the hole diameter.

**Substrate Printing**    In flatpack applications only, the top side of the substrate may be used for circuit patterns. The bottom is metalized to provide a solderable surface for mechanical attachment. In TO or DIP applications, both sides of the substrate may be used for circuitry with top and bottom connections via the riser pins. However, chip components and flywires may be attached only to the top surface while resistors may be printed on the top or bottom surface, but not on both.

**Conductors**    The following guidelines are recommended for the design of thick-film conductors:
1. *Length.* Make conductor lengths as short as possible to minimize circuit resistance.
2. *Width and spacing.* The minimum conductor width is 0.025 cm; this is equivalent to two openings of a 200-mesh screen. A width of 0.05 cm is preferred. The spacing between conductors or pads must be at least 0.025 cm. A spacing of at least 0.05 cm is preferred.

**Conductor Routing**    Conductors should be run in straight horizontal and vertical lines wherever possible. Angles and curves are permissible but will print sawtooth edges owing to nonalignment of artwork with the screen mesh.

**Edge Conductors**    Do not run conductors around the edge of a substrate. Edge printing requires special fixturization and printing equipment. Use riser pins or spring clips to make connections from one side of the substrate to the other.

**Edge Clearances**    Keep conductors at least 0.05 cm away from the substrate edge.

**Crossovers**    Crossovers require extra printing and firing operations with concomitant increased cost and reduced yield. If there are only a few crossovers in the circuit, use a jumper wire over a local insulating overglaze to make the connection (Fig. 24-18). If the circuit has a large number of crossovers, the connections may be made with a screen-printed conductor over a multilayer dielectric. See Fig. 24-19 for dimensioning.

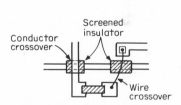

**Fig. 24-18**   Crossover methods.

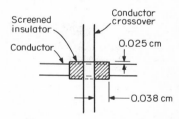

**Fig. 24-19**   Insulator dimensions.

**Parallel Crossovers**    For closely spaced parallel crossovers, the screened dielectric should be stepped to prevent the crossover conductors from running together at the edges (Fig. 24-20).

**Pad Diameters**    Pads surrounding feed-through holes should be at least 0.083 cm larger than the hole diameter with a minimum of 0.15 cm diameter (Fig. 24-21). The area covering the hole may be blocked out. The pad must cover at least 180° of the area around the hole.

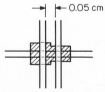

**Fig. 24-20**   Parallel conductor crossovers.

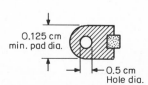

**Fig. 24-21**   Hole and pad sizes.

**Resistor Orientation**    Always orient resistors vertically or horizontally, never at an angle. Try to locate resistors so that all trimming may be done from one or two directions. This will minimize trimming tooling and setup requirements.

**Resistor Dimensioning**    The minimum length or width dimension of any resistor is 0.05 cm; the best results are achieved with resistors which are at least 0.1 cm square (Fig. 24-22). Small resistors tend to have a wider spread in value than larger resistors and are more difficult to trim.

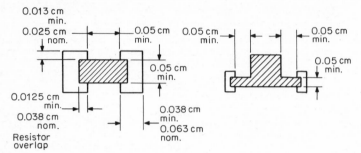

**Fig. 24-22**    Resistor size and overlap allowance.

**Conductor Interface**    Allow resistors to overlap conductor electrodes by at least 0.010 cm on each side; an overlap of 0.015 cm is better (Fig. 24-22). The larger overlap permits easier registration of the resistor/conductor screen patterns and gives a better ohmic contact.

**Trimming Clearance**    When using the air abrasive method of resistor trimming, allow at least 0.076-cm clearance along the trim side of the resistor for the abrasive tool; a spacing of 0.125 cm is preferred. If components are too closely spaced, the overspray from the abrasive tool may cut into adjacent resistors or conductors.

**Closed Loops**    Avoid closed loops. A circuit design may call for a resistor-conductor loop, or for resistors connected in parallel. To make measuring and trimming of the resistors possible, the loop must be broken until resistor tests and adjustments are complete. It may then be closed by means of a flywire jumper (Fig. 24-23).

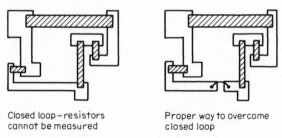

Closed loop – resistors
cannot be measured

Proper way to overcome
closed loop

**Fig. 24-23**    Overcoming closed resistor loops.

**Common Conductors**    When two resistors of the same sheet resistivity have a common conductor, they may be printed as a single pattern completely overlapping the common conductor (Fig. 24-24). If the resistivities are different, leave a space of at least 0.025 cm between the two resistors.

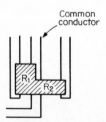

Common
conductor

**Fig. 24-24**    Resistors with common conductor and same sheet resistivity.

**Chip Component Pads**   The mounting surface of many chip components also serves as an electrical connection. On transistor chips, it is usually the collector; on diodes, it may be the anode or the cathode. These chips are eutectic bonded (Fig. 24-25) or glued to the pad with an electrically conductive epoxy.

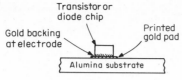

**Fig. 24-25**   Eutectic bonding.

The minimum mounting pad area must be at least 0.0125 cm larger on each side than the chip body. If printed resistors, or additional chips, are added to the same pad, a minimum separation of 0.076 cm must be maintained between components.

**Capacitor Mounting Pads**   Chip capacitors generally used in hybrid circuits have a rectangular ceramic body with metalized ends. The metalization is usually silver, palladium-silver, or palladium-gold. The electrical connection is made by soldering or by gluing with conductive epoxy. The mounting pads should be at least 0.025 cm larger on each side than the metalized area. The pads should be at least 0.05 cm apart to prevent shorting in case the epoxy flows under the chip body. It is permissible to run conductors under the body of large capacitor chips, but the conductor must be covered with a protective overglaze to prevent shorting.

**Pads for Lead-Attached Components**   Most components used in hybrid circuits are of chip form. Occasionally, however, it is necessary to include a lead-attached component in the hybrid package. The best way to attach the leads to the substrate pads is by soldering, although sometimes conductive epoxy may be used.

The substrate pads should be at least three times the lead diameter or width in order to get a good solder fillet. If the substrate pad extends to other circuit components, a glass solder barrier should be placed across the pad to prevent solder wicking (Fig. 24-26).

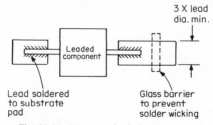

**Fig. 24-26**   Mounting leaded components.

**Flywire Interconnections**   Although flip-chip and beam-lead semiconductor devices have been around for a number of years, chip and wire construction still remains by far the most popular method of hybrid fabrication. It is easy to design, inspect, and repair, and it has a proven track record of reliability.

**Wire Lengths**   Wire interconnections must be kept as short as possible. Maximum lengths should not exceed 0.125 cm for chip-to-substrate connections and 0.25 cm for substrate-to-substrate connections.

**Crossovers**   Wires should not cross over exposed conductors or circuit components. If a wire crosses over a conductor, the conductor should be overglazed to prevent shorting.

**Chip-to-Chip Connections**   Chip-to-chip flywire connections should always be made via an intermediate bonding pad. Direct connections are subject to breakage and should be avoided.

**Substrate Pads**   Allow an area approximately 0.025 by 0.025 cm for each wire termination point. In cases where eutectic die bonding and flywire bonding are to be done on the same pad, provision must be made for preventing the eutectic material from wicking onto the wire bonding surface. This may be accomplished with a glass solder barrier or by designing a pad with an L-shaped addition for the wire bond (Fig. 24-27).

**solution**   The first step in the design process is to make some basic decisions regarding construction of the hybrid circuit. All semiconductor components are readily available as gold-backed silicon chips with aluminum metalization on the top surface. This type of chip lends itself nicely to standard chip-and-wire thermocompression bonding to gold thick-film conductors.

The resistors are all of normal value and can be screen printed. We shall need three resistive inks: a 100-$\Omega$/square ink for $R_1$, $R_4$, and $R_5$; a 10 k$\Omega$/square ink for $R_2$; and a 100-k$\Omega$/square ink for $R_3$. We assume that this circuit is being manufactured in large quantities. If we were making only a few test circuits, it would be easier and cheaper to substitute thick- or thin-film resistor chips for $R_2$ and $R_3$ than it would be to screen-print them.

The only resistor that dissipates any substantial power is $R_2$. The minimum area requirement for $R_2$, based on 3.87 W/cm², is:

$$\text{Minimum area} = \frac{0.030 \text{ W}}{3.87 \text{ W/cm}^2} = 0.0078 \text{ cm}^2$$

The smallest resistor size we can tolerate for $R_2$ is 30 × 40 mils, which should not present a problem.

We assume that the resistor pastes available have the geometry characteristics shown in Figs. 24-28 to 24-30. Using these design curves, and the method described in Sec. 24.4, we obtain the following results:

| Resistor | $N$ | $L$, mils | $W$, mils | $R$, $\Omega$ | Design limits Max., $\Omega$ | Min., $\Omega$ | Resistive paste, $\Omega$/square | Manufacturing limits, $\Omega$ Max. | Min. |
|---|---|---|---|---|---|---|---|---|---|
| $R_1$ | 1.50 | 60 | 40 | 220 | 242 | 110 | 100 | 192 | 128 |
| $R_2$ | 0.75 | 30 | 40 | 7500 | 825,000 | 3750 | 10,000 | 7300 | 5300 |
| $R_3$ | 1.50 | 60 | 40 | 330,000 | 363,000 | 165,000 | 100,000 | 200,000 | 132,000 |
| $R_4$ | 1.25 | 50 | 40 | 180 | 198 | 90 | 100 | 156 | 102 |
| $R_5$ | 2.75 | 110 | 40 | 550 | 605 | 275 | 100 | 413 | 275 |

Note that all these sizes are based on a standard 40-mil width according to our design curves. If we had a family of design curves available for each ink showing sheet resistivity at various widths, we could calculate any number of alternate solutions.

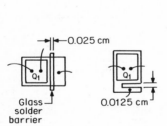

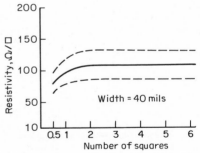

**Fig. 24-27**   Pad area designs for die and wire bonding.

**Fig. 24-28**   Resistor design curve, 100-$\Omega$/ square ink on gold termination.

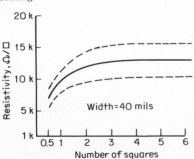

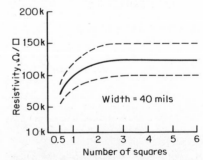

**Fig. 24-29**   Resistor design curve, 10-k$\Omega$/ square ink on gold termination.

**Fig. 24-30**   Resistor design curve, 100-k$\Omega$/ square ink on gold termination.

The next step is to determine the substrate size and package configuration. Since all components have been defined, we do not have to estimate sizes. We set up the calculation as follows:

| Component | Power, mW | Size, cm | Area, cm² |
|---|---|---|---|
| $Q_1$ | 30 | $0.063 \times 0.63$ | 0.004 |
| $Q_2$ | 25 | $0.063 \times 0.063$ | 0.004 |
| $Q_3$ | 25 | $0.063 \times 0.063$ | 0.004 |
| $D_1$ | | $0.05 \times 0.05$ | 0.0025 |
| $R_1$ | 15 | $0.15 \times 0.1$ | 0.015 |
| $R_2$ | 30 | $0.075 \times 0.1$ | 0.0075 |
| $R_3$ | 5 | $0.075 \times 0.1$ | 0.0075 |
| $R_4$ | 15 | $0.125 \times 0.1$ | 0.0125 |
| $R_5$ | | $0.275 \times 0.1$ | 0.0275 |
| Totals | 145 mW | | 0.0845 cm² |

Our total substrate area requirement is:

$$\text{Area} = 5 \times 0.0845 = 0.4225 \text{ cm}^2$$

The circuit will probably fit into a 0.9525 by 0.9525 cm 14-lead flatpack which has a substrate area of 0.5 cm² (0.707 by 0.707 cm) and a power-handling capability of 300 mW.

The final step is to draw the layout using the design rules set forth in the previous paragraphs.

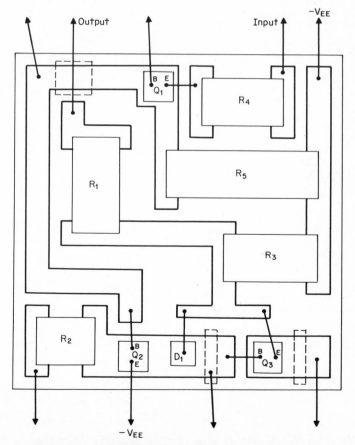

**Fig. 24-31**  Hybrid circuit layout of analog switch driver (Problem 24.8).

This is a trial-and-error process, and any number of solutions are possible. One feasible solution is shown in Fig. 24-31. Note the application of the following design rules and considerations:

1. All resistors overlap their electrodes by 0.025 cm. The area of overlap is not considered in calculations of resistor power and size.

2. All semiconductor dice are eutectically bonded to gold pads which extend at least 0.0125 cm beyond the die in each direction. The eutectic bond serves as a collector connection for the transistors and as a cathode connection for the diode.

3. All conductors are 0.05 cm wide and run horizontally and vertically on the substrate.

4. Flywire lengths do not exceed 0.125 cm.

5. Overglazing is provided as an insulator where a flywire crosses over a conductor, and as a solder barrier to prevent eutectic material from wicking onto flywire bonding pads.

## 24.10   BIBLIOGRAPHY

Designing with Beryllia, National Beryllia Corp., 1976.

Dryden, W. G.: "Achieving Consistent Results with Thick-Film," *Solid State Technology,* September 1971.

Harper, C. A., et al.: *Handbook of Thick Film Hybrid Microelectronics,* McGraw-Hill, New York, 1974.

Holmes, P. J., and Loasby, R. G.: *Handbook of Thick-Film Technology,* Electrochemical Publications, Ltd., 1976.

Isaaak, H.: "Voltage Coefficient of Thick-Film Resistors," International Hybrid Microelectronics Symposium, 1970.

Keister, F., and Auda, D.: "Thick Film Hybrid Design: The Right Way," *EDN,* June 10, 1968.

Lane, G. D.: "Designing with High Reliability Thick-Film Resistors," Second Symposium on Hybrid Microelectronics, October 1967.

Ost, R., Nash, D., and Tramantana, P.: "Automation Applied to Establishing Thick-Film Resistor Characteristics," International Hybrid Microelectronics Symposium, 1970.

Sargent, J.: "Thick Film Hybrid Microcircuits," Industrial and Scientific Conference Management, Inc., 1975.

"Some Design Layout Rules for Thick-Film Circuits," *Circuits Manufacturing,* September 1970.

Stern, L.: *Fundamentals of Integrated Circuits,* Hayden, New York, 1968.

# Appendix

**A. The Greek Alphabet**

| Name | Large | Small |
|---|---|---|
| Alpha | A | $\alpha$ |
| Beta | B | $\beta$ |
| Gamma | $\Gamma$ | $\gamma$ |
| Delta | $\Delta$ | $\delta$ |
| Epsilon | E | $\epsilon$ |
| Zeta | Z | $\zeta$ |
| Eta | H | $\eta$ |
| Theta | $\Theta$ | $\theta$ |
| Iota | I | $\iota$ |
| Kappa | K | $\kappa$ |
| Lambda | $\Lambda$ | $\lambda$ |
| Mu | M | $\mu$ |
| Nu | N | $\nu$ |
| Xi | $\Xi$ | $\xi$ |
| Omicron | O | $o$ |
| Pi | $\Pi$ | $\pi$ |
| Rho | P | $\rho$ |
| Sigma | $\Sigma$ | $\sigma$ |
| Tau | T | $\tau$ |
| Upsilon | Y | $\upsilon$ |
| Phi | $\Phi$ | $\phi$ |
| Chi | X | $\chi$ |
| Psi | $\Psi$ | $\psi$ |
| Omega | $\Omega$ | $\omega$ |

## B. Probable Values of General Physical Constants

| Constant | Symbol | Value |
|---|---|---|
| Electronic charge | $q$ | $1.602 \times 10^{-19}$ C |
| Electronic mass | $m$ | $9.109 \times 10^{-31}$ kg |
| Ratio of charge to mass of an electron | $q/m$ | $1.759 \times 10^{11}$ C/kg |
| Mass of atom of unit atomic weight (hypothetical) | ..... | $1.660 \times 10^{-27}$ kg |
| Mass of proton | $m_p$ | $1.673 \times 10^{-27}$ kg |
| Ratio of proton to electron mass | $m_p/m$ | $1.837 \times 10^3$ |
| Planck's constant | $h$ | $6.626 \times 10^{-34}$ J-s |
| Boltzmann constant | $\bar{k}$ | $1.381 \times 10^{-23}$ J/K |
|  | $k$ | $8.620 \times 10^{-5}$ eV/K |
| Stefan-Boltzmann constant | $\sigma$ | $5.670 \times 10^{-8}$ W/(m²)(K⁴) |
| Avogadro's number | $N_A$ | $6.023 \times 10^{23}$ molecules/mol |
| Gas constant | $R$ | $8.314$ J/(deg)(mol) |
| Velocity of light | $c$ | $2.998 \times 10^8$ m/s |
| Faraday's constant | $F$ | $9.649 \times 10^3$ C/mol |
| Volume per mole | $V_o$ | $2.241 \times 10^{-2}$ m³ |
| Acceleration of gravity | $g$ | $9.807$ m/s² |
| Permeability of free space | $\mu_o$ | $1.257 \times 10^{-6}$ H/m |
| Permittivity of free space | $\epsilon_o$ | $8.849 \times 10^{-12}$ F/m |

° E. A. Mechtly, "The International System of Units: Physical Constants and Conversion Factors," National Aeronautics and Space Administration, NASA SP-7012, Washington, D.C., 1964.

From J. Millman and C. Halkias, "Integrated Electronics: Analog Digital Circuits and Systems," McGraw-Hill Book Company, 1972.

## C. Conversion Factors and Prefixes

| | | | | |
|---|---|---|---|---|
| 1 ampere (A) | = 1 C/s | | 1 lumen per square foot | = 1 ft-candle (fc) |
| 1 angstrom unit (Å) | = $10^{-10}$ m | | mega (M) | = $\times 10^6$ |
| | = $10^{-4}$ $\mu$m | | 1 meter (m) | = 39.37 in |
| 1 atmosphere pressure | = 760 mm Hg | | micro ($\mu$) | = $\times 10^{-6}$ |
| 1 coulomb (C) | = 1 A-s | | 1 micron | = $10^{-6}$ m |
| 1 electronvolt (eV) | = $1.60 \times 10^{-19}$ J | | | = 1 $\mu$m |
| 1 farad (F) | = 1 C/V | | 1 mil | = $10^{-3}$ in |
| 1 foot (ft) | = 0.305 m | | | = 25 $\mu$m |
| 1 gram-calorie | = 4.185 J | | 1 mile | = 5,280 ft |
| giga (G) | = $\times 10^9$ | | | = 1.609 km |
| 1 henry (H) | = 1 V-s/A | | milli (m) | = $\times 10^{-3}$ |
| 1 hertz (Hz) | = 1 cycle/s | | nano (n) | = $\times 10^{-9}$ |
| 1 inch (in) | = 2.54 cm | | 1 newton (N) | = 1 kg-m/s² |
| 1 joule (J) | = $10^7$ ergs | | pico (p) | = $\times 10^{-12}$ |
| | = 1 W-s | | 1 pound (lb) | = 453.6 g |
| | = $6.25 \times 10^{18}$ eV | | 1 tesla (T) | = 1 Wb/m² |
| | = 1 N-m | | 1 ton | = 2,000 lb |
| | = 1 C-V | | 1 volt (V) | = 1 W/A |
| kilo (k) | = $\times 10^3$ | | 1 watt (W) | = 1 J/s |
| 1 kilogram (kg) | = 2.205 lb | | 1 weber (Wb) | = 1 V-s |
| 1 kilometer (km) | = 0.622 mi | | 1 weber per square | |
| 1 lumen | = 0.0016 W | | meter (Wb/m²) | = $10^4$ gauss |
| | (at 0.55 $\mu$m) | | | |

From J. Millman and C. Halkias, "Integrated Electronics: Analog Digital Circuits and Systems," McGraw-Hill Book Company, 1972.

## D. Decibel Table

| Db | Current and voltage ratio Gain | Loss | Power ratio Gain | Loss | Db | Current and voltage ratio Gain | Loss | Power ratio Gain | Loss |
|---|---|---|---|---|---|---|---|---|---|
| 0.1 | 1.01 | 0.989 | 1.02 | 0.977 | 8.0 | 2.51 | 0.398 | 6.31 | 0.158 |
| 0.2 | 1.02 | 0.977 | 1.05 | 0.955 | 8.5 | 2.66 | 0.376 | 7.08 | 0.141 |
| 0.3 | 1.03 | 0.966 | 1.07 | 0.933 | 9.0 | 2.82 | 0.355 | 7.94 | 0.126 |
| 0.4 | 1.05 | 0.955 | 1.10 | 0.912 | 9.5 | 2.98 | 0.335 | 8.91 | 0.112 |
| 0.5 | 1.06 | 0.944 | 1.12 | 0.891 | 10.0 | 3.16 | 0.316 | 10.00 | 0.100 |
| 0.6 | 1.07 | 0.933 | 1.15 | 0.871 | 11.0 | 3.55 | 0.282 | 12.6 | 0.079 |
| 0.7 | 1.08 | 0.923 | 1.17 | 0.851 | 12.0 | 3.98 | 0.251 | 15.8 | 0.063 |
| 0.8 | 1.10 | 0.912 | 1.20 | 0.832 | 13.0 | 4.47 | 0.224 | 19.9 | 0.050 |
| 0.9 | 1.11 | 0.902 | 1.23 | 0.813 | 14.0 | 5.01 | 0.199 | 25.1 | 0.040 |
| 1.0 | 1.12 | 0.891 | 1.26 | 0.794 | 15.0 | 5.62 | 0.178 | 31.6 | 0.032 |
| 1.1 | 1.13 | 0.881 | 1.29 | 0.776 | 16.0 | 6.31 | 0.158 | 39.8 | 0.025 |
| 1.2 | 1.15 | 0.871 | 1.32 | 0.759 | 17.0 | 7.08 | 0.141 | 50.1 | 0.020 |
| 1.3 | 1.16 | 0.861 | 1.35 | 0.741 | 18.0 | 7.94 | 0.126 | 63.1 | 0.016 |
| 1.4 | 1.17 | 0.851 | 1.38 | 0.724 | 19.0 | 8.91 | 0.112 | 79.4 | 0.013 |
| 1.5 | 1.19 | 0.841 | 1.41 | 0.708 | 20.0 | 10.00 | 0.100 | 100.0 | 0.010 |
| 1.6 | 1.20 | 0.832 | 1.44 | 0.692 | 25.0 | 17.8 | 0.056 | $3.16 \times 10^2$ | $3.16 \times 10^{-3}$ |
| 1.7 | 1.22 | 0.822 | 1.48 | 0.676 | 30.0 | 31.6 | 0.032 | $10^3$ | $10^{-3}$ |
| 1.8 | 1.23 | 0.813 | 1.51 | 0.661 | 35.0 | 56.2 | 0.018 | $3.16 \times 10^3$ | $3.16 \times 10^{-4}$ |
| 1.9 | 1.24 | 0.803 | 1.55 | 0.646 | 40.0 | 100.0 | 0.010 | $10^4$ | $10^{-4}$ |
| 2.0 | 1.26 | 0.794 | 1.58 | 0.631 | 45.0 | 177.8 | 0.006 | $3.16 \times 10^4$ | $3.16 \times 10^{-5}$ |
| 2.2 | 1.29 | 0.776 | 1.66 | 0.603 | 50.0 | 316 | 0.003 | $10^5$ | $10^{-5}$ |
| 2.4 | 1.32 | 0.759 | 1.74 | 0.575 | 55.0 | 562 | 0.002 | $3.16 \times 10^5$ | $3.16 \times 10^{-6}$ |
| 2.6 | 1.35 | 0.741 | 1.82 | 0.550 | 60.0 | 1,000 | 0.001 | $10^6$ | $10^{-6}$ |
| 2.8 | 1.38 | 0.724 | 1.90 | 0.525 | 65.0 | 1,770 | 0.0006 | $3.16 \times 10^6$ | $3.16 \times 10^{-7}$ |
| 3.0 | 1.41 | 0.708 | 1.99 | 0.501 | 70.0 | 3,160 | 0.0003 | $10^7$ | $10^{-7}$ |
| 3.2 | 1.44 | 0.692 | 2.09 | 0.479 | 75.0 | 5,620 | 0.0002 | $3.16 \times 10^7$ | $3.16 \times 10^{-8}$ |
| 3.4 | 1.48 | 0.676 | 2.19 | 0.457 | 80.0 | 10,000 | 0.0001 | $10^8$ | $10^{-8}$ |
| 3.6 | 1.51 | 0.661 | 2.29 | 0.436 | 85.0 | 17,800 | 0.00006 | $3.16 \times 10^8$ | $3.16 \times 10^{-9}$ |
| 3.8 | 1.55 | 0.646 | 2.40 | 0.417 | 90.0 | 31,600 | 0.00003 | $10^9$ | $10^{-9}$ |
| 4.0 | 1.58 | 0.631 | 2.51 | 0.398 | 95.0 | 56,200 | 0.00002 | $3.16 \times 10^9$ | $3.16 \times 10^{-10}$ |
| 4.2 | 1.62 | 0.617 | 2.63 | 0.380 | 100.0 | 100,000 | 0.00001 | $10^{10}$ | $10^{-10}$ |
| 4.4 | 1.66 | 0.603 | 2.75 | 0.363 | 105.0 | 178,000 | 0.000006 | $3.16 \times 10^{10}$ | $3.16 \times 10^{-11}$ |
| 4.6 | 1.70 | 0.589 | 2.88 | 0.347 | 110.0 | 316,000 | 0.000003 | $10^{11}$ | $10^{-11}$ |
| 4.8 | 1.74 | 0.575 | 3.02 | 0.331 | 115.0 | 562,000 | 0.000002 | $3.16 \times 10^{11}$ | $3.16 \times 10^{-12}$ |
| 5.0 | 1.78 | 0.562 | 3.16 | 0.316 | 120.0 | 1,000,000 | 0.000001 | $10^{12}$ | $10^{-12}$ |
| 5.5 | 1.88 | 0.531 | 3.55 | 0.282 | 130.0 | $3.16 \times 10^6$ | $3.16 \times 10^{-7}$ | $10^{13}$ | $10^{-13}$ |
| 6.0 | 1.99 | 0.501 | 3.98 | 0.251 | 140.0 | $10^7$ | $10^{-7}$ | $10^{14}$ | $10^{-14}$ |
| 6.5 | 2.11 | 0.473 | 4.47 | 0.224 | 150.0 | $3.16 \times 10^7$ | $3.16 \times 10^{-8}$ | $10^{15}$ | $10^{-15}$ |
| 7.0 | 2.24 | 0.447 | 5.01 | 0.199 | 160.0 | $10^8$ | $10^{-8}$ | $10^{16}$ | $10^{-16}$ |
| 7.5 | 2.37 | 0.422 | 5.62 | 0.178 | 170.0 | $3.16 \times 10^8$ | $3.16 \times 10^{-9}$ | $10^{17}$ | $10^{-17}$ |

From F. E. Terman, "Radio Engineers' Handbook," McGraw-Hill Book Company, 1943.

## E. Natural Sines and Cosines

NOTE.—For cosines use right-hand column of degrees and lower line of tenths.

| Deg | °0.0 | °0.1 | °0.2 | °0.3 | °0.4 | °0.5 | °0.6 | °0.7 | °0.8 | °0.9 | |
|---|---|---|---|---|---|---|---|---|---|---|---|
| 0° | 0.0000 | 0.0017 | 0.0035 | 0.0052 | 0.0070 | 0.0087 | 0.0105 | 0.0122 | 0.0140 | 0.0157 | 89 |
| 1 | 0.0175 | 0.0192 | 0.0209 | 0.0227 | 0.0244 | 0.0262 | 0.0279 | 0.0297 | 0.0314 | 0.0332 | 88 |
| 2 | 0.0349 | 0.0366 | 0.0384 | 0.0401 | 0.0419 | 0.0436 | 0.0454 | 0.0471 | 0.0488 | 0.0506 | 87 |
| 3 | 0.0523 | 0.0541 | 0.0558 | 0.0576 | 0.0593 | 0.0610 | 0.0628 | 0.0645 | 0.0663 | 0.0680 | 86 |
| 4 | 0.0698 | 0.0715 | 0.0732 | 0.0750 | 0.0767 | 0.0785 | 0.0802 | 0.0819 | 0.0837 | 0.0854 | 85 |
| 5 | 0.0872 | 0.0889 | 0.0906 | 0.0924 | 0.0941 | 0.0958 | 0.0976 | 0.0993 | 0.1011 | 0.1028 | 84 |
| 6 | 0.1045 | 0.1063 | 0.1080 | 0.1097 | 0.1115 | 0.1132 | 0.1149 | 0.1167 | 0.1184 | 0.1201 | 83 |
| 7 | 0.1219 | 0.1236 | 0.1253 | 0.1271 | 0.1288 | 0.1305 | 0.1323 | 0.1340 | 0.1357 | 0.1374 | 82 |
| 8 | 0.1392 | 0.1409 | 0.1426 | 0.1444 | 0.1461 | 0.1478 | 0.1495 | 0.1513 | 0.1530 | 0.1547 | 81 |
| 9 | 0.1564 | 0.1582 | 0.1599 | 0.1616 | 0.1633 | 0.1650 | 0.1668 | 0.1685 | 0.1702 | 0.1719 | 80° |
| 10° | 0.1736 | 0.1754 | 0.1771 | 0.1788 | 0.1805 | 0.1822 | 0.1840 | 0.1857 | 0.1874 | 0.1891 | 79 |
| 11 | 0.1908 | 0.1925 | 0.1942 | 0.1959 | 0.1977 | 0.1994 | 0.2011 | 0.2028 | 0.2045 | 0.2062 | 78 |
| 12 | 0.2079 | 0.2096 | 0.2113 | 0.2130 | 0.2147 | 0.2164 | 0.2181 | 0.2198 | 0.2215 | 0.2232 | 77 |
| 13 | 0.2250 | 0.2267 | 0.2284 | 0.2300 | 0.2317 | 0.2334 | 0.2351 | 0.2368 | 0.2385 | 0.2402 | 76 |
| 14 | 0.2419 | 0.2436 | 0.2453 | 0.2470 | 0.2487 | 0.2504 | 0.2521 | 0.2538 | 0.2554 | 0.2571 | 75 |
| 15 | 0.2588 | 0.2605 | 0.2622 | 0.2639 | 0.2656 | 0.2672 | 0.2689 | 0.2706 | 0.2723 | 0.2740 | 74 |
| 16 | 0.2756 | 0.2773 | 0.2790 | 0.2807 | 0.2823 | 0.2840 | 0.2857 | 0.2874 | 0.2890 | 0.2907 | 73 |
| 17 | 0.2924 | 0.2940 | 0.2957 | 0.2974 | 0.2990 | 0.3007 | 0.3024 | 0.3040 | 0.3057 | 0.3074 | 72 |
| 18 | 0.3090 | 0.3107 | 0.3123 | 0.3140 | 0.3156 | 0.3173 | 0.3190 | 0.3206 | 0.3223 | 0.3239 | 71 |
| 19 | 0.3256 | 0.3272 | 0.3289 | 0.3305 | 0.3322 | 0.3338 | 0.3355 | 0.3371 | 0.3387 | 0.3404 | 70° |
| 20° | 0.3420 | 0.3437 | 0.3453 | 0.3469 | 0.3486 | 0.3502 | 0.3518 | 0.3535 | 0.3551 | 0.3567 | 69 |
| 21 | 0.3584 | 0.3600 | 0.3616 | 0.3633 | 0.3649 | 0.3665 | 0.3681 | 0.3697 | 0.3714 | 0.3730 | 68 |
| 22 | 0.3746 | 0.3762 | 0.3778 | 0.3795 | 0.3811 | 0.3827 | 0.3843 | 0.3859 | 0.3875 | 0.3891 | 67 |
| 23 | 0.3907 | 0.3923 | 0.3939 | 0.3955 | 0.3971 | 0.3987 | 0.4003 | 0.4019 | 0.4035 | 0.4051 | 66 |
| 24 | 0.4067 | 0.4083 | 0.4099 | 0.4115 | 0.4131 | 0.4147 | 0.4163 | 0.4179 | 0.4195 | 0.4210 | 65 |
| 25 | 0.4226 | 0.4242 | 0.4258 | 0.4274 | 0.4289 | 0.4305 | 0.4321 | 0.4337 | 0.4352 | 0.4368 | 64 |
| 26 | 0.4384 | 0.4399 | 0.4415 | 0.4431 | 0.4446 | 0.4462 | 0.4478 | 0.4493 | 0.4509 | 0.4524 | 63 |
| 27 | 0.4540 | 0.4555 | 0.4571 | 0.4586 | 0.4602 | 0.4617 | 0.4633 | 0.4648 | 0.4664 | 0.4679 | 62 |
| 28 | 0.4695 | 0.4710 | 0.4726 | 0.4741 | 0.4756 | 0.4772 | 0.4787 | 0.4802 | 0.4818 | 0.4833 | 61 |
| 29 | 0.4848 | 0.4863 | 0.4879 | 0.4894 | 0.4909 | 0.4924 | 0.4939 | 0.4955 | 0.4970 | 0.4985 | 60° |
| 30° | 0.5000 | 0.5015 | 0.5030 | 0.5045 | 0.5060 | 0.5075 | 0.5090 | 0.5105 | 0.5120 | 0.5135 | 59 |
| 31 | 0.5150 | 0.5165 | 0.5180 | 0.5195 | 0.5210 | 0.5225 | 0.5240 | 0.5255 | 0.5270 | 0.5284 | 58 |
| 32 | 0.5299 | 0.5314 | 0.5329 | 0.5344 | 0.5358 | 0.5373 | 0.5388 | 0.5402 | 0.5417 | 0.5432 | 57 |
| 33 | 0.5446 | 0.5461 | 0.5476 | 0.5490 | 0.5505 | 0.5519 | 0.5534 | 0.5548 | 0.5563 | 0.5577 | 56 |
| 34 | 0.5592 | 0.5606 | 0.5621 | 0.5635 | 0.5650 | 0.5664 | 0.5678 | 0.5693 | 0.5707 | 0.5721 | 55 |
| 35 | 0.5736 | 0.5750 | 0.5764 | 0.5779 | 0.5793 | 0.5807 | 0.5821 | 0.5835 | 0.5850 | 0.5864 | 54 |
| 36 | 0.5878 | 0.5892 | 0.5906 | 0.5920 | 0.5934 | 0.5948 | 0.5962 | 0.5976 | 0.5990 | 0.6004 | 53 |
| 37 | 0.6018 | 0.6032 | 0.6046 | 0.6060 | 0.6074 | 0.6088 | 0.6101 | 0.6115 | 0.6129 | 0.6143 | 52 |
| 38 | 0.6157 | 0.6170 | 0.6184 | 0.6198 | 0.6211 | 0.6225 | 0.6239 | 0.6252 | 0.6266 | 0.6280 | 51 |
| 39 | 0.6293 | 0.6307 | 0.6320 | 0.6334 | 0.6347 | 0.6361 | 0.6374 | 0.6388 | 0.6401 | 0.6414 | 50° |
| 40° | 0.6428 | 0.6441 | 0.6455 | 0.6468 | 0.6481 | 0.6494 | 0.6508 | 0.6521 | 0.6534 | 0.6547 | 49 |
| 41 | 0.6561 | 0.6574 | 0.6587 | 0.6600 | 0.6613 | 0.6626 | 0.6639 | 0.6652 | 0.6665 | 0.6678 | 48 |
| 42 | 0.6691 | 0.6704 | 0.6717 | 0.6730 | 0.6743 | 0.6756 | 0.6769 | 0.6782 | 0.6794 | 0.6807 | 47 |
| 43 | 0.6820 | 0.6833 | 0.6845 | 0.6858 | 0.6871 | 0.6884 | 0.6896 | 0.6909 | 0.6921 | 0.6934 | 46 |
| 44 | 0.6947 | 0.6959 | 0.6972 | 0.6984 | 0.6997 | 0.7009 | 0.7022 | 0.7034 | 0.7046 | 0.7059 | 45 |
| | °1.0 | °0.9 | °0.8 | °0.7 | °0.6 | °0.5 | °0.4 | °0.3 | °0.2 | °0.1 | Deg |

From "Standard Handbook for Electrical Engineers," 7th ed.

## E. Natural Sines and Cosines (*Concluded*)

| Deg | °0.0 | °0.1 | °0.2 | °0.3 | °0.4 | °0.5 | °0.6 | °0.7 | °0.8 | °0.9 | |
|---|---|---|---|---|---|---|---|---|---|---|---|
| 45 | 0.7071 | 0.7083 | 0.7096 | 0.7108 | 0.7120 | 0.7133 | 0.7145 | 0.7157 | 0.7169 | 0.7181 | 44 |
| 46 | 0.7193 | 0.7206 | 0.7218 | 0.7230 | 0.7242 | 0.7254 | 0.7266 | 0.7278 | 0.7290 | 0.7302 | 43 |
| 47 | 0.7314 | 0.7325 | 0.7337 | 0.7349 | 0.7361 | 0.7373 | 0.7385 | 0.7396 | 0.7408 | 0.7420 | 42 |
| 48 | 0.7431 | 0.7443 | 0.7455 | 0.7466 | 0.7478 | 0.7490 | 0.7501 | 0.7513 | 0.7524 | 0.7536 | 41 |
| 49 | 0.7547 | 0.7559 | 0.7570 | 0.7581 | 0.7593 | 0.7604 | 0.7615 | 0.7627 | 0.7638 | 0.7649 | 40° |
| 50° | 0.7660 | 0.7672 | 0.7683 | 0.7694 | 0.7705 | 0.7716 | 0.7727 | 0.7738 | 0.7749 | 0.7760 | 39 |
| 51 | 0.7771 | 0.7782 | 0.7793 | 0.7804 | 0.7815 | 0.7826 | 0.7837 | 0.7848 | 0.7859 | 0.7869 | 38 |
| 52 | 0.7880 | 0.7891 | 0.7902 | 0.7912 | 0.7923 | 0.7934 | 0.7944 | 0.7955 | 0.7965 | 0.7976 | 37 |
| 53 | 0.7986 | 0.7997 | 0.8007 | 0.8018 | 0.8028 | 0.8039 | 0.8049 | 0.8059 | 0.8070 | 0.8080 | 36 |
| 54 | 0.8090 | 0.8100 | 0.8111 | 0.8121 | 0.8131 | 0.8141 | 0.8151 | 0.8161 | 0.8171 | 0.8181 | 35 |
| 55 | 0.8192 | 0.8202 | 0.8211 | 0.8221 | 0.8231 | 0.8241 | 0.8251 | 0.8261 | 0.8271 | 0.8281 | 34 |
| 56 | 0.8290 | 0.8300 | 0.8310 | 0.8320 | 0.8329 | 0.8339 | 0.8348 | 0.8358 | 0.8368 | 0.8377 | 33 |
| 57 | 0.8387 | 0.8396 | 0.8406 | 0.8415 | 0.8425 | 0.8434 | 0.8443 | 0.8453 | 0.8462 | 0.8471 | 32 |
| 58 | 0.8480 | 0.8490 | 0.8499 | 0.8508 | 0.8517 | 0.8526 | 0.8536 | 0.8545 | 0.8554 | 0.8563 | 31 |
| 59 | 0.8572 | 0.8581 | 0.8590 | 0.8599 | 0.8607 | 0.8616 | 0.8625 | 0.8634 | 0.8643 | 0.8652 | 30° |
| 60° | 0.8660 | 0.8669 | 0.8678 | 0.8686 | 0.8695 | 0.8704 | 0.8712 | 0.8721 | 0.8729 | 0.8738 | 29 |
| 61 | 0.8746 | 0.8755 | 0.8763 | 0.8771 | 0.8780 | 0.8788 | 0.8796 | 0.8805 | 0.8813 | 0.8821 | 28 |
| 62 | 0.8829 | 0.8838 | 0.8846 | 0.8854 | 0.8862 | 0.8870 | 0.8878 | 0.8886 | 0.8894 | 0.8902 | 27 |
| 63 | 0.8910 | 0.8918 | 0.8926 | 0.8934 | 0.8942 | 0.8949 | 0.8957 | 0.8965 | 0.8973 | 0.8980 | 26 |
| 64 | 0.8988 | 0.8996 | 0.9003 | 0.9011 | 0.9018 | 0.9026 | 0.9033 | 0.9041 | 0.9048 | 0.9056 | 25 |
| 65 | 0.9063 | 0.9070 | 0.9078 | 0.9085 | 0.9092 | 0.9100 | 0.9107 | 0.9114 | 0.9121 | 0.9128 | 24 |
| 66 | 0.9135 | 0.9143 | 0.9150 | 0.9157 | 0.9164 | 0.9171 | 0.9178 | 0.9184 | 0.9191 | 0.9198 | 23 |
| 67 | 0.9205 | 0.9212 | 0.9219 | 0.9225 | 0.9232 | 0.9239 | 0.9245 | 0.9252 | 0.9259 | 0.9265 | 22 |
| 68 | 0.9272 | 0.9278 | 0.9285 | 0.9291 | 0.9298 | 0.9304 | 0.9311 | 0.9317 | 0.9323 | 0.9330 | 21 |
| 69 | 0.9336 | 0.9342 | 0.9348 | 0.9354 | 0.9361 | 0.9367 | 0.9373 | 0.9379 | 0.9385 | 0.9391 | 20° |
| 70° | 0.9397 | 0.9403 | 0.9409 | 0.9415 | 0.9421 | 0.9426 | 0.9432 | 0.9438 | 0.9444 | 0.9449 | 19 |
| 71 | 0.9455 | 0.9461 | 0.9466 | 0.9472 | 0.9478 | 0.9483 | 0.9489 | 0.9494 | 0.9500 | 0.9505 | 18 |
| 72 | 0.9511 | 0.9516 | 0.9521 | 0.9527 | 0.9532 | 0.9537 | 0.9542 | 0.9548 | 0.9553 | 0.9558 | 17 |
| 73 | 0.9563 | 0.9568 | 0.9573 | 0.9578 | 0.9583 | 0.9588 | 0.9593 | 0.9598 | 0.9603 | 0.9608 | 16 |
| 74 | 0.9613 | 0.9617 | 0.9622 | 0.9627 | 0.9632 | 0.9636 | 0.9641 | 0.9646 | 0.9650 | 0.9655 | 15 |
| 75 | 0.9659 | 0.9664 | 0.9668 | 0.9673 | 0.9677 | 0.9681 | 0.9686 | 0.9690 | 0.9694 | 0.9699 | 14 |
| 76 | 0.9703 | 0.9707 | 0.9711 | 0.9715 | 0.9720 | 0.9724 | 0.9728 | 0.9732 | 0.9736 | 0.9740 | 13 |
| 77 | 0.9744 | 0.9748 | 0.9751 | 0.9755 | 0.9759 | 0.9763 | 0.9767 | 0.9770 | 0.9774 | 0.9778 | 12 |
| 78 | 0.9781 | 0.9785 | 0.9789 | 0.9792 | 0.9796 | 0.9799 | 0.9803 | 0.9806 | 0.9810 | 0.9813 | 11 |
| 79 | 0.9816 | 0.9820 | 0.9823 | 0.9826 | 0.9829 | 0.9833 | 0.9836 | 0.9839 | 0.9842 | 0.9845 | 10° |
| 80° | 0.9848 | 0.9851 | 0.9854 | 0.9857 | 0.9860 | 0.9863 | 0.9866 | 0.9869 | 0.9871 | 0.9874 | 9 |
| 81 | 0.9877 | 0.9880 | 0.9882 | 0.9885 | 0.9888 | 0.9890 | 0.9893 | 0.9895 | 0.9898 | 0.9900 | 8 |
| 82 | 0.9903 | 0.9905 | 0.9907 | 0.9910 | 0.9912 | 0.9914 | 0.9917 | 0.9919 | 0.9921 | 0.9923 | 7 |
| 83 | 0.9925 | 0.9928 | 0.9930 | 0.9932 | 0.9934 | 0.9936 | 0.9938 | 0.9940 | 0.9942 | 0.9943 | 6 |
| 84 | 0.9945 | 0.9947 | 0.9949 | 0.9951 | 0.9952 | 0.9954 | 0.9956 | 0.9957 | 0.9959 | 0.9960 | 5 |
| 85 | 0.9962 | 0.9963 | 0.9965 | 0.9966 | 0.9968 | 0.9969 | 0.9971 | 0.9972 | 0.9973 | 0.9974 | 4 |
| 86 | 0.9976 | 0.9977 | 0.9978 | 0.9979 | 0.9980 | 0.9981 | 0.9982 | 0.9983 | 0.9984 | 0.9985 | 3 |
| 87 | 0.9986 | 0.9987 | 0.9988 | 0.9989 | 0.9990 | 0.9990 | 0.9991 | 0.9992 | 0.9993 | 0.9993 | 2 |
| 88 | 0.9994 | 0.9995 | 0.9995 | 0.9996 | 0.9996 | 0.9997 | 0.9997 | 0.9997 | 0.9998 | 0.9998 | 1 |
| 89 | 0.9998 | 0.9999 | 0.9999 | 0.9999 | 0.9999 | 1.000 | 1.000 | 1.000 | 1.000 | 1.000 | 0° |
| | °1.0 | °0.9 | °0.8 | °0.7 | °0.6 | °0.5 | °0.4 | °0.3 | °0.2 | °0.1 | Deg |

From "Standard Handbook for Electrical Engineers," 7th ed.

## F. Natural Tangents and Cotangents

NOTE.—For cotangents use right-hand column of degrees and lower line of tenths.

| Deg | °0.0 | °0.1 | °0.2 | °0.3 | °0.4 | °0.5 | °0.6 | °0.7 | °0.8 | °0.9 | |
|---|---|---|---|---|---|---|---|---|---|---|---|
| 0° | 0.0000 | 0.0017 | 0.0035 | 0.0052 | 0.0070 | 0.0087 | 0.0105 | 0.0122 | 0.0140 | 0.0157 | 89 |
| 1 | 0.0175 | 0.0192 | 0.0209 | 0.0227 | 0.0244 | 0.0262 | 0.0279 | 0.0297 | 0.0314 | 0.0332 | 88 |
| 2 | 0.0349 | 0.0367 | 0.0384 | 0.0402 | 0.0419 | 0.0437 | 0.0454 | 0.0472 | 0.0489 | 0.0537 | 87 |
| 3 | 0.0524 | 0.0542 | 0.0559 | 0.0577 | 0.0594 | 0.0612 | 0.0629 | 0.0647 | 0.0664 | 0.0682 | 86 |
| 4 | 0.0699 | 0.0717 | 0.0734 | 0.0752 | 0.0769 | 0.0787 | 0.0805 | 0.0822 | 0.0840 | 0.0857 | 85 |
| 5 | 0.0875 | 0.0892 | 0.0910 | 0.0928 | 0.0945 | 0.0963 | 0.0981 | 0.0998 | 0.1016 | 0.1033 | 84 |
| 6 | 0.1051 | 0.1069 | 0.1086 | 0.1104 | 0.1122 | 0.1139 | 0.1157 | 0.1175 | 0.1192 | 0.1210 | 83 |
| 7 | 0.1228 | 0.1246 | 0.1263 | 0.1281 | 0.1299 | 0.1317 | 0.1334 | 0.1352 | 0.1370 | 0.1388 | 82 |
| 8 | 0.1405 | 0.1423 | 0.1441 | 0.1459 | 0.1477 | 0.1495 | 0.1512 | 0.1530 | 0.1548 | 0.1566 | 81 |
| 9 | 0.1584 | 0.1602 | 0.1620 | 0.1638 | 0.1655 | 0.1673 | 0.1691 | 0.1709 | 0.1727 | 0.1745 | 80° |
| 10° | 0.1763 | 0.1781 | 0.1799 | 0.1817 | 0.1835 | 0.1853 | 0.1871 | 0.1890 | 0.1908 | 0.1926 | 79 |
| 11 | 0.1944 | 0.1962 | 0.1980 | 0.1998 | 0.2016 | 0.2035 | 0.2053 | 0.2071 | 0.2089 | 0.2107 | 78 |
| 12 | 0.2126 | 0.2144 | 0.2162 | 0.2180 | 0.2199 | 0.2217 | 0.2235 | 0.2254 | 0.2272 | 0.2290 | 77 |
| 13 | 0.2309 | 0.2327 | 0.2345 | 0.2364 | 0.2382 | 0.2401 | 0.2419 | 0.2438 | 0.2456 | 0.2475 | 76 |
| 14 | 0.2493 | 0.2512 | 0.2530 | 0.2549 | 0.2568 | 0.2586 | 0.2605 | 0.2623 | 0.2643 | 0.2661 | 75 |
| 15 | 0.2679 | 0.2698 | 0.2717 | 0.2736 | 0.2754 | 0.2773 | 0.2792 | 0.2811 | 0.2830 | 0.2849 | 74 |
| 16 | 0.2867 | 0.2886 | 0.2905 | 0.2924 | 0.2943 | 0.2962 | 0.2981 | 0.3000 | 0.3019 | 0.3038 | 73 |
| 17 | 0.3057 | 0.3076 | 0.3096 | 0.3115 | 0.3134 | 0.3153 | 0.3172 | 0.3191 | 0.3211 | 0.3230 | 72 |
| 18 | 0.3249 | 0.3269 | 0.3288 | 0.3307 | 0.3327 | 0.3346 | 0.3365 | 0.3385 | 0.3404 | 0.3424 | 71 |
| 19 | 0.3443 | 0.3463 | 0.3482 | 0.3502 | 0.3522 | 0.3541 | 0.3561 | 0.3581 | 0.3600 | 0.3620 | 70° |
| 20° | 0.3640 | 0.3659 | 0.3679 | 0.3699 | 0.3719 | 0.3739 | 0.3759 | 0.3779 | 0.3799 | 0.3819 | 69 |
| 21 | 0.3339 | 0.3859 | 0.3879 | 0.3899 | 0.3919 | 0.3939 | 0.3959 | 0.3979 | 0.4000 | 0.4020 | 68 |
| 22 | 0.4040 | 0.4061 | 0.4081 | 0.4101 | 0.4122 | 0.4142 | 0.4163 | 0.4183 | 0.4204 | 0.4224 | 67 |
| 23 | 0.4245 | 0.4265 | 0.4286 | 0.4307 | 0.4327 | 0.4348 | 0.4369 | 0.4390 | 0.4411 | 0.4431 | 66 |
| 24 | 0.4452 | 0.4473 | 0.4494 | 0.4515 | 0.4536 | 0.4557 | 0.4578 | 0.4599 | 0.4621 | 0.4642 | 65 |
| 25 | 0.4663 | 0.4684 | 0.4706 | 0.4727 | 0.4748 | 0.4770 | 0.4791 | 0.4813 | 0.4834 | 0.4856 | 64 |
| 26 | 0.4877 | 0.4899 | 0.4921 | 0.4942 | 0.4964 | 0.4986 | 0.5008 | 0.5029 | 0.5051 | 0.5073 | 63 |
| 27 | 0.5095 | 0.5117 | 0.5139 | 0.5161 | 0.5184 | 0.5206 | 0.5228 | 0.5250 | 0.5272 | 0.5295 | 62 |
| 28 | 0.5317 | 0.5340 | 0.5362 | 0.5384 | 0.5407 | 0.5430 | 0.5452 | 0.5475 | 0.5498 | 0.5520 | 61 |
| 29 | 0.5543 | 0.5566 | 0.5589 | 0.5612 | 0.5635 | 0.5658 | 0.5681 | 0.5704 | 0.5727 | 0.5750 | 60° |
| 30° | 0.5774 | 0.5797 | 0.5820 | 0.5844 | 0.5867 | 0.5890 | 0.5914 | 0.5938 | 0.5961 | 0.5985 | 59 |
| 31 | 0.6009 | 0.6032 | 0.6056 | 0.6080 | 0.6104 | 0.6128 | 0.6152 | 0.6176 | 0.6200 | 0.6224 | 58 |
| 32 | 0.6249 | 0.6273 | 0.6297 | 0.6322 | 0.6346 | 0.6371 | 0.6395 | 0.6420 | 0.6445 | 0.6469 | 57 |
| 33 | 0.6494 | 0.6519 | 0.6544 | 0.6569 | 0.6594 | 0.6619 | 0.6644 | 0.6669 | 0.6694 | 0.6720 | 56 |
| 34 | 0.6745 | 0.6771 | 0.6796 | 0.6822 | 0.6847 | 0.6873 | 0.6899 | 0.6924 | 0.6950 | 0.6976 | 55 |
| 35 | 0.7002 | 0.7028 | 0.7054 | 0.7080 | 0.7107 | 0.7133 | 0.7159 | 0.7186 | 0.7212 | 0.7239 | 54 |
| 36 | 0.7265 | 0.7292 | 0.7319 | 0.7346 | 0.7373 | 0.7400 | 0.7427 | 0.7454 | 0.7481 | 0.7508 | 53 |
| 37 | 0.7536 | 0.7563 | 0.7590 | 0.7618 | 0.7646 | 0.7673 | 0.7701 | 0.7729 | 0.7757 | 0.7785 | 52 |
| 38 | 0.7813 | 0.7841 | 0.7869 | 0.7898 | 0.7926 | 0.7954 | 0.7983 | 0.8012 | 0.8040 | 0.8069 | 51 |
| 39 | 0.8098 | 0.8127 | 0.8156 | 0.8185 | 0.8214 | 0.8243 | 0.8273 | 0.8302 | 0.8332 | 0.8361 | 50° |
| 40° | 0.8391 | 0.8421 | 0.8451 | 0.8481 | 0.8511 | 0.8541 | 0.8571 | 0.8601 | 0.8632 | 0.8662 | 49 |
| 41 | 0.8693 | 0.8724 | 0.8754 | 0.8785 | 0.8816 | 0.8847 | 0.8878 | 0.8910 | 0.8941 | 0.8972 | 48 |
| 42 | 0.9004 | 0.9036 | 0.9067 | 0.9099 | 0.9131 | 0.9163 | 0.9195 | 0.9228 | 0.9260 | 0.9293 | 47 |
| 43 | 0.9325 | 0.9358 | 0.9391 | 0.9424 | 0.9457 | 0.9490 | 0.9523 | 0.9556 | 0.9590 | 0.9623 | 46 |
| 44 | 0.9657 | 0.9691 | 0.9725 | 0.9759 | 0.9793 | 0.9827 | 0.9861 | 0.9896 | 0.9930 | 0.9965 | 45 |
| | °1.0 | °0.9 | °0.8 | °0.7 | °0.6 | °0.5 | °0.4 | °0.3 | °0.2 | °0.1 | Deg |

From "Standard Handbook for Electrical Engineers," 7th ed.

## F. Natural Tangents and Cotangents (*Concluded*)

| Deg | °0.0 | °0.1 | °0.2 | °0.3 | °0.4 | °0.5 | °0.6 | °0.7 | °0.8 | °0.9 | |
|---|---|---|---|---|---|---|---|---|---|---|---|
| 45 | 1.0000 | 1.0035 | 1.0070 | 1.0105 | 1.0141 | 1.0176 | 1.0212 | 1.0247 | 1.0283 | 1.0319 | 44 |
| 46 | 1.0355 | 1.0392 | 1.0428 | 1.0464 | 1.0501 | 1.0538 | 1.0575 | 1.0612 | 1.0649 | 1.0686 | 43 |
| 47 | 1.0724 | 1.0761 | 1.0799 | 1.0837 | 1.0875 | 1.0913 | 1.0951 | 1.0990 | 1.1028 | 1.1067 | 42 |
| 48 | 1.1106 | 1.1145 | 1.1184 | 1.1224 | 1.1263 | 1.1303 | 1.1343 | 1.1383 | 1.1423 | 1.1463 | 41 |
| 49 | 1.1504 | 1.1544 | 1.1585 | 1.1626 | 1.1667 | 1.1708 | 1.1750 | 1.1792 | 1.1833 | 1.1875 | 40° |
| 50° | 1.1918 | 1.1960 | 1.2002 | 1.2045 | 1.2088 | 1.2131 | 1.2174 | 1.2218 | 1.2261 | 1.2305 | 39 |
| 51 | 1.2349 | 1.2393 | 1.2437 | 1.2482 | 1.2527 | 1.2572 | 1.2617 | 1.2662 | 1.2708 | 1.2753 | 38 |
| 52 | 1.2799 | 1.2846 | 1.2892 | 1.2938 | 1.2985 | 1.3032 | 1.3079 | 1.3127 | 1.3175 | 1.3222 | 37 |
| 53 | 1.3270 | 1.3319 | 1.3367 | 1.3416 | 1.3465 | 1.3514 | 1.3564 | 1.3613 | 1.3663 | 1.3713 | 36 |
| 54 | 1.3764 | 1.3814 | 1.3865 | 1.3916 | 1.3968 | 1.4019 | 1.4071 | 1.4124 | 1.4176 | 1.4229 | 35 |
| 55 | 1.4281 | 1.4335 | 1.4388 | 1.4442 | 1.4496 | 1.4550 | 1.4605 | 1.4659 | 1.4715 | 1.4770 | 34 |
| 56 | 1.4826 | 1.4882 | 1.4938 | 1.4994 | 1.5051 | 1.5108 | 1.5166 | 1.5224 | 1.5282 | 1.5340 | 33 |
| 57 | 1.5399 | 1.5458 | 1.5517 | 1.5577 | 1.5637 | 1.5697 | 1.5757 | 1.5818 | 1.5880 | 1.5941 | 32 |
| 58 | 1.6003 | 1.6066 | 1.6128 | 1.6191 | 1.6255 | 1.6319 | 1.6383 | 1.6447 | 1.6512 | 1.6577 | 31 |
| 59 | 1.6643 | 1.6709 | 1.6775 | 1.6842 | 1.6909 | 1.6977 | 1.7045 | 1.7113 | 1.7182 | 1.7251 | 30° |
| 60° | 1.7321 | 1.7391 | 1.7461 | 1.7532 | 1.7603 | 1.7675 | 1.7747 | 1.7820 | 1.7893 | 1.7966 | 29 |
| 61 | 1.8040 | 1.8115 | 1.8190 | 1.8265 | 1.8341 | 1.8418 | 1.8495 | 1.8572 | 1.8650 | 1.8728 | 28 |
| 62 | 1.8807 | 1.8887 | 1.8967 | 1.9047 | 1.9128 | 1.9210 | 1.9292 | 1.9375 | 1.9458 | 1.9542 | 27 |
| 63 | 1.9626 | 1.9711 | 1.9797 | 1.9883 | 1.9970 | 2.0057 | 2.0145 | 2.0233 | 2.0323 | 2.0413 | 26 |
| 64 | 2.0503 | 2.0594 | 2.0686 | 2.0778 | 2.0872 | 2.0965 | 2.1060 | 2.1155 | 2.1251 | 2.1348 | 25 |
| 65 | 2.1445 | 2.1543 | 2.1642 | 2.1742 | 2.1842 | 2.1943 | 2.2045 | 2.2148 | 2.2251 | 2.2355 | 24 |
| 66 | 2.2460 | 2.2566 | 2.2673 | 2.2781 | 2.2889 | 2.2998 | 2.3109 | 2.3220 | 2.3332 | 2.3445 | 23 |
| 67 | 2.3559 | 2.3673 | 2.3789 | 2.3906 | 2.4023 | 2.4142 | 2.4262 | 2.4383 | 2.4504 | 2.4627 | 22 |
| 68 | 2.4751 | 2.4876 | 2.5002 | 2.5129 | 2.5257 | 2.5386 | 2.5517 | 2.5649 | 2.5782 | 2.5916 | 21 |
| 69 | 2.6051 | 2.6187 | 2.6325 | 2.6464 | 2.6605 | 2.6746 | 2.6889 | 2.7034 | 2.7179 | 2.7326 | 20° |
| 70° | 2.7475 | 2.7625 | 2.7776 | 2.7929 | 2.8083 | 2.8239 | 2.8397 | 2.8556 | 2.8716 | 2.8878 | 19 |
| 71 | 2.9042 | 2.9208 | 2.9375 | 2.9544 | 2.9714 | 2.9887 | 3.0061 | 3.0237 | 3.0415 | 3.0595 | 18 |
| 72 | 3.0777 | 3.0961 | 3.1146 | 3.1334 | 3.1524 | 3.1716 | 3.1910 | 3.2106 | 3.2305 | 3.2506 | 17 |
| 73 | 3.2709 | 3.2914 | 3.3122 | 3.3332 | 3.3544 | 3.3759 | 3.3977 | 3.4197 | 3.4420 | 3.4646 | 16 |
| 74 | 3.4874 | 3.5105 | 3.5339 | 3.5576 | 3.5816 | 3.6059 | 3.6305 | 3.6554 | 3.6806 | 3.7062 | 15 |
| 75 | 3.7321 | 3.7583 | 3.7848 | 3.8118 | 3.8391 | 3.8667 | 3.8947 | 3.9232 | 3.9520 | 3.9812 | 14 |
| 76 | 4.0108 | 4.0408 | 4.0713 | 4.1022 | 4.1335 | 4.1653 | 4.1976 | 4.2303 | 4.2635 | 4.2972 | 13 |
| 77 | 4.3315 | 4.3662 | 4.4015 | 4.4374 | 4.4737 | 4.5107 | 4.5483 | 4.5864 | 4.6252 | 4.6646 | 12 |
| 78 | 4.7046 | 4.7453 | 4.7867 | 4.8288 | 4.8716 | 4.9152 | 4.9594 | 5.0045 | 5.0504 | 5.0970 | 11 |
| 79 | 5.1446 | 5.1929 | 5.2422 | 5.2924 | 5.3435 | 5.3955 | 5.4486 | 5.5026 | 5.5578 | 5.6140 | 10° |
| 80° | 5.6713 | 5.7297 | 5.7894 | 5.8502 | 5.9124 | 5.9758 | 6.0405 | 6.1066 | 6.1742 | 6.2433 | 9 |
| 81 | 6.3138 | 6.3859 | 6.4596 | 6.5350 | 6.6122 | 6.6912 | 6.7720 | 6.8548 | 6.9395 | 7.0264 | 8 |
| 82 | 7.1154 | 7.2066 | 7.3002 | 7.3962 | 7.4947 | 7.5958 | 7.6996 | 7.8062 | 7.9158 | 8.0285 | 7 |
| 83 | 8.1443 | 8.2636 | 8.3863 | 8.5126 | 8.6427 | 8.7769 | 8.9152 | 9.0579 | 9.2052 | 9.3572 | 6 |
| 84 | 9.5144 | 9.677 | 9.845 | 10.02 | 10.20 | 10.39 | 10.58 | 10.78 | 10.99 | 11.20 | 5 |
| 85 | 11.43 | 11.66 | 11.91 | 12.16 | 12.43 | 12.71 | 13.00 | 13.30 | 13.62 | 13.95 | 4 |
| 86 | 14.30 | 14.67 | 15.06 | 15.46 | 15.89 | 16.35 | 16.83 | 17.34 | 17.89 | 18.46 | 3 |
| 87 | 19.08 | 19.74 | 20.45 | 21.20 | 22.02 | 22.90 | 23.86 | 24.90 | 26.03 | 27.27 | 2 |
| 88 | 28.64 | 30.14 | 31.82 | 33.69 | 35.80 | 38.19 | 40.92 | 44.07 | 47.74 | 52.08 | 1 |
| 89 | 57.29 | 63.66 | 71.62 | 81.85 | 94.49 | 114.6 | 143.2 | 191.0 | 286.5 | 573.0 | 0° |
| | °1.0 | °0.9 | °0.8 | °0.7 | °0.6 | °0.5 | °0.4 | °0.3 | °0.2 | °0.1 | Deg |

From "Standard Handbook for Electrical Engineers," 7th ed.

## G. Common Logarithms of Numbers

| N | 0 | 1 | 2 | 3 | 4 | 5 | 6 | 7 | 8 | 9 |
|---|---|---|---|---|---|---|---|---|---|---|
| 10 | 0000 | 0043 | 0086 | 0128 | 0170 | 0212 | 0253 | 0294 | 0334 | 0374 |
| 11 | 0414 | 0453 | 0492 | 0531 | 0569 | 0607 | 0645 | 0682 | 0719 | 0755 |
| 12 | 0792 | 0828 | 0864 | 0899 | 0934 | 0969 | 1004 | 1038 | 1072 | 1106 |
| 13 | 1139 | 1173 | 1206 | 1239 | 1271 | 1303 | 1335 | 1367 | 1399 | 1430 |
| 14 | 1461 | 1492 | 1523 | 1553 | 1584 | 1614 | 1644 | 1673 | 1703 | 1732 |
| 15 | 1761 | 1790 | 1818 | 1847 | 1875 | 1903 | 1931 | 1959 | 1987 | 2014 |
| 16 | 2041 | 2068 | 2095 | 2122 | 2148 | 2175 | 2201 | 2227 | 2253 | 2279 |
| 17 | 2304 | 2330 | 2355 | 2380 | 2405 | 2430 | 2455 | 2480 | 2504 | 2529 |
| 18 | 2553 | 2577 | 2601 | 2625 | 2648 | 2672 | 2695 | 2718 | 2742 | 2765 |
| 19 | 2788 | 2810 | 2833 | 2856 | 2878 | 2900 | 2923 | 2945 | 2967 | 2989 |
| 20 | 3010 | 3032 | 3054 | 3075 | 3096 | 3118 | 3139 | 3160 | 3181 | 3201 |
| 21 | 3222 | 3243 | 3263 | 3284 | 3304 | 3324 | 3345 | 3365 | 3385 | 3404 |
| 22 | 3424 | 3444 | 3464 | 3483 | 3502 | 3522 | 3541 | 3560 | 3579 | 3598 |
| 23 | 3617 | 3636 | 3655 | 3674 | 3692 | 3711 | 3729 | 3747 | 3766 | 3784 |
| 24 | 3802 | 3820 | 3838 | 3856 | 3874 | 3892 | 3909 | 3927 | 3945 | 3962 |
| 25 | 3979 | 3997 | 4014 | 4031 | 4048 | 4065 | 4082 | 4099 | 4116 | 4133 |
| 26 | 4150 | 4166 | 4183 | 4200 | 4216 | 4232 | 4249 | 4265 | 4281 | 4298 |
| 27 | 4314 | 4330 | 4346 | 4362 | 4378 | 4393 | 4409 | 4425 | 4440 | 4456 |
| 28 | 4472 | 4487 | 4502 | 4518 | 4533 | 4548 | 4564 | 4579 | 4594 | 4609 |
| 29 | 4624 | 4639 | 4654 | 4669 | 4683 | 4698 | 4713 | 4728 | 4742 | 4757 |
| 30 | 4771 | 4786 | 4800 | 4814 | 4829 | 4843 | 4857 | 4871 | 4886 | 4900 |
| 31 | 4914 | 4928 | 4942 | 4955 | 4969 | 4983 | 4997 | 5011 | 5024 | 5038 |
| 32 | 5051 | 5065 | 5079 | 5092 | 5105 | 5119 | 5132 | 5145 | 5159 | 5172 |
| 33 | 5185 | 5198 | 5211 | 5224 | 5237 | 5250 | 5263 | 5276 | 5289 | 5302 |
| 34 | 5315 | 5328 | 5340 | 5353 | 5366 | 5378 | 5391 | 5403 | 5416 | 5428 |
| 35 | 5441 | 5453 | 5465 | 5478 | 5490 | 5502 | 5514 | 5527 | 5539 | 5551 |
| 36 | 5563 | 5575 | 5587 | 5599 | 5611 | 5623 | 5635 | 5647 | 5658 | 5670 |
| 37 | 5682 | 5694 | 5705 | 5717 | 5729 | 5740 | 5752 | 5763 | 5775 | 5786 |
| 38 | 5798 | 5809 | 5821 | 5832 | 5843 | 5855 | 5866 | 5877 | 5888 | 5899 |
| 39 | 5911 | 5922 | 5933 | 5944 | 5955 | 5966 | 5977 | 5988 | 5999 | 6010 |
| 40 | 6021 | 6031 | 6042 | 6053 | 6064 | 6075 | 6085 | 6096 | 6107 | 6117 |
| 41 | 6128 | 6138 | 6149 | 6160 | 6170 | 6180 | 6191 | 6201 | 6212 | 6222 |
| 42 | 6232 | 6243 | 6253 | 6263 | 6274 | 6284 | 6294 | 6304 | 6314 | 6325 |
| 43 | 6335 | 6345 | 6355 | 6365 | 6375 | 6385 | 6395 | 6405 | 6415 | 6425 |
| 44 | 6435 | 6444 | 6454 | 6464 | 6474 | 6484 | 6493 | 6503 | 6513 | 6522 |
| 45 | 6532 | 6542 | 6551 | 6561 | 6571 | 6580 | 6590 | 6599 | 6609 | 6618 |
| 46 | 6628 | 6637 | 6646 | 6656 | 6665 | 6675 | 6684 | 6693 | 6702 | 6712 |
| 47 | 6721 | 6730 | 6739 | 6749 | 6758 | 6767 | 6776 | 6785 | 6794 | 6803 |
| 48 | 6812 | 6821 | 6830 | 6839 | 6848 | 6857 | 6866 | 6875 | 6884 | 6893 |
| 49 | 6902 | 6911 | 6920 | 6928 | 6937 | 6946 | 6955 | 6964 | 6972 | 6981 |
| 50 | 6990 | 6998 | 7007 | 7016 | 7024 | 7033 | 7042 | 7050 | 7059 | 7067 |
| 51 | 7076 | 7084 | 7093 | 7101 | 7110 | 7118 | 7126 | 7135 | 7143 | 7152 |
| 52 | 7160 | 7168 | 7177 | 7185 | 7193 | 7202 | 7210 | 7218 | 7226 | 7235 |
| 53 | 7243 | 7251 | 7259 | 7267 | 7275 | 7284 | 7292 | 7300 | 7308 | 7316 |
| 54 | 7324 | 7332 | 7340 | 7348 | 7356 | 7364 | 7372 | 7380 | 7388 | 7396 |

From "Standard Handbook for Electrical Engineers," 7th ed.

## G. Common Logarithms of Numbers (*Concluded*)

| N | 0 | 1 | 2 | 3 | 4 | 5 | 6 | 7 | 8 | 9 |
|---|---|---|---|---|---|---|---|---|---|---|
| 55 | 7404 | 7412 | 7419 | 7427 | 7435 | 7443 | 7451 | 7459 | 7466 | 7474 |
| 56 | 7482 | 7490 | 7497 | 7505 | 7513 | 7520 | 7528 | 7536 | 7543 | 7551 |
| 57 | 7559 | 7566 | 7574 | 7582 | 7589 | 7597 | 7604 | 7612 | 7619 | 7627 |
| 58 | 7634 | 7642 | 7649 | 7657 | 7664 | 7672 | 7679 | 7686 | 7694 | 7701 |
| 59 | 7709 | 7716 | 7723 | 7731 | 7738 | 7745 | 7752 | 7760 | 7767 | 7774 |
| 60 | 7782 | 7789 | 7796 | 7803 | 7810 | 7818 | 7825 | 7832 | 7839 | 7846 |
| 61 | 7853 | 7860 | 7868 | 7875 | 7882 | 7889 | 7896 | 7903 | 7910 | 7917 |
| 62 | 7924 | 7931 | 7938 | 7945 | 7952 | 7959 | 7966 | 7973 | 7980 | 7987 |
| 63 | 7993 | 8000 | 8007 | 8014 | 8021 | 8028 | 8035 | 8041 | 8048 | 8055 |
| 64 | 8062 | 8069 | 8075 | 8082 | 8089 | 8096 | 8102 | 8109 | 8116 | 8122 |
| 65 | 8129 | 8136 | 8142 | 8149 | 8156 | 8162 | 8169 | 8176 | 8182 | 8189 |
| 66 | 8195 | 8202 | 8209 | 8215 | 8222 | 8228 | 8235 | 8241 | 8248 | 8254 |
| 67 | 8261 | 8267 | 8274 | 8280 | 8287 | 8293 | 8299 | 8306 | 8312 | 8319 |
| 68 | 8325 | 8331 | 8338 | 8344 | 8351 | 8357 | 8363 | 8370 | 8376 | 8382 |
| 69 | 8388 | 8395 | 8401 | 8407 | 8414 | 8420 | 8426 | 8432 | 8439 | 8445 |
| 70 | 8451 | 8457 | 8463 | 8470 | 8476 | 8482 | 8488 | 8494 | 8500 | 8506 |
| 71 | 8513 | 8519 | 8525 | 8531 | 8537 | 8543 | 8549 | 8555 | 8561 | 8567 |
| 72 | 8573 | 8579 | 8585 | 8591 | 8597 | 8603 | 8609 | 8615 | 8621 | 8627 |
| 73 | 8633 | 8639 | 8645 | 8651 | 8657 | 8663 | 8669 | 8675 | 8681 | 8686 |
| 74 | 8692 | 8698 | 8704 | 8710 | 8716 | 8722 | 8727 | 8733 | 8739 | 8745 |
| 75 | 8751 | 8756 | 8762 | 8768 | 8774 | 8779 | 8785 | 8791 | 8797 | 8802 |
| 76 | 8808 | 8814 | 8820 | 8825 | 8831 | 8837 | 8842 | 8848 | 8854 | 8859 |
| 77 | 8865 | 8871 | 8876 | 8882 | 8887 | 8893 | 8899 | 8904 | 8910 | 8915 |
| 78 | 8921 | 8927 | 8932 | 8938 | 8943 | 8949 | 8954 | 8960 | 8965 | 8971 |
| 79 | 8976 | 8982 | 8987 | 8993 | 8998 | 9004 | 9009 | 9015 | 9020 | 9025 |
| 80 | 9031 | 9036 | 9042 | 9047 | 9053 | 9058 | 9063 | 9069 | 9074 | 9079 |
| 81 | 9085 | 9090 | 9096 | 9101 | 9106 | 9112 | 9117 | 9122 | 9128 | 9133 |
| 82 | 9138 | 9143 | 9149 | 9154 | 9159 | 9165 | 9170 | 9175 | 9180 | 9186 |
| 83 | 9191 | 9196 | 9201 | 9206 | 9212 | 9217 | 9222 | 9227 | 9232 | 9238 |
| 84 | 9243 | 9248 | 9253 | 9258 | 9263 | 9269 | 9274 | 9279 | 9284 | 9289 |
| 85 | 9294 | 9299 | 9304 | 9309 | 9315 | 9320 | 9325 | 9330 | 9335 | 9340 |
| 86 | 9345 | 9350 | 9355 | 9360 | 9365 | 9370 | 9375 | 9380 | 9385 | 9390 |
| 87 | 9395 | 9400 | 9405 | 9410 | 9415 | 9420 | 9425 | 9430 | 9435 | 9440 |
| 88 | 9445 | 9450 | 9455 | 9460 | 9465 | 9469 | 9474 | 9479 | 9484 | 9489 |
| 89 | 9494 | 9499 | 9504 | 9509 | 9513 | 9518 | 9523 | 9528 | 9533 | 9538 |
| 90 | 9542 | 9547 | 9552 | 9557 | 9562 | 9566 | 9571 | 9576 | 9581 | 9586 |
| 91 | 9590 | 9595 | 9600 | 9605 | 9609 | 9614 | 9619 | 9624 | 9628 | 9633 |
| 92 | 9638 | 9643 | 9647 | 9652 | 9657 | 9661 | 9666 | 9671 | 9675 | 9680 |
| 93 | 9685 | 9689 | 9694 | 9699 | 9703 | 9708 | 9713 | 9717 | 9722 | 9727 |
| 94 | 9731 | 9736 | 9741 | 9745 | 9750 | 9754 | 9759 | 9763 | 9768 | 9773 |
| 95 | 9777 | 9782 | 9786 | 9791 | 9795 | 9800 | 9805 | 9809 | 9814 | 9818 |
| 96 | 9823 | 9827 | 9832 | 9836 | 9841 | 9845 | 9850 | 9854 | 9859 | 9863 |
| 97 | 9868 | 9872 | 9877 | 9881 | 9886 | 9890 | 9894 | 9899 | 9903 | 9908 |
| 98 | 9912 | 9917 | 9921 | 9926 | 9930 | 9934 | 9939 | 9943 | 9948 | 9952 |
| 99 | 9956 | 9961 | 9965 | 9969 | 9974 | 9978 | 9983 | 9987 | 9991 | 9996 |

From "Standard Handbook for Electrical Engineers," 7th ed.

## H. Natural, Napierian, or Hyperbolic Logarithms

| N | 0 | 1 | 2 | 3 | 4 | 5 | 6 | 7 | 8 | 9 |
|---|---|---|---|---|---|---|---|---|---|---|
| 0 | — ∞ | 0.0000 | 0.6931 | 1.0986 | 1.3863 | 1.6094 | 1.7918 | 1.9459 | 2.0794 | 2.1972 |
| 10 | 2.3026 | 2.3979 | 2.4849 | 2.5649 | 2.6391 | 2.7081 | 2.7726 | 2.8332 | 2.8904 | 2.9444 |
| 20 | 2.9957 | 3.0445 | 3.0910 | 3.1355 | 3.1781 | 3.2189 | 3.2581 | 3.2958 | 3.3322 | 3.3673 |
| 30 | 3.4012 | 3.4340 | 3.4657 | 3.4965 | 3.5264 | 3.5553 | 3.5835 | 3.6109 | 3.6376 | 3.6636 |
| 40 | 3.6889 | 3.7136 | 3.7377 | 3.7612 | 3.7842 | 3.8067 | 3.8286 | 3.8501 | 3.8712 | 3.8918 |
| 50 | 3.9120 | 3.9318 | 3.9512 | 3.9703 | 3.9890 | 4.0073 | 4.0254 | 4.0431 | 4.0604 | 4.0775 |
| 60 | 4.0943 | 4.1109 | 4.1271 | 4.1431 | 4.1589 | 4.1744 | 4.1897 | 4.2047 | 4.2195 | 4.2341 |
| 70 | 4.2485 | 4.2627 | 4.2767 | 4.2905 | 4.3041 | 4.3175 | 4.3307 | 4.3438 | 4.3567 | 4.3694 |
| 80 | 4.3820 | 4.3944 | 4.4067 | 4.4188 | 4.4308 | 4.4427 | 4.4543 | 4.4659 | 4.4773 | 4.4886 |
| 90 | 4.4998 | 4.5109 | 4.5218 | 4.5326 | 4.5433 | 4.5539 | 4.5643 | 4.5747 | 4.5850 | 4.5951 |
| 100 | 4.6052 | 4.6151 | 4.6250 | 4.6347 | 4.6444 | 4.6540 | 4.6634 | 4.6728 | 4.6821 | 4.6913 |
| 110 | 4.7005 | 4.7095 | 4.7185 | 4.7274 | 4.7362 | 4.7449 | 4.7536 | 4.7622 | 4.7707 | 4.7791 |
| 120 | 4.7875 | 4.7958 | 4.8040 | 4.8122 | 4.8203 | 4.8283 | 4.8363 | 4.8442 | 4.8520 | 4.8598 |
| 130 | 4.8675 | 4.8752 | 4.8828 | 4.8903 | 4.8978 | 4.9053 | 4.9127 | 4.9200 | 4.9273 | 4.9345 |
| 140 | 4.9416 | 4.9488 | 4.9558 | 4.9628 | 4.9698 | 4.9767 | 4.9836 | 4.9904 | 4.9972 | 5.0039 |
| 150 | 5.0106 | 5.0173 | 5.0239 | 5.0304 | 5.0370 | 5.0434 | 5.0499 | 5.0562 | 5.0626 | 5.0689 |
| 160 | 5.0752 | 5.0814 | 5.0876 | 5.0938 | 5.0999 | 5.1059 | 5.1120 | 5.1180 | 5.1240 | 5.1299 |
| 170 | 5.1358 | 5.1417 | 5.1475 | 5.1533 | 5.1591 | 5.1648 | 5.1705 | 5.1761 | 5.1818 | 5.1874 |
| 180 | 5.1930 | 5.1985 | 5.2040 | 5.2095 | 5.2149 | 5.2204 | 5.2257 | 5.2311 | 5.2364 | 5.2417 |
| 190 | 5.2470 | 5.2523 | 5.2575 | 5.2627 | 5.2679 | 5.2730 | 5.2781 | 5.2832 | 5.2883 | 5.2933 |
| 200 | 5.2983 | 5.3033 | 5.3083 | 5.3132 | 5.3181 | 5.3230 | 5.3279 | 5.3327 | 5.3375 | 5.3423 |
| 210 | 5.3471 | 5.3519 | 5.3566 | 5.3613 | 5.3660 | 5.3706 | 5.3753 | 5.3799 | 5.3845 | 5.3891 |
| 220 | 5.3936 | 5.3982 | 5.4027 | 5.4072 | 5.4116 | 5.4161 | 5.4205 | 5.4250 | 5.4293 | 5.4337 |
| 230 | 5.4381 | 5.4424 | 5.4467 | 5.4510 | 5.4553 | 5.4596 | 5.4638 | 5.4681 | 5.4723 | 5.4765 |
| 240 | 5.4806 | 5.4848 | 5.4889 | 5.4931 | 5.4972 | 5.5013 | 5.5053 | 5.5094 | 5.5134 | 5.5175 |
| 250 | 5.5215 | 5.5255 | 5.5294 | 5.5334 | 5.5373 | 5.5413 | 5.5452 | 5.5491 | 5.5530 | 5.5568 |
| 260 | 5.5607 | 5.5645 | 5.5683 | 5.5722 | 5.5759 | 5.5797 | 5.5835 | 5.5872 | 5.5910 | 5.5947 |
| 270 | 5.5984 | 5.6021 | 5.6058 | 5.6095 | 5.6131 | 5.6168 | 5.6204 | 5.6240 | 5.6276 | 5.6312 |
| 280 | 5.6348 | 5.6384 | 5.6419 | 5.6454 | 5.6490 | 5.6525 | 5.6560 | 5.6595 | 5.6630 | 5.6664 |
| 290 | 5.6699 | 5.6733 | 5.6768 | 5.6802 | 5.6836 | 5.6870 | 5.6904 | 5.6937 | 5.6971 | 5.7004 |
| 300 | 5.7038 | 5.7071 | 5.7104 | 5.7137 | 5.7170 | 5.7203 | 5.7236 | 5.7268 | 5.7301 | 5.7333 |
| 310 | 5.7366 | 5.7398 | 5.7430 | 5.7462 | 5.7494 | 5.7526 | 5.7557 | 5.7589 | 5.7621 | 5.7652 |
| 320 | 5.7683 | 5.7714 | 5.7746 | 5.7777 | 5.7807 | 5.7838 | 5.7869 | 5.7900 | 5.7930 | 5.7961 |
| 330 | 5.7991 | 5.8021 | 5.8051 | 5.8081 | 5.8111 | 5.8141 | 5.8171 | 5.8201 | 5.8230 | 5.8260 |
| 340 | 5.8289 | 5.8319 | 5.8348 | 5.8377 | 5.8406 | 5.8435 | 5.8464 | 5.8493 | 5.8522 | 5.8551 |
| 350 | 5.8579 | 5.8608 | 5.8636 | 5.8665 | 5.8693 | 5.8721 | 5.8749 | 5.8777 | 5.8805 | 5.8833 |
| 360 | 5.8861 | 5.8889 | 5.8916 | 5.8944 | 5.8972 | 5.8999 | 5.9026 | 5.9054 | 5.9081 | 5.9108 |
| 370 | 5.9135 | 5.9162 | 5.9189 | 5.9216 | 5.9243 | 5.9269 | 5.9296 | 5.9322 | 5.9349 | 5.9375 |
| 380 | 5.9402 | 5.9428 | 5.9454 | 5.9480 | 5.9506 | 5.9532 | 5.9558 | 5.9584 | 5.9610 | 5.9636 |
| 390 | 5.9661 | 5.9687 | 5.9713 | 5.9738 | 5.9764 | 5.9789 | 5.9814 | 5.9839 | 5.9865 | 5.9890 |
| 400 | 5.9915 | 5.9940 | 5.9965 | 5.9989 | 6.0014 | 6.0039 | 6.0064 | 6.0088 | 6.0113 | 6.0137 |
| 410 | 6.0162 | 6.0186 | 6.0210 | 6.0234 | 6.0259 | 6.0283 | 6.0307 | 6.0331 | 6.0355 | 6.0379 |
| 420 | 6.0403 | 6.0426 | 6.0450 | 6.0474 | 6.0497 | 6.0521 | 6.0544 | 6.0568 | 6.0591 | 6.0615 |
| 430 | 6.0638 | 6.0661 | 6.0684 | 6.0707 | 6.0730 | 6.0753 | 6.0776 | 6.0799 | 6.0822 | 6.0845 |
| 440 | 6.0868 | 6.0890 | 6.0913 | 6.0936 | 6.0958 | 0.0981 | 6.1003 | 6.1026 | 6.1048 | 6.1070 |
| 450 | 6.1092 | 6.1115 | 6.1137 | 6.1159 | 6.1181 | 6.1203 | 6.1225 | 6.1247 | 6.1269 | 6.1291 |
| 460 | 6.1312 | 6.1334 | 6.1356 | 6.1377 | 6.1399 | 6.1420 | 6.1442 | 6.1463 | 6.1485 | 6.1506 |
| 470 | 6.1527 | 6.1549 | 6.1570 | 6.1591 | 6.1612 | 6.1633 | 6.1654 | 6.1675 | 6.1696 | 6.1717 |
| 480 | 6.1738 | 6.1759 | 6.1779 | 6.1800 | 6.1821 | 6.1841 | 6.1862 | 6.1883 | 6.1903 | 6.1924 |
| 490 | 6.1944 | 6.1964 | 6.1985 | 6.2005 | 6.2025 | 6.2046 | 6.2066 | 6.2086 | 6.2106 | 6.2126 |

From "Standard Handbook for Electrical Engineers," 7th ed.

| | n | n × 2.3026 |
|---|---|---|
| NOTE 1: Moving the decimal point n places to the right (or left) in the number is equivalent to adding (or subtracting) n times 2.3026. | 1 | 2.3026 = 0.6974−3 |
| | 2 | 4.6052 = 0.3948−5 |
| | 3 | 6.9078 = 0.0922−7 |
| NOTE 2: | 4 | 9.2103 = 0.7897−10 |
| $\log_e x = 2.3026 \log_{10} x$ | 5 | 11.5129 = 0.4871−12 |
| $\log_{10} x = 0.4343 \log_e x$ | 6 | 13.8155 = 0.1845−14 |
| $\log_e 10 = 2.3026$ | 7 | 16.1181 = 0.8819−17 |
| $\log_{10} e = 0.4343$ | 8 | 18.4207 = 0.5793−19 |
| | 9 | 20.7233 = 0.2767−21 |

## H. Natural, Napierian, or Hyperbolic Logarithms (*Concluded*)

| N | 0 | 1 | 2 | 3 | 4 | 5 | 6 | 7 | 8 | 9 |
|---|---|---|---|---|---|---|---|---|---|---|
| 500 | 6.2146 | 6.2166 | 6.2186 | 6.2206 | 6.2226 | 6.2246 | 6.2265 | 6.2285 | 6.2305 | 6.2324 |
| 510 | 6.2344 | 6.2364 | 6.2383 | 6.2403 | 6.2422 | 6.2442 | 6.2461 | 6.2480 | 6.2500 | 6.2519 |
| 520 | 6.2538 | 6.2558 | 6.2577 | 6.2596 | 6.2615 | 6.2634 | 6.2653 | 6.2672 | 6.2691 | 6.2710 |
| 530 | 6.2729 | 6.2748 | 6.2766 | 6.2785 | 6.2804 | 6.2823 | 6.2841 | 6.2860 | 6.2879 | 6.2897 |
| 540 | 6.2916 | 6.2934 | 6.2953 | 6.2971 | 6.2989 | 6.3008 | 6.3026 | 6.3044 | 6.3063 | 6.3081 |
| 550 | 6.3099 | 6.3117 | 6.3135 | 6.3154 | 6.3172 | 6.3190 | 6.3208 | 6.3226 | 6.3244 | 6.3261 |
| 560 | 6.3279 | 6.3297 | 6.3315 | 6.3333 | 6.3351 | 6.3368 | 6.3386 | 6.3404 | 6.3421 | 6.3439 |
| 570 | 6.3456 | 6.3474 | 6.3491 | 6.3509 | 6.3256 | 6.3544 | 6.3561 | 6.3578 | 6.3596 | 6.3613 |
| 580 | 6.3630 | 6.3648 | 6.3665 | 6.3682 | 6.3699 | 6.3716 | 6.3733 | 6.3750 | 6.3767 | 6.3784 |
| 590 | 6.3801 | 6.3818 | 6.3835 | 6.3852 | 6.3869 | 6.3886 | 6.3902 | 6.3919 | 6.3936 | 6.3953 |
| 600 | 6.3969 | 6.3986 | 6.4003 | 6.4019 | 6.4036 | 6.4052 | 6.4069 | 6.4085 | 6.4102 | 6.4118 |
| 610 | 6.4135 | 6.4151 | 6.4167 | 6.4184 | 6.4200 | 6.4216 | 6.4232 | 6.4249 | 6.4265 | 6.4281 |
| 620 | 6.4297 | 6.4313 | 6.4329 | 6.4345 | 6.4362 | 6.4378 | 6.4394 | 6.4409 | 6.4425 | 6.4441 |
| 630 | 6.4457 | 6.4473 | 6.4489 | 6.4505 | 6.4520 | 6.4536 | 6.4552 | 6.4568 | 6.4583 | 6.4599 |
| 640 | 6.4615 | 6.4630 | 6.4646 | 6.4661 | 6.4677 | 6.4693 | 6.4708 | 6.4723 | 6.4739 | 6.4754 |
| 650 | 6.4770 | 6.4785 | 6.4800 | 6.4816 | 6.4831 | 6.4846 | 6.4862 | 6.4877 | 6.4892 | 6.4907 |
| 660 | 6.4922 | 6.4938 | 6.4953 | 6.4968 | 6.4983 | 6.4998 | 6.5013 | 6.5028 | 6.5043 | 6.5058 |
| 670 | 6.5073 | 6.5088 | 6.5103 | 6.5117 | 6.5132 | 6.5147 | 6.5162 | 6.5177 | 6.5191 | 6.5206 |
| 680 | 6.5221 | 6.5236 | 6.5250 | 6.5265 | 6.5280 | 6.5294 | 6.5309 | 6.5323 | 6.5338 | 6.5352 |
| 690 | 6.5367 | 6.5381 | 6.5396 | 6.5410 | 6.5425 | 6.5439 | 6.5453 | 6.5468 | 6.5482 | 6.5497 |
| 700 | 6.5511 | 6.5525 | 6.5539 | 6.5554 | 6.5568 | 6.5582 | 6.5596 | 6.5610 | 6.5624 | 6.5639 |
| 710 | 6.5653 | 6.5667 | 6.5681 | 6.5695 | 6.5709 | 6.5723 | 6.5737 | 6.5751 | 6.5765 | 6.5779 |
| 720 | 6.5793 | 6.5806 | 6.5820 | 6.5834 | 6.5848 | 6.5862 | 6.5876 | 6.5889 | 6.5903 | 6.5917 |
| 730 | 6.5930 | 6.5944 | 6.5958 | 6.5971 | 6.5985 | 6.5999 | 6.6012 | 6.6026 | 6.6039 | 6.6053 |
| 740 | 6.6067 | 6.6080 | 6.6093 | 6.6107 | 6.6120 | 6.6134 | 6.6147 | 6.6161 | 6.6174 | 6.6187 |
| 750 | 6.6201 | 6.6214 | 6.6227 | 6.6241 | 6.6254 | 6.6267 | 6.6280 | 6.6294 | 6.6307 | 6.6320 |
| 760 | 6.6333 | 6.6346 | 6.6359 | 6.6373 | 6.6386 | 6.6399 | 6.6412 | 6.6425 | 6.6438 | 6.6451 |
| 770 | 6.6464 | 6.6477 | 6.6490 | 6.6503 | 6.6516 | 6.6529 | 6.6542 | 6.6554 | 6.6567 | 6.6580 |
| 780 | 6.6593 | 6.6606 | 6.6619 | 6.6631 | 6.6644 | 6.6657 | 6.6670 | 6.6682 | 6.6695 | 6.6708 |
| 790 | 6.6720 | 6.6733 | 6.6746 | 6.6758 | 6.6771 | 6.6783 | 6.6796 | 6.6809 | 6.6821 | 6.6834 |
| 800 | 6.6846 | 6.6859 | 6.6871 | 6.6884 | 6.6896 | 6.6908 | 6.6921 | 6.6933 | 6.6946 | 6.6958 |
| 810 | 6.6970 | 6.6983 | 6.6995 | 6.7007 | 6.7020 | 6.7032 | 6.7044 | 6.7056 | 6.7069 | 6.7081 |
| 820 | 6.7093 | 6.7105 | 6.7117 | 6.7130 | 6.7142 | 6.7154 | 6.7166 | 6.7178 | 6.7190 | 6.7202 |
| 830 | 6.7214 | 6.7226 | 6.7238 | 6.7250 | 6.7262 | 6.7274 | 6.7286 | 6.7298 | 6.7310 | 6.7322 |
| 840 | 6.7334 | 6.7346 | 6.7358 | 6.7370 | 6.7382 | 6.7393 | 6.7405 | 6.7417 | 6.6429 | 6.7441 |
| 850 | 6.7452 | 6.7464 | 6.7476 | 6.7488 | 6.7499 | 6.7511 | 6.7523 | 6.7534 | 6.7546 | 6.7558 |
| 860 | 6.7569 | 6.7581 | 6.7593 | 6.7604 | 6.7616 | 6.7627 | 6.7639 | 6.7650 | 6.7662 | 6.7673 |
| 870 | 6.7685 | 6.7696 | 6.7708 | 6.7719 | 6.7731 | 6.7742 | 6.7754 | 6.7765 | 6.7776 | 6.7788 |
| 880 | 6.7799 | 6.7811 | 6.7822 | 6.7833 | 6.7845 | 6.7856 | 6.7867 | 5.7878 | 6.7890 | 6.7901 |
| 890 | 6.7912 | 6.7923 | 6.7935 | 6.7946 | 6.7957 | 6.7968 | 6.7979 | 6.7991 | 6.8002 | 6.8013 |
| 900 | 6.8024 | 6.8035 | 6.8046 | 6.8057 | 6.8068 | 6.8079 | 6.8090 | 6.8101 | 6.8112 | 6.8123 |
| 910 | 6.8134 | 6.8145 | 6.8156 | 6.8167 | 6.8178 | 6.8189 | 6.8200 | 6.8211 | 6.8222 | 6.8233 |
| 920 | 6.8244 | 6.8255 | 6.8265 | 6.8276 | 6.8287 | 6.8298 | 6.8309 | 6.8320 | 6.8330 | 6.8341 |
| 930 | 6.8352 | 6.8363 | 6.8373 | 6.8384 | 6.8395 | 6.8405 | 6.8416 | 6.8427 | 6.8437 | 6.8448 |
| 940 | 6.8459 | 6.8469 | 6.8480 | 6.8491 | 6.8501 | 6.8512 | 6.8522 | 6.8533 | 6.8544 | 8.8554 |
| 950 | 6.8565 | 6.8575 | 6.8586 | 6.8596 | 6.8607 | 6.8617 | 6.8628 | 6.8638 | 6.8648 | 6.8659 |
| 960 | 6.8669 | 6.8680 | 6.8690 | 6.8701 | 6.8711 | 6.8721 | 6.8732 | 6.8742 | 6.8752 | 6.8763 |
| 970 | 6.8773 | 6.8783 | 6.8794 | 6.8804 | 6.8814 | 6.8824 | 6.8835 | 6.8845 | 6.8855 | 6.8865 |
| 980 | 6.8876 | 6.8886 | 6.8896 | 6.8906 | 6.8916 | 6.8926 | 6.8937 | 6.8947 | 6.8957 | 6.8967 |
| 990 | 6.8977 | 6.8987 | 6.8997 | 6.9007 | 6.9017 | 6.9027 | 6.9037 | 6.9047 | 6.9057 | 6.9068 |

From "Standard Handbook for Electrical Engineers," 7th ed.

## I. Degrees and Minutes Expressed in Radians

| Degrees | | | | | | Hundredths | | | | Minutes | |
|---|---|---|---|---|---|---|---|---|---|---|---|
| 1° | 0.0175 | 61° | 1.0647 | 121° | 2.1118 | 0°.01 | 0.0002 | 0°.51 | 0.0089 | 1′ | 0.0003 |
| 2 | 0.0349 | 62 | 1.0821 | 122 | 2.1293 | 0 .02 | 0.0003 | 0 .52 | 0.0091 | 2′ | 0.0006 |
| 3 | 0.0524 | 63 | 1.0996 | 123 | 2.1468 | 0 .03 | 0.0005 | 0 .53 | 0.0093 | 3′ | 0.0009 |
| 4 | 0.0698 | 64 | 1.1170 | 124 | 2.1642 | 0 .04 | 0.0007 | 0 .54 | 0.0094 | 4′ | 0.0012 |
| 5° | 0.0873 | 65° | 1.1345 | 125° | 2.1817 | 0 .05 | 0.0009 | 0 .55 | 0.0096 | 5′ | 0.0015 |
| 6 | 0.1047 | 66 | 1.1519 | 126 | 2.1991 | 0 .06 | 0.0010 | 0 .56 | 0.0098 | 6′ | 0.0017 |
| 7 | 0.1222 | 67 | 1.1694 | 127 | 2.2166 | 0 .07 | 0.0012 | 0 .57 | 0.0099 | 7′ | 0.0020 |
| 8 | 0.1396 | 68 | 1.1868 | 128 | 2.2340 | 0 .08 | 0.0014 | 0 .58 | 0.0101 | 8′ | 0.0023 |
| 9 | 0.1571 | 69 | 1.2043 | 129 | 2.2515 | 0 .09 | 0.0016 | 0 .59 | 0.0103 | 9′ | 0.0026 |
| 10° | 0.1745 | 70° | 1.2217 | 130° | 2.2689 | 0°.10 | 0.0017 | 0°.60 | 0.0105 | 10′ | 0.0029 |
| 11 | 0.1920 | 71 | 1.2392 | 131 | 2.2864 | 0 .11 | 0.0019 | 0 .61 | 0.0106 | 11′ | 0.0032 |
| 12 | 0.2094 | 72 | 1.2566 | 132 | 2.3038 | 0 .12 | 0.0021 | 0 .62 | 0.0108 | 12′ | 0.0035 |
| 13 | 0.2269 | 73 | 1.2741 | 133 | 2.3213 | 0 .13 | 0.0023 | 0 .63 | 0.0110 | 13′ | 0.0038 |
| 14 | 0.2443 | 74 | 1.2915 | 134 | 2.3387 | 0 .14 | 0.0024 | 0 .64 | 0.0112 | 14′ | 0.0041 |
| 15° | 0.2618 | 75° | 1.3090 | 135° | 2.3562 | 0 .15 | 0.0026 | 0 .65 | 0.0113 | 15′ | 0.0044 |
| 16 | 0.2793 | 76 | 1.3265 | 136 | 2.3736 | 0 .16 | 0.0028 | 0 .66 | 0.0115 | 16′ | 0.0047 |
| 17 | 0.2967 | 77 | 1.3439 | 137 | 2.3911 | 0 .17 | 0.0030 | 0 .67 | 0.0117 | 17′ | 0.0049 |
| 18 | 0.3142 | 78 | 1.3614 | 138 | 2.4086 | 0 .18 | 0.0031 | 0 .68 | 0.0119 | 18′ | 0.0052 |
| 19 | 0.3316 | 79 | 1.3788 | 139 | 2.4260 | 0 .19 | 0.0033 | 0 .69 | 0.0120 | 19′ | 0.0055 |
| 20° | 0.3491 | 80° | 1.3963 | 140° | 2.4435 | 0°.20 | 0.0035 | 0°.70 | 0.0122 | 20′ | 0.0058 |
| 21 | 0.3665 | 81 | 1.4137 | 141 | 2.4609 | 0 .21 | 0.0037 | 0 .71 | 0.0124 | 21′ | 0.0061 |
| 22 | 0.3840 | 82 | 1.4312 | 142 | 2.4784 | 0 .22 | 0.0038 | 0 .72 | 0.0126 | 22′ | 0.0064 |
| 23 | 0.4014 | 83 | 1.4486 | 143 | 2.4958 | 0 .23 | 0.0040 | 0 .73 | 0.0127 | 23′ | 0.0067 |
| 24 | 0.4189 | 84 | 1.4661 | 144 | 2.5133 | 0 .24 | 0.0042 | 0 .74 | 0.0129 | 24′ | 0.0070 |
| 25° | 0.4363 | 85° | 1.4835 | 145° | 2.5307 | 0 .25 | 0.0044 | 0 .75 | 0.0131 | 25′ | 0.0073 |
| 26 | 0.4538 | 86 | 1.5010 | 146 | 2.5482 | 0 .26 | 0.0045 | 0 .76 | 0.0133 | 26′ | 0.0076 |
| 27 | 0.4712 | 87 | 1.5184 | 147 | 2.5656 | 0 .27 | 0.0047 | 0 .77 | 0.0134 | 27′ | 0.0079 |
| 28 | 0.4887 | 88 | 1.5359 | 148 | 2.5831 | 0 .28 | 0.0049 | 0 .78 | 0.0136 | 28′ | 0.0081 |
| 29 | 0.5061 | 89 | 1.5533 | 149 | 2.6005 | 0 .29 | 0.0051 | 0 .79 | 0.0138 | 29′ | 0.0084 |
| 30° | 0.5236 | 90° | 1.5708 | 150° | 2.6180 | 0°.30 | 0.0052 | 0°.80 | 0.0140 | 30′ | 0.0087 |
| 31 | 0.5411 | 91 | 1.5882 | 151 | 2.6354 | 0 .31 | 0.0054 | 0 .81 | 0.0141 | 31′ | 0.0090 |
| 32 | 0.5585 | 92 | 1.6057 | 152 | 2.6529 | 0 .32 | 0.0056 | 0 .82 | 0.0143 | 32′ | 0.0093 |
| 33 | 0.5760 | 93 | 1.6232 | 153 | 2.6704 | 0 .33 | 0.0058 | 0 .83 | 0.0145 | 33′ | 0.0096 |
| 34 | 0.5934 | 94 | 1.6406 | 154 | 2.6878 | 0 .34 | 0.0059 | 0 .84 | 0.0147 | 34′ | 0.0099 |
| 35° | 0.6109 | 95° | 1.6581 | 155° | 2.7053 | 0 .35 | 0.0061 | 0 .85 | 0.0148 | 35′ | 0.0102 |
| 36 | 0.6283 | 96 | 1.6755 | 156 | 2.7227 | 0 .36 | 0.0063 | 0 .86 | 0.0150 | 36′ | 0.0105 |
| 37 | 0.6458 | 97 | 1.6930 | 157 | 2.7402 | 0 .37 | 0.0065 | 0 .87 | 0.0152 | 37′ | 0.0108 |
| 38 | 0.6632 | 98 | 1.7104 | 158 | 2.7576 | 0 .38 | 0.0066 | 0 .88 | 0.0154 | 38′ | 0.0111 |
| 39 | 0.6807 | 99 | 1.7279 | 159 | 2.7751 | 0 .39 | 0.0068 | 0 .89 | 0.0155 | 39′ | 0.0113 |
| 40° | 0.6981 | 100° | 1.7453 | 160° | 2.7925 | 0°.40 | 0.0070 | 0°.90 | 0.0157 | 40′ | 0.0116 |
| 41 | 0.7156 | 101 | 1.7628 | 161 | 2.8100 | 0 .41 | 0.0072 | 0 .91 | 0.0159 | 41′ | 0.0119 |
| 42 | 0.7330 | 102 | 1.7802 | 162 | 2.8274 | 0 .42 | 0.0073 | 0 .92 | 0.0161 | 42′ | 0.0122 |
| 43 | 0.7505 | 103 | 1.7977 | 163 | 2.8449 | 0 .43 | 0.0075 | 0 .93 | 0.0162 | 43′ | 0.0125 |
| 44 | 0.7679 | 104 | 1.8151 | 164 | 2.8623 | 0 .44 | 0.0077 | 0 .94 | 0.0164 | 44′ | 0.0128 |
| 45° | 0.7854 | 105° | 1.8326 | 165° | 2.8798 | 0 .45 | 0.0079 | 0 .95 | 0.0166 | 45′ | 0.0131 |
| 46 | 0.8029 | 106 | 1.8500 | 166 | 2.8972 | 0 .46 | 0.0080 | 0 .96 | 0.0168 | 46′ | 0.0134 |
| 47 | 0.8203 | 107 | 1.8675 | 167 | 2.9147 | 0 .47 | 0.0082 | 0 .97 | 0.0169 | 47′ | 0.0137 |
| 48 | 0.8378 | 108 | 1.8850 | 168 | 2.9322 | 0 .48 | 0.0084 | 0 .98 | 0.0171 | 48′ | 0.0140 |
| 49 | 0.8552 | 109 | 1.9024 | 169 | 2.9496 | 0 .49 | 0.0086 | 0 .99 | 0.0173 | 49′ | 0.0143 |
| 50° | 0.8727 | 110° | 1.9199 | 170° | 2.9671 | 0°.50 | 0.0087 | 1°.00 | 0.0175 | 50′ | 0.0145 |
| 51 | 0.8901 | 111 | 1.9373 | 171 | 2.9845 | .... | ....... | .... | ....... | 51′ | 0.0148 |
| 52 | 0.9076 | 112 | 1.9548 | 172 | 3.0020 | .... | ....... | .... | ....... | 52′ | 0.0151 |
| 53 | 0.9250 | 113 | 1.9722 | 173 | 3.0194 | .... | ....... | .... | ....... | 53′ | 0.0154 |
| 54 | 0.9425 | 114 | 1.9897 | 174 | 3.0369 | .... | ....... | .... | ....... | 54′ | 0.0157 |
| 55° | 0.9599 | 115° | 2.0071 | 175° | 3.0543 | .... | ....... | .... | ....... | 55′ | 0.0160 |
| 56 | 0.9774 | 116 | 2.0246 | 176 | 3.0718 | .... | ....... | .... | ....... | 56′ | 0.0163 |
| 57 | 0.9948 | 117 | 2.0420 | 177 | 3.0892 | .... | ....... | .... | ....... | 57′ | 0.0166 |
| 58 | 1.0123 | 118 | 2.0595 | 178 | 3.1067 | .... | ....... | .... | ....... | 58′ | 0.0169 |
| 59 | 1.0297 | 119 | 2.0769 | 179 | 3.1241 | .... | ....... | .... | ....... | 59′ | 0.0172 |
| 60° | 1.0472 | 120° | 2.0944 | 180° | 3.1416 | .... | ....... | .... | ....... | 60′ | 0.0175 |

Arc 1° = 0.01745.   Arc 1′ = 0.0002909.   Arc 1″ = 0.000004848.
1 radian = 57°.296 = 57° 17′.75 = 57° 17′ 44″.81.
From Lionel S. Marks, "Mechanical Engineers' Handbook."

## J. Radians Expressed in Degrees

| rad | deg | rad | deg | rad | deg | rad | deg | rad | deg |
|---|---|---|---|---|---|---|---|---|---|
| 0.01 | 0°.57 | 0.64 | 36°.67 | 1.27 | 72°.77 | 1.90 | 108°.86 | 2.53 | 144°.96 |
| 0.02 | 1°.15 | 0.65 | 37°.24 | 1.28 | 73°.34 | 1.91 | 109°.43 | 2.54 | 145°.53 |
| 0.03 | 1°.72 | 0.66 | 37°.82 | 1.29 | 73°.91 | 1.92 | 110°.01 | 2.55 | 146°.10 |
| 0.04 | 2°.29 | 0.67 | 38°.39 | 1.30 | 74°.48 | 1.93 | 110°.58 | 2.56 | 146°.68 |
| 0.05 | 2°.86 | 0.68 | 38°.96 | 1.31 | 75°.06 | 1.94 | 111°.15 | 2.57 | 147°.25 |
| 0.06 | 3°.44 | 0.69 | 39°.53 | 1.32 | 75°.63 | 1.95 | 111°.73 | 2.58 | 147°.82 |
| 0.07 | 4°.01 | 0.70 | 40°.11 | 1.33 | 76°.20 | 1.96 | 112°.30 | 2.59 | 148°.40 |
| 0.08 | 4°.58 | 0.71 | 40°.68 | 1.34 | 76°.78 | 1.97 | 112°.87 | 2.60 | 148°.97 |
| 0.09 | 5°.16 | 0.72 | 41°.25 | 1.35 | 77°.35 | 1.98 | 113°.45 | 2.61 | 149°.54 |
| 0.10 | 5°.73 | 0.73 | 41°.83 | 1.36 | 77°.92 | 1.99 | 114°.02 | 2.62 | 150°.11 |
| 0.11 | 6°.30 | 0.74 | 42°.40 | 1.37 | 78°.50 | 2.00 | 114°.59 | 2.63 | 150°.69 |
| 0.12 | 6°.88 | 0.75 | 42°.97 | 1.38 | 79°.07 | 2.01 | 115°.16 | 2.64 | 151°.26 |
| 0.13 | 7°.45 | 0.76 | 43°.54 | 1.39 | 79°.64 | 2.02 | 115°.74 | 2.65 | 151°.83 |
| 0.14 | 8°.02 | 0.77 | 44°.12 | 1.40 | 80°.21 | 2.03 | 116°.31 | 2.66 | 152°.41 |
| 0.15 | 8°.59 | 0.78 | 44°.69 | 1.41 | 80°.79 | 2.04 | 116°.88 | 2.67 | 152°.98 |
| 0.16 | 9°.17 | 0.79 | 45°.26 | 1.42 | 81°.36 | 2.05 | 117°.46 | 2.68 | 153°.55 |
| 0.17 | 9°.74 | 0.80 | 45°.84 | 1.43 | 81°.93 | 2.06 | 118°.03 | 2.69 | 154°.13 |
| 0.18 | 10°.31 | 0.81 | 46°.41 | 1.44 | 82°.51 | 2.07 | 118°.60 | 2.70 | 154°.70 |
| 0.19 | 10°.89 | 0.82 | 46°.98 | 1.45 | 83°.08 | 2.08 | 119°.18 | 2.71 | 155°.27 |
| 0.20 | 11°.46 | 0.83 | 47°.56 | 1.46 | 83°.65 | 2.09 | 119°.75 | 2.72 | 155°.84 |
| 0.21 | 12°.03 | 0.84 | 48°.13 | 1.47 | 84°.22 | 2.10 | 120°.32 | 2.73 | 156°.42 |
| 0.22 | 12°.61 | 0.85 | 48°.70 | 1.48 | 84°.80 | 2.11 | 120°.89 | 2.74 | 156°.99 |
| 0.23 | 13°.18 | 0.86 | 49°.27 | 1.49 | 85°.37 | 2.12 | 121°.47 | 2.75 | 157°.56 |
| 0.24 | 13°.75 | 0.87 | 49°.85 | 1.50 | 85°.94 | 2.13 | 122°.04 | 2.76 | 158°.14 |
| 0.25 | 14°.32 | 0.88 | 50°.42 | 1.51 | 86°.52 | 2.14 | 122°.61 | 2.77 | 158°.71 |
| 0.26 | 14°.90 | 0.89 | 50°.99 | 1.52 | 87°.09 | 2.15 | 123°.19 | 2.78 | 159°.28 |
| 0.27 | 15°.47 | 0.90 | 51°.57 | 1.53 | 87°.66 | 2.16 | 123°.76 | 2.79 | 159°.86 |
| 0.28 | 16°.04 | 0.91 | 52°.14 | 1.54 | 88°.24 | 2.17 | 124°.33 | 2.80 | 160°.43 |
| 0.29 | 16°.62 | 0.92 | 52°.71 | 1.55 | 88°.81 | 2.18 | 124°.90 | 2.81 | 161°.00 |
| 0.30 | 17°.19 | 0.93 | 53°.29 | 1.56 | 89°.38 | 2.19 | 125°.48 | 2.82 | 161°.57 |
| 0.31 | 17°.76 | 0.94 | 53°.86 | 1.57 | 89°.95 | 2.20 | 126°.05 | 2.83 | 162°.15 |
| 0.32 | 18°.33 | 0.95 | 54°.43 | 1.58 | 90°.53 | 2.21 | 126°.62 | 2.84 | 162°.72 |
| 0.33 | 18°.91 | 0.96 | 55°.00 | 1.59 | 91°.10 | 2.22 | 127°.20 | 2.85 | 163°.29 |
| 0.34 | 19°.48 | 0.97 | 55°.58 | 1.60 | 91°.67 | 2.23 | 127°.77 | 2.86 | 163°.87 |
| 0.35 | 20°.05 | 0.98 | 56°.15 | 1.61 | 92°.25 | 2.24 | 128°.34 | 2.87 | 164°.44 |
| 0.36 | 20°.63 | 0.99 | 56°.72 | 1.62 | 92°.82 | 2.25 | 128°.92 | 2.88 | 165°.01 |
| 0.37 | 21°.20 | 1.00 | 57°.30 | 1.63 | 93°.39 | 2.26 | 129°.49 | 2.89 | 165°.58 |
| 0.38 | 21°.77 | 1.01 | 57°.87 | 1.64 | 93°.97 | 2.27 | 130°.06 | 2.90 | 166°.16 |
| 0.39 | 22°.35 | 1.02 | 58°.44 | 1.65 | 94°.54 | 2.28 | 130°.63 | 2.91 | 166°.73 |
| 0.40 | 22°.92 | 1.03 | 59°.01 | 1.66 | 95°.11 | 2.29 | 131°.21 | 2.92 | 167°.30 |
| 0.41 | 23°.49 | 1.04 | 59°.59 | 1.67 | 95°.68 | 2.30 | 131°.78 | 2.93 | 167°.88 |
| 0.42 | 24°.06 | 1.05 | 60°.16 | 1.68 | 96°.26 | 2.31 | 132°.35 | 2.94 | 168°.45 |
| 0.43 | 24°.64 | 1.06 | 60°.73 | 1.69 | 96°.83 | 2.32 | 132°.93 | 2.95 | 169°.02 |
| 0.44 | 25°.21 | 1.07 | 61°.31 | 1.70 | 97°.40 | 2.33 | 133°.50 | 2.96 | 169°.60 |
| 0.45 | 25°.78 | 1.08 | 61°.88 | 1.71 | 97°.98 | 2.34 | 134°.07 | 2.97 | 170°.17 |
| 0.46 | 26°.36 | 1.09 | 62°.45 | 1.72 | 98°.55 | 2.35 | 134°.65 | 2.98 | 170°.74 |
| 0.47 | 26°.93 | 1.10 | 63°.03 | 1.73 | 99°.12 | 2.36 | 135°.22 | 2.99 | 171°.31 |
| 0.48 | 27°.50 | 1.11 | 63°.60 | 1.74 | 99°.69 | 2.37 | 135°.79 | 3.00 | 171°.89 |
| 0.49 | 28°.07 | 1.12 | 64°.17 | 1.75 | 100°.27 | 2.38 | 136°.36 | 3.01 | 172°.46 |
| 0.50 | 28°.65 | 1.13 | 64°.74 | 1.76 | 100°.84 | 2.39 | 136°.94 | 3.02 | 173°.03 |
| 0.51 | 29°.22 | 1.14 | 65°.32 | 1.77 | 101°.41 | 2.40 | 137°.51 | 3.03 | 173°.61 |
| 0.52 | 29°.79 | 1.15 | 65°.89 | 1.78 | 101°.99 | 2.41 | 138°.08 | 3.04 | 174°.18 |
| 0.53 | 30°.37 | 1.16 | 66°.46 | 1.79 | 102°.56 | 2.42 | 138°.66 | 3.05 | 174°.75 |
| 0.54 | 30°.94 | 1.17 | 67°.04 | 1.80 | 103°.13 | 2.43 | 139°.23 | 3.06 | 175°.33 |
| 0.55 | 31°.51 | 1.18 | 67°.61 | 1.81 | 103°.71 | 2.44 | 139°.80 | 3.07 | 175°.90 |
| 0.56 | 32°.09 | 1.19 | 68°.18 | 1.82 | 104°.28 | 2.45 | 140°.37 | 3.08 | 176°.47 |
| 0.57 | 32°.66 | 1.20 | 68°.75 | 1.83 | 104°.85 | 2.46 | 140°.95 | 3.09 | 177°.04 |
| 0.58 | 33°.23 | 1.21 | 69°.33 | 1.84 | 105°.42 | 2.47 | 141°.52 | 3.10 | 177°.62 |
| 0.59 | 33°.80 | 1.22 | 69°.90 | 1.85 | 106°.00 | 2.48 | 142°.09 | 3.11 | 178°.19 |
| 0.60 | 34°.38 | 1.23 | 70°.47 | 1.86 | 106°.57 | 2.49 | 142°.67 | 3.12 | 178°.76 |
| 0.61 | 34°.95 | 1.24 | 71°.05 | 1.87 | 107°.14 | 2.50 | 143°.24 | 3.13 | 179°.34 |
| 0.62 | 35°.52 | 1.25 | 71°.62 | 1.88 | 107°.72 | 2.51 | 143°.81 | 3.14 | 179°.91 |
| 0.63 | 36°.10 | 1.26 | 72°.19 | 1.89 | 108°.29 | 2.52 | 144°.39 | 3.15 | 180°.48 |

### Interpolation

| | |
|---|---|
| 0.0002 | 0°.01 |
| 0.0004 | 0°.02 |
| 0.0006 | 0°.03 |
| 0.0008 | 0°.05 |
| 0.0010 | 0°.06 |
| 0.0012 | 0°.07 |
| 0.0014 | 0°.08 |
| 0.0016 | 0°.09 |
| 0.0018 | 0°.10 |
| 0.0020 | 0°.11 |
| 0.0022 | 0°.13 |
| 0.0024 | 0°.14 |
| 0.0026 | 0°.15 |
| 0.0028 | 0°.16 |
| 0.0030 | 0°.17 |
| 0.0032 | 0°.18 |
| 0.0034 | 0°.19 |
| 0.0036 | 0°.21 |
| 0.0038 | 0°.22 |
| 0.0040 | 0°.23 |
| 0.0042 | 0°.24 |
| 0.0044 | 0°.25 |
| 0.0046 | 0°.26 |
| 0.0048 | 0°.28 |
| 0.0050 | 0°.29 |
| 0.0052 | 0°.30 |
| 0.0054 | 0°.31 |
| 0.0056 | 0°.32 |
| 0.0058 | 0°.33 |
| 0.0060 | 0°.34 |
| 0.0062 | 0°.36 |
| 0.0064 | 0°.37 |
| 0.0066 | 0°.38 |
| 0.0068 | 0°.39 |
| 0.0070 | 0°.40 |
| 0.0072 | 0°.41 |
| 0.0074 | 0°.42 |
| 0.0076 | 0°.44 |
| 0.0078 | 0°.45 |
| 0.0080 | 0°.46 |
| 0.0082 | 0°.47 |
| 0.0084 | 0°.48 |
| 0.0086 | 0°.49 |
| 0.0088 | 0°.50 |
| 0.0090 | 0°.52 |
| 0.0092 | 0°.53 |
| 0.0094 | 0°.54 |
| 0.0096 | 0°.55 |
| 0.0098 | 0°.56 |

### Multiples of $\pi$

| | | |
|---|---|---|
| 1 | 3.1416 | 180° |
| 2 | 6.2832 | 360° |
| 3 | 9.4248 | 540° |
| 4 | 12.5664 | 720° |
| 5 | 15.7080 | 900° |
| 6 | 18.8496 | 1080° |
| 7 | 21.9911 | 1260° |
| 8 | 25.1327 | 1440° |
| 9 | 28.2743 | 1620° |
| 10 | 31.4159 | 1800° |

From Lionel S. Marks, "Mechanical Engineers' Handbook."

## K. Exponentials

$$(e^n \text{ and } e^{-n})$$

| $n$ | $e^n$ | Diff. | $n$ | $e^n$ | Diff. | $n$ | $e^n$ | $n$ | $e^{-n}$ | Diff. | $n$ | $e^{-n}$ | $n$ | $e^{-n}$ |
|---|---|---|---|---|---|---|---|---|---|---|---|---|---|---|
| 0.00 | 1.000 | 10 | 0.50 | 1.649 | 16 | 1.0 | 2.718† | 0.00 | 1.000 | -10 | 0.50 | 0.607 | 1.0 | 0.368 |
| 0.01 | 1.010 | 10 | 0.51 | 1.665 | 17 | 1.1 | 3.004 | 0.01 | 0.990 | -10 | 0.51 | 0.600 | 1.1 | 0.333 |
| 0.02 | 1.020 | 10 | 0.52 | 1.682 | 17 | 1.2 | 3.320 | 0.02 | 0.980 | -10 | 0.52 | 0.595 | 1.2 | 0.301 |
| 0.03 | 1.030 | 11 | 0.53 | 1.699 | 17 | 1.3 | 3.669 | 0.03 | 0.970 | -9 | 0.53 | 0.589 | 1.3 | 0.273 |
| 0.04 | 1.041 | 10 | 0.54 | 1.716 | 17 | 1.4 | 4.055 | 0.04 | 0.961 | -10 | 0.54 | 0.583 | 1.4 | 0.247 |
| 0.05 | 1.051 | 11 | 0.55 | 1.733 | 18 | 1.5 | 4.482 | 0.05 | 0.951 | -9 | 0.55 | 0.577 | 1.5 | 0.223 |
| 0.06 | 1.062 | 11 | 0.56 | 1.751 | 17 | 1.6 | 4.953 | 0.06 | 0.942 | -10 | 0.56 | 0.571 | 1.6 | 0.202 |
| 0.07 | 1.073 | 10 | 0.57 | 1.768 | 18 | 1.7 | 5.474 | 0.07 | 0.932 | -9 | 0.57 | 0.566 | 1.7 | 0.183 |
| 0.08 | 1.083 | 11 | 0.58 | 1.786 | 18 | 1.8 | 6.050 | 0.08 | 0.923 | -9 | 0.58 | 0.560 | 1.8 | 0.165 |
| 0.09 | 1.094 | 11 | 0.59 | 1.804 | 18 | 1.9 | 6.686 | 0.09 | 0.914 | -9 | 0.59 | 0.554 | 1.9 | 0.150 |
| 0.10 | 1.105 | 11 | 0.60 | 1.822 | 18 | 2.0 | 7.389 | 0.10 | 0.905 | -9 | 0.60 | 0.549 | 2.0 | 0.135 |
| 0.11 | 1.116 | 11 | 0.61 | 1.840 | 19 | 2.1 | 8.166 | 0.11 | 0.896 | -9 | 0.61 | 0.543 | 2.1 | 0.122 |
| 0.12 | 1.127 | 12 | 0.62 | 1.859 | 19 | 2.2 | 9.025 | 0.12 | 0.887 | -9 | 0.62 | 0.538 | 2.2 | 0.111 |
| 0.13 | 1.139 | 11 | 0.63 | 1.878 | 18 | 2.3 | 9.974 | 0.13 | 0.878 | -9 | 0.63 | 0.533 | 2.3 | 0.100 |
| 0.14 | 1.150 | 12 | 0.64 | 1.896 | 20 | 2.4 | 11.02 | 0.14 | 0.869 | -8 | 0.64 | 0.527 | 2.4 | 0.0907 |
| 0.15 | 1.162 | 12 | 0.65 | 1.916 | 19 | 2.5 | 12.18 | 0.15 | 0.861 | -9 | 0.65 | 0.522 | 2.5 | 0.0821 |
| 0.16 | 1.174 | 11 | 0.66 | 1.935 | 19 | 2.6 | 13.46 | 0.16 | 0.852 | -8 | 0.66 | 0.517 | 2.6 | 0.0743 |
| 0.17 | 1.185 | 12 | 0.67 | 1.954 | 20 | 2.7 | 14.88 | 0.17 | 0.844 | -9 | 0.67 | 0.512 | 2.7 | 0.0672 |
| 0.18 | 1.197 | 12 | 0.68 | 1.974 | 20 | 2.8 | 16.44 | 0.18 | 0.835 | -8 | 0.68 | 0.507 | 2.8 | 0.0608 |
| 0.19 | 1.209 | 12 | 0.69 | 1.994 | 20 | 2.9 | 18.17 | 0.19 | 0.827 | -8 | 0.69 | 0.502 | 2.9 | 0.0550 |
| 0.20 | 1.221 | 13 | 0.70 | 2.014 | 20 | 3.0 | 20.09 | 0.20 | 0.819 | -8 | 0.70 | 0.497 | 3.0 | 0.0498 |
| 0.21 | 1.234 | 12 | 0.71 | 2.034 | 20 | 3.1 | 22.20 | 0.21 | 0.811 | -8 | 0.71 | 0.492 | 3.1 | 0.0450 |
| 0.22 | 1.246 | 13 | 0.72 | 2.054 | 21 | 3.2 | 24.53 | 0.22 | 0.803 | -8 | 0.72 | 0.487 | 3.2 | 0.0408 |
| 0.23 | 1.259 | 12 | 0.73 | 2.075 | 21 | 3.3 | 27.11 | 0.23 | 0.795 | -8 | 0.73 | 0.482 | 3.3 | 0.0369 |
| 0.24 | 1.271 | 13 | 0.74 | 2.096 | 21 | 3.4 | 29.96 | 0.24 | 0.787 | -8 | 0.74 | 0.477 | 3.4 | 0.0334 |
| 0.25 | 1.284 | 13 | 0.75 | 2.117 | 21 | 3.5 | 33.12 | 0.25 | 0.779 | -8 | 0.75 | 0.472 | 3.5 | 0.0302 |
| 0.26 | 1.297 | 13 | 0.76 | 2.138 | 22 | 3.6 | 36.60 | 0.26 | 0.771 | -8 | 0.76 | 0.468 | 3.6 | 0.0273 |
| 0.27 | 1.310 | 13 | 0.77 | 2.160 | 21 | 3.7 | 40.45 | 0.27 | 0.763 | -7 | 0.77 | 0.463 | 3.7 | 0.0247 |
| 0.28 | 1.323 | 13 | 0.78 | 2.181 | 22 | 3.8 | 44.70 | 0.28 | 0.756 | -8 | 0.78 | 0.458 | 3.8 | 0.0224 |
| 0.29 | 1.336 | 14 | 0.79 | 2.203 | 23 | 3.9 | 49.40 | 0.29 | 0.748 | -7 | 0.79 | 0.454 | 3.9 | 0.0202 |
| 0.30 | 1.350 | 13 | 0.80 | 2.226 | 22 | 4.0 | 54.60 | 0.30 | 0.741 | -8 | 0.80 | 0.449 | 4.0 | 0.0183 |
| 0.31 | 1.363 | 14 | 0.81 | 2.248 | 22 | 4.1 | 60.34 | 0.31 | 0.733 | -7 | 0.81 | 0.445 | 4.1 | 0.0166 |
| 0.32 | 1.377 | 14 | 0.82 | 2.270 | 23 | 4.2 | 66.69 | 0.32 | 0.726 | -7 | 0.82 | 0.440 | 4.2 | 0.0150 |
| 0.33 | 1.391 | 14 | 0.83 | 2.293 | 23 | 4.3 | 73.70 | 0.33 | 0.719 | -7 | 0.83 | 0.436 | 4.3 | 0.0136 |
| 0.34 | 1.405 | 14 | 0.84 | 2.316 | 24 | 4.4 | 81.45 | 0.34 | 0.712 | -7 | 0.84 | 0.432 | 4.4 | 0.0123 |
| 0.35 | 1.419 | 14 | 0.85 | 2.340 | 23 | 4.5 | 90.02 | 0.35 | 0.705 | -7 | 0.85 | 0.427 | 4.5 | 0.0111 |
| 0.36 | 1.433 | 15 | 0.86 | 2.363 | 24 | | | 0.36 | 0.698 | -7 | 0.86 | 0.423 | | |
| 0.37 | 1.448 | 14 | 0.87 | 2.387 | 24 | 5.0 | 148.4 | 0.37 | 0.691 | -7 | 0.87 | 0.419 | 5.0 | 0.00674 |
| 0.38 | 1.462 | 15 | 0.88 | 2.411 | 24 | 6.0 | 403.4 | 0.38 | 0.684 | -7 | 0.88 | 0.415 | 6.0 | 0.00248 |
| 0.39 | 1.477 | 15 | 0.89 | 2.435 | 25 | 7.0 | 1097. | 0.39 | 0.677 | -7 | 0.89 | 0.411 | 7.0 | 0.000912 |
| 0.40 | 1.492 | 15 | 0.90 | 2.460 | 24 | 8.0 | 2981. | 0.40 | 0.670 | -6 | 0.90 | 0.407 | 8.0 | 0.000335 |
| 0.41 | 1.507 | 15 | 0.91 | 2.484 | 25 | 9.0 | 8103. | 0.41 | 0.664 | -7 | 0.91 | 0.403 | 9.0 | 0.000123 |
| 0.42 | 1.522 | 15 | 0.92 | 2.509 | 26 | 10.0 | 22026. | 0.42 | 0.657 | -6 | 0.92 | 0.399 | 10.0 | 0.000045 |
| 0.43 | 1.537 | 16 | 0.93 | 2.535 | 25 | $\pi/2$ | 4.810 | 0.43 | 0.651 | -7 | 0.93 | 0.395 | | |
| 0.44 | 1.553 | 15 | 0.94 | 2.560 | 26 | $2\pi/2$ | 23.14 | 0.44 | 0.644 | -7 | 0.94 | 0.391 | | |
| 0.45 | 1.568 | 16 | 0.95 | 2.586 | 26 | $3\pi/2$ | 111.3 | 0.45 | 0.638 | -7 | 0.95 | 0.387 | $\pi/2$ | 0.208 |
| 0.46 | 1.584 | 16 | 0.96 | 2.612 | 26 | $4\pi/2$ | 535.5 | 0.46 | 0.631 | -7 | 0.96 | 0.383 | $2\pi/2$ | 0.0432 |
| 0.47 | 1.600 | 16 | 0.97 | 2.638 | 26 | $5\pi/2$ | 2576 | 0.47 | 0.625 | -6 | 0.97 | 0.379 | $3\pi/2$ | 0.00898 |
| 0.48 | 1.616 | 16 | 0.98 | 2.664 | 27 | $6\pi/2$ | 12392 | 0.48 | 0.619 | -6 | 0.98 | 0.375 | $4\pi/2$ | 0.00187 |
| 0.49 | 1.632 | 17 | 0.99 | 2.691 | 27 | $7\pi/2$ | 59610 | 0.49 | 0.613 | -6 | 0.99 | 0.372 | $5\pi/2$ | 0.000388 |
| 0.50 | 1.649 | | 1.00 | 2.718 | | $8\pi/2$ | 286751 | 0.50 | 0.607 | | 1.00 | 0.368 | $6\pi/2$ | 0.000081 |

$e = 2.71828$.   $1/e = 0.367879$.   $\log_{10} e = 0.4343$.   $1/(0.4343) = 2.3026$.
From Lionel S. Marks, "Mechanical Engineers' Handbook."
† NOTE: Do not interpolate in this column.

## L. Trigonometric Relations

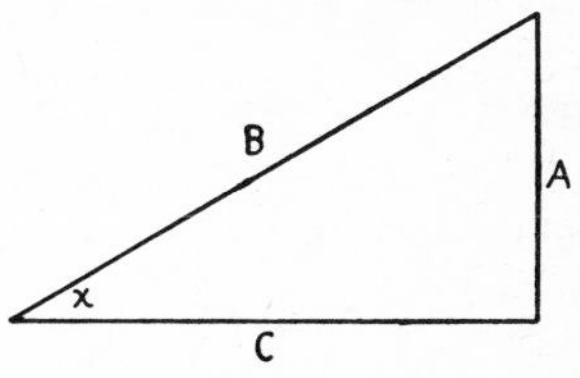

$\sin x = \dfrac{A}{B}$

$\cos x = \dfrac{C}{B}$

$\tan x = \dfrac{A}{C} = \dfrac{\sin x}{\cos x}$

$\sec x = \dfrac{B}{C} = \dfrac{1}{\cos x}$

$\csc x = \dfrac{B}{A} = \dfrac{1}{\sin x}$

$\cot x = \dfrac{C}{A} = \dfrac{1}{\tan x} = \dfrac{\cos x}{\sin x}$

Fig. 1.

$\sin(-x) = -\sin x$
$\cos(-x) = \cos x$
$\tan(-x) = -\tan x$

$\sin\left(\dfrac{\pi}{2} - x\right) = \cos x$ $\qquad$ $\sin\left(\dfrac{\pi}{2} + x\right) = \cos x$

$\cos\left(\dfrac{\pi}{2} - x\right) = \sin x$ $\qquad$ $\cos\left(\dfrac{\pi}{2} + x\right) = -\sin x$

$\tan\left(\dfrac{\pi}{2} - x\right) = \cot x$ $\qquad$ $\tan\left(\dfrac{\pi}{2} + x\right) = -\cot x$

$\sin(\pi - x) = \sin x$ $\qquad$ $\sin(\pi + x) = -\sin x$
$\cos(\pi - x) = -\cos x$ $\qquad$ $\cos(\pi + x) = -\cos x$
$\tan(\pi - x) = -\tan x$ $\qquad$ $\tan(\pi + x) = \tan x$

$\left.\begin{array}{l} \sin(x + 2\pi n) = \sin x \\ \cos(x + 2\pi n) = \cos x \\ \tan(x + 2\pi n) = \tan x \end{array}\right\}$ ($n$ a positive or negative integer)

$\sin(x + y) = \sin x \cos y + \cos x \sin y$
$\sin(x - y) = \sin x \cos y - \cos x \sin y$
$\cos(x + y) = \cos x \cos y - \sin x \sin y$
$\cos(x - y) = \cos x \cos y + \sin x \sin y$

$\tan(x + y) = \dfrac{\tan x + \tan y}{1 - \tan x \tan y}$

$\tan(x - y) = \dfrac{\tan x - \tan y}{1 + \tan x \tan y}$

$\sin 2x = 2 \sin x \cos x$ $\qquad$ $\cos 2x = 2 \cos^2 x - 1 = 1 - 2 \sin^2 x$

$\tan 2x = \dfrac{2 \tan x}{1 - \tan^2 x}$ $\qquad$ $\cot 2x = \dfrac{(\cot^2 x - 1)}{2 \cot x}$

$\sin 3x = 3 \sin x - 4 \sin^3 x$ $\qquad$ $\cos 3x = 4 \cos^3 x - 3 \cos x$

$\sin \dfrac{1}{2} x = \left[\dfrac{1}{2}(1 - \cos x)\right]^{1/2} = \dfrac{1}{2}(1 + \sin x)^{1/2} - \dfrac{1}{2}(1 - \sin x)^{1/2}$

$\cos \dfrac{1}{2} x = \left[\dfrac{1}{2}(1 + \cos x)\right]^{1/2} = \dfrac{1}{2}(1 + \sin x)^{1/2} + \dfrac{1}{2}(1 - \sin x)^{1/2}$

$\tan \dfrac{1}{2} x = \dfrac{(1 - \cos x)^{1/2}}{(1 + \cos x)^{1/2}} = \dfrac{1 - \cos x}{\sin x} = \dfrac{\sin x}{1 + \cos x}$

$\sin^2 x = \dfrac{1}{2}(1 - \cos 2x)$ $\qquad$ $\cos^2 x = \dfrac{1}{2}(1 + \cos 2x)$

$\sin^3 x = \dfrac{1}{4}(3 \sin x - \sin 3x)$ $\qquad$ $\cos^3 x = \dfrac{1}{4}(\cos 3x + 3 \cos x)$

## L. Trigonometric Relations (*Concluded*)

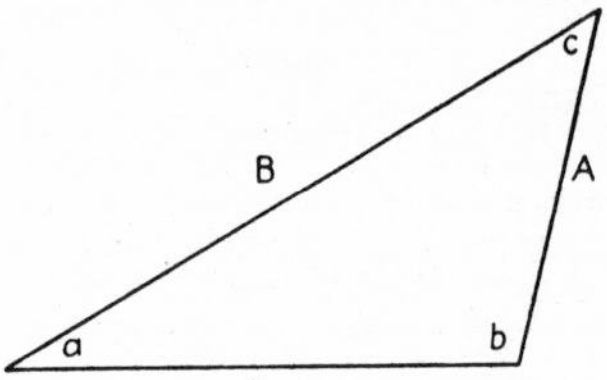

$$\sin x \sin y = \frac{1}{2} \cos (x - y) - \frac{1}{2} \cos (x + y)$$

$$\cos x \cos y = \frac{1}{2} \cos (x - y) + \frac{1}{2} \cos (x + y)$$

$$\sin x \cos y = \frac{1}{2} \sin (x - y) + \frac{1}{2} \sin (x + y)$$

$$\sin x + \sin y = 2 \sin \frac{1}{2} (x + y) \cos \frac{1}{2} (x - y)$$

$$\sin x - \sin y = 2 \cos \frac{1}{2} (x + y) \sin \frac{1}{2} (x - y)$$

$$\cos x + \cos y = 2 \cos \frac{1}{2} (x + y) \cos \frac{1}{2} (x - y)$$

$$\cos x - \cos y = - 2 \sin \frac{1}{2} (x + y) \sin \frac{1}{2} (x - y)$$

$$\tan x + \tan y = \frac{\sin (x + y)}{\cos x \cos y} \qquad \tan x - \tan y = \frac{\sin (x - y)}{\cos x \cos y}$$

$$\sin^2 x - \sin^2 y = \sin (x + y) \sin (x - y)$$
$$\cos^2 x - \cos^2 y = - \sin (x + y) \sin (x - y)$$
$$\cos^2 x - \sin^2 y = \cos (x + y) \cos (x - y)$$
$$\sin^2 x + \cos^2 x = 1 \qquad\qquad \sec^2 x - \tan^2 x = 1$$

FIG. 2.

In any triangle (Fig. 2):

$$\frac{A}{\sin a} = \frac{B}{\sin b} = \frac{C}{\sin c} \qquad\qquad \text{(law of sines)}$$

$$A^2 = B^2 + C^2 - 2BC \cos a \qquad\qquad \text{(law of cosines)}$$

$$\frac{A + B}{A - B} = \frac{\tan \frac{1}{2} (a + b)}{\tan \frac{1}{2} (a - b)} = \frac{\sin a + \sin b}{\sin a - \sin b} \qquad \text{(law of tangents)}$$

$$a + b + c = 180° \qquad\qquad A = B \cos c + C \cos b$$
$$B = C \cos a + A \cos c \qquad\qquad C = A \cos b + B \cos a$$

## M. Properties of Hyperbolic Functions

$$\sinh x = \frac{\epsilon^x - \epsilon^{-x}}{2} = x + \frac{x^3}{3!} + \frac{x^5}{5!} + \cdots$$

$$\cosh x = \frac{\epsilon^x + \epsilon^{-x}}{2} = 1 + \frac{x^2}{2!} + \frac{x^4}{4!} + \cdots$$

$$\tanh x = \frac{\sinh x}{\cosh x}$$

$\sinh (-x) = -\sinh x$

$\cosh (-x) = \cosh x$

$\cosh^2 x = 1 + \sinh^2 x$

$$\frac{d\,(\sinh x)}{dx} = \cosh x$$

$$\frac{d\,(\cosh x)}{dx} = \sinh x$$

$\int \sinh x \cdot dx = \cosh x$

$\int \cosh x \cdot dx = \sinh x$

$\sinh (x \pm j\pi) = -\sinh x$

$\cosh (x \pm j\pi) = -\cosh x$

$\sinh (x \pm j2\pi) = \sinh x$

$\cosh (x \pm j2\pi) = \cosh x$

## N. Copper-Wire Table, Standard Annealed Copper American Wire Gage (B & S)

| Gage No. | Diameter, mils at 20°C | Cross section at 20°C — Circular mils | Cross section at 20°C — Sq in. | Ohms per 1,000 ft at 20°C (= 68°F) | Lb. per 1,000 ft | Ft per lb | Ft per ohm at 20°C (= 68°F) | Ohms per lb at 20°C (= 68°F) | Gage No. |
|---|---|---|---|---|---|---|---|---|---|
| 0000 | 460.0 | 211,600.0 | 0.1662 | 0.04901 | 640.5 | 1.561 | 20,400.0 | 0.00007652 | 0000 |
| 000 | 409.6 | 167,800.0 | 0.1318 | 0.06180 | 507.9 | 1.968 | 16,180.0 | 0.0001217 | 000 |
| 00 | 364.8 | 133,100.0 | 0.1045 | 0.07793 | 402.8 | 2.482 | 12,830.0 | 0.0001935 | 00 |
| 0 | 324.9 | 105,500.0 | 0.08289 | 0.09827 | 319.5 | 3.130 | 10,180.0 | 0.0003076 | 0 |
| 1 | 289.3 | 83,690.0 | 0.06573 | 0.1239 | 253.3 | 3.947 | 8,070.0 | 0.0004891 | 1 |
| 2 | 257.6 | 66,370.0 | 0.05213 | 0.1563 | 200.9 | 4.977 | 6,400.0 | 0.0007778 | 2 |
| 3 | 229.4 | 53,640.0 | 0.04134 | 0.1970 | 159.3 | 6.276 | 5,075.0 | 0.001237 | 3 |
| 4 | 204.3 | 41,740.0 | 0.03278 | 0.2485 | 126.4 | 7.914 | 4,025.0 | 0.001966 | 4 |
| 5 | 181.9 | 33,100.0 | 0.02600 | 0.3133 | 100.2 | 9.980 | 3,192.0 | 0.003127 | 5 |
| 6 | 162.0 | 26,250.0 | 0.02062 | 0.3951 | 79.46 | 12.58 | 2,531.0 | 0.004972 | 6 |
| 7 | 144.3 | 20,820.0 | 0.01635 | 0.4982 | 62.02 | 15.87 | 2,007.0 | 0.007905 | 7 |
| 8 | 128.5 | 16,510.0 | 0.01297 | 0.6282 | 49.98 | 20.01 | 1,592.0 | 0.01257 | 8 |
| 9 | 114.4 | 13,090.0 | 0.01028 | 0.7921 | 39.63 | 25.23 | 1,262.0 | 0.01999 | 9 |
| 10 | 101.9 | 10,380.0 | 0.008155 | 0.9989 | 31.43 | 31.82 | 1,001.0 | 0.03178 | 10 |
| 11 | 90.74 | 8,234.0 | 0.006467 | 1.260 | 24.92 | 40.12 | 794.0 | 0.05053 | 11 |
| 12 | 80.81 | 6,530.0 | 0.005129 | 1.588 | 19.77 | 50.59 | 629.6 | 0.08035 | 12 |
| 13 | 71.96 | 5,178.0 | 0.004067 | 2.003 | 15.68 | 63.80 | 499.3 | 0.1278 | 13 |
| 14 | 64.08 | 4,107.0 | 0.003225 | 2.525 | 12.43 | 80.44 | 396.0 | 0.2032 | 14 |
| 15 | 57.07 | 3,257.0 | 0.002558 | 3.184 | 9.858 | 101.4 | 314.0 | 0.3230 | 15 |
| 16 | 50.82 | 2,583.0 | 0.002028 | 4.016 | 7.818 | 127.9 | 249.0 | 0.5136 | 16 |

| | | | | | | | | | |
|---|---|---|---|---|---|---|---|---|---|
| 17 | 45.26 | 2,048.0 | 0.001609 | 5.064 | 6.200 | 161.3 | 197.5 | 0.8167 | 17 |
| 18 | 40.30 | 1,624.0 | 0.001276 | 6.385 | 4.917 | 203.4 | 156.6 | 1.299 | 18 |
| 19 | 35.89 | 1,288.0 | 0.001012 | 8.051 | 3.899 | 256.5 | 124.2 | 2.065 | 19 |
| 20 | 31.96 | 1,022.0 | 0.0008023 | 10.15 | 3.092 | 323.4 | 98.50 | 3.283 | 20 |
| 21 | 28.46 | 810.1 | 0.0006363 | 12.80 | 2.452 | 407.8 | 78.11 | 5.221 | 21 |
| 22 | 25.35 | 642.4 | 0.0005046 | 16.14 | 1.945 | 514.2 | 61.95 | 8.301 | 22 |
| 23 | 22.57 | 509.5 | 0.0004002 | 20.36 | 1.542 | 648.4 | 49.13 | 13.20 | 23 |
| 24 | 20.10 | 404.0 | 0.0003173 | 25.67 | 1.223 | 817.7 | 38.96 | 20.99 | 24 |
| 25 | 17.90 | 320.4 | 0.0002517 | 32.37 | 0.9699 | 1,031.0 | 30.90 | 33.37 | 25 |
| 26 | 15.94 | 254.1 | 0.0001996 | 40.81 | 0.7692 | 1,300.0 | 24.50 | 53.06 | 26 |
| 27 | 14.20 | 201.5 | 0.0001583 | 51.47 | 0.6100 | 1,639.0 | 19.43 | 84.37 | 27 |
| 28 | 12.64 | 159.8 | 0.0001255 | 64.90 | 0.4837 | 2,067.0 | 15.41 | 134.2 | 28 |
| 29 | 11.26 | 126.7 | 0.00009953 | 81.83 | 0.3836 | 2,607.0 | 12.22 | 213.3 | 29 |
| 30 | 10.03 | 100.5 | 0.00007894 | 103.2 | 0.3042 | 3,287.0 | 9.691 | 339.2 | 30 |
| 31 | 8.928 | 79.70 | 0.00006260 | 130.1 | 0.2413 | 4,145.0 | 7.685 | 539.3 | 31 |
| 32 | 7.950 | 63.21 | 0.00004964 | 164.1 | 0.1913 | 5,227.0 | 6.095 | 857.6 | 32 |
| 33 | 7.080 | 50.13 | 0.00003937 | 206.9 | 0.1517 | 6,591.0 | 4.833 | 1,364.0 | 33 |
| 34 | 6.305 | 39.75 | 0.00003122 | 260.9 | 0.1203 | 8,310.0 | 3.833 | 2,168.0 | 34 |
| 35 | 5.615 | 31.52 | 0.00002476 | 329.0 | 0.09542 | 10,480.0 | 3.040 | 3,448.0 | 35 |
| 36 | 5.000 | 25.00 | 0.00001964 | 414.8 | 0.07568 | 13,210.0 | 2.411 | 5,482.0 | 36 |
| 37 | 4.453 | 19.83 | 0.00001557 | 523.1 | 0.06001 | 16,660.0 | 1.912 | 8,717.0 | 37 |
| 38 | 3.965 | 15.72 | 0.00001235 | 659.6 | 0.04759 | 21,010.0 | 1.516 | 13,860.0 | 38 |
| 39 | 3.531 | 12.47 | 0.000009793 | 831.8 | 0.03774 | 26,500.0 | 1.202 | 22,040.0 | 39 |
| 40 | 3.145 | 9.888 | 0.000007766 | 1,049.0 | 0.02993 | 33,410.0 | 0.9534 | 35,040.0 | 40 |

From F. E. Terman, "Radio Engineers' Handbook," McGraw-Hill Book Company, 1943.

## O. SI Unit Prefixes

| Multiples | Submultiples |
|---|---|
| $10^{12}$ = tera = T | $10^{-1}$ = deci = d |
| $10^{9}$ = giga = G | $10^{-2}$ = centi = c |
| $10^{6}$ = mega = M | $10^{-3}$ = milli = m |
| $10^{3}$ = kilo = k | $10^{-6}$ = micro = $\mu$ |
| $10^{2}$ = hecto = h | $10^{-9}$ = nano = n |
| 10 = deca = da | $10^{-12}$ = pico = p |
| | $10^{-15}$ = femto = f |
| | $10^{-18}$ = atto = a |

## P. SI Base Units

| Quantity | Base unit | Symbol |
|---|---|---|
| Length | meter | m |
| Mass | kilogram | kg |
| Time | second | s |
| Electric current | ampere | A |
| Thermodynamic temp. | kelvin | K |
| Amount of substance | mole | mol |
| Luminous intensity | candela | cd |

## Q. The International System (SI) of Units for Electricity

| Quantity Name | Symbol | SI unit Name | Unit | Conversions | Base units (m, kg, s, A) |
|---|---|---|---|---|---|
| Electric current | I | ampere | A | A | A |
| Electric current density | J, (S) | ampere per square meter | A/m² | | $m^{-2} \cdot A$ |
| Electric charge, quantity of electricity | Q | coulomb | C | $1\,C = 1\,A \cdot s$ | $s \cdot A$ |
| Electric potential | V, $\varphi$ | | | | |
| Electrical potential difference, electric tension (voltage) | U, (V) | volt | V | $1\,V = 1\,W/A$<br>Link to base units:<br>$1\,V$ = energy per unity of charge<br>$= 1\,J/C = 1\,N \cdot m/C$<br>$= 1\,kg \cdot m^2/(A \cdot s^3)$ | $m^2 \cdot kg \cdot s^{-3} \cdot A^{-1}$ |
| Electric source voltage, electromotive force | E | | | | |
| Power* | P | watt | W | $1\,W = 1\,V \cdot A = 1\,J/s$ | $m^2 \cdot kg \cdot s^{-3}$ |
| Energy | W | joule | J | $1\,J = 1\,W \cdot s = 1\,n \cdot m$ | $m^2 \cdot kg \cdot s^{-2}$ |
| Impedance | Z | | | | |
| Resistance | R | ohm | Ω | $1\,\Omega = 1\,V/A$ | $m^2 \cdot kg \cdot s^{-3} \cdot A^{-2}$ |
| Reactance | X | | | | |
| Admittance | Y | | | | |
| Conductance | G | siemen | S | $1\,S = 1\,A/V = 1/\Omega$<br>$Y = 1/Z$<br>$G = 1/R \quad B = 1/X$ | |
| Susceptance | B | | | | |
| Specific resistance, resistivity | $\rho$ | ohm-meter | $\Omega \cdot m$ | | $m^3 \cdot kg \cdot s^{-3} \cdot A^{-2}$ |
| Specific conductance, conductivity | $\gamma$, $\sigma$ | siemens per meter | S/m | $\gamma = 1/\rho$ | $m^{-3} \cdot kg^{-1} \cdot s^3 \cdot A^2$ |
| Electric flux ($= \int D_n dS$) | $\Psi$ | coulomb | C | | $s \cdot A$ |

| Quantity | | SI unit | | | Base units |
| Name | Symbol | Name | Unit | Conversions | (m, kg, s, A) |
|---|---|---|---|---|---|
| Electric field strength | E | volt per meter | V/m | 1 V/m = 1 N/C | $m \cdot kg \cdot s^{-3} \cdot A^{-1}$ |
| Dielectric displacement | D | coulomb per square meter | $C/m^2$ | | $m^{-2} \cdot s \cdot A$ |
| Electric polarization | P, (D) | | | $P = D - \epsilon_0 E$ | |
| Electric dipole moment | p, $P_e$ | coulomb-meter | C · m | | M · s · A |
| Capacitance | C | farad | F | 1 F = 1 C/V <br> C = Q/U | $m^{-2} \cdot kg^{-1} \cdot s^4 \cdot A^2$ |
| Permittivity, dielectric constant | $\epsilon$ | farad per meter | F/m | $\epsilon = D/E$ | $m^{-3} \cdot kg^{-1} \cdot s^4 \cdot A^2$ |
| Permittivity of vacuum, electric constant | $\epsilon_0$ | | | $\epsilon_0 = 10^{-9}/(36 \cdot \pi)F/m$ | |
| Relative permittivity | $\epsilon_r$ | (dimensionless) | — | $\epsilon_r = \epsilon/\epsilon_0$ | — |
| Relative electric susceptibility | $\chi, \chi_e$ | (dimensionless) | — | $\chi = \epsilon_r - 1$ | — |
| Magnetic flux ($= \int B_n dS$) | $\Phi$ | weber | Wb | 1 Wb = 1 V · s | $m^2 \cdot kg \cdot s^{-2} \cdot A$ |
| Magnetic field strength | H | ampere per meter | A/m | 1 A/m = 1 N/Wb | $m^{-1} \cdot A$ |
| Magnetic induction | B | weber per square meter (tesla) | Wb/m² (T) | | $kg \cdot s^{-2} \cdot A^{-1}$ |
| Magnetic polarization | J | | | $J = B - \mu_0 H$ | |
| Magnetic dipole moment | j | weber-meter | Wb · m | | $m^3 \cdot kg \cdot s^{-2} \cdot A^{-1}$ |
| Self-inductance | L | henry | H | 1 H = 1 V · s/A = 1 Wb/A | $m^2 \cdot kg \cdot s^{-2} \cdot A^{-2}$ |
| Mutual inductance | M, $L_{12}$ | | | | |
| Permeability magnetic induction constant | $\mu$ | henry per meter | H/m | $\mu = B/H$ | $m \cdot kg \cdot s^{-2} \cdot A^{-2}$ |
| Permeability of vacuum, magnetic induction constant | $\mu_0$ | | | $\mu_0 = 4\pi \cdot 10^{-7} H/m$ | |
| Relative permeability | $\mu_r$ | (dimensionless) | — | $\mu_r = \mu/\mu_0$ | — |
| Relative magnetic permeability | $\kappa_r$ | (dimensionless) | — | $\kappa_r = \mu_r - 1$ | — |
| Phase displacement | $\varphi$ | radian | rad | | — |
| Frequency | f, $\nu$ | hertz | Hz | 1 Hz = 1/s | $s^{-1}$ |
| Angular frequency, angular velocity | $\omega$ | radian per second | rad/s | $\omega = 2\pi \cdot f$ | $s^{-1}$ |

*In electric power technology, active power (symbol P) is expressed in the unit watt (W): apparent power (symbol S or $P_s$) in volt-ampere (V · A) and reactive (positive or negative) power (symbol Q and $P_q$) in var (var).

## R. ASTAP Language Statements with Examples of Common Variations

(On pages A-22 and A-23)

# LANGUAGE STATEMENTS

MODEL DESCRIPTION
  comments
  MODEL model name (disposition) (node-node-....-node)

  ELEMENTS
    comments
    element name, from node-to-node = element value
    ................

  parameter name = parameter value
  ................

  reference name = MODEL model name (node-node-....-node) (changes)

  mutual name, element 1 - element 2 = mutual inductance value

FUNCTIONS
  TABLE table name, X, Y, Y, Y, ..............., X, Y
  EQUATION equation name (dummy argument list) = (FORTRAN-like expression)
  ................

  DISTRIBUTION distribution name (type) X, Y, Y, Y, ......, X, Y
  ................

  STATISTICAL TABLE table name (DEP = DISTRIBUTION name) X, Y1, Y2, X, Y1, Y2
    OR
  STATISTICAL TABLE table name (IND = DISTRIBUTION name) X1, X2, Y, X1, X2, Y
FEATURES
  GROUND = (node name)
  GLOBAL = (entry, entry,...., entry)

  PORTS = (node-node, node-node,...... node-node)
  NORMALIZE = S (value 1, value 2,........ )
EXECUTION CONTROLS
  comments
  ANALYZE model name (mode of analysis)

  comments

# FURTHER EXPLANATION

disposition
RETAIN
REPLACE
DELETE
PRINT

element value
  numeric constant
  TABLE table name (argument)
  EQUATION equation name (argument list)
    OR
  (FORTRAN-like expression)

  DISTRIBUTION distribution name (nominal value, tolerance %)
    OR
  DISTRIBUTION distribution name (tol 0%, nominal value, tol 1%)
    OR
  DISTRIBUTION distribution name (value 0, nominal value, value 1)
  DISTRIBUTION distribution name (value 0, value 1)
  STATISTICAL TABLE table name (argument, DISTRIBUTION distribution name)
    OR
  STATISTICAL TABLE table name (argument, parameter)
    OR
  STATISTICAL TABLE table name (argument)
  COMPLEX MAGPH (real argument, imaginary argument)
  COMPLEX MAGPH (magnitude, phase)
  FOURIER (argument 1, argument 2)

parameter value
  All element value forms except COMPLEX MAGPH and FOURIER
  plus
  COMPLEX EXP (FORTRAN-like complex expression)

Dummy arguments must not begin with letters
  I
  J
  K
  L
  M
  N

# EXAMPLES

Key letters for elements
R
C
L
G
E
M

MODEL 2N 2369 (RETAIN) (B-E-C)
MODEL ACTIVE FILTER ( )
MODEL CRYSTAL (1-4)

RBASE, BASE-X = .05
JBE, B-E = TABLE EMITTER DIODE (VBE)
CJUNCTION, 34-37 = EQUATION CAP (VR84, ILCOIL)
JE, B-4 = (.182 + DLOG (3/(JET + .1)/20)
C2, G-A = (DIODEQ (2.5E − 13, 38.1, VJE)
RC, V4-VN = D GAUSSIAN (4.5, 10%)
RIN, 1-2 = D NORMAL (.055 * PX * .020 * PX * PRN, 2%)
RIN, 1-2 = D NORMAL (−10%, 4.5, 10%)
C2, G-A = D DIST2 (−2%, 10, 5%)
RES, C-D = D8 (4.5, 5.0, 5.5)

REXT, C-A = D (0, 1)
JET3, B-E = ST DIODE (VJET3, D GAUSSIAN)

J1, V1 = STABLE SBD (VJ1, PT)
J2, V2 = STABLE SBD (VJ2, PT)
JE, B-E = STATISTICAL TABLE JE (VJE)
E1, GND-1 = COMPLEX (1, 0)
JSOURCE, N2-N1 = COMPLEX MAGPH (5, 90)
EIN, 1-2 = FOURIER (PFTAB, PHARMS)

Key letters for parameters
A
B
F
H
I
K
O
P
J
W
X
Y
Z

EIN, 1-2 = FOURIER (PFTAB, PHARM)
PFTAB = TABLE F (TIME)
PHARMS = 15

PZOUT = COMPLEX EXP (VRL/IRL)

T1 = MODEL JCTRAN (2-3-GND)
Q1 = MODEL 2N 2369 (1-2-3) (RE = .001)
M12, L1-L2 = 99
MUTUAL, LPRIMARY-LSECONDARY = TABLE MUTABLE (ILPRIMARY)

TABLE EIN, 0, −4, 2, −4, 8.25, 4, 30, 4, 36.25, −4, 50, −.4
EQUATION CAP (A, B, C) = (A + B + C)
Q1 (A, B, C, D, E, F, G, H) = (A1/(B-C)** D + E*F*(G + H)
Q DIODE (A,B) = (DIODEQ (A, 32, B))
DISTRIBUTION BETA (DENSITY) 2, 0, 4, 1, 10, 0
D BETA, 2, 0, 4, 1, 10, 0
D ALPHA (CUMULATIVE) 6, 0.8, 2, 9, 8, 1, 1
D2 (GAUSSIAN) MEAN = 0, SIGMA = 1, MIN = −2, MAX = 2.5
D3 (NORMAL) MIN = 4, MEAN = 20, MAX = −36.4, SIGMA = 9.1

STATISTICAL TABLE RB (DEP) 0, .179, .181, .1, .156, .163, ...., .15, .089, .092
STABLE SBD (IND = D GAUSSIAN) −.5, −.5, −.001, ...., .535, .565, 2.0

GROUND = (0)
GLOBAL = (P4)
GLOBAL = (ALPHA, BETA, PR, TJE)
PORTS = (GND-1, GND-4)

ANALYZE DIGITAL LOGIC INVERTER (TRANSIENT)
ANALYZE OSCILLATOR (D/O)
ANALYZE OSCILLATOR (AC)

RUN CONTROLS
   run control name = numeric constant
   OR
   run control name
   INITIAL CONDITIONS
OR
OPERATING POINT
   V element name = numeric constant
   I element name = numeric constant

```
START TIME = −20
TOPOLOGY
CASES = 1000
INITIAL CONDITIONS
    VCE = −3
    IL1, T4, CELL 3 = 5.5
    PCAP = 5
    IRL = 0.5
                                        ELEMENTS
                                          RX, B-BB = (3.0 + PX1)
                                          PX1 = TABLE PX1
                                          JE, BB-E = TABLE JE (WE)
OPERATING POINT                           LC, CC1 = (5.65 + 0.31 * ILC^2)
    VJE = 63
    ILC = 2.55
```

OUTPUTS
   PRINT variable list        variable list
                              variable (rename), variable (rename), . . .

   PLOT (plot options) variable list VS independent variable      plot options
              LABEL = (label text)
              COMMON SCALES or COMMON SCALES (min, max)
              INTERVAL = x        variable list
              DIVISIONS = n       variable (rename) (min, max) (variable option), . . . .
              BODE             independent variable
              REAL             variable option
              IMAG             REAL
              MAG             IMAG
              PHASE           MAG
              LOGM           PHASE
              ENVELOPE       LOGM

```
PRINT      VC1, VC2, CEMIT.Q1, CCOL.Q1
PRINT      EIN, N2, VRL2 (IN-PHASE-OUTPUT)
PRINT      DJE/DVJE, DJCNL,NDVR1
PLOT       VC1 (BASE VOLTS), VC2 (OUTPUT VOLTS)
PRINT, PLOT (LABEL = (INPUT AND OUTPUT VS TIME), COMMON SCALES) VC1, VC2
PLOT (INTERVAL = .01) JE (0, 100) VS VJE (0, 1.0)
PLOT, PRINT N1, N2, N3, N4
PLOT (REAL) PALPHA, PBETA
PLOT PALPHA (REAL), PALPHA (IMAG), PBETA (REAL), PBETA (IMAG)
```

HISTOGRAM (histogram options) variable list      histogram options
                              SAMPLE = (T1, T2,.....)
                              INTERVAL = class interval value
                              DIVISIONS = number of class intervals

SCATTERGRAM (sample option) variable list VS independent variable      variable list

TEST variable (min, max), variable (min, max), . . .      same as for PLOT statement

```
HISTOGRAM SAMPLE = (0, 30, 50) VRL2
PRINT, HIST (SAMPLE = (3., 4., 5.), DIVISIONS = 100) VLOAD

SCATTERGRAM PTONDIN VS PIDBG
SCAT (SAMPLE = (10, 20)) VC1, VS EIN
TEST IRLOAD (4, 12), POWER (30, 50), VC1 (1.5, 2.1)
```

RERUNS
   parametric name = (constant 1, constant 2, .....)

```
            ELEMENTS
              R1, 1-2 = (P1)
              P1 = 5
              P2 = 35

            ANALYZE.........
            RERUNS
              P1 = (10, 20, 30)
              P2 = (30, 25)
```

UTILITY CONTROLS
   PRINT NOTES
   PRINT USER DIRECTORY
   PRINT PROGRAM DIRECTORY
   PRINT ALL DIRECTORIES
   PRINT USER LIBRARY
   PRINT PROGRAM LIBRARY
   PRINT ALL LIBRARIES
   COMPRESS USER LIBRARY
   COMPRESS PROGRAM LIBRARY
END

# Index